AF478556

Supersymmetry

Supersymmetry:
Theory, Experiment, and Cosmology

P. Binétruy
AstroParticule et Cosmologie
Université Paris 7

OXFORD
UNIVERSITY PRESS

OXFORD
UNIVERSITY PRESS

Great Clarendon Street, Oxford OX2 6DP

Oxford University Press is a department of the University of Oxford.
It furthers the University's objective of excellence in research, scholarship,
and education by publishing worldwide in

Oxford New York

Auckland Cape Town Dar es Salaam Hong Kong Karachi
Kuala Lumpur Madrid Melbourne Mexico City Nairobi
New Delhi Shanghai Taipei Toronto

With offices in

Argentina Austria Brazil Chile Czech Republic France Greece
Guatemala Hungary Italy Japan Poland Portugal Singapore
South Korea Switzerland Thailand Turkey Ukraine Vietnam

Oxford is a registered trade mark of Oxford University Press
in the UK and in certain other countries

Published in the United States
by Oxford University Press Inc., New York

British Library Cataloguing in Publication Data

Data available

Library of Congress Cataloging in Publication Data

Binétruy, P.
Supersymmetry : theory, experiment, and cosmology / P. Binétruy.
p. cm.
Includes index.
ISBN-13: 978-0-19-850954-7 (alk. paper)
1. Supersymmetry. I. Title.
QC174.17.S9B56 2006
539.7′25—dc22
2006017203

Typeset by Newgen Imaging Systems (P) Ltd., Chennai, India
Printed in Great Britain
on acid-free paper by
Biddles Ltd, King's Lynn, Norfolk

ISBN 978–0–19–850954–7 (Hbk)

3 5 7 9 10 8 6 4 2

Contents

Introduction

The turn of the century witnesses a somewhat unprecedented situation in the study of fundamental interactions. On one hand, we have a theory, the Standard Model, which has been tested successfully to a few per mil: present data shows no deviation from this theory. On the other hand, we have a symmetry, supersymmetry, which seems to be the necessary ingredient to discuss the questions left aside by the Standard Model: diversity of masse scales, vacuum energy, etc. Supersymmetry is obviously not realized in the spectrum of fundamental particles. If it is realized at some deeper level, it must therefore be spontaneously broken. But even though it might be hidden in this way, there must be a level of precision to which experiments should be able to test deviations from the Standard Model. Clearly, this has not happened yet. But the ideas developed within the supersymmetric framework have lead to a general picture which seems supported by recent experimental data: gauge coupling unification, small neutrino masses, relatively light Higgs, a Universe with a density close to the critical density, etc.

It is the purpose of this book to give the tools necessary to discuss these issues. Because fundamental interactions provide, at some deeper level, a general picture of our own world, these issues may be solved at different levels: consistency of the theory, experimental discoveries, observation of the Universe. It would be preposterous to say at this point which will be the most decisive approach. It is therefore important to develop as much as possible a common language and to be able to follow the progress of each path: this will ease the way to a very enriching and most probably fruitful exchange between the different communities involved in the search for supersymmetry. This is why this book is conceived for readers with different backgrounds and varied interests: it may provide an introduction to the concepts and methods of supersymmetry for theorists; it can also be used as an introduction to the phenomenology of supersymmetric models for high energy experimentalists involved in supersymmetry searches; finally it targets the community of cosmologists involved in unravelling the properties of the early Universe. Of course supersymmetry, as we understand it now, may not be found in the end but one may be confident that the theory that will emerge eventually will feed upon the concepts, the methods and the results developed in a supersymmetric context.

Regarding experimental results, the present times represent a turning point. The precision tests at the LEP collider have successfully confronted the Standard Model and have started constraining the bulk of the parameter space of supersymmetric models. If supersymmetry is realized as we think now, the discovery of supersymmetric particles should wait for the turning on of the LHC collider. This does not mean that there is nothing to expect besides: searches at the Tevatron collider and precise measurements in B factories have been confronted with supersymmetry. Moreover, astrophysics and cosmology will most probably provide another perspective to study supersymmetric models: searches for dark matter will reach their maturity and cosmology which has become a quantitative science will provide a unique window on the very high mass scale regime of the theory.

On a parallel track, fundamental theories are being developed. It is probable that string theories provide, as they did from the beginning, the logical framework for supersymmetry. They are still under construction but it is important to be aware of the general picture they present us with. For example, recent ideas about extra spacetime dimensions have enriched the phenomenology of high energy colliders. Therefore, one chapter of this book (Chapter 10) presents in a non-technical way the general string and brane picture.

Roadmap

The text is organized in such a way that it can be read using different tracks depending on the interests of the reader. It can provide: (i) a theoretical introduction to supersymmetry; (ii) a presentation of supersymmetric models for high energy experimentalists; and (iii) an introduction to supersymmetry emphasizing its rôle in the early Universe. The sections that should be read in each case are summarized in the following table.

Theoretical Introduction	High Energy Experimentalist	Astrophysicist or Cosmologist
1	1.1, 1.2, 1.3.1	1.1, 1.2, 1.3.1
2	2.1–2.3	2.1–2.3
3 and App. B, C	3	3
4	—	—
5.1–5.4	5	5, esp. 5.5
D.1 and 6	D.1 and 6.1–6.8	6.1–6.4
7.1–7.4	7	7.4, 7.5
8	8.1	8.1
9	9	9.1, 9.4.1
10	10 (no box)	10.1–10.4.2 (no box)
—	—	11
12	12.1	12.2

In the course of the general text, some comments intended solely for readers who are following the "Theoretical introduction" track are put between square brackets: [...].

One should note that some very elementary knowledge of quantum field theory is assumed. Appendix A provides a sketch of the basic notions which are needed, which might prove useful to some readers to refresh their memory. It also describes some more advanced notions which will be necessary to the more theoretically oriented reader, and the bulk of the appendix is a presentation of the Standard Model of electroweak unification. The reader is urged to browse through it in order to get acquainted with the notation used throughout the book, as well as to make sure that he (she) feels at ease with the notions discussed there.

In parallel, Appendix D provides a brief introduction to the Standard Model of cosmological evolution which should prove useful to the non-expert to follow the chapters or sections on cosmology.

Appendices B and C introduce the notion of superfield and superspace. They lie on the "Theoretical introduction" track.

Finally, Appendix E provides a summary of the renormalization group equations which are scattered through the main text.

References and index

In addition to the general bibliography, a set of a few key references is presented at the end of each chapter that may provide further reading material.

Use of this book in teaching

It is difficult to discuss the interest of supersymmetry in the context of extensions of the Standard Model of electroweak interactions without having in mind the details of this Model. This is why a rather extended presentation of this Standard Model is presented in Appendix A. Depending on the origin and level of the students, it may provide the basis for some introductory lectures. Indeed, this represents the bulk of a half-semester course which I have given several times on the Standard Model.

Then, according to the nature of the audience, one of the three tracks described in the table above, or a mixture of them, may be followed. Each chapter is followed by exercises that develop some technical points or introduce applications not included in the main text. Exercises with a larger scope are called "Problems." They may provide material for more extended homework. The purpose of most exercises is not to challenge the students but rather to help them work out some details as well as to open their perspectives. This is why hints of solutions, or sometimes detailed solutions, are provided. They should also be of much help to the student who is studying this book by her(him)self.

Acknowledgments

This book owes much to my collaborators on the subject. It is through their own perception of the field that I have acquired some familiarity with the many methods and tools necessary to grasp the central ideas. Let me mention in particular Mary K. Gaillard, I. Hinchliffe, G. Girardi, R. Grimm, P. Ramond, S. Lavignac, and E. Dudas.

It was based originally on a set of lectures given in 1995 at Laboratoire de l'Accélérateur Linéaire in Orsay and has evolved through lectures at Ecole Normale Supérieure for several years. Let me thank all those who attended these various lectures and contributed one way or another to their improvement, in particular B. Delamotte, D. Langlois and Y. Mambrini. And Mireille Calvet for typing expertly some of these lectures.

I wish to extend my gratitude to all my colleagues in the Groupement de Recherche (GDR turned to Euro-GDR) SUSY who have shared over the years the excitement of working on the various aspects of supersymmetry. This has been a very enriching experience. Thanks to all, especially J.-F. Grivaz.

A preliminary version was used in class by several colleagues, in particular M.K. Gaillard and G. Kane, whom I thank for their many suggestions. In particular, I am especially indebted to Mary K. for her many major and minor corrections and her invaluable comments.

This book owes to much people but also to places, through which I have carried my ever growing manuscript over the years. I may mention the Institute of Fundamental Physics in Santa Barbara (several times!), the Aspen Center for Physics (several times as well), the University of Michigan, the Perimeter Institute and the Laboratoire d'Annecy-le-Vieux de Physique Théorique. Not to forget my own lab for more than ten years, the Laboratoire de Physique Théorique d'Orsay!

Very warm thanks to Sonke Adlung, at Oxford University Press, whose patience with the progress of the manuscript has been immense. His repeated encouragements have been the main reason for my not dropping the seemingly impossible task, especially in the last few years when I was becoming increasingly busy.

Finally, I would like to pay tribute to my family and my friends for their heroic patience over all these weekends, holidays, and vacations which were supposed to see the book completed in a few weeks and then days. This book is dedicated to them: they are the only ones who will not need to read it but will be satisfied with it being there on the shelf.

Pierre Binétruy
Paris, December 2005

1

The problems of the Standard Model

The Standard Model is a very satisfying theory of fundamental interactions: it has been superbly confirmed by experimental data, especially precision tests at the LEP collider. But it does not answer all the questions that one may raise about fundamental interactions. It is therefore believed to be the low energy limit of some deeper theory. It is the purpose of this chapter to make this statement more explicit. We will see that most of the questions which remain to be answered require new physics at much larger mass scales than the typical Standard Model scale. Such masses, more precisely quantum fluctuations associated with these heavy degrees of freedom, tend to destabilize the Standard Model mass scales in contradiction with experimental data. Supersymmetry provides the narrow path out of this trap by keeping these fluctuations under control. Readers who do not need any incentive to study supersymmetry may proceed directly to Chapter 2 and come back to this one if they ever lose faith.

1.1 General discussion

1.1.1 Full agreement with experimental data

Strictly speaking, the Standard Model of electroweak interactions has no problem with data and has been tested, in particular at LEP, to better than a percent level[1]. This means that it has been tested as a quantum theory. For example, the top quark quantum fluctuations allowed the LEP collaborations to make a rather precise estimate of the top quark mass even before it was discovered at the Tevatron collider. Similarly, there are now some indications in the context of the Standard Model that the Higgs particle is light [270].

Let us list in parallel the successes of the Standard Model from a theoretical point of view:

- The abelian gauge symmetry of quantum electrodynamics $U(1)_{\mathrm{QED}}$ is ensured without any tuning of parameters.
- The chiral nature of the Standard Model, *i.e.* the property that the left and right chiralities of its fermions have different quantum numbers, forbids any mass term: mass generation arises through symmetry breaking. Thus fermion masses are at most of the order of the electroweak scale.

---

[1] A review of the Standard Model and its precision tests is given in Appendix Appendix A.

- The most general Lagrangian consistent with the gauge symmetry and the field content of the Standard Model conserves baryon and lepton number, as observed in nature.

One introduces a Higgs field with the same quantum numbers as the lepton doublets (or more precisely their charge conjugates). There is however a fundamental difference between them: quantum statistics. The leptons are fermions and should be counted as matter whereas the Higgs, as a boson, participates in the interactions.

The only "smudge" in this picture is the recent realization, coming in particular from the deficit of solar as well as atmospheric neutrinos, that neutrinos have a mass. This leads in most cases[2] to a minor modification of the Standard Model as it was proposed initially: the introduction of right-handed neutrinos. One may note immediately that, following the relation

$$q = t^3 + \frac{y}{2}$$

between charge q, weak isospin t^3, and hypercharge y, the right-handed neutrino N_R has vanishing quantum numbers under $SU(3) \times SU(2) \times U(1)$, and is therefore a good indicator of physics beyond the Standard Model.

Let us be more precise about this statement. One of the remarkable properties of the Standard Model is that, given its field content, the Lagrangian is the most general compatible with the gauge symmetry $SU(3) \times SU(2) \times U(1)$. If we do not want to include an extra symmetry, we must keep this principle once we introduce the right-handed neutrino. Then, besides the Yukawa term $\lambda_Y \phi \bar{N}_R \nu_L$ (ϕ is the Higgs field and we adopt a different notation for ν_L and N_R since they may not form the two helicity eigenstates of a single Dirac fermion) which gives a Dirac mass term to the neutrino after spontaneous symmetry breaking, we must include as well a Majorana mass term $M\bar{N}_L^c N_R$, where $N_L^c = C \left(\overline{N_R} \right)^T$ (see Appendix B). A Majorana mass term $\bar{\nu}_R^c \nu_L$ for the left-handed neutrino ($\nu_R^c = C \left(\overline{\nu_L} \right)^T$) is forbidden by the gauge symmetry and may neither be generated by a Yukawa term if we do not extend the Higgs sector[2].

One thus obtains the following mass matrix, written for a single neutrino species,

$$
\begin{array}{cc}
\nu_L & N_L^c
\end{array}
$$
$$
\begin{pmatrix} 0 & m \\ m & M \end{pmatrix}
\begin{array}{c}
\nu_R^c \\
N_R
\end{array}
$$

The nondiagonal entry is fixed by $SU(2) \times U(1)$ spontaneous breaking: the scale m is a typical electroweak scale (a few hundred GeV) if the Yukawa coupling λ_Y is of order 1. The nonvanishing diagonal entry M remains unconstrained by the electroweak gauge symmetry: it may therefore be very large. In the case $M \gg m$, one obtains

---

[2] That is, if one does not extend the Higgs sector beyond doublets. Otherwise, a Majorana mass term for the left-handed neutrino may arise from a Yukawa coupling to a isosinglet or isovector scalar field. The Zee model [385] is an example of such a possibility.

two eigenvalues: $m_1 \sim M$ and $m_2 \sim m^2/M$. This is the famous seesaw mechanism [177, 383]. If we interpret the latter eigenvalue as the small neutrino mass observed ($m_\nu \sim 10^{-1}$ to 10^{-2} eV), this gives an indication that new physics appears at a scale $M \sim 10^{14}$ GeV, under the form of a superheavy neutrino (the second eigenstate).

1.1.2 Theoretical issues

The Standard Model is in any case not completely satisfactory from a theoretical point of view. For example, one may list the free parameters:

- three gauge couplings
- two parameters in the Higgs sector: m and λ
- nine quark (u, d, c, s, t, b) and charged lepton (e, μ, τ) masses
- three mixing angles and one CP-violating phase for the quark system
- the QCD parameter θ (coupling of the $F^a_{\mu\nu}\widetilde{F}^{a\mu\nu}$ term)

which amounts to 19 free parameters. One may wish fewer parameters for a fundamental theory.

Indeed, extensions of the Standard Model are proposed which relate some of these parameters. For example, grand unified theories unify the three gauge couplings at a scale M_U of order 10^{16} GeV. This works only approximately in a nonsupersymmetric framework. They also classify the quark and lepton fields in larger representations, which induces relations among the mass parameters. For example, grand unified theories predict with success the ratio m_b/m_τ.

The theory is also determined by the quantum numbers of each field. There are open questions in this respect: why are all the electric charges a multiple of $e/3$, where $-e$ is the electron charge? This is often called the problem of the quantization of charge. Why are the quantum numbers of quarks and leptons such that all anomalies cancel? Again, grand unified theories give a first answer to these problems by including the electric charge among the nonabelian gauge symmetry generators.

The gap observed in quark and lepton masses when one goes from one family to another is not satisfactorily explained in the Standard Model. This is often referred to as the problem of mass or flavor problem. The Standard Model of electroweak interactions clearly establishes the breaking of the $SU(2) \times U(1)$ gauge symmetry as the origin of mass. This is summarized in the formula:

$$m_f = \lambda_f \langle \phi \rangle \tag{1.1}$$

where the mass m_f of a fermion f is expressed in terms of the vacuum expectation value (vev) of the Higgs field and of its Yukawa coupling λ_f to this field.

The vacuum expectation value of the scalar field $\langle \phi \rangle \equiv v/\sqrt{2}$ is fixed by the low energy effective Fermi theory:

$$v = \left(\frac{1}{G_F \sqrt{2}} \right)^{1/2} = 246 \text{ GeV}, \tag{1.2}$$

where G_F is the Fermi constant. Thus (1.1) means that the only fermion with a "natural" mass scale is the top quark of mass $m_t = 175$ GeV: the corresponding

Yukawa coupling is of order 1. But it does not account for the diversity of quark and lepton masses, especially from one family of quarks and leptons to another. This question must be addressed by a theory of Yukawa couplings yet to come. If complete, this theory should also predict the number of families.

Similarly for mixing angle and phases. It is well-known that, in the quark sector, mass eigenstates do not coincide with interaction eigenstates (*i.e.* the fields with definite quantum numbers under $SU(3) \times SU(2) \times U(1)$). This results in mixing angles and relative phases among the different quarks. The mixing angles are found to be very different in size. Again, this diversity has yet to be explained in the framework of a more fundamental theory. As for phases, the Standard Model provides a single one, the Cabibbo–Kobayashi–Maskawa phase, and thus a unique source of CP violation. This, however, fails to explain the baryon number of the Universe. Similarly, the strong CP problem seems to indicate that we are missing in the Standard Model some important ingredient, whether the axion or some other solution.

The final question is how to treat gravity. Because Newton's constant is dimensionful ($[G_N] = M^{-2}$), the theory of gravity is nonrenormalizable. Quantum effects become important at a scale $(\hbar c/G_N)^{1/2} = M_P \sim 1.22 \times 10^{19}$ GeV, known as the Planck scale. How does one obtain a quantum theory of gravity, that is how does one put gravity on the same level as the theory of the other fundamental interactions? Most probably, answering such a question requires some drastic changes, such as the ones proposed in the string approach where the fundamental objects are no longer pointlike.

We will return to these questions in later chapters. For the time being, we will address the question of the coexistence at the quantum level of two vastly different scales: the electroweak scale M_W and the Planck scale M_P (or for that matter any of the superheavy scales we have discussed above: $M \sim 10^{14}$ GeV, $M_U \sim 10^{16}$ GeV, etc.).

1.2   Naturalness and the problem of hierarchy

The central question that we will address in this section is the existence of quadratic divergences associated with the presence of a fundamental scalar field, such as the Higgs field in the Standard Model. Before dealing with this, we must have a short presentation of the notion of effective theory.

1.2.1   Effective theories

In the modern point of view, a given theory (*e.g.* the Standard Model) is always the effective theory of a more complete underlying theory, which adequately describes physics at a energy scale higher than a threshold M. This threshold is physical in the sense that the complete physical spectrum includes particles with a mass of order M. For example, in the case of the seesaw mechanism for neutrino masses, discussed in Section 1.1.1, there is a neutrino field of mass M.

The description in terms of an effective theory, restricted to the light states, is obviously valid only up to the scale M. The heavy fields (of mass M or larger) regulate the theory and therefore the scale M acts as a cut-off Λ on loop momenta.

In quantum field theory, the renormalization procedure allows us to deal with infinities, *i.e.* contributions that diverge when the cut-off is sent to infinity. However, the cut-offs that we consider here are physical and thus cannot be sent to arbitrary values. There is then the possibility that the corrections due to the heavy fields (of mass M) destabilize the low energy theory. As we will see in the next section, this is indeed a possibility when we are working with fundamental scalars.

In some theories, we may infer some upper bound on the physical cut-off Λ (which we identify from now on with the scale of new physics M) from the value of the low-energy parameters. We will discuss briefly the three standard methods used (unitarity, triviality, and vacuum stability) and illustrate them on the example of a complex scalar field.

More precisely, we consider, as in the Standard Model, a complex scalar field Φ with Lagrangian

$$\mathcal{L} = \partial^\mu \Phi^\dagger \partial_\mu \Phi - V\left(\Phi^\dagger \Phi\right)$$
$$V\left(\Phi^\dagger \Phi\right) = -m^2 \Phi^\dagger \Phi + \lambda \left(\Phi^\dagger \Phi\right)^2. \tag{1.3}$$

The minimization of this potential gives the background value $\langle \Phi^\dagger \Phi \rangle = v^2/2$ with

$$v^2 \equiv m^2/\lambda. \tag{1.4}$$

We thus parametrize Φ as:

$$\Phi = \begin{pmatrix} \phi^+ \\ \frac{1}{\sqrt{2}}(v + h + i\phi^0) \end{pmatrix}. \tag{1.5}$$

The fields ϕ^+ and ϕ^0 are Goldstone bosons whereas the mass of the h field is

$$m_h^2 = 2m^2 = 2\lambda v^2. \tag{1.6}$$

This scalar field may have extra couplings to gauge fields or the top quark for example.

Unitarity ([268, 282])[3]

Unitarity of the S-matrix, which is a consequence of the conservation of probabilities at the quantum level, imposes some constraints on scattering cross-sections, especially on their high-energy behavior. This is usually expressed in terms of partial-wave expansion: if $\mathcal{M}(s, \theta)$ is the amplitude for a $2 \to 2$ scattering process with center of mass energy $\sqrt{s}$ and diffusion angle θ, one defines the Jth partial wave as:

$$a_J(s) = \frac{1}{32\pi} \int d\cos\theta \, P_J(\cos\theta)\mathcal{M}(s, \theta), \tag{1.7}$$

where P_J is the Jth Legendre polynomial. The constraint coming from unitarity reads

$$\text{Im } a_J \geq |a_J|^2 = \left(\text{Re } a_J\right)^2 + \left(\text{Im } a_J\right)^2, \tag{1.8}$$

from which we obtain

$$\left(\text{Re } a_J\right)^2 \leq \text{Im } a_J \left(1 - \text{Im } a_J\right). \tag{1.9}$$

---

[3]For a nice introduction to the Standard Model from the point of view of unitarity, see the lectures of [252].

Since the right-hand side of this equation is bounded by 1/4, it implies

$$|\text{Re } a_J| \leq \frac{1}{2}. \tag{1.10}$$

Such limits were considered in the context of the Fermi model of weak interactions to introduce an intermediate vector boson (see Section A.3 of Appendix Appendix A). They may also be applied to the physics of the Standard Model. For example, in the absence of a fundamental Higgs field, the $J = 0$ tree level amplitude for $W_L^+ W_L^- \rightarrow Z_L Z_L$ ($W_L^\pm$, Z_L are the longitudinal components of $W^\pm$ and Z respectively) simply reads $a_0(s) = G_F \sqrt{2}\, s/(16\pi)$. The tree level unitarity constraint (1.10) imposes that new physics (the fundamental Higgs in the Standard Model) appears at a scale $\Lambda < \Lambda_U = \sqrt{8\pi/(G_F \sqrt{2})} \sim 1.2$ TeV.

If we include a Higgs doublet (1.5), we may use the equivalence theorem [73, 268] to identify $\phi^\pm$ with $W_L^\pm$ and ϕ^0 with Z_L. Then a_0 receives an extra contribution coming from the Higgs field

$$a_0(s) = \frac{G_F \sqrt{2}\, s}{16\pi} - \frac{G_F \sqrt{2}\, s}{16\pi} \frac{s}{s - m_h^2} + O\left(\frac{M_W^2}{s}\right) \xrightarrow{s \gg m_h^2} -m_h^2 \frac{G_F \sqrt{2}}{16\pi} \tag{1.11}$$

and the unitarity constraint (1.10) gives a constraint on the Higgs mass. It turns out that the most stringent constraint comes from the mixed zero-isospin channel $2W_L^+ W_L^- + Z_L Z_L$ and reads, in terms of the electroweak breaking scale $v = \left(G_F \sqrt{2}\right)^{-1/2}$,

$$m_h < \sqrt{\frac{16\pi}{5}}\, v = 780 \text{ GeV}. \tag{1.12}$$

Triviality ([61, 277])

In the renormalization group approach, the scalar self-coupling λ is turned into a running coupling $\lambda(\mu)$ varying with the momentum scale μ characteristic of the process considered. The study of one-loop radiative corrections allows us to compute to lowest order (in λ) the evolution of $\lambda(\mu)$ with the scale μ, *i.e.* its beta function:

$$\mu \frac{d\lambda}{d\mu} = \frac{3}{2\pi^2} \lambda^2 + \cdots \tag{1.13}$$

where the dots refer to higher order contributions[4].

We see that the coupling $\lambda(\mu)$ is monotonically increasing. If we want the theory described by the Lagrangian (1.3) to make sense all the way up to the scale Λ, we must impose that $\lambda(\mu) < \infty$ for scales $\mu < \Lambda$. If Λ is known, this imposes some bound on the value of λ at low energy, say $\lambda(v)$. For example, if we send Λ to infinity, this imposes $\lambda(v) = 0$. This is why a theory described by an action (1.3) which would be

---

[4]In the case of the Standard Model, there are further one-loop contributions, mainly due to the couplings of the Higgs doublet to the gauge fields and the top. In the case of a large Higgs mass (*i.e.* from (1.6), large λ), the term given here is the dominant one.

valid at *all* energy scales is known as *trivial*, *i.e.* is a free field theory in the infrared (low energy) regime. In practice, this only means that at some scale Λ smaller than the scale $\Lambda_{\mathrm{Landau}}$ where the coupling would explode (known as the Landau pole, see below), some new physics appears.

The exact value of the Landau pole requires a nonperturbative computation since the running coupling explodes at this scale. Complete calculations show that it is not unreasonable to use the one-loop result (1.13) to obtain an order of magnitude for the Landau pole. Thus, solving for λ the differential equation (1.13),

$$\lambda^{-1}(\mu) = \lambda^{-1} - \frac{3}{2\pi^2} \ln \frac{\mu}{v}, \tag{1.14}$$

where $\lambda \equiv \lambda(v)$, one obtains, using $\lambda^{-1}(\Lambda_{\mathrm{Landau}}) = 0$,

$$\Lambda_{\mathrm{Landau}} \sim v \, e^{2\pi^2/(3\lambda)}. \tag{1.15}$$

Since λ can be expressed in terms of m_h itself through (1.6), this is used to put an upper bound, a triviality bound, on the scale of new physics. Keeping the same level of approximation as before, this gives roughly

$$\Lambda < v \, e^{4\pi^2 v^2/(3m_h^2)} \equiv \Lambda_T(m_h). \tag{1.16}$$

The right-hand side is a monotonically decreasing function of m_h. Alternatively, one may say that, for a given value of Λ, the Higgs mass is bounded by

$$m_h^2 < \frac{4\pi^2 v^2}{3\ln(\Lambda/v)}, \tag{1.17}$$

a decreasing function of Λ.

The reader should be reminded that our expressions are only rough estimates which give the general trend and that they can be refined. In any case, such limits should be taken with a grain of salt: obviously, in the case where Λ is not much greater than m_h, the Higgs mass receives new corrections arising from effective operators (scaling like some negative powers of Λ) which will induce possibly large corrections to the triviality bound.

Vacuum stability

An ever-increasing $\lambda(\mu)$ coupling leads to the constraints just discussed. An ever-decreasing $\lambda(\mu)$ coupling leads to difficulties of another kind: as soon as the quartic coupling $\lambda(\mu)$ turns negative, the potential (1.3) becomes unbounded from below at large values of the field ϕ and the theory suffers from an instability. Notwithstanding considerations regarding the cosmological stability of the false vacuum, this instability is a sign that some new physics will take charge.

This situation is faced in particular when the scalar field that we have considered is light (λ is small) but has other interactions that contribute to lower $\lambda(\mu)$ as μ increases. For example, if the Standard Model Higgs is light, the dominant term comes from the top Yukawa interaction:

$$\mu \frac{d\lambda}{d\mu} = -\frac{3}{8\pi^2} \lambda_t^4 + \cdots \tag{1.18}$$

If we neglect gauge interactions, as we did in (1.18), then the top Yukawa coupling $\lambda_t(\mu)$ does not run and (1.18) is solved as:

$$\lambda(\mu) = \lambda - \frac{3}{8\pi^2}\lambda_t^4 \ln\frac{\mu}{v}. \tag{1.19}$$

Defining the scale Λ_U at which the instability appears by the condition $\lambda(\Lambda_U) = 0$, one obtains

$$\Lambda_U = v\, e^{8\pi^2\lambda/(3\lambda_t^4)}. \tag{1.20}$$

Then the scale of new physics Λ is constrained by

$$\Lambda < \Lambda_U = v\, e^{\pi^2 m_h^2 v^2/(3m_t^4)} \tag{1.21}$$

where we have used (1.6) and $m_t^2 = \lambda_t^2 v^2/2$. Alternatively, for a given value of Λ, we have

$$m_h^2 > \frac{3m_t^4}{\pi^2 v^2}\ln\frac{\Lambda}{v}. \tag{1.22}$$

Again, this formula just shows a trend (for example that the vacuum stability bound increases with Λ, as well as with the top mass). For large enough Λ, it is not possible to neglect the running of λ_t due to gauge interactions: $\lambda_t(\mu)$ increases with μ, which lowers the scale Λ_U where the instability appears. This strengthens the bound on m_h.

1.2.2   The concept of naturalness

The presence of fundamental scalar fields leads to the well-known problem [186, 347, 359] of quadratic divergences as soon as one introduces a finite cut-off Λ in the theory. Indeed a diagram of the type given in Fig. 1.1 generically gives a contribution

$$\delta m^2 = \lambda \int^\Lambda \frac{d^4 k}{(2\pi)^4}\frac{1}{k^2} \sim \frac{\lambda}{16\pi^2}\int^\Lambda dk^2,$$

which is of order $\lambda\Lambda^2/16\pi^2$, to the scalar mass-squared m^2.

Let us denote by m_0 the bare mass (which, in this context, is the mass of the scalar field in the absence of underlying physics); we obtain at the one-loop level a scalar mass-squared

$$m^2 = m_0^2 + \alpha\lambda\frac{\Lambda^2}{16\pi^2}$$

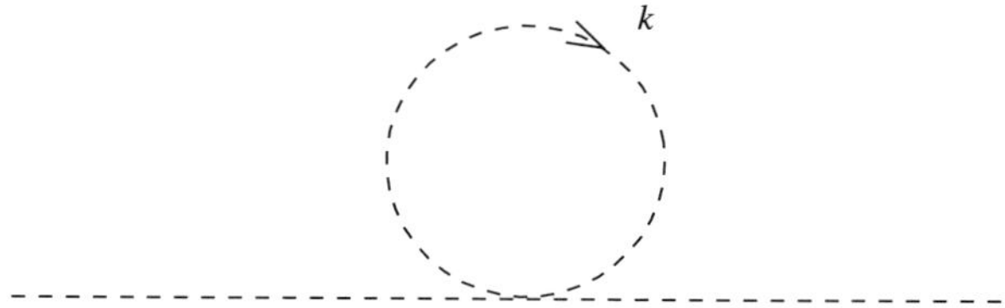

Fig. 1.1 The scalar one-loop diagram giving rise to a quadratic divergence.

where α is a positive or negative number of order one. Taking Λ as a fundamental mass unit,

$$\frac{m_0^2}{\Lambda^2} = \frac{m^2}{\Lambda^2} - \alpha \frac{\lambda}{16\pi^2}. \tag{1.23}$$

Plugging typical numbers, say $m \sim 100$ GeV and $\Lambda \sim M_P \sim 10^{19}$ GeV, we see that m_0^2/Λ^2 must be adjusted to more than 30 orders of magnitude. This is to most people an intolerable fine tuning.

Let us be more precise in the case of the Standard Model. Then, the Higgs mass receives the following one-loop corrections:

$$\delta m_h^2 = \frac{3\Lambda^2}{8\pi^2 v^2} \left[\left(4m_t^2 - 2M_W^2 - M_Z^2 - m_h^2\right) + O\left(\log \frac{\Lambda}{\mu}\right) \right], \tag{1.24}$$

where we recognize the contribution of the diagram of Fig. 1.1, proportional to $\lambda \sim m_h^2/v^2$, as well as the contribution of the top quark loop, proportional to $\lambda_t^2 \sim m_t^2/v^2$. The latter is leading in the case of a light Higgs. One may also note that the one-loop quadratically divergent contribution vanishes if we have the following relation between the masses:

$$4m_t^2 = 2M_W^2 + M_Z^2 + m_h^2 \tag{1.25}$$

This relation, known as the Veltman condition [352], is obviously not ensured to hold at higher orders.

One may define the amount f of fine tuning discussed above by

$$\frac{\delta m_h^2}{m_h^2} \equiv \frac{1}{f}. \tag{1.26}$$

Indeed, if $\delta m_h^2 = 100 m_h^2$, then one needs to fine tune the Higgs bare mass m_0^2 to the per cent level in order to recover the right physical Higgs mass m_h^2. This amount of fine-tuning is represented on a plot (m_h, Λ) [257] in Fig. 1.2 for values of Λ smaller than 100 TeV. The regions forbidden by the triviality and the vacuum stability bounds discussed in the preceding sections are also presented. One may note that, in the region corresponding to the Veltman condition (1.25), there is less need for fine tuning: this region does not extend, however, to very large values of Λ because of the higher order contributions.

In any case, it is clear that fine tunings larger than the per cent level are necessary as soon as the scale of new physics Λ is larger than 100 TeV.

Such a fine tuning goes against the prejudice that the observable properties of a theory (masses, charges,...) are stable under small variations of the fundamental parameters (the bare parameters). One talks of the naturalness of a theory to describe such behavior.

Let us see how this operates in quantum electrodynamics, which is the archetype of a natural theory. Since QED is characterized by a dimensionless coupling, one may wonder what is the fundamental scale. However, we are again in the situation of a "trivial" theory with an ever increasing coupling (at least perturbatively) and thus

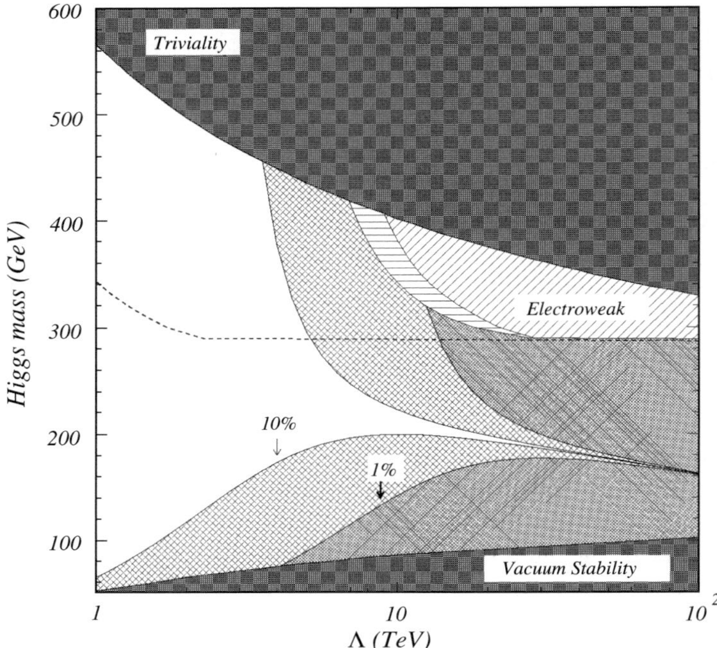

Fig. 1.2 Plot in the m_h–Λ plane showing the triviality (dark region at top) and stability (dark region at bottom) constraints, as well as the tuning contours. The darkly hatched region marked "1%" represents tunings of greater than 1 part in 100; the "10%" region means greater than 1 part in 10. The empty region has less than 1 part in 10 fine tuning [257].

a Landau pole: the corresponding mass scale provides a dynamical scale for QED. Again, we may use the one-loop beta function

$$\mu\frac{de^2}{d\mu} = \frac{N_f}{6\pi^2}e^4, \tag{1.27}$$

where N_f is the number of fermions, to obtain an order of magnitude for this scale. Solving for $e(\mu)$

$$\frac{1}{e^2} - \frac{1}{e^2(\mu)} = \frac{N_f}{6\pi^2}\ln\frac{\mu}{m_e}$$

we obtain $(e^2(\Lambda_{\mathrm{Landau}}) \to \infty)$[5]

$$\Lambda_{\mathrm{Landau}} \sim m_e\, e^{6\pi^2/(N_f e^2)} \gg M_P. \tag{1.28}$$

Nevertheless, QED is a natural theory because its parameters such as the electron mass m_e, are naturally small: they are protected from important radiative corrections by a symmetry.

[5]The fact that the Landau pole lies beyond the Planck scale explains why the original result of [267] was never considered as a severe problem for quantum electrodynamics.

For example, the electron mass is protected by the chiral symmetry: in the limit $m_e \to 0$, the symmetry of the system is enhanced to include invariance under $\psi_e \to e^{i\alpha\gamma_5}\psi_e$ where ψ_e is the Dirac spinor describing the electron. The presence of this symmetry imposes that the corrections to the electron mass are themselves proportional to m_e. It follows that the coefficient of proportionality is dimensionless and thus behaves like $\log \Lambda$.

This leads to the formulation of naturalness by 't Hooft [349]:

A theory is natural if, for all its parameters p which are small with respect to the fundamental scale Λ, the limit $p \to 0$ corresponds to an enhancement of the symmetry of the system.

1.2.3   The case of the scalar field

Let us return to the case of a complex scalar field with self-coupling λ which was discussed in the previous section.

The fundamental high-energy scale of the theory is given again by its Landau pole $\Lambda_{\mathrm{Landau}}$ given in (1.15). But not all parameters are naturally small at this scale:

- λ may be small because $\lambda = 0$ corresponds to an enhancement of the symmetry (conservation of the number of Φ particles).
- m^2 is not naturally small because $m^2 = 0$ does not correspond to any symmetry enhancement at the quantum level. To be more precise, in the limit $\lambda \to 0$ and $m^2 \to 0$, the symmetry of the system is larger: $\Phi(x) \to \Phi(x)+C$, with C constant. This could be the symmetry of a theory which would appear at a scale Λ_{nat}. At low energy this symmetry would be broken by effects described by a parameter ε: $\lambda = 0(\varepsilon)$, $m^2 = 0(\varepsilon)\,\Lambda_{\mathrm{nat}}^2$. Thus

$$\Lambda_{\mathrm{nat}} \sim \frac{m}{\sqrt{\lambda}}.$$

In other words, a Φ^4 theory is natural for a fundamental scale of order $m/\sqrt{\lambda}$. If one applies this to the Standard Model, one obtains $m/\sqrt{\lambda} \sim v$. Thus, strictly speaking, the Standard Model is only natural up to a scale no larger than the TeV.

1.2.4   The case of asymptotically free gauge theories. Technicolor

Contrary to scalar field theories, asymptotically free gauge theories provide good candidates for natural theories. In the limit of vanishing gauge coupling (free theory!), the conservation of the number of gauge bosons enhances the symmetry. One can easily verify that there is no severe problem of fine-tuning.

Denote by Λ the fundamental scale, where the gauge coupling is, in some sense, the bare coupling g_0. Then keeping for simplicity only the first term in the beta function

$$\mu\,\frac{dg}{d\mu} = -\frac{b}{16\pi^2}g^3 + O(g^5), \tag{1.29}$$

we have

$$\frac{1}{g^2(\mu)} = \frac{1}{g_0^2} + \frac{b}{8\pi^2}\,\log\frac{\mu}{\Lambda} \tag{1.30}$$

and the coupling explodes for $\mu = \Lambda_{IR}$ given by

$$\frac{\Lambda_{IR}}{\Lambda} = e^{-8\pi^2/(bg_0^2)}. \tag{1.31}$$

At this scale, gauge interactions become strong and typically generate physical masses of this order. Taking the example of $SU(3)$ ($b = 11$), one obtains $\Lambda_{\mathrm{IR}}/\Lambda \sim 10^{-19}$ for $g_0^2 = 0.16$: no need to adjust g_0^2 intolerably! This is precisely the property which is used in technicolor models to overcome the naturalness problem: scalars are bound states and $m_{\mathrm{scalar}}^2 \sim \Lambda_{\mathrm{IR}}^2 \ll \Lambda^2$.

1.3 Supersymmetry as a solution to the problem of naturalness

We have seen that:

(i) setting the mass of a scalar field to zero does not enhance the symmetry;
(ii) setting the mass of a fermion field to zero enhances the symmetry (chiral symmetry).

The idea is therefore to relate under a new symmetry a scalar field with a fermion field. This symmetry – supersymmetry – must be such that the masses of the scalar and of the fermion fields be equal. It is therefore related, in some sense to be defined, to the invariance under the Poincaré group since it connects representations of different spin. In such a scheme, the relation $m_s/\Lambda \ll 1$ is natural because $m_f/\Lambda \ll 1$ is natural and because the scalar mass m_s is related to the fermion mass m_f.

Then, the contribution of fermions to the quadratic divergence cancels the contribution of bosons.

1.3.1 An explicit example: the Wess–Zumino model

We will check this result explicitly, on a model known as the Wess–Zumino model. This model might seem at first rather contrived. We will learn how to construct it and to understand the beauty of it (see Chapter 3). For the time being, it will be an opportunity to familiarize ourselves with some typical supersymmetric interactions.

The model contains:

- a complex scalar field (two degrees of freedom): $\phi = (A + iB)/\sqrt{2}$;
- a fermion field described by a Majorana spinor Ψ (two degrees of freedom): $\Psi^c = C\bar{\Psi}^T = \Psi$.

The Lagrangian decomposes into

(i) a kinetic term

$$\mathcal{L}_k = \partial^\mu \phi^* \partial_\mu \phi + \frac{i}{2}\,\bar{\Psi}\slashed{\partial}\Psi = \frac{1}{2}\,\partial^\mu A \partial_\mu A + \frac{1}{2}\,\partial^\mu B \partial_\mu B + \frac{i}{2}\bar{\Psi}\slashed{\partial}\Psi; \tag{1.32}$$

(ii) an interaction term which is expressed in terms of a single function $W(\phi)$ known as the superpotential

$$\mathcal{L}_i = -\left|\frac{dW}{d\phi}\right|^2 - \frac{1}{2}\left(\frac{d^2W}{d\phi^2}\,\bar{\Psi}_R\Psi_L + \frac{d^2W^*}{d\phi^{*2}}\,\bar{\Psi}_L\Psi_R\right). \tag{1.33}$$

Explicitly

$$W(\phi) = \frac{1}{2}m\phi^2 + \frac{1}{3}\lambda\phi^3, \tag{1.34}$$

$$\mathcal{L}_i = -|m\phi + \lambda\phi^2|^2 - \frac{1}{2}\left[m\left(\bar{\Psi}_R\Psi_L + \bar{\Psi}_L\Psi_R\right) + 2\lambda\left(\phi\,\bar{\Psi}_R\Psi_L + \phi^*\,\bar{\Psi}_L\Psi_R\right)\right]$$

$$= -\frac{1}{2}m^2\left(A^2 + B^2\right) - \frac{m\lambda}{\sqrt{2}}\,A\left(A^2 + B^2\right) - \frac{\lambda^2}{4}\left(A^2 + B^2\right)^2$$

$$-\frac{1}{2}m\,\bar{\Psi}\Psi - \frac{\lambda}{\sqrt{2}}\,\bar{\Psi}\left(A - iB\,\gamma_5\right)\Psi. \tag{1.35}$$

The self-energy diagrams for the scalar field A which contribute at one loop to the quadratic divergence are given in Fig. 1.3. This gives respectively:

$$(1)\quad -\frac{i\lambda^2}{4}\,4\times 3\int\frac{d^4k}{(2\pi)^4}\,\frac{i}{k^2 - m^2} = 3\lambda^2\int\frac{d^4k}{(2\pi)^4}\,\frac{1}{k^2 - m^2}$$

$$(2)\quad -\frac{i\lambda^2}{2}\,2\int\frac{d^4k}{(2\pi)^4}\,\frac{i}{k^2 - m^2} = \lambda^2\int\frac{d^4k}{(2\pi)^4}\,\frac{1}{k^2 - m^2}$$

$$(3)\;(-)\left(-\frac{i\lambda}{\sqrt{2}}\right)^2 2\int\frac{d^4k}{(2\pi)^4}\,\mathrm{Tr}\left[\frac{i}{\slashed{k} - m}\,\frac{i}{(\slashed{k} - \slashed{p} - m)}\right]$$

$$= -\lambda^2\int\frac{d^4k}{(2\pi)^4}\,\frac{\mathrm{Tr}(\slashed{k} + m)(\slashed{k} - \slashed{p} + m)}{(k^2 - m^2)[(k - p)^2 - m^2]}.$$

Using $\mathrm{Tr}(\slashed{k} + m)(\slashed{k} - \slashed{p} + m) = 4\left[k.(k - p) + m^2\right] = 2\left[(k^2 - m^2) + ((k - p)^2 - m^2)\right.$ $\left. -p^2 + 4m^2\right]$, one finds a total contribution

$$2\lambda^2\left\{\int\frac{d^4k}{(2\pi)^4}\,\frac{1}{k^2 - m^2} - \int\frac{d^4k}{(2\pi)^4}\,\frac{1}{(k - p)^2 - m^2}\right.$$

$$\left.+\int\frac{d^4k}{(2\pi)^4}\,\frac{p^2 - 4m^2}{(k^2 - m^2)[(p - k)^2 - m^2]}\right\},$$

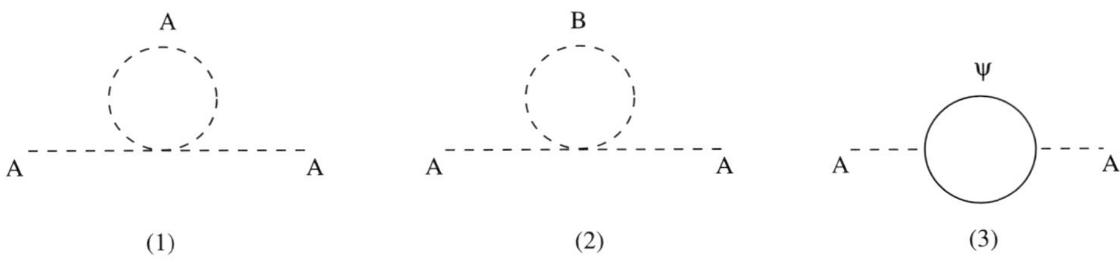

Fig. 1.3 Quadratically divergent self-energy diagrams for the A scalar field.

which shows that the quadratic divergences (still present in the first two terms) cancel[6].

One may show that the cancellation is even larger and that logarithmic divergences are only present in wave function renormalization. There are no mass counterterms and, more generally, the parameters of the superpotential are not renormalized (no finite or infinite quantum corrections). This is an example of the famous nonrenormalization theorems that have made the success of supersymmetry.

[The rest of this chapter is not necessary for reading the next chapters and may be dealt with when reaching Chapter 5.]

1.3.2   A graphic method

Let us check it for the mass by using a graphic method which is particularly instructive. We are going to show that the properties of chirality of the theory forbid any mass counterterm for fermions (and by supersymmetry for bosons). The mass term for fermions is written

$$\mathcal{L}_m = -\tfrac{1}{2}m \left(\bar{\Psi}_R \Psi_L + \bar{\Psi}_L \Psi_R \right)$$

which we depict graphically by:

Similarly for the Yukawa coupling

$$\mathcal{L}_y = -\lambda \left(\Phi\, \bar{\Psi}_R \Psi_L + \Phi^*\, \bar{\Psi}_L \Psi_R \right)$$

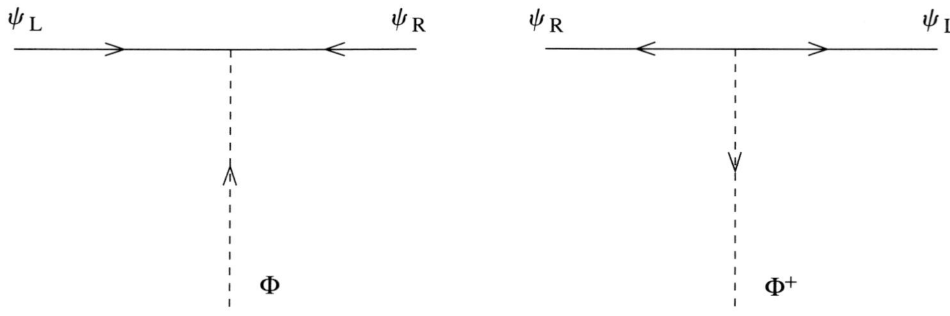

---

[6]We are supposing here that the integrals are properly regularized and thus that there exists a regularization procedure that respects supersymmetry.

The convention is that all arrows are incoming or outgoing at the vertex. The scalar propagator (associated with the contraction $\phi^*\phi$) is denoted:

Indeed, we can define a direction on the propagator of the scalar field, a direction conserved by interactions, because we may associate it by supersymmetry with the chirality of its fermionic partner.

At the one-loop level, the self-energy of fermions is given by the diagram:

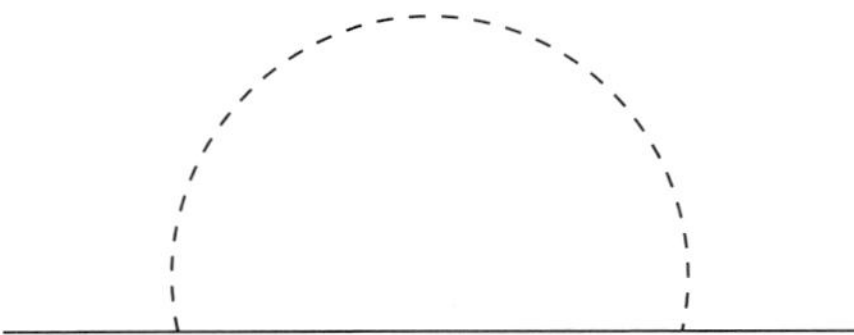

Let us represent chiralities in the case corresponding to a mass counterterm:

In either diagram, the two vertices impose contradictory directions for the arrow of the scalar propagator. Hence, we cannot write a contribution at one loop: there is no mass counterterm for fermions. This is a direct consequence of the chiral nature of the interactions.

Let us note that:

(i) One can easily write a contribution for wave function renormalization $\left(\bar{\Psi}_L i\partial\!\!\!/\, \Psi_L + \bar{\Psi}_R i\partial\!\!\!/\, \Psi_R\right)$:

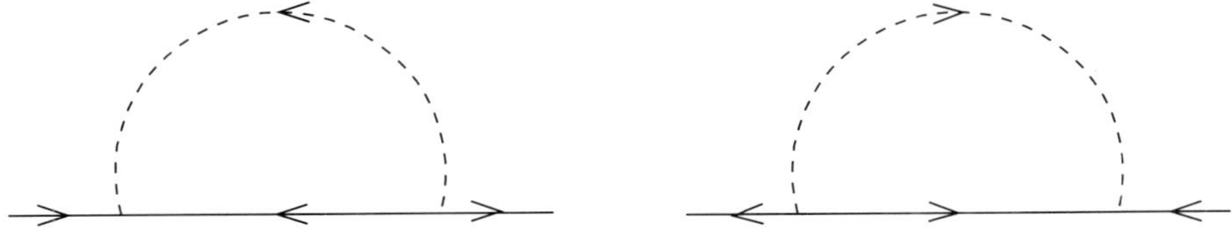

(ii) the argument was given in terms of the complex field ϕ because this field is reminiscent, in its interactions, of the chirality of the spinor field associated. We could have argued in terms of the real fields A and B:

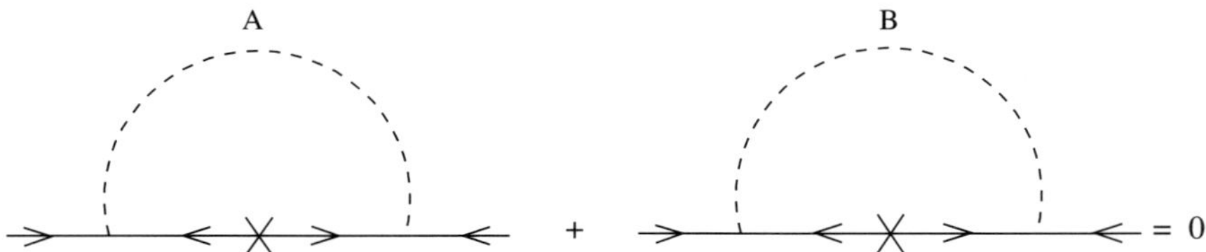

The $i^2 = -1$ in the pseudoscalar contribution plays the rôle of the minus sign in the fermion loop, which is central in the cancellation of quadratic divergences. We check here the importance of the fact that supersymmetry involves a *complex* scalar field $((A + iB)/\sqrt{2})$.

1.3.3   Soft breaking of supersymmetry

Nonrenormalization theorems are characteristic of global supersymmetry. But nature is obviously nonsupersymmetric and if supersymmetry be, it must be broken. However, for what concerns us here – the cancellation of quadratic divergences, supersymmetry is not strictly speaking necessary.

Let us imagine for example that, in the Wess–Zumino model, we modify the mass squared of scalar A by δm_A^2 or B by δm_B^2:

$$\delta\mathcal{L}_{\mathrm{SB}} = -\tfrac{1}{2}\delta m_A^2\, A^2 - \tfrac{1}{2}\delta m_B^2\, B^2. \tag{1.36}$$

This is obviously not compatible with supersymmetry, as can be seen from the mass spectrum. The scalar contributions become

$$3\lambda^2 \int \frac{d^4 k}{(2\pi)^4} \frac{1}{k^2 - m^2 - \delta m_A^2}$$

$$= 3\lambda^2 \int \frac{d^4 k}{(2\pi)^4} \frac{1}{k^2 - m^2}\left[1 + \frac{\delta m_A^2}{k^2 - m^2}\right] + \text{finite}$$

$$\lambda^2 \int \frac{d^4 k}{(2\pi)^4} \frac{1}{k^2 - m^2 - \delta m_B^2}$$

$$= \lambda^2 \int \frac{d^4 k}{(2\pi)^4} \frac{1}{k^2 - m^2}\left[1 + \frac{\delta m_B^2}{k^2 - m^2}\right] + \text{finite}$$

and thus the cancellation of quadratic divergences (but not of logarithmic divergences) follows without problem. One represents generically these terms breaking supersymmetry by

$$\delta\mathcal{L}_{\text{SB}} = -\delta M^2 \phi^* \phi - \delta M'^2 \left(\phi^2 + \phi^{*2}\right)$$
$$= -\tfrac{1}{2}\left(\delta M^2 + 2\delta M'^2\right) A^2 - \tfrac{1}{2}\left(\delta M^2 - 2\delta M'^2\right) B^2. \tag{1.37}$$

Let us note that $\delta M'^2$ destroys the symmetry between A and B (hence as well the cancellations between A and B just described).

On the other hand, a mass term for fermions

$$\delta\mathcal{L} = -\tfrac{1}{2}\delta m \,\bar{\Psi}\Psi \tag{1.38}$$

generates quadratic divergences:

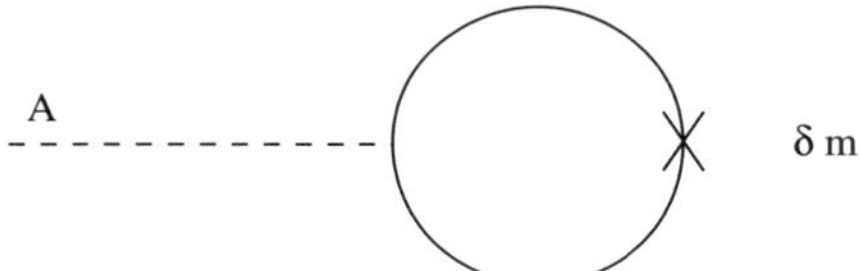

The reason is that, in a supersymmetric theory, there is a cancellation between the following tadpole diagrams (see Exercise 2):

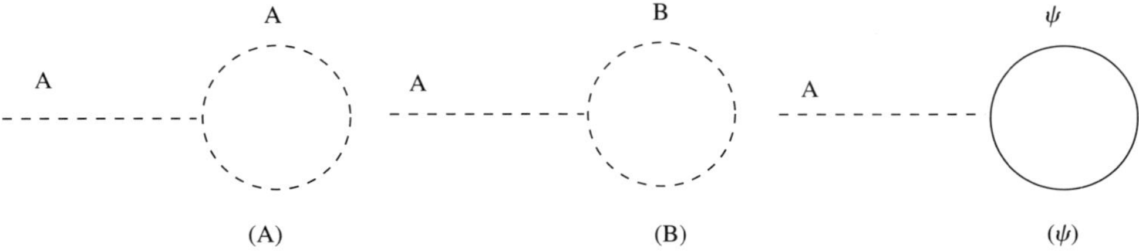

This cancellation is destroyed by the mass term above. It might appear surprising that one may shift scalar but not fermion masses. But one should not forget that it is the relation between fermion masses and other couplings (for example of $A(A^2 + B^2)$) in (1.35) which is responsible for the supersymmetric cancellation of quadratic divergences.

Another possibility is a modification of the couplings of the form:

$$\delta\mathcal{L}_{\text{SB}} = \mathcal{A}\left(\phi^3 + \phi^{*3}\right) = \frac{\mathcal{A}}{\sqrt{2}}\left(A^3 - 3AB^2\right), \tag{1.39}$$

known as the A-term. Indeed such a term provides only cubic scalar interactions which cannot lead to quadratically divergent two-point functions. And the combination between A^3 and AB^2 is such that the contributions to tadpole diagrams (A) and (B) above cancel. On the other hand, a term such as

$$\delta\mathcal{L} = \rho\, A^3 \tag{1.40}$$

would lead to quadratic divergences.

The only other possibility lies in the gauge sector, if present: it is a mass term for the gauginos, supersymmetric partners of the gauge fields,

$$\delta\mathcal{L}_{\mathrm{SB}} = -\tfrac{1}{2} M_\lambda\, \bar\lambda\lambda. \tag{1.41}$$

As we will see in Chapter 5, one can show that mass terms for scalars and pseudoscalars, A-terms, and mass terms for gauginos represent the only possible terms breaking supersymmetry without generating quadratic divergences. One expresses this property by saying that, under these conditions, supersymmetry is *softly* broken. The terms just discussed are the only ones which may be induced through a soft breaking of supersymmetry. A mass term for fermions corresponds on the other hand to a hard breaking of supersymmetry.

Further reading

- G. 't Hooft in *Recent Developments in Gauge Theories*, Cargèse 1979, NATO ASI series B, Vol. 59, Plenum Press, p. 135–157.

Exercises

Exercise 1 Majorana and Dirac masses.

The purpose of this first exercise is to familiarize the reader with the notions of Majorana, Weyl, and Dirac spinor. This is done here on the example of the neutrino, as in Section 1.1.1, but will prove to be useful in the rest of the book when we deal with supersymmetric particles.

Consider a spinor Ψ. One defines its chirality eigenstates

$$\Psi_L = \frac{1-\gamma_5}{2}\Psi, \quad \Psi_R = \frac{1+\gamma_5}{2}\Psi, \tag{1.42}$$

with $\gamma_5 = \begin{pmatrix} -I & 0 \\ 0 & I \end{pmatrix}$ and its charge conjugate as

$$\Psi^c \equiv C\overline{\Psi}^T$$

where C satisfies $C\gamma_\mu^T C^{-1} = -\gamma_\mu$. In the representation chosen, $C^T = C^\dagger = -C$ and $CC^\dagger = C^\dagger C = 1$.

(a) Show that

$$\left(\Psi^c\right)_L (x) = C\overline{\left(\Psi_R\right)}^T (x),$$

$$\left(\Psi^c\right)_R (x) = C\overline{\left(\Psi_L\right)}^T (x).$$

(b) Consider a Weyl neutrino ν_L of left-handed chirality and its charge conjugate ν_R^c of opposite chirality. The following mass term

$$\mathcal{L}_m = -\tfrac{1}{2}m_M\,\bar{\nu}_L\,\nu_R^c - \tfrac{1}{2}m_M\,\bar{\nu}_R^c\,\nu_L,$$

is called a Majorana mass term. Write it in terms of the single field $\nu_M \equiv \nu_L + \nu_R^c$, which is called a Majorana neutrino. Can such a term arise in the Standard Model from gauge symmetry breaking (by a Higgs doublet)?

(c) One introduces a neutral lepton described by a Weyl spinor N_R of right-handed chirality. What are its quantum numbers under $SU(2) \times U(1)$? Introduce the Dirac neutrino $\nu \equiv \nu_L + N_R$. Rewrite the Dirac mass term

$$\mathcal{L}_m = -\tfrac{1}{2}m_D\,\bar{\nu}\nu$$

in terms of ν_L and N_R. Can it be obtained by gauge symmetry breaking (with a Higgs doublet) in the context of the Standard Model?

Hints:

(a) $C\gamma_5^T C^{-1} = \gamma_5$, $\gamma_5 = \gamma_5^T = \gamma_5^\dagger$, hence $\overline{\Psi}_R^{\,T} = \frac{1-\gamma_5}{2}\overline{\Psi}^{\,T}$ and $C\overline{\Psi}_R^{\,T} = \frac{1-\gamma_5}{2}C\overline{\Psi}^{\,T}$.

(b) $\mathcal{L}_m = -\tfrac{1}{2}m_M\,\bar{\nu}_M\,\nu_M$, $m_M \sim \langle\phi\rangle$ if ϕ is an isovector ($t_3 = 1/2$ for ν_L as well as $\bar{\nu}_R^c$).

(c) N_R is a gauge singlet; $\mathcal{L}_m = -\tfrac{1}{2}m_D\,\bar{\nu}_L\,N_R + \text{h.c.}$; $m_D \sim \langle\phi\rangle$ if ϕ is an isodoublet (as in the Standard Model).

<u>Exercise 2</u> Show that there is a cancellation between the tadpole diagrams (A), (B) and (Ψ) in the last section.

Hints:

(A) $\dfrac{3m\lambda}{\sqrt{2}} \displaystyle\int \dfrac{d^4k}{(2\pi)^4}\, \dfrac{1}{k^2 - m^2}$

(B) $\dfrac{m\lambda}{\sqrt{2}} \displaystyle\int \dfrac{d^4k}{(2\pi)^4}\, \dfrac{1}{k^2 - m^2}$

(Ψ) $(-)\dfrac{4m\lambda}{\sqrt{2}} \displaystyle\int \dfrac{d^4k}{(2\pi)^4}\, \dfrac{1}{k^2 - m^2}.$

2

The singular rôle of supersymmetry

This chapter presents basic notions about supersymmetry and supersymmetry breaking. In particular, it underlines the way supersymmetry is a unique symmetry among spacetime symmetries.

2.1 Why supersymmetry?

Supersymmetry is a symmetry between boson and fermion fields. Why should we try to search for such a symmetry? One very fundamental reason is the following.

Bosons are the mediators of interaction: their statistics allows for a coherent superposition and thus for a macroscopic force, such as the Coulomb force. On the other hand, fermions are the constituents of matter: their statistics is translated at the macroscopic level into the additive character of matter. It is certainly a very fundamental question to ask whether there exists a symmetry which unifies matter and radiation.

The educated reader already knows that the present models of supersymmetry applied to particle physics do not fulfill this goal. Should this be taken as an indication that they are not yet complete? But let us put this aside and try to identify the general properties of a symmetry between bosons and fermions.

First, the generators of such a symmetry must carry a spinorial index, since they correspond to the transformation of an integer spin field into a spinor field. They are thus not commuting with Lorentz transformations. This is much in contrast with the behavior of generators of any internal symmetry, such as $SU(3)$. In this sense, supersymmetry is necessarily a spacetime symmetry.

The first attempts were faced with a no-go theorem due to Coleman and Mandula [80] which we could rephrase in a non-technical way as:
The only conserved charges which transform as tensors under the Lorentz group are:

- P_μ, *the generators of translations;*
- $M_{\mu\nu}$, *the generators of Lorentz transformations.*

We recall (see Section A.1 of Appendix Appendix A) that the charges obtained in the Noether procedure may be interpreted as the generators of the transformations they are associated with. The Coleman–Mandula theorem implies that the charges associated with symmetries besides those of Poincaré group are necessarily Lorentz scalars.

Let us illustrate the physical limitations behind this no-go theorem on an example due to E. Witten [374].

We consider two free scalar fields of vanishing mass ϕ_1 and ϕ_2. They are described by the Lagrangian

$$\mathcal{L} = \tfrac{1}{2}\partial^\mu \phi_1 \partial_\mu \phi_1 + \tfrac{1}{2}\partial^\mu \phi_2 \partial_\mu \phi_2 \tag{2.1}$$

and the corresponding equation of motion

$$\Box \phi_1 = 0, \qquad \Box \phi_2 = 0, \tag{2.2}$$

implies the conservation $\partial^\mu J_\mu = 0$ of the current

$$J_\mu = \phi_1 \partial_\mu \phi_2 - \phi_2 \partial_\mu \phi_1. \tag{2.3}$$

But since one also has $\Box \partial^\alpha \phi_{1,2} = 0 = \Box \partial^\alpha \partial^\beta \phi_{1,2}$, the currents

$$J_\mu^\rho = \partial^\rho \phi_1 \partial_\mu \phi_2 - \phi_2 \partial_\mu \partial^\rho \phi_1$$
$$J_\mu^{\rho\sigma} = \partial^\rho \partial^\sigma \phi_1 \partial_\mu \phi_2 - \phi_2 \partial_\mu \partial^\rho \partial^\sigma \phi_1 \tag{2.4}$$

are also conserved:

$$\partial^\mu J_\mu^\rho = 0, \qquad \partial^\mu J_\mu^{\rho\sigma} = 0. \tag{2.5}$$

Thus, following the Noether procedure, the charges

$$Q^\rho = \int d^3x\, J_0^\rho, \qquad Q^{\rho\sigma} = \int d^3x\, J_0^{\rho\sigma} \tag{2.6}$$

are conserved.

This does not mean much because ϕ_1 and ϕ_2 are free fields. To the question "can one write an interaction term between ϕ_1 and ϕ_2 such that Q^ρ and $Q^{\rho\sigma}$ remain conserved?", Coleman and Mandula answer "no". Indeed, following Witten, let us suppose that it is possible. Then the value of the traceless part of $Q^{\rho\sigma}$ in a one (massless) particle state $|p\rangle$ of momentum p is

$$\langle p|Q^{\rho\sigma}|p\rangle = A\left(p^\rho p^\sigma - \tfrac{1}{4}p^2 g^{\rho\sigma}\right) \tag{2.7}$$

where we have imposed the tracelessness condition $g_{\rho\sigma}Q^{\rho\sigma} = 0$.

If we consider an interaction between two such massless particles on-shell: $|p_1\, p_2\rangle \rightarrow |p_3\, p_4\rangle$. the tensorial charge in the initial state is $A(p_1^\rho p_1^\sigma + p_2^\rho p_2^\sigma)$ and in the final state $A(p_3^\rho p_3^\sigma + p_4^\rho p_4^\sigma)$. Thus, if the charge is conserved in interactions, one obtains the condition:

$$p_1^\rho p_1^\sigma + p_2^\rho p_2^\sigma = p_3^\rho p_3^\sigma + p_4^\rho p_4^\sigma. \tag{2.8}$$

This yields, in the center of mass of the reaction ($\mathbf{p_1} = -\mathbf{p_2}, \mathbf{p_3} = -\mathbf{p_4}$, $E_1 = E_2 = E_3 = E_4$), $(\mathbf{p_1}^2)^2 = (\mathbf{p_1} \cdot \mathbf{p_3})^2$ or

$$E_1^4(1 - \cos^2\theta) = 0. \tag{2.9}$$

The scattering angle θ takes specific values: $\theta = 0$ or π. This is only compatible with the analyticity of the scattering amplitudes if the particles did not interact in the first place ($\theta = 0$).

This example, which may be generalized to particles with spin [239], illustrates a very general fact. In a two-particle collision, conservation of P_μ and $M_{\mu\nu}$ leaves a scattering angle undetermined. If there existed another tensorial charge, scattering would only be possible for discrete values of the angle, in contradiction with observation, as well as the principles of quantum field theory (themselves based on observation!).

The obstruction seems rather general. But the hypotheses of Coleman and Mandula were too restrictive. More specifically, they considered the following algebra

$$
\begin{aligned}
&[P_\mu, P_\nu] = 0 \\
&[P_\mu, M_{\rho\sigma}] = i\,(\eta_{\mu\rho}\, P_\sigma - \eta_{\mu\sigma}\, P_\rho) \\
&[M_{\mu\nu}, M_{\rho\sigma}] = i\,(\eta_{\nu\rho}\, M_{\mu\sigma} - \eta_{\nu\sigma}\, M_{\mu\rho} - \eta_{\mu\rho}\, M_{\nu\sigma} + \eta_{\mu\sigma}\, M_{\nu\rho}) \\
&[T^a, T^b] = i\, C^{abc}\, T^c
\end{aligned}
\tag{2.10}
$$

where the last line corresponds to a Lie algebra associated with the internal symmetry. They conclude that the generators T_a are Lorentz scalars:

$$
[T^a, P_\mu] = [T^a, M_{\mu\nu}] = 0.
\tag{2.11}
$$

But we are looking here for a symmetry whose charges/generators are operators acting on the Hilbert space which replace a fermion field by a boson field. They can be written as

$$
Q = \sum_{ij} \int d^3k \int d^3k'\; f_{ij}(k, k') a^\dagger_{B_j}(k') a_{F_i}(k)
\tag{2.12}
$$

where $a_{F_i}(k)$ annihilates a fermion of type i and momentum k and $a^\dagger_{B_j}(k')$ creates a boson of type j and momentum k'. Due to the (anti)commutation rules of bosons and fermions, Q must have anticommutation relations[1].

This is what allows us to evade the conclusions of Coleman and Mandula. In order to see that, let us consider a second example with a complex scalar field ϕ and a fermion of left chirality Ψ_L. The free field Lagrangian is

$$
\mathcal{L} = \partial^\mu \phi^* \partial_\mu \phi + \bar{\Psi}_L\, i\gamma^\mu \partial_\mu \Psi_L.
\tag{2.13}
$$

One easily checks, using the equations of motion

$$
\Box \phi = 0, \qquad \gamma^\mu \partial_\mu \Psi_L = 0,
\tag{2.14}
$$

that the current

$$
J_{r\mu} = \partial^\rho \phi^* \left(\gamma_\rho \gamma_\mu \Psi_L\right)_r
\tag{2.15}
$$

---

[1] In mathematical terms, we must consider *graded* Lie algebras instead of the usual Lie algebras whose definitions only involve commutators.

is conserved. Indeed, using the Dirac equation,

$$\partial^\mu J_{r\mu} = (\partial^\mu \partial^\rho \phi^*)\,(\gamma_\rho \gamma_\mu \Psi_L)_r$$
$$= (\partial^\mu \partial^\rho \phi^*)\,\left(\tfrac{1}{2}\{\gamma_\rho, \gamma_\mu\}\Psi_L\right)_r = \Box\phi^*\,\Psi_{L\,r} = 0. \tag{2.16}$$

Thus $Q_r = \int d^3x\, J_{r0}$ is conserved.

The surprise is that it is now possible to write an interaction term which leaves intact the conservation of Q_r. Indeed, consider

$$\mathcal{L}_{int} = -\lambda^2 |\phi|^4 - \lambda \left(\phi\,\overline{\Psi^c}_R \Psi_L + \phi^* \overline{\Psi}_L \Psi^c_R\right). \tag{2.17}$$

Then $J_{r\mu}$ receives an extra contribution

$$J_{r\mu} = \partial^\rho \phi^*\,(\gamma_\rho \gamma_\mu \Psi_L)_r + i\lambda \phi^{*2}\,(\gamma_\mu \Psi^c_R)_r \tag{2.18}$$

but is still conserved. And thus Q_r conservation follows.

Although the Coleman–Mandula no-go theorem can be evaded, it imposes severe constraints on the possible ways out.

For example, the anticommutator $\{Q_r, Q_s\}$ is an operator which is a symmetric combination of two spin $1/2$ objects. It is thus of bosonic type and obeys commutation relations. Moreover, being a symmetric combination of two Lorentz spinors, it is a vector under Lorentz transformations ($[(1/2) \otimes (1/2)]_s = (1)$). Coleman–Mandula restrictions then apply to it and impose it to be proportional to the generator of translations P_μ.

One may push this analysis [215] further and show (see for example [342]) that such constraints basically fix the algebra of fermionic generators[2]. In this sense, supersymmetry is unique.

2.2   The supersymmetry algebra

Let us write the supersymmetry charge Q_r which, by convention, we choose to be a Majorana spinor[3]. The basic supersymmetry algebra reads

$$\{Q_r, \bar{Q}_s\} = 2\gamma^\mu_{rs}\, P_\mu \qquad \text{(I)}$$

$$[Q_r, P^\mu] = 0 \qquad \text{(II)}$$

$$[Q_r, M^{\mu\nu}] = i\sigma^{\mu\nu}_{rs}\, Q_s \qquad \text{(III)}$$

---

[2] [We will return to this in Chapter 4.]

[3] As stressed in Appendix B, a Majorana spinor has four real components in the Majorana basis where the gamma matrices are purely imaginary. Useful relations involving Majorana spinors may be found in equations (B.40) and (B.41) of this Appendix.

where

$$\bar{Q}_r = \left(Q^T \gamma^0\right)_r \tag{2.19}$$

and

$$\sigma^{\mu\nu} = \tfrac{1}{4}[\gamma^\mu, \gamma^\nu]. \tag{2.20}$$

The relation (III) simply means that Q_r transforms as a spinor in spacetime rotations.

Before going further let us note that one may define N such supersymmetry charges: Q^i, $i = 1, \ldots, N$. In this case, (I) is replaced by:

$$\{Q_r^i, \bar{Q}_s^j\} = 2\delta^{ij}\, \gamma_{rs}^\mu\, P_\mu. \tag{2.21}$$

We will return to this more general case in Chapter 4, and show that interesting new terms may appear.

The supersymmetry algebra has several important consequences. First from (II) one infers

$$[Q_r, P^\mu P_\mu] = 0. \tag{2.22}$$

Let us then consider two states $|b\rangle$ and $|f\rangle$ of respective mass m_b and m_f which are obtained from one another by supersymmetry:

$$Q_r |b\rangle = |f\rangle.$$

Then since $P^\mu P_\mu |b\rangle = m_b^2 |b\rangle$ and $P^\mu P_\mu |f\rangle = m_f^2 |f\rangle$,

$$\begin{aligned}
P^\mu P_\mu\, Q_r |b\rangle = P^\mu P_\mu |f\rangle &= m_f^2 |f\rangle \\
= Q_r\, P^\mu P_\mu |b\rangle &= m_b^2 Q_r |b\rangle = m_b^2 |f\rangle,
\end{aligned}$$

hence $m_b = m_f$.

If supersymmetry is explicit in the particle spectrum, bosons and fermions which are supersymmetric partners must have equal mass. However, we do not observe in nature a boson field with the same mass as the electron. We conclude that, if the Hamiltonian describing the fundamental interactions is supersymmetric, supersymmetry is not explicit in the spectrum: it must be spontaneously broken.

Secondly, we may obtain from (I) an expression for the Hamiltonian of the system. Indeed, (I) reads

$$\{Q_r, Q_t\}\gamma_{ts}^0 = 2\gamma_{rs}^\mu\, P_\mu.$$

Contracting with γ_{sr}^0, one obtains

$$\sum_{r,t} \{Q_r, Q_t\}\left(\gamma^{02}\right)_{tr} = 2\, Tr\left(\gamma^0 \gamma^\mu\right) P_\mu.$$

Using $\gamma^{02} = \mathbb{1}$ and $\mathrm{Tr}(\gamma^0\gamma^\mu) = 4g^{0\mu}$, one obtains

$$\sum_r Q_r^2 = 4P^0 = 4P_0$$

thus

$$H = \frac{1}{4}\sum_r Q_r^2. \tag{2.23}$$

In a (globally) supersymmetric theory, the energy of the states is positive. This has some important consequences for supersymmetry breaking as we now see.

2.3 Supersymmetry breaking

Let $|\Omega\rangle$ be the vacuum of the theory. If supersymmetry is a symmetry of the vacuum, that is if there is no spontaneous breaking of supersymmetry,

$$Q_r|\Omega\rangle = 0. \tag{2.24}$$

In other words, the vacuum is invariant under e^{iQ}. Then

$$H|\Omega\rangle = \frac{1}{4}\sum_r Q_r^2|\Omega\rangle = 0$$

and the energy of the vacuum vanishes.

Reciprocally, if the vacuum energy vanishes,

$$\langle\Omega|H|\Omega\rangle = 0 = \frac{1}{4}\sum_r \langle\Omega|Q_rQ_r|\Omega\rangle = \frac{1}{4}\sum_r \parallel Q_r|\Omega\rangle\parallel^2$$

and

$$Q_r|\Omega\rangle = 0,$$

i.e. supersymmetry is a symmetry of the vacuum.

We thus have proven the following statement: supersymmetry is spontaneously broken (*i.e.* it is not a symmetry of the vacuum) if and only if

$$\langle\Omega|H|\Omega\rangle \neq 0. \tag{2.25}$$

Thus *vacuum energy is the order parameter for global supersymmetry breaking*.

The statement obviously differs from the corresponding one for gauge symmetry breaking. Indeed let us consider a field ϕ, which is nonsinglet under a gauge group G, and its potential $V(\phi)$. We denote by $\langle\phi\rangle$ the value of ϕ at the minimum of $V(\phi)$. Being space and time independent, it minimizes the kinetic energy (which vanishes) and by definition the potential energy. Thus, the vacuum energy is given by $V(\langle\phi\rangle)$.

(i) If $\langle\phi\rangle = 0$ and $V(\langle\phi\rangle) = 0$, neither gauge symmetry, nor supersymmetry is broken.

(ii) If $\langle\phi\rangle = 0$ and $V(\langle\phi\rangle) \neq 0$, supersymmetry is spontaneously broken but not gauge symmetry.

(iii) If $\langle\phi\rangle \neq 0$ and $V(\langle\phi\rangle) = 0$, on the contrary gauge symmetry is spontaneously broken but not supersymmetry.

(iv) If $\langle\phi\rangle \neq 0$ and $V(\langle\phi\rangle) \neq 0$, both supersymmetry and gauge symmetry are spontaneously broken.

An important consequence of having vacuum energy as the order parameter of supersymmetry breaking is that supersymmetry may be broken in a finite volume. Indeed, we will discuss in the next section supersymmetry breaking in quantum mechanical systems, *i.e.* systems with only "time" dependence. This puts again supersymmetry on a fundamentally different status from an ordinary continuous symmetry. It is well-known that there is in the latter case no possibility of spontaneous symmetry breaking in a finite volume. Through superposition, the groundstate can always be made invariant under the symmetry because mixing between states is always possible in finite volume. It is only in the infinite volume limit that one may define unmixed ground states with spontaneously broken symmetry.

If supersymmetry is a symmetry of the vacuum for any finite value of the volume, then, obviously (the limit of zero being zero), in the infinite volume limit the vacuum energy vanishes and supersymmetry is conserved. The converse is not true however: it may well be that supersymmetry is spontaneously broken for any finite value of the volume and restored in the infinite volume limit.

Let us also stress that the criterion (2.25) is a criterion of spontaneous breaking of *global* symmetry. When we make supersymmetry local, we need to include gravity. Indeed, the commutation relation (I) imposes to include "local" translations, which are nothing but the local reparametrizations of spacetime which play a central rôle in general relativity. The local version of supersymmetry is called supergravity, and the inclusion of gravity modifies the criterion for spontaneous supersymmetry breaking. In fact, we will see in Chapter 6 that local supersymmetry may be spontaneously broken even when $\langle\Omega|H|\Omega\rangle = 0$. This is welcome news since $\langle\Omega|H|\Omega\rangle$ is interpreted as the cosmological constant which is observed to be either zero or very small.

Let us return to the pairing of boson–fermion states discussed in the previous section. In the supersymmetric case, all states are paired except for the vacuum state (since $Q_r|\Omega\rangle = 0$). Once supersymmetry is spontaneously broken, the supersymmetry algebra remains valid but $Q_r|\Omega\rangle \neq 0$. We conclude that all states should be paired. What is then the fermionic state associated with the bosonic state $|b\rangle$ since there is no longer a supersymmetric partner $|f\rangle$? It turns out that, as in any spontaneous breaking of a global symmetry, there appears in the spectrum a massless Goldstone excitation. Because of the fermionic nature of supersymmetry, it is a Goldstone fermion, usually referred to as the Goldstino. This massless fermion, which we note $\tilde{G}$, can be created from the vacuum by the supersymmetry current:

$$\langle\Omega|J_{r\mu}|\tilde{G}_s\rangle \propto (\gamma_\mu)_{rs}. \tag{2.26}$$

It is described by a Majorana spinor.

Then, any bosonic state $|b\rangle$ is paired to (*i.e.* has the same energy as) the fermionic state $|b\tilde{G}\rangle$ (since $\tilde{G}$ is massless). Since $\tilde{G}$ has two helicity states $\tilde{G}\uparrow$ and $\tilde{G}\downarrow$, the number of bosonic and fermionic zero modes does not seem to match. It is only because we are missing the bosonic state $|b\tilde{G}\uparrow\tilde{G}\downarrow\rangle$ which completes the energy level.

For example, the vacuum state $|\Omega\rangle$ is now paired with the bosonic state $|\tilde{G}\uparrow\tilde{G}\downarrow\rangle$ and the fermionic states $|\tilde{G}\uparrow\rangle$ and $|\tilde{G}\downarrow\rangle$. The rôle of this massless Goldstino field is central in the discussions of supersymmetry breaking.

2.4 Supersymmetric quantum mechanics

We turn, in this section, to a topic which may seem to lie somewhat out of our main track. It provides, however, a nice illustration of the notions introduced in the previous sections and it has a wide range of applications. Moreover, some of the methods developed here will be used to discuss later the central issue of supersymmetry breaking.

It proves to be useful to advocate the methods of supersymmetry to study one-dimensional potential problems in quantum mechanics. Moreover, since finite volume supersymmetric systems may be assimilated to quantum mechanical systems with a finite number of degrees of freedom, such a study will allow us to discuss the issue of supersymmetry breaking in finite volume. Before being more explicit, let us review a few facts of one-dimensional quantum mechanics.

In one-dimensional potential problems, there is a deep connection between the bound state wave function and the potential. Consider the ground state (of zero energy) wave function $\psi_0(x)$ of the Schrödinger equation with potential $V_1(x)$

$$H_1\psi_0(x) = -\frac{\hbar^2}{2m}\,\partial_x^2\,\psi_0(x) + V_1(x)\,\psi_0(x) = 0. \tag{2.27}$$

Then obviously

$$V_1(x) = \frac{\hbar^2}{2m}\,\frac{\psi_0''(x)}{\psi_0(x)} \tag{2.28}$$

and the potential can be reconstructed if the wave function has no node. One can in fact factorize the Hamiltonian in the following form $H_1 = A^+A$ with

$$A = \frac{p}{\sqrt{2m}} + iw(x), \qquad A^+ = \frac{p}{\sqrt{2m}} - iw(x) \tag{2.29}$$

where $p = i\hbar\partial_x$, and the identification

$$V_1(x) = w^2(x) - \frac{\hbar}{\sqrt{2m}}\,w'(x). \tag{2.30}$$

The "superpotential" $w(x)$ is given in terms of the ground state wave function as

$$w(x) = -\frac{\hbar}{\sqrt{2m}}\,\frac{\psi_0'(x)}{\psi_0(x)}. \tag{2.31}$$

In other words, the ground state wave function satisfies $A\psi_0(x) = 0$, since this implies $H_1\psi_0(x) = 0$.

There is obviously a companion Hamiltonian $H_2 = AA^+$:

$$H_2 = -\frac{\hbar^2}{2m}\,\partial_x^2 + V_2(x), \tag{2.32}$$

with

$$V_2(x) = w^2(x) + \frac{\hbar}{\sqrt{2m}}\,w'(x). \tag{2.33}$$

Note that the energy eigenstates of the two systems are related. If

$$H_1|\psi_n^{(1)}\rangle = E_n^{(1)}\,|\psi_n^{(1)}\rangle$$

then

$$H_2 A|\psi_n^{(1)}\rangle = AH_1|\psi_n^{(1)}\rangle = E_n^{(1)}\,A|\psi_n^{(1)}\rangle.$$

Hence, apart from the ground state for which $A|\psi_0\rangle = 0$, the energy eigenstates of the two systems are in one-to-one correspondence (see Fig. 2.1).

One may wonder why we end up with a dissymmetry when H_1 and H_2 seem to play symmetric roles. The reason is that the zero energy ground states of H_1 and H_2 satisfy, respectively, $A|\psi_0^{(1)}\rangle = 0$ and $A^+|\psi_0^{(2)}\rangle = 0$ and thus can be written as

$$\psi_0^{(1)}(x) = N^{(1)}\exp\left[-\frac{\sqrt{2m}}{\hbar}\int_0^x w(y)dy\right], \tag{2.34}$$

$$\psi_0^{(2)}(x) = N^{(2)}\exp\left[+\frac{\sqrt{2m}}{\hbar}\int_0^x w(y)dy\right] \tag{2.35}$$

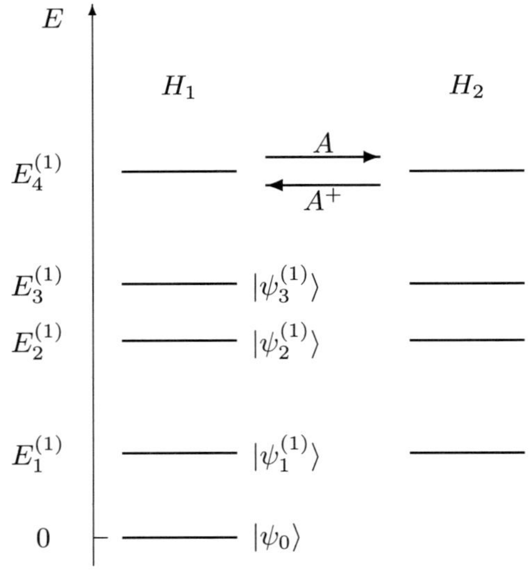

Fig. 2.1 Energy spectrum of the two companion Hamiltonians H_1 and H_2, respectively (2.27) and (2.32).

where $N^{(1)}$, $N^{(2)}$ are normalization constants. Obviously, only one of them (if any) is a normalizable state. We made the assumption that it is $\psi_0^{(1)}(x)$ and we labelled it simply $\psi_0(x)$. We will return later to the case where none of these states is normalizable.

Let us now make explicit the presence of the supersymmetry algebra. Define

$$Q_1 = \begin{pmatrix} 0 & A^+ \\ A & 0 \end{pmatrix} = \frac{p}{\sqrt{2m}}\sigma^1 + w(x)\sigma^2,$$

$$Q_2 = \begin{pmatrix} 0 & -iA^+ \\ iA & 0 \end{pmatrix} = \frac{p}{\sqrt{2m}}\sigma^2 - w(x)\sigma^1. \tag{2.36}$$

One may easily check the algebra

$$\{Q_1, Q_1\} = \{Q_2, Q_2\} = 2H, \quad \{Q_1, Q_2\} = 0, \quad [Q_1, H] = 0 = [Q_2, H], \tag{2.37}$$

with

$$H = \begin{pmatrix} A^+A & 0 \\ 0 & AA^+ \end{pmatrix} = \begin{pmatrix} H_1 & 0 \\ 0 & H_2 \end{pmatrix}$$

$$= \frac{p^2}{2m} + w^2(x) - \sigma^3 \frac{\hbar}{\sqrt{2m}}w'(x) \tag{2.38}$$

One recognizes the supersymmetry algebra, this time with a single dimension (time).

The Hamiltonian (2.38) is found for example in the following one-dimensional physical system: an electron moving along the z-axis in a x-dependent magnetic field aligned along the z-axis. Indeed, the corresponding Hamiltonian reads:

$$H = \frac{(\mathbf{p} - e\mathbf{A})^2}{2m} + \frac{ie}{2m}\text{div }\mathbf{A} + \frac{|e|\hbar}{2m}\sigma.\mathbf{B}, \tag{2.39}$$

with

$$p_x = p_y = 0, p_z = p,$$

$$A_x = A_z = 0, A_y = -\sqrt{\frac{2m}{e^2}}w(x). \tag{2.40}$$

The state of the system is described by a two-component spinor $\Psi = \begin{pmatrix} \alpha \\ \beta \end{pmatrix}$, $\begin{pmatrix} 1 \\ 0 \end{pmatrix}$ corresponding to spin up and $\begin{pmatrix} 0 \\ 1 \end{pmatrix}$ to spin down along the z-axis. Since $[\sigma^3, H] = 0$, the spin is conserved in interactions. Note that in the one-dimensional models that we consider, the notion of spin is purely artificial (there is no rotation group). In fact, fermions may be bosonized.

Since the Q's obey anticommutation relations, one may interpret them, as well as A and A^+, as fermion type operators. We may call the eigenstates of H_1, which include the ground state $|\psi_0\rangle$, the bosonic states of the system. The eigenstates of H_2 are obtained from the previous ones by the action of A and are thus interpreted as the fermion degrees of freedom. The ground state of the total system is described by

$$\Psi_0(x) = \begin{pmatrix} \psi_0(x) \\ 0 \end{pmatrix}.$$

We can go further and write a Lagrangian formulation of the system. There is a single spacetime variable – the time t – and the Lagrangian reads [78, 328]:

$$L = \frac{1}{2} m \left(\frac{dx(t)}{dt} \right)^2 - [W'(x)]^2 + i\hbar\psi^\dagger \frac{d\psi}{dt} - \frac{\hbar}{\sqrt{2m}} W''(x) \left[\psi^\dagger, \psi \right]. \tag{2.41}$$

We have introduced a fermion variable ψ in L because, in this form, L is invariant under a supersymmetry transformation which relates x and ψ (see Exercise 2). We need an explicit matrix representation of the fermion variable ψ which realizes the standard anticommutation relations:

$$\{\psi(t), \psi(t)\} = \{\psi^\dagger(t), \psi^\dagger(t)\} = 0, \quad \{\psi(t), \psi^\dagger(t)\} = 1. \tag{2.42}$$

One may simply choose

$$\psi(t) = \begin{pmatrix} 0 & 1 \\ 0 & 0 \end{pmatrix}, \quad \psi^\dagger(t) = \begin{pmatrix} 0 & 0 \\ 1 & 0 \end{pmatrix}. \tag{2.43}$$

The corresponding Hamiltonian reads

$$H = \frac{p^2}{2m} + [W'(x)]^2 + \frac{\hbar}{\sqrt{2m}} W''(x) \, [\psi^\dagger, \psi]. \tag{2.44}$$

which coincides with (2.38) if one uses $[\psi^\dagger, \psi] = -\sigma^3$ and one makes the identification[4]:

$$w(x) = W'(x). \tag{2.45}$$

It is interesting to note that the fermion number operator

$$N_F = \psi^\dagger\psi = \begin{pmatrix} 0 & 0 \\ 0 & 1 \end{pmatrix} = \frac{1}{2} \left(1 - \sigma^3 \right) \tag{2.46}$$

is conserved through interactions, as can be checked using the ψ equation of motion. Once again, one finds that states of the form $\begin{pmatrix} \alpha \\ 0 \end{pmatrix}$ are of the boson type whereas states of the form $\begin{pmatrix} 0 \\ \beta \end{pmatrix}$ are of the fermion type. For future reference, we note that $\sigma^3 = (-)^{N_F}$.

Let us now use such models to discuss the issue of spontaneous supersymmetry breaking, following the seminal paper by [371]. We consider systems described by the Hamiltonian (2.38). We assume that $|w(x)| \to \infty$ when $|x| \to \infty$ in order to have a discrete spectrum.

---

[4]Note that, by analogy of (2.41) with equation (1.33) of Chapter 1, what we should call the superpotential in this section is the function $W(x)$, not $w(x)$.

In the weak coupling limit $\hbar \to 0$, $V = w^2$. Hence the number of supersymmetric vacua is the number of zeros of w. Let us then consider corrections of order $\hbar$. Assuming that w has a simple zero at $x = x_0$, we write

$$w(x) = \sqrt{\frac{m}{2}}\,\omega(x - x_0) + O\left[(x - x_0)^2\right].$$
(2.47)

Then the Hamiltonian (2.38) reads

$$H = \frac{1}{2m}p^2 + \frac{1}{2}m\omega^2(x - x_0)^2 - \hbar\frac{\omega}{2}\sigma^3.$$
(2.48)

One recognizes in the first two terms (bosonic in the sense of the Hamiltonian (2.44)) the Hamiltonian of a harmonic oscillator with frequency ω: the zero point energy is $\hbar\omega/2$. The third term has eigenvalues $\mp\hbar\omega/2$. The ground state thus corresponds to the eigenvalue $\sigma^3 = +1$ (a boson state) and has vanishing energy[5]: $\hbar\omega/2 - \hbar\omega/2 = 0$. One may check that the cancellation persists to higher orders in $(x - x_0)$. One can also note that the fermion state ($\sigma^3 = -1$) has energy $\hbar\omega/2 + \hbar\omega/2 = \hbar\omega$.

Let us now consider the exact spectrum of the theory and not rely on a perturbative analysis. A supersymmetric ground state satisfies:

$$Q_1|\psi\rangle = Q_2|\psi\rangle = 0.$$
(2.49)

Since $H = Q_1^2 = Q_2^2$, it suffices that $Q_1|\psi\rangle = 0$. Multiplying the first of equations (2.36) by σ^1, one obtains

$$\frac{\hbar}{\sqrt{2m}}\frac{\partial\psi}{\partial x} = -w(x)\sigma^3\psi.$$
(2.50)

The solution of this first order differential equation is

$$\psi(x) = \exp\left[-\int_0^x dy\,\frac{\sqrt{2m}}{\hbar}w(y)\sigma^3\right]\psi(0).$$
(2.51)

For this to be the ground state of the theory with quantized energy levels, it has to be normalizable, which imposes that $w(x)$ has opposite signs when $x \to +\infty$ and $x \to -\infty$. An equivalent statement is to require that w has an odd number of zeroes.

Let us take for example $w(x) = \lambda x^3$. Then $\int_0^x dy\, w(y) = \lambda x^4/4$, and the choice $\sigma^3\psi(0) = +\,\psi(0)$ corresponds to a normalizable wave function. Note that this corresponds to the spectrum described earlier, with bosonic degrees of freedom (including the vacuum) associated with $\sigma^3 = +1$.

On the other hand, if w has an even number of zeros, then it is not possible to write a supersymmetric vacuum which is a normalizable state: supersymmetry is spontaneously broken. If w has no zero, the perturbative analysis already told us so. But if w has zeros (in even number) this contradicts the perturbative analysis above: supersymmetry is broken at a nonperturbative level. One speaks of *dynamical breaking of supersymmetry*. We will study in more detail this type of breaking in Chapter 8.

---

[5] This thus results from a cancellation between the bosonic terms and the fermionic terms in (2.44).

2.5   Witten index

It is a remarkable fact that the issue of spontaneous supersymmetry breaking was settled in the preceding section by discussing the behavior of a function at infinity or the number of its zeroes. Such a property is, to some extent, stable under continuous deformations of the parameters of the theory: it is, in some sense, topological. This has been formalized by E. Witten [373] and connected with higher mathematics – the [14, 15] index formula. Let us have a closer look at this from the physics point of view.

The problem is whether one can find a supersymmetric ground state, *i.e.* a normalizable state of vanishing energy. This question can be settled in a finite volume: if the energy of the state vanishes for all finite values of the volume, it vanishes as well in the infinite volume limit. Since finite volume field theory amounts to quantum mechanics with a finite number of degrees of freedom, the framework described in the preceding section is sufficient to address the problem.

For example, the spectrum of the theory has necessarily – if there is a supersymmetric ground state – the form shown in Fig. 2.1. Apart from the ground state, bosonic and fermionic levels are paired. If we deform continuously the parameters of the theory, levels move up or down, they might coalesce, reach zero energy or on the contrary emerge from zero energy, but they remain paired. Thus the difference between the number of bosons n_B and the number of fermions n_F at a given level remains unchanged. This is of particular importance at the zero energy level where we expect to find the ground state. Witten thus defines the index (a topological notion) as:

$$I = n_B^{(E=0)} - n_F^{(E=0)}. \tag{2.52}$$

Introducing the operator

$$\Gamma_F = (-)^{N_F} \tag{2.53}$$

which commutes with the Hamiltonian and is equal to $+1$ for a boson and -1 for a fermion, we have

$$I = \mathrm{Tr}\Gamma_F = \mathrm{Tr}(-)^{N_F} \tag{2.54}$$

where Tr means that we sum over all states. Clearly, if $I \neq 0$, there exists one or more states with vanishing energy: supersymmetry is a symmetry of the vacuum. If one is looking for spontaneous symmetry breaking, one should thus restrict one's attention to theories with vanishing Witten index. However, $I = 0$ is obviously only a necessary condition.

Let us take a few examples for the sake of illustration. First, consider as "superpotential" $w(x)$ a monotonic function such that $w(x_0) = 0$. Then, we can deform the parameters of the theory in order to have

$$w(x) = \lambda(x - x_0). \tag{2.55}$$

If $\lambda > 0$, then an analysis similar to the one of the preceding section that led to (2.51) shows that the corresponding state is a boson ($\sigma^3|\psi\rangle = +|\psi\rangle$). If $\lambda < 0$, it is a fermion ($\sigma^3|\psi\rangle = -|\psi\rangle$). In any case, $I \neq 0$ and there is no chance to find the spontaneous breaking of supersymmetry. On the other hand, if $w(x) = \lambda(x-x_1)(x-x_2)$ with $x_1 \neq x_2$, then we find one boson state (by analogy with the preceding case, this corresponds to the zero where w is increasing) and one fermion state (corresponding to the zero where w is decreasing). Thus $I = 0$: one may have spontaneous supersymmetry breaking. We see that I is directly related to the parity of the number of zeroes of $w(x)$ (or just as well to the comparative sign between $w(+\infty)$ and $w(-\infty)$).

As another example, we may consider the Wess–Zumino model discussed in Chapter 1. To compute the Witten index, one may use perturbation theory since one is allowed to change continuously the parameters. The potential of the model is given by equation (1.35) in Chapter 1:

$$V = |m\phi + \lambda\phi^2|^2. \tag{2.56}$$

Assuming m large and λ small in order to stay within perturbation theory, we have two ground states $\phi_0 = 0$ and $\phi_0 = -m/\lambda$. On the other hand, we have seen that all fields (bosons and fermions) are massive, of mass m. We thus have two bosonic states of vanishing energy, the two vacua, and $I = 2$. This remains true when one reaches the nonperturbative regime as well as when m goes to zero[6].

The previous example should make clear which deformations of the theory are permissible. If we add a term of order ϕ^6 to the potential (2.56), we obviously change the asymptotic behavior of the energy and there is no reason for the Witten index to remain unchanged: states may be brought in or sent out to infinity. Thus, the Witten index is only invariant under changes which do not modify the asymptotic (large field) regime: adding a new coupling or setting an existing one to zero may require some caution.

We end this section by showing how the Witten index can be formally computed using functional methods. One needs to introduce a regulator β and write the regulated Witten index as

$$I(\beta) = \mathrm{Tr}(-)^{N_F} e^{-\beta H/\hbar}. \tag{2.57}$$

We will let the regulator β go to zero at the end of the computation.

If we consider the general system (2.38) with the two companion Hamiltonians H_1 and H_2, then we have:

$$I(\beta) = \mathrm{Tr}\left(e^{-\beta H_1/\hbar} - e^{-\beta H_2/\hbar}\right). \tag{2.58}$$

This may be written explicitly as

$$
\begin{aligned}
I(\beta) &= \mathrm{Tr}\ \sigma^3 \int \frac{[dpdx]}{2\pi\hbar} e^{-\beta[p^2/(2m)+w^2-\sigma^3\hbar w'/\sqrt{2m}]/\hbar} \\
&= \int \frac{[dpdx]}{\pi\hbar} e^{-\beta[p^2/(2m)+w^2]/\hbar} \sinh\left[\frac{\beta}{\sqrt{2m}}w'\right] \\
&\sim \int \frac{[dpdx]}{\pi\hbar} e^{-\beta[p^2/(2m)+w^2]/\hbar} \frac{\beta}{\sqrt{2m}}w',
\end{aligned} \tag{2.59}
$$

where, in the last line, we have used an expansion in β since we will send it to zero.

If we take $w(x) = \lambda x^{2n+1}$, we obtain $I = 1$ whereas for $w(x) = \lambda x^{2n}$ (n integer), the integrand is odd under $x \to -x$ and $I = 0$. We recover that it is only for an odd number of zeros (here coinciding) that the Witten index vanishes, which allows for the possibility of spontaneous supersymmetry breaking.

---

[6]In this limit, the fermion becomes massless. However, because $I \neq 0$, this is not a Goldstino field.

Further reading

- E. Witten, *Introduction to Supersymmetry*, in Proceedings of the International School of Subnuclear Physics, Erice 1981, ed. by A. Zichichi, Plenum Press, 1983.
- F. Cooper, A. Khare and U. Sukhatme, *Supersymmetry and quantum mechanics*, Phys. Rep. C 251 (1995) 267–385.

Exercises

Exercise 1

(a) Show that, for every spinor field Ψ,

$$\left(\overline{\Psi^c}_R \Psi_L\right) \Psi_L = 0.$$

(b) Write the equations of motion for the theory described by the Lagrangians (2.13) and (2.17).

(c) Deduce from these equations of motion that the current $J_{r\mu}$ defined in (2.18) is conserved.

Hints:

(a) Ψ_L is expressed in terms of two independent spinor components, Ψ^c_R is expressed in terms of the same components; hence the left-hand side is a cubic monomial of only two independent spinor components: two are necessarily identical and their product vanishes (because spinor components anticommute: $\Psi_r \Psi_r = -\Psi_r \Psi_r = 0$).

(b)
$$\Box \phi^* = -2\lambda^2 \phi^{*2}\phi - \lambda \overline{\Psi^c}_R \Psi_L,$$
$$i\gamma^\mu \partial_\mu \Psi_L = 2\lambda \phi^* \Psi^c_R,$$
$$i\gamma^\mu \partial_\mu \Psi^c_R = 2\lambda \phi \Psi_L.$$

(c) Straightforward when using (a).

Exercise 2 Using the Euler equations for $x(t)$ and $\psi(t)$ derived from the Lagrangian L in (2.41), show that L is invariant (up to a total derivative) under the supersymmetry transformations:

$$\delta x = \varepsilon^\dagger \psi + \psi^\dagger \varepsilon,$$

$$\delta \psi = \varepsilon \left(\alpha \frac{dx}{dt} + \beta W(x)\right), \tag{2.60}$$

$$\delta \psi^\dagger = \varepsilon^\dagger \left(\alpha^* \frac{dx}{dt} + \beta^* W(x)\right), \tag{2.61}$$

where ε is a spinor, parameter of the supersymmetry transformation, and α, β are complex numbers.

Hints: The equations of motion read:

$$m\frac{d^2x}{dt^2} + 2W'W'' + \frac{\hbar}{\sqrt{2m}}W''''[\psi^\dagger,\psi] = 0,$$

$$-i\hbar\frac{d\psi}{dt} + 2\frac{\hbar}{\sqrt{2m}}W''\psi = 0,$$

$$-i\hbar\frac{d\psi^\dagger}{dt} - 2\frac{\hbar}{\sqrt{2m}}W''\psi^\dagger = 0.$$

Using these equations,

$$\delta L = \frac{d}{dt}\left[m\frac{dx}{dt}\left(\varepsilon^\dagger\psi + \psi^\dagger\varepsilon\right) + i\hbar\left(a\frac{dx}{dt} + \beta W\right)\psi^\dagger\varepsilon\right]. \tag{2.62}$$

Compare the transformation (2.60) in one dimension with the supersymmetry transformations in four dimensions, e.g. (3.26) of Chapter 3 with $F = -(dW/d\phi)^*$.

<u>Exercise 3</u> Consider a quantum mechanical model with "superpotential" $w(x) = \lambda(x^2 + a^2)$. For $a^2 > 0$, supersymmetry is broken at tree level since, at the classical level $(\hbar \to 0)$, $V = w^2(x) > 0$. For $a^2 < 0$, w has an even number of zeros and the arguments of Section 2.4 show that supersymmetry is spontaneously broken. The purpose of this exercise is to show that there exists a change of parameters which interpolates between the two situations.

(a) Define $Q_\pm = (Q_1 \pm iQ_2)/2$. Express $Q_\pm$ in terms of A and A^+. Show that

$$Q_+^2 = Q_-^2 = 0$$
$$Q_+Q_- + Q_-Q_+ = H. \tag{2.63}$$

(b) Explain why the zero energy eigenstates are the states $|\chi\rangle$ such that $Q_+|\chi\rangle = 0$ but $|\chi\rangle \neq Q_+|\psi\rangle$ for any $|\psi\rangle$.

(c) Determine $P_\pm$ such that
$$\tilde{Q}_\pm = e^{P_\pm x}Q_\pm e^{-P_\pm x}$$

correspond to the "superpotential" $\tilde{w}(x) = \lambda(x^2 - a^2)$.

(d) Using (c) and the criterion defined in (b), conclude that, if H corresponding to $w(x) = \lambda(x^2 + a^2)$ has no zero-energy eigenstates, then $\tilde{H}$ corresponding to $\tilde{w}(x) = \lambda(x^2 - a^2)$ has no zero-energy eigenstates.

Hints: See [373], Section 3.

<u>Exercise 4</u> Show that the cancellation of the ground state energy still holds when one considers terms of order $(x - x_0)^2$ in (2.47).

Hints: See [339], equation (14).

<u>Exercise 5</u> *We have studied in Section 2.4 only cases where the spectrum is discrete. Supersymmetric quantum mechanical methods prove to be useful also with continuum spectra. We illustrate this by studying the relations that exist between the reflection and transmission coefficients of the two companion Hamiltonians H_1 and H_2.*

In order to have a continuum spectrum, we must assume that the potential remains finite as $x \to -\infty$ or $x \to +\infty$. We thus define $w(x \to \pm\infty) \equiv w_\pm$.

(a) One considers an incident plane wave e^{ikx} of energy E coming from $-\infty$. The scattering on the potentials $V_{1,2}$ induces reflection and transmission with respective coefficients $R_{1,2}$ and $T_{1,2}$. One thus has:

$$\psi_{1,2}(k, x \to -\infty) \to e^{ikx} + R_{1,2} e^{-ikx},$$
$$\psi_{1,2}(k', x \to +\infty) \to T_{1,2} e^{ik'x}. \tag{2.64}$$

Writing the "supersymmetry" relations between the wave functions of the two Hamiltonian, show that

$$R_1 = \left(\frac{w_- + ik}{w_- - ik} \right) R_2 \ , \quad T_1 = \left(\frac{w_+ - ik'}{w_- - ik} \right) T_2 \tag{2.65}$$

and express k and k' in terms of E, w_+ and w_-.

(b) Deduce that the companion potentials have identical reflection and transmission probabilities. In which case is $T_1 = T_2$?

(c) Infer that, if one of the potentials is constant, the other one is necessarily reflectionless. Using $w(x) = \mu \tanh \alpha x$, show that the potential $V(x) = \lambda \, \mathrm{sech}^2 \alpha x$ is reflectionless.

Hint: See Cooper *et al.*, (1995), p. 278ff.

Exercise 6 The methods of supersymmetric quantum mechanics are useful to discuss the problem of localization of four-dimensional gravity in models with extra dimensions. In the five-dimensional model of Randall and Sundrum [319], the Kaluza–Klein modes of the five-dimensional graviton (the mediator of five-dimensional gravity) obey a standard Schrödinger equation (2.27):

$$H\psi = -\partial_x^2 \psi(x) + V(x)\psi(x),$$

where the coordinate x parametrizes the fifth dimension (a factor $\hbar^2/(2m)$ has been absorbed) and

$$V(x) = \frac{15}{4(|x| + R)^2} - \frac{3}{R} \delta(x), \tag{2.66}$$

where R is a constant (R is called the AdS_5 curvature radius in this model).

(a) The potential $V(x)$ is called the volcano potential. Draw it to understand this name.

(b) Identify the function $w(x)$ corresponding to this potential.

(c) Deduce from it the form of the ground state wave function (Kaluza–Klein zero mode of the five-dimensional graviton). Show that it is localized around $x = 0$. This zero mode is identified as the four-dimensional graviton field: four-dimensional gravity is thus localized around $x = 0$ in the fifth dimension.

Hints: (b) $w(x) = (3/2) (2\theta(x) - 1)1/(|x| + R)$; (c) $\psi_0(x) \sim 1/(|x| + R)^{3/2}$.

3
Basic supermultiplets

We present here, in a hopefully nontechnical fashion, the two basic supermultiplets: (spin 0/spin $\frac{1}{2}$) and (spin 1/spin $\frac{1}{2}$). The former involves a complex scalar field and a spinor field chosen to be real (Majorana) or to have a given helicity (Weyl). It is called a **chiral supermultiplet**. The latter involves a real vector field and a Majorana spinor field. It is called a **vector supermultiplet**. If it is associated with a gauge symmetry, the vector is a gauge field and its supersymmetric partner is called a gaugino. We will review the form of the supersymmetry transformations for both multiplets and insist on the need for auxiliary fields. Moreover, in both cases, we will discuss the issue of supersymmetry breaking.

This chapter is self-contained. The more theoretically oriented reader may, however, find it useful to read in parallel Appendix C where the use of superspace allows for a more systematic approach. Appendix C uses two-component spinors and it might be a useful exercise to translate its results back into four-component spinors as used in this chapter.

3.1 Chiral supermultiplet

3.1.1 A first look

We consider here, as in the Wess–Zumino model of Chapter 1, a chiral supermultiplet which consists of:

- a complex scalar field $\phi(x) = (A(x) + iB(x))/\sqrt{2}$;
- a Majorana spinor field $\Psi(x)$.

Obviously, the scalar field consists of two real degrees of freedom: $A(x)$ and $B(x)$. On the other hand, the Majorana spinor has four real components. Its equation of motion – the Dirac equation – fixes two of these degrees of freedom. To see this, consider for simplicity a planar wave of energy $k^0 > 0$ described by the spinor $u(k)$ which satisfies

$$(\not{k} - m)\ u(k) = 0. \tag{3.1}$$

In the center of mass, $k_\mu = (m, 0, 0, 0)$ and (3.1) becomes

$$\left(\gamma^0 - 1\right)\ u(k) = 0.$$

In the Dirac representation $\gamma^0 - 1 = \begin{pmatrix} 0 & 0 \\ 0 & -2 \end{pmatrix}$. We are thus left with two degrees of freedom.

Thus, the number of bosonic and fermionic degrees of freedom of the chiral supermultiplet seem to be only equal on-shell (that is, using the equations of motion).

Let us now consider the free field Lagrangian

$$\mathcal{L} = \partial^\mu \phi^* \partial_\mu \phi - m^2 \phi^* \phi + \frac{1}{2} \bar\Psi \left(i\gamma^\mu \partial_\mu - m \right) \Psi. \tag{3.2}$$

It is invariant under the infinitesimal supersymmetry transformation

$$\delta A = \bar\varepsilon \Psi, \quad \delta B = i\bar\varepsilon \gamma_5 \Psi \tag{3.3}$$

$$\delta \Psi_r = - \left[i\gamma^\mu \partial_\mu \left(A + iB\gamma_5 \right) + m \left(A + iB\gamma_5 \right) \right]_{rs} \varepsilon_s \tag{3.4}$$

where ε_s is the (Majorana) spinor parameter of the transformation[1].

There is, however, an undesirable problem with the transformations (3.3–3.4). According to the supersymmetry algebra (I), one has

$$\left[\bar\varepsilon_1 Q, \bar\varepsilon_2 Q \right] = \left[\bar\varepsilon_1 Q, \bar Q \varepsilon_2 \right] = \bar\varepsilon_{1r} \, \varepsilon_{2s} \left\{ Q_r, \bar Q_s \right\}$$
$$= 2 \left(\bar\varepsilon_1 \gamma_\mu \varepsilon_2 \right) P^\mu \tag{3.5}$$

where we have used the relation $\bar\varepsilon_2 Q = \bar Q \varepsilon_2$ (see Appendix B equation (B.40)) valid for Majorana spinors. The commutators of two supersymmetry transformations of parameters ε_1 and ε_2 is a translation of vector $a_\mu = 2(\bar\varepsilon_1 \gamma_\mu \varepsilon_2)$ and one should have

$$\left[\delta_1, \delta_2 \right] \Psi_r = 2 \left(\bar\varepsilon_1 \gamma_\mu \varepsilon_2 \right) \left(i\partial^\mu \Psi_r \right) \tag{3.6}$$

whereas the transformations (3.3)–(3.4) yield

$$\left[\delta_1, \delta_2 \right] \Psi_r = 2i \left(\bar\varepsilon_1 \gamma_\mu \varepsilon_2 \right) \left[\partial^\mu \Psi + \frac{i}{2} \gamma^\mu \left(i\gamma^\nu \partial_\nu - m \right) \Psi \right]_r. \tag{3.7}$$

This coincides with (3.6) only when we make use of the Dirac equation. In other words, the algebra of supersymmetry only closes on-shell in this formulation.

3.1.2   Auxiliary fields

Luckily, by introducing auxiliary fields, J. Wess and B. Zumino (1974b) have provided us with a formulation of supersymmetry where the algebra closes off-shell. Following them, let us add a complex scalar field $F(x) = (F_1(x) + iF_2(x))/\sqrt{2}$ and consider the Lagrangian

$$\mathcal{L} = \partial^\mu \phi^* \partial_\mu \phi + \frac{1}{2} \bar\Psi \left(i\gamma^\mu \partial_\mu - m \right) \Psi + \mathcal{L}_{\text{aux}}$$
$$\mathcal{L}_{\text{aux}} = F^* F + m \left(F\phi + F^* \phi^* \right). \tag{3.8}$$

Since F has no kinetic term, there is no dynamical degree of freedom associated with it. One may solve for it using its equation of motion:

$$F = -m\phi^*. \tag{3.9}$$

Hence $\mathcal{L}_{\text{aux}} = -m^2 \phi^* \phi$ and one recovers the Lagrangian (3.2): the theory is identical to the previous one on-shell (that is, making use of the equations of motion).

---

[1] Let us note here the canonical dimension of ε. Since scalar fields (A, B) have canonical dimension 1 and spin $1/2$ fields (Ψ) have canonical dimension $3/2$, ε has dimension $-1/2$.

But, in this formulation, the supersymmetry transformations read:

$$\delta_S A = \bar{\varepsilon}\Psi, \quad \delta_S B = i\bar{\varepsilon}\gamma_5\Psi$$

$$\delta_S \Psi = \left[-i\gamma^\mu \partial_\mu \left(A + iB\gamma_5\right) + F_1 - iF_2\gamma_5\right]\varepsilon,$$

$$\delta_S F_1 = -i\bar{\varepsilon}\gamma^\mu \partial_\mu \Psi, \quad \delta_S F_2 = -\bar{\varepsilon}\gamma_5\gamma^\mu \partial_\mu \Psi. \tag{3.10}$$

Note that the first two coincide with (3.3)–(3.4) when one makes use of the equation of motion (3.9). The novel feature is that one recovers (3.6) (and similar relations for ϕ and F), **without** making use of any equation of motion: in this formulation the supersymmetry algebra closes off-shell.

It is not so surprising that we now have an off-shell formulation of supersymmetry: with the introduction of auxiliary fields, the number of off-shell bosonic degrees of freedom (four: A, B, F_1, F_2) equals the number of off-shell fermionic degrees of freedom (four: Ψ_r). We will see that this generalizes to all types of multiplets.

The supersymmetry transformations (3.10) call for several comments.

Let us first look for signs of spontaneous supersymmetry breaking. Since supersymmetry is global (*i.e.* the parameter ε is constant), we should look for a Goldstone field. As recalled in Section A.2.1 of Appendix Appendix A, in the case of a continuous symmetry which is spontaneously broken through the vacuum expectation value $v \neq 0$ of a scalar field, a constant term in the transformation law, *i.e.*

$$\delta\phi = \alpha v + \cdots \tag{3.11}$$

where α is the parameter of the transformation, characterizes ϕ as a Goldstone boson. In our case, supersymmetry is of a fermionic nature and, as discussed in Chapter 2, we expect the Goldstone field to be a fermion, the Goldstino, namely Ψ in our example. A look at (3.10) shows that a vacuum value for ϕ (*i.e.* A or B) does not generate a constant term in the transformation law of Ψ whereas a constant value for the auxiliary field does. For example, if $\langle F_1 \rangle \neq 0$ (this is obviously not the case in the free field theory that we consider here), then

$$\delta\psi = \langle F_1 \rangle \varepsilon + \cdots \tag{3.12}$$

Thus a non-vanishing ground state value for the auxiliary field is a signature of spontaneous supersymmetry breakdown.

If a continuous bosonic symmetry is local, the finite version of (3.11) shows that one can choose the parameter $\alpha(x)$ locally in such a way that it cancels the field $\phi(x)$: this field disappears from the spectrum. This is the Higgs mechanism. We will see in Chapter 6 that a similar phenomenon, the super-Higgs mechanism, occurs when supersymmetry is local: the Goldstone fermion disappears from the spectrum.

Finally, one should note that, since we are dealing with global supersymmetry, both δF_1 and δF_2 are total derivatives. This is a characteristic property of auxiliary fields, which will allow us to construct easily invariant action terms.

3.1.3   Interacting theory

There is no difficulty in adding interactions. As we have seen earlier, this is precisely what makes supersymmetry special. Indeed, the Lagrangian

$$\mathcal{L} = \partial^\mu \phi^* \partial_\mu \phi + \tfrac{1}{2}\bar{\Psi}\left(i\gamma^\mu\partial_\mu - m\right)\Psi - \lambda\left(\phi\bar{\Psi}_R\Psi_L + \phi^*\bar{\Psi}_L\Psi_R\right)$$
$$+ F^*F + F\left(m\phi + \lambda\phi^2\right) + F^*\left(m\phi^* + \lambda\phi^{*2}\right) \tag{3.13}$$

is invariant under the supersymmetry transformations (3.10).

Solving for F,

$$F = -\left(m\phi^* + \lambda\phi^{*2}\right) \tag{3.14}$$

yields

$$\mathcal{L} = \partial^\mu \phi^* \partial_\mu \phi + \tfrac{1}{2}\bar{\Psi}\left(i\gamma^\mu\partial_\mu - m\right)\Psi - \lambda\left(\phi\bar{\Psi}_R\Psi_L + \phi^*\bar{\Psi}_L\psi_R\right) - V(\phi) \tag{3.15}$$

with

$$V(\phi) = \left|m\phi + \lambda\phi^2\right|^2 . \tag{3.16}$$

We may introduce the function

$$W(\phi) = \tfrac{1}{2}m\phi^2 + \tfrac{1}{3}\lambda\phi^3 \tag{3.17}$$

which is analytic in the field ϕ and is called the **superpotential**. All interaction terms involve the superpotential and its derivatives

$$\frac{dW}{d\phi} = m\phi + \lambda\phi^2, \qquad \frac{d^2W}{d\phi^2} = m + 2\lambda\phi. \tag{3.18}$$

Let us write for the sake of completeness the Lagrangian describing n such super-multiplets (ϕ_i, ψ_i, $i = 1,\ldots,n$) with a general superpotential $W(\phi_i)$, analytic in the fields ϕ_i:

$$\mathcal{L} = \sum_i \partial^\mu \phi_i^* \partial_\mu \phi_i + \frac{1}{2}\sum_i \bar{\Psi}_i i\gamma^\mu\partial_\mu\Psi_i$$

$$-\frac{1}{2}\sum_{ij}\left[\frac{\partial^2 W}{\partial\phi_i\partial\phi_j}\bar{\Psi}_{iR}\Psi_{jL} + \frac{\partial^2 W^*}{\partial\phi_i^*\partial\phi_j^*}\bar{\Psi}_{iL}\Psi_{jR}\right]$$

$$+\sum_i\left[F_i^*F_i + F_i\frac{\partial W}{\partial\phi_i} + F_i^*\frac{\partial W^*}{\partial\phi_i^*}\right]. \tag{3.19}$$

Solving for F_i yields the scalar potential

$$V(\phi_i) = \sum_i\left|\frac{\partial W}{\partial\phi_i}\right|^2 = \sum_i |F_i|^2 . \tag{3.20}$$

Let us stress that the analyticity of the superpotential W (it depends on the fields ϕ_i but not on ϕ_i^*) is a crucial ingredient for supersymmetry. As for supersymmetry breaking, we see that supersymmetry is broken spontaneously if and only if $\langle \partial W/\partial\phi_i \rangle \neq 0$, that is

$$\langle F_i \rangle \neq 0. \tag{3.21}$$

It is this criterion – an auxiliary field acquires a non-zero vacuum expectation value (vev) – which will remain the basic one when we move to local supersymmetry. Indeed, as discussed above, it is associated with the presence of a Goldstone fermion, and is a necessary condition for the super-Higgs mechanism to take place.

3.1.4 An example of F-term spontaneous supersymmetry breaking: O'Raifeartaigh mechanism

It is easy to see that, with a single field ϕ and a polynomial superpotential, it is not possible to break supersymmetry spontaneously: since ϕ is complex, $dW/d\phi = 0$ always has solutions, which are the ground states of the theory and $\langle V \rangle = 0$.

The simplest example of F-term supersymmetry breaking involves three scalar fields and was devised by [300]. Let us indeed consider the superpotential

$$W(A, X, Y) = m\,YA + \lambda X \left(A^2 - M^2 \right) \tag{3.22}$$

where $\lambda > 0$, m and M are real parameters. The corresponding potential is given by (3.20). It reads

$$V = m^2\,|A|^2 \quad + \lambda^2\,|A^2 - M^2|^2 + |mY + 2\lambda AX|^2$$

$$= \quad |F_Y|^2 + \quad |F_X|^2 \quad + \quad |F_A|^2.$$

If $M \neq 0$, one cannot set the three terms separately to zero. Thus $\langle V \rangle \neq 0$ and supersymmetry is spontaneously broken.

Let us see this in more detail. One may always choose $\langle Y \rangle$ and $\langle X \rangle$ in order to set $\langle F_A \rangle = 0$. Indeed, there is a direction in field space ($mY + 2\lambda AX = 0$) where the potential is flat for a fixed value of A: this is the first example that we encounter of a flat direction of the scalar potential, a characteristic of supersymmetric models that will have important phenomenological and cosmological consequences.

Once this is done, one is left with a potential depending solely on A. One finds the following ground states:

- if $M^2 \leq m^2/(2\lambda^2)$, $\quad \langle A \rangle = 0 \quad$ and $\quad \langle V \rangle = \lambda^2 M^4 = F_X^2$
- if $M^2 \geq m^2/(2\lambda^2)$, $\quad \langle A^2 \rangle = M^2 - m^2/(2\lambda^2)$
$$\text{and } \langle V \rangle = m^2 \left(M^2 - m^2/(4\lambda^2) \right).$$

One may check that the spectrum is no longer supersymmetric. Take the first case $\langle A \rangle = 0$; we will take the opportunity of the flat direction to compute the spectrum at $\langle X \rangle = 0$. Note that X and Ψ_X are massless: Ψ_X is the Goldstone fermion associated with the spontaneous breaking ($\langle F_X \rangle \neq 0$). On the other hand, the masses of the (A, Ψ_A) and (Y, Ψ_Y) supermultiplets are no longer supersymmetric: one finds $\sqrt{m^2 - 2\lambda^2 M^2}$ for Re A, $\sqrt{m^2 + 2\lambda^2 M^2}$ for Im A and m for Y. The only nonvanishing fermion mass term mixes Ψ_A and Ψ_Y. The corresponding eigenvectors are $(\Psi_A \pm \Psi_Y)/\sqrt{2}$ with mass $\pm m$ (as in any fermion mass, the sign can be redefined through phase redefinitions).

If one computes the supertrace of squared masses defined as

$$\text{STr } M^2 = \sum_{J=0}^{1/2} (-1)^{2J} (2J + 1) M_J^2 = \sum M_{\text{boson}}^2 - 2 \sum M_{\text{fermion}}^2 \tag{3.23}$$

one finds

$$\text{STr } M^2 = 2m^2 + m^2 - 2\lambda^2 M^2 + m^2 + 2\lambda^2 M^2 - 2(2m^2) = 0. \tag{3.24}$$

This relation is characteristic of F-type breaking. The nonrenormalization theorems discussed in Chapter 1 ensure that it does not receive divergent contributions in higher orders. Such a relation poses however some severe phenomenological problems: it means that the average boson mass squared coincides with the average fermion mass squared whereas limits on supersymmetric particles (mostly scalars) tend to show that, on average, bosons are much heavier than fermions (mostly the quarks and leptons that we know).

[We may check on this model the criterion for supersymmetry breaking based on the use of the Witten index I (see Section 2.5 of Chapter 2). It is easiest to compute I in the perturbative regime where m and λM^2 are finite whereas $\lambda \ll 1$. Since $\lambda M^2 \ll m^2/(2\lambda)$ in this regime, the minimum is at $\langle A \rangle = 0$ and $\langle V \rangle = (\lambda M^2)^2$. This vacuum energy must be added to all states: for example, the fermionic state with one Goldstino Ψ_X has the same energy $\langle V \rangle$ since the Goldstino is massless[2]. There is therefore no state with vanishing energy and the Witten index I is zero, as in any situation of spontaneous supersymmetry breaking.]

3.1.5   Chiral supermultiplet with a Weyl spinor

When we are to apply supersymmetry to the Standard Model of electroweak interactions, we will not encounter Majorana spinors but spinors of definite chirality or Weyl spinors: for example, the right-handed electron e_R has specific quantum numbers, different from those of e_L; this is indeed one of the characteristics of the Standard Model. A Weyl spinor has two on-shell degrees of freedom, just like the Majorana spinor: out of the four degrees of freedom of a Dirac spinor, two are projected out by performing the chirality projection. This is the right number of on-shell fermionic degrees of freedom to match the two degrees of freedom of a complex scalar. Indeed, the free field Lagrangian

$$\mathcal{L} = \partial^\mu \phi^* \partial_\mu \phi + \bar{\Psi}_L i\gamma^\mu \partial_\mu \Psi_L - \frac{1}{2} m \left(\bar{\Psi^c}_R \Psi_L + \bar{\Psi}_L \Psi^c_R \right)$$
$$+ F^* F + m \left(F\phi + F^* \phi^* \right), \tag{3.25}$$

---

[2] As discussed in Section 2.3 of Chapter 2, this fermion state is degenerate with the vacuum and with the bosonic state which consists of two Goldstinos of opposite helicities.

is invariant under the supersymmetry transformation

$$\delta_S \phi = \sqrt{2}\,\bar{\varepsilon}\Psi_L,$$

$$\delta_S \Psi_L = \frac{1-\gamma_5}{2}\,[F - i\gamma^\mu\partial_\mu\phi]\,\varepsilon\sqrt{2},$$

$$\delta_S F = -i\sqrt{2}\,\bar{\varepsilon}\gamma^\mu\partial_\mu\Psi_L \tag{3.26}$$

where ε is the *Majorana* spinor of the transformation.

This supersymmetry transformation is similar to the one found for a Majorana spinor (3.10), except for the chirality projector $L \equiv (1-\gamma_5)/2$ in the transformation law of the fermion field: this ensures that it remains left-handed under a supersymmetry transformation.

By taking the hermitian conjugate of (3.26), one may obtain, using formulas (B.36) of Appendix B,

$$\delta_S \phi^* = \sqrt{2}\,\bar{\varepsilon}\Psi_R^c,$$

$$\delta_S \Psi_R^c = \frac{1+\gamma_5}{2}\,[F^* - i\gamma^\mu\partial_\mu\phi^*]\,\varepsilon\sqrt{2},$$

$$\delta_S F^* = -i\sqrt{2}\,\bar{\varepsilon}\gamma^\mu\partial_\mu\Psi_R^c \tag{3.27}$$

where, as usual, $\Psi_R^c = C\left(\overline{\Psi_L}\right)^T$.

It is then straightforward to show that one may include interactions to our original free field Lagrangian, much in the way of (3.13) but with a special attention to chiralities:

$$\mathcal{L} = \partial^\mu\phi^*\partial_\mu\phi + \bar{\Psi}_L i\gamma^\mu\partial_\mu\Psi_L - \tfrac{1}{2}m\left(\overline{\Psi}_R^c\Psi_L + \bar{\Psi}_L\Psi_R^c\right)$$

$$-\lambda\left(\phi\overline{\Psi}_R^c\Psi_L + \phi^*\bar{\Psi}_L\Psi_R^c\right) + F^*F + F\left(m\phi + \lambda\phi^2\right) + F^*\left(m\phi^* + \lambda\phi^{*2}\right). \tag{3.28}$$

One recognizes the first and second derivatives of the superpotential (3.17). The mass term is a Majorana mass term and (once one solves for F) the interaction term is precisely the one which was given in (2.17) of Chapter 2 when we first discussed supersymmetric interactions as a way of evading Coleman–Mandula no-go theorem.

Now, if we look closely at (3.27), we realize that, just as (ϕ, Ψ_L, F) transformed as (3.26) with a left-handed chirality projector, (ϕ^*, Ψ_R^c, F^*) transform with a right-handed chirality projector. By convention, the first transformation law (3.26) is the one of a chiral supermultiplet, whereas a set of fields $(\check{\phi}, \check{\Psi}_R, \check{F})$ which transforms like (3.27) forms an antichiral supermultiplet:

$$\delta_S \check{\phi} = \sqrt{2}\,\bar{\varepsilon}\check{\Psi}_R,$$

$$\delta_S \check{\Psi}_R = \frac{1+\gamma_5}{2}\,[\check{F} - i\gamma^\mu\partial_\mu\check{\phi}]\,\varepsilon\sqrt{2},$$

$$\delta_S \check{F} = -i\sqrt{2}\,\bar{\varepsilon}\gamma^\mu\partial_\mu\check{\Psi}_R. \tag{3.29}$$

Obviously then, $(\check{\phi}^*,\ \check{\Psi}^c_L,\ \check{F}^*)$ transforms as a chiral multiplet. When we come to the Standard Model, we will use the relation just found to put all the fermion fields into chiral supermultiplets. For example, denoting the complex scalar which is the supersymmetric partner of e_R by $\tilde{e}_R$, we will describe their supersymmetric interactions in terms of $\tilde{e}^*_R$ and e^c_L, members of the associate chiral supermultiplet.

For future use, we may write the general interaction of a set of chiral supermultiplets with Weyl spinor fields $(\phi_i,\ \Psi_{iL},\ F_i)$[3] (to be compared with the Majorana case (3.19)[4]):

$$\mathcal{L} = \sum_i \partial^\mu \phi^{i*} \partial_\mu \phi_i + \sum_i \overline{\Psi}^i_L i\gamma^\mu \partial_\mu \Psi_{iL}$$

$$- \frac{1}{2} \sum_{ij} \left[\frac{\partial^2 W}{\partial \phi_i \partial \phi_j} \overline{\Psi^c}_{iR} \Psi_{jL} + \frac{\partial^2 W^*}{\partial \phi^{*i} \partial \phi^{*j}} \overline{\Psi}^i_L \Psi^{cj}_R \right]$$

$$+ \sum_i \left[F^{*i} F_i + F_i \frac{\partial W}{\partial \phi_i} + F^{*i} \frac{\partial W^*}{\partial \phi^{*i}} \right]. \tag{3.30}$$

Let us take this opportunity to introduce the general notion of an F-term. Since the kinetic terms of the free field theory are invariant under supersymmetry, the terms depending on W should be invariant on their own. Indeed we have, using (3.26) and (3.27) $(\varepsilon = \varepsilon^c)$

$$\delta_S \left[\sum_i F_i \frac{\partial W}{\partial \phi^i} - \frac{1}{2} \sum_{ij} \frac{\partial^2 W}{\partial \phi^i \partial \phi^j} \overline{\Psi^c}_{iR} \Psi_{jL} \right] = \partial_\mu \left[-i\sqrt{2} \sum_i \bar{\varepsilon} \gamma^\mu \Psi_{iL} \frac{\partial W}{\partial \phi_i} \right]. \tag{3.31}$$

This transformation law is actually a consequence of the fact that this combination may be considered as the auxiliary field of a composite chiral supermultiplet[5]. We thus note, for any analytic function W of the fields ϕ_i,

$$[W(\phi)]_F \equiv \sum_i F_i \frac{\partial W}{\partial \phi^i} - \frac{1}{2} \sum_{ij} \frac{\partial^2 W}{\partial \phi^i \partial \phi^j} \overline{\Psi^c}_{iR} \Psi_{jL}. \tag{3.32}$$

We end this section by discussing some subtleties associated with the derivation of the supersymmetric current. We start with the Lagrangian (3.28) which we write, using the notation just introduced,

$$\mathcal{L} = \partial^\mu \phi^* \partial_\mu \phi + \overline{\Psi}_L i\gamma^\mu \partial_\mu \Psi_L + F^* F + [W(\phi)]_F + [W^*(\phi^*)]_F. \tag{3.33}$$

---

[3]In the following equation, we have adopted the convention to put lower index i for fields whose supersymmetry transformation follows the chiral supermultiplet transformation rule (3.26), such as $\phi_i, \Psi_{iL}, \ldots$, and upper index i to fields whose supersymmetry transformation follows the antichiral supermultiplet transformation rule (3.29), such as $\phi^{*i}, \Psi^{ci}_R, \ldots$.

[4]Using (B.34) of Appendix B, one shows that for, for a Majorana spinor $(\Psi = \Psi^c)$, $\overline{\Psi}_R i\gamma^\mu \partial_\mu \Psi_R = \overline{\Psi}_L i\gamma^\mu \partial_\mu \Psi_L$, up to a total derivative. Hence the different normalization of the fermion kinetic term in the two Lagrangians.

[5][See Section C.2.2 of Appendix C for details.]

The Ncether current $J_\mu^{(N)}$ associated with the supersymmetry transformations (3.26) and (3.27) reads

$$\bar{\epsilon}J_\mu^{(N)} = \sum_X \frac{\delta\mathcal{L}}{\delta\partial^\mu X}\,\delta_s X\,, \tag{3.34}$$

$$J_\mu^{(N)} = \sqrt{2}\,\partial_\mu\phi^*\Psi_L + \sqrt{2}\,\partial^\rho\phi\left(g_{\rho\mu} + \gamma_\rho\gamma_\mu\right)\Psi_R^c - i\sqrt{2}F\gamma_\mu\Psi_R^c. \tag{3.35}$$

This is, however, *not* the supersymmetry current J_μ because the supersymmetry transformations are not an invariance of the Lagrangian (3.33). They transform it into a total derivative

$$\delta_s\mathcal{L} = \partial^\mu\left(\bar{\epsilon}K_\mu\right)\,,$$

$$K_\mu = \sqrt{2}\,\partial^\rho\phi^*\left(g_{\rho\mu} - \gamma_\rho\gamma_\mu\right)\Psi_L + \sqrt{2}\,\partial^\mu\phi\Psi_R^c - i\sqrt{2}\,F\gamma_\mu\Psi_R^c$$
$$-i\sqrt{2}\,\gamma^\mu\Psi_L\frac{\partial W}{\partial\phi} - i\sqrt{2}\,\gamma^\mu\Psi_R^c\frac{\partial W^*}{\partial\phi^*}\,, \tag{3.36}$$

where we have used (3.31). The supersymmetry current is thus

$$J_\mu \equiv J_\mu^{(N)} - K_\mu$$
$$= \sqrt{2}\,\partial^\rho\phi\gamma_\rho\gamma_\mu\Psi_R^c + i\sqrt{2}\,\frac{\partial W}{\partial\phi}\gamma^\mu\Psi_L \tag{3.37}$$
$$+\sqrt{2}\,\partial^\rho\phi^*\gamma_\rho\gamma_\mu\Psi_L + i\sqrt{2}\,\frac{\partial W^*}{\partial\phi^*}\gamma^\mu\Psi_R^c.$$

We recognize (up to a factor $\sqrt{2}$) the current constructed in equation (2.18) of Chapter 2.

3.2   Vector supermultiplet and gauge interactions

We will be much briefer in this case since the discussion is parallel to the previous one.

3.2.1   Vector supermultiplet

The off-shell formulation of a vector supermultiplet uses a real vector field A_μ, a Majorana spinor λ and a real auxiliary pseudoscalar field D. This makes $3+1$ bosonic degrees of freedom and 4 fermionic degrees of freedom in the off-shell formulation; on-shell, the auxiliary field is no longer independent and we have two bosonic and two fermionic degrees of freedom. The free field Lagrangian reads:

$$\mathcal{L}_V = -\tfrac{1}{4}F^{\mu\nu}F_{\mu\nu} + \tfrac{1}{2}\bar{\lambda}i\gamma^\mu\partial_\mu\lambda + \tfrac{1}{2}D^2 \tag{3.38}$$

where $F_{\mu\nu} = \partial_\mu A_\nu - \partial_\mu A_\nu$. It is invariant under the abelian gauge transformation:

$$\delta_g A_\mu = -\frac{1}{g}\partial_\mu \theta,$$
$$\delta_g \lambda_r = 0,$$
$$\delta_g D = 0. \tag{3.39}$$

And it is invariant under the supersymmetry transformation[6]

$$\delta_s A_\mu = \bar{\varepsilon}\gamma_\mu \gamma_5 \lambda,$$
$$\delta_s \lambda_r = -D\varepsilon_r + \frac{1}{2}\left(\sigma^{\mu\nu}\gamma_5\varepsilon\right)_r F_{\mu\nu},$$
$$\delta_s D = -i\bar{\varepsilon}\gamma^\mu \gamma_5 \partial_\mu \lambda. \tag{3.40}$$

Once again, if $\langle D \rangle \neq 0$, the supersymmetry transformation of the spinor field includes a constant term, which is the trademark of a Goldstone fermion or Goldstino. Then, the supersymmetry transformation of the auxiliary field D is a total derivative. This means that the following Lagrangian

$$\mathcal{L}_{\mathrm{FI}} = -\xi D \tag{3.41}$$

leads by itself to a supersymmetric term in the action[7]. This is the so-called Fayet-Iliopoulos [145] term. If the vector field is an abelian gauge field, it turns out that this term is gauge invariant and may be added to the Lagrangian (3.38).

Now that we have introduced auxiliary D fields, we may present the concept of a D-term (which somewhat parallels the F-term introduced in the previous section). It turns out[8] that the kinetic terms of a chiral supermultiplet may be understood (up to a total derivative) as the D auxiliary field of a composite supermultiplet which we note $\phi^\dagger \phi$:

$$\left(\phi^\dagger \phi\right)_D = \partial^\mu \phi^* \partial_\mu \phi + \bar{\Psi}_L i\gamma^\mu \partial_\mu \Psi_L + F^* F + \text{ total derivative.} \tag{3.42}$$

This is indeed the reason why this provides a supersymmetric invariant action (as an auxiliary field, it transforms into a total derivative). The vector component (known as the Kähler connection) of this composite supermultiplet is $A_\mu = -i\phi^*\partial_\mu\phi + i\phi\partial_\mu\phi^* - \bar{\Psi}_L\gamma_\mu\Psi_L$. Such a construction can be generalized[9] to any real function $K(\phi^\dagger, \phi)$ (called a Kähler potential). We thus see that full supersymmetric actions may easily be constructed as F components (*cf.* (3.32)) or D components (*cf.* (3.42)) of some composite supermultiplets.

---

[6]The γ_5 in the transformation law of A_μ might seem surprising. However, as can be seen from (B.41) of Appendix B, it ensures the reality of $\delta_s A_\mu$. Following a standard convention [362], one may redefine λ_L ($\hat{\lambda}_L = i\lambda_L$) and λ_R ($\hat{\lambda}_R = -i\lambda_R$) in such a way that $\bar{\varepsilon}\gamma_\mu\gamma_5\lambda = \bar{\varepsilon}\gamma_\mu\lambda_R - \bar{\varepsilon}\gamma_\mu\lambda_L = i\bar{\varepsilon}\gamma_\mu(\hat{\lambda}_R + \hat{\lambda}_L) = i\bar{\varepsilon}\gamma_\mu\hat{\lambda}$. We will refrain from doing so because this introduces spurious i's in gaugino interactions.

[7]A similar term of the form $aF +$ h.c. is in principle possible if the corresponding scalar field ϕ is a gauge singlet. However, as can be seen from (3.19), it corresponds to a linear term in the superpotential which can be absorbed through a redefinition of the field ϕ.

[8][See Section 3.2.2 for details.]

[9][See Section 3.2.4.]

3.2.2 Coupling of a chiral supermultiplet to an abelian gauge supermultiplet

We will consider here the case of a chiral supermultiplet with a Weyl spinor (see Exercise 5 for the case of Majorana spinors). The following Lagrangian may be shown to be invariant under supersymmetry transformations:

$$\mathcal{L} = D^\mu \phi^* D_\mu \phi + \bar{\Psi}_L i\gamma^\mu D_\mu \Psi_L + F^* F$$

$$+ gq\left(D\phi^* \phi + \sqrt{2}\bar{\lambda}\Psi_L\, \phi^* + \sqrt{2}\bar{\lambda}\Psi^c_R\, \phi \right) \tag{3.43}$$

where $D_\mu = \partial_\mu - igqA_\mu$ is the covariant derivative, g is the gauge coupling and q the $U(1)$ charge. More precisely, it is invariant under (3.40) and a variation of (3.10) where one replaces derivatives by covariant derivatives and one completes the auxiliary field transformation:

$$\delta_S \phi = \sqrt{2}\,\bar{\varepsilon}\Psi_L,$$

$$\delta_S \Psi_L = \frac{1 - \gamma_5}{2}\,[F - i\gamma^\mu D_\mu \phi]\,\varepsilon\sqrt{2},$$

$$\delta_S F = -i\sqrt{2}\,\bar{\varepsilon}\gamma^\mu D_\mu \Psi_L - 2gq\phi\,\bar{\varepsilon}\frac{1 + \gamma_5}{2}\lambda. \tag{3.44}$$

If we add $\mathcal{L}_V$ given by (3.38) and possibly $\mathcal{L}_{\mathrm{FI}}$ in (3.41), we have a supersymmetric theory of a chiral supermultiplet coupled with an abelian gauge supermultiplet. The novel feature is that gauge interactions give a contribution to the scalar potential through the D auxiliary field. Indeed solving for D yields

$$D = -gq\phi^*\phi + \xi$$

and the scalar potential reads

$$V(\phi) = \tfrac{1}{2}\left(gq\phi^*\phi - \xi\right)^2 = \tfrac{1}{2}D^2. \tag{3.45}$$

If there are several such chiral supermultiplets with scalar fields ϕ_i of charge q_i, the potential reads:

$$V(\phi_i) = \frac{1}{2}\left(g\sum_i q_i \phi^{*i}\phi_i - \xi\right)^2. \tag{3.46}$$

If we apply this to the abelian $U(1)_Y$ hypercharge symmetry of the Standard Model, then since the charges y_i of the quarks and leptons (and of their scalar supersymmetric partners) have both signs, the minimum of the potential corresponds to zero energy and supersymmetry is not spontaneously broken. One is therefore tempted to introduce another abelian symmetry under which all charges q_i have the sign opposite to the sign of ξ. Obviously, the minimum of the potential corresponds to all scalar fields vanishing and $\langle V \rangle = \xi^2/2$: supersymmetry is spontaneously broken. And, by expanding the potential (3.46), we check that the scalar squared masses are simply $-gq_i\xi > 0$: if ξ is large enough, they can be made much larger than typical quark and lepton masses. However, such an abelian gauge symmetry has potential problems with quantum anomalies[10].

We will see in Section 6.3.2 of Chapter 6 that the phenomenological problems that we have encountered both with F-type breaking and D-type breaking leads one to isolate the supersymmetry-breaking sector from the sector of quarks, leptons and their supersymmetric partners.

3.2.3   Nonabelian gauge symmetries

The previous considerations are easily generalized to nonabelian gauge theories. Let us consider a gauge group G with coupling constant g (G is a simple Lie group) and structure constants C^{abc} (the index a runs over the adjoint representation of the group): the generators of the group satisfy

$$[t^a, t^b] = iC^{abc}t^c. \tag{3.47}$$

The gauge supermultiplet is simply $(A^a_\mu, \lambda^a, D^a)$, where one introduces a gaugino λ^a and a real auxiliary D^a for each gauge vector field A^a_μ. The straightforward generalization of (3.38) reads

$$\mathcal{L}_V = -\tfrac{1}{4}F^{a\mu\nu}F^a_{\mu\nu} + \tfrac{1}{2}\bar{\lambda}^a i\gamma^\mu D_\mu \lambda^a + \tfrac{1}{2}D^a D^a, \tag{3.48}$$

where

$$F^a_{\mu\nu} = \partial_\mu A^a_\nu - \partial_\nu A^a_\mu + gC^{abc}A^b_\mu A^c_\nu$$

$$D_\mu \lambda^a = \partial_\mu \lambda^a + gC^{abc}A^b_\mu \lambda^c \tag{3.49}$$

are, respectively, the covariant field strength and the covariant derivative of the gaugino field[11]. This action is invariant under the supersymmetry transformation (3.40) where a superscript a should be added to all fields.

Likewise, the coupling of chiral supermultiplets to this gauge supermultiplet follows the same lines as before. Consider a set of chiral supermultiplets (ϕ_i, Ψ_i, F_i) transforming under the gauge group G in a representation with hermitian matrices $(t^a)_i{}^j$:

$$\begin{aligned}
\delta_g \phi_i &= -i\alpha^a (t^a)_i{}^j \phi_j, \\
\delta_g \phi^{*i} &= i\alpha^a \phi^{j*} (t^a)_j{}^i,
\end{aligned} \tag{3.51}$$

---

[10]Remember that the triangle anomaly diagram with three abelian gauge bosons is proportional to $\sum_i q_i^3$.

[11]The corresponding infinitesimal gauge transformations of the gauge supermultiplet are:

$$\delta_g A^a_\mu = -\frac{1}{g}\partial_\mu \alpha^a + C^{abc}A^b_\mu \alpha^c,$$

$$\delta_g \lambda^a = C^{abc}\lambda^b \alpha^c,$$

$$\delta_g D^a = C^{abc}D^b \alpha^c. \tag{3.50}$$

and similarly, respectively, for Ψ_{iL}, F_i and Ψ_R^{ci}, F^{*i}. Then, the Lagrangian describing the coupled gauge and chiral supermultiplets reads

$$\mathcal{L} = \mathcal{L}_V + \sum_i D^\mu \phi^{*i} D_\mu \phi_i + \sum_i \overline{\Psi}_L^i i\gamma^\mu D_\mu \Psi_{iL}$$

$$-\frac{1}{2}\sum_{ij}\left[\frac{\partial^2 W}{\partial\phi_i \partial\phi_j}\overline{\Psi^c}_{iR}\Psi_{jL} + \frac{\partial^2 W^*}{\partial\phi^{*i}\partial\phi^{*j}}\overline{\Psi}_L^i \Psi_R^{cj}\right]$$

$$+\sum_i\left[F^{*i}F_i + F_i\frac{\partial W}{\partial\phi_i} + F^{*i}\frac{\partial W^*}{\partial\phi^{*i}}\right]$$

$$+g\sqrt{2}\left[\bar{\phi}^{*i}\bar{\lambda}^a\left(t^a\right)_i{}^j\Psi_{jL} + \overline{\Psi}_L^i\lambda^a\left(t^a\right)_i{}^j\phi_j\right] + gD^a\phi^{*i}\left(t^a\right)_i{}^j\phi_j \qquad (3.52)$$

where $\mathcal{L}_V$ is given in (3.48); the rest of the first three lines corresponds to (3.19) where we have just replaced the ordinary derivatives by covariant derivatives:

$$D_\mu\phi_i = \partial_\mu\phi_i - igA_\mu^a\left(t^a\right)_i{}^j\phi_j$$
$$D_\mu\phi^{*i} = \partial_\mu\phi^{*i} + igA_\mu^a\phi^{*j}\left(t^a\right)_j{}^i$$
$$D_\mu\Psi_{iL} = \partial_\mu\Psi_{iL} - igA_\mu^a\left(t^a\right)_i{}^j\Psi_{jL}$$
$$D_\mu\Psi_R^{ci} = \partial_\mu\Psi_R^{ci} + igA_\mu^a\Psi_R^{cj}\left(t^a\right)_j{}^i. \qquad (3.53)$$

The last line of (3.52) should be compared with (3.43) where the abelian symmetry allows a much lighter (and thus transparent) notation.

One can solve for the auxiliary fields F_i or F^{*i} and D^a to obtain the scalar potential. One obtains

$$V\left(\phi_i, \phi^{*i}\right) = \sum_i F^{*i}F_i + \frac{1}{2}\sum_a \left(D^a\right)^2 + \frac{1}{2}\sum_m \left(D^m\right)^2$$

$$= \sum_i\left|\frac{\partial W}{\partial\phi_i}\right|^2 + \frac{1}{2}\sum_a g_a^2 \left[\phi^{*i}\left(t^a\right)_i{}^j\phi_j\right]^2$$

$$+\frac{1}{2}\sum_m\left[g_m\sum_i q_i^m\phi^{*i}\phi_i - \xi^m\right]^2. \qquad (3.54)$$

In this last equation, we have been slightly more general than previously:

- We have included the possibility of having a gauge group G which is a product of simple groups: for all generators t^a corresponding to the same simple group factor, the coupling g_a is identical. For example, in the case of the Standard Model where $G|_{\text{nonabel.}} = SU(3) \times SU(2)$, all the generators of SU(3) have coupling g_s and all generators of $SU(2)$ have coupling g.
- We have included in the last line the possibility of having a certain number of abelian $U(1)$ gauge groups, labelled by m (q_i^m is the charge of the field ϕ_i under $U(1)_m$), with possible Fayet–Iliopoulos terms (labelled ξ^m for $U(1)_m$). Let us note that a Fayet–Iliopoulos term (3.41) is not allowed for a nonabelian symmetry because it is not gauge invariant[12].

---

[12] Compare the gauge transformations of the auxiliary fields in (3.39) and (3.50).

The general scalar potential (3.54) with its F-terms and D-terms will be the central object of study when we discuss gauge or supersymmetry breaking.

Exercises

<u>Exercise 1</u>

(a) Show that, under the supersymmetry transformation (3.3), (3.4), we have:

$$[\delta_1, \delta_2]\, A = 2i\, \bar\varepsilon_1 \gamma^\mu \varepsilon_2\, \partial_\mu A$$
$$[\delta_1, \delta_2]\, B = 2i\, \bar\varepsilon_1 \gamma^\mu \varepsilon_2\, \partial_\mu B$$

i.e. supersymmetry transformations on scalar fields close off-shell.

(b) Prove (3.7) using Fierz transformations (equation (B.29) of Appendix B).

Hints:

(a) Use (B.40) of Appendix B. The fact that supersymmetry transformations close off-shell on scalar fields indicates that there is no need to introduce auxiliary fields in the supersymmetry transformations of scalar fields.

(b) Use equation (B.29) of Appendix B to prove:

$$\varepsilon_2 \bar\varepsilon_1 - \varepsilon_1 \bar\varepsilon_2 - (\gamma_5 \varepsilon_2)(\bar\varepsilon_1 \gamma_5) + (\gamma_5 \varepsilon_1)(\bar\varepsilon_2 \gamma_5) = -\gamma_\mu \left(\bar\varepsilon_1 \gamma^\mu \epsilon_2\right).$$

<u>Exercise 2</u> Compute the spectrum for the O'Raifeartaigh model of Section 3.1.4 in the case $\langle A \rangle = 0$ but for any value of $\langle X \rangle$ along the flat direction. Check that STr $M^2 = 0$.

Hint: This time, the A and Y supermultiplets mix in the mass matrices. The eigenvalues are for the system (Ψ_A, Ψ_X): $\sqrt{m^2 + \lambda^2 X^2} \pm \lambda X$, for the system (Re A, Re X):

$$m^2 + \lambda^2(2X^2 - M^2) \pm \lambda\sqrt{\lambda^2(2X^2 - M^2)^2 + 4m^2 X^2}$$

and for the system (Re A, Re X):

$$m^2 + \lambda^2(2X^2 + M^2) \pm \lambda\sqrt{\lambda^2(2X^2 + M^2)^2 + 4m^2 X^2}.$$

<u>Exercise 3</u> *We generalize here the O'Raifeartaigh model to make explicit the structure that leads to supersymmetry breaking.*

Let us consider a set of N supermultiplets (A_i, Ψ_{A_i}) and P supermultiplets (X_m, Ψ_{X_m}) whose interactions are described by the superpotential:

$$W(A, X) = \sum_{m=1}^{P} X_m f_m(A), \tag{3.55}$$

where f_m are P analytic functions of the variables A_i.

(a) Write the scalar potential.
(b) Show that, when $N < P$, supersymmetry is spontaneously broken in the generic case. The O'Raifeartaigh model (3.22) corresponds to $N = 1$ and $P = 2$.
[(c) Give an R-symmetry that would make natural the choice (3.55).]

Hints:

(a) $V = \sum_m |f_m(A)|^2 + \sum_i |\sum_m X_m \partial f_m / \partial A_i|^2$.

(b) Take $\langle X_m \rangle = 0$ or any point on the flat directions defined by $\sum_m X_m \partial f_m / \partial A_i = 0$. Then $\langle V \rangle = 0$ requires $f_m(A) = 0$, *i.e.* P conditions for N fields. If $N < P$, this is not possible in the general case: it would require specific values of the parameters.

(c) $R(A_i) = 0$, $R(X_m) = 2$.

Exercise 4 Prove (3.27) from (3.26) using the formulas (B.36) of Appendix B.

Hint: To obtain $\delta_S \Psi^c_R$, introduce a spinor Ψ_1 and use $\left(\bar{\Psi}_1 \delta_S \Psi_L \right)^* = \bar{\Psi}^c_1 \delta_S \Psi^c_R$.

Exercise 5 *We write in this exercise the supersymmetric version of quantum electrodynamics as first introduced by [363].*

If we want to couple a chiral supermultiplet with a Majorana spinor to an abelian gauge supermultiplet, we face the problem that a gauge (phase) transformation is incompatible with the reality imposed by the Majorana condition. We must thus introduce *two* Majorana spinors Ψ_1 and Ψ_2 in order to form a complex Majorana spinor $\Psi_\pm = (\Psi_1 \pm i\Psi_2)/\sqrt{2}$ which undergoes a phase transformation under the abelian symmetry.

We thus start from the Lagrangian (3.43) with two chiral supermultiplets $(\phi_\pm, \Psi_\pm, F_\pm)$ of charge $\pm q$:

$$\mathcal{L} = D^\mu \phi^*_+ D_\mu \phi_+ + D^\mu \phi^*_- D_\mu \phi_- + \tfrac{1}{2} \overline{\Psi}_+ i\gamma^\mu D_\mu \Psi_+$$

$$+ \tfrac{1}{2} \overline{\Psi}_- i\gamma^\mu D_\mu \Psi_- + F^*_+ F_+ + F^*_- F_-$$

$$+ gq \left(D\phi^*_+ \phi_+ + \sqrt{2}\, \bar{\lambda}\Psi_{+L}\, \phi^*_+ + \sqrt{2}\, \bar{\lambda}\Psi^c_{+R}\, \phi_+ \right) \tag{3.56}$$

$$- gq \left(D\phi^*_- \phi_- + \sqrt{2}\, \bar{\lambda}\Psi_{-L}\, \phi^*_- + \sqrt{2}\, \bar{\lambda}\Psi^c_{-R}\, \phi_- \right) + \mathcal{L}_V \,,$$

where $\mathcal{L}_V$ is given in (3.38). This Lagrangian is invariant under (3.40) and the following transformations:

$$\delta_S \phi_\pm = \sqrt{2}\, \bar{\varepsilon}\Psi_{\pm L},$$

$$\delta_S \Psi_{\pm L} = \frac{1 - \gamma_5}{2} \left[F_\pm - i\gamma^\mu D_\mu \phi_\pm \right] \varepsilon \sqrt{2},$$

$$\delta_S F_\pm = -i\sqrt{2}\, \bar{\varepsilon}\gamma^\mu D_\mu \Psi_{\pm L} \mp 2gq\phi_\pm\, \bar{\varepsilon}\frac{1 + \gamma_5}{2}\lambda. \tag{3.57}$$

Introducing, as for the fermion fields,

$$\phi_\pm \equiv \frac{1}{\sqrt{2}} (\phi_1 \pm i\phi_2) \equiv \frac{1}{\sqrt{2}} \left(\frac{A_1 + iB_1}{\sqrt{2}} \pm i\frac{A_2 + iB_2}{\sqrt{2}} \right) ,$$

$$F_\pm \equiv \frac{1}{\sqrt{2}} (\mathcal{F}_1 \pm i\mathcal{F}_2) \equiv \frac{1}{\sqrt{2}} \left(\frac{F_1 + iG_1}{\sqrt{2}} \pm i\frac{F_2 + iG_2}{\sqrt{2}} \right) ,$$

show that the action (3.56) is written

$$\mathcal{L} = -\tfrac{1}{4}F^{\mu\nu}F_{\mu\nu} + \tfrac{1}{2}\bar{\lambda}i\gamma^\mu\partial_\mu\lambda + \tfrac{1}{2}D^2$$
$$+\tfrac{1}{2}\left(\partial^\mu A_1\partial_\mu A_1 + \partial^\mu B_1\partial_\mu B_1 + \partial^\mu A_2\partial_\mu A_2 + \partial^\mu B_2\partial_\mu B_2\right) - \tfrac{1}{2}gqD(A_1B_2 - A_2B_1)$$
$$+\tfrac{1}{2}\left(\bar{\Psi}_1i\gamma^\mu\partial_\mu\Psi_1 + \bar{\Psi}_2i\gamma^\mu\partial_\mu\Psi_2\right) + \tfrac{1}{2}\left(F_1^2 + G_1^2 + F_2^2 + G_2^2\right)$$
$$-gqA^\mu\left(A_1\partial_\mu A_2 - A_2\partial_\mu A_1 + B_1\partial_\mu B_2 - B_2\partial_\mu B_1 - i\bar{\Psi}_1\gamma_\mu\Psi_2\right)$$
$$+gq\left[\bar{\lambda}\left(B_1 - iA_1\gamma_5\right)\Psi_2 - \bar{\lambda}\left(B_2 - iA_2\gamma_5\right)\Psi_1\right]. \tag{3.58}$$

The supersymmetry transformations corresponding to (3.57) are

$$\delta_s A_i = \bar{\varepsilon}\Psi_i, \qquad \delta_s B_i = i\bar{\varepsilon}\gamma_5\Psi_i, \quad i = 1,2,$$
$$\delta_s \Psi_1 = \left[-i\gamma^\mu D_\mu\left(A_1 + iB_1\gamma_5\right) + F_1 - iG_1\gamma_5 - igq\gamma^\mu A_\mu\left(A_2 + iB_2\gamma_5\right)\right]\varepsilon,$$
$$\delta_s \Psi_2 = \left[-i\gamma^\mu D_\mu\left(A_2 + iB_2\gamma_5\right) + F_2 - iG_2\gamma_5 + igq\gamma^\mu A_\mu\left(A_1 + iB_1\gamma_5\right)\right]\varepsilon,$$
$$\delta_s F_1 = -i\bar{\varepsilon}\gamma^\mu D_\mu\Psi_1 - igq\,\bar{\varepsilon}\left(A_2 + iB_2\gamma_5\right)\lambda - igq\,\bar{\varepsilon}\gamma^\mu A_\mu\Psi_2,$$
$$\delta_s F_2 = -i\bar{\varepsilon}\gamma^\mu D_\mu\Psi_2 + igq\,\bar{\varepsilon}\left(A_1 + iB_1\gamma_5\right)\lambda + igq\,\bar{\varepsilon}\gamma^\mu A_\mu\Psi_1,$$
$$\delta_s G_1 = -\bar{\varepsilon}\gamma_5\gamma^\mu D_\mu\Psi_1 - igq\,\bar{\varepsilon}\left(B_2 - iA_2\gamma_5\right)\lambda + gq\,\bar{\varepsilon}\gamma^\mu\gamma_5 A_\mu\Psi_2,$$
$$\delta_s G_2 = -\bar{\varepsilon}\gamma_5\gamma^\mu D_\mu\Psi_2 + igq\,\bar{\varepsilon}\left(B_1 - iA_1\gamma_5\right)\lambda - gq\,\bar{\varepsilon}\gamma^\mu\gamma_5 A_\mu\Psi_1.$$

Note that, out of the two Majorana spinors Ψ_1 and Ψ_2, one can form a Dirac spinor which describes the electron in this super-QED theory.

Hint: $D_\mu\phi_\pm = \partial_\mu\phi_\pm \mp igqA_\mu\phi_\pm \cdots$

Exercise 6

(a) Check that $D_\mu\lambda^a$ defined in (3.49) transforms as a covariant derivative *i.e.* transforms as λ^a in (3.50).

(b) Check that the derivatives (3.53) transform as covariant derivatives, *i.e.* transform as in (3.51).

Hints:

(a) Use the relation $C^{abe}C^{cde} + C^{bce}C^{ade} + C^{cae}C^{bde} = 0$ which expresses nothing but the commutation relations (3.47) in the adjoint representation, where $(T^a)^{bc} = -iC^{abc}$.

(b) Note that Hermitian conjugation raises or lowers the indices; thus, for example from the first equation in (3.51), one obtains

$$\delta_g\phi^{*i} = i\alpha^a\,(t^{a*})^i{}_j\phi^{*j} = i\alpha^a\,(t^a)_j{}^i\phi^{*j}$$

where we used the fact that the matrix t^a is hermitian: $t^{a*} = t^{aT}$.

4

The supersymmetry algebra and its representations

Now that we have acquired some familiarity with supersymmetry and the techniques involved, we may discuss more thoroughly the supersymmetry algebra and its representations. As an example of application, we will then discuss in more details the construction of the representations of $N = 2$ supersymmetry. Of great importance is the appearance of short representations of supersymmetry, associated with the concept of BPS states in soliton physics. Their study allows us to introduce such notions as moduli, moduli space, etc. important when we discuss supersymmetry breaking or string and brane theory. This chapter is intended for the more theoretically oriented reader[1].

4.1 Supersymmetry algebra

We use two-component notation (see Appendix B). The basic commutator of $N = 1$ supersymmetry algebra

$$\{Q_r, \bar{Q}_s\} = 2\gamma^\mu_{rs}\, P_\mu$$

simply reads

$$\{Q_\alpha, \bar{Q}_{\dot\alpha}\} = 2\sigma^\mu_{\alpha\dot\alpha}\, P_\mu$$

$$\{Q_\alpha, Q_\beta\} = \{\bar{Q}_{\dot\alpha}, \bar{Q}_{\dot\beta}\} = 0. \tag{4.1}$$

Indeed

$$\left\{ \begin{pmatrix} Q_\alpha \\ \bar{Q}^{\dot\alpha} \end{pmatrix}_r , \left(Q^\beta\, \bar{Q}_{\dot\beta}\right)_s \right\} = \begin{pmatrix} \{Q_\alpha, Q^\beta\} & \{Q_\alpha, \bar{Q}_{\dot\beta}\} \\ \{\bar{Q}^{\dot\alpha}, Q^\beta\} & \{\bar{Q}^{\dot\alpha}, \bar{Q}_{\dot\beta}\} \end{pmatrix}_{rs}$$

$$= \begin{pmatrix} 0 & 2\sigma^\mu_{\alpha\dot\beta}\, P_\mu \\ 2\bar{\sigma}^{\mu\dot\alpha\beta}\, P_\mu & 0 \end{pmatrix}_{rs}$$

$$= 2\left(\gamma^\mu\right)_{rs}\, P_\mu.$$

In the case of N supersymmetric (Majorana) charges Q^i_r, that is $Q_{i\alpha}$, $\bar{Q}^{i\dot\alpha}$, $i = 1,\ldots,N$, one may use, as in Chapter 2, Coleman–Mandula restrictions to

---

[1]Since this chapter is on the *Theorist* track we will assume in Section 4.3 familiarity with the superspace techniques developed in Appendix C. We also use the Van der Waerden notation for spinors (see Appendix B).

infer the general form of the supersymmetry algebra, thus following Haag, Łopuszanski and Sohnius [215]. Since $\{Q_{i\alpha}, \bar{Q}^{j}_{\dot\beta}\}$ belongs to the representation $(1/2, 0) \times (0, 1/2) = (1/2, 1/2)$, it is proportional to P_μ times a matrix M^j_i. The latter matrix may be shown to be hermitian and positive definite and one may redefine the supersymmetry charges so that M^j_i be proportional to δ^j_i. In other words,

$$\{Q_{i\alpha}, \bar{Q}^{j}_{\dot\beta}\} = 2\delta^j_i\, \sigma^\mu_{\alpha\dot\beta}\, P_\mu. \tag{4.2}$$

There is, however, something new with the anticommutator $\{Q_{i\alpha}, Q_{j\beta}\}$. In all generality, it belongs to the representation $(1/2, 0) \times (1/2, 0) = (1, 0) + (0, 0)$ and, accordingly, we may write

$$\{Q_{i\alpha}, Q_{j\beta}\} = (\sigma^{\mu\nu})_\alpha{}^\gamma\, \varepsilon_{\gamma\beta}\, M_{\mu\nu}\, Y_{ij} + \varepsilon_{\alpha\beta}\, Z_{ij} \tag{4.3}$$

where $(\sigma^{\mu\nu})_\alpha{}^\beta = \frac{1}{4}(\sigma^\mu\bar\sigma^\nu - \sigma^\nu\bar\sigma^\mu)_\alpha{}^\beta$, Y is symmetric ($Y_{ij} = +Y_{ji}$), and Z is antisymmetric ($Z_{ij} = -Z_{ji}$). Before going further, let us note that $[Q_{i\alpha}, P_\mu]$ belongs to $(1/2, 0) \times (1/2, 1/2) = (0, 1/2) + (1, 1/2)$ and write therefore $((Q_{i\alpha})^\dagger = \bar{Q}^i_{\dot\alpha})$

$$[Q_{i\alpha}, P_\mu] = X_{ij}\, \sigma_{\mu\alpha\dot\beta}\, \bar{Q}^{j\dot\beta}$$

$$[\bar{Q}^{i\dot\alpha}, P_\mu] = (X_{ij})^*\, \bar\sigma_\mu^{\dot\alpha\beta}\, Q_{j\beta}.$$

Using the Jacobi identity[2]

$$[[Q_{i\alpha}, P_\mu], P_\nu] + [[P_\mu, P_\nu], Q_{i\alpha}] + [[P_\nu, Q_{i\alpha}], P_\mu] = 0$$

and $[P_\mu, P_\nu] = 0$, one obtains $XX^* = 0$.

Then consider the other Jacobi identity

$$[\{Q_{i\alpha}, Q_{j\beta}\}, P_\mu] - \{[Q_{j\beta}, P_\mu], Q_{i\alpha}\} + \{[P_\mu, Q_{i\alpha}], Q_{j\beta}\} = 0. \tag{4.5}$$

Contracting with $\varepsilon^{\alpha\beta}$ yields[3] $X_{ij} = X_{ji}$. Hence $XX^\dagger = 0$ and $X_{ij} = 0$. We thus have

$$[Q_{i\alpha}, P_\mu] = 0 = \left[\bar{Q}^j_{\dot\alpha}, P_\mu\right] \tag{4.6}$$

i.e. the N supersymmetry charges commute with the generator of translations (we have already seen that this implies that the fields in a supermultiplet have a common

---

[2]The Jacobi identity familiar in Lie algebras

$$[[T_1, T_2], T_3] + [[T_2, T_3], T_1] + [[T_3, T_1], T_2] = 0$$

reads in the case of the graded Lie algebras that we consider

$$[[G_1, G_2\}, G_3\} \pm [[G_2, G_3\}, G_1\} \pm [[G_3, G_1\}, G_2\} = 0 \tag{4.4}$$

where $[G_i, G_j\}$ is an anticommutator if G_i and G_j are both fermionic operators, a commutator otherwise. The signs are $+$ or $-$ depending on whether the number of permutations of fermionic operators is even or odd, *e.g.*

$$[\{F_1, F_2\}, B] - \{[F_2, B], F_1\} + \{[B, F_1], F_2\} = 0.$$

[3]Z_{ij} being an internal generator commutes with P_μ.

invariant mass since $[Q_{i\alpha}, P^{\mu}P_{\mu}] = 0)$. Also, once we have proven that $X_{ij} = 0$, it follows from (4.5) that $Y_{ij} = 0$. Hence

$$\{Q_{i\alpha}, Q_{j\beta}\} = \varepsilon_{\alpha\beta} \, Z_{ij}$$
$$\left\{\bar{Q}^i_{\dot\alpha}, \bar{Q}^j_{\dot\beta}\right\} = \varepsilon_{\dot\alpha\dot\beta} \, (Z_{ij})^* . \tag{4.7}$$

The complex constants Z_{ij} are called central charges because they are shown to commute with all generators (see Exercise 3). Note that the central charges have the dimension of a mass.

The supersymmetry charges usually also transform under an internal compact symmetry group G which is a product of a semisimple group and an abelian factor. Denoting by B_r the hermitian generators of this group, which satisfy

$$[B_r, B_s] = iC_{rst} B_t \tag{4.8}$$

we have

$$[Q_{i\alpha}, B_r] = (b_r)_i{}^j \, Q_{j\alpha}$$
$$\left[B_r, \bar{Q}^i_{\dot\alpha}\right] = (b_r^*)^i{}_j \, \bar{Q}^j_{\dot\alpha}. \tag{4.9}$$

The Jacobi identity

$$\left[\left\{Q_{i\alpha}, \bar{Q}^j_{\dot\beta}\right\}, B_r\right] - \left\{\left[\bar{Q}^j_{\dot\beta}, B_r\right], Q_{i\alpha}\right\} + \left\{[B_r, Q_{i\alpha}], \bar{Q}^j_{\dot\beta}\right\} = 0$$

yields $(b_r^*)^j{}_i = (b_r)_i{}^j$. Hence the matrices B_r are hermitian and the largest possible internal symmetry group which acts on the supersymmetry generators is $U(N)$.

In the case of $N = 1$ supersymmetry, one recovers (4.1) since Z, being antisymmetric, vanishes ($Z = -Z$): there are no nonvanishing central charges. Also, by rescaling the generators B_r by $(b_r)_1{}^1$, one obtains from (4.9)

$$[Q_\alpha, B_r] = Q_\alpha$$
$$[\bar{Q}_{\dot\alpha}, B_r] = -\bar{Q}_{\dot\alpha}.$$

There is only one independent combination R of the generators B_r which acts nontrivially on the supersymmetry charges

$$[Q_\alpha, R] = Q_\alpha$$
$$[\bar{Q}_{\dot\alpha}, R] = -\bar{Q}_{\dot\alpha}. \tag{4.10}$$

Thus $N = 1$ supersymmetry possesses an internal global $U(1)$ symmetry known as the *R-symmetry*: Q_α has R-charge $+1$ and $\bar{Q}_{\dot\alpha}$ has R-charge -1. We will see later the important rôle played by this symmetry (see Section C.2.3 of Appendix C and Chapter 8).

4.2 Supermultiplet of currents

The supersymmetry current plays an important rôle when we discuss the issue of supersymmetry breaking. A key property is that it belongs itself to a supermultiplet, known as the supermultiplet of currents. This is related to the fact that supersymmetry is a spacetime symmetry: since two supersymmetries give a translation, the supersymmetry transformation of the supersymmetry current should give the spacetime translation current, *i.e.* the energy–momentum tensor.

To be more explicit, the supersymmetry transformation of the supersymmetry current $J_{r\mu}$ reads (as in (A.17) of Appendix Appendix A)

$$\delta J_{r\mu} = i\bar{\epsilon}_s \left\{ Q_s, J_{r\mu} \right\}. \tag{4.11}$$

Integrating the zeroth component over space, we obtain

$$\int d^3x \; \delta J_{r0} = i\bar{\epsilon}_s \left\{ Q_s, Q_r \right\} = -i \left\{ Q_r, \bar{Q}_s \right\} \epsilon_s$$

$$= -2i \left(\gamma^\nu \epsilon\right)_r P_\nu = \int d^3x \left[-2i \left(\gamma^\nu \epsilon\right)_r \Theta_{0\nu} \right]. \tag{4.12}$$

We deduce that the transformation of the supersymmetry current includes the following term:

$$\delta J_{r\mu} = -2i \left(\gamma^\nu \epsilon\right)_r \Theta_{\mu\nu} + \cdots \tag{4.13}$$

This shows that the supersymmetry current belongs to the same multiplet as the energy–momentum tensor. A simple count of the number of fermionic $(J_{r\mu})$ and bosonic $(\Theta_{\mu\nu})$ degrees of freedom shows that one is missing some bosonic components. One can then note that the algebra for the R-symmetry, which can be written with Majorana spinors as

$$[Q_r, R] = -\gamma^5_{rs} Q_s, \tag{4.14}$$

gives, along the same reasoning as before, the supersymmetry transformation of the R-current:

$$\delta J^R_\mu = -i\bar{\epsilon}_r \gamma^5_{rs} J_{s\mu} + \cdots \tag{4.15}$$

Indeed, the currents $(\Theta_{\mu\nu}, J_{r\mu}, J^R_\mu)$ form the supermultiplet of currents.

Let us note that the trace of the energy–momentum tensor transforms under supersymmetry into $\gamma^\mu J^R_\mu$: from (4.13), we deduce

$$\delta \left(\gamma^\mu J^R_\mu\right) = -2i \left(\gamma^\mu \gamma^\nu \epsilon\right) \Theta_{\mu\nu} + \cdots = -i \left\{\gamma^\mu, \gamma^\nu\right\} \epsilon \, \Theta_{\mu\nu} + \cdots$$
$$= -2i\epsilon \, \Theta^\mu{}_\mu + \cdots \tag{4.16}$$

As is recalled in Section A.5.1 of Appendix Appendix A, the tracelessness of the energy–momentum tensor ensures conformal symmetry. In a supersymmetric context, the two algebraic constraints

$$\Theta^\mu{}_\mu = 0, \quad \gamma^\mu J^R_\mu = 0 \tag{4.17}$$

lead to the superconformal algebra.

We first note that $(\Theta_{\mu\nu}, J_{r\mu}, J^R_\mu)$ describe an equal number of bosonic and fermionic degrees of freedom. The symmetric tensor $\Theta_{\mu\nu}$ has 10 independent components minus 4 for current conservation $(\partial^\mu \Theta_{\mu\nu} = 0)$ minus 1 for the tracelessness condition, that is five degrees of freedom. Similarly, $J_{r\mu}$ accounts for $16-4-4=8$ and J_μ^R for $4-1=3$. The full supersymmetry transformations read

$$\delta J_\mu^R = -i\bar\epsilon_r \gamma_{rs}^5 J_{s\mu},$$

$$\delta J_{r\mu} = -2i\left(\gamma^\nu \epsilon\right)_r \Theta_{\mu\nu} - \left(\gamma^\nu \gamma_5 \epsilon\right)_r \partial_\nu J_\mu^R - \frac{i}{2}\epsilon_{\mu\nu\rho\sigma}\left(\gamma^\nu \epsilon\right)_r \partial^\rho J^{R\sigma},$$

$$\delta\Theta_{\mu\nu} = -\frac{1}{4}\left(\bar\epsilon\,\sigma_{\mu\rho}\partial^\rho J_\nu + \bar\epsilon\,\sigma_{\nu\rho}\partial^\rho J_\mu\right). \tag{4.18}$$

Introducing the currents (see (A.222) and (A.225) in Appendix Appendix A)

$$D_\mu = x^\rho \Theta_{\mu\rho}, \quad K_{\mu\nu} = 2x_\mu x^\rho \Theta_{\rho\nu} - x^2 \Theta_{\lambda\mu}, \quad S_{r\mu} = ix_\nu \gamma_{rs}^\nu J_{s\mu}, \tag{4.19}$$

and the corresponding charges

$$D = \int d^3x\, D_0, \quad K_\mu = \int d^3x\, K_{\mu 0}, \quad S_r = \int d^3x\, S_{r0}, \tag{4.20}$$

we may derive the following (anti)commutation relations[4]

$$
\begin{aligned}
[Q, M_{\mu\nu}] &= i\sigma_{\mu\nu}\, Q, & [S, M_{\mu\nu}] &= i\sigma_{\mu\nu}\, S,\\
[Q, D] &= \frac{i}{2}Q, & [S, D] &= -\frac{i}{2}S,\\
[Q, P_\mu] &= 0, & [S, P_\mu] &= \gamma_\mu Q,\\
[Q, K_\mu] &= \gamma_\mu S, & [S, K_\mu] &= 0,\\
[Q, R] &= -\gamma_5 Q, & [S, R] &= \gamma_5 S,\\
\{Q, \bar Q\} &= 2\gamma^\mu P_\mu, & \{S, \bar S\} &= 2\gamma^\mu K_\mu,
\end{aligned}
\tag{4.21}
$$

$$\{S, \bar Q\} = 2iD + 2i\sigma^{\mu\nu}M_{\mu\nu} - 3\gamma_5 R. \tag{4.22}$$

This, together with (2.10) of Chapter 2 and (A.230) of Appendix Appendix A, gives the algebra of conformal supersymmetry.

4.3 Representations of the supersymmetry algebra

Remember that, in the case of the Poincaré algebra, representations are classified according to the mass operator

$$P^2 = P^\mu\, P_\mu$$

and the spin operator which is built out of the Pauli–Ljubanski operator

$$W^\mu = -\frac{1}{2}\,\varepsilon^{\mu\nu\rho\sigma}\, P_\nu\, M_{\rho\sigma}. \tag{4.23}$$

W^2 has eigenvalues $-m^2 j(j+1)$ for massive states of spin $j = 0, \frac{1}{2}, 1, \ldots$ and $W_\mu = \lambda P_\mu$ for massless states of helicity λ.

Of the two "Casimir" operators, when we go supersymmetric, P^2 remains unchanged since it commutes with supersymmetry, whereas W^2 includes some extra contributions.

---

[4]We do not include the commutators of R with the generators of the algebra of the conformal group (P_μ, $M_{\mu\nu}$, D and K_μ) since they all vanish.

4.3.1   Massless states, in the absence of central charges

We choose a special frame where $P_\mu = (E, 0, 0, E)$. Then $\{Q_{i\alpha}, \bar{Q}^j_{\dot\alpha}\} = 2\delta^j_i\, \sigma^\mu_{\alpha\dot\alpha}\, P_\mu$ reads:

$$\{Q_{i1}, \bar{Q}^j_{\dot1}\} = 4\,\delta^j_i\, E$$
$$\{Q_{i2}, \bar{Q}^j_{\dot2}\} = 0. \tag{4.24}$$

Defining

$$a_i \equiv \frac{Q_{i1}}{2\sqrt{E}}, \qquad a^{j\dagger} \equiv \frac{\bar{Q}^j_{\dot1}}{2\sqrt{E}},$$

we have from (4.24)

$$\{a_i, a^{j\dagger}\} = \delta^j_i, \qquad \{a_i, a_j\} = 0 = \{a^{i\dagger}, a^{j\dagger}\}.$$

Note that $a^{j\dagger}$ is in the representation $(0, 1/2) \equiv 2_R$ and carries helicity[5] $\lambda = +1/2$, whereas a_i is in the representation $(1/2, 0) \equiv 2_L$ and carries helicity $\lambda = -1/2$.

Let us consider a state $|\Omega_0\rangle \equiv |E, \lambda_0\rangle$, of energy E and helicity λ_0, "annihilated" by the $a_i : a_i|\Omega_0\rangle = 0$. By successive applications of the N "creation" operators $a^{j\dagger}$, $j = 1 \cdots N$, one obtains the states

$$a^{j\dagger}|\Omega_0\rangle = |E, \lambda_0 + \frac{1}{2}, j\rangle : N \text{ states}$$

$$a^{j_1\dagger}\, a^{j_2\dagger}|\Omega_0\rangle = |E, \lambda_0 + 1; j_1 j_2\rangle : \frac{1}{2}\, N(N-1) \text{ states}$$

$$a^{j_1\dagger} \cdots a^{j_k\dagger}|\Omega_0\rangle = |E, \lambda_0 + \frac{k}{2}; j_1 \cdots j_k\rangle : \binom{N}{k} = \frac{N!}{k!(N-k)!} \text{ states}$$

$$a^{1\dagger} \cdots a^{N\dagger}|\Omega_0\rangle = |E, \lambda_0 + \frac{N}{2}\rangle : \binom{N}{N} = 1 \text{ state.}$$

We thus have constructed a set of $\sum\limits_{k=0}^{N} \binom{N}{k} = 2^N$ states which form a representation of supersymmetry[6].

Let us consider several examples:

- $N = 4$ $\qquad\qquad$ $\lambda_0 = -1$ $\qquad$ vector supermultiplet

λ	-1	$-1/2$	0	$+1/2$	$+1$
no. of states	1	4	6	4	1

Note that for $N > 4$, there is at least one field of spin 3/2 which does not allow renormalizable couplings: hence, renormalizable theories are found only for $N \leq 4$. Also for $N > 8$, there is at least one field of spin 5/2; such fields are not known to have consistent couplings to gravity. Hence consistent theories of (super)gravity require $N \leq 8$.

---

[5] For massless states, $W_\mu = \lambda P_\mu$.

[6] Since, for any state $|\Phi\rangle$, $0 = \langle\Phi|\{Q_{i2}, \bar{Q}^i_{\dot2}\}|\Phi\rangle = |Q_{i2}|\Phi\rangle|^2 + |\bar{Q}^i_{\dot2}|\Phi\rangle|^2$, we have $Q_{i2}|\Phi\rangle = \bar{Q}^i_{\dot2}|\Phi\rangle = 0$.

- $N = 2$ $\lambda_0 = -\frac{1}{2}$ hypermultiplet

λ	$-1/2$	0	$1/2$
no. of states	1	2	1

All the preceding supermultiplets meet the following CPT theorem requirement: for every state of helicity λ, one should have a parity reflected state of helicity $-\lambda$. They are CPT self-conjugate.

- $N = 2$ $\lambda_0 = -1$

λ	-1	$-1/2$	0
no. of states	1	2	1

This supermultiplets does not satisfy the requirement above. One must add to it a supermultiplet built from $|\Omega_0'\rangle = |E, \lambda_0 = 0\rangle$

λ	0	$1/2$	1
no. of states	1	2	1

In total, one obtains a $N = 2$ vector supermultiplet

λ	-1	$-1/2$	0	$1/2$	1
no. of states	1	2	2	2	1

Hence a non-CPT self-conjugate supermultiplet has 2^{N+1} helicity states.

- $N = 1$, $\lambda_0 = -\frac{1}{2}$

Similarly, one obtains by including the CPT conjugate ($\lambda_0 = 0$)

λ	$-1/2$	0	$1/2$
no. of states	1	2	1

which describes the chiral supermultiplet.

- $N = 1$, $\lambda_0 = -1, \frac{1}{2}$ (to include the CPT conjugate)

λ	-1	$-1/2$	$1/2$	1
no. of states	1	1	1	1

which describes the vector supermultiplet.

We should note that all the degrees of freedom discussed here are on-shell degrees of freedom.

4.3.2   Massive states, in the absence of central charges

In this case, one chooses a rest frame where $P_\mu = (m, 0, 0, 0)$ and $W^2 = -m^2 \mathbf{J}^2$, where the Lorentz generators $(J_i) = (M_{23}, M_{31}, M_{12})$ form a $SO(3) \sim SU(2)$ algebra.

The only nontrivial anticommutator relation is now

$$\{Q_{i\alpha}, \bar{Q}^j_{\dot\beta}\} = 2m\, \delta^j_i\, \delta_{\alpha\dot\beta}$$

which defines $2N$ "annihilation" and "creation" operators $a_{i\alpha} = Q_{i\alpha}/\sqrt{2m}$ and $a^{j\dagger}_{\dot\beta} = \bar{Q}^j_{\dot\beta}/\sqrt{2m}$ satisfying

$$\{a_{i\alpha}, a^{j\dagger}_{\dot\beta}\} = \delta_{ij}\, \delta_{\alpha\dot\beta}.$$

Under the spin group $SU(2)$, the creation operators behave like a doublet

$$a^{j\dagger}_{\dot{1}} \sim \bar{Q}^j_{\dot{1}} \qquad J = 1/2 \;\; J_3 = -1/2,$$
$$a^{j\dagger}_{\dot{2}} \sim \bar{Q}^j_{\dot{2}} \qquad J = 1/2 \;\; J_3 = +1/2.$$

In a way similar to the preceding case, one may therefore construct 2^{2N} states out of a state $|\Omega_0\rangle$ of given spin j_0.

Let us take the case $N = 1$ for simplicity and consider a state $|\Omega_0\rangle$ of total spin j_0 annihilated by the Q_α. One may therefore form

$$|\Omega_0\rangle \qquad \text{spin } j_0$$
$$a^\dagger_{\dot\beta}|\Omega_0\rangle \qquad \text{spin } j_0 \pm \frac{1}{2} \qquad \left(\text{if } j_0 \geq \frac{1}{2}\right)$$
$$a^\dagger_{\dot\beta}\, a^\dagger_{\dot\gamma}|\Omega_0\rangle = -\frac{1}{2}\, \varepsilon_{\dot\beta\dot\gamma}\left(a^\dagger_{\dot\alpha}\, a^{\dagger\dot\alpha}\right)|\Omega_0\rangle \quad \text{spin } j_0.$$

In the last line, we have used the fact that the indices $\dot\beta$ and $\dot\gamma$ are necessarily distinct and that $a^\dagger_{\dot\alpha}\, a^{\dagger\dot\alpha}$ is a scalar entity (spin 0) made out of two spin $\frac{1}{2}$ objects.

We have for example constructed:

- a massive chiral supermultiplet $(j_0 = 0)$

j	0	1/2
no. of states	2	1
no. of degrees of freedom	2	2

- a massive vector supermultiplet $(j_0 = \frac{1}{2})$

j	0	1/2	1
no. of states	1	2	1
no. of degrees of freedom	1	4	3

Let us note that the latter field content is compatible with the supersymmetric version of the Higgs mechanism. With a massless Higgs chiral superfield (two scalar and two spin 1/2 degrees of freedom) and a massless vector superfield (two spin 1/2 and two spin 1 degrees of freedom), we can make a massive vector superfield where one scalar component provides the missing longitudinal degree of freedom of the vector particle.

In the case of general N, starting again from a state $|\Omega_0\rangle$ of total spin j_0, we construct a state of maximal spin $j = j_0 + N/2$ in the form of $a_{\frac{1}{2}}^{1\dagger}\cdots a_{\frac{1}{2}}^{N\dagger}|\Omega_0\rangle$. Thus, in the absence of central charges, massive multiplets of $N > 1$ supersymmetry always contain fields of spin larger or equal to 1.

4.3.3 Massive states in the presence of central charges

In this case, the supersymmetry charges cannot be straightforwardly interpreted as creation and annihilation operators. We consider here only the case N even. Through a unitary transformation [388], one first puts the antisymmetric matrix Z_{ij} into the form

$$Z = \begin{pmatrix} 0 & D \\ -D & 0 \end{pmatrix}$$

where $D = \begin{pmatrix} z_1 & & \\ & z_r & \\ & & z_{N/2} \end{pmatrix}$, with all eigenvalues real and positive.

The index i of the supersymmetry charges is then substituted to (ar) where $a = 1,2$ and $r = 1,\ldots,N/2$. The supersymmetry algebra (4.2, 4.7) reads

$$\{Q_{\alpha ar}, \bar{Q}_{\dot\beta}^{bs}\} = 2\,\delta_a^b\,\delta_r^s\,\sigma^\mu_{\alpha\dot\beta}\,P_\mu,$$

$$\{Q_{\alpha ar}, Q_{\beta bs}\} = \varepsilon_{\alpha\beta}\,\varepsilon_{ab}\,\delta_{rs}\,z_r, \tag{4.25}$$

$$\{\bar{Q}_{\dot\alpha}^{ar}, \bar{Q}_{\dot\beta}^{bs}\} = \varepsilon_{\dot\alpha\dot\beta}\,\varepsilon^{ab}\,\delta^{rs}\,z_r$$

where $[\varepsilon_{ab}] = [\varepsilon^{ab}] = i\sigma_2$.

One easily checks that massless representations are only compatible with $z_r = 0$ (see Exercise 6).

For massive representations, one works again in the frame where $P_\mu = (m,0,0,0)$. Introducing the combinations

$$A_{ar}^{\pm} = \frac{1}{\sqrt{2}}\ \left(Q_{\alpha 1 r} \pm \bar{Q}^{\dot\alpha 2r}\right), \tag{4.26}$$

the only nontrivial anticommutation relations are

$$\left\{A_{ar}^{\pm}, \left(A_{\beta s}^{\pm}\right)^{\dagger}\right\} = \delta_{\alpha\beta}\,\delta_{rs}\,(2m \mp z_r). \tag{4.27}$$

Hence necessarily $z_r \leq 2m$.

- When $z_r < 2m$, the multiplicities are the same as in the case of no central charge, i.e. $2^{2N}(2j+1)$. One describes them as long multiplets.

- When $z_r = 2m$ for some or all z_r, the multiplicity is decreased. For example, if there are n_0 central charges that saturate the bound, *i.e.* such that $z_r = 2m$, the multiplicity is decreased to $2^{2(N-n_0)}(2j + 1)$ and one speaks of short multiplets. Such states are called BPS-saturated states because of their connection with BPS monopoles, as we will see in the Section 4.5.

As an example, we may take for $N = 2$ supersymmetry $n_0 = 1$, $j = 0$ which corresponds to the hypermultiplet constructed in the next section.

4.4 Multiplets of $N = 2$ supersymmetry

Extended supersymmetries play an important rôle in the understanding of the nonperturbative aspects of supersymmetric theories, as well in supersymmetric models arising from higher dimensional theories, such as string theories. In this section, we construct the basic $N = 2$ supermultiplets [147, 327] by combining the $N = 1$ supermultiplets that we are now acquainted with.

4.4.1 Vector supermultiplet

We have encountered the $N = 2$ vector supermultiplet in Section 4.3.1 above. It is easy to check from the field content that it can be decomposed into a $N = 1$ vector and $N = 1$ chiral supermultiplets:

λ	-1	$-1/2$	0	$1/2$	1
$N = 2$ vector	1	2	2	2	1
$N = 1$ vector	1	1	0	1	1
$N = 1$ chiral	0	1	2	1	0

We may thus try to construct an $N = 2$ supersymmetric action out of a $N = 1$ vector supermultiplet $(A^a_\mu, \lambda^a, D^a)$ and a $N = 1$ chiral supermultiplet (ϕ^a, ψ^a, F^a) in the adjoint representation. We use equations (C.81) and (C.84) of Appendix C to write the action

$$
S = \int d^4x \left[\int d^2\theta \; \frac{1}{4} \mathrm{Tr}\,(W^\alpha W_\alpha) + \text{h.c.} \right] + \int d^4x \int d^4\theta \; \Phi^\dagger e^{2gV} \Phi
$$

$$
= \int d^4x \mathrm{Tr}\left(-\frac{1}{4} F_{\mu\nu} F^{\mu\nu} + i\lambda\sigma^\mu D_\mu \bar\lambda + \frac{1}{2} D^2 + D^\mu \phi^\dagger D_\mu \phi + i\psi\sigma^\mu D_\mu \bar\psi + F^\dagger F \right)
$$

$$
- g\,\mathrm{Tr}\left(D\left[\phi^\dagger, \phi\right] + \sqrt{2}\psi\left[\lambda, \phi^\dagger\right] - \sqrt{2}\bar\psi\left[\bar\lambda, \phi\right] \right). \tag{4.28}
$$

We recall that the generators in the adjoint representation can be written as

$$
(T^a)^{bc} = -iC^{abc} \tag{4.29}
$$

where the C^{abc} are the structure constants defined, for any representation t^a as

$$
\left[t^a, t^b\right] = iC^{abc}t^c. \tag{4.30}
$$

Hence the covariant derivatives for $\lambda \equiv \lambda^a T^a$, $\phi \equiv \phi^a T^a$, $\psi \equiv \psi^a T^a$ read

$$
\begin{aligned}
D_\mu \lambda &= \partial_\mu \lambda - ig\,[A_\mu, \lambda] \\
D_\mu \phi &= \partial_\mu \phi - ig\,[A_\mu, \phi] \\
D_\mu \psi &= \partial_\mu \psi - ig\,[A_\mu, \psi]
\end{aligned}
\tag{4.31}
$$

and the field strength $F_{\mu\nu} \equiv F^a_{\mu\nu} T^a$

$$
F_{\mu\nu} = \partial_\mu A_\nu - \partial_\nu A_\mu - ig\,[A_\mu, A_\nu].
\tag{4.32}
$$

Similarly, we have noted in (4.28) the auxiliary fields $D \equiv D^a T^a$ and $F \equiv F^a T^a$.

We have chosen the normalizations of the vector and chiral actions in order to normalize in the same way the λ and ψ fermion kinetic terms. In this way, the action (4.28) has an $O(2)$ symmetry which may be made more visible by writing $\lambda \equiv \lambda^1$ and $\psi \equiv \lambda^2$.[7]

This is, however, not the $U(2)$ symmetry that we expect (see Section 4.1 above). In order to make the latter symmetry explicit, we must introduce symplectic Majorana spinors. They are four-component spinors defined from two-component spinors λ^i_α, $i = 1, 2$, as

$$
\lambda^i \equiv \begin{pmatrix} -i\epsilon^{ij} \lambda^j_\alpha \\ \bar{\lambda}^{\dot{\alpha} i} \end{pmatrix},
\tag{4.33}
$$

where ϵ^{ij} is the antisymmetric tensor ($\epsilon^{12} = 1$). They thus satisfy the condition

$$
\lambda^i = i\epsilon^{ij} \gamma_5 C \bar{\lambda}^{jT},
\tag{4.34}
$$

where, as usual, C is the charge conjugation matrix, defined in (B.30) of Appendix B.

We may now rewrite the action (4.28) in an explicitly $U(2)$ symmetric way. In order to write later the supersymmetry transformations in a compact way, we redefine the auxiliary fields as

$$
F \equiv (P^1 - iP^2)/\sqrt{2}, \quad D \equiv -P^3 + g\,[\phi^\dagger, \phi],
\tag{4.35}
$$

to form a triplet P^r, $r = 1, 2, 3$ of $U(2)$. Then, the action (4.28) reads

$$
\mathcal{S} = \int d^4x\, \mathrm{Tr}\left[-\frac{1}{4} F_{\mu\nu} F^{\mu\nu} + \frac{i}{2} \bar{\lambda}^i \gamma^\mu D_\mu \lambda^i + \frac{1}{2} D^\mu A D_\mu A + \frac{1}{2} D^\mu B D_\mu B + \frac{1}{2} \sum_{r=1}^{3} (P^r)^2 \right]
$$

$$
+ \frac{1}{2} g^2\, \mathrm{Tr}\left([A, B]^2\right) - \frac{i}{2} g\, \mathrm{Tr}\left(\bar{\lambda}^i \gamma_5 [\lambda^i, A] + i\bar{\lambda}^i [\lambda^i, B]\right).
\tag{4.36}
$$

where, as usual, we have denoted the scalar field $\phi = (A + iB)/\sqrt{2}$.

---

[7]Thus the interaction term in (4.28) involves for example $\left(\lambda^{1a} \lambda^{2b} - \lambda^{2a} \lambda^{1b}\right) C^{abc}$ which is $O(2)$ invariant.

This action is invariant under the following $N = 2$ supersymmetry transformations:

$$\begin{aligned}
\delta_s A_\mu &= \bar{\xi}^i \gamma_\mu \gamma_5 \lambda^i, \\
\delta_s A &= i\, \bar{\xi}^i \lambda^i, \\
\delta_s B &= -\bar{\xi}^i \gamma_5 \lambda^i, \\
\delta_s \lambda^i &= i\sigma^{\mu\nu} \gamma_5 \xi^i F_{\mu\nu} - \gamma^\mu \xi^i D_\mu A - i\gamma^\mu \gamma_5 \xi^i D_\mu B - ig\xi^i [A, B] + (P^r \sigma^r)^{ij} \gamma_5 \xi^j, \\
\delta_s P^r &= -i\bar{\xi}^i (\sigma^r)^{ij} \gamma^\mu \gamma_5 D_\mu \lambda^j + ig\bar{\xi}^i (\sigma^r)^{ij} \left[A + iB\gamma_5, \lambda^j \right],
\end{aligned} \tag{4.37}$$

where the transformation parameter ξ^i, $i = 1, 2$, is a symplectic Majorana spinor:

$$\xi^i \equiv \begin{pmatrix} -i\epsilon^{ij} \eta^j_\alpha \\ \bar{\eta}^{\dot\alpha i} \end{pmatrix}. \tag{4.38}$$

The transformation with parameter $\eta^1_\alpha, \bar{\eta}^{1\dot\alpha}$ is simply the $N = 1$ supersymmetry transformation that acts between the components of the chiral and vector supermultiplets (see Exercise 7).

4.4.2 Hypermultiplet

Matter may be described by hypermultiplets which involve on-shell two complex scalar fields ϕ^i, $i = 1, 2$, forming a doublet of $U(2)$, and one Dirac fermion $\Psi \equiv \begin{pmatrix} \psi^1_\alpha \\ \bar{\psi}^{2\dot\alpha} \end{pmatrix}$.
Off-shell, this is complemented by two complex auxiliary fields F^i, $i = 1, 2$, and the supersymmetry transformation reads

$$\begin{aligned}
\delta_s \phi^i &= \sqrt{2}\, \bar{\xi}^i \Psi, \\
\delta_s \Psi &= -i\sqrt{2}\, \xi^i F^i - i\sqrt{2}\gamma^\mu \xi^i\, \partial_\mu \phi^i, \\
\delta_s F^i &= \sqrt{2}\, \bar{\xi}^i \gamma^\mu \partial_\mu \Psi,
\end{aligned} \tag{4.39}$$

where the transformation parameter ξ^i, $i = 1, 2$, is again the symplectic Majorana spinor of (4.38). Under the $N = 1$ supersymmetry transformation with parameter $\eta^1_\alpha, \bar{\eta}^{1\dot\alpha}$, $(\phi^1, \psi^1_\alpha, F^2)$ is a chiral supermultiplet whereas $(i\phi^2, \bar{\psi}^{2\dot\alpha}, -iF^1)$ is antichiral. The combination of two supersymmetry transformations takes the form

$$[\delta_1, \delta_2] = 2\bar{\xi}^k_{(1)} \gamma^\mu \xi^k_{(2)}\, i\partial_\mu + \bar{\xi}^k_{(1)} \xi^k_{(2)}\, \delta_z, \tag{4.40}$$

where

$$\begin{aligned}
\delta_z \phi^i &= 2i\, F^i \\
\delta_z \Psi &= 2i\, \gamma^\mu \partial_\mu \Psi \\
\delta_z F^i &= 2i\, \Box \phi^i,
\end{aligned} \tag{4.41}$$

corresponds to a central charge:

$$[\delta_s, \delta_z] = 0. \tag{4.42}$$

The invariant Lagrangian is simply

$$\mathcal{L} = \partial^\mu \phi^{i*} \partial_\mu \phi^i + F^{i*} F^i + i\bar{\Psi}\gamma^\mu \partial_\mu \Psi - m \left[i\phi^{i*} F^i - iF^{i*} \phi^i + \bar{\Psi}\Psi \right]. \tag{4.43}$$

We see that the supersymmetry algebra closes on shell only in the case of massless fields ($F^i = 0$, $\gamma^\mu \partial_\mu \Psi = 0$, $\Box \phi^i = 0$). In the case of nonvanishing mass, one must take into account the central charge. We note that

$$\delta_z^2 = -4\Box \tag{4.44}$$

is the operator equivalent of the condition $z^2 = 4m^2$, or $z = 2m$ found at the end of the previous section to obtain short multiplets such as the hypermultiplet.

The $U(2)$ symmetry is not compatible with any self-interactions of the hypermultiplet. The only allowed interactions are gauge interactions. If we couple the hypermultiplet with the vector supermultiplet introduced in the previous subsection, the Lagrangian (4.43) becomes, for $m = 0$,

$$\mathcal{L} = D^\mu \phi^{i\dagger} D_\mu \phi^i + F^{i\dagger} F^i + i\bar{\Psi}\gamma^\mu D_\mu \Psi$$

$$+ g\sqrt{2}\,\phi^{i\dagger}\bar{\lambda}^i \Psi + g\sqrt{2}\bar{\Psi}\lambda^i \phi^i - g\bar{\Psi}(A - i\gamma_5 B)\Psi$$

$$- g\phi^{i\dagger}(\sigma^r)^{ij}\,P^r \phi^j - g^2 \phi^{i\dagger}\left(A^2 + B^2\right)\phi^i. \tag{4.45}$$

The supersymmetry and central charge transformations (4.39) and (4.41) of the hypermultiplet then include extra terms involving the gauge supermultiplet.

4.5 BPS states

Solitons, which can be defined as a (partially) localized and nondispersive concentration of energy will play an increasingly important rôle in the remainder of this book. Since they provide a beautiful illustration of the uses of short supermultiplets we will take the opportunity to present some key notions using as examples kinks and monopoles. This will allow us to illustrate important new notions such as BPS states, moduli space, zero modes, duality, and their connection with supersymmetry.

4.5.1 Introduction: The kink solution and domain walls

We start by illustrating some of the basic methods used in this section by considering the simpler case of domain walls. Domain walls appear in models where a discrete symmetry is broken: two regions of space which correspond to two different ground states are separated by a sheet-like structure where energy is concentrated (corresponding to the energy barrier between the two ground states). We take as an example a *real* scalar field whose potential energy $V(\phi)$, being invariant under the reflection $\phi \leftrightarrow -\phi$, has minima at $\phi = \pm\,\phi_0$ ($\phi_0 > 0$): when we need to be more explicit, we choose $V(\phi) = \frac{1}{2}\lambda(\phi^2 - \phi_0^2)^2$. We will assume that the domain wall is flat and parallel to the y–z plane. The problem becomes $1 + 1$ dimensional, *i.e.* depends only on time t and on the spatial coordinate x and we can write the scalar field action

$$S = \int dt \int_{-\infty}^{+\infty} \left\{ \frac{1}{2}\left(\frac{\partial\phi}{\partial t}\right)^2 - \frac{1}{2}\left(\frac{\partial\phi}{\partial x}\right)^2 - V(\phi) \right\} dx. \tag{4.46}$$

The corresponding energy per unit area reads:

$$E = \int_{-\infty}^{+\infty} \left[\frac{1}{2}\left(\frac{\partial\phi}{\partial t}\right)^2 + \frac{1}{2}\left(\frac{\partial\phi}{\partial x}\right)^2 + V(\phi) \right] dx. \tag{4.47}$$

Finiteness requires $|\phi| \to |\phi_0|$ for $x \to \pm\infty$. We will be focussing our attention on the kink (resp. antikink) solution where $\phi \to \pm\phi_0$ (resp. $\phi \to \mp\phi_0$) as $x \to \pm\infty$. This twisted boundary condition stabilizes the solution: changing for example from $-\phi_0$ to $+\phi_0$ all the way to $x \to -\infty$ in a finite time would require an infinite amount of energy. In other words

$$T \equiv \int_{-\infty}^{+\infty} dx \frac{\partial\phi}{\partial x} = \phi(+\infty) - \phi(-\infty) \tag{4.48}$$

may be interpreted as a topological charge: being the integral of a total divergence, it vanishes for topologically trivial backgrounds[8].

One may try to solve the static equation of motion

$$\frac{d^2\phi}{dx^2} = \frac{dV}{d\phi} \tag{4.49}$$

which is second order. [47] has introduced a trick which allows us to identify a first order equation and simplifies the task of finding an explicit solution in more complicated situations. One may write (4.47) for the static solution $\phi(x)$

$$E = \int_{-\infty}^{+\infty} \frac{1}{2}\left[\frac{d\phi}{dx} \pm \sqrt{2V(\phi)}\right]^2 dx \mp \int_{\phi(-\infty)}^{\phi(+\infty)} \sqrt{2V(\phi)}\, d\phi \tag{4.50}$$

and conclude that

$$E \geq \left| \int_{\phi(-\infty)}^{\phi(+\infty)} \sqrt{2V(\phi)}\, d\phi \right| \tag{4.51}$$

the equality being obtained for

$$\frac{d\phi}{dx} = \mp\sqrt{2V(\phi)} \tag{4.52}$$

where the sign is chosen in order to obtain the absolute value in (4.51) and depends on the asymptotic values of ϕ: the choice $\phi(\pm\infty) = \pm\phi_0$ yields a plus sign.

We have thus obtained a bound on the energy which depends only on the asymptotic values of the field. Moreover, any solution of (4.52) minimizes E and thus the

---

[8]More precisely, we may define a current $j^\mu = \varepsilon^{\mu\nu}\partial_\nu\phi$ where $\varepsilon^{\mu\nu}$ is the antisymmetric tensor with two indices ($\mu = 0$ for t and 1 for x). This current is conserved by construction: $\partial_\mu j^\mu = \varepsilon^{\mu\nu}\partial_\mu\partial_\nu\phi = 0$. Thus the topological charge T is conserved: $T = \int dx\, j^0$.

action S in the static case: it is therefore a solution of the equation of motion (4.49) as can be checked directly. In the specific case considered, one finds

$$\phi(x) = \phi_0 \tanh\left[m(x - x_0)/\sqrt{2}\right] \tag{4.53}$$

where we have introduced the mass of the ϕ excitation: $m = \sqrt{2\lambda}\phi_0$. The energy (rest mass) per unit area is obtained directly from the boundary value (4.51):

$$E = \frac{4}{3}\sqrt{\lambda}\,\phi_0^3 = \frac{\sqrt{2}}{3}\frac{m^3}{\lambda}. \tag{4.54}$$

Note that, for a given value of the elementary excitation mass, the soliton rest mass is inversely proportional to the coupling λ: the smaller the coupling, the larger the mass. This is to be contrasted with the standard situation of the Higgs mechanism where masses are proportional to couplings.

The kink is thus characterized by the unique collective coordinate x_0 which fixes the position of the domain wall (*i.e.* where energy is localized). One calls such a collective coordinate a modulus and the space of solutions of minimal energy (for a given topological charge) is referred to as the moduli space: in this case, since $x_0 \in \mathbb{R}$, it is isomorphic to $\mathbb{R}$.

A boost in the x direction may move this position to any arbitrary value. Thus, one may turn the static solution into a dynamical one by letting x_0 be time-dependent: $x_0(t)$. This motion in moduli space corresponds to an adiabatic evolution through the space of static solutions. Of course, other time dependent configurations of the system may be considered but this motion dominates at long wavelength over all other excitations: it corresponds to a massless fluctuation around the kink solution, what we will call later a zero mode. If we want to quantize around the kink background, we must first factor out this type of mode [316].

Replacing $\phi(x,t)$ by $\phi(x)$ in (4.53) with $x_0 = x_0(t)$, one obtains for the action (4.46)

$$S = \frac{4}{3}\sqrt{\lambda}\,\phi_0^3 \int dt \left[\frac{1}{2}\left(\frac{dx_0}{dt}\right)^2 - 1\right] \tag{4.55}$$

which is, up to an additive constant, the action of a nonrelativistic particle.

The analysis has been conducted until now at the classical level: for example, the estimate of the rest mass of the soliton has been purely classical. We will now see that one can go further in the context of supersymmetry. Indeed, if we turn supersymmetric, we must replace the action (4.46) by a supersymmetric two-dimensional action. For simplicity, in what follows, we simply ignore the two extra spatial dimensions associated with the domain wall. The action reads:

$$\mathcal{S} = \int d^2x \left[\frac{1}{2}\left(\partial_\mu\phi\right)^2 + \frac{1}{2}\bar{\psi}i\partial\!\!\!/\psi - \frac{1}{2}W'^2(\phi) - \frac{1}{2}W''(\phi)\bar{\psi}\psi\right] \tag{4.56}$$

where $\mu = 0, 1$ correspond to the variables $x^0 \equiv t$ and $x^1 \equiv x$, respectively, and ψ is a Majorana fermion in two dimensions. The scalar potential is, as usual, written in terms of the superpotential $V(\phi) = W'^2(\phi)/2$. We recover the previous potential for $W(\phi) = \sqrt{\lambda}(\phi\phi_0^2 - \phi^3/3)$.

One can then define [368] (see Problem 1) a supersymmetry charge with components Q_+, Q_- satisfying

$$Q_+^2 = P_+, \quad Q_-^2 = P_-$$
$$Q_+Q_- + Q_-Q_+ = Z \tag{4.57}$$

where $P_\pm \equiv P_0 \pm P_1$ and

$$Z = \int dx \, 2W'(\phi)\frac{\partial \phi}{\partial x} = \int dx \frac{\partial}{\partial x}[2W(\phi)] . \tag{4.58}$$

Note that Z is of a topological nature, much in the way of the charge T introduced in (4.48). It vanishes for topologically trivial backgrounds but is nonzero for the kink: for $W(\phi) = \sqrt{\lambda}(\phi\phi_0^2 - \phi^3/3)$, we have $Z = 8\sqrt{\lambda}\phi_0^3/3$.

Now, using the algebra (4.57), one has

$$P_+ + P_- = (Q_+ \pm Q_-)^2 \mp Z \ge |Z|. \tag{4.59}$$

Thus at rest, since $P_+ = P_- = E$, we find

$$E \ge \frac{1}{2}\,|Z| = \left|\int \sqrt{2V(\phi)}\,d\phi\right| \tag{4.60}$$

just as in (4.51). This inequality is saturated for all states $|\alpha\rangle$ satisfying

$$(Q_+ \pm Q_-)\,|\alpha\rangle = 0. \tag{4.61}$$

The kink solution satisfies this equation with a $-$ sign. The orthogonal combination is broken by the kink solution: indeed, the domain wall obviously breaks translational invariance in the x direction.

We now turn to truly four-dimensional examples. In this case, the rôle of the topological charge is played by the electric or magnetic charge. Indeed, the electric charge can be written as a surface integral: using Gauss' law, one may write for the electric charge

$$Q_e = \int d^3x \, \partial_i E_i = \int dS_i E_i, \tag{4.62}$$

which can be compared with the two-dimensional case just discussed, equation (4.58). A discussion of the magnetic charge requires some background on magnetic monopoles which we now present.

4.5.2   Monopoles and the Dirac quantization condition

Magnetic monopoles appear naturally when one tries to introduce sources that respect the electric–magnetic duality observed in Maxwell equations in the vacuum. We start by reviewing the argument: this notion of duality will play an important rôle in subsequent chapters (see Chapter 8).

Maxwell equations in the vacuum

$$\partial_\mu F^{\mu\nu} = 0, \quad \partial_\mu \tilde{F}^{\mu\nu} = 0, \quad \tilde{F}_{\mu\nu} \equiv \tfrac{1}{2}\epsilon_{\mu\nu\rho\sigma}F^{\rho\sigma} \tag{4.63}$$

are invariant under the duality transformation

$$F^{\mu\nu} \to \tilde{F}^{\mu\nu}. \tag{4.64}$$

This may be rephrased in terms of the electric $F^{0i} = -E_i$ and magnetic $F^{ij} = -\epsilon_{ijk}B_k$ fields[9]:

$$\mathbf{E} \to \mathbf{B}, \quad \mathbf{B} \to -\mathbf{E}, \tag{4.65}$$

which may be generalized to the continuous transformation

$$\mathbf{E}' + i\mathbf{B}' = e^{i\theta}\left(\mathbf{E} + i\mathbf{B}\right). \tag{4.66}$$

However, introducing charged matter through the current j_E^μ

$$\partial_\mu F^{\mu\nu} = j_E^\nu, \quad \partial_\mu \tilde{F}^{\mu\nu} = 0 \tag{4.67}$$

introduces an imbalance between electric and magnetic. At the classical level, the obvious solution is to introduce magnetic sources through a magnetic current j_M^ν in the right-hand side of the second equation. This is not as straightforward in the quantum theory because the wave equation for a charge particle involves the vector potential A_μ which is ill-defined once one introduces magnetic charges; indeed, if the vector potential $\mathbf{A}$ is uniquely defined, we can surround any magnetic charge Q_m by a closed surface S and write

$$Q_m = \int_S \mathbf{B}.d\mathbf{S} = \int_S \nabla \wedge \mathbf{A}.d\mathbf{S} = 0. \tag{4.68}$$

However, the description of the electromagnetic field by the potential A_μ is redundant: it is the famous gauge invariance $A'_\mu = A_\mu - \partial_\mu\alpha$. Since A_μ and A'_μ describe the same physical reality, we may use either of them in different parts of space, under the condition that one makes sure that there is full compatibility in overlapping regions. This is precisely what one does in the presence of magnetic charges since $\mathbf{A}$ cannot be uniquely defined.

Let us consider a magnetic monopole Q_m. As shown on Fig. 4.1, we draw a closed surface around the monopole and consider a plane going through the monopole. The vector potential is $\mathbf{A}^{(1)}$ above the plane and $\mathbf{A}^{(2)}$ below: in order that the electromagnetic field be uniquely defined, we must have $\mathbf{A}^{(1)} - \mathbf{A}^{(2)} = \nabla\alpha$. The plane divides the surface between an upper hemisphere $S^{(1)}$ and a lower sphere $S^{(2)}$: $S^{(1)} \cap S^{(2)}$ is a curve C parametrized by an angle ϕ $(0 \le \phi \le 2\pi)$. Then

$$Q_m = \int_{S^{(1)}\cup S^{(2)}} \mathbf{B}.d\mathbf{S} = \int_{S^{(1)}} \mathbf{B}.d\mathbf{S} + \int_{S^{(2)}} \mathbf{B}.d\mathbf{S}$$

$$= \int_C \left(\mathbf{A}^{(1)} - \mathbf{A}^{(2)}\right).d\mathbf{l} = \alpha(2\pi) - \alpha(0). \tag{4.69}$$

---

[9]Note that the square of this transformation corresponds to charge conjugation.

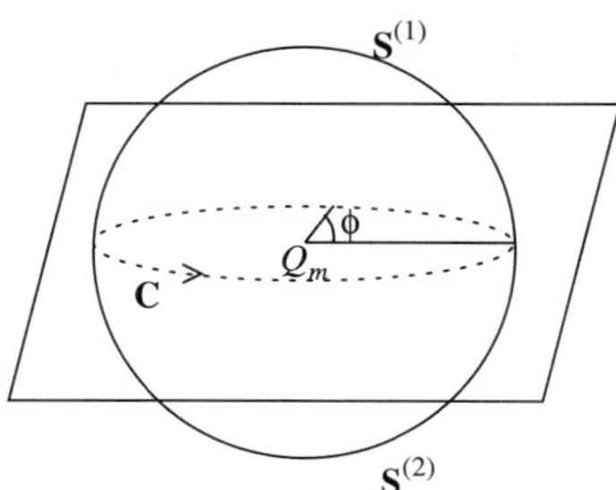

Fig. 4.1 Closed surface around a magnetic monopole

Under the gauge transformation, the wave function corresponding to a field of electric charge Q_e transforms as

$$\psi'(x) = e^{iQ_e\alpha/\hbar}\psi(x). \tag{4.70}$$

Requiring such wave functions to be single-valued (as $\phi \to \phi + 2\pi$) implies that

$$Q_e Q_m = 2n\pi\hbar, \quad n \in \mathbb{Z}, \tag{4.71}$$

which is the famous Dirac quantization condition [118, 381]: the presence of a single monopole imposes that all electric charges are multiples of a given unit of charge $(2\pi\hbar/Q_m)$. This gives a rationale for the problem of the quantization of charge mentioned in Chapter 1. We will encounter a seemingly different reason when we discuss grand unified theories in Chapter 9 but we will see that, at a deeper level, the two solutions coincide.

Until now, we have only considered electric charges and magnetic monopoles. One may envisage the presence of *dyons* [241], *i.e.* objects with both electric charge Q_e and magnetic charge Q_m. If we have two dyons, the Dirac quantization condition becomes [332, 390]:

$$Q_{e1}Q_{m2} - Q_{e2}Q_{m1} = 2n\pi\hbar, \quad n \in \mathbb{Z}. \tag{4.72}$$

We deduce that, if the theory has electrons of charges $(-e, 0)$, then the quantization reads, for any dyon field (Q_e, Q_m): $Q_m = 2n\pi\hbar/e$. Let us consider for example the dyon $(Q_e, 2\pi\hbar/e)$ and assume that we have CP invariance. Under CP, the electric charge is odd whereas the magnetic charge is even. We thus also have a dyon $(-Q_e, 2\pi\hbar/e)$. Applying the quantization condition (4.72) to this pair, we obtain $Q_e = ne/2$: such dyons must have integer or half-integer charge.

Moreover, applying again the quantization condition to $(Q_e, 2\pi\hbar/e)$ and $(Q'_e, 2\pi\hbar/e)$, we obtain $Q_e - Q'_e = ne$. Hence all of them must have integer charges, or have half-integer charges.

However, if there is a source of CP violation, we may expect that there will be a departure from these quantized values. Witten [370] has in particular stressed the importance of adding the CP violating term

$$\delta\mathcal{L} = -\theta\frac{e^2}{32\pi^2}F^{\mu\nu}\tilde{F}_{\mu\nu}, \tag{4.73}$$

which gives a contribution of order θ to the dyon mass. We will return in more details to this issue in Section 4.5.7.

4.5.3 Monopoles in the Georgi–Glashow model

In order to be able to consider the magnetic charge as a topological charge, we need to unify the electromagnetic $U(1)$ symmetry as part of a larger nonabelian symmetry. The simplest case is the model of [180] (see also Exercise 9 of Appendix Appendix A) with a $SU(2)$ gauge symmetry. We note that, in such cases, the monopole is a solution of the field equations and therefore must exist, in the context of these theories. From now on, we set $\hbar = 1$.

The Lagrangian of the Georgi–Glashow model reads

$$\mathcal{L} = -\frac{1}{4}F^a_{\mu\nu}F^{a\mu\nu} - \frac{\theta}{32\pi^2}g^2 F^a_{\mu\nu}\tilde{F}^{a\mu\nu} + \frac{1}{2}D^\mu\phi^a D_\mu\phi^a - V(\phi), \tag{4.74}$$

where $(a = 1, 2, 3)$

$$F^a_{\mu\nu} = \partial_\mu A^a_\nu - \partial_\nu A^a_\mu + g\epsilon^{abc}A^b_\mu A^c_\nu \tag{4.75}$$

$$D_\mu\phi^a = \partial_\mu\phi^a + g\epsilon^{abc}A^b_\mu\phi^c, \tag{4.76}$$

and

$$V(\phi) = \frac{\lambda}{4}\left(\phi^a\phi^a - a^2\right)^2. \tag{4.77}$$

The corresponding Euler equations are

$$D_\mu F^{a\mu\nu} = -g\epsilon^{abc}\phi^b D^\nu\phi^c$$
$$(D^\mu D_\mu\phi)^a = -\lambda\phi^a\left(\phi^b\phi^b - a^2\right), \tag{4.78}$$

where $D_\mu F^{a\mu\nu}$ is defined in (A.57) of Appendix Appendix A (as is well known, the θ term does not contribute to the equation of motion thanks to the Bianchi identity $D_\mu\tilde{F}^{a\mu\nu} = 0$).

The symmetric energy–momentum tensor simply reads:

$$\Theta^{\mu\nu} = -F^{a\mu\rho}F^{a\nu}{}_\rho + D^\mu\phi^a D^\nu\phi^a - g^{\mu\nu}\mathcal{L}. \tag{4.79}$$

If we want to find the ground state, we must minimize the energy density

$$\Theta^{00} = \frac{1}{2}\left(E^a_i E^a_i + B^a_i B^a_i + D_0\phi^a D_0\phi^a + D_i\phi^a D_i\phi^a\right) + V(\phi), \tag{4.80}$$

with standard definitions for the electric and magnetic fields (see the preceding section): $E^a_i = -F^{a0i}$, $B^a_i = -\frac{1}{2}\epsilon_{ijk}F^{ajk}$. Hence the vacuum corresponds to a vanishing gauge field and a constant scalar field value that minimizes the potential: $\langle\phi^a\phi^a\rangle \equiv a^2$, which defines a two-sphere S^2_{vac}. For example, $\phi^3 = a \neq 0$ and $\phi^1 = \phi^2 = 0$ breaks $SU(2)$ to $U(1)$ and gives a mass ag to the fields $W^\pm_\mu \equiv \left(A^1_\mu \mp iA^2_\mu\right)/\sqrt{2}$, where the index $\pm$ refers to the charge under the $U(1)$ symmetry. The field $A_\mu \equiv A^3_\mu$ remains massless.

We are interested in finite energy configurations. Obviously, for the energy to remain finite, each term in (4.80) must fall off faster than $1/r^2$ at infinity. On the surface at infinity, the two-sphere S^2_∞, we must recover the vacuum. Hence a finite energy configuration defines a mapping $\phi^a: S^2_\infty \to S^2_{\text{vac}}$. Such a map is characterized by an integer, the winding number n, which counts the number of times one wraps one sphere around the other. This can be expressed as

$$n = \frac{1}{8\pi} \int_{S^2_\infty} dS_i \epsilon_{ijk} \epsilon^{abc} \hat{\phi}^a \partial_j \hat{\phi}^b \partial_k \hat{\phi}^c \tag{4.81}$$

where $\hat{\phi}^a \equiv \phi^a/a$. Note that this expression is invariant under a small change $\delta\hat{\phi}^a$ of the unit vector $\hat{\phi}^a$: it is a topological invariant[10].

We may express this in a more mathematical way. We first note that the vector $\hat{\phi}^a$ which describes the vacuum parametrizes the coset group $G/H = SU(2)/U(1)$ in the breaking of $G = SU(2)$ to $H = U(1)$ (see Section A.2.3 of Appendix Appendix A). We have thus defined a mapping from S^2_∞ to $SU(2)/U(1)$. The group of inequivalent mappings (*i.e.* mappings which cannot be deformed continuously into one another) from the k-sphere S^k to a manifold $\mathcal{M}$ (such as G/H) is called its kth homotopy group and noted $\pi_k(\mathcal{M})$. We have just established that $\pi_2(SU(2)/U(1))$ is isomorphic to $\mathbb{Z}$.

We must have in particular

$$D_i \hat{\phi}^a = \partial_i \hat{\phi}^a + g\epsilon^{abc} A_i^b \hat{\phi}^c \sim 0 \tag{4.82}$$

on the surface at infinity. This is solved by

$$A_i^a \sim -\frac{1}{g}\epsilon^{abc} \hat{\phi}^b \partial_i \hat{\phi}^c + \hat{\phi}^a A_i \tag{4.83}$$

where A_i is a $SU(2)$ invariant potential. We then deduce from (4.82) that, at large distances, the field strength is aligned along the scalar field[11], that is along the direction of unbroken $U(1)$ symmetry:

$$F_{ij}^a = \hat{\phi}^a F_{ij} \tag{4.84}$$

with

$$F_{ij} = \partial_i A_j - \partial_j A_i - \frac{1}{g}\epsilon^{abc} \hat{\phi}^a \partial_i \hat{\phi}^b \partial_j \hat{\phi}^c. \tag{4.85}$$

The magnetic charge corresponding to this field configuration is expressed in terms of the winding number n given in (4.81) as:

$$Q_m = \int_{S^2_\infty} B_i dS_i = -\frac{1}{2}\int_{S^2_\infty} \epsilon_{ijk} dS_i F_{jk} = \frac{1}{2g}\int_{S^2_\infty} dS_i \epsilon_{ijk} \epsilon^{abc} \hat{\phi}^a \partial_j \hat{\phi}^b \partial_k \hat{\phi}^c = \frac{4\pi n}{g}. \tag{4.86}$$

which corresponds to the Dirac quantization condition (4.71) since, in this theory, fields in the fundamental of $SU(2)$ (not introduced here) would have an electric charge $Q_e = g/2$. We conclude that the finite energy configurations of the Georgi–Glashow model are the magnetic monopoles.

---

[10]Since $\hat{\phi}^a$ is a unit vector, $\delta\hat{\phi}^a$, and its spatial derivatives all lie in the plane perpendicular to $\hat{\phi}^a$: thus $\delta n = 0$.

[11]Using (A.51) of Appendix Appendix A and $(T^a)^{bc} = -i\epsilon^{abc}$ in the adjoint representation of $SU(2)$, we have

$$0 = [D_i, D_j]^{ab} \hat{\phi}^b = -g\epsilon^{abc} \hat{\phi}^b F_{ij}^c = -g\left(\hat{\phi} \wedge \mathbf{F}_{ij}\right)^a.$$

More generally, if the symmetry group of a theory is G and the unbroken subgroup H, then monopoles exist if the second homotopy group $\pi_2(G/H)$ is nontrivial, *i.e.* not reduced to the identity. If G is simply connected (*i.e.* any loop can be continuously deformed into a point: $\pi_1(G)$ is trivial) such as $SU(n)$ or any simple compact connected Lie group besides $SO(n)$, then

$$\pi_2(G/H) = \pi_1(H). \tag{4.87}$$

For example, if $H = U(1)$, $\pi_1(U(1)) = \mathbb{Z}$, then monopoles exist and are characterized by a integer n. This is what happens with the 't Hooft–Polyakov model just discussed or the breaking of the grand unified group $SU(5)$ (see Chapter 9). If $H = SO(n)$ ($n \geq 3$), $\pi_1(SO(n)) = \mathbb{Z}_2$, there is a single type of monopole corresponding to the nontrivial element -1 of $\mathbb{Z}_2$. If G is not simply connected, one may find useful the following theorem: $\pi_2(G/H)$ is isomorphic to the kernel of the homomorphism of $\pi_1(H)$ into $\pi_1(G)$ (a loop in H is naturally a loop in G).

4.5.4 BPS monopoles

We note that, according to (4.84), we may write the electric and magnetic charges with respect to $U(1)$ as

$$Q_e = \frac{1}{a} \int dS_i \; E_i^a \phi^a$$

$$Q_m = \frac{1}{a} \int dS_i \; B_i^a \phi^a. \tag{4.88}$$

Following [47], we may derive a lower bound for the energy of the stationary configurations that we have just identified. For simplicity, we work in the temporal gauge[12] $A_0^a = 0$. We obtain for the energy density (4.80)

$$\mathcal{E} = \mathcal{T} + \mathcal{V},$$

$$\mathcal{T} = \frac{1}{2} \int d^3x \; [\partial_0 A_i^a \partial_0 A_i^a + \partial_0 \phi^a \partial_0 \phi^a], \tag{4.90}$$

$$\mathcal{V} = \frac{1}{2} \int d^3x \; [(B_i^a \mp D_i\phi^a)(B_i^a \mp D_i\phi^a) + 2V(\phi)] \pm \int d^3x \; B_i^a D_i\phi^a,$$

where we have separated the contribution which is quadratic in time derivatives ("kinetic energy" $\mathcal{T}$) from the one which does not involve time derivatives ("potential" $\mathcal{V}$). Since, according to (4.88), $aQ_m = \int d^3x \; B_i^a D_i\phi^a$, one obtains (choosing the upper, respectively lower, sign for positive, respectively negative, magnetic charge):

$$\mathcal{E} \geq a\,|Q_m|. \tag{4.91}$$

---

[12] Accordingly, we must impose the Gauss law constraint, which corresponds to ensuring the A^0 equation of motion (4.78):

$$D_i\left(\partial_0 A_i^a\right) + g\epsilon^{abc}\phi^b \partial_0 \phi^c = 0. \tag{4.89}$$

The bound is saturated for states satisfying the Bogomol'nyi conditions

$$E_i^a = 0 \ , \ \ D_0\phi^a = 0 \tag{4.92}$$

$$B_i^a = \pm\, D_i\phi^a \tag{4.93}$$

in the limit of vanishing potential $V(\phi)$. This limit is known as the Prasad–Sommerfield limit [315] and the states which saturate the bound are thus known as BPS states. The upper (lower) sign corresponds to the monopole (antimonopole) solution.

As we will emphasize later, the Prasad–Sommerfield condition of vanishing potential is unstable under quantum fluctuations unless we are in a supersymmetric context.

The Bogomol'nyi equations can be solved using the following static and spherically symmetric ansatz:

$$\phi^a(\mathbf{x}) = aH(\xi)\frac{x^a}{r}, \quad A_0^a(\mathbf{x}) = 0,$$

$$A_i^a(\mathbf{x}) = \epsilon^{aij}\frac{x^j}{gr^2}\,(1 - K(\xi)) \tag{4.94}$$

where $r = |\mathbf{x}|$ and H and K are functions of the rescaled radial variable $\xi = agr$. Plugging this ansatz into the Bogomol'nyi equation (4.93) for the (anti)monopole yields

$$H'(\xi) = \mp\frac{1}{\xi^2}(K^2 - 1)$$

$$K'(\xi) = \mp\, KH, \tag{4.95}$$

the boundary conditions being $H \to 0$, $K \to 1$ as $\xi \to 0$ and $H \to 1$, $K \to 0$ as $\xi \to \infty$. The solution corresponding to the monopole (upper sign in the preceding equation) reads:

$$H(\xi) = \coth\xi - \frac{1}{\xi}, \quad K(\xi) = \frac{\xi}{\sinh\xi}. \tag{4.96}$$

4.5.5 Moduli space

Before quantizing the monopole, we need to identify its collective coordinates. The monopole solution which has been constructed in the previous subsection sits at the origin but the translation invariance of the equations implies that a solution with the same energy can be constructed at any point in space. Thus the spatial coordinates of the monopole center of mass provide three collective coordinates.

Since the inclusion of these degrees of freedom does not change the (potential) energy of the monopole, they are zero modes, or moduli, of the system and their quantization requires a special treatment. Just as for the kink, the standard way [316] is to start with the classical solution (4.94) which we denote $\phi_{\mathrm{cl}}^a(\mathbf{x})$, $A_{i\,\mathrm{cl}}^a(\mathbf{x})$ and to translate this solution to a time-dependent position $\mathbf{X}(t)$ to form the ansatz:

$$\phi^a(\mathbf{x}, t) = \phi_{\mathrm{cl}}^a\left(\mathbf{x} - \mathbf{X}(t)\right), \quad A_i^a(\mathbf{x}, t) = A_{i\,\mathrm{cl}}^a\left(\mathbf{x} - \mathbf{X}(t)\right). \tag{4.97}$$

This represents, as long as the position $\mathbf{X}(t)$ is a slowly moving function, the low energy ansatz obtained by integrating out the massive modes (*i.e.* ignoring in the semiclassical approach contributions of order $\hbar\omega_i$ where ω_i are the nonvanishing frequencies associated with the eigenvalues of the double derivatives of the potential energy, evaluated around the monopole solution). Substituting this ansatz into the action yields

$$S = \int dt \mathcal{L} = \int dt \, (\mathcal{T} - \mathcal{V}) = \frac{4\pi a}{g} \int dt \left[\frac{1}{2} \left(\frac{d\mathbf{X}(t)}{dt} \right)^2 - 1 \right]. \tag{4.98}$$

This is the moduli space approximation: the motion in moduli space is made at constant potential energy since moduli are associated with the zero modes of the system. One remains with a kinetic energy associated with the slow motion in moduli space (here in position space).

In order to search for all the moduli fields, we adopt a more general strategy. We start with the BPS monopole solution satisfying $B_i^a = D_i\phi^a$ and deform it ($A_i^a + \delta A_i^a$, $\phi^a + \delta\phi^a$) while keeping the potential energy fixed. Then, to first order, the Bogomol'nyi equation yields:

$$-\epsilon_{ijk} D^j \delta A^{ak} = D_i \delta\phi^a + g\epsilon^{abc} \delta A_i^b \phi^c, \tag{4.99}$$

where D_i is the covariant derivative in the adjoint representation ($D_i X^a = \partial_i X^a + g\epsilon^{abc} A_i^b X^c$).

This has to be complemented by the Gauss law constraint (4.89): writing for a general modulus Z, as in (4.97),

$$\phi^a(\mathbf{x}, t) = \phi_{\text{cl}}^a(\mathbf{x}, Z(t)), \quad A_i^a(\mathbf{x}, t) = A_{i\,\text{cl}}^a(\mathbf{x}, Z(t)), \tag{4.100}$$

the constraint reads

$$D_i(\delta A_i^a) + g\epsilon^{abc} \phi^b \delta\phi^c = 0. \tag{4.101}$$

A general deformation is a combination of spacetime translation and gauge transformation ($\delta A_i^a = -(1/g)D_i\alpha^a$, $\delta\phi^a = C^{abc}\alpha^b\phi^c$). We have already considered moduli associated with spacetime translations. The only remaining modulus corresponds to the gauge transformation $\exp[i\chi(t)T^a\phi^a(\mathbf{x})/a]$. We note that, since ϕ^a is nontrivial at spatial infinity, this gauge transformation does not approach the identity: in the standard language of gauge theories, this is a *large* gauge transformation which relates two distinct physical points (by comparison, a *small* gauge transformation, *i.e.* one that approaches the identity at spatial infinity, corresponds to a mere redundancy in the description of physical states). Writing thus $\alpha^a(x) = -\chi(t)\phi^a(\mathbf{x})/a$, the new modulus corresponds to the deformation[13]

$$\delta A_i^a = \frac{1}{ga} D_i[\chi(t)\phi^a(\mathbf{x})], \quad \delta\phi^a = 0. \tag{4.102}$$

It is straightforward to check that this satisfies the linearized Bogomol'nyi constraint (4.99) (using the original equation $B_i^a = D_i\phi^a$) and the Gauss law constraint (4.101)

---

[13] In order to keep $A_0^a = 0$ we must complement the ansatz (4.100) with $A_0^a = -\dfrac{1}{ga}\dot\chi(t)\phi^a(\mathbf{x})$.

(using the scalar field equation of motion (4.78) with vanishing scalar potential). We see that the new coordinate is conjugate to the electric charge (just as the space coordinates are conjugate to momentum): motion in moduli space generates momentum and charge for the monopoles, which thus become dyons. We note that there are no moduli corresponding to angular momentum: classical motion in moduli space does not generate intrinsic angular momentum for the monopoles.

Since the unbroken $U(1)$ gauge group is compact, χ has a compact range (typically $0 \leq \chi < 2\pi$). Thus the corresponding space is the circle S^1 and the full moduli space for a single monopole[14] is $\mathbb{R}^3 \times S^1$. In the moduli space approximation, we now have

$$S = \frac{4\pi a}{g} \int dt \left[\frac{1}{2} \left(\frac{d\mathbf{X}(t)}{dt} \right)^2 + \frac{1}{2a^2 g^2} \left(\frac{d\chi(t)}{dt} \right)^2 - 1 \right]. \tag{4.103}$$

We note that the corresponding plane waves are simply $e^{i\mathbf{P}\cdot\mathbf{X}} e^{in_e\chi}$, with n_e integer. Charge is simply $Q_e = -ig\partial_\chi$: such plane waves correspond to dyons with charge $Q_e = n_e g$. We may also infer their masses in this approximation: since the corresponding Hamiltonian is

$$\mathcal{H} = \frac{1}{2} \frac{g}{4\pi a} \mathbf{P}^2 + \frac{1}{2} \frac{ag^3}{4\pi} P_\chi^2 + \frac{4\pi a}{g}, \tag{4.104}$$

where $\mathbf{P}$ and P_χ are the canonical momenta associated with $\mathbf{X}$ and χ, respectively, the mass of the dyons is simply

$$M = \frac{4\pi a}{g} + \frac{1}{2} \frac{n_e^2 ag^3}{4\pi} = \frac{4\pi a}{g} \left[1 + \frac{1}{2} \frac{n_e^2 g^4}{16\pi^2} \right]. \tag{4.105}$$

It can be showed [81] that the classical solutions of the equations of motion satisfy the general Bogomol'nyi-type bound $M \geq a\sqrt{Q_e^2 + Q_m^2}$. We see that the dyons saturate this bound: indeed, (4.105) is the expansion for $g \ll 1$ of the formula

$$M = a\sqrt{Q_e^2 + Q_m^2} = a\,|Q_e + iQ_m|, \tag{4.106}$$

since $Q_m = 4\pi/g$ and $Q_e = n_e g$.

4.5.6 Supersymmetry

The picture that emerged in the last section is likely to suffer some undesirable modifications once one includes quantum corrections. First, as already noted, radiative corrections should generate a nonvanishing potential, thus contradicting the Prasad–Sommerfield condition of vanishing potential. Similarly, the mass bounds should be corrected by quantum fluctuations. Obviously, this makes such bounds less interesting.

We now know enough about supersymmetry to understand that it might provide the extra ingredient required to control quantum corrections. Indeed, both vanishing potentials and masses are protected by supersymmetry. Since we already have vector ($W_\mu^\pm$ and A_μ) and scalar (monopole) degrees of freedom, the natural set up, once one includes fermionic degrees of freedom, is $N = 2$ supersymmetry.

---

[14]More generally, the moduli space of n monopoles is $4n$-dimensional [357].

We thus consider a $N = 2$ vector supermultiplet, whose action is written in (4.36) (we will set $P^r = 0$ and thus consider the on-shell action). We note that the condition of vanishing potential requires $[A, B] = 0$, which may be satisfied for nonzero vacuum expectation values of A and B. Supersymmetry ensures that this holds at the quantum level.

Choosing the gauge symmetry group $SU(2)$, as in the Georgi–Glashow model, we may set the vacuum expectation value for B to zero. A nonzero value for A (say $\langle A^3 \rangle = a \neq 0$) breaks $SU(2)$ to $U(1)$. The supersymmetry current for the model considered reads (see (3.37) in Chapter 3 and (C.73) in Appendix C):

$$J_{\mu i} = \sigma^{\rho\sigma} F_{\rho\sigma} \gamma_\mu \gamma_5 \lambda_i + \frac{1}{\sqrt{2}} D^\rho A \gamma_\rho \gamma_\mu \lambda_i + \frac{1}{\sqrt{2}} D^\rho B \gamma_\rho \gamma_\mu \gamma_5 \lambda_i + g \gamma_\mu \gamma_5 [A, B] \lambda_i. \quad (4.107)$$

One may then deduce the supersymmetry charge anticommutation relations, with special attention to the boundary terms. Introducing

$$U = \int d^3x \; \partial_i \left(A^a F_{0i}^a + B^a \frac{1}{2} \epsilon_{ijk} F_{jk}^a \right)$$

$$V = \int d^3x \; \partial_i \left(A^a \frac{1}{2} \epsilon_{ijk} F_{jk}^a + B^a F_{0i}^a \right), \quad (4.108)$$

one obtains [369]

$$\{Q_{ri}, \bar{Q}_{sj}\} = \delta_{ij} \gamma_{rs}^\mu P_\mu + \epsilon_{ij} \left(\delta_{rs} U + \gamma_{rs}^5 V \right). \quad (4.109)$$

Since we set $B = 0$, we have, using (4.88),

$$U = aQ_e, \quad V = aQ_m. \quad (4.110)$$

We see that, in (4.108), U and V play the rôle of central charges. We thus expect mass bounds similar to the ones derived in subsection 4.3.3. Indeed, if we work in the frame where $P_\mu = (M, 0, 0, 0)$, then (4.109) reads

$$\left\{Q_{ri}, Q_{sj}^\dagger\right\} = M \delta_{ij} \delta_{rs} + a \epsilon_{ij} \left[\left(Q_e + i\gamma^5 Q_m\right) \gamma^0 \right]_{rs}. \quad (4.111)$$

Since the left-hand side is positive definite, the eigenvalues of the right-hand side are positive. But $\left[\left(Q_e + i\gamma^5 Q_m\right) \gamma^0 \right]^2 = \left(Q_e + i\gamma^5 Q_m\right) \left(Q_e - i\gamma^5 Q_m\right) = Q_e^2 + Q_m^2$. The eigenvalues are then $M \pm \sqrt{Q_e^2 + Q_m^2}$ and one obtains the constraint

$$M \geq a \sqrt{Q_e^2 + Q_m^2}. \quad (4.112)$$

Obviously states which saturate the bound are annihilated by at least one of the supersymmetry generators.

We note that all states in the theory saturate the bound: for example, the photon A_μ is massless whereas $W_\mu^\pm$ has mass ga; and dyon masses are given by (4.106). This explains what could have been seen as a puzzle otherwise. Indeed, we started with the eight states of the (non-CPT self-conjugate) $N = 2$ vector supermultiplet: A^μ, λ^i, A, B. Symmetry breaking gives mass to some of these states but does not provide the degrees of freedom necessary to form a (long) massive $N = 2$ vector supermultiplet, which has 16 degrees of freedom. But these eight states form a short massive $N = 2$ supermultiplet, in presence of the central charges.

This also explains why the Bogomol'nyi bound remains unaffected by radiative corrections or nonperturbative effects in this context. If it did, this would mean that the mass spectrum departs from the one fixed by the central charges. The massive states would then fall into long supermultiplets but we are lacking the necessary extra degrees of freedom.

4.5.7   Montonen–Olive duality

Since magnetic monopoles carry magnetic charge, they generate a magnetic field which should be described by a gauge theory. The coupling associated with this theory should satisfy, according to (4.86),

$$g_m = \frac{4\pi}{g}. \tag{4.113}$$

A key property of this relation is that it exchanges strong and weak coupling regimes: if the "electric" gauge symmetry is weakly coupled, then the "magnetic" gauge symmetry is strongly coupled, and vice versa. This implies that the fundamental degrees of freedom of both theories are not simultaneously accessible: in the electric description, the degrees of freedom of magnetic charge $q_m \equiv Q_m/g_m = \pm 1$ (*i.e.* the magnetic monopoles) arise as solitons; conversely, the gauge fields of charge $q_e \equiv Q_e/g = \pm 1$ should arise as solitons of the magnetic theory. It might seem surprising to find here topological solitons with spin. It is well-known [223, 236] that, in the context of the quantum theory, monopoles convert isospin into spin.

Making full use of the mass spectrum derived above, *i.e.* $M = a\,|Q_e + iQ_m|$, we may represent the different fields on the lattice (q_e, q_m). The fundamental fields of the electric gauge theory are located at points $(0,0)$ and $(\pm 1, 0)$ whereas the fundamental fields of the magnetic gauge theory are found at $(0,0)$ and $(0, \pm 1)$. The sites off the axis correspond to dyon states. The electric–magnetic duality encapsuled in (4.113) appears in this diagram as the symmetry associated with a rotation of $\pi/2$. Such a duality was conjectured by [288].

Besides the mass spectrum, a nontrivial check is provided by monopole interactions. The force between two (positively charged) monopoles is velocity-dependent [281]: it vanishes at rest. Duality would imply that the static W^+-W^+ force vanishes as well. It turns out that the standard Coulomb repulsion is cancelled by the force resulting from Higgs exchange (note that, in the Prasad–Sommerfield limit, the Higgs is massless).

Obviously, it is crucial to the Montonen–Olive duality conjecture that the mass formula does not receive contributions from radiative corrections or nonperturbative effects. This is why it has to be considered in the supersymmetric context. We have already discussed what $N = 2$ supersymmetry brings forth. It is only when one goes to $N = 4$ supersymmetry that one finds monopole of spin 1 and thus a full realization of Montonen–Olive duality [301][15]. This goes beyond the scope of this chapter and we refer the reader to the existing reviews on the subject (see for example [222]).

---

[15]Moreover the beta function of the $N = 4$ symmetric Yang–Mills theory vanishes, which settles the question of whether the couplings that we have been using are running or not.

Associated with the electric–magnetic duality, there is an interesting $SL(2, \mathbb{Z})$ symmetry. To make it explicit, we must pay more attention to the θ-term (4.73). In the Prasad–Sommerfield limit ($V = 0$), the Georgi–Glashow Lagrangian (4.74) may be written

$$\mathcal{L} = -\frac{1}{32\pi} \text{Im}\tau \left(F^{a\mu\nu} + i\tilde{F}^{a\mu\nu} \right) \left(F^a_{\mu\nu} + i\tilde{F}^a_{\mu\nu} \right) + \frac{1}{2} D^\mu \Phi^a D_\mu \Phi^a, \tag{4.114}$$

where we have rescaled the gauge fields by the coupling constant and

$$\tau = \frac{\theta}{2\pi} + \frac{4\pi i}{g^2}. \tag{4.115}$$

Since there is periodicity under $\theta \to \theta + 2\pi$, the transformation $\tau \to \tau + 1$ should be a symmetry. Moreover, when $\theta = 0$, the duality invariance ((4.113) with $g_m = g$) reads $\tau \to -1/\tau$. These two transformations generate the group $SL(2, Z)$ of projective transformations:

$$\tau \to \frac{a\tau + b}{c\tau + d}, \quad a, b, c, d \in \mathbb{Z}, \ ad - bc = 1. \tag{4.116}$$

It has been noted by [370] that, once one restores a nonzero vacuum angle θ, the monopole charge is shifted by an amount proportional to θ. To see this, we introduce the charge N associated with the dyon collective coordinate χ introduced earlier. Following (4.102), we have

$$\begin{aligned}
N &= \int d^3x \, \frac{1}{ga} \frac{\delta\mathcal{L}}{\delta\partial_0 A^a_i} D_i\phi^a \\
&= \frac{1}{ga} \int d^3x D_i\phi^a F^a_{0i} + \frac{\theta g}{8\pi^2 a} \int d^3x D_i\phi^a \frac{1}{2}\epsilon_{ijk} F^a_{jk} \\
&= \frac{1}{ga} \int d^3x \partial_i \left(\phi^a F^a_{0i} \right) + \frac{\theta g}{8\pi^2 a} \int d^3x \partial_i \left(\phi^a \frac{1}{2}\epsilon_{ijk} F^a_{jk} \right).
\end{aligned} \tag{4.117}$$

This gives, using (4.88),

$$N = \frac{1}{g} Q_e + \frac{\theta g}{8\pi^2} Q_m. \tag{4.118}$$

Because n is associated with a gauge transformation about ϕ^a, and a $SO(3) \sim SU(2)/\mathbb{Z}_2$ rotation of 2π is identity, we have the condition $e^{2\pi N} = 1$. Hence N is an integer n_e and the charge is ($Q_m = 4\pi/g$ for a monopole)[16]

$$Q_e = n_e g - \frac{\theta g}{2\pi} n_m. \tag{4.119}$$

---

[16]In the moduli space approximation, this gives $Q = -ig\partial_\chi - (g\theta/2\pi)n_m$.

We see that the BPS mass formula for a dyon (4.106) is now given by

$$M = a\,|n_e + n_m \tau|,\tag{4.120}$$

and the $SL(2, Z)$ transformation acts on this integers as

$$(n_m, n_e) \rightarrow (n_m, n_e)\begin{pmatrix} a & b \\ c & d \end{pmatrix}^{-1}.\tag{4.121}$$

Further reading

- M. Sohnius, *Introducing supersymmetry*, Phys. Rep. C 128 (1985) 39–204.
- J. A Harvey, *Magnetic monopoles, Duality, and Supersymmetry*, 1995 Trieste Summer School in High Energy Physics and Cosmology, p. 66–125.

Exercises

<u>Exercise 1</u> Consider the quantity $T^{\mu\nu} = \xi^\alpha \sigma^{\mu\nu}{}_\alpha{}^\beta \xi_\beta$, where $\sigma^{\mu\nu}{}_\alpha{}^\beta$ is defined in (B.24) of Appendix B. Show that $(\sigma^{\mu\nu}\epsilon)_{\alpha\gamma} = (\sigma^{\mu\nu}\epsilon)_{\gamma\alpha}$. Infer that $T_{\mu\nu}$ belongs to the representation (1,0) of the Lorentz group.

   This is what allowed us to write the decomposition (4.3).

Hint: $T_{\mu\nu} = \xi^\alpha \left(\sigma^{\mu\nu}\epsilon\right)_{\alpha\gamma} \xi^\gamma$ and $[(1/2, 0) \times (1/2, 0)]_s = (1, 0)$.

<u>Exercise 2</u> Use a Jacobi identity to prove that the matrices b_r defined in (4.9) provide a representation of the internal symmetry algebra:

$$[b_r, b_s] = i C_{rst} b_t.$$

Hints: Use $[[B_r, B_s], Q_{i\alpha}] + [[B_s, Q_{i\alpha}], B_r] + [[Q_{i\alpha}, B_r], B_s] = 0$.

<u>Exercise 3</u> *We show in this exercise that the central charges Z_{ij} introduced in Section 4.1 commute with all generators of the supersymmetry algebra; hence their name.*

1. Prove that the subalgebra spanned by the Z_{ij} is invariant:

$$[Z_{ij}, B_r] = (b_r)_i{}^k Z_{kj} + (b_r)_j{}^k Z_{ik}.\tag{4.122}$$

2. One decomposes the Z_{ij} on the internal symmetry generators:

$$Z_{ij} = \sum_r a^r_{ij} B_r.\tag{4.123}$$

Use the relevant Jacobi identity to prove the following relation:

$$\sum_r a^r_{ij} (b_r^*)^k{}_l = 0.\tag{4.124}$$

3. Using in (4.124) the fact that B_r is hermitian, prove that

$$[Z_{ij}, Q_{k\alpha}] = 0, \quad [Z_{ij}, \bar{Q}^k_{\dot\alpha}] = 0,$$
$$[Z_{ij}, Z_{lm}] = 0.$$

(4.125)

4. Deduce that

$$[Z_{ij}, B_r] = 0.$$

(4.126)

Hence Z_{ij} commutes with all generators.

5. Prove the following relation:

$$\sum_k (b_r)_i{}^k a^s_{kj} + (b_r)_j{}^k a^s_{ik} = 0.$$

(4.127)

This relation shows that the $N \times N$ antisymmetric matrix a^s is invariant under the internal symmetry group G. This imposes a symplectic structure on the semisimple part of the corresponding algebra.

Using the fact that b_r is hermitian, equation (4.127) can be written as:

$$\sum_k (b_r)_i{}^k a^s_{kj} = -\sum_k a^s_{ik} (b^*_r)^k{}_j.$$

(4.128)

This is known as the intertwining relation since it shows that the matrix a^s intertwines the representation b_r with its conjugate $-b^*_r$.

Hints:

1. Use the Jacobi identity $[\{Q_{i\alpha}, Q_{j\beta}\}, B_r] - \{[Q_{j\beta}, B_r], Q_{i\alpha}\} + \{[B_r, Q_{i\alpha}], Q_{j\beta}\} = 0$.
2. $[\{Q_{i\alpha}, Q_{j\beta}\}, \bar{Q}^k_{\dot\gamma}] + [\{Q_{j\beta}, \bar{Q}^k_{\dot\gamma}\}, Q_{i\alpha}] + [\{\bar{Q}^k_{\dot\gamma}, Q_{i\alpha}\}, Q_{j\beta}] = 0$.
3. From (4.124), $\sum_r a^r_{ij} (b_r)_l{}^k = 0$; and from (4.122),

$$[Z_{ij}, Z_{lm}] = \sum_r (b_r)_i{}^k a^r_{lm} Z_{kj} + (b_r)_j{}^k a^r_{lm} Z_{ik} = 0.$$

4. The subalgebra spanned by the Z_{ij} is an invariant abelian subalgebra. It must therefore belong to the abelian factor since the internal symmetry group G is the product of a semisimple group (no abelian invariant subgroup) and an abelian factor.
5. Write explicitly (4.122) using (4.126).

<u>Exercise 4</u> In the construction of Section 4.3.1, show that there are 2^{N-1} bosonic states (integer helicity) and 2^{N-1} fermionic states (half integer helicity).

Hint:

$$(1 \pm 1)^N = \sum_{k=0}^{[N/2]} \binom{N}{2k} \pm \sum_{k=0}^{[N/2]-1} \binom{N}{2k+1},$$

$$\text{hence} \quad \sum_{k=0}^{[N/2]} \binom{N}{2k} = \sum_{k=0}^{[N/2]-1} \binom{N}{2k+1} = 2^{N-1}.$$

<u>Exercise 5</u> Write equation (III) of Section 2.2 of Chapter 2 in two-component notation:

$$[Q_\alpha, M^{\mu\nu}] = i\,(\sigma^{\mu\nu})_\alpha{}^\beta Q_\beta,$$

$$[\bar{Q}^{\dot\alpha}, M^{\mu\nu}] = i\,(\bar{\sigma}^{\mu\nu})^{\dot\alpha}{}_{\dot\beta}\bar{Q}^{\dot\beta}. \tag{4.129}$$

Deduce the spin components of $\bar{Q}_{\dot{1}}$ and $\bar{Q}_{\dot{2}}$, i.e. the values of $J_3 \equiv M_{12}$ and J, as indicated in Section 4.3.2.

Hint: Use (B.23) and (B.24) of Appendix B, as well as the explicit forms of σ_μ and $\bar{\sigma}_\mu$.

<u>Exercise 6</u> Prove that massless representations are only compatible with vanishing central charges. For this purpose, choose the frame of Section 4.3.1 where $P_\mu = (E, 0, 0, E)$ and show that Q_{2ar} annihilates any state $|\Phi\rangle$.

Hint: See the footnote 6 of Section 4.3.1; if Q_{2ar} annihilates any state $|\Phi\rangle$, then $z_r = 0$ from (4.25).

<u>Exercise 7</u> Show that the $N = 2$ supersymmetry transformations (4.37) include the $N = 1$ supersymmetry transformations of the vector and chiral supermultiplets of parameters $\eta^1_\alpha, \bar{\eta}^{1\dot\alpha}$ (given in equation (C.99) of Appendix C).

Hint: Using the fact that all fields belong to the adjoint representation and writing $\lambda \equiv \lambda^1$, $\psi \equiv \lambda^2$, $\phi \equiv (A+iB)/\sqrt{2}$, $F \equiv (F_1+iF_2))/\sqrt{2}$, we obtain from the transformation law (C.99) of Appendix C (with parameter $\eta \equiv \eta^1$):

$$\delta_s A = \eta^1\lambda^2 + \bar{\eta}^1\bar{\lambda}^2,$$
$$\delta_s B = -i\left(\eta^1\lambda^2 - \bar{\eta}^1\bar{\lambda}^2\right),$$
$$\delta_s \lambda^2_\alpha = \sqrt{2}\eta^1_\alpha F - i(\sigma^\mu\bar{\eta}^1)_\alpha D_\mu A + (\sigma^\mu\bar{\eta}^1)_\alpha D_\mu B,$$
$$\delta_s F_1 = -i\eta^1\sigma^\mu D_\mu\bar{\lambda}^2 - i\bar{\eta}^1\bar{\sigma}^\mu D_\mu\lambda^2 - g\left[A, \eta^1\lambda^1 - \bar{\eta}^1\bar{\lambda}^1\right] + ig\left[B, \eta^1\lambda^1 + \bar{\eta}^1\bar{\lambda}^1\right],$$
$$\delta_s F_2 = \eta^1\sigma^\mu D_\mu\bar{\lambda}^2 - \bar{\eta}^1\bar{\sigma}^\mu D_\mu\lambda^2 - ig\left[A, \eta^1\lambda^1 + \bar{\eta}^1\bar{\lambda}^1\right] - g\left[B, \eta^1\lambda^1 - \bar{\eta}^1\bar{\lambda}^1\right],$$
$$\delta_s A_\mu = \eta^1\sigma_\mu\bar{\lambda}^1 - \bar{\eta}^1\bar{\sigma}_\mu\lambda^1$$
$$\delta_s \lambda^1_\alpha = -\eta^1_\alpha D - i\left(\sigma^{\mu\nu}\eta^1\right)_\alpha F_{\mu\nu},$$
$$\delta_s D = i\eta^1\sigma^\mu D_\mu\bar{\lambda}^1 + i\bar{\eta}^1\bar{\sigma}^\mu D_\mu\lambda^1,$$

which corresponds to the transformations (4.37) of parameter η^1, once one uses the four-component symplectic Majorana notation for spinors and the form (4.35) for the auxiliary fields.

<u>Exercise 8</u> Consider in electromagnetism the energy density $\frac{1}{2}(\mathbf{E}^2 + \mathbf{B}^2)$ and Poynting vector $\frac{1}{4\pi}\mathbf{E} \wedge \mathbf{B}$. By expressing them in terms of $\mathbf{E} + i\mathbf{B}$, show that they are invariant under the duality transformation (4.66).

Hint: $|\mathbf{E} + i\mathbf{B}|^2 = (\mathbf{E}^2 + \mathbf{B}^2)$, $(\mathbf{E} + i\mathbf{B})^* \wedge (\mathbf{E} + i\mathbf{B}) = 2i\,\mathbf{E} \wedge \mathbf{B}$.

<u>Exercise 9</u> Compute the winding number n in (4.81) when $\hat{\phi}^a \to \hat{r}^a$ as $r \to \infty$ ($\hat{r}^a$ is the unit vector in space).

Hint: $n = 1$.

<u>Problem 1</u> *We study in this problem the two-dimensional supersymmetric model of Section 4.5.1. Two-dimensional supersymmetry plays also an important rôle in string models because the string, being a one-dimensional object, follows a two-dimensional world-sheet in its motion: string theory can be understood as a two-dimensional quantum theory on the world-sheet (see Section 10.1 of Chapter 10).*

1. We first discuss spinors in $d = 2$ dimensions. Since the spacetime dimension falls in the class $d = 2$ mod 8, we can impose at the same time the Majorana and Weyl conditions. Moreover, the Lorentz group is simply $SO(2) \sim U(1)$ (or rather $SO(1,1)$ in the Lorentz case that we consider here). Thus its representations are characterized by a single quantum number. The spinor representation has dimension 2; it can be decomposed into spinors of chirality ± 1: a Majorana–Weyl spinor is thus a one-dimensional object and chirality is the additive quantum number just mentioned.

   More precisely, we consider the two-dimensional gamma matrices:

   $$\rho^0 = \begin{pmatrix} 0 & -i \\ i & 0 \end{pmatrix}, \quad \rho^1 = \begin{pmatrix} 0 & i \\ i & 0 \end{pmatrix}, \tag{4.130}$$

   which satisfy

   $$\{\rho^\mu, \rho^\nu\} = 2g^{\mu\nu} \tag{4.131}$$

   (the metric being $(+, -)$). We define also the two-dimensional analog of the γ_5 matrix:

   $$\rho^3 = -\rho^0 \rho^1 = \begin{pmatrix} -1 & 0 \\ 0 & 1 \end{pmatrix}. \tag{4.132}$$

   Then a two-dimensional spinor is written $\psi = \begin{pmatrix} \psi_- \\ \psi_+ \end{pmatrix}$ where $\psi_\pm$ is the $\pm$ chirality component:

   $$\rho^3 \begin{pmatrix} \psi_- \\ \psi_+ \end{pmatrix} = \begin{pmatrix} -\psi_- \\ \psi_+ \end{pmatrix}. \tag{4.133}$$

   (a) Show that $C = -\rho^0$ is the matrix that intertwines ρ_μ with $-\rho_\mu^T$:

   $$C \rho^{\mu T} C^{-1} = -\rho^\mu. \tag{4.134}$$

   One defines as usual $\bar{\psi} = \psi^\dagger \rho^0$ and $\psi^c = C\bar{\psi}^T$. Show that the Majorana condition $\psi^c = \psi$ is simply a reality condition with the gamma matrix basis chosen here.

   (b) Show that, for Majorana spinors ψ and χ,

   $$\begin{aligned} \bar{\psi}\chi &= \bar{\chi}\psi, \\ \bar{\psi}\rho^\mu \chi &= -\bar{\chi}\rho^\mu \psi, \\ \bar{\psi}\rho^\mu \rho^\nu \chi &= \bar{\chi}\rho^\nu \rho^\mu \psi, \end{aligned} \tag{4.135}$$

   which are the analogs of the four-dimensional formulas (B.40) of Appendix B.

2. We consider the action

$$S = \int d^2x \left[\frac{1}{2} (\partial_\mu \phi)^2 + \frac{1}{2} \bar{\psi} i \partial\!\!\!/ \psi + \frac{1}{2} F^2 + F W'(\phi) - \frac{1}{2} W''(\phi) \bar{\psi}\psi \right], \qquad (4.136)$$

where ϕ, F are real scalar fields, ψ is a Majorana spinor and $W = W(\phi)$, $W' = dW/d\phi$. Solving for the auxiliary field F (no kinetic term), one recovers the action (4.56).

(a) Show that S is invariant under the global supersymmetry transformations (compare with the four-dimensional analog (3.10) of Chapter 3)

$$\begin{aligned}
\delta\phi &= \bar{\varepsilon}\psi, \\
\delta\psi &= -i\rho^\mu \varepsilon \partial_\mu \phi + F\varepsilon, \\
\delta F &= -i\bar{\varepsilon}\rho^\mu \partial_\mu \psi,
\end{aligned} \qquad (4.137)$$

where ε is an anticommuting Majorana spinor.

(b) How many off-shell and on-shell bosonic degrees of freedom in the theory? How many fermionic?

(c) Determine the Noether current j_μ associated with the supersymmetry transformation:

$$j_\mu = \partial^\nu \phi \rho_\nu \rho_\mu \psi + iW' \rho_\mu \psi. \qquad (4.138)$$

3. (a) Using the chiral decomposition introduced in 1 for ψ and j_μ, show that the chiral components of the supersymmetry charge are respectively:

$$Q_+ = \int dx \, [\partial_+ \phi \, \psi_+ - W'(\phi)\psi_-]$$

$$Q_- = \int dx \, [\partial_- \phi \, \psi_- + W'(\phi)\psi_+] \qquad (4.139)$$

where $\partial_\pm \equiv \partial_0 \pm \partial_1$.

(b) Show that the energy–momentum tensor can be written as

$$T_{\mu\nu} = \partial_\mu \phi \partial_\nu \phi - \frac{1}{2} g_{\mu\nu} \partial^\sigma \phi \partial_\sigma \phi + \frac{1}{2} W'^2 g_{\mu\nu} + \frac{i}{2} \bar{\psi}\rho_\mu \partial_\nu \psi - g_{\mu\nu}$$

$$\times \left[\frac{1}{2} \bar{\psi} i \gamma^\sigma \partial_\sigma \psi - \frac{1}{2} W'' \bar{\psi}\psi \right]. \qquad (4.140)$$

Use the equations of motion to check that it is conserved: $\partial^\mu T_{\mu\nu} = 0$.

(c) Use the canonical commutation relations

$$\begin{aligned}
[\phi(t,x), \partial_0\phi(t,y)] &= i\delta(x-y) \\
\{\psi_+(t,x), \psi_+(t,y)\} &= \delta(x-y) \\
\{\psi_-(t,x), \psi_-(t,y)\} &= \delta(x-y)
\end{aligned} \qquad (4.141)$$

in order to check how the different fields behave under the translation generator $P_\mu = \int dx \, T_{0\mu}$:

$$[\phi(x), P_\mu] = i\partial_\mu\phi(x), \quad [\psi_+(x), P_\mu] = i\partial_\mu\psi_+(x), \quad [\psi_-(x), P_\mu] = i\partial_\mu\psi_-(x). \qquad (4.142)$$

4. (a) Prove that

$$Q_\pm^2 = P_\pm \equiv P_0 \pm P_1. \tag{4.143}$$

(b) Show that

$$\{Q_+, Q_-\} = \int dx\, 2W'\left[\phi(x)\right] \partial_x \phi(x) \equiv Z. \tag{4.144}$$

(c) Compute Z for the kink state when $V(\phi) = W'^2(\phi)/2 = \lambda(\phi^2 - \phi_0^2)^2/2$.

(d) Expressing $Q_+ \pm Q_-$ for a static solution $(\partial_0 \phi = 0)$, show that one of these combinations vanishes for the kink (antikink), *i.e.* for a solution of (4.52).

Hints: 1(b) Remember that spinors anticommute: hence transposition of a quadratic combination of spinor fields introduces a minus sign.

2(a,c) $\delta\bar\psi = i\bar\varepsilon \rho^\mu \partial_\mu \phi + F\bar\varepsilon$.

In order to compute the Noether current, it is advisable to introduce a space-time dependent $\varepsilon(x)$ spinor parameter and to make use of the relation

$$\delta S = \int d^2 x\, \partial_\mu \bar\varepsilon j^\mu.$$

This allows us to take full account of the anticommuting nature of ε. One finds the following variation for the Lagrangian $\mathcal{L}$:

$$\mathcal{L} = \partial_\mu \left[\partial^\mu \phi \bar\varepsilon \psi - \frac{1}{2}\bar\varepsilon \rho^\nu \rho^\mu \psi \partial_\nu \phi - \frac{i}{2}(F + 2W')\bar\varepsilon \rho^\mu \psi \right]$$
$$+ \partial_\mu \bar\varepsilon \left[\partial_\nu \phi \rho^\nu \rho^\mu \psi + iW' \rho^\mu \psi \right].$$

(b) Off-shell, two scalar (ϕ, F) and two spinor $(\psi_\pm)$ degrees of freedom. On-shell, one must take into account the special nature of $1{+}1$ dimensional spacetimes. Indeed, the free field equations of motion $\partial^\mu \partial_\mu \phi = 0$ and $i\rho^\mu \partial_\mu \psi = 0$ are solved by $\phi = \phi_+(x+t) + \phi_-(x-t)$, $\psi_+ = \psi_+(x+t)$, $\psi_- = \psi_-(x-t)$. Hence one scalar and one spinor left-moving degrees of freedom. And the same for right-movers.

3(c)

$$P_0 = \int dx \left[\frac{1}{2}(\partial_0 \phi)^2 + \frac{1}{2}(\partial_1 \phi)^2 + \frac{1}{2}W'^2 + \frac{i}{2}\psi_+ \partial_1 \psi_+ - \frac{i}{2}\psi_- \partial_1 \psi_- + iW'' \psi_+ \psi_- \right]$$
$$P_1 = \int dx \left[\partial_0 \phi\, \partial_1 \phi + \frac{i}{2}\psi_+ \partial_1 \psi_+ + \frac{i}{2}\psi_- \partial_1 \psi_- \right].$$

To compute $[\psi_\pm(x), P_0]$, use the equations of motion:

$$\partial_\mp \psi_\pm = \pm W'' \psi_\mp.$$

4 (a) $P_\pm = \int dx \left[\frac{1}{2}(\partial_\pm \phi)^2 + \frac{1}{2}W'^2 \pm i\psi_\pm \partial_x \psi_\pm + iW'' \psi_+ \psi_- \right].$

(b) Note that the commutator $[\partial_+ \phi(x), \partial_- \phi(y)]$ at equal time vanishes.

(c) $Z = 2\left| \int_{-\phi_0}^{+\phi_0} d\phi \sqrt{2V(\phi)} \right| = 8\sqrt{\lambda}\phi_0^3/3.$

(d) For a static solution, $Q_+ \pm Q_- = \int dx (\partial_x \phi \pm W')(\psi_+ \mp \psi_-).$

5
The minimal supersymmetric model

As a part of the general effort to implement supersymmetry at the level of fundamental interactions, the minimal supersymmetric extension of the Standard Model is motivated and described here at tree level. The presentation of its phenomenology requires a discussion of radiative corrections which play here a central rôle ; it is therefore postponed to Chapter 7. Of great interest for astrophysics is the lightest supersymmetric particle (LSP) if it is stable: as a weakly interacting massive particle (WIMP), it provides a candidate for dark matter. The last section of this chapter is devoted to these considerations.

5.1 Why double the number of fundamental fields?

5.1.1 First attempts

If supersymmetry rules the microscopic world of elementary particles, then one should be able to classify the known particles in supersymmetric multiplets or supermultiplets. As is customary (*cf.* baryons in the $SU(3)$ classification), some elements of the puzzle, *i.e.* some components of the supermultiplets might be missing and experimental effort should be looking for them.

In the years 1976–1977, P. Fayet undertook such an effort, to come to the conclusion that half of the jigsaw puzzle is presently missing. This is such a sweeping statement (although not such an implausible one: after all, the prediction of antiparticles which also led to an almost doubling of the spectrum of known particles, was later confirmed by experiment) that it is worth following P. Fayet's early efforts to put known particles in supermultiplets.

Obviously, the mass spectrum is not a reliable tool since supersymmetry is (at best) spontaneously broken and thus not visible in the particle spectrum. However, photon and neutrino are a boson–fermion pair of massless particles and it is tempting to associate their charged electroweak partners $W^{\pm}$ and $e^{\pm}$. Indeed a model could be built along these lines [146].

This does not, however, solve the problem at hand: what to do with the other two neutrinos? how to relate the colored particles: the bosons (gluons) are in a different representation of $SU(3)$ than the fermions (quarks)? etc.

Table 5.1

BOSONS	FERMIONS
gauge field	*gaugino*
Higgs	*Higgsino*
squark	quark
slepton	lepton

5.1.2 Squarks, sleptons, gauginos, higgsinos, and the Higgs sector

These considerations lead P. Fayet [148, 149] to double the number of fields and introduce a supersymmetric partner to each field of the Standard Model, as shown in Table 5.1[1]. Obviously, this is a very large step backward in our goal of unifying matter and radiation: we end up introducing new forms of matter and of radiation.

There is an extra complication: we need to enrich the gauge symmetry breaking sector of the theory and to introduce two Higgs doublets. We will give two reasons for this.

First, we must pay attention to the gauge anomalies of the new theory. In the Standard Model, the compensation of chiral anomalies is ensured by the field content: as recalled at the end of Section A.6 of Appendix Appendix A, the dangerous triangle diagrams of Figure A.11, which jeopardize the local gauge symmetry at the quantum level, add up to zero when one sums over all possible fermions (three colors of quarks, leptons) in the loop. In the theory that we consider now, we have introduced new fermions: gauginos and Higgsino. As far as chiral anomalies are concerned, gauginos are not a problem since they couple vectorially. On the other hand, a unique Higsino introduces anomalies: for example, a triangle diagram with three $U(1)_Y$ vector fields attached to a Higgsino loop is proportional to $y_H^3 = (+1)^3 = +1$ and is nonvanishing. Introducing a second Higgs doublet of opposite hypercharge solves the problem: $y_{H_1}^3 + y_{H_2}^3 = (-1)^3 + (+1)^3 = 0$.

This second doublet is also needed for the purpose of breaking correctly the $SU(2) \times U(1)$ gauge symmetry. We have seen in Chapter 4 that a massive vector supermultiplet consists of one real scalar field, one massive vector field and two Majorana spinors. This can be understood as well by reference to the Higgs mechanism: we start in the Hamiltonian with a massless vector supermultiplet (one vector and one Majorana spinor) and a chiral supermultiplet (two real scalars and one Majorana spinor); one of the scalars provides the longitudinal massive vector degree of freedom and we find in the spectrum precisely the massive vector supermultiplet just described.

The Z^0 and $W^\pm$ supermultiplets thus require three longitudinal degrees of freedom Z_L^0, $W_L^\pm$ and three real scalars, respectively H^0 and $H^\pm$. A single complex doublet provides four degrees of freedom, which is clearly insufficient. Two complex doublets H_1 and H_2 have a total of eight degrees of freedom: the six above plus a scalar h^0 and a pseudoscalar A^0 which form a chiral supermultiplet (with a Majorana fermion). Thus a supersymmetric version of the Higgs mechanism for breaking gauge invariance requires the presence of two complex doublets.

---

[1]One uses the denomination photino, wino, zino, and gluino for the supersymmetric partner of the photon, W, Z, or gluon.

It will prove to be useful in what follows to remember this supersymmetric limit in which $m_{H^0} \to M_Z$, $m_{H^\pm} \to M_W$ and $m_{h^0} \to m_{A^0}$.

5.2 Model building

5.2.1 Chiral supermultiplets and their supersymmetric couplings

The Standard Model is chiral, in the sense that left-handed and right-handed components transform differently under the gauge interactions. When discussing supersymmetry, it is however more convenient to describe all fermions by spinors of the same chirality, say left-handed. If necessary, we will use charge conjugation in order to do so; for example we will trade u_R for u_L^c. One thus introduces, for each generation of quarks and leptons:

$$(SU(3),\ SU(2),\ Y)$$

$$Q = \begin{pmatrix} U \\ D \end{pmatrix} \quad \in \quad \left(\mathbf{3},\ \mathbf{2}\ ;\ \frac{1}{3}\right)$$

$$L = \begin{pmatrix} N \\ E \end{pmatrix} \quad \in \quad (\mathbf{1},\ \mathbf{2}\ ;\ -1)$$

$$U^c \quad \in \quad \left(\overline{\mathbf{3}},\ \mathbf{1}\ ;\ -\frac{4}{3}\right)$$

$$D^c \quad \in \quad \left(\overline{\mathbf{3}},\ \mathbf{1}\ ;\ \frac{2}{3}\right)$$

$$E^c \quad \in \quad (\mathbf{1},\ \mathbf{1}\ ;\ 2)$$

where the capital letter indicates the superfield. For example U contains the left-handed quark u_L and the squark field $\widetilde{u}_L$; whereas U^c contains the left-handed antiquark u_L^c and the antisquark $\widetilde{u}_R^*$ (respectively, charge conjugates of the right-handed quark u_R and of the squark $\widetilde{u}_R$). Note that $\widetilde{u}_L$ and $\widetilde{u}_R$ are completely independent scalar fields and that the index refers to the chirality of their supersymmetric partners.

Moreover we have just seen that we must introduce two Higgs doublets:

$$H_1 = \begin{pmatrix} H_1^0 \\ H_1^- \end{pmatrix} \quad \in \quad (\mathbf{1},\ \mathbf{2}\ ;\ -1)$$

$$H_2 = \begin{pmatrix} H_2^+ \\ H_2^0 \end{pmatrix} \quad \in \quad (\mathbf{1},\ \mathbf{2}\ ;\ 1)\ .$$

Since by definition, the Minimal SuperSymmetric Model (MSSM) has a minimal field content, the former fields form the minimal set that we are looking for.

Their supersymmetric couplings are described by the superpotential, which is an analytic function of the (super)fields. We may form quadratic terms using the Higgs fields only

$$W^{(2)} = -\mu\, H_1 \cdot H_2 = \mu\, H_2 \cdot H_1 \tag{5.1}$$

where $H_1 \cdot H_2 \equiv \varepsilon_{ij} \, H_1^i \, H_2^j$ selects the combination which is invariant under $SU(2)_L$ ($\varepsilon_{12} = -\varepsilon_{21} = 1$). The parameter μ will turn out to be the only dimensionful supersymmetric parameter[2] (cubic terms in the superpotential have dimensionless couplings). If the fundamental scale of the underlying theory is very large, unification scale M_U of order 10^{16} GeV or Planck scale, one should explain why the scale μ is so much smaller (as we will see, a superheavy μ would destabilize the electroweak vacuum). This is known as the μ-problem.

Cubic terms in the superpotential yield the Yukawa couplings of the Standard Model

$$W^{(3)} = \lambda_d \, Q \cdot H_1 D^c + \lambda_u \, Q \cdot H_2 U^c + \lambda_e \, L \cdot H_1 E^c \tag{5.2}$$

with notation similar to (5.1) and color indices suppressed. Indeed, plugging this superpotential into equation (3.30) of Chapter 3 gives (beware of signs which arise from the ϵ_{ij} contractions denoted with a dot)

$$m_d = -\lambda_d \, \langle H_1^0 \rangle, \quad m_u = \lambda_u \, \langle H_2^0 \rangle, \quad m_e = -\lambda_e \, \langle H_1^0 \rangle. \tag{5.3}$$

Let us stress that hypercharge conservation requires to use the Higgs doublet H_2 in the up-type quark coupling because the "chiral" nature of the superpotential forbids to use H_1^*. We thus need H_1 to give nonvanishing mass to down-type quarks and charged leptons, and H_2 to give mass to up-type quarks[3].

A very nice property of the Standard Model lies in the fact that its Yukawa couplings form the most general set of couplings compatible with gauge invariance and renormalizability. This is unfortunately not so in the case of its supersymmetric extensions, because there are more scalar fields (the squarks and sleptons) and more fermion fields (the Higgsinos). Indeed the following superpotential terms are allowed by gauge invariance and renormalizability (which allows terms up to dimension three in the superpotential):

$$L \cdot L \, E^c, \quad Q \cdot L \, D^c, \quad U^c D^c D^c. \tag{5.4}$$

They are potentially dangerous because they violate lepton or baryon number.

We will define the MSSM as the model with minimal field *and coupling* content. Whereas the couplings in (5.2) are necessary in order to give quarks and leptons a mass, the latter couplings are not. We therefore assume that they are absent in the MSSM. But, before going further, we will give a rationale for such an absence.

5.2.2   *R*-parity

The presence of the couplings (5.4) in the theory leads through the exchange of a squark or a slepton to effective four-fermion interactions in the low energy regime which violate lepton number (for the first two) or baryon number (for the third one). This was clearly undesirable, especially at a time when one thought that squarks and

---

[2]Beware that there is no general agreement on the convention for the sign of μ, *i.e.* on the sign in front of (5.1). This is quite unfortunate because this sign turns out to play an important rôle in the physical consequences, as we will see later. It is therefore important when comparing results to check the authors' convention on the sign of μ.

[3]This is why one often finds in the literature the notation: $H_1 \equiv H_d$ and $H_2 \equiv H_u$.

Table 5.2

Field	B	L	S	$3B + L + 2S$
quark	1/3	0	1/2	2
squark	1/3	0	0	1
lepton	0	1	1/2	2
slepton	0	1	0	1

sleptons might be fairly light. In order to avoid this, P. Fayet [148, 149] postulated the existence of a discrete symmetry, known as R-parity. The fields of the Standard Model have R-parity $+1$ and their supersymmetric partners R-parity -1.

As may be checked from Table 5.2, this can be summarized as

$$R = (-1)^{3B+L+2S} \tag{5.5}$$

which shows the connection of R-parity with baryon (B) and lepton (L) numbers. Such a parity obviously forbids the single exchange of a squark or a slepton between ordinary fermions

At the level of the superpotential, Q, L, U^c, D^c and E^c have R-parity -1 whereas H_1 and H_2 have R-parity $+1$ (in each case, the R-parity of the corresponding scalar field). Then obviously terms in (5.2) are invariant whereas terms in (5.4) are not.

This R-parity may find its origin in a R-symmetry, a continuous $U(1)$ symmetry which treats differently a particle and its supersymmetric partner, *i.e.* a continuous $U(1)$ symmetry which does not commute with supersymmetry (see Chapter 4 for details). Gaugino fields are for example charged under such a symmetry. Supersymmetry breaking induces gaugino masses and thus breaks this continuous symmetry down to a discrete symmetry, R-parity. We will see later that R-symmetries often play a central rôle in supersymmetric theories.

For the time being, we remain at a phenomenological level. The assumption of R-parity has some remarkable consequences:

(i) Supersymmetric partners are produced by pairs: since the initial state is formed of ordinary matter, it has R-parity $+1$. A final state such as slepton–antislepton or squark–antisquark pair has also R-parity $(-1)^2 = +1$.

(ii) The lightest supersymmetric particle (LSP) is stable: since it has $R = -1$, it cannot decay into ordinary matter; since it is the lightest, it cannot decay into supersymmetric matter. We will see in Section 5.5 that this provides us in many cases with an excellent candidate for dark matter.

Even though R-parity might be well motivated, there is nothing sacred with it, all the more so with the existing limits on the supersymmetric spectrum. Indeed, a single-sfermion exchange amplitude is typically of order $\lambda_{\cancel{R}}^2/m_{\tilde{f}}^2$ where $\lambda_{\cancel{R}}$ is a R-violating coupling and $m_{\tilde{f}}$ the sfermion mass. It is thus possible to include R-parity violations under the condition that the R-violating coupling $\lambda_{\cancel{R}}$ is not too large[4]. We will study in detail this possibility in Section 7.6 of Chapter 7. Let us note here only that R-parity

---

[4]Similarly, Higgs exchange between electrons has an amplitude of order $\lambda_e^2/m_h^2 \sim m_e^2/(v^2 m_h^2) \ll 1/v^2 \sim G_F$, where λ_e is the electron Yukawa coupling. It is therefore a minor correction to weak amplitudes.

violation is connected with B or L violations. As long as both B and L are not violated simultaneously, there is no problem with proton decay: proton decay channels (*e.g.* $\mathrm{p} \to \pi^0 \mathrm{e}^+$, $\mathrm{p} \to \mathrm{K}^0 \mu^+$, $\mathrm{p} \to \mathrm{K}^+ \nu_\tau$) are forbidden by either B or L conservation.

5.2.3 Soft supersymmetry breaking terms

We have until now discussed the supersymmetric couplings. Supersymmetry breaking will obviously generate new couplings. If we do not want to restrict ourselves to a given scenario of supersymmetry breaking, we may just require that this mechanism is chosen such that it does not generate quadratic divergences: otherwise the *raison d'être* of supersymmetry is lost. One says that supersymmetry is only broken softly. [The analysis at the end of Chapter 1 indicates which terms are allowed by this condition of soft supersymmetry breaking that we impose.] It turns out that the allowed terms have been classified by Girardello and Grisaru [188] and are remarkably simple. Soft supersymmetry breaking terms are of three kinds:

- Scalar mass terms. These terms are of two different forms: for a complex scalar field ϕ

$$\delta\mathcal{L}_{\mathrm{SB}} = \delta m^2 \phi^* \phi + \delta m'^2 \left(\phi^2 + \phi^{*2}\right)$$
$$= \left(\delta m^2 + 2\delta m'^2\right)(\mathrm{Re}\ \phi)^2 + \left(\delta m^2 - 2\delta m'^2\right)(\mathrm{Im}\ \phi)^2. \tag{5.6}$$

  Thus one term (δm^2) treats scalar $(\mathrm{Re}\ \phi)$ and pseudoscalar $(\mathrm{Im}\ \phi)$ of the super-multiplet equivalently whereas the other term $(\delta m'^2)$ introduces a gap between them.

- A-terms. These terms are (the real part of) cubic analytic functions:

$$\delta\mathcal{L}_{\mathrm{SB}} = -\mathcal{A}\left(\phi^3 + \phi^{*3}\right). \tag{5.7}$$

  If the term ϕ^3 is allowed by the gauge symmetry, then in all generality, it is also allowed in the superpotential. For reasons that will be more clear in the case of supergravity, it has become customary to write $\mathcal{A} \equiv A\lambda$ where λ is the corresponding superpotential coupling. In other words,

$$\delta\mathcal{L}_{\mathrm{SB}} = -A\lambda\left(\phi^3 + \phi^{*3}\right). \tag{5.8}$$

- Gaugino masses.

$$\delta\mathcal{L}_{\mathrm{SB}} = -\frac{1}{2}M_\lambda \bar{\lambda}\lambda. \tag{5.9}$$

5.3 The Minimal SuperSymmetric Model (MSSM)

We study in this section the archetype of supersymmetric models, the Minimal SuperSymmetric Model (MSSM). This is the linearly realized supersymmetric version of the Standard Model with the minimal number of fields and the minimum number of couplings. From the discussion of the previous sections, this means that the field content is one of vector supermultiplets (gauge fields, gauginos) associated with the gauge symmetry $SU(3) \times SU(2) \times U(1)$ and chiral supermultiplets describing quarks, leptons, the two Higgs doublets and their supersymmetric partners (squarks, slep-tons, Higgsinos). As for the couplings, our minimality requirement leads us to impose

R-parity; all other cubic couplings (5.2) are necessary in order to provide a mass for all quarks and leptons. Soft supersymmetry breaking terms generate masses for squarks, sleptons, and gauginos.

Obviously, the qualifier "minimal" is a misnomer. Besides the 19 parameters of the Standard Model[5], one counts 105 new parameters: five real parameters and three CP-violating phases in the gaugino–Higgsino sector, 21 masses, 36 real mixing angles and 40 CP-violating phases in the squark and slepton sector. Overall, this makes 124 parameters for this not so minimal MSSM! Underlying physics (grand unification, family symmetries, string theory, etc.) usually provides additional relations between these parameters. In the case of the minimal supergravity model introduced in Chapter 6, we will be left with only five extra parameters besides those of the Standard Model. For the time being, we will consider the MSSM as a low energy model which provides the framework for many different underlying theories. We will start in the next chapter to unravel the high energy dynamics which may constrain further the model.

In our description of the MSSM, we begin by studying the central issue of gauge symmetry breaking and thus the Higgs sector.

5.3.1   Gauge symmetry breaking and the Higgs sector

In order to discuss gauge symmetry breaking, we should write the full scalar potential and discuss its minimization. Scalars include not only the Higgs fields but also squarks and sleptons. But squarks and/or sleptons (other than the sneutrino) getting a nonzero vacuum expectation value lead to the spontaneous breaking of charge or color. And a sneutrino getting a nonzero vacuum expectation value leads to the spontaneous breaking of R-parity. We will exclude such possibilities. This obviously leads to constraints on the parameters: we will return to this in Chapter 7. For the time being, we assume that such constraints are satisfied and set all scalar fields besides the Higgs fields to their vanishing vacuum expectation values.

The supersymmetric part of the potential consists of F-terms and D-terms. Since the only term of the superpotential which involves H_1 or H_2 solely is (5.1), and since

$$\sum_i \left| \frac{\partial W}{\partial H_1^i} \right|^2 = \sum_{ijk} |\mu|^2 \, \varepsilon_{ij} \, H_2^{j*} \, \varepsilon_{ik} \, H_2^k = \sum_j |\mu|^2 \, H_2^{j*} \, H_2^j = |\mu|^2 \, H_2^\dagger H_2$$

and similarly for H_2, we have

$$V_F = |\mu|^2 \left(H_1^\dagger H_1 + H_2^\dagger H_2 \right). \tag{5.10}$$

The D term involves contributions from the $SU(2)$ as well as $U(1)$ gauge vector supermultiplets:

$$V_D = \frac{1}{2} \, g^2 \left(H_1^\dagger \frac{\vec{\tau}}{2} H_1 + H_2^\dagger \frac{\vec{\tau}}{2} H_2 \right)^2 + \frac{1}{2} \left(\frac{g'}{2} \right)^2 \left(-H_1^\dagger H_1 + H_2^\dagger H_2 \right)^2 \tag{5.11}$$

where g and $g'/2$ are, respectively, the couplings of $SU(2)$ and $U(1)$.

---

[5]The two scalar potential parameters m and λ are replaced by the two real parameters in μ (real and imaginary part).

The sum $V_F + V_D$ represents the full supersymmetric contribution to the scalar potential. We may make at this point two remarks:

- The quadratic couplings in (5.10) are obviously positive, which does not allow at this stage gauge symmetry breaking. It is the supersymmetry breaking terms that will induce gauge symmetry breaking.
- The quartic couplings in (5.11) are of order of the gauge couplings squared, and thus small. In the Standard Model, a small quartic coupling λ is associated with a rather light Higgs field ($m^2 \sim \lambda v^2 \sim g^2 v^2 \sim M_W^2$).

As discussed above, spontaneous supersymmetry breaking induces soft supersymmetry breaking terms. Gauge invariance forbids any cubic A-terms since one cannot form a gauge singlet out of three fields H_1 and/or H_2. Thus supersymmetry breaking induces the following terms:

$$V_{\text{SB}} = m_{H_1}^2 \, H_1^\dagger H_1 + m_{H_2}^2 \, H_2^\dagger H_2 + (B_\mu \, H_1 \cdot H_2 + \text{h.c.}) \tag{5.12}$$

where we assume B_μ to be real through a redefinition of the relative phase of H_1 and H_2.

One may check that the minimization of the complete scalar potential $V \equiv V_F + V_D + V_{\text{SB}}$ leads to $\langle H_1^\pm \rangle = 0 = \langle H_2^\pm \rangle$, *i.e.* , irrespective of the parameters[6] (μ, $m_{H_1}^2$, $m_{H_2}^2$, B_μ and the gauge couplings) charge is conserved in the Higgs sector. We will leave the proof of this result as Exercise 2 and we will restrict from now on our attention to the neutral scalars. The potential V then reads:

$$V(H_1^0, H_2^0) = m_1^2 |H_1^0|^2 + m_2^2 |H_2^0|^2 + B_\mu \left(H_1^0 H_2^0 + H_1^{0*} H_2^{0*} \right)$$
$$+ \frac{g^2 + g'^2}{8} \left(|H_1^0|^2 - |H_2^0|^2 \right)^2 \tag{5.13}$$

where we have introduced the mass parameters

$$m_1^2 \equiv m_{H_1}^2 + |\mu|^2$$
$$m_2^2 \equiv m_{H_2}^2 + |\mu|^2. \tag{5.14}$$

Thus, apart from the gauge couplings, V depends on three parameters : m_1^2, m_2^2 and B_μ. These parameters must satisfy two sets of conditions:

(i) Stability conditions: the potential should be stable (*i.e.* be bounded from below) in all directions of field space. Since the coefficients of the quartic terms are obviously positive, a nontrivial condition arises only in the direction of field space where these terms vanish: $H_2^0 = H_1^{0*} e^{i\varphi}$, with φ an undetermined phase. Since

$$V \left(H_1^0, H_1^{0*} e^{i\varphi} \right) = \left(m_1^2 + m_2^2 + 2 B_\mu \cos\varphi \right) |H_1^0|^2, \tag{5.15}$$

the stability condition reads $m_1^2 + m_2^2 + 2 B_\mu \cos\varphi > 0$, that is

$$m_1^2 + m_2^2 > 2|B_\mu|. \tag{5.16}$$

---

[6]This property is specific to the MSSM and not found in immediate generalizations of this model.

(ii) Gauge symmetry breaking condition: $H_1^0 = H_2^0 = 0$ should not be a minimum. This translates into the condition

$$\left| \begin{array}{cc} \dfrac{\partial^2 V}{\partial H_1^0 \, \partial H_1^{0*}} & \dfrac{\partial^2 V}{\partial H_1^0 \, \partial H_2^0} \\[2ex] \dfrac{\partial^2 V}{\partial H_2^{0*} \, \partial H_1^{0*}} & \dfrac{\partial^2 V}{\partial H_2^{0*} \, \partial H_2^0} \end{array} \right|_{H_1^0 = H_2^0 = 0} < 0 \tag{5.17}$$

or

$$m_1^2 \, m_2^2 < |B_\mu|^2. \tag{5.18}$$

We note immediately that $m_1^2 = m_2^2$ is not compatible with (5.16) and (5.18) simultaneously. This requires that the soft masses $m_{H_1}^2$ and $m_{H_2}^2$ differ. We will see in Section 6.5 of Chapter 6 that this may be induced by radiative corrections: top quark loops affect the H_2 propagator but not H_1 to first order (*cf.* (5.2)).

Mass eigenstates are obtained by performing independent rotations in the pseudoscalar and the scalar sectors:

$$\begin{pmatrix} \mathrm{Im}\, H_1^0 \\[1ex] \mathrm{Im}\, H_2^0 \end{pmatrix} = \frac{1}{\sqrt{2}} \begin{pmatrix} \cos\beta & \sin\beta \\[1ex] -\sin\beta & \cos\beta \end{pmatrix} \begin{pmatrix} Z_L^0 \\[1ex] A^0 \end{pmatrix}, \tag{5.19}$$

$$\begin{pmatrix} \mathrm{Re}\, H_1^0 \\[1ex] \mathrm{Re}\, H_2^0 \end{pmatrix} = \frac{1}{\sqrt{2}} \begin{pmatrix} \cos\alpha & -\sin\alpha \\[1ex] \sin\alpha & \cos\alpha \end{pmatrix} \begin{pmatrix} H^0 \\[1ex] h^0 \end{pmatrix}, \tag{5.20}$$

where the massless pseudoscalar field Z_L^0 provides the longitudinal degree of freedom of the massive Z^0 gauge boson. The three other fields are physical scalar fields; by convention h^0 is the lightest scalar: $m_{h^0} < m_{H^0}$. If we reintroduce the charged Higgs fields $H_1^\pm$ and $H_2^\pm$, one finds that the same rotation as in (5.19) leads to the charged mass eigenstates:

$$\begin{pmatrix} H_1^\pm \\[1ex] H_2^\pm \end{pmatrix} = \begin{pmatrix} \cos\beta & \sin\beta \\[1ex] -\sin\beta & \cos\beta \end{pmatrix} \begin{pmatrix} W_L^\pm \\[1ex] H^\pm \end{pmatrix}. \tag{5.21}$$

Again, the massless eigenstates $W_L^\pm$ are the longitudinal degrees of freedom of the massive $W^\pm$ and $H^\pm$ is the physical charged Higgs.

One may replace the three parameters m_1^2, m_2^2 and B_μ by the more physical set: the vacuum expectation values

$$\begin{aligned} v_1 &\equiv \langle H_1^0 \rangle, \\ v_2 &\equiv \langle H_2^0 \rangle, \end{aligned} \tag{5.22}$$

which can be chosen to be real through a $SU(2) \times U(1)$ rotation, and the mass of the lightest scalar m_{h^0} or of the pseudoscalar m_{A^0}. But, a combination of v_1 and v_2, namely $v^2/2 \equiv v_1^2 + v_2^2$, yields the Z^0 mass:

$$M_z^2 = \frac{1}{2}\left(g^2 + g'^2\right)\left(v_1^2 + v_2^2\right) = \frac{1}{4}\left(g^2 + g'^2\right)v^2 \tag{5.23}$$

(v corresponds to the vev of the Higgs field in the Standard Model; *cf.* (A.131) of Appendix Appendix A: $v = 246$ GeV).

Hence, we remain with two new parameters:

$$\tan\beta \equiv \frac{v_2}{v_1} \tag{5.24}$$

and m_{h^0} or m_{A^0}. It turns out that the angle β thus defined is precisely the one that appears in the pseudoscalar (5.19) or charged (5.21) scalar rotation matrix. This angle β plays a central rôle in supersymmetric models.

Obviously one may express the new set of parameters in terms of the previous one; for example:

$$\sin 2\beta = -2\,\frac{B_\mu}{m_1^2 + m_2^2}, \tag{5.25}$$

$$m_{A^0}^2 = m_1^2 + m_2^2. \tag{5.26}$$

Following (5.25), we note that, had we forgotten the B_μ term in (5.12), we would have found $\sin 2\beta = 0$, *i.e.* $\langle H_1^0 \rangle = 0$ or $\langle H_2^0 \rangle = 0$: up-type quarks or down-type quarks would be massless.

Another relation of importance for the future discussion on the issue of fine-tuning is

$$\frac{1}{2}\,M_z^2 = \frac{m_1^2 - m_2^2 \tan^2\beta}{\tan^2\beta - 1}. \tag{5.27}$$

Of course, all physical Higgs masses and mixing angles can be expressed, at tree level, in terms of $\tan\beta$ and m_{h^0} (or m_{A^0}). For instance, we have[7]

$$m_{H^0}^2 = M_z^2 \cos^2 2\beta\,\frac{M_z^2 - m_{h^0}^2}{M_z^2 \cos^2 2\beta - m_{h^0}^2}, \tag{5.28}$$

$$m_{A^0}^2 = m_{h^0}^2\,\frac{M_z^2 - m_{h^0}^2}{M_z^2 \cos^2 2\beta - m_{h^0}^2}, \tag{5.29}$$

$$m_{H^0}^2 + m_{h^0}^2 = m_{A^0}^2 + M_z^2, \tag{5.30}$$

$$m_{H^\pm}^2 = m_{A^0}^2 + M_W^2, \tag{5.31}$$

$$\sin 2\alpha = -\sin 2\beta\,\frac{m_{H^0}^2 + m_{h^0}^2}{m_{H^0}^2 - m_{h^0}^2}, \tag{5.32}$$

$$\tan 2\alpha = \tan 2\beta\,\frac{m_{A^0}^2 + M_z^2}{m_{A^0}^2 - M_z^2}. \tag{5.33}$$

---

[7]Note the two interesting limits: when $m_{h^0}^2 \to M_z^2 \cos^2 2\beta$, all other three masses become infinite; when $m_{h^0}^2 \to 0$, then $m_{H^0} \to M_z$, $m_{A^0} \to 0$, $m_{H^\pm} \to M_W$ and $\alpha \to -\beta$: one recovers the supersymmetry limit discussed at the end of Section 5.1.2.

The stability and gauge symmetry breaking conditions lead to the following bound:

$$m_{h^0}^2 < M_z^2 \cos^2 2\beta < M_z^2. \tag{5.34}$$

As expected, the lightest scalar is found to be relatively light. Such a bound has led to the hope that the lightest Higgs of the MSSM could be detected or ruled out by the LEP collider at CERN. We will now see, however, that top/stop radiative corrections modify the bound (5.34).

We will have a general discussion of the radiative corrections to the Higgs potential in Chapter 7 but a simple argument [22] may be used to obtain the bulk of the one-loop contributions to the lightest Higgs mass. It rests on an analysis of the standard Higgs quartic coupling λ. We will make the simplifying hypothesis that the only relevant degrees of freedom are the Higgs scalar (assimilated to the lightest scalar h^0), the top quark and its supersymmetric partner, the stop (of mass $m_{\tilde{t}}$).

For mass scales μ larger than $m_{\tilde{t}}$, the model is supersymmetric and, as we have seen in (5.13) the quartic coupling $\lambda(\mu)$ is of the order of the gauge coupling squared g^2. For scales $m_t < \mu < m_{\tilde{t}}$, the effective theory is no longer supersymmetric (the stop decouples) and one typically recovers the Standard Model. The evolution of λ is governed by the Standard Model renormalization group equation, which is dominated by the top Yukawa coupling λ_t (of order 1, and thus much larger than λ: $\lambda(m_{\tilde{t}}) \sim g^2$):

$$\mu \frac{d\lambda}{d\mu} \sim -\frac{3}{8\pi^2} \lambda_t^4. \tag{5.35}$$

This term is simply due to a diagram with four scalars attached to a top quark loop.

The fact that the top quark is heavy, *i.e.* that λ_t is of order one, thus leads to a sharp increase in the magnitude of λ when one goes down in scale from $m_{\tilde{t}}$ to m_t. For scales $M_z < \mu < m_t$, the top has decoupled and the evolution of $\lambda(\mu)$ is milder.

The global effect on λ is thus of order $(3\lambda_t^4/16\pi^2) \ln(m_{\tilde{t}}^2/m_t^2)$. This gives for the Higgs mass ($m_h^2 = 2\lambda v^2$ from (A.138) of Appendix Appendix A)

$$\delta m_h^2 \sim \frac{3\lambda_t^4}{8\pi^2} v^2 \ln \frac{m_{\tilde{t}}^2}{m_t^2}$$

$$\sim \frac{3g^2}{8\pi^2} \frac{m_t^4}{M_W^2} \ln \frac{m_{\tilde{t}}^2}{m_t^2} \tag{5.36}$$

where we have used $M_W^2 = \frac{1}{4}g^2 v^2$ and $m_t = -\lambda_t v/\sqrt{2}$ (*cf.* (A.141) of Appendix Appendix A).

We see that the effect is negligible if m_t is small with respect to M_W (this is why these corrections were assumed to be small at an early period when one expected the top to have a typical quark mass) as well as if $m_t \sim m_{\tilde{t}}$ (approximate supersymmetry in the top sector). We will give in Chapter 7 a more detailed account of radiative corrections, but (5.36) represents the bulk of these corrections. If one allows a generous 185 GeV for the top mass and 1 TeV for the stop mass, this gives an upper limit of 130 GeV for the lightest Higgs of the MSSM.

Table 5.3 Couplings of fermions to the neutral scalars.

f	λ_{hff}	λ_{Hff}	λ_{Aff}
u	$\dfrac{gm_u}{2M_W}\dfrac{\cos\alpha}{\sin\beta}$	$\dfrac{gm_u}{2M_W}\dfrac{\sin\alpha}{\sin\beta}$	$\dfrac{gm_u}{2M_W}\cot\beta$
d	$\dfrac{gm_d}{2M_W}\dfrac{\sin\alpha}{\cos\beta}$	$\dfrac{gm_d}{2M_W}\dfrac{\cos\alpha}{\cos\beta}$	$\dfrac{gm_d}{2M_W}\tan\beta$
e	$\dfrac{gm_e}{2M_W}\dfrac{\sin\alpha}{\cos\beta}$	$\dfrac{gm_e}{2M_W}\dfrac{\cos\alpha}{\cos\beta}$	$\dfrac{gm_e}{2M_W}\tan\beta$

To conclude this section, we give here for future reference the couplings of the fermion fields to the neutral Higgs. They are obtained straightforwardly from the superpotential (5.2) using the decomposition (5.19)–(5.20):

$$\mathcal{L} = \sum_{f=u,d,e} \lambda_{hff}\, h^0 \bar{f}f - \lambda_{Hff}\, H^0 \bar{f}f + i\lambda_{Aff}\, A^0 \bar{f}\gamma^5 f, \tag{5.37}$$

where the couplings are given in Table 5.3.

5.3.2 The gaugino–Higgsino sector

The coupling of the Higgs fields to the gauge fields induces by supersymmetry a Higgsino –Higgs–gaugino coupling. This in turn generates, once the Higgs field is set to its vacuum expectation value, a mixed Higgsino –gaugino mass term. Mass eigenstates are therefore mixed Higgsino –gaugino states. Since color is not broken, charged (resp. neutral) fields mix among themselves: the mass eigenstates are called charginos (resp. neutralinos).

In this respect, the gluino $\widetilde{g}$, supersymmetric partner of the gluon, stands alone: color conservation prevents it from mixing with Higgsino fields. Only soft supersymmetry-breaking terms generate a mass term:

$$\mathcal{L}_{\mathrm{SB}} = -\frac{1}{2}\, M_3\, \overline{\lambda_g}\, \lambda_g. \tag{5.38}$$

As for the chargino mass matrix, supersymmetric contributions arise from:

- the term $-\mu H_1 \cdot H_2 = \mu\left(H_1^-\, H_2^+ - H_1^0\, H_2^0\right)$ in the superpotential which yields a fermion term

$$\mathcal{L} = -\mu\, \overline{\Psi^c_{H_2^+}}_R\, \Psi_{H_1^-\,L} + \mu\, \overline{\Psi^c_{H_2^0}}_R\, \Psi_{H_1^0\,L} + \text{h.c.} \tag{5.39}$$

- the coupling $g\overline{\lambda_-}\,\Psi_{H_2^-}\,H_2^0$ which is related by supersymmetry to the gauge coupling $gW^+ H_2^- H_2^0$ gives a term $(\langle H_2^0\rangle = \sqrt{v_1^2 + v_2^2}\,\sin\beta = M_W\sqrt{2}\sin\beta/g)$

$$\mathcal{L} = M_W\sqrt{2}\sin\beta\, \overline{\lambda_+}_R\, \Psi_{H_2^+\,L} + \text{h.c.} \tag{5.40}$$

Thus the chargino mass matrix receives a supersymmetric contribution

$$M_{\text{SUSY}} = \begin{pmatrix} 0 & -M_w\sqrt{2}\sin\beta \\ -M_w\sqrt{2}\cos\beta & \mu \end{pmatrix} \begin{matrix} \lambda_- \\ \Psi_{H_1^-} \end{matrix} \qquad \begin{matrix} \lambda_+ \qquad\qquad \Psi_{H_2^+} \end{matrix} . \tag{5.41}$$

On the other hand, soft supersymmetry breaking only generates a gaugino mass (a Higgsino mass would generate quadratic divergences)

$$M_{\text{SB}} = \begin{pmatrix} M_2 & 0 \\ 0 & 0 \end{pmatrix}. \tag{5.42}$$

In total, the chargino mass term reads

$$\mathcal{L}_c = -\begin{pmatrix} \overline{\lambda_+}_R & \overline{\Psi^c_{H_1^-}}_R \end{pmatrix} \mathcal{M}_c \begin{pmatrix} \lambda_{+L} \\ \Psi_{H_2^+ L} \end{pmatrix} + \text{h.c.} \tag{5.43}$$

with

$$\mathcal{M}_c = \begin{pmatrix} M_2 & -M_w\sqrt{2}\sin\beta \\ -M_w\sqrt{2}\cos\beta & \mu \end{pmatrix}. \tag{5.44}$$

The mass eigenstates are traditionally written $\chi_1^\pm$, $\chi_2^\pm$ with $m_{\chi_1^\pm} \le m_{\chi_2^\pm}$. More precisely, one defines

$$\begin{pmatrix} \chi_{1L}^+ \\ \chi_{2L}^+ \end{pmatrix} = Z_L \begin{pmatrix} \lambda_{+L} \\ \Psi_{H_2^+ L} \end{pmatrix}, \qquad \begin{pmatrix} \chi_{1R}^+ \\ \chi_{2R}^+ \end{pmatrix} = Z_R \begin{pmatrix} \lambda_{+R} \\ \Psi^c_{H_1^- R} \end{pmatrix}, \tag{5.45}$$

such that $\mathcal{L}_c = -m_{\chi_1^\pm}\overline{\chi_1^+}_R\chi_{1L}^+ - m_{\chi_2^\pm}\overline{\chi_2^+}_R\chi_{2L}^+ + \text{h.c.}$, *i.e.*

$$Z_R\mathcal{M}_c Z_L^\dagger = \begin{pmatrix} m_{\chi_1^\pm} & 0 \\ 0 & m_{\chi_2^\pm} \end{pmatrix}. \tag{5.46}$$

One has

$$m^2_{\chi_{1,2}^\pm} = \tfrac{1}{2}[\mu^2 + M_2^2 + 2M_w^2$$

$$\mp \sqrt{(\mu^2 - M_2^2)^2 + 4M_w^2(\mu^2 + 2\mu M_2\sin\beta + M_2^2) + 4M_w^4\cos^2 2\beta}], \tag{5.47}$$

and[8] $Z_{L,R} = \begin{pmatrix} \cos\phi_{L,R} & \sin\phi_{L,R} \\ -\sin\phi_{L,R} & \cos\phi_{L,R} \end{pmatrix}$ with

$$\tan 2\phi_L = 2M_w\sqrt{2}\frac{\mu\cos\beta + M_2\sin\beta}{\mu^2 - M_2^2 - 2M_w^2\cos 2\beta},$$

$$\tan 2\phi_R = 2M_w\sqrt{2}\frac{\mu\sin\beta + M_2\cos\beta}{\mu^2 - M_2^2 + 2M_w^2\cos 2\beta}. \tag{5.48}$$

---

[8]If $\det\mathcal{M}_c < 0$, write Z_L as $\begin{pmatrix} \cos\phi_L & \sin\phi_L \\ \sin\phi_L & -\cos\phi_L \end{pmatrix}$ to have positive mass eigenvalues.

Note that, in the case where the scales associated with supersymmetry are large ($M_W \ll M_2, |\mu|$), one of the charginos is predominantly wino whereas the other one is mostly Higgsino (see Exercise 5).

Similarly, the neutralino mass term reads (with obvious notations, λ_B, resp. λ_3, is the gaugino associated with the $U(1)_Y$ gauge field B_μ, resp. the $SU(2)_L$ gauge field A_μ^3)

$$\mathcal{L}_n = -\frac{1}{2} \left(\overline{\lambda_B}_R \ \overline{\lambda_3}_R \ \overline{\Psi^c_{H_1^0}}_R \ \overline{\Psi^c_{H_2^0}}_R \right) \mathcal{M}_n \begin{pmatrix} \lambda_{B_L} \\ \lambda_{3_L} \\ \Psi_{H_1^0 L} \\ \Psi_{H_2^0 L} \end{pmatrix} + \text{h.c.} \tag{5.49}$$

with

$$\mathcal{M}_n = \begin{pmatrix} M_1 & 0 & M_z \cos\beta \sin\theta_W & -M_z \sin\beta \sin\theta_W \\ 0 & M_2 & -M_z \cos\beta \cos\theta_W & M_z \sin\beta \cos\theta_W \\ M_z \cos\beta \sin\theta_W & -M_z \cos\beta \cos\theta_W & 0 & -\mu \\ -M_z \sin\beta \sin\theta_W & +M_z \sin\beta \cos\theta_W & -\mu & 0 \end{pmatrix}$$
$$\tag{5.50}$$

where M_1 and M_2 are, respectively, the $U(1)_Y$ and $SU(2)_L$ soft supersymmetry breaking gaugino mass terms. The eigenstates are written χ_1^0, χ_2^0, χ_3^0, χ_4^0 in increasing mass order (we will often denote the lightest neutralino simply by χ^0 or χ and other neutralinos by χ'). More precisely, one writes

$$\begin{pmatrix} \chi_{1_L}^0 \\ \chi_{2_L}^0 \\ \chi_{3_L}^0 \\ \chi_{4_L}^0 \end{pmatrix} = N \begin{pmatrix} \lambda_{B_L} \\ \lambda_{3_L} \\ \Psi_{H_1^0 L} \\ \Psi_{H_2^0 L} \end{pmatrix}, \quad \begin{pmatrix} \chi_{1_R}^0 \\ \chi_{2_R}^0 \\ \chi_{3_R}^0 \\ \chi_{4_R}^0 \end{pmatrix} = N^* \begin{pmatrix} \lambda_{B_R} \\ \lambda_{3_R} \\ \Psi^c_{H_1^0 R} \\ \Psi^c_{H_2^0 R} \end{pmatrix}, \tag{5.51}$$

where N is a unitary matrix (the second equation derives from charge conjugation: $\lambda = \lambda^c$) which satisfies:

$$N^* \mathcal{M}_n N^{-1} = \text{diag}\left(m_{\chi_1^0}, m_{\chi_2^0}, m_{\chi_3^0}, m_{\chi_4^0} \right). \tag{5.52}$$

By construction, the neutralinos are Majorana spinors.

We note that, in the limit $M_z \ll M_1 < M_2 < |\mu|$, the lightest neutralino is mostly bino (λ_B), the next to lightest mostly wino (λ_3) and the heaviest are Higgsinos ($\Psi_{H_1^0} \pm \Psi_{H_2^0}$) (see Exercise 5).

We see that, at tree level, the -ino sector depends on two supersymmetric parameters (μ, $\tan\beta$) and three soft supersymmetry breaking mass terms (M_1, M_2, M_3). We will see in the next chapter that the assumption of gaugino mass universality at the gauge unification scale yields relations among the latter three parameters.

5.3.3 The squark and slepton sector

Assuming that the squark mass matrices can be diagonalized (in family space indexed by $i \in \{1, 2, 3\}$) simultaneously with those of the corresponding quarks, we have

$$
\widetilde{M}^2_{u_i} = \begin{pmatrix} \widetilde{m}^2_{u_i LL} & \widetilde{m}^2_{u_i LR} \\ \widetilde{m}^2_{u_i RL} & \widetilde{m}^2_{u_i RR} \end{pmatrix}
$$

$$
= \begin{pmatrix} m^2_{Q_i} + m^2_{u_i} + \frac{1}{6}(4M^2_W - M^2_Z)\cos 2\beta & m_{u_i}(A^*_{u_i} - \mu \cot\beta) \\ m_{u_i}(A_{u_i} - \mu^* \cot\beta) & m^2_{U_i} + m^2_{u_i} + \frac{2}{3}(-M^2_W + M^2_Z)\cos 2\beta \end{pmatrix},
\tag{5.53}
$$

$$
\widetilde{M}^2_{d_i} = \begin{pmatrix} \widetilde{m}^2_{d_i LL} & \widetilde{m}^2_{d_i LR} \\ \widetilde{m}^2_{d_i RL} & \widetilde{m}^2_{d_i RR} \end{pmatrix}
$$

$$
= \begin{pmatrix} m^2_{Q_i} + m^2_{d_i} - \frac{1}{6}(2M^2_W + M^2_Z)\cos 2\beta & m_{d_i}(A^*_{d_i} - \mu \tan\beta) \\ m_{d_i}(A_{d_i} - \mu^* \tan\beta) & m^2_{D_i} + m^2_{d_i} + \frac{1}{3}(M^2_W - M^2_Z)\cos 2\beta \end{pmatrix},
\tag{5.54}
$$

where $m_{u_i} = \lambda_{u_i} v_2$, $m_{d_i} = -\lambda_{d_i} v_1$ are the fermion masses. Also $m^2_{Q_i}$, $m^2_{U_i}$ and $m^2_{D_i}$ are the soft masses, respectively, for the $SU(2)_L$ scalar doublet Q_i and singlets U_i and D_i, and A_{u_i}, A_{d_i} are the trilinear soft terms (defined as in (5.8)):

$$
\mathcal{L}_{\text{soft}} = -m^2_{Q_i}\left(\widetilde{u}^*_{i_L}\,\widetilde{u}_{i_L} + \widetilde{d}^*_{i_L}\,\widetilde{d}_{i_L}\right) - m^2_{U_i}\,\widetilde{u}^*_{i_R}\,\widetilde{u}_{i_R} - m^2_{D_i}\,\widetilde{d}^*_{i_R}\,\widetilde{d}_{i_R}
$$

$$
- \left(A_{d_i}\lambda_{d_i}\,\widetilde{q}_{i_L}\cdot H_1\,\widetilde{d}^*_{i_R} + A_{u_i}\lambda_{u_i}\,\widetilde{q}_{i_L}\cdot H_2\,\widetilde{u}^*_{i_R} + \text{h.c.}\right).
\tag{5.55}
$$

We will identify in turn the origin of each term. This will provide a good illustration of the different interactions obtained earlier. For simplicity, we consider only up-type squarks. One finds in the entries of the mass-squared matrix (5.53):

- Supersymmetric mass terms arising from the superpotential after gauge symmetry breaking. For example, the terms

$$
-\mu H_1 \cdot H_2 + \lambda_u Q \cdot H_2 U^c
$$

in the superpotential (5.2) yield a term $|\lambda_u Q \cdot H_2|^2$ in $|dW/dU^c|^2$ and $|\lambda_u U U^c - \mu H^0_1|^2$ in $|dW/dH^0_2|^2$. After gauge symmetry breaking ($\langle H^0_1 \rangle = v_1$, $\langle H^0_2 \rangle = v_2$), this yields quadratic terms in the scalar potential, respectively, $\lambda^2_u v^2_2 \widetilde{u}_L \widetilde{u}^*_L = m^2_u \widetilde{u}_L \widetilde{u}^*_L$ and, arising from the cross-product in the second case, $-\mu^* \lambda_u v_1 \widetilde{u}_L \widetilde{u}^*_R + \text{h.c.} = -\mu^* m_u \cot\beta + \text{h.c.}$

- Supersymmetric mass terms arising from the D-terms after gauge symmetry breaking. The D-term reads

$$V_D = \frac{1}{2}\left(g\sum_{i,j}\phi^{*i}t_i^{3j}\phi_j\right)^2 + \frac{1}{2}\left(\frac{g'}{2}\sum_i\phi^{*i}y_i\phi_i\right)^2 + \cdots$$

$$= \tfrac{1}{2}g^2\left(\tfrac{1}{2}\tilde{u}_L^*\tilde{u}_L + \tfrac{1}{2}H_1^{0*}H_1^0 - \tfrac{1}{2}H_2^{0*}H_2^0 + \cdots\right)^2$$

$$+ \tfrac{1}{2}g'^2\left(\tfrac{1}{6}\tilde{u}_L^*\tilde{u}_L - \tfrac{1}{2}H_1^{0*}H_1^0 + \tfrac{1}{2}H_2^{0*}H_2^0 + \cdots\right)^2. \tag{5.56}$$

After symmetry breaking ($\langle H_1^0\rangle = v_1$, $\langle H_2^0\rangle = v_2$), the cross-terms yield:

$$\frac{1}{2}(v_1^2 - v_2^2)\left(\frac{1}{2}g^2 - \frac{1}{6}g'^2\right)\tilde{u}_L^*\tilde{u}_L,$$

or more generally

$$\tfrac{1}{2}(v_1^2 - v_2^2)\left(g^2t_i^3 - g'^2y_i/2\right)\phi^{*i}\phi_i,$$

which can be expressed in terms of M_z, M_w and β using (5.23) and $\cos 2\beta = (v_1^2 - v_2^2)/(v_1^2 + v_2^2)$:

$$\left[q_i M_w^2 - (y_i/2)M_z^2\right]\cos 2\beta \, \phi^{*i}\phi_i.$$

- Soft supersymmetry-breaking scalar mass terms, respectively, m_Q^2 and m_U^2 for $\tilde{u}_L$ and $\tilde{u}_R$.
- Mass terms arising from the soft-supersymmetry-breaking A-terms in (5.55) after gauge symmetry breaking.

Similarly for charged leptons, one has

$$\widetilde{M}_{e_i}^2 = \begin{pmatrix} \tilde{m}_{e_iLL}^2 & \tilde{m}_{e_iLR}^2 \\ \tilde{m}_{e_iLR}^2 & \tilde{m}_{e_iRR}^2 \end{pmatrix}$$

$$= \begin{pmatrix} m_{L_i}^2 + m_{e_i}^2 + \tfrac{1}{2}(M_z^2 - 2M_w^2)\cos 2\beta & m_{e_i}(A_{e_i}^* - \mu\tan\beta) \\ m_{e_i}(A_{e_i} - \mu^*\tan\beta) & m_{E_i}^2 + m_{e_i}^2 + (M_w^2 - M_z^2)\cos 2\beta \end{pmatrix}. \tag{5.57}$$

We note that, in all three cases (u, d, e-type sfermions), the nondiagonal terms are proportional to the corresponding fermion mass. They are nonnegligible only in the case of the third family. Thus, for $f = t, b$ or τ, the mass eigenstates are

$$\begin{pmatrix} \tilde{f}_1 \\ \tilde{f}_2 \end{pmatrix} = \begin{pmatrix} \cos\theta_f & \sin\theta_f \\ -\sin\theta_f & \cos\theta_f \end{pmatrix} = \begin{pmatrix} \tilde{f}_L \\ \tilde{f}_R \end{pmatrix}, \tag{5.58}$$

with corresponding eigenvalues $\tilde{m}_{f1}^2 < \tilde{m}_{f2}^2$ and $0 < \theta_f < \pi$.

5.4 Baryon and lepton number

Realizing supersymmetry by doubling the number of fundamental fields has the immediate consequence of losing a remarkable property of the Standard Model: renormalizability and $SU(3) \times SU(2) \times U(1)$ gauge symmetry impose automatic baryon (B) and lepton (L) number conservation. Indeed, it is possible to generalize baryon and lepton number to the supersymmetric particles (squarks, sleptons) but the introduction of these new fields allows us to write new renormalizable couplings which are B or L violating. These couplings were written in (5.4) and, if all are present, they lead to rapid proton decay.

In order to prevent this, one may choose to impose B or L global symmetries. This is not a favored solution of the problem because global symmetries, if not protected by a local symmetry, tend to be broken by gravitational interactions. One prefers to obtain them as a consequence of a broken local symmetry.

In Section 5.2.2, we have discussed the introduction of R-parity which may arise from the breaking of a continuous R-symmetry. Such a R-parity allows us to set to zero all renormalizable interactions breaking B or L number (*cf.* (5.4)) since Q, L, U^c, D^c and E^c have R-parity -1 whereas H_1 and H_2 have R-parity $+1$.

The question must however also be addressed at the level of nonrenormalizable interactions. Since a supersymmetric generalization of the Standard Model usually arises as a low energy effective theory of a more fundamental theory, nonrenormalizable terms are expected, with coefficients of the order of the mass scale M of the underlying theory to some power. This power is fixed by the dimension of the corresponding term. We will see later explicit examples in the context of grand unification or string theory.

A thorough discussion therefore requires to list all the possible terms according to their dimensions. Since we are working in a supersymmetric set up, we write them as F-terms or D-terms. We recall a few facts from Chapter 3:

- an F-term of a product of superfields with total dimension n is a term in the Lagrangian with dimension $n + 1$;
- a D-term of a product of superfields and possibly their hermitian conjugates, with total dimension n, is a term in the Lagrangian with dimension $n + 2$.

Then, imposing R-parity leaves us the possibility of writing the following gauge invariant terms at dimension 5:

$$(LLH_2H_2)_F, \ (QQU^cD^c)_F, \ (QU^cLE^c)_F, \ (QQQL)_F, \ (U^cU^cD^cE^c)_F,$$
$$\left(LE^cH_2^\dagger\right)_D, \ \left(QD^cH_2^\dagger\right)_D, \ \left(QU^cH_1^\dagger\right)_D,$$

where $SU(2)$ and $SU(3)$ indices are contracted in order for the corresponding term to be singlet under these groups. Each term has a coupling of order M^{-1}. It is already clear at this stage that R-parity is not sufficient to prevent baryon nor lepton violation: the terms $(QQQL)_F$ and $(U^cU^cD^cE^c)_F$ violate both. It is thus important that either the fundamental scale M be superheavy (as in grand unification) or extra symmetries be imposed.

5.5 The LSP and dark matter

We have seen above that models with R-parity include a stable particle: the lightest supersymmetric particle (LSP). We first show why this particle is an excellent candidate to account for cold dark matter. We then discuss its nature as well as its properties.

5.5.1 The LSP as a WIMP

We will not recall here the arguments that lead to the need for dark matter. It suffices to say that the total matter content ρ_M of the energy density in the Universe is believed to be

$$\Omega_M \equiv \frac{\rho_M}{\rho_c} \sim 0.3, \tag{5.59}$$

where $\rho_c = 3H_0^2 m_P^2 \simeq 10^{-26} \text{kg/m}^3$ is the critical density (corresponding to a spatially flat spacetime)[9]. On the other hand, the density of baryonic matter ρ_B is limited to be

$$\Omega_B \equiv \frac{\rho_B}{\rho_c} \leq 0.02. \tag{5.60}$$

This leaves a lot missing, under the form of nonbaryonic matter or of a more exotic component.

Let us see under which conditions a given particle of mass m_X might provide the right amount of dark matter. We will suppose that this particle is neutral and colorless, otherwise it would have some observable effects through scattering on matter. There are two competing effects to modify the abundance of this species: annihilation and expansion of the Universe. Indeed, the faster is the dilution associated with the expansion, the least effective is the annihilation because the particles recede from one another. This is summarized in the following Boltzmann equation which gives the evolution with time of the particle number density n_X:

$$\frac{dn_X}{dt} + 3Hn_X = -\langle \sigma_{\mathrm{ann}} v \rangle \left(n_X^2 - n_X^{(\mathrm{eq})2} \right), \tag{5.61}$$

where $\langle \sigma_{\mathrm{ann}} v \rangle$ is the thermal average of the $X\bar{X}$ annihilation cross-section times the relative velocity of the two particles annihilating, $n_X^{(\mathrm{eq})}$ is the equilibrium density, and H the Hubble parameter. When the temperature drops below the mass m_X, the annihilation rate becomes smaller than the expansion rate and there is a freezing of the number of particles in a covolume. We study this freezing for a general species in Section D.3.3 of Appendix D, to which we refer the reader. We find in the case of cold relics (*i.e.* relics which are nonrelativistic at the time of freezing) that the freezing temperature T_f is given in terms of the variable $x_f = m_X/(kT_f)$ by equation (D.73) of this appendix:

$$x_f = \ln \left(0.038 \frac{M_P m_X \langle \sigma_{\mathrm{ann}} v \rangle}{\hbar^2} \frac{g_X}{g_*^{1/2}} \right), \tag{5.62}$$

where g_X is the number of internal degrees of freedom of the particle and g_* is the total number of relativistic degrees of freedom present in the Universe at the time of decoupling. One checks that x_f is a quantity of order one (more precisely 20).

---

[9]See Appendix D for a more detailed treatment.

The relic density of particles is given by

$$\Omega_x \equiv \frac{m_x n_x(t_0)}{\rho_c} = 40 \sqrt{\frac{\pi}{5} \frac{s_0}{k}} \frac{\hbar^3}{H_0^2 M_P^3} \frac{g_*^{1/2}}{g_s} \frac{x_f}{\langle \sigma_{\rm ann} v \rangle} \tag{5.63}$$

where s_0 is the present entropy density and g_s an effective number of degrees of freedom which enters in the expression of the entropy (see equation (D.64) of Appendix D). Putting the explicit values, one finds

$$\Omega_x h_0^2 \sim \frac{1.07 \times 10^9 \ {\rm GeV}^{-1}}{g_*^{1/2} M_P} \frac{x_f}{\langle \sigma_{\rm ann} v \rangle}. \tag{5.64}$$

We note for further reference that the smaller the annihilation cross-section is, the larger is the relic density. We find $\Omega_x \sim (100 \ {\rm TeV})^{-2} (\langle \sigma_{\rm ann} v \rangle)^{-1} \sim 10^{-13} \ {\rm barn}/\langle \sigma_{\rm ann} v \rangle$ (in units where $\hbar = c = 1$, 1 barn $= 2.5 \times 10^3 \ {\rm GeV}^{-2}$).

Thus Ω_x will be of order 1 (more precisely a fraction of 1) if $\langle \sigma_{\rm ann} v \rangle$ is of the order of a picobarn to a femtobarn, which is a typical size for an electroweak process. Also writing dimensionally

$$\langle \sigma_{\rm ann} v \rangle \sim \frac{\alpha^2}{m_x^2}, \tag{5.65}$$

where α is a generic coupling strength, we find that Ω_x is of order 1 for a mass $m_x \sim \alpha \times 100 \ {\rm TeV}$, *i.e.* in the TeV range. This is why one is searching for a weakly interacting massive particle (WIMP).

As we have seen earlier, in supersymmetric models with R-parity, the lightest supersymmetric particle (LSP) is stable and provides a good candidate: supersymmetric particles are massive (as it results from negative searches) and many of them are weakly interacting (gauginos, Higgsinos and sleptons). In the following section, we will discuss the possible nature of the LSP.

The simple analysis that we presented is somewhat more involved when the WIMP appears within a system of particles with which it can annihilate, as is the case in the supersymmetric set up that we study: such coannihilations are particularly helpful to decrease the relic density in the case where the WIMP is almost degenerate with some of the other particles [40, 207].

To illustrate this, let us consider a system of two Majorana particles X_1 and X_2 of masses $m_1 < m_2$. Their abundance is determined by the following reactions:

$$\text{annihilation} \ \ X_i X_j \to f \bar{f}, \quad i = 1, 2 \tag{5.66}$$

$$\text{scattering} \ \ X_2 f \to X_1 f, \tag{5.67}$$

$$\text{decay} \ \ X_2 \to X_1 f \bar{f}, \tag{5.68}$$

where f and $\bar{f}$ stand for light Standard Model particles (we have assumed the conservation of a X-parity). The Boltzmann equations describing the evolution with time of their respective number densities n_1 and n_2 read [40] (compare with (5.61)):

$$\frac{dn_1}{dt} + 3Hn_1 = \sum_f \left[-\langle \sigma^{11}_{\text{ann}} v \rangle \left(n_1^2 - n_1^{(\text{eq})2} \right) - \langle \sigma^{12}_{\text{ann}} v \rangle \left(n_1 n_2 - n_1^{(\text{eq})} n_2^{(\text{eq})} \right) \right.$$

$$\left. + \langle \sigma^{2\to1}_{\text{scat}} v \rangle n_2 n_f^{(\text{eq})} - \langle \sigma^{1\to2}_{\text{scat}} v \rangle n_1 n_f^{(\text{eq})} + \Gamma^{21} \left(n_2 - n_2^{(\text{eq})} \right) \right] \quad (5.69)$$

$$\frac{dn_2}{dt} + 3Hn_2 = \sum_f \left[-\langle \sigma^{22}_{\text{ann}} v \rangle \left(n_2^2 - n_2^{(\text{eq})2} \right) - \langle \sigma^{12}_{\text{ann}} v \rangle \left(n_1 n_2 - n_1^{(\text{eq})} n_2^{(\text{eq})} \right) \right.$$

$$\left. - \langle \sigma^{2\to1}_{\text{scat}} v \rangle n_2 n_f^{(\text{eq})} + \langle \sigma^{1\to2}_{\text{scat}} v \rangle n_1 n_f^{(\text{eq})} - \Gamma^{21} \left(n_2 - n_2^{(\text{eq})} \right) \right] \quad (5.70)$$

with obvious notation: $\sigma^{ij}_{\text{ann}} = \sigma \left(X_i X_j \to f\overline{f} \right)$, $\sigma^{i\to j}_{\text{scat}} = \sigma \left(X_i f \to X_j f \right)$ and $\Gamma^{ij} = \Gamma \left(X_i \to X_j f\overline{f} \right)$. Since all the X_2 which survive annihilation eventually decay into X_1, the relic density will be the present value of $n_X \equiv n_1 + n_2$. Summing the previous two equaticns, we obtain the evolution equation for n_X:

$$\frac{dn_X}{dt} + 3Hn_X = -\sum_{i,j} \langle \sigma^{ij}_{\text{ann}} v \rangle \left(n_i n_j - n_i^{(\text{eq})} n_j^{(\text{eq})} \right). \quad (5.71)$$

This analysis can straightforwardly be generalized to a system of N particles X_i ($i = 1, \ldots, N$) with g_i internal degrees of freedom and masses m_i ($m_1 \le m_2 \le \cdots \le m_N$). Equation (5.71) applies to $n_X \equiv \sum_{i=1}^{N} n_i$ and allows us to determine the relic density of the only stable particle of this system, *i.e.* X_1.

To simplify (5.71), one notes [207] that the reaction rates are much faster for (5.67) and (5.68) than for (5.66): this ensures that the relative number densities of each species remain at their equilibrium value: $n_i/n_X \sim n_i^{(\text{eq})}/n_X^{(\text{eq})}$. Then, (5.71) simply reads, just as (5.61),

$$\frac{dn_X}{dt} + 3Hn_X = -\langle \sigma^{\text{eff}}_{\text{ann}} v \rangle \left(n_X^2 - n_X^{(\text{eq})2} \right), \quad (5.72)$$

where we have defined

$$\sigma^{\text{eff}}_{\text{ann}} \equiv \sum_{i,j=1}^{N} \sigma^{ij}_{\text{ann}} \frac{n_i^{(\text{eq})}}{n_X^{(\text{eq})}} \frac{n_j^{(\text{eq})}}{n_X^{(\text{eq})}}, \quad (5.73)$$

with (see (D.60) of Appendix D)

$$\frac{n_i^{(\text{eq})}}{n_X^{(\text{eq})}} = \frac{g_i}{g_{\text{eff}}} \left(\frac{m_i}{m_1} \right)^{3/2} e^{(1-m_i/m_1)x_1}, \quad g_{\text{eff}} = \sum_{i=1}^{N} g_i \left(\frac{m_i}{m_1} \right)^{3/2} e^{(1-m_i/m_1)x_1}, \quad (5.74)$$

where $x_1 = m_1/(kT)$.

The decoupling temperature T_f and relic density Ω_{X_1} are now given by

$$x_{1f} \equiv \frac{m_1}{kT_f} = \ln\left(0.038 \frac{M_P m_1 \langle \sigma_{\text{ann}}^{\text{eff}} v \rangle}{\hbar^2} \frac{g_{\text{eff}}}{g_*^{1/2}}\right),$$

$$\Omega_{X_1} h_0^2 \sim \frac{1.07 \times 10^9 \text{ GeV}^{-1}}{g_*^{1/2} M_P} \frac{x_{1f}}{\langle \sigma_{\text{ann}}^{\text{eff}} v \rangle}. \tag{5.75}$$

We see that, in the case of near degeneracy in mass, and equal degrees of freedom and cross-sections, $g_{\text{eff}} \sim N g_1$ and $\Omega_{X_1}/\Omega_X = x_{1f}/x_f \sim 1 + (\ln N)/x_f \sim 1 + (\ln N)/20$, where x_f and Ω_X are the quantities computed ignoring coannihilations (*i.e.* using (5.62) and (5.63)). A more important effect arises if the coannihilators X_i ($i \geq 2$) have a larger annihilation cross-section: this increases $\langle \sigma_{\text{ann}}^{\text{eff}} v \rangle$ with respect to $\langle \sigma_{\text{ann}} v \rangle$ and may decrease Ω_{X_1} with respect to Ω_X by several orders of magnitude.

Another possible deviation from the simple analysis presented earlier arises when one approaches poles in the annihilation cross-section [207]. Indeed, if the exchange of a particle of mass m_A and width Γ_A is possible in the s-channel (in supersymmetric models, this often turns out to be the pseudoscalar A, hence the notation), the estimate (5.65) is more correctly written

$$\langle \sigma_{\text{ann}} v \rangle \sim \frac{4\alpha^2 s}{\left(s - m_A^2\right)^2 + \Gamma_A^2 m_A^2}, \tag{5.76}$$

where $s = 4m_X^2$ in the limit of zero velocity. We see that, close to the pole *i.e.* for $m_A^2 \sim 4m_X^2$, the cross-section reaches the value $\langle \sigma_{\text{ann}} v \rangle|_{\text{pole}} = 4\alpha^2/\Gamma_A^2$. If the resonance is narrow, this might lead to a significant increase in the cross-section and decrease in the relic density.

5.5.2 Nature of the LSP

If the LSP had nonzero electric charge or color, it would have condensed with baryonic matter and should be present today in anomalous heavy isotopes. The absence of such isotopes puts very stringent bounds on such particles, which therefore could not play the rôle of dark matter.

A possible candidate would be a sneutrino. Direct searches (see Figure 5.3) exclude a LSP sneutrino in a mass window between 25 GeV and 5 TeV [28]. But lighter sneutrinos have been excluded by searches at LEP and, on the other side of the allowed spectrum, a mass larger than several TeV represents a rather unnatural value for a LSP. In the context of the theories that we consider here, this discards the sneutrino as a LSP.

We are thus left[10] with the lightest neutralino $\chi_1^0 \equiv \chi$ which, as we have seen in Section 5.3.2, is a combination of gauginos and Higgsinos.

---

[10]Another possibility is that the lightest supersymmetric particle is the supersymmetric partner of the graviton, the gravitino (see Chapter 6).

We list below for further use the main interactions of the lightest neutralino (for a more complete treatment, see [242]). First, the coupling to the neutral vectors is given by:

$$\mathcal{L}_{\chi\chi Z} = -\frac{1}{4}\frac{g}{\cos\theta_w}\left(|N_{13}|^2 - |N_{14}|^2\right)\bar{\chi}\gamma^\mu\gamma^5\chi \, Z_\mu, \qquad (5.77)$$

or more generally

$$\mathcal{L}_{\chi\chi_m^0 Z} = \frac{1}{4}\frac{g}{\cos\theta_w}Z_\mu\left[\bar{\chi}\gamma^\mu L\chi_m^0\left(N_{13}N_{m3}^* - N_{14}N_{m4}^*\right) - \bar{\chi}\gamma^\mu R\chi_m^0\left(N_{13}^*N_{m3} - N_{14}^*N_{m4}\right)\right], \qquad (5.78)$$

where the N matrix is defined in (5.51), $m = 1,\dots,4$, and $L, R = (1 \mp \gamma_5)/2$. We note that such interactions probe the Higgsino content of the neutralino LSP. The coupling to charginos and W reads:

$$\begin{aligned}\mathcal{L}_{\chi\chi_r^\pm W^\mp} = -g\bar{\chi}\gamma^\mu &\left[\left(N_{12}Z_{Lr1}^* - \frac{1}{\sqrt{2}}N_{14}Z_{Lr2}^*\right) L\right.\\ &\left.+ \left(N_{12}^*Z_{Rr1}^* + \frac{1}{\sqrt{2}}N_{13}^*Z_{Rr2}^*\right) R\right] \chi_r^+ W_\mu^- + \text{h.c.} \qquad (5.79)\end{aligned}$$

where the Z_L, Z_R matrices are defined in (5.45).

Next, the couplings to the neutral Higgs system can be summarized in

$$\begin{aligned}\mathcal{L}_{\chi\chi H} = \frac{g}{2}(N_{12} - \tan\theta_w N_{11}) &\left[(\cos\alpha N_{13} - \sin\alpha N_{14})H^0 - (\sin\alpha N_{13} + \cos\alpha N_{14})h^0\right.\\ &\left.+i(\sin\beta N_{13} - \cos\beta N_{14})A^0\right]\bar{\chi}R\chi + \text{h.c.} \qquad (5.80)\end{aligned}$$

We see that such couplings vanish when the lightest neutralino is purely gaugino ($N_{13} = N_{14} = 0$) or purely Higgsino ($N_{11} = N_{12} = 0$). Moreover, if the N matrix elements are real, the coupling of χ to h^0 and H^0 (resp. A^0) is purely scalar (resp. pseudoscalar).

Finally, the couplings to quarks and squarks are given by

$$\mathcal{L}_{\chi q\tilde{q}} = -\bar{q}_{iL}\chi\left(X_i\tilde{q}_{iL} + Z_i^{q*}\tilde{q}_{iR}\right) - \bar{q}_{iR}\chi\left(Y_i^*\tilde{q}_{iR} + Z_i^q\tilde{q}_{iL}\right) + \text{h.c.} \qquad (5.81)$$

where

$$\begin{aligned}X_i &= -g\sqrt{2}\left[t_i^3 N_{12} + (y_i/2)\tan\theta_w N_{11}\right],\\ Y_i &= g\sqrt{2}\, q_i\, N_{11}\tan\theta_w, \qquad\qquad\qquad\qquad (5.82)\\ Z_i^u &= \frac{g}{\sqrt{2}}\frac{m_{u_i}N_{14}^*}{M_w\sin\beta}, \qquad Z_i^d = \frac{g}{\sqrt{2}}\frac{m_{d_i}N_{13}^*}{M_w\cos\beta},\end{aligned}$$

where $i = 1, 2, 3$ is a family index (we have neglected quark and squark mixings).

5.5.3 LSP annihilation

We assume from now on that the LSP is the lightest neutralino χ. We first give in Table 5.4 the main annihilation channels for a neutralino at rest. We note that the

Table 5.4 Main annihilation channels of the lightest neutralino χ at rest, with the indication of the type of exchanged particle (and the corresponding channel). The couplings λ_{Aff} are given in Table 5.3.

Final state	Exchanged		Amplitude $\propto$
	s	t, u	
$f\bar{f}$		$\tilde{f}$	$g^2 \left[1 + \left(\dfrac{m_{\tilde{f}}}{m_\chi} \right)^2 - \left(\dfrac{m_f}{m_\chi} \right)^2 \right]^{-1}$
$f\bar{f}$	A		$g\lambda_{Aff} \dfrac{(N_{12} - \tan\theta_W N_{11})(\sin\beta N_{13} - \cos\beta N_{14})}{4 - m_A^2/m_\chi^2 + i\Gamma_A m_A/m_\chi^2}$
$f\bar{f}$	Z		$\dfrac{g^2}{\cos^2\theta_W} \dfrac{m_f m_\chi}{M_Z^2} \left(|N_{13}|^2 - |N_{14}|^2 \right)$
W^+W^-		$\chi_r^\pm$	$g^2 \sqrt{1 - \dfrac{M_W^2}{m_\chi^2}} \dfrac{\left| N_{12} Z_{Lr1}^* - \frac{1}{\sqrt{2}} N_{14} Z_{Lr2}^* \right|^2 + \left| N_{12} Z_{Rr1} + \frac{1}{\sqrt{2}} N_{13} Z_{Rr2} \right|^2}{1 + m_{\chi_r^\pm}^2/m_\chi^2 - M_W^2/m_\chi^2}$
ZZ		χ_m^0	$\dfrac{g^2}{\cos^2\theta_W} \sqrt{1 - \dfrac{M_Z^2}{m_\chi^2}} \dfrac{\left(N_{13} N_{m3}^* - N_{14} N_{m4}^* \right)^2}{1 + m_{\chi_m^0}^2/m_\chi^2 - M_Z^2/m_\chi^2}$

amplitudes for most channels behave dominantly as m_χ^2/M_{SUSY} where M_{SUSY} is a supersymmetric mass, except for the Z exchange (which dominates when it is open). The Higgsino fraction N_{13} or N_{14} plays a significant rôle: it enhances the annihilation cross-section and thus tends to decrease the relic density for a Higgsino-like LSP. Finally, the annihilation amplitude into $f\bar{f}$ is proportional to the fermion mass and thus important only for the third family.

Other annihilation channels at rest include $W^\pm H^\mp$ (through the exchange of A in the s-channel or charginos in the t, u channels), ZH or AH and Zh or Ah (through the exchange of Z or A in the s-channel or neutralinos in the t, u channels). They turn out to be less relevant in explicit computations, except when one considers light scalars. We also note that decays to photons or gluons arise only at the one loop level.

As is clear from the considerations of Section 5.5.1, the neutralino χ is not at rest at the time of freezing. Typically, its average velocity is given by $\langle v^2 \rangle = 6kT/m = 6/x$. One may then expand the averaged cross-section as $\langle \sigma v \rangle = a + b\langle v^2 \rangle + \cdots$. The term a is the S wave contribution already discussed, whereas the term b contains contributions from S and P waves. New contributions appear when one considers the decay of a neutralino LSP which is not at rest (for example s-channel exchange of Z^0, h^0 and H^0 for the decay into W^+W^- pairs), as well as new final states: $Z^0 A^0$, $A^0 A^0$, $h^0 h^0$, $H^0 H^0$, $H^0 h^0$ and $H^+ H^-$.

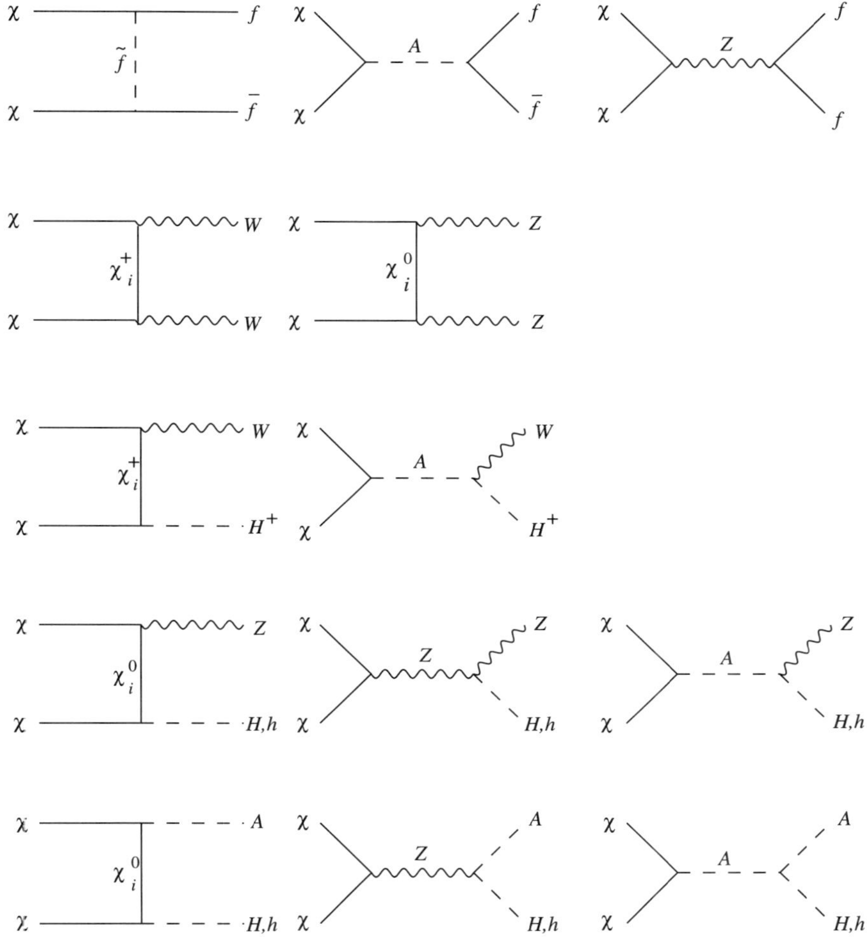

Fig. 5.1 Main annihilation channels of the lightest neutralino χ at rest.

5.5.4 LSP interaction with matter

It is important to be able to determine as precisely as possible the interaction of the LSP neutralino with nonrelativistic matter in order to make predictions for its direct detection as well as its capture by material bodies.

Since we are considering matter in the nonrelativistic limit, we first recall that in this limit, solutions of the Dirac equation have the form[11]: $\Psi = \begin{pmatrix} \varphi \\ 0 \end{pmatrix}$ where φ is a two-component spinor field. Then, $\bar{\Psi}\Psi$ and $\bar{\Psi}\gamma^{\mu}\Psi$ yield the scalar $\varphi^{\dagger}\varphi$ whereas

---

[11] We use here the Dirac representation for gamma matrices: $\gamma^0 = \beta = \begin{pmatrix} 1 & 0 \\ 0 & -1 \end{pmatrix}$, $\gamma^i = \beta\alpha^i$ with $\alpha^i = \begin{pmatrix} 0 & \sigma^i \\ \sigma^i & 0 \end{pmatrix}$ and $\gamma^5 = \begin{pmatrix} 0 & 1 \\ 1 & 0 \end{pmatrix}$. For example, $\bar{\Psi}\gamma^0\Psi = \Psi^{\dagger}\Psi = \varphi^{\dagger}\varphi$ whereas $\bar{\Psi}\gamma^i\Psi = \Psi^{\dagger}\alpha^i\Psi = 0$.

$\bar{\Psi}\gamma^\mu\gamma^5\Psi$ and $\bar{\Psi}\sigma^{\mu\nu}\Psi$ yield the vector $\varphi^\dagger\sigma^i\varphi$ and $\bar{\Psi}\gamma^5\Psi$ vanishes in the nonrelativistic limit.

Because of the Majorana nature of the neutralino (which implies $\bar{\chi}\gamma^\mu\chi = 0$ and $\bar{\chi}\sigma^{\mu\nu}\chi = 0$, see (B.40) of Appendix B), the most general interaction is described at the level of quarks, by the effective four-fermion Lagrangian:

$$\mathcal{L} = \sum_i \left[d_i \, \bar{\chi}\gamma^\mu\gamma^5\chi \, \bar{q}_i\gamma_\mu\gamma^5 q_i + f_i \, \bar{\chi}\chi \, \bar{q}_i q_i \right]. \tag{5.83}$$

One is thus left with the axial-vector (also called spin dependent) interaction with coefficient d_i and the scalar interaction (also called spin independent) with coefficient f_i.

From the interactions derived in Section 5.5.2, we see that (*cf.* Fig. 5.2) scalar (resp. spin-dependent) interactions arise through the exchange of H, h (resp. Z) in the t channel. The exchange of a sfermion in the s or t channels leads through Fierz reordering to both types of interactions.

It turns out that the scalar contribution is dynamically reduced with respect to the spin-dependent one. As we will see in the next section, this is compensated experimentally by the fact that it behaves coherently and the effects of all the nucleons in a nucleus add up. For the time being, we will identify the origin of the dynamical reduction at the level of a single nucleon.

We first consider squark exchange. A large contribution of this type would require the lighter squarks of the third family and thus the implementation of sfermion mixing. More precisely, the contributions to the couplings d_i and f_i in (5.83) read

$$d_i\big|_{\tilde{q}\text{ exch.}} = \frac{1}{4}\sum_{r=1}^{2} \frac{a_{ir}^2 + b_{ir}^2}{\tilde{m}_{q_i r}^2 - (m_\chi + m_{q_i})^2}, \tag{5.84}$$

$$f_i\big|_{\tilde{q}\text{ exch.}} = -\frac{1}{4}\sum_{r=1}^{2} \frac{a_{ir}^2 - b_{ir}^2}{\tilde{m}_{q_i r}^2 - (m_\chi + m_{q_i})^2}. \tag{5.85}$$

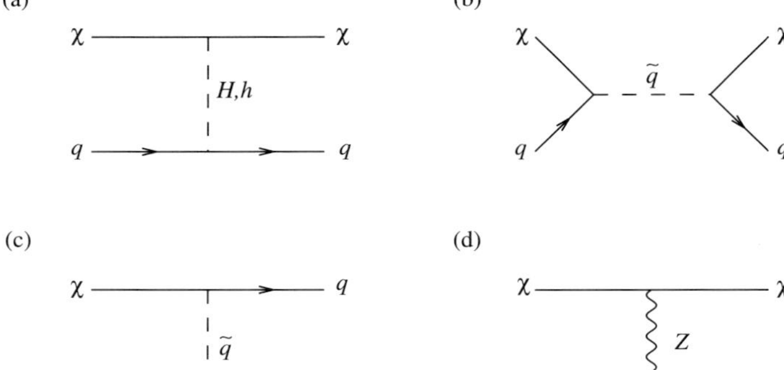

Fig. 5.2 Scalar (a), (b), (c) and spin-dependent (b), (c), (d) interactions of the lightest neutralino χ with matter.

The couplings a_{ir} and b_{ir} describe scalar and pseudoscalar LSP–squark–quark interactions:

$$\mathcal{L}_{\chi q\tilde{q}} = \sum_{i}\sum_{r=1}^{2} \bar{q}_{ir}\left(a_{ir} + b_{ir}\gamma^5\right)\chi\tilde{q}_{ir} + \text{ h.c.,} \tag{5.86}$$

where the index r refers to the two mass eigenstates in the case of sfermion mixing
(see (5.58)). They can be obtained from (5.81) and read explicitly

$$a_{i1} = -\tfrac{1}{2}\left[\cos\theta_{q_i}(X_i + Z_i^q) + \sin\theta_{q_i}(Y_i^* + Z_i^{q*})\right],$$

$$b_{i1} = -\tfrac{1}{2}\left[\cos\theta_{q_i}(X_i - Z_i^q) + \sin\theta_{q_i}(Y_i^* - Z_i^{q*})\right], \tag{5.87}$$

with $\cos\theta_{q_i}$ replaced by $-\sin\theta_{q_i}$ and $\sin\theta_{q_i}$ replaced by $\cos\theta_{q_i}$ for $r = 2$.

Realizing that $|a_{i_r}| = |b_{i_r}|$ in the chiral limit (Z_i^q goes to zero with the fermion
mass), we see that f_i vanishes in this limit. This is true also for the contribution
coming from H, h exchange since their couplings to quarks are proportional to the
mass, see Table 5.3. The scalar contribution is thus naturally disfavored for the light
quarks that are abundant in nucleons. Heavy quarks couple to nucleons only through
a loop [123, 124]. The combined effects lead to a dynamical suppression of the scalar
interaction. We may note that, going to large values of $\tan\beta$ may significantly increase
the cross-section because, as can be seen from Table 5.3, it enhances the couplings of
H, h to down-type quarks [123].

Regarding the spin-dependent cross-section, we infer from (5.77) that a large contribution from Z exchange requires a large Higgsino component, which is disfavored
if we want a substantial relic abundance.

To give an order of magnitude, in models where the LSP is mostly a bino, one
typically obtains spin-dependent cross-sections $\sigma_{\chi-p}^{SD}$ in the 10^{-7} to 10^{-5} pb and scalar
cross-sections $\sigma_{\chi-p}^{SI}$ in the range 10^{-10} to 10^{-7} pb range.

5.5.5 Direct detection

Goodman and Witten (1985) were the first to propose to use bolometers to detect dark
matter candidates. Typically, the event rate R is given by the number $\mathcal{N}_N = N_A/A$ of
nuclei in the target (N_A is Avogadro's number and A the atomic number) times the
flux $F = \rho_\chi v_\chi/m_\chi$ (ρ_χ is the local WIMP density and v_χ the average WIMP velocity)
times the cross-section σ. Putting typical numbers, this gives

$$R = 4.7 \text{ evts.kg}^{-1}.\text{day}^{-1}\,\frac{1}{A}\left(\frac{\rho_\chi}{0.3\text{GeV/cm}^3}\right)\left(\frac{v_\chi}{300\text{km/s}}\right)\left(\frac{100\text{ GeV}}{m_\chi}\right)\left(\frac{\sigma}{1\text{ pb}}\right). \tag{5.88}$$

If the invariant amplitude $\mathcal{M}$ is a constant at low energy, the cross-section reads

$$\sigma = \frac{\mu^2}{\pi}|\mathcal{M}|^2 \tag{5.89}$$

where μ is the reduced mass ($\mu^{-1} = M_A^{-1} + m_\chi^{-1}$). For the processes discussed in
the preceding section, $\mathcal{M}$ is typically of the form g^2/M^2, where g is a gauge coupling and

M a heavy scale (Z mass, Higgs mass or squark mass) which is larger than say $100\,\text{GeV}$. This gives, in the most optimistic case, a cross-section of the order of a picobarn, and thus rates of a few events per kg and per day. This is typically the rate obtained by present day experiments, which thus start to put some limits on the supersymmetric parameter space. This is also the rate of radioactive or cosmic ray background, which urges the need to perform these experiments in underground laboratories.

For a target nucleus of mass $M_A \sim A\,\text{GeV}$, the momentum transferred is $|\mathbf{q}^2| = 2\mu^2 v_\chi^2 (1 - \cos\theta^*)$ where θ^* is the scattering angle in the center of mass frame. The recoil energy is then

$$Q \equiv \frac{|\mathbf{q}^2|}{2M_A} = \frac{m_\chi^2 M_A}{(m_\chi + M_A)^2} v_\chi^2 \, (1 - \cos\theta^*). \tag{5.90}$$

For example, if $m_\chi \ll M_A$,

$$Q \sim \left(\frac{m_\chi}{1\,\text{GeV}}\right)^2 \left(\frac{v_\chi}{300\,\text{km/s}}\right)^2 \frac{1}{A} \,\text{keV}. \tag{5.91}$$

Hence, one expects an energy deposited in the detector in the range of a few keV.

We now discuss separately the spin-dependent and spin-independent contributions to the cross-section. As alluded to above, the scalar interaction adds coherently among the nucleons: this means that the invariant amplitude is proportional to A and the cross-section (5.89) scales like $\mu^2 A^2$. This represents a large enhancement factor if one uses heavy nuclei. Typical examples are germanium ($Z = 32, A = 76$) or xenon ($Z = 54, A = 136$).

More precisely the scalar differential cross-section reads

$$\frac{d\sigma^{\text{SI}}}{d|\mathbf{q}|^2} = \frac{1}{\pi v_\chi^2} \left[Z f_p + (A - Z) f_n \right]^2 F^2(Q), \tag{5.92}$$

where f_p and f_n are the effective neutralino couplings to proton and neutron respectively [18] and $F(Q)$ is the nuclear form factor.

For the spin-dependent contribution, ones introduces the nucleonic matrix element

$$\langle N | \bar{q}_i \gamma_\mu \gamma^5 q_i | N \rangle \equiv 2 s_\mu \Delta q_i^{(N)} \tag{5.93}$$

where s_μ is the spin vector of the nucleon N. The experimental values of the constants thus introduced are $\Delta u^{(p)} = \Delta d^{(n)} = 0.78$, $\Delta d^{(p)} = \Delta u^{(n)} = -0.5$ and $\Delta s^{(p)} = \Delta s^{(n)} = -0.16$. Then, the relevant part of (5.83) reads

$$\mathcal{L}_{\text{SD}} = 2\sqrt{2}\,\bar{\chi}\gamma^\mu \gamma_5 \chi \, \left[a_p \bar{p} s_\mu p + a_n \bar{n} s_\mu n \right],$$

$$a_p = \frac{1}{\sqrt{2}} \sum_{i=u,d,s} d_i \Delta q_i^{(p)}, \qquad a_n = \frac{1}{\sqrt{2}} \sum_{i=u,d,s} d_i \Delta q_i^{(n)}. \tag{5.94}$$

Then the spin-dependent differential cross-section reads

$$\frac{d\sigma^{\text{SD}}}{d|\mathbf{q}|^2} = \frac{8}{\pi v_\chi^2} \lambda^2 J(J+1) \frac{S(|\mathbf{q}|)}{S(0)}, \tag{5.95}$$

where $\mathcal{S}(|\mathbf{q}|)/S(0)$ is the nuclear spin form factor, J is the total spin of the nucleus, and λ is a nucleonic matrix element that basically describes the fraction of the total spin that is due to the spin of the nucleons:

$$\lambda = \frac{1}{J}\left[a_p\langle S_p\rangle + a_n\langle S_n\rangle\right], \tag{5.96}$$

where $\langle S_{p,n}\rangle$ is the expectation value of the proton (neutron) group spin content of the nucleus.

To recapitulate, the differential detection rate is calculated to be

$$\frac{dR}{dQ} = \frac{4}{\pi^{3/2}}\frac{\rho_\chi}{m_\chi v_\chi}T(Q)\left\{[Zf_p + (A-Z)f_n]^2\,F^2(Q)\right.$$

$$\left. +8\lambda^2 J(J+1)\frac{S(|\mathbf{q}|)}{S(0)}\right\}, \tag{5.97}$$

where

$$T(Q) \equiv \frac{\sqrt{\pi}v_\chi}{2}\int_{v_{\min}}^{\infty}\frac{f_\chi(v)}{v}dv \tag{5.98}$$

integrates over the neutralino velocity distribution $f_\chi(v)$ ($v_{\min} = \mu^{-1}\sqrt{M_A Q/2}$).

For the sake of illustration, let us take a simple Maxwellian distribution of velocities in the halo (isothermal model):

$$f_\chi(v) = \frac{4v^2}{v_\chi^3\sqrt{\pi}}e^{-v^2/v_\chi^2}. \tag{5.99}$$

Then (5.98) gives simply

$$T(Q) = \exp\left(-\frac{v_{\min}^2}{v_\chi^2}\right) = \exp\left(-\frac{QM_A}{2\mu^2 v_\chi^2}\right). \tag{5.100}$$

The total rate R is, in this case, easily computed. Assuming an experimental cut-off E_T for the recoil energy measured,

$$R = \int_{E_T}^{\infty}\frac{dR}{dQ}dQ \sim \frac{\rho_\chi}{m_\chi}\exp\left(-\frac{E_T M_A}{2\mu^2 v_\chi^2}\right), \tag{5.101}$$

where we have made explicit in the last term the dependence in the neutralino density and mass. If an experiment puts an upper limit on R, this translates into a upper limit on ρ_χ of the order of

$$m_\chi\exp\left[\frac{E_T}{2M_A v_\chi^2}\left(1+\frac{M_A}{m_\chi}\right)^2\right].$$

This behaves linearly with m_χ for large m_χ and grows exponentially for small m_χ. Thus, exclusion plots typically look like Fig. 5.3: assuming a given value of ρ_χ leads to the exclusion of neutralino masses in a range $[m_\chi^{\min}, m_\chi^{\max}]$.

Since the rate depends on the velocity of the incident WIMP, one expects an annual modulation due to the motion of the Earth in the Galactic frame. More precisely, the

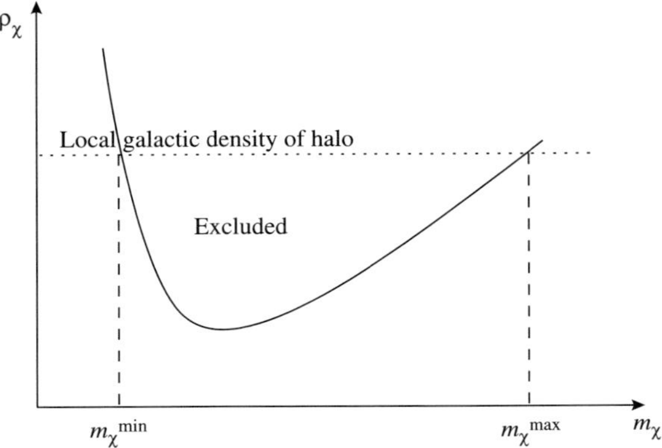

Fig. 5.3 Typical form of the exclusion plot for a WIMP in the plane local density versus mass.

Earth motion is a superposition of its rotation around the Sun and of the rotation of the Sun around the Galactic Center (with velocity $v_\odot = 220$ km/s):

$$v_\oplus = v_\odot \left[1.05 + 0.07 \cos\left(2\pi(t - t_p)/\mathrm{year}\right)\right], \tag{5.102}$$

where t is measured in days and $t_p \sim 153$ days (June 2) is the day when the direction of the Earth's motion around the Sun matches the direction of the Sun's motion around the Galactic Center [164]. The numerical coefficients reflect the $60°$ angle between the two planes of rotation. The yearly modulation comes from the fact that, in the lab (Earth) frame, one must shift the velocity distribution: $f(\mathbf{v} + \mathbf{v}_\oplus)$.

5.5.6 Indirect detection

WIMPs annihilate and their annihilation products can be detected. The variety of these particles (gamma rays, neutrinos, positrons, antiprotons, antinuclei) leads to expect to extract rich information from correlated observations. Because the annihilation rate is proportional to the square of dark matter density, one increases the expected flux by looking at places where the dark matter accumulates. Typical places are the center of the Earth or the Sun: WIMPs loose energy by scattering with the nucleons of matter and sink to the center. Or the Galactic Center: the density profile of the dark matter halo is expected to grow as a power law ($\rho(r) \propto r^{-\gamma}$); this effect may even be enhanced by adiabatic accretion on the central massive black hole leading to a central "spike" [201].

WIMPs which have accumulated in the Earth or the Sun annihilate into ordinary particles. Most of the decay products of these particles will be absorbed, except for high energy neutrinos (with typical energy $\frac{1}{3}$ to $\frac{1}{2}$ the WIMP mass) which pass through the Earth or the Sun. Thus energetic neutrinos are a signature for dark matter. Interactions of such neutrinos in the Earth give energetic muons which are detected in large underwater (or ice) detectors such as AMANDA, ANTARES, or NESTOR.

The capture rate of neutralinos is proportional to the nucleus–neutralino elastic scattering cross-section, and thus depends on $\sigma_{\chi-p,n}^{\mathrm{SI,SD}}$. In the case of the Earth, scalar interactions are favored because of the zero spin of Fe^{56}. For the Sun, the spin of hydrogen allows spin-dependent interactions and capture is more efficient.

On general grounds, the annihilation cross-section of neutralinos at rest (a good approximation in the case of neutralinos trapped in the center of a celestial body) into massless fermions vanishes [195]. Indeed, Fermi statistics imposes that the spin of the two neutralinos in the initial state (S-wave) are opposite. In the final state, the required spin flip is ensured by a mass insertion; hence the cross-section is proportional to the mass of the fermions. Direct annihilation into neutrinos is thus strongly disfavored and neutrinos are produced as secondary particles: $\chi\chi \to t\bar{t}, b\bar{b}, c\bar{c}, \tau^+\tau^-, W^+W^-, ZZ \to X\nu$. The more energetic the neutrinos are, the better their conversion into muons and the longer the distance they cover. Hence energetic neutrinos are more easy to detect.

Annihilation of neutralinos in the halo also gives monoenergetic photons (through the one-loop processes $\chi\chi \to \gamma\gamma, \gamma Z$) but also a continuous spectrum of photons through the decay of annihilation products (mostly from the decay of π^0 produced in hadronization). Generally speaking, the annihilation flux is proportional to (i) the number of annihilations per second in a unit volume $\langle\sigma_{\rm ann}v\rangle\rho^2(r)/M^2$ for dark matter particles of mass M and annihilation cross-section $\sigma_{\rm ann}$, (ii) the spectrum dN_i/dE of secondary particles of type i. Performing an integration over the line of sight, one obtains the observed flux of secondary particles of type i in a direction making an angle ψ with the direction of the galactic center:

$$\Phi_i(E,\psi) = \frac{\langle\sigma_{\rm ann}v\rangle}{4\pi M^2}\frac{dN_i}{dE}\int_{\rm line\ of\ sight} ds\ \rho^2(r) \qquad (5.103)$$

where the galactocentric coordinate is expressed in terms of the coordinate s along the line of sight and the angle ψ as: $r^2 = s^2 + R_0^2 - 2sR_0\cos\psi$ ($R_0 \sim 8$ kpc is the solar distance from the galactic center).

There is thus a strong dependence on the density profile of dark matter $\rho(r)$ which is poorly known. It is usually parametrized as

$$\rho(r) = \frac{\rho_0}{(r/R)^\gamma[1 + (r/R)^\alpha]^{(\beta-\gamma)/\alpha}}, \qquad (5.104)$$

where R is a characteristic length and α, β, and γ are parameters. Table 5.5 gives their values obtained in typical models based on N-body simulations. Because of the strong model dependence of the astrophysics input in the computation of the flux, it proves useful to factorize it and to introduce $\bar{J}(\Delta\Omega)$, the average of the function

$$J(\psi) = \int_{\rm line\ of\ sight} \frac{ds}{8.5\ {\rm kpc}} \left(\frac{\rho(r)}{0.3\ {\rm GeV/cm}^3}\right)^2 \qquad (5.105)$$

Table 5.5 Values of the parameters used to parametrize dark matter profiles for typical models: Navarro–Frenk–White (NFW) [293], Moore et al. [289] and isothermal [31].

Model	α	β	γ	R (kpc)	$\bar{J}(10^{-3})$
NFW	1.0	3.0	1.0	20	1.35×10^3
Moore	1.5	3.0	1.5	28	1.54×10^5
Isothermal	2.0	2.0	0	3.5	2.87×10^1

over a region of solid angle $\Delta\Omega$ centered on $\psi = 0$. The values of $\bar{J}(10^{-3})$ are shown in Table 5.5 for various models: they show that the prediction for the astrophysical factor based on N-body simulations varies by several orders of magnitude!

Then the flux coming from a solid angle $\Delta\Omega$ reads

$$\Phi_i(E, \Delta\Omega) = 5.6 \times 10^{-12} \frac{dN_i}{dE} \left(\frac{\langle\sigma_{\rm ann}v\rangle}{1\ {\rm pb}}\right) \left(\frac{1\ {\rm TeV}}{M}\right)^2 \bar{J}(\Delta\Omega)\Delta\Omega\ {\rm cm}^{-2}{\rm s}^{-1}.$$

$$(5.106)$$

The typical energy of gamma rays from neutralino annihilations is in the GeV–TeV range. Since the interaction length of photons with matter is of the order of 30 g cm^{-2} (to be compared with the thickness of the atmosphere 1030 g cm^{-2}), only space missions like GLAST, which follows EGRET launched in 1991, will be able to provide a significant test for supersymmetric models. However ground Cherenkov telescopes such as H.E.S.S, MAGIC, and VERITAS, which measure the Cherenkov light emitted by the particles produced in the cosmic gamma ray shower, show some sensitivity to models with a large neutralino mass.

The annihilation of neutralinos can also produce positrons. Moreover, the secondary electrons and positrons propagate in the galactic magnetic field, producing synchrotron radiation. Typical frequencies are of the order of a few hundred MHz. The corresponding observations provide significant constraints on the positron flux expected. Other annihilation products include antiprotons and antideuterons. Given the uncertainties on the astrophysical quantities, the level at which these particles are expected in a given supersymmetric model is difficult to predict. On the other hand, ratios of fluxes corresponding to different species are less sensitive to the details of halo models.

5.5.7 Dark matter codes

Public codes exist which provide, for a given supersymmetric model, relic abundances, as well as cross-sections, expected fluxes, etc. We refer the interested reader to the webpage of two of these codes:

DARKSUSY [200]: http://www.physto.se/edsjo/darksusy/
MicrOMEGAs [29]: http://wwwlapp.in2p3.fr/lapth/micromegas

5.6 Nonminimal models

We end this chapter with a discussion of the simplest extension of the MSSM, obtained by adding a gauge singlet S and all its renormalizable interaction [129,295]. This model is called the next-to-minimal supersymmetric model (NMSSM, sometimes also referred to as the (M+1)SSM). We will see in Chapters 9 and 10 that singlets naturally appear in the context of grand unified models or string models.

The most general superpotential compatible with the symmetries is:

$$W = \tfrac{1}{2}\mu_S S^2 + \tfrac{1}{6}\kappa_S S^3 + \lambda_S S H_2 \cdot H_1 + W^{(2)} + W^{(3)}, \tag{5.107}$$

where $W^{(2)}$ and $W^{(3)}$ are the standard MSSM quadratic and cubic terms given in equations (5.1) and (5.2).

One of the motivations for the NMSSM is that it provides a framework to account for a dynamical origin of the μ term. Indeed, let us assume the existence of a discrete symmetry that prohibits any quadratic term in the superpotential: $\mu = \mu_S = 0$. A rationale behind this assumption is that the typical mass scale of the fundamental theory underlying this effective low energy theory is very large: any dimensionful parameter should be fixed by this superheavy scale or vanishing. The low energy theory then has only dimensionless parameters such as κ_S, λ_S and the Yukawa couplings, and its mass scale appears only through symmetry breaking. Indeed, if S acquires a vacuum expectation value, then we obtain an effective μ-term with

$$\mu = \lambda_S \langle S \rangle. \tag{5.108}$$

Correspondingly, soft terms read

$$V_{\text{SB}}|_{\text{NMSSM}} = V_{\text{SB}}|_{\text{MSSM}} + \left(\tfrac{1}{6} A_\kappa \kappa_S S^3 + A_\lambda \lambda_S S H_2 \cdot H_1 + \text{h.c.}\right), \tag{5.109}$$

which generates an effective B_μ term: $B_\mu = A_\lambda \lambda_S \langle S \rangle$.

The spectrum of the NMSSM is the one of the MSSM plus one scalar, one pseudoscalar, and one Weyl fermion (often referred to as the singlino). The scalar mixes with h^0 and H^0, the pseudoscalar with A^0. The singlino mixes with the other neutralinos: the neutralino mass matrix $\mathcal{M}_n$ is now five-dimensional. In most of the parameter space, these mixings are small and the fields associated with the S superfield basically decouple. Otherwise, one may encounter large differences with the phenomenology of the MSSM [135].

Further reading

- H. Haber and G. Kane, *The search for supersymmetry: probing physics beyond the Standard Model*, Physics Reports **117** (1985) 75.

- G. Jungman, M. Kamionkowski and K. Griest, *Supersymmetric dark matter*, Physics Reports **267** (1996) 195–373.

Exercises

<u>Exercise 1</u> Explain in detail why the MSSM has 105 real parameters besides the 19 of the Standard Model, *i.e.* 124 real parameters in total.

Hints: See Chapter 12, Section 12.1.

<u>Exercise 2</u> Show that the potential $V \equiv V_F + V_D + V_{\rm SB}$ given by (5.10)-(5.12) has only charge-conserving minima: $\langle H_1^\pm \rangle = 0 = \langle H_2^\pm \rangle$.

<u>Exercise 3</u> Compute the mass matrices corresponding to the scalar potential of the MSSM (Section 5.3.1) for the neutral scalars (Re H_i^0) and pseudoscalars (Im H_i^0). Deduce (5.28)–(5.30) and (5.32)–(5.33).

Hints:

$$M_{\rm sc}^2 = \begin{pmatrix} m_1^2 + \dfrac{g^2 + g'^2}{4}(3v_1^2 - v_2^2) & B_\mu - \dfrac{g^2 + g'^2}{2}v_1 v_2 \\[2ex] B_\mu - \dfrac{g^2 + g'^2}{2}v_1 v_2 & m_2^2 + \dfrac{g^2 + g'^2}{4}(3v_2^2 - v_2^1) \end{pmatrix}$$

$$M_{\rm ps}^2 = \begin{pmatrix} m_1^2 + \dfrac{g^2 + g'^2}{4}(v_1^2 - v_2^2) & -B_\mu \\[2ex] -B_\mu & m_2^2 + \dfrac{g^2 + g'^2}{4}(v_2^2 - v_2^1) \end{pmatrix}.$$

<u>Exercise 4</u> Using the Lagrangians presented in Chapter 3, prove (5.43)–(5.44) where $\lambda_\pm \equiv (\lambda^1 \mp \lambda^2)/\sqrt{2}$ is the supersymmetric partner of $W_\mu^\pm \equiv (A_\mu^1 \mp A_\mu^2)/\sqrt{2}$.

Hints: Use equations (3.30) and (3.52) of Chapter 3. Note that, because λ^1 and λ^2 are Majorana spinors and because charge conjugation implies hermitian conjugation, one has $\lambda_\pm^c = \lambda_\mp$. Also $\overline{\lambda_\pm} = (\overline{\lambda^1} \pm i\overline{\lambda^2})/\sqrt{2}$.

<u>Exercise 5</u> Compute the spectrum of charginos and neutralinos in the limit case where $M_W, M_Z \ll M_1 < M_2 < |\mu|$.

Hints: To leading order in $M_{W,Z}^2/\mu$,

$$m_{\chi_1^\pm} = M_2 - M_W^2 \frac{M_2 + \mu \sin 2\beta}{\mu^2 - M_2^2}, \quad m_{\chi_2^\pm} = |\mu| + M_W^2 \frac{|\mu| + M_2{\rm sgn}(\mu)\sin 2\beta}{\mu^2 - M_2^2},$$

$$m_{\chi_1^0} = M_1 - M_Z^2 \sin^2\theta_w \frac{M_1 + \mu\sin 2\beta}{\mu^2 - M_1^2}, \quad m_{\chi_2^0} = M_2 - M_Z^2 \cos^2\theta_w \frac{M_2 + \mu\sin 2\beta}{\mu^2 - M_2^2},$$

$$m_{\chi_{3,4}^0} = |\mu| + M_Z^2(1 \mp {\rm sgn}(\mu)\sin 2\beta)\frac{\left(|\mu| \pm M_1 \cos^2\theta_w \pm M_2 \sin^2\theta_w\right)}{2(|\mu| \pm M_1)(|\mu| \pm M_2)}. \tag{5.110}$$

<u>Exercise 6</u> Check the terms in the d-type squark mass matrix (5.54) or slepton mass matrix (5.57).

<u>Exercise 7</u> *We wish to study the evolution with time of the number N of neutralinos in the center of a celestial body under the influence of capture and annihilation.*

The annihilation rate in the center of the celestial body is

$$\Gamma_{\text{ann}} = \tfrac{1}{2} C_{\text{ann}} N^2, \quad C_{\text{ann}} = \langle \sigma_{\text{ann}} v \rangle V_2 / V_1^2, \tag{5.111}$$

where the quantities V_j are effective volumes: $V_j = \left[\frac{3 M_P^2 T}{(2 j m_\chi \rho)} \right]^{3/2}$ (ρ is the core density and M_P the Planck mass).[12]

(a) Write in terms of Γ_{ann} and the capture rate C the differential equation expressing the evolution of N with respect to time and solve it.

(b) Express then Γ_{ann} in terms of time, C, and the critical time $\tau \equiv (CC_{\text{ann}})^{-1/2}$.

(c) Compare the values of τ for the Earth and for the Sun, given that the capture rate in the Earth is some 10^{-9} smaller than in the Sun. If the age of the solar system is larger than the critical time, show that the annihilation rate is only determined by the elastic scattering cross-section.

Hints:

(a) $\dot{N} = C - 2\Gamma_{\text{ann}} = C - C_{\text{ann}} N^2$.

(b) $\Gamma_{\text{ann}} = \dfrac{1}{2} C \tanh^2(t/\tau)$.

(c) $\tau_\oplus < \tau_\odot$; $\Gamma_{\text{ann}} = \dfrac{1}{2} C$ and C is fixed by the scattering cross-section.

Exercise 8 Show that in the NMSSM model of Section 5.6.1 with a discrete symmetry prohibiting quadratic terms in the superpotential ($\mu = \mu_S = 0$), the lightest Higgs mass satisfies the upper bound:

$$m_h^2 \leq M_z^2 + \left(\tfrac{1}{2} \lambda_S^2 v^2 - M_z^2 \right) \sin^2 2\beta. \tag{5.112}$$

---

[12] For the Sun [206], $V_j = [j m_\chi / 10 \text{ GeV}]^{-3/2}\ 6.5 \times 10^{28}\ \text{cm}^3$ and for the Earth [204], $V_j = [j m_\chi / 10 \text{ GeV}]^{-3/2}\ 2.0 \times 10^{25} \text{cm}^3$.

6

Supergravity

There are two directions one might take to generalize supersymmetry and its models as we have presented them until this point: (a) to try to incorporate gravity as the fourth fundamental interaction, or (b) by using our experience with gauge theories, to try to make supersymmetry local. It turns out that these two attempts converge: local supersymmetry makes it necessary to introduce a graviton field (and its supersymmetric partner the gravitino) and therefore gravity. Thus the theory of local supersymmetry is called supergravity.

Let us note immediately that this is not yet a full quantum theory of gravity: just as gravity, supergravity is nonrenormalizable; its coupling constant κ is still the dimensionful Newton's constant $\kappa \sim G_N^{1/2}$ and interactions scale like powers of κ. It can thus be understood as an effective theory of a full-fledged quantum theory of gravity whose fundamental scale is typically the Planck scale $M_P \equiv \sqrt{\hbar c / G_N} \sim 1.22 \times 10^{19}$ GeV. We will return to these considerations at the beginning of Chapter 10, when we discuss string theory, which represents, to date, the most elaborate attempt to write a quantum theory of gravity coupled with matter.

For the time being, we will content ourselves with this nonrenormalizable theory of supergravity, which still represents major progress since it allows us to consider the theory up to scales a few orders of magnitude smaller than the Planck scale. We recall that it was precisely one of the motivations for supersymmetry to push the appearance of new physics to superheavy scales, the hope being that the theory becomes simpler. In this chapter, we will thus renormalize the low energy supersymmetric models up to scales close to the Planck scale. This leads to a new way of envisaging model building: a few assumptions on masses and couplings at this high energy scale, to be discussed later, allow us to restrict the parameter space of the MSSM. This gives in particular the so-called *minimal supergravity model* which has provided a benchmark for discussing the phenomenology of supersymmetry at high energy colliders, especially hadron colliders.

6.1   Local supersymmetry is supergravity

Let us start from global supersymmetry and make the supersymmetry transformations local. In other words, the parameter ε of the transformation is made spacetime dependent: $\varepsilon(x)$. Then following (3.5) of Chapter 3, we have

$$[\bar{\varepsilon}_1(x)Q, \bar{\varepsilon}_2(x)Q] = 2\bar{\varepsilon}_1(x)\gamma^\mu \varepsilon_2(x)\, P_\mu. \tag{6.1}$$

Hence the commutator of two local supersymmetry transformations is a spacetime dependent translation, what is known as a spacetime general coordinate transformation. It is precisely by making the relativity theory invariant under local coordinate transformations that Einstein has derived the theory of general relativity. Thus one expects, as immediately stressed by Wess and Zumino [364] in the first article on four-dimensional supersymmetry, that local supersymmetry will lead to a theory of gravity. This theory is known as supergravity.

We expect to find in the particle spectrum a graviton field (of spin 2) and its supersymmetric partner, a field of spin 3/2, the gravitino. The gravitino plays a fundamental rôle in supergravity, as we will now see.

Let us first recall how one makes a gauge symmetry local in order to follow the same steps with supersymmetry (see Section A.1.3. of Appendix Appendix A for more details). The simple action

$$S_0 = \int d^4x \; \bar{\Psi}(x) \; i\gamma^\mu \partial_\mu \; \Psi(x)$$

is not invariant under the *local* transformation: $\Psi(x) \to e^{i\theta(x)} \; \Psi(x)$. Infinitesimally

$$\delta S_0 = - \int d^4x \; \partial_\mu \theta(x) \; J^\mu(x) \tag{6.2}$$

where $J^\mu(x)$ is the Noether current: $J^\mu(x) = \bar{\Psi}(x)\gamma^\mu\Psi(x)$. The solution is to introduce a vector field $A_\mu(x)$ which transforms as:

$$\delta A_\mu(x) = -\frac{1}{g}\partial_\mu \theta(x) \tag{6.3}$$

and add a term $S_1 = -g \int d^4x A_\mu(x) \; J^\mu(x)$ in order that the full action

$$S = S_0 + S_1 = \int d^4x \; \bar{\Psi}(x) \; i\gamma^\mu \left(\partial_\mu + igA_\mu(x) \right) \Psi(x) \tag{6.4}$$

be invariant under local gauge transformations. Finally one includes a dynamics for the gauge field by adding a kinetic term

$$S' = \int d^4x \left[-\frac{1}{4} \; F^{\mu\nu}(x) \; F_{\mu\nu}(x) \right]. \tag{6.5}$$

Let us now consider supersymmetry transformations. Since the parameter of the transformation carries a spinor index r ($\varepsilon_r(x)$), the equivalent of (6.3) necessarily reads

$$\delta\psi_r^\mu(x) \sim \partial^\mu \varepsilon_r(x). \tag{6.6}$$

There is therefore a need for a field carrying a spin 1 **and** a spin 1/2 index, which amounts to a spin 3/2 (and spin 1/2) field: the gravitino field. As above, we follow the Noether procedure to couple this field to matter and then give this field a dynamics.

Let us start, as an example, with the kinetic terms of the Wess–Zumino model:

$$S_0 = \int d^4x \left[\partial^\mu \phi^* \partial_\mu \phi + \frac{i}{2} \, \bar\Psi \gamma^\mu \partial_\mu \Psi \right] \tag{6.7}$$

where $\phi = (A + iB)/\sqrt{2}$ and Ψ is a Majorana spinor. We have seen that S_0 is invariant under the *global* transformation (3.3), (3.4) of Chapter 3:

$$\delta A = \bar\varepsilon \Psi,$$
$$\delta B = i\bar\varepsilon \gamma_5 \Psi,$$
$$\delta \Psi_r = -i \left(\gamma^\mu \left[\partial_\mu (A + i\gamma_5 B) \right] \varepsilon \right)_r, \tag{6.8}$$

the canonical dimension of the Majorana spinor ε_r being $-1/2$. If we make ε_r spacetime dependent, S_0 is no longer invariant:

$$\delta S_0 = \int d^4x \, \partial_\mu \bar\varepsilon \, \gamma^\rho \gamma^\mu \left[\partial_\rho (A + i\gamma_5 B) \right] \Psi \equiv \int d^4x \, \partial_\mu \bar\varepsilon_r \, J_r^\mu(x). \tag{6.9}$$

We thus introduce a new field $\psi_r^\mu(x)$ with vector and spinor indices and a new term in the action

$$S_1 = -\frac{\kappa}{2} \int d^4x \, \bar\psi_\mu \gamma^\rho \gamma^\mu \left[\partial_\rho (A + i\gamma_5 B) \right] \Psi. \tag{6.10}$$

Since $\psi_{\mu r}$ is a spinor field, it has canonical dimension $\frac{3}{2}$, just as Ψ; and the coupling κ has dimension -1. In other words, the theory has a dimensionful coupling and is obviously nonrenormalizable. The supersymmetry transformation of ψ_μ^r is chosen in order to cancel (6.9):

$$\delta \psi_r^\mu(x) = \frac{2}{\kappa} \, \partial^\mu \varepsilon_r(x). \tag{6.11}$$

Note that dimensions match.

However, contrary to the previous case, the story does not end here since $S_0 + S_1$ is not invariant under local supersymmetry. The transformation of S_1 yields for example the following terms quadratic in A:

$$\delta S_1 = i\kappa \int d^4x \, \bar\psi_\mu(x) \gamma_\nu \varepsilon(x) \left[\partial^\mu A \partial^\nu A - \frac{1}{2} \, g^{\mu\nu} \partial^\rho A \partial_\rho A + \cdots \right] \tag{6.12}$$

where one notices between the brackets the energy–momentum tensor $\Theta^{\mu\nu}$ for the scalar field. Remembering that $\Theta^{\mu\nu}$ is obtained by varying the action with respect to the metric

$$\delta S = \int d^4x \, \frac{1}{2} \, \delta g_{\mu\nu} \, \Theta^{\mu\nu}, \tag{6.13}$$

one introduces a local metric $g_{\mu\nu}(x)$ such that, under supersymmetry,

$$\delta g_{\mu\nu}(x) = -i\kappa \left(\bar\psi_\mu \gamma_\nu \varepsilon(x) + \bar\psi_\nu \gamma_\mu \varepsilon(x) \right). \tag{6.14}$$

The corresponding spin 2 field – the graviton – is obviously the supersymmetric partner of the gravitino, a spin 3/2 field, and the constant κ is nothing but the gravitational coupling; κ^{-1} is the reduced Planck scale:

$$\kappa^{-1} = \sqrt{\frac{\hbar c}{8\pi G_N}} = \frac{M_P}{\sqrt{8\pi}} \equiv m_P = 2.4 \times 10^{18} \text{ GeV}. \tag{6.15}$$

We could continue the procedure and show for example that the gravitino transformation law receives other contributions besides (6.11). It turns out however that we have basically found the supersymmetry transformations of the (spin 2, spin 3/2) system on shell, in the absence of matter:

$$\delta g_{\mu\nu}(x) = -i\kappa \left(\bar{\psi}_\mu \gamma_\nu \varepsilon(x) + \bar{\psi}_\nu \gamma_\mu \varepsilon(x) \right) \tag{6.16}$$

$$\delta \psi^\mu(x) = \frac{2}{\kappa} \, \mathcal{D}^\mu \varepsilon(x). \tag{6.17}$$

The only modification is that, because we work in curved spacetime, we have written a covariant derivative in the second equation: $\mathcal{D}^\mu \varepsilon(x) = \partial^\mu \varepsilon(x) + \frac{1}{2} \omega_\mu{}^{ab} \sigma_{ab} \varepsilon(x)$ where $\omega_\mu{}^{ab}$ is a spin connection (see Appendix D). The Lagrangian invariant under the transformations (6.16), (6.17) is simply:

$$S = \int d^4x \sqrt{g} \left[-\frac{1}{2\kappa^2} R - \frac{i}{2} \epsilon^{\mu\nu\rho\sigma} \bar{\psi}_\mu \gamma_5 \gamma_\nu \partial_\rho \psi_\sigma \right] \tag{6.18}$$

where $g \equiv \det g_{\mu\nu}$. The first term involves the curvature tensor built out of the metric $g_{\mu\nu}$: its variation with respect to the metric yields the Einstein tensor $R_{\mu\nu} - \frac{1}{2} g_{\mu\nu} R$ present in the Einstein equation. The second term is a kinetic term for the spin 3/2 field, the so-called Rarita-Schwinger [321] action (see Section 6.4). Thus the action (6.18) provides a dynamics for the graviton–gravitino system.

As with the other supermultiplets that we have considered the off-shell description of the gravity supermultiplet – which consists of the graviton and the gravitino – involves the introduction of auxiliary fields. They consist in this case of a complex scalar field $M = (M_1 + iM_2)/\sqrt{2}$ and a real vector field b^μ. For example, whereas the supersymmetry transformation of the graviton field (6.16) remains unmodified, the gravitino transformation (6.17) includes extra terms of the form $(\gamma^\mu \gamma^\nu \varepsilon) b_\nu$, $(\gamma^\mu \varepsilon) M_1$ or $(\gamma^\mu \gamma_5 \varepsilon) M_2$.

6.2   Coupling of matter to supergravity

We will just describe here the coupling of chiral supermultiplets $\Phi^i = (\phi^i, \Psi^i, F^i)$ to supergravity[1]. This coupling is summarized by three basic functions:

- The superpotential which is an analytic function of the scalar fields $W(\phi^i)$: it determines the self-interactions of the scalar fields as well as their Yukawa interactions to fermions.

---

[1] Their complex conjugates are written $\bar{\Phi}^{\bar{\imath}} = (\bar{\phi}^{\bar{\imath}}, \bar{\Psi}^{\bar{\imath}}, \bar{F}^{\bar{\imath}})$.

- The Kähler potential $K(\phi^i, \bar{\phi}^{\bar{\jmath}})$, a real function of the scalar fields, determines in particular the kinetic term of scalar fields:

$$\mathcal{L}_{\text{kin}} = g_{i\bar{\jmath}}(\phi^i, \bar{\phi}^{\bar{\jmath}}) \, \partial^\mu \phi^i \partial_\mu \bar{\phi}^{\bar{\jmath}} \tag{6.19}$$

where the function $g_{i\bar{\jmath}}(\phi^i, \bar{\phi}^{\bar{\jmath}})$ called Kähler metric is defined as

$$g_{i\bar{\jmath}} = \frac{\partial^2 K}{\partial \phi^i \partial \bar{\phi}^{\bar{\jmath}}}. \tag{6.20}$$

This is interpreted in a geometric way: the scalar fields ϕ^i, $\bar{\phi}^{\bar{\jmath}}$ parametrize a complex manifold and the Kähler metric is the metric on this manifold. One speaks of a flat Kähler metric if $g_{i\bar{\jmath}} = \delta_{ij}$. If the metric is nonflat, the nonnormalized kinetic term is obviously not renormalizable. Nonrenormalizable interaction terms are also generated besides the kinetic term, for example a four-fermion interaction: they are proportional to derivatives of the Kähler metric. [A more technical introduction to Kähler potential is given in Appendix C.]

The complete supergravity Lagrangian is invariant under the Kähler transformation:

$$K\left(\phi^i, \bar{\phi}^{\bar{\jmath}}\right) \rightarrow K\left(\phi^i, \bar{\phi}^{\bar{\jmath}}\right) + F\left(\phi^i\right) + \bar{F}\left(\bar{\phi}^{\bar{\imath}}\right), \tag{6.21}$$

$$W\left(\phi^i\right) \rightarrow e^{-\kappa^2 F(\phi^i)} \, W\left(\phi^i\right), \tag{6.22}$$

where $F\left(\phi^i\right)$ is an analytic function of the scalar fields ϕ^i. For example, the Kähler metric is invariant under (6.21). The specific choice $\kappa^2 F\left(\phi^i\right) = \ln W\left(\phi^i\right)$ shows that the supergravity Lagrangian only depends on the combination [91]:

$$\mathcal{G} = \kappa^2 K + \ln |W|^2. \tag{6.23}$$

We refrain from using this function in what follows because the Kähler invariance of the Lagrangian is no longer explicit.

Finally, we note that the Kähler transformation acts, in a way similar to a R-symmetry, as a chiral $U(1)$ rotation on the individual fermion fields. More precisely, one has:

$$\Psi_L \rightarrow e^{-iw\text{Im}F/2}\Psi_L, \quad \Psi_R \rightarrow e^{+iw\text{Im}F/2}\Psi_R \tag{6.24}$$

where the chiral weight is $w = 1$ for the gravitino and gaugino fields, $w = -1$ for a fermion in a chiral supermultiplet [2].

- A kinetic function $f(\phi^i)$ for the gauge fields which determines the gauge field kinetic term:

$$\mathcal{L}_{\text{kin}} = -\frac{1}{4}\text{Re}f_{ab}(\phi^i) \, F^{a\mu\nu}F^b_{\mu\nu} + \frac{1}{4}\text{Im}f_{ab}(\phi^i) \, F^{a\mu\nu}\tilde{F}^b_{\mu\nu}. \tag{6.25}$$

---

[2]We have assumed here that all scalar fields have vanishing chiral weights. [More precisely, using the superfield language of Appendix C, one should note that the θ variables have chiral weight $w = 1$, i.e. transform as $\theta^\alpha \rightarrow e^{-i\text{Im}F/2}\theta^\alpha$. Thus, if a scalar fields has chiral weight w_0, its fermion partner has weight $w_0 - 1$.]

For complete expressions of the action describing matter coupled to supergravity, we refer the reader to the references at the end of the chapter. We will concentrate here on the scalar potential and will therefore write explicitly the nonderivative terms involving scalar fields. They read:

$$\frac{1}{\sqrt{g}}\mathcal{L} = -\frac{1}{3}M\bar{M} + F^i g_{i\bar{j}}\bar{F}^{\bar{j}} + e^{\kappa^2 K/2}\left\{-\kappa M\bar{W} - \kappa\bar{M}W\right.$$

$$\left. + F^k\left(\frac{\partial W}{\partial\phi^k} + \kappa^2\frac{\partial K}{\partial\phi^k}W\right) + \bar{F}^{\bar{k}}\left(\frac{\partial\bar{W}}{\partial\bar{\phi}^{\bar{k}}} + \kappa^2\frac{\partial K}{\partial\bar{\phi}^{\bar{k}}}\bar{W}\right)\right\} \tag{6.26}$$

$$+ \frac{1}{2}\mathrm{Re}f_{ab}D^a D^b + \frac{1}{2}D^a\left[K_i\left(t^a\phi\right)^i + \left(\bar{\phi}t^a\right)^{\bar{i}}K_{\bar{i}}\right],$$

where $\kappa = \sqrt{8\pi}/M_P = 1/m_p$ and we suppressed the summation over repeated indices. We recognize the auxiliary fields M of the gravity supermultiplet, F of the matter supermultiplets and D of the gauge supermultiplets.

It is worth recalling that, whereas the superpotential W has canonical mass dimension 3, the Kähler potential has dimension 2 (so that the Kähler metric (6.20) is dimensionless); the auxiliary fields F^i, $\bar{F}^{\bar{j}}$ and M have dimension 2.

In the limit of vanishing gravitational coupling ($\kappa \to 0$), flat Kähler metric and trivial kinetic function ($f = 1/g^2$), one recovers the global supersymmetry Lagrangian (3.52) of Chapter 3:

$$\mathcal{L} = F^i\bar{F}^{\bar{i}} + F^i\frac{\partial W}{\partial\phi^i} + \bar{F}^{\bar{i}}\frac{\partial\bar{W}}{\partial\bar{\phi}^{\bar{i}}} + \frac{1}{2g^2}D^a D^a + D^a\left(\bar{\phi}t^a\phi\right). \tag{6.27}$$

Solving for the auxiliary fields

$$F^i = -\frac{\partial\bar{W}}{\partial\bar{\phi}^{\bar{i}}}, \quad D^a = -g^2\bar{\phi}t^a\phi, \tag{6.28}$$

we obtained the scalar potential

$$V = -\mathcal{L} = \sum_i\left|\frac{\partial W}{\partial\phi^i}\right|^2 + \frac{1}{2}g^2\sum_a\left(\bar{\phi}t^a\phi\right)^2. \tag{6.29}$$

We follow the same steps in the case of supergravity. The auxiliary field equations of motion read:

$$M = -3\kappa\, e^{\kappa^2 K/2}W, \tag{6.30}$$

$$F^i = -g^{i\bar{j}}\, e^{\kappa^2 K/2}D_{\bar{j}}\bar{W}, \tag{6.31}$$

$$\bar{F}^{\bar{j}} = -g^{\bar{j}k}\, e^{\kappa^2 K/2}D_k W, \tag{6.32}$$

$$D^a = (\mathrm{Re}f)_{ab}^{-1}K_i\left(t^a\phi\right)^i, \tag{6.33}$$

where we have used the inverse Kähler metric $g^{i\bar{j}}$ ($g^{i\bar{j}}g_{\bar{j}k} = \delta^i_k$) and we have introduced the notation[3]

---

[3]We have also used the invariance of the Kähler potential under gauge transformations, *i.e.* $K_i\left(t^a\phi\right)^i - \left(\bar{\phi}\,t^a\right)^{\bar{i}}K_{\bar{i}} = 0.$

$$D_i W = \frac{\partial W}{\partial \phi^i} + \kappa^2 \frac{\partial K}{\partial \phi^i} W, \tag{6.34}$$

$$\bar{D}_{\bar{\imath}} \bar{W} = \frac{\partial \bar{W}}{\partial \bar{\phi}^{\bar{\imath}}} + \kappa^2 \frac{\partial K}{\partial \bar{\phi}^{\bar{\imath}}} \bar{W}. \tag{6.35}$$

The scalar potential thus reads:

$$V = -\frac{1}{\sqrt{g}} \mathcal{L} = e^{\kappa^2 K} \left[D_i W g^{i\bar{\jmath}} D_{\bar{\jmath}} \bar{W} - 3\kappa^2 \, |W|^2 \right] + \frac{1}{2} \, (\mathrm{Re} f)_{ab}^{-1} \, K_i \, (t^a \phi)^i \, (\bar{\phi} t^a)^{\bar{\jmath}} \, K_{\bar{\jmath}}$$

$$= F^i g_{i\bar{\jmath}} \bar{F}^{\bar{\jmath}} - \frac{1}{3} M \bar{M} + \frac{1}{2} \, (\mathrm{Re} f)_{ab} \, D^a D^b. \tag{6.36}$$

We note that, contrary to (6.29), the scalar potential is no longer positive definite.
If we use the generalized Kähler potential (6.23)

$$V = \kappa^2 e^{\mathcal{G}} \left[\mathcal{G}_i \mathcal{G}^{i\bar{\jmath}} \mathcal{G}_{\bar{\jmath}} - 3 \right] + \frac{1}{2} \, (\mathrm{Re} f)_{ab} \, D^a D^b, \tag{6.37}$$

where $\mathcal{G}_i = \partial \mathcal{G}/\partial \phi^i$, $\mathcal{G}_{\bar{\imath}} = \partial \mathcal{G}/\partial \bar{\phi}^{\bar{\imath}}$, ... and $\mathcal{G}^{i\bar{\jmath}} \mathcal{G}_{\bar{\jmath}k} = \delta^i_k$.

6.3 Supersymmetry breaking

We have seen in Chapter 3 that global supersymmetry is broken if and only if
the vacuum energy vanishes: vacuum energy is the order parameter for global
supersymmetry breaking. When we turn to local supersymmetry, gravity comes into
the game and strongly modifies the issue. As we have just seen, the scalar potential is
for example no longer positive definite. We will show however that one can trade the
vacuum energy for another order parameter.

6.3.1 Criterion for spontaneous local supersymmetry breaking

In order to see that, assume that the minimum of the potential energy corresponds to
a value $V_0 \neq 0$: global supersymmetry is spontaneously broken through this nonvanishing vacuum energy. The full action now includes a "cosmological constant" term:

$$\mathcal{S}_0 = \int d^4 x \sqrt{g} \, [-V_0]. \tag{6.38}$$

Unlike the global case, this term involves some nontrivial dynamics because of the presence of the determinant g of the metric $g_{\mu\nu}$: for example, V_0 appears on the left-hand
side of the Einstein equations (obtained by the variation of the full action with respect
to the metric) as a contribution to the energy–momentum tensor. We recall that $\sqrt{g}$
appears in (6.38) to ensure reparametrization invariance: under a reparametrization
$x^\mu \to x'^\mu = x'^\mu(x^\rho)$, the metric transforms as

$$g'_{\mu\nu} = g_{\rho\sigma} \frac{\partial x^\rho \partial x^\sigma}{\partial x'^\mu \partial x'^\nu} \quad \Rightarrow \quad \sqrt{g'} = \sqrt{g} \, \det \left(\frac{\partial x^\rho}{\partial x'^\mu} \right).$$

Thus $d^4 x' \sqrt{g'} = d^4 x \sqrt{g}$.

Obviously, a term such as (6.38) is non invariant under the supersymmetry transformations (6.16)–(6.17). It must appear in connection with other dynamical terms. On
the other hand, there exists a cosmological constant term which is perfectly compatible

with these supersymmetry transformations:

$$\tilde{\mathcal{S}}_0 = \int d^4x \sqrt{g} \left[3\frac{m_{3/2}^2}{\kappa^2} - m_{3/2}\bar{\psi}_\mu \sigma^{\mu\nu}\psi_\nu \right]. \qquad (6.39)$$

There are several comments to be made on this (locally) supersymmetric term in the action.

First, the term quadratic in the gravitino field looks very much like a mass term:

$$-m_{3/2}\bar{\psi}_\mu \sigma^{\mu\nu}\psi_\nu = -\tfrac{1}{4}m_{3/2}\bar{\psi}_\mu \left[\gamma^\mu, \gamma^\nu\right]\psi_\nu = \tfrac{1}{4}m_{3/2}\bar{\psi}_\mu \left\{\gamma^\mu, \gamma^\nu\right\}\psi_\nu = \tfrac{1}{2}m_{3/2}\bar{\psi}^\mu \psi_\mu,$$

where we have used the gauge condition (see next Section) $\gamma_\mu\psi^\mu = 0$. But although this resembles a gravitino mass term, this is not sufficient to identify it as such: if there is only the term (6.39) besides (6.18), supersymmetry is conserved and the gravitino is massless!

Second, the cosmological constant term in (6.39) corresponds to a negative vacuum energy: $E_0 = -3m_{3/2}^2/\kappa^2$. The corresponding spacetime is no longer Minkowski but it is known as an anti-de Sitter space[4]. [This is the first time, and not the last one, that we encounter this type of space. It suffices to say that (anti)de Sitter space can be described geometrically as the surface of equation

$$\left(x^5\right)^2 \mp \eta_{\mu\nu}x^\mu x^\nu = R^2, \qquad (6.40)$$

in the five-dimensional space with metric

$$ds^2 = \eta_{\mu\nu}dx^\mu dx^\nu \mp \left(dx^5\right)^2, \qquad (6.41)$$

with upper sign for de Sitter, lower sign for anti-de Sitter space.

The spacetime symmetries of such anti-de Sitter spaces are not described by the Poincaré group $O(1,3)$ of translations and Lorentz transformations (as in Minkowski spacetime), but rather by the group $O(2,3)$.[5]]

In anti-de Sitter space, the definition of mass from quadratic terms in the action is subtle. Indeed, if one redefines the covariant derivatives and Riemann tensor *in the field equations* to account for the symmetries specific of such spacetime, the two terms coming from the action (6.39) are absorbed [348]. This solves the puzzle of the seemingly spurious mass term discussed above.

However, if we combine (6.38) with (6.39) and assume cancellation of the vacuum energy:

$$-V_0 + \frac{3m_{3/2}^2}{\kappa^2} = 0 \qquad (6.42)$$

then we are back in Minkowski space and

$$\mathcal{S}_0 + \tilde{\mathcal{S}}_0 = \int d^4x \left[\frac{i}{2}m_{3/2}\bar{\psi}_\mu \sigma^{\mu\nu}\psi_\nu\right] \qquad (6.43)$$

---

[4]The de Sitter space familiar in the inflation scenario would correspond to a positive vacuum energy.

[5]The group $O(m,n)$ is the group of linear transformations that leave invariant a metric with m positive and n negative diagonal components. Clearly, $m = 2$ and $n = 3$ for the metric (6.41).

now allows us to interpret $m_{3/2}$ as the mass of the gravitino. From (6.42),

$$m_{3/2} = \kappa \sqrt{\frac{V_0}{3}}. \tag{6.44}$$

We have thus traded the vacuum energy for the gravitino mass: the criterion for the spontaneous breaking of local supersymmetry is a nonvanishing gravitino mass. The physical meaning is obvious: since the graviton stays massless, a massive gravitino is a clear sign of supersymmetry breaking in the spectrum.

This is quite similar to the situation in spontaneous gauge symmetry breaking: the criterion is a non zero mass for the gauge vector field. One can in fact push the analogy further and speak of a *super-Higgs mechanism*. Indeed, the longitudinal degree of freedom of the massive vector field is provided by the Goldstone particle associated with the spontaneous breaking of the continuous gauge symmetry. Similarly, the missing degrees of freedom for the massive gravitino are provided by the massless Goldstone fermion or Goldstino that was necessarily present in the spectrum when we discussed the spontaneous breaking of global supersymmetry. The two degrees of freedom of a massless gravitino and the two degrees of freedom of the Goldstino make up the four degrees of freedom of the massive gravitino. Just as in the gauge case, the Goldstino thus does not appear in the physical spectrum of the theory when supersymmetry is local.

The previous analysis was obviously sketchy since it did not include the dynamics of supersymmetry breaking. One may take a more realistic approach by using the explicit coupling of matter to supergravity discussed in the preceding Section. This is relevant if supersymmetry is broken by the F-component of some chiral supermultiplet.

Besides the scalar potential (6.36), the complete Lagrangian includes a term quadratic in the gravitino field:

$$\mathcal{L} = \frac{i}{2}\kappa^2 e^{\kappa^2 K/2} W(\phi)\, \bar{\psi}_\mu \sigma^{\mu\nu} \psi_\nu \;+\; \text{h.c.} \tag{6.45}$$

If $\langle e^{\kappa^2 K/2} W(\phi) \rangle \neq 0$ in the ground state, then supersymmetry is spontaneously broken:

$$m_{3/2} = \kappa^2 \langle e^{\kappa^2 K/2} W \rangle = \frac{\kappa}{3}\langle M \rangle \tag{6.46}$$

where we have used (6.30).

Assuming a vanishing vacuum energy $\langle V \rangle = 0$, we have the following relation from (6.36):

$$\sqrt{\langle D_i W g^{i\bar{\jmath}} D_{\bar{\jmath}} \bar{W} \rangle} = \kappa\sqrt{3}\langle |W| \rangle \tag{6.47}$$

or in terms of the auxiliary fields

$$\sqrt{\langle F_i g^{i\bar{\jmath}} F_{\bar{\jmath}} \rangle} = \frac{1}{\sqrt{3}}\langle |M| \rangle. \tag{6.48}$$

We can thus write

$$|m_{3/2}| = \frac{\kappa}{\sqrt{3}}\sqrt{\langle F^i g_{i\bar{\jmath}} \bar{F}^{\bar{\jmath}} \rangle}, \tag{6.49}$$

which relates directly the gravitino mass to the measure of supersymmetry breaking, *i.e.* the vacuum value of an auxiliary field. As discussed in Section 3.1.2 of Chapter 3, it is also this nonvanishing vacuum value which generates a field independent term in the Goldstino supersymmetry transformation.

It is customary to introduce a supersymmetry breaking scale M_{SB} and write on dimensional grounds $\langle F^i g_{i\bar{j}} \bar{F}^{\bar{j}} \rangle \equiv M_{SB}^4$. Then,

$$m_{3/2} = \frac{M_{SB}^2}{m_P \sqrt{3}}. \qquad (6.50)$$

6.3.2 Hidden sector

When discussing the issue of supersymmetry breaking in Chapter 3, we have encountered difficulties when there is a direct coupling between the supersymmetry-breaking fields and the quark and lepton fields. In the case of F-term breaking (O'Raifeartaigh model of Section 3.1.4), the spectrum obtained is unrealistic (STr $M^2 = 0$). In the case of D-term breaking (see Section 3.2.2), one faces quantum anomalies.

The idea is therefore to somewhat decouple the sector of supersymmetry breaking from the observable sector of quarks, leptons and gauge interactions. One refers to the former as a *hidden sector*. Obviously, one cannot totally "hide" this sector since the information of supersymmetry breaking must reach the observable sector. But, in the context of supergravity, gravity itself provides a necessary coupling between the two sectors. As we will see, the gravitational coupling between the two sectors tends to soften the divergences: whereas supersymmetry breaking effects are of order $\langle F \rangle \sim M_{SB}^2 \sim m_{3/2} m_{Pl}$ in the hidden sector, they are damped by the gravitational coupling $\kappa \sim m_{Pl}^{-1}$ in the observable sector and one expects there effects of order $m_{3/2}$. We will see below that indeed soft supersymmetry breaking terms are of the order of the gravitino mass (or smaller if more powers of κ are involved). This means that, barring accidental cancellations, one expects STr $M^2 \equiv \text{Tr} M_0^2 - 2\text{Tr} M_{1/2}^2$ to be of order $m_{3/2}^2$. For example, for a single (gauge singlet) chiral superfield z coupled to supergravity, the supertrace, including terms up to spin $3/2$, reads [209]

$$\mathcal{S}\text{Tr} M^2 \equiv \sum_{J=0}^{3/2} (-)^{2J}(2J+1)\,\text{Tr} M_J^2 = 2e^{\mathcal{G}} \frac{\mathcal{G}_z \mathcal{G}_{\bar{z}}}{\mathcal{G}_{z\bar{z}}} \frac{R_{z\bar{z}}}{\mathcal{G}_{z\bar{z}}}\bigg|_{\text{min}}, \qquad (6.51)$$

where $R_{z\bar{z}}$ is the curvature of the Kähler manifold: $R_{z\bar{z}} = -\partial_z \partial_{\bar{z}} \ln \mathcal{G}_{z\bar{z}}$ [see (C.57) of Appendix C]. Then, for a flat Kähler manifold ($\mathcal{G}_{z\bar{z}} = 1$ hence $R_{z\bar{z}} = 0$), we have, for $z = (A+iB)/\sqrt{2}$ (the corresponding fermion field is the Goldstino; it disappears from the physical spectrum):

$$m_A^2 + m_B^2 = 4m_{3/2}^2. \qquad (6.52)$$

We have seen in Chapter 3, after equation (3.23) that STr M^2 measures approximately the difference between an average supersymmetric particle mass and an average quark/lepton mass. As such, it should be of the order of the $(\text{TeV})^2$. This explains why a gravitino mass of a few hundred GeV or a TeV is often quoted as a preferred value (although there exists viable models with a superlight or superheavy gravitino).

We now illustrate these ideas on a very simple model, known as the Polonyi [313] model, which has provided for a long time a guideline for this type of gravity-mediated supersymmetry breaking.

The hidden sector consists of a single chiral supermultiplet with scalar component denoted by z and a superpotential

$$W_h(z) = \mu m_P(z + \beta). \tag{6.53}$$

Since $F_z = dW_h/dz = \mu m_P \neq 0$, supersymmetry is broken (and arguments above suggest to take μ in the 100 GeV to 1 TeV range); β is a real constant of order m_P to be fixed below. One then adds observable supermultiplets, with scalar field components denoted by ϕ^i. If we assume that the complete potential has the form

$$\mathcal{W}(z, \phi^i) = W_h(z) + W_o(\phi^i), \tag{6.54}$$

then the scalar potential reads in the limit of global supersymmetry (*i.e.* for vanishing gravitational coupling: $\kappa \to 0$)

$$\mathcal{V}(z, \phi^i) = \left| \frac{dW_h}{dz} \right|^2 + \sum_i \left| \frac{\partial W_o}{\partial \phi^i} \right|^2. \tag{6.55}$$

Hence, the two sectors are completely decoupled. If we restore a nonvanishing κ, we see that the two sectors are coupled only gravitationally.

In order to discuss such couplings, we must specify the Kähler potential; for simplicity, we will choose a flat Kähler metric:

$$\mathcal{K}(z, \bar{z}, \phi^i, \bar{\phi}^{\bar{\imath}}) = \bar{z}z + \phi^i \bar{\phi}^{\bar{\imath}}. \tag{6.56}$$

The scalar potential (6.36) then reads

$$\mathcal{V}(z, \phi^i) = e^{\kappa^2 \left(|z|^2 + \phi^i \bar{\phi}^{\bar{\imath}} \right)} \left[\left| \frac{dW_h}{dz} + \kappa^2 z^*(W_h + W_o) \right|^2 + \left| \frac{\partial W_o}{\partial \phi^i} + \kappa^2 \bar{\phi}^{\bar{\imath}}(W_h + W_o) \right|^2 \right.$$
$$\left. - 3\kappa^2 \left| W_h + W_o \right|^2 \right]. \tag{6.57}$$

Since only z acquires a vacuum expectation value of order m_{Pl}, it is consistent to consider first this field alone by setting the ϕ^i to zero. Requiring a vanishing cosmological constant at the minimum $\langle z \rangle$ of $\mathcal{V}(z, 0)$ tunes the constant β:

$$\kappa\beta = \pm 2 - \sqrt{3}$$
$$\kappa\langle z \rangle = \sqrt{3} \mp 1 \tag{6.58}$$

and the gravitino mass reads from (6.46):

$$m_{3/2} = \mu e^{2 \mp \sqrt{3}}. \tag{6.59}$$

We note that

$$\left\langle e^{\kappa^2 |z|^2/2} \left(\frac{dW_h}{dz} + \kappa^2 z^* W_h \right) \right\rangle = \kappa\sqrt{3}\langle W_h e^{\kappa^2 |z|^2/2} \rangle = \frac{\sqrt{3}}{\kappa} m_{3/2}. \tag{6.60}$$

Hence, we can write

$$\mathcal{V}(\langle z \rangle, \phi^i) = e^{\kappa^2 |\phi^i|^2 / 2} \left\{ \left| \sqrt{3}\kappa^{-1} m_{3/2} + \kappa(\sqrt{3} \mp 1) W \right|^2 \right.$$
$$+ \sum_i \left| \frac{\partial W}{\partial \phi^i} + \kappa^2 \bar{\phi}^{\bar{i}} W + m_{3/2} \bar{\phi}^{\bar{i}} \right|^2$$
$$\left. -3 \left| \kappa^{-1} m_{3/2} + \kappa W \right|^2 \right\} \tag{6.61}$$

where

$$W(\phi^i) \equiv e^{\kappa^2 |\langle z \rangle|^2 / 2} W_o(\phi^i). \tag{6.62}$$

In the limit $\kappa \to 0$, finite $m_{3/2}$ and finite $\kappa\langle z \rangle$, we have

$$\mathcal{V}(\langle z \rangle, \phi^i) \sim \sum_i \left| \frac{\partial W}{\partial \phi^i} \right|^2 + m_{3/2}^2 \sum_i |\phi^i|^2$$
$$+ m_{3/2} \left[\sum_i \phi^i \frac{\partial W}{\partial \phi^i} + (A - 3) W + \text{h.c.} \right], \tag{6.63}$$

with

$$A = 3 \mp \sqrt{3}. \tag{6.64}$$

We recognize:

- in the first term the standard scalar potential of global supersymmetry;
- in the second term, scalar masses of order $m_{3/2}$;
- in the third term, analytic terms in ϕ^i plus their hermitian conjugates. The constant A has been introduced because for a cubic potential $W(\phi) = \lambda \phi^3$, these terms simply read $A m_{3/2} \lambda \left(\phi^3 + \phi^{*3} \right)$. This is a standard A-term of the type discussed in Section 5.2.3 of Chapter 5.

We see that we have generated through gravitational interactions soft supersymmetry-breaking terms in the observable sector precisely of the order of the gravitino mass $m_{3/2}$: scalar masses[6] and A-terms. One may wonder about the last type of soft terms, gaugino masses. It turns out that the full supergravity Lagrangian includes a term:

$$\mathcal{L}|_{\text{quad } \lambda} = \text{Re} \left[f(z, \phi^i) \right] \frac{i}{4} \bar{\lambda} \gamma^\mu \partial_\mu \lambda - \frac{1}{4} \frac{\partial f}{\partial z} F_z \bar{\lambda}_R \lambda_L + \text{h.c.} \tag{6.65}$$

where the gauge kinetic function $f(z, \phi^i)$ is defined in (6.25). Using (6.32), we find $\langle F_z \rangle$ of order $m_{3/2} \kappa^{-1}$. Thus, we obtain gaugino masses of order $m_{3/2}$ as well if the gauge kinetic function depends on z.

---

[6]The other type of analytic mass term in equation (5.6) of Chapter 5 may be recovered from the last line in (6.63) if $W(\phi)$ includes a quadratic term $\rho\phi^2$: then $\delta m'^2 = (A - 1)\rho m_{3/2}$.

6.4   The gravitino and Goldstino fields

It is time to pause and to discuss in some more detail the physics associated with this spin 3/2 field which provides the supersymmetric partner of the graviton: the gravitino.

We first consider the massless gravitino described by the Majorana spin-vector $\psi_{r\mu}$ (r spinor index and μ vector index). The free field equation of motion simply reads in flat space:

$$\epsilon^{\lambda\mu\nu\rho}\gamma_5\gamma_\mu\partial_\nu\psi_\rho = 0, \tag{6.66}$$

which is invariant under the gauge transformation $\psi_\mu \rightarrow \psi_\mu + \partial_\mu\eta$, with η a Majorana spinor. Equation (6.66) is known as the Rarita–Schwinger [321] equation.

It can also be written, using (B.21) of Appendix B:

$$i\gamma^{[\mu}\gamma^\nu\gamma^{\rho]}\partial_\nu\psi_\rho = 0, \tag{6.67}$$

which has the advantage of being suitable for generalization to higher dimensions (see Chapter 10).

We make the standard gauge choice of the harmonic gauge:

$$\gamma_\rho\psi^\rho = 0. \tag{6.68}$$

This leaves us with a residual gauge symmetry $\psi_\mu \rightarrow \psi_\mu + \partial_\mu\eta$ with $\gamma^\rho\partial_\rho\eta = 0$.

With this gauge choice, (6.67) reads $\gamma^\nu\partial_\nu\psi^\mu - \gamma^\mu\partial_\nu\psi^\nu = 0$. Contracting with γ_μ gives

$$\partial_\nu\psi^\nu = 0 \tag{6.69}$$

and thus

$$\gamma^\rho\partial_\rho\psi_\mu = 0 \tag{6.70}$$

which is the Dirac equation for a massless fermion field.

The spin-vector $\psi_{r\mu}$ belongs a priori to the representation $(1/2) \otimes (1) = (1/2) \oplus (3/2)$. The condition $\gamma^\rho\psi_\rho = 0$ above selects the $(3/2)$ representation: it is an irreducibility condition. To see this, we decompose ψ_μ in momentum (p) space using the complete set provided by the vector (e.g. photon) polarization states[7] $\varepsilon_\mu^\pm$ associated with helicity ± 1, the momentum p_μ and a conjugate momentum $\bar{p}_\mu$ such that $\bar{p}^2 = p^2 = 0$ and $p^\mu\bar{p}_\mu = 1$:

$$\psi_{r\mu} = \varepsilon_\mu^+ u_r^+ + \varepsilon_\mu^- u_r^- + p_\mu u_r + \bar{p}_\mu v_r, \tag{6.71}$$

where $u^\pm$, u and v are spinor fields. From (6.69), we have $p^\rho\psi_\rho = 0$ and thus $v = 0$, whereas (6.70) yields $\gamma^\rho\partial_\rho u^\pm = \gamma^\rho\partial_\rho u = 0$. We then use the residual gauge symmetry (see the comment following equation (6.68)) to absorb the term $p_\mu u_r$:

$$\psi_{r\mu} = \varepsilon_\mu^+ u_r^+ + \varepsilon_\mu^- u_r^-. \tag{6.72}$$

Then the irreducibility condition (6.68) yields:

$$0 = \gamma^\rho\varepsilon_\rho^-\gamma^\sigma\psi_\sigma = \gamma^\rho\varepsilon_\rho^-\gamma^\sigma\varepsilon_\sigma^+ u^+ = \left(1 - \frac{p_i\Sigma^i}{|p|}\right)u^+, \tag{6.73}$$

---

[7] They satisfy $\varepsilon_\mu^\pm p^\mu = 0$ (transverse polarization) and $\varepsilon_\mu^+ \varepsilon^{\mu+} = 0 = \varepsilon_\mu^- \varepsilon^{\mu-}$.

where Σ^i is the spin matrix which takes the form $\begin{pmatrix} \sigma^i & 0 \\ 0 & \sigma^i \end{pmatrix}$ in the basis described in footnote 11 of Chapter 5. Hence u^+ has helicity $+1/2$ and $\varepsilon_\mu^+ u^+$ represents the component of helicity $+3/2$; similarly $\varepsilon_\mu^- u^-$ is the component of helicity $-3/2$.

Let us recapitulate the counting of the number of on-shell degrees of freedom for a massless gravitino: it has the degrees of freedom of a massless vector field (2) times those of a Majorana spinor (2) but the gauge condition $\gamma^\rho \psi_\rho = 0$ kills those of the spin $1/2$ part (2); we are left with $2 \times 2 - 2 = 2$ degrees of freedom.

This should be put in parallel with the counting of on-shell degrees of freedom of a graviton. The graviton field is obtained as a fluctuation of the metric $g_{\mu\nu} = \eta_{\mu\nu} + h_{\mu\nu}$ and is thus described by the symmetric traceless tensor $h_{\mu\nu}$ which has $(4 \times 5)/2 - 1$ (for the trace), that is nine independent components. The gauge freedom $\delta h_{\mu\nu} = \partial_\mu \xi_\nu + \partial_\nu \xi_\mu$ is partly fixed by four gauge conditions which specify $\partial^\mu h_{\mu\nu}$; but there is, as usual, a residual gauge invariance corresponding to ξ_μ satisfying $\partial_\mu \xi^\mu = 0$ and $\Box \xi^\mu = 0$, which eliminates another three degrees of freedom (four massless scalars minus one condition). We are left with two on-shell degrees of freedom for the graviton.

It is thus perfectly consistent to have a gravity supermultiplet with only the graviton and the gravitino. As we have seen in Chapter 3 for other supermultiplets, the counting of *off-shell* degrees of freedom, on the other hand, imposes the presence of auxiliary fields, as described at the end of Section 6.1.

We now turn to the massive gravitino case. The equation of motion reads:

$$-\frac{i}{2}\epsilon^{\mu\nu\rho\sigma}\gamma_5\gamma_\nu\partial_\rho\psi_\sigma - m_{3/2}\sigma^{\mu\rho}\psi_\rho = 0. \tag{6.74}$$

Contracting with ∂_μ yields, in the case of nonvanishing gravitino mass,

$$[\gamma^\sigma\partial_\sigma,\ \gamma^\rho]\ \psi_\rho = 0. \tag{6.75}$$

Then, contracting (6.74) with γ_μ and using the equation just obtained gives, again for $m_{3/2} \neq 0$,

$$\gamma^\rho\psi_\rho = 0. \tag{6.76}$$

Thus the gauge condition of the massless case is obtained as a consequence of the equations of motion (a similar situation is found for the case of a massive vector field, see the comment following equation (A.95) of Appendix Appendix A). One obtains:

$$\partial^\rho\psi_\rho = \frac{1}{2}\{\gamma^\sigma\partial_\sigma,\gamma^\rho\}\psi_\rho = -\frac{1}{2}[\gamma^\sigma\partial_\sigma,\gamma^\rho]\psi_\rho = 0 \tag{6.77}$$

from (6.75) and (6.76). Also the equation of motion simply reads:

$$i\epsilon^{\mu\nu\rho\sigma}\gamma_5\gamma_\nu\partial_\rho\psi_\sigma = m_{3/2}\psi_\mu. \tag{6.78}$$

Contracting this with $\gamma_\mu\gamma_\lambda$, one finally obtains

$$[\gamma^\rho\partial_\rho - m_{3/2}]\psi_\lambda = 0 \tag{6.79}$$

which is the standard Dirac equation for a massive spinor field.

Whereas the MSSM was strictly speaking a low energy theory, the inclusion of gravity, even at the level of the nonrenormalizable theory that is supergravity, allows us to consider supersymmetric models up to scales close to the Planck scale. The hope is that, as one renormalizes the model up to higher scales, it becomes simpler, reflecting a symmetry principle that becomes transparent at a high fundamental scale (we will study for example in later chapters grand unification theories or string theories which provide explicit examples of such paradigms). In the remaining sections, we will therefore use renormalization group techniques to bring the low energy supersymmetric models discussed in the previous chapter to higher scales, and before considering more elaborate high energy models, we will make some simplifying assumptions at a scale close to the Planck scale to introduce a toy model known as *minimal supergravity* which will represent our first attempt at writing a simple high energy paradigm.

6.5 Radiative breaking of $SU(2) \times U(1)$

We start by studying the breaking of the electroweak symmetry. We have seen in Chapter 5, equation (5.13) that, in the Higgs sector, the scalar potential

$$V = m_1^2 |H_1^0|^2 + m_2^2 |H_2^0|^2 + B_\mu \left(H_1^0 H_2^0 + H_1^{0*} H_2^{0*} \right)$$
$$+ \frac{g^2 + g'^2}{8} \left(|H_1^0|^2 - |H_2^0|^2 \right)^2$$

has a nontrivial minimum under the conditions

$$m_1^2 + m_2^2 > 2|B_\mu|$$
$$m_1^2 m_2^2 < |B_\mu|^2$$

which are not compatible with $m_1^2 = m_2^2$. Luckily enough, because H_1 and H_2 couple to different fields, radiative corrections give a different treatment to H_1 and H_2, even if supersymmetry breaking yields the same contribution of order $m_{3/2}$. Let us see in more detail how this arises.

The main difference between H_1 and H_2 is that H_2 couples to the top quark. Restricting therefore our attention to the heavy quark loops, we must consider the diagrams of Fig. 6.1 which contribute to the renormalization of the soft scalar mass $m_{H_2}^2 \equiv m_2^2 - |\mu|^2$.

The first three diagrams come from the term $\lambda_t Q_3 \cdot H_2 T^c$ where $Q_3 = \binom{T}{B}$ in the superpotential. Each individual diagram contributes a quadratic divergence – which is cancelled in the sum, since supersymmetry is spontaneously broken– and a logarithmic divergence. The fourth diagram arises from an A-term proportional to the same term $\lambda_t Q_3 \cdot H_2 T^c$ in the superpotential: $A_t \lambda_t \widetilde{q}_{3L} \cdot H_2 \widetilde{t}_R^* + h.c.$ and the last term arises from the Higgs gauge interactions. One obtains (see Appendix E for a more complete formula)

$$8\pi^2 \frac{dm_{H_2}^2}{dt} = 3\lambda_t^2 \left(m_{H_2}^2 + m_T^2 + m_Q^2 + A_t^2 \right) - 3g^2 M_2^2 \tag{6.80}$$

where $t = \ln(\mu/\mu_0)$, μ being the renormalization scale (not to be confused with the supersymmetric parameter μ!) This should be solved with the boundary condition

$$m_{H_2}^2(\mu = m_P) = m_{3/2}^2.$$

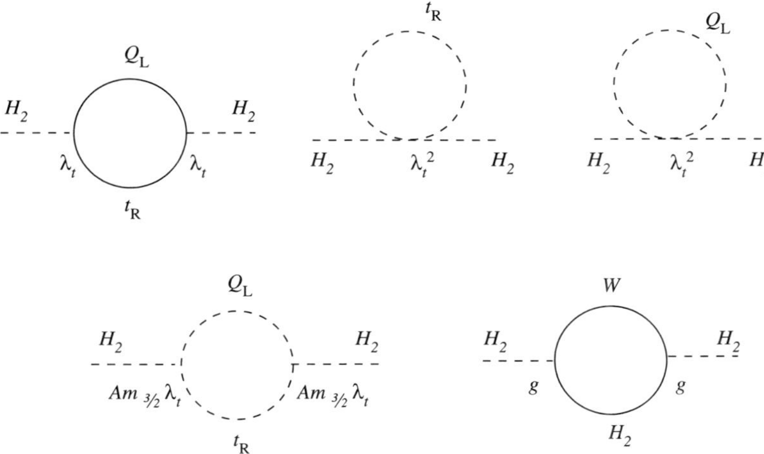

Fig. 6.1 One-loop diagrams contributing to the renormalization of $m_{H_2}^2$. Q, t_R solid (resp. dashed) lines are quark (resp. squark) lines, solid W or H_2 lines are gaugino or Higgsino lines.

We see that the term proportional to λ_t tends to decrease $m_{H_2}^2$ as μ decreases. One should remember the relation (5.27) of Chapter 5 which we rewrite

$$\frac{1}{2} M_z^2 = \frac{m_{H_1}^2 - m_{H_2}^2 \tan^2 \beta}{\tan^2 \beta - 1} - |\mu|^2 > 0. \tag{6.81}$$

This gives the condition that the soft supersymmetry breaking parameters $m_{H_1}^2$ and $m_{H_2}^2$ must fulfill at low energy in order to have the right breaking: the smaller $m_{H_2}^2$ is (and possibly negative), the easier it is to fulfill.

Similar equations are obtained for the evolution of m_T^2 and m_Q^2

$$8\pi^2 \frac{dm_T^2}{dt} = 2\lambda_t^2 \left(m_{H_2}^2 + m_T^2 + m_Q^2 + A_t^2 \right) - \frac{16}{3} g_3^2 \, M_3^2 \tag{6.82}$$

$$8\pi^2 \frac{dm_Q^2}{dt} = \lambda_t^2 \left(m_{H_2}^2 + m_T^2 + m_Q^2 + A_t^2 \right) - \frac{16}{3} g_3^2 \, M_3^2. \tag{6.83}$$

If we can neglect the effect of the gaugino masses M_2, M_3, we see that, as μ decreases, all three $m_{H_2}^2$, m_T^2 and m_Q^2 decrease but $m_{H_2}^2$ faster than the other two: hence $SU(2) \times U(1)$ is broken before there is a danger that $SU(3)$ gets broken (through $m_T^2 < 0$ or $m_Q^2 < 0$). If the effect of M_3 is nonnegligible, it will tend to slow down the evolution of m_T^2, m_Q^2, which in turn enhances the effect on $m_{H_2}^2$.

To get a crude estimate of the scale μ_{EW} at which the electroweak symmetry is broken, we may neglect the running of all quantities besides m_{H_2}. The leading contribution to the renormalization of m_{H_2} is then, from (6.80),

$$\delta m_{H_2}^2 \sim -\frac{3\lambda_t^2}{8\pi^2} m_Q^2 \, \ln \frac{\mu}{m_P}. \tag{6.84}$$

If we start with a boundary condition where $m_{H_2}^2(m_P) \sim m_Q^2$, then the scale μ_{EW} at which $m_{H_2}^2 + \delta m_{H_2}^2$ becomes negative is then

$$\mu_{EW} \sim m_P\, e^{-\mathcal{O}(1)\times 4\pi/\lambda_t^2}. \tag{6.85}$$

Hence, a large Yukawa coupling induces a scale for electroweak symmetry breaking which is exponentially small compared to the Planck scale.

In principle, however, a large top mass is not compulsory: at the time when the top mass was thought to be around 40 GeV, electroweak symmetry breaking was shown to be consistent with such a mass, although this involved some fine-tuning of the gaugino mass [48, 230].

6.6 Gaugino masses

In a similar way, one should let the gaugino masses evolve from a scale of order m_{Pl} where they are typically equal to $m_{3/2}$ to a low energy scale. Their renormalization group evolution is particularly simple:

$$\frac{dM_i}{dt} = -\frac{\alpha_i}{2\pi}\, b_i\, M_i \tag{6.86}$$

where α_i is the corresponding gauge coupling constant which evolves as:

$$\frac{d\alpha_i}{dt} = -\frac{b_i}{2\pi}\, \alpha_i^2 \tag{6.87}$$

and b_i is thus the beta function coefficient at one loop (see Appendix E). One obtains

$$\frac{d(M_i/\alpha_i)}{dt} = 0 \tag{6.88}$$

which shows that M_i/α_i is not renormalized.

Thus, if we assume equality of gauge couplings at some unification scale M_U

$$\alpha_1(M_U) = \alpha_2(M_U) = \alpha_3(M_U) \tag{6.89}$$

and correspondingly (confusing M_U and m_P for that purpose) equality of gaugino masses at this scale[8]

$$M_1(M_U) = M_2(M_U) = M_3(M_U) = M_{1/2}, \tag{6.90}$$

where $M_{1/2}$ is a constant of the order of the gravitino mass, the gaugino masses and gauge couplings *at low energy* are related:

$$\frac{M_1}{\alpha_1} = \frac{M_2}{\alpha_2} = \frac{M_3}{\alpha_3}. \tag{6.91}$$

In other words, the gaugino mass terms M_1 and M_2 which appear in the chargino/neutralino mass matrices can be expressed in terms of the single gluino mass M_3. We have the following relations (see last footnote)

$$M_3 = \frac{\alpha_3}{\alpha}\sin^2\theta_W\, M_2 = \frac{3}{5}\frac{\alpha_3}{\alpha}\cos^2\theta_W\, M_1. \tag{6.92}$$

Hence, at the electroweak scale, $M_3 : M_2 : M_1 \sim 7 : 2 : 1$.

---

[8]Note that, as usual, α_1 corresponds to the normalization of the $U(1)$ hypercharge imposed by grand unification: the $U(1)_Y$ gauge coupling used traditionally in the Standard Model is g' such that $\alpha' \equiv g'^2/(4\pi) = 3\alpha_1/5$ (see Chapter 9, equations (9.26) and (9.30)).

We have seen that the equation of evolutions for the scalar masses are rather more complicated. If we assume that scalar masses are smaller than gaugino masses at all scales, the gaugino loop contribution dominates in equations such as (6.82–6.83) and the renormalization group evolutions are easy to solve. One finds, to a good approximation,

$$M_3^2(M_Z) \simeq 9.8\ M_{1/2}^2,$$

$$m_Q^2(M_Z) \simeq m_Q^2(M_U) + 8.3\ M_{1/2}^2,$$

$$m_{U,D}^2(M_Z) \simeq m_{U,D}^2(M_U) + 8\ M_{1/2}^2, \tag{6.93}$$

$$m_L^2(M_Z) \simeq m_L^2(M_U) + 0.7\ M_{1/2}^2,$$

$$m_E^2(M_Z) \simeq m_E^2(M_U) + 0.23\ M_{1/2}^2.$$

6.7 Scalar masses

There is however a severe problem connected with scalar masses arising from supersymmetry breaking: the problem of flavor changing neutral currents (FCNC). The origin of these FCNC is similar to the Standard Model but there is, in general, no equivalent to the GIM cancellation mechanism (see Section A.3.4 of Appendix Appendix A).

Let us recall that, in general, the nondiagonal nature of Yukawa couplings leads, after gauge symmetry breaking, to nondiagonal quark and lepton mass matrices. One thus needs to diagonalize them through some matrices $V^{q,\ell}$.

Similarly, we have seen earlier that supersymmetry breaking leads to nondiagonal squark and slepton mass matrices, which may be diagonalized through some matrices $\widetilde{V}^{q,\ell}$.

As long as the two issues – supersymmetry breaking and Yukawa couplings – are not connected, there is no reason to relate $V^{q,\ell}$ with $\widetilde{V}^{q,\ell}$, *i.e.* the rotations among quarks and leptons and the rotation of squarks and sleptons. This leads to undesirable FCNC's. Let us consider for example the squark–quark–gluino coupling: $\widetilde{q}^\dagger \lambda_g q$, where q and $\widetilde{q}$ are, respectively, the quark and squark fields which are interaction eigenstates (*i.e.* the original fields in the interaction Lagrangian).

If one wants to express this coupling in terms of the mass eigenstates one needs to make the rotations just introduced: $\widetilde{q}^\dagger \widetilde{V}^{q\dagger} \lambda_g V^q q$. In other words, we have a flavor mixing matrix

$$W \equiv \widetilde{V}^{q\dagger}\ V^q.$$

In general, $W \neq 1$, which generates FCNC. Let us consider for example the famous $s\bar{d} \to \bar{s}d$ amplitude which contributes to the K_L–K_S mass difference Δm_K (see Fig. A.6 of Appendix Appendix A). As long as $W \neq 1$, there is a new contribution coming from the following box diagram involving gluinos and squarks as shown in Fig. 6.2. The corresponding contribution reads typically

$$\frac{\Delta m_K}{m_K} \sim \alpha_3^2 \sum_{ij} W_{di}^\dagger\ W_{is}\ W_{dj}^\dagger\ W_{js}\ f\left(\frac{\widetilde{m}_i^2}{M_3^2}, \frac{\widetilde{m}_j^2}{M_3^2}\right) \tag{6.94}$$

where $\widetilde{m}_i$ is the $\widetilde{u}_i$ squark mass. For a generic W matrix, this is much too large.

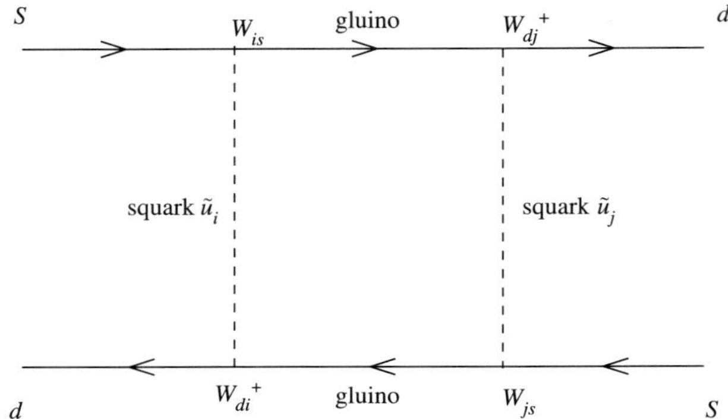

Fig. 6.2 Box diagram of supersymmetric contribution to $s\bar{d} \to \bar{s}d$ amplitude.

We will return in detail to this problem in Chapter 12. Let us say here only that there are several ways of accounting for it:

(i) Universality. To first order, all squarks and sleptons are degenerate. One can then choose $\widetilde{V}^q = V^q$, in which case $W = 1$ and the supersymmetric box diagram vanishes. Of course, one can only make this assumption at a given scale, of order m_{Pl} or M_U for supergravity models, and renormalization to low energies will tend to generate nondiagonal terms in the mass matrices and thus FCNC at low energy.

(ii) Effective supersymmetry [79]. FCNC constraints are only crucial for the first two families, whereas the fine-tuning associated with the hierarchy problem concerns mostly the third family in the radiative breaking scenario (since the breaking of $SU(2) \times U(1)$ and thus the scale M_Z arises through the top Yukawa coupling). A possible solution is therefore to make the squarks of the first two families heavy (TeV scale), whereas the third family squarks remain relatively light in order to avoid an unwanted fine-tuning. In this case, because of the squark propagators, the function $f(\widetilde{m}_i^2/M_3^2, \widetilde{m}_j^2/M_3^2)$ is very small.

(iii) Quark–squark alignment. If one "aligns" the squark flavors along the quark flavors, that is if the same matrix may diagonalize quark and squark mass matrices, then $V^q = \widetilde{V}^q$ and $W = 1$. This obviously requires a theory of flavor. It may indeed be obtained by imposing flavor symmetries known as horizontal symmetries (in contrast with the ordinary gauge symmetries of the Standard Model which are vertical, *i.e.* act within a given family of quarks and leptons).

6.8 The minimal supergravity model

In the case of the minimal supergravity model, one therefore assumes, besides (6.90), that all scalar masses are equal at unification scale, as well as all A-terms:

$$m_{H_1}(M_U) = m_{H_2}(M_U) = m_0 \tag{6.95}$$

$$m_Q(M_U) = m_U(M_U) = m_E(M_U) = m_0 \tag{6.96}$$

$$m_D(M_U) = m_L(M_U) = m_0 \tag{6.97}$$

$$A_{QH_2U^c}(M_U) = \cdots = A_0 \tag{6.98}$$

where m_0 and A_0 are typically of order $m_{3/2}$.

The original motivation was that gravity is the messenger of supersymmetry breaking and that gravity is "flavor blind". For example, the Polonyi model which illustrated in Section 6.3.2 the simplest example of gravitationally hidden sector yields universal soft terms. However, this turns out to be often an undue simplification: in many explicit models of supersymmetry breaking in a gravitationally hidden sector (such as those inspired by superstring theories), the soft supersymmetry breaking terms are not flavor blind. It remains however true that this property is recovered in some special limits: the minimal supergravity model thus addresses such cases.

As stressed earlier, such a universality of scalar masses is not stable under quantum fluctuations and renormalization down to low energies leads to nonuniversalities and thus FCNCs.

The attractiveness of the minimal supergravity model is its small number of parameters, besides those of the Standard Model:

$$m_0, \quad M_{1/2}, \quad \tan\beta, \quad A_0, \quad \text{sign } \mu.$$

Let us note in particular that $|\mu|$ is determined through the condition (6.81) of $SU(2) \times U(1)$ breaking. The low energy parameters m_1^2 and m_2^2 may be expressed in terms of the high energy parameters m_0, A_0 and $M_{1/2}$, which gives for example for $\tan\beta = 1.65$

$$M_z^2 = 5.9\ m_0^2 + 0.1\ A_0^2 - 0.3\ A_0\ M_{1/2} + 15\ M_{1/2}^2 - 2|\mu|^2. \tag{6.99}$$

This already shows that, for small $\tan\beta$, the minimal supergravity model requires some level of fine-tuning: as the experimental limits on squark, slepton and gaugino masses are raised, this pushes higher m_0, $M_{1/2}$ and thus (6.99) requires some fine-tuning among large parameters. We return to this issue in Section 6.10. We also see that this requires to have $|\mu| > M_{1/2}$.

Generally speaking for the minimal supergravity model, when $|\mu| > M_{1/2}$, the LSP $\widetilde{\chi}_1^0$ is predominantly a bino $\widetilde{B}$ with mass $M_1 = M_2\alpha_1/\alpha_2 = \frac{5}{3}\ M_2 g'^2/g^2 = \frac{5}{3}\tan^2\theta_W\ M_2$ and the lightest chargino–neutralino spectrum is approximately: $m_{\widetilde{\chi}_1^0} \sim \frac{1}{2}\ m_{\widetilde{\chi}_1^\pm} \sim \frac{1}{2}\ m_{\widetilde{\chi}_2^0} \sim \frac{1}{6}\ M_3$ where M_3 is the gluino mass.

The remaining sections, which lie in the "Theoretical Introduction" track, are somewhat technical, although they deal with important issues: dynamical determination of key parameters, and fine tuning. The phenomenology oriented readers may have interest to skip these sections and return to them at a later stage.

6.9 Infrared fixed points, quasi-infrared fixed points and focus points

If the fundamental theory underlying the Standard Model lies at a scale which is close to the Planck scale, this means that, by studying today the Standard Model, we are looking at the deep infrared regime of the theory. The renormalization group flow might thus lead us to some specific corners of the parameter space. We will illustrate in this section this possibility with a few examples.

6.9.1   The top quark and the infrared fixed point

Among the Yukawa couplings, the top coupling is the only one which is of order 1. In other words, it is the only one which is of the order of some of the gauge couplings, say the strong coupling g_3. Is this a coincidence? Pendleton and Ross [306] showed that this might be expected in the infrared regime of the theory because of the presence of a fixed point in the low energy evolution of the renormalization group equations. Before being more specific, let us note that fixing dynamically the top quark Yukawa coupling would allow us to determine the key parameter $\tan\beta$ through the relation:

$$\sin\beta = \frac{\tan\beta}{\sqrt{1+\tan^2\beta}} = \frac{\sqrt{2}}{\lambda_t}\frac{m_t}{v}, \tag{6.100}$$

where we have used the relation $m_t = \lambda_t v_2$.

Let us indeed consider the renormalization group equation for the top Yukawa coupling, given in equation (E.13) of Appendix E. We will for simplicity neglect all Yukawa couplings but the top quark one and all gauge couplings but the QCD one. The relevant renormalization group equations are then:

$$\frac{\mu}{\lambda_t}\frac{d\lambda_t}{d\mu} = \frac{1}{16\pi^2}\left(6\lambda_t^2 - \frac{16}{3}g_3^2\right) \tag{6.101}$$

$$\frac{\mu}{g_3}\frac{dg_3}{d\mu} = -\frac{3}{16\pi^2}g_3^2.$$

We thus deduce the following equation for the ratio of these two couplings:

$$\mu\frac{d\ln\left(\lambda_t^2/g_3^2\right)}{d\mu} = \frac{3g_3^2}{4\pi^2}\left(\frac{\lambda_t^2}{g_3^2} - \frac{7}{18}\right). \tag{6.102}$$

It is easy to check that the ratio

$$\frac{\lambda_t^2}{g_3^2} = \frac{7}{18} \tag{6.103}$$

is an infrared fixed point: as one goes down in scale, one is attracted to this value. This is the Pendleton–Ross fixed point, or quasifixed point, since the top Yukawa coupling continues to evolve: it only tracks closer and closer the strong gauge coupling as one goes down in energy.

It was, however, emphasized by Hill [224] that, as large as the 16 orders of magnitude between the Planck scale and the top or W mass might seem to be, they are not sufficient to allow the top Yukawa coupling to significantly approach its quasi-fixed point value given by (6.103). Since we can solve exactly the system of equations (6.101) [224, 231], we can check this explicitly. The reader who is only interested in the result may proceed directly to equation (6.111). For the sake of completeness, we will reinstate all the gauge couplings. Then, introducing the standard notation

$$Y_t \equiv \frac{\lambda_t^2}{16\pi^2}, \qquad Y_b \equiv \frac{\lambda_b^2}{16\pi^2}, \tag{6.104}$$

we may write the renormalization equation for the top Yukawa coupling as

$$\mu\frac{dY_t^{-1}}{d\mu} = Y_t^{-1}\left(\frac{32}{3}\frac{g_3^2}{16\pi^2} + 6\frac{g_2^2}{16\pi^2} + \frac{25}{15}\frac{g_1^2}{16\pi^2}\right) - 12. \tag{6.105}$$

The linear equation is readily solved by writing it in the form (using (E.2)–(E.4) of Appendix E)

$$\frac{\mu}{Y_t^{-1}}\frac{dY_t^{-1}}{d\mu} = -\frac{32}{9}\frac{\mu}{g_3}\frac{dg_3}{d\mu} + 6\frac{\mu}{g_2}\frac{dg_2}{d\mu} + \frac{26}{99}\frac{\mu}{g_1}\frac{dg_1}{d\mu}. \tag{6.106}$$

The solution is $Y_t(\mu) = \alpha_t^{-1}E_t(\mu)$ with α_t constant and

$$E_t(\mu) = \left(\frac{g_3(\mu)}{g_3(\mu_0)}\right)^{32/9}\left(\frac{g_2(\mu)}{g_2(\mu_0)}\right)^{-6}\left(\frac{g_1(\mu)}{g_1(\mu_0)}\right)^{-26/99}. \tag{6.107}$$

The solution to the full equation is solved by the method of variation of the constant. Namely, we let α_t depend on μ. Then the solution satisfies the equation (6.105) if $\mu\, d\alpha_t/d\mu = -12E_t(\mu)$ which is readily solved. Since $\alpha_t(\mu_0) = Y_t^{-1}(\mu_0)$, we may finally write the solution of (6.105) as [231]

$$Y_t(\mu) = \frac{Y_t(\mu_0)E_t(\mu)}{1 - 12Y_t(\mu_0)F_t(\mu)}, \tag{6.108}$$

with

$$F_t(\mu) \equiv \int_{\mu_0}^{\mu} E_t(\mu')\frac{d\mu'}{\mu'}. \tag{6.109}$$

We note that, for large values of $Y_t(\mu_0)$[9], that is for large values of the top Yukawa coupling at the fundamental scale (the grand unified scale M_U or the Planck scale M_P), the low energy value becomes independent of this high energy value:

$$Y_t(\mu) = -\frac{1}{12}\frac{E_t(\mu)}{F_t(\mu)}. \tag{6.110}$$

Is this the Pendleton–Ross fixed point? Not quite. In order to compare the two, let us disregard again the evolution of g_1 and g_2. Then, using the explicit form:

$$\frac{g_3(\mu)}{g_3(\mu_0)} = \left(1 + \frac{3}{8\pi^2}g_3^2(\mu_0)\ln\frac{\mu}{\mu_0}\right)^{-1/2},$$

we may compute explicitly $F_t(\mu)$ and we obtain for (6.110)

$$Y_t(\mu) = \frac{7}{18}\frac{g_3^2(\mu)}{16\pi^2}\frac{1}{\left[1 - \left(\frac{g_3(\mu_0)}{g_3(\mu)}\right)^{14/9}\right]}, \tag{6.111}$$

which coincides with the Pendleton–Ross fixed point (6.103) only in the limit where $g_3(\mu)/g_3(\mu_0) \gg 1$. As we stressed earlier, there is often not enough energy span for the top coupling to fall into the fixed point: for example $g_3(M_Z)/g_3(M_U) \sim 1.7$. However, the solution (6.111) is independent of the initial conditions. By a slight variation over the previous denomination, it has become known in the literature as the quasi-infrared

---

[9]Strictly speaking large values of $12Y_t(\mu_0)F_t(\mu)$, which leaves us some margin before we enter the nonperturbative regime for the top coupling $Y_t(\mu_0) > 1$.

fixed point, probably to stress that, although the behavior has become universal, the couplings have not completely fallen yet into the Pendleton–Ross (quasi-)fixed point.

If the value of the top Yukawa coupling is large at some superheavy scale such as the grand unification scale, then we just saw that we can estimate the value of $\tan\beta$ through (6.100). The top mass which is present in this equation is the running mass $m_t(m_t)$. This should be distinguished from the physical top mass M_t, which receives two important corrections at one-loop: QCD gluon corrections[10] and stop/gluino corrections. Taking these corrections into account [69], one obtains a value of $\tan\beta$ of the order of 1.5. As we will see in Chapter 7, such a value is not favored by data because it gives too light a Higgs. For $\tan\beta \geq 30$, the bottom Yukawa coupling $\lambda_b = \lambda_t(m_b/m_t)\tan\beta$ cannot be neglected and the previous analysis must be changed: it turns out that, for values of $\tan\beta$ larger than 30, m_t is decreasing with $\tan\beta$.

6.9.2 Focus points

Besides fixed points, there is the possibility of renormalization group trajectories focussing towards a given point: the evolution does not stop there and, as in the focussing of light, the trajectories diverge again beyond the focus point. It has been pointed out [153–155] that such focus points could be of interest if they correspond to the electroweak scale. Then, the low energy physics would be less dependent on the boundary values of parameters at high energy. This puts in a different perspective the discussion of naturalness which we will undertake in the next section. We will therefore discuss in some details the nature of such focus points.

The set of renormalization group equations that one has to solve at the one loop level is presented in Appendix E. They turn out to have a very specific structure which allows us to solve them, at least formally. We start by solving for the gauge couplings from which we can infer the evolution of the gaugino masses (see (6.88) above). This allows us to solve for the A-terms. In parallel, we may obtain the evolution of the Yukawa couplings following the methods developed in the previous section (we will apply it in Section 9.3.3 of Chapter 9 to obtain the evolution of λ_b).

Finally, there remains mainly to solve the evolution equations of the soft scalar squared masses, *i.e.* equations (E.17)–(E.23) of Appendix E. They turn out to involve only the following combinations

$$
\begin{aligned}
X_t &\equiv m_Q^2 + m_T^2 + m_{H_2}^2, \\
X_b &\equiv m_Q^2 + m_B^2 + m_{H_1}^2, \\
X_\tau &\equiv m_L^2 + m_E^2 + m_{H_1}^2.
\end{aligned}
\tag{6.112}
$$

The evolution equations can thus be transformed into a set of equations for these variables, and by adding to each mass squared m_i^2 ($i = H_1, H_2, B, \ldots$) an appropriate combination of X_t, X_b and X_τ, we obtain a new variable $\widehat{m}_i$ whose evolution depends only on gauge couplings and gaugino masses. We will illustrate this below.

---

[10] In the relevant renormalization scheme ($\overline{DR}$), described in Section E.8 of Appendix E,

$$
\left.\frac{M_t - m_t(m_t)}{m_t(m_t)}\right|_{\text{QCD}} = \frac{5g_3^2}{12\pi^2}.
$$

See Section 9.3.3 of Chapter 9 for a discussion of similar corrections in the case of the bottom quark.

Now, from the point of view of electroweak symmetry breaking and naturalness, we will be especially interested in the Higgs soft masses, in particular $m_{H_2}^2$. Consider two distinct solutions $(m_i^{(1)2})$ and $(m_i^{(2)2})$ of the renormalization group equations, with the same boundary conditions for the A-terms and gaugino masses, but different ones for the scalar masses. Then, given the structure of the renormalization group equations described above, the differences $(\Delta m_i^2 \equiv m_i^{(1)2} - m_i^{(2)2})$ satisfy a linear system which is easy to solve. If there is a scale $\mu_{i_0 F}$ for which $\Delta m_{i_0}^2(\mu_{i_0 F}) = 0$, then all solutions will converge at this scale to the same value for the particular mass-squared $m_{i_0}^2$. If this happens for $m_{H_2}^2$ at a scale close to the electroweak scale, then the low energy value of $m_{H_2}^2$ is insensitive to its boundary value at the superheavy scale.

Let us show this analytically. For simplicity, we will neglect the lepton evolution and thus the X_τ variable, as well as the electroweak gauge interactions. Then, one obtains from equations (E.17)–(E.23) of Appendix E the evolution equation for X_t and X_b:

$$8\pi^2 \frac{d}{dt}\begin{pmatrix} X_t \\ X_b \end{pmatrix} = \begin{pmatrix} 6\lambda_t^2 & \lambda_b^2 \\ \lambda_t^2 & 6\lambda_b^2 \end{pmatrix}\left[\begin{pmatrix} X_t \\ X_b \end{pmatrix} + \begin{pmatrix} A_t^2 \\ A_b^2 \end{pmatrix}\right] - \frac{32}{3}\begin{pmatrix} 1 \\ 1 \end{pmatrix} g_3^2 M_3^2, \tag{6.113}$$

where $t = \ln\mu$. In order to define the variables $\widehat{m}_i$, it is convenient to introduce the following combinations:

$$Z_t = -\frac{6}{35}X_t + \frac{1}{35}X_b \qquad X_t = -(6Z_t + Z_b)$$

$$Z_b = \frac{1}{35}X_t - \frac{6}{35}X_b \qquad X_b = -(Z_t + 6Z_b). \tag{6.114}$$

Then the following combinations satisfy the simple equations[11]:

$$\widehat{m}_{H_1}^2 \equiv m_{H_1}^2 + 3Z_b, \qquad 8\pi^2 d\widehat{m}_{H_1}^2/dt = \frac{32}{7}g_3^2 M_3^2,$$

$$\widehat{m}_{H_2}^2 \equiv m_{H_2}^2 + 3Z_t, \qquad 8\pi^2 d\widehat{m}_{H_2}^2/dt = \frac{32}{7}g_3^2 M_3^2, \tag{6.116}$$

$$\widehat{m}_Q^2 \equiv m_Q^2 + Z_t + Z_b, \qquad 8\pi^2 d\widehat{m}_Q^2/dt = -\frac{16}{7}g_3^2 M_3^2,$$

$$\widehat{m}_T^2 \equiv m_T^2 + 2Z_t, \qquad 8\pi^2 d\widehat{m}_T^2/dt = -\frac{16}{7}g_3^2 M_3^2,$$

$$\widehat{m}_B^2 \equiv m_B^2 + 2Z_b, \qquad 8\pi^2 d\widehat{m}_B^2/dt = -\frac{16}{7}g_3^2 M_3^2.$$

[11] In the case of universal boundary conditions as in (6.95)–(6.97), we have

$$X_t = X_b = 3m_0^2, \qquad Z_t = Z_b = -\frac{3}{7}m_0^2$$

$$\widehat{m}_{H_1}^2 = \widehat{m}_{H_2}^2 = -\frac{2}{7}m_0^2, \qquad \widehat{m}_Q^2 = \widehat{m}_T^2 = \widehat{m}_B^2 = \frac{1}{7}m_0^2. \tag{6.115}$$

If we have two sets of solutions (1) and (2) to these equations, which differ only by the scalar mass boundary conditions, then defining as above $\Delta Z_{t,b} \equiv Z_{t,b}^{(1)} - Z_{t,b}^{(2)}, \ldots,$ we obtain from (6.113), using the notation (6.104)

$$\frac{d}{dt}\Delta Z_t = 2Y_t\left(6\Delta Z_t + \Delta Z_b\right),$$

$$\frac{d}{dt}\Delta Z_b = 2Y_b\left(\Delta Z_t + 6\Delta Z_b\right), \tag{6.117}$$

whereas the combinations $\Delta\widehat{m}_i^2$ corresponding to (6.116) do not evolve.

The solution to (6.117) reads (see Exercise 3)

$$\Delta Z_t(t) = \Delta Z_t(t_0) + Y_t(t)e^{2\int_0^t dt_1 g_3^2(t_1)/(3\pi^2)}$$

$$\times \left\{ 2\left(6\Delta Z_t(t_0) + \Delta Z_b(t_0)\right) \int_0^t dt_1 e^{-2\int_0^{t_1} dt_2 g_3^2(t_2)/(3\pi^2)} \right. \tag{6.118}$$

$$-20\left(\Delta Z_t(t_0) - \Delta Z_b(t_0)\right) \int_0^t dt_1 Y_b(t_1)e^{-2\int_0^{t_1} dt_2 Y(t_2)}$$

$$\left. \times \int_0^{t_1} dt_2 e^{2\int_0^{t_2} dt_3\left[Y(t_3)-g_3^2(t_3)/(3\pi^2)\right]} \right\},$$

with $Y \equiv Y_b + Y_t$, and similarly for $\Delta Z_b(t)$ with the exchange $b \leftrightarrow t$. The explicit solution for $Y_t(t)$ was given in (6.108). The solution for $Y_b(t)$ may be found in Chapter 9, equation (9.77).

We deduce from (6.116) that the focus scale $t_F = \ln\mu_F$ for $m_{H_2}^2$ satisfies the condition

$$\Delta m_{H_2}^2(\mu_F) = 0 = \Delta\widehat{m}_{H_2}^2(\mu_0) - 3\Delta Z_t(\mu_F), \tag{6.119}$$

where we have used the fact that $\Delta\widehat{m}_{H_2}^2$ is not renormalized. In the case of universal boundary conditions (6.115), an overall factor Δm_0^2 drops out of this equation, which reads using (6.118),

$$Y_t(t_F)e^{2\int_0^{t_F} dt g_3^2(t)/(3\pi^2)} \int_0^{t_F} dt e^{-2\int_0^t dt_1 g_3^2(t_1)/(3\pi^2)} = -\frac{1}{18}. \tag{6.120}$$

We note that this depends only on the value of the top Yukawa coupling at the focus scale $Y_t(t_F)$. It turns out that, for $m_t \sim 174$ GeV and $\tan\beta \sim 5$, this scale is the electroweak weak scale [154]. For larger values of $\tan\beta$, we see from (6.100) that Y_t, or λ_t, is fixed by the top mass. Since the condition (6.100) does not depend on the bottom Yukawa coupling, one deduces that it stills corresponds to a focus at the electroweak scale.

6.10   The issue of fine tuning

We have already stressed that, as the scale of supersymmetry breaking becomes higher (because of the nondiscovery of supersymmetric partners), the parameters of the theory get more fine tuned. The basic relation is equation (6.81) which we rewrite here:

$$\frac{1}{2}M_Z^2 = \frac{1}{\tan^2\beta - 1}m_{H_1}^2 - \frac{\tan^2\beta}{\tan^2\beta - 1}m_{H_2}^2 - |\mu|^2. \tag{6.121}$$

It expresses the scale of electroweak symmetry breaking in terms of the parameters of the low energy theory (say at $\mu = M_z$). In the approach that we adopt in this chapter and the following ones, the relevant parameters are rather the ones at the high energy scale ($\mu_0 = m_P$ or M_U or the string scale). One should therefore express $m^2_{H_{1,2}}$ and $|\mu|^2$ in terms of the high energy parameters. This is what we did in (6.99) for the minimal supergravity model. More generally it takes the following form:

$$M_z^2 = \sum_i C_i \, m_i^2(\mu_0) + \sum_{ij} C_{ij} \, m_i(\mu_0) \, m_j(\mu_0), \qquad (6.122)$$

where m_i represents a generic parameter of the softly broken supersymmetric Lagrangian (scalar mass, gaugino mass, A-term or $|\mu|$). Obviously, the parameters corresponding to the largest coefficients C_i or C_{ij} will have the most impact on the discussion of fine tuning: a small variation of these parameters induces large corrections which must be compensated by other terms in the sum (6.122). Moreover, the nonobservation of supersymmetric partners imposes lower limits on these parameters which are often much higher than M_z: the sum on the right-hand side thus equals M_z at the expense of delicately balancing large numbers one against another.

It turns out that, quite generically, it is the gluino mass M_3 which appears in the sum with the largest coefficient. We are going to show this analytically by using the methods developed in the previous sections. We make the following simplifying assumptions: we neglect in the renormalization group equations terms proportional to g_1^2, g_2^2, λ_b^2 and λ_τ^2; we also set the A-terms to zero. Then from (E.5), (E.17), and (E.18) of Appendix E, $m^2_{H_1}$ is not renormalized whereas we have

$$\frac{1}{\mu}\frac{d\mu}{dt} = 3Y_t, \qquad \frac{dm^2_{H_2}}{dt} = 6Y_t X_t, \qquad (6.123)$$

where Y_t and X_t are defined in (6.104) and (6.112), and $t = \ln \mu/\mu_0$. The solution for Y_t is written in (6.108). The equation of evolution for X_t is given in (6.113). We see that it involves a term in $g_3^2(t)M_3^2(t) = g_3^6(t)M_3^2(0)/g_3^4(0)$ (we have used (6.88)): it is this contribution that induces a term proportional to $M_3^2(0)$. Solving (6.113) for X_t (see Exercise 4), one obtains, in the limit $\lambda_t(0) \gg g_3(0)$:

$$m^2_{H_2}(t) = \frac{1}{2}\left(m^2_{H_2}(0) - m^2_Q(0) - m^2_T(0)\right)$$

$$-M_3^2(0)\frac{391\gamma_3^4 - 1400 + 1009\gamma_3^{-14/9}}{1575\left(1 - \gamma_3^{-14/9}\right)},$$

$$\mu(t) = \mu(0)\left[1 + \frac{36}{7}\frac{8\pi^2 Y_t(0)}{g_3^2(0)}\left(\gamma_3^{14/9} - 1\right)\right]^{-1/4}, \qquad (6.124)$$

where $\gamma_3 \equiv g_3(t)/g_3(0) \sim 1.7$. Replacing these into (6.121), this gives

$$M_z^2 = \frac{1}{\tan^2\beta - 1} 2m_{H_1}^2(0) - \frac{\tan^2\beta}{\tan^2\beta - 1} m_{H_2}^2(0)$$

$$+ \frac{\tan^2\beta}{\tan^2\beta - 1} \left(m_Q^2(0) + m_T^2(0) \right)$$

$$+ \frac{\tan^2\beta}{\tan^2\beta - 1} 5.2 M_3^2(0) - 1.1 \frac{g_3(0)}{\lambda_t(0)} |\mu(0)|^2. \tag{6.125}$$

We see that the fine tuning gets larger for values of $\tan\beta$ close to one (the coefficients C_i behave as $(\tan^2\beta - 1)^{-1}$). On the other hand, the fine tuning becomes independent of $\tan\beta$ for values significantly larger than 1 (but not too large, otherwise the λ_b coupling may not be neglected).

Our approximate result may be compared with complete computations. For example, for $\tan\beta = 2.5$, one obtains [246]

$$M_z^2 = 0.38\ m_{H_1}^2(0) - 1.42\ m_{H_2}^2(0) + 0.96 \left(m_Q^2(0) + m_T^2(0) \right)$$

$$+7.2\ M_3^2(0) - 0.24\ M_2^2(0) + 0.01\ M_1^2(0) - 1.74\ |\mu(0)|^2$$

$$+0.18 A_t(0)^2 - 0.68 A_t(0)M_3(0) - 0.14 A_t(0)M_2(0) - 0.02 A_t(0)M_1(0)$$

$$+0.5\ M_2(0)M_3(0) + 0.06\ M_1(0)M_3(0) + 0.01\ M_1(0)M_2(0). \tag{6.126}$$

The large coefficient of the $M_3^2(0)$ term (and the nondiscovery so far of the gluino, which sets a lower limit for M_3) imposes to envisage a cancellation among large numbers, in order to obtain the right scale M_z. We note on the other hand that neither M_1 nor M_2 play a significant rôle in this discussion about fine tuning, except obviously if they are related to M_3 as in the minimal supergravity model.

Several authors [23, 128] have tried to quantify the amount of fine tuning in the following way: writing $M_z^2 = M_z^2(m_1 \cdots m_n)$ as above, they define

$$\Delta_{\mathrm{max}} = \max_i \left| \frac{m_i}{M_z^2} \frac{\partial M_z^2}{\partial m_i} \right|. \tag{6.127}$$

One obtains, at the tree level, $\Delta_{\mathrm{max}} > 100$ for $\tan\beta = 1.65$ or $\Delta_{\mathrm{max}} > 40$ for $\tan\beta < 4$. But including one-loop corrections [25] stabilizes the model and the fine-tuning is only $\Delta_{\mathrm{max}} > 8$ for $\tan\beta < 4$.

One may note that correlations between the different parameters arising from a given supersymmetry breaking mechanism may decrease the degree of fine tuning. Obviously, a correlation between M_3 and μ would serve this purpose but might be difficult to realize. Otherwise, one may envisage correlations between M_3 and m_{H_2} or between the gaugino masses (which would then imply $M_3 < M_1$ or M_2).

If we return to the Higgs mass, we may note an apparent paradox. As the experimental limits on the lightest Higgs mass are larger than $M_z \cos 2\beta$, one must take into account loop corrections to the mass (all the more as $\tan\beta$ is closer to 1).

These corrections given in equation (5.36) of Chapter 5, depend logarithmically on the stop mass. They may be increased by exponentially increasing the stop mass, *i.e.* by increasing accordingly the values of the soft parameters, in particular $M_3(0)$ [246]. This is obviously at the expense of increasing the amount of fine tuning.

6.11   The μ problem

We have seen in Chapter 5 that the only mass scale of the low energy supersymmetric theory is the μ parameter: all other mass scales are provided by supersymmetry breaking. If the energy scale of the underlying fundamental theory is superheavy (grand unification, string or Planck scale), it is difficult to understand why μ is not of this order. This has been called the μ problem [250].

Of course, a symmetry could ensure that μ vanishes at the large scale. But this has to be broken at low energy, otherwise the low energy theory has a $U(1)$ Peccei–Quinn symmetry [303] which is broken by the vacuum expectation values of H_1 and H_2, leading to an unwanted axion. Moreover, in the simplest supergravity models, the soft parameter B_μ turns out to be proportional to μ and $B_\mu = 0$ implies $\langle H_1^0 \rangle = 0$ or $\langle H_2^0 \rangle = 0$ (see (5.25) of Chapter 5).

It is thus important to find at low energy a dynamical origin for the μ term. One possibility is the one discussed in Section 5.6 of Chapter 5: a cubic term $\lambda S H_2 \cdot H_1$ gives rise to the μ term when the singlet field S acquires a vacuum expectation value at low energy ($\mu = \lambda \langle S \rangle$). The μ term may also be generated by radiative corrections or by nonrenormalizable interactions.

To date, the most elegant solution has been given by Giudice and Masiero [190] who have shown that, in a supersymmetric theory with no dimensional parameters (besides the fundamental scale), the μ parameter may be generated by supersymmetry breaking. Of course, the fundamental problem of how to generate a scale of supersymmetry breaking much smaller than the fundamental scale remains open. We will return to it in Chapter 8.

To illustrate the Giudice–Masiero mechanism, we consider a supergravity model along the lines of Section 6.3.2: the hidden sector consists of N_h chiral superfields with scalar component z^p and the observable sector of N_o chiral superfields with scalar component ϕ^i (and their conjugates). We now make the assumption that there is no mass scale in the observable sector: observable fields have only renormalizable couplings among themselves and gravitational couplings (suppressed by powers of κ) with the hidden sector. Introducing the dimensionless hidden fields $\xi^p \equiv \kappa z^p$, we thus write the superpotential

$$W = \kappa^{-2} W^{(1)}(\xi^p) + \kappa^{-1} W^{(2)}(\xi^p) + W^{(3)}(\xi^p, \phi^i), \tag{6.128}$$

$$W^{(3)}(\xi^p, \phi^i) = \frac{1}{3!} \sum_{ijk} \lambda_{ijk}(\xi^p) \phi^i \phi^j \phi^k,$$

and the Kähler potential

$$K = \kappa^{-2} K^{(0)}(\xi^p, \bar{\xi}^{\bar p}) + \kappa^{-1} K^{(1)}(\xi^p, \bar{\xi}^{\bar p}) + K^{(2)}(\xi^p, \bar{\xi}^{\bar p}, \phi^i, \bar{\phi}^{\bar i}), \tag{6.129}$$

$$K^{(2)}(\xi^p, \bar{\xi}^{\bar p}, \phi^i, \bar{\phi}^{\bar i}) = \sum_i \phi^i \bar{\phi}^{\bar i} + \frac{1}{2} \sum_{ij} \left[\alpha_{ij}(\xi^p, \bar{\xi}^{\bar p}) \phi^i \phi^j + \text{h.c.} \right],$$

where the index in parenthesis gives the mass dimension of the function (for simplicity, we have assumed a flat Kähler metric for the observable fields). As in Section 6.3.2, the hidden sector involves mass scales (M_{SB}) such that, in the low energy limit $\kappa \to 0$, the gravitino mass

$$m_{3/2} = e^{K^{(0)}/2} W^{(1)} \tag{6.130}$$

is a low energy scale.

Precisely, in the low energy limit $\kappa \to 0$, the theory in the observable sector is described by a supersymmetric Lagrangian and soft supersymmetry breaking terms whose magnitude is fixed by the scale $m_{3/2}$. More precisely, the effective superpotential $\hat{W}$ of the low energy effective theory reads

$$\hat{W}(\phi^i) = \hat{W}_3 + \left[1 - \sum_{\bar{q}} Y^{\bar{q}} \frac{\partial}{\partial \bar{\xi}^{\bar{q}}} \right] m_{3/2} \hat{W}_2,$$

$$\hat{W}_3 = W^{(3)} e^{K^{(0)}/2}, \quad \hat{W}_2 = \frac{1}{2} \sum_{i,j} \alpha_{ij}(\xi^p, \bar{\xi}^{\bar{p}}) \, \phi^i \phi^j, \tag{6.131}$$

$$Y^{\bar{q}} = g^{(0)\bar{q}p} \left(\frac{1}{W^{(1)}} \frac{\partial W^{(1)}}{\partial \xi^p} + \frac{\partial K^{(0)}}{\partial \xi^p} \right),$$

where $g^{(0)\,\bar{q}p}$ is the inverse Kähler metric associated with the Kähler potential $K^{(0)}$. We note that it is holomorphic in the low energy dynamical fields ϕ^i but it depends on the vacuum expectation values of the hidden sector (superheavy) fields ξ^p and $\bar{\xi}^{\bar{p}}$. We see that the terms quadratic in the observable fields in the Kähler potential $K^{(2)}(\xi^p, \bar{\xi}^{\bar{p}}, \phi^i, \bar{\phi}^{\bar{i}})$ induce quadratic terms ($\hat{W}^2$) in the effective superpotential with scale fixed by the supersymmetry breaking scale $m_{3/2}$. One of them may be the μ term.

The low energy scalar interactions are described by the Lagrangian (see Exercise 5)

$$\mathcal{L}_{\text{eff}} = \sum_i \left| \frac{\partial \hat{W}}{\partial \phi^i} \right|^2 + m_{3/2}^2 \sum_i |\phi^i|^2$$

$$+ m_{3/2} \left\{ \sum_i \phi^i \frac{\partial \hat{W}}{\partial \phi^i} + \left[Y^p \left(\frac{\partial K^{(0)}}{\partial \xi^p} + \frac{1}{W^{(3)}} \frac{\partial W^{(3)}}{\partial \xi^p} \right) - 3 \right] \hat{W}_3 \right.$$

$$\left. + \left(Y^p \frac{\partial}{\partial \xi^p} + Y^{\bar{p}} \frac{\partial}{\partial \bar{\xi}^{\bar{p}}} - Y^p Y^{\bar{q}} \frac{\partial^2}{\partial \xi^p \partial \bar{\xi}^{\bar{q}}} \right) m_{3/2} \hat{W}_2 + \text{h.c.} \right\}. \tag{6.132}$$

We see that soft supersymmetry breaking terms cubic (A-terms) and quadratic (like the B_μ term) are naturally generated. The scale is fixed, as in the simpler example of (6.63), by the gravitino mass but this time the coefficients may take very diverse values, depending on the specific form of the hidden sector.

6.12   No-scale models

In the context of electroweak symmetry radiative breaking, we have seen that the electroweak symmetry breaking scale comes out naturally small compared to the Planck scale. This however does not determine the gravitino mass which fixes the masses of the supersymmetric partners. Indeed, only dimensionless quantities such as mass ratios (*e.g.* gaugino mass over $m_{3/2}$) or A-terms play a rôle in the determination of the electroweak scale. However, if we do not want to destabilize the electroweak vacuum through radiative corrections, the overall supersymmetric mass should be a low energy scale. It is thus desirable to look for a determination of the scale $m_{3/2}$ through low energy physics.

Obviously, in this case, the gravitino mass should remain undetermined by the hidden sector: in other words, it should correspond to a flat direction of the scalar potential. Let us get some orientation with a hidden sector reduced to a single field z [92]. For the time being, we ignore low energy observable fields. The scalar potential may be written from (6.37)

$$V(z, \bar{z}) = 9\ e^{4\mathcal{G}/3}\mathcal{G}_{z\bar{z}}^{-1}\ \partial_z\partial_{\bar{z}}e^{-\mathcal{G}/3}. \tag{6.133}$$

It vanishes for $\partial_z\partial_{\bar{z}}e^{-\mathcal{G}/3} = 0$, *i.e.*

$$\mathcal{G}(z, \bar{z}) = K(z, \bar{z}) = -3\ln\left(f(z) + \bar{f}(\bar{z})\right). \tag{6.134}$$

One notes that the corresponding curvature, defined in (6.51), is proportional to the Kähler metric, *i.e.* $R_{z\bar{z}} = -\frac{2}{3}\mathcal{G}_{z\bar{z}}$. This property defines an Einstein–Kähler manifold.

One may redefine the coordinates on the Kähler manifold by performing the replacement $f(z) \to z$. Then,

$$K(z, \bar{z}) = -3\ln\left(z + \bar{z}\right). \tag{6.135}$$

The corresponding Lagrangian is restricted to a nontrivial kinetic term:

$$\mathcal{L} = 3\frac{1}{(z + \bar{z})^2}\partial^\mu z\partial_\mu\bar{z}. \tag{6.136}$$

In order to discuss the symmetries associated with such a Lagrangian, we define $z = (y + 1)/(y - 1)$ such that $\mathcal{L} = 3\ \partial^\mu y\partial_\mu\bar{y}/\left(1 - |y|^2\right)^2$, which is invariant under:

$$y \to \frac{ay + b}{\bar{b}y + \bar{a}}, \quad |a|^2 - |b|^2 = 1. \tag{6.137}$$

This defines the noncompact group $SU(1, 1)$ [72, 92, 131, 132][12]

In terms of the variable z, this $SU(1, 1)$ symmetry reads

$$z \to \frac{\alpha z + i\beta}{i\gamma z + \delta}, \quad \alpha\delta + \beta\gamma = 1, \ \alpha, \beta, \gamma, \delta \in \mathbb{R}. \tag{6.138}$$

---

[12]A representation of this group is provided by the matrices $\begin{pmatrix} a & b \\ \bar{b} & \bar{a} \end{pmatrix}$ which, acting on vectors $\begin{pmatrix} y_1 \\ y_2 \end{pmatrix}$, leave invariant the quadratic form $|y_1|^2 - |y_2|^2$ (to be compared with $|y_1|^2 + |y_2|^2$ for the compact group $SO(2)$).

This includes:

(i) imaginary translations, forming a noncompact $\tilde{U}(1)_a$ group: $z \to z + i\alpha$;

(ii) dilatations: $z \to \beta^2 z$;

(iii) conformal transformations: $z \to (z + i\tan\theta)/(i\tan\theta + 1)$.

The complete supergravity Lagrangian is invariant under this symmetry, except the gravitino mass term (6.45):

$$\mathcal{L} = \frac{i}{2} e^{\mathcal{G}/2} \bar{\psi}_\mu \sigma^{\mu\nu} \psi_\nu + \text{h.c.} \tag{6.139}$$

This is invariant under $\tilde{U}_a(1)$ but not under dilatations since it has dimension $d = 3$ (see Section A.5.1 of Appendix Appendix A), nor on conformal transformations (under which the gravitino and the Goldstino transform differently). This term thus breaks $SU(1,1)$ down to the noncompact $\tilde{U}_a(1)$.

Because, by construction, the tree level scalar potential is flat, we must rely on quantum fluctuations to determine $m^2_{3/2} = e^{\mathcal{G}}$: since supersymmetry is broken, these radiative corrections should lift the flat direction. If only light fields (with a mass of order M_W or μ_{EW} as defined in (6.85)) contribute, then one expects $m_{3/2}$ to be of the same order. On the other hand, if superheavy supermultiplets of mass of order M_U (with mass splittings of order $m_{3/2}$) contribute, then one expects a vacuum energy $(M_U^2 + m^2_{3/2})^2 - M_U^4$ of order $m^2_{3/2} M_U^2$. This would destabilize the gravitino mass towards M_U.

We thus need to find models where superheavy matter supermultiplets do not feel supersymmetry breaking. We construct in what follows such models [130]. Let us consider for this purpose the following generalized Kähler potential

$$\mathcal{G} = -3\ln\left[z + \bar{z} + h\left(\phi^i, \bar{\phi}^{\bar{i}}\right)\right], \tag{6.140}$$

where h is a real function of n chiral supermultiplets ($i = 1, \dots, n$). We note ϕ^I, $I = 0, \dots, n$ the scalar fields $\phi^0 \equiv z$ and $\phi^{I=i} \equiv \phi^i$. The scalar potential then reads, from (6.37), setting $\kappa = 1$,

$$V = e^{\mathcal{G}} \left[\mathcal{G}_I \mathcal{G}^{I\bar{J}} \mathcal{G}_{\bar{J}} - 3\right] + \frac{1}{2} \left(\text{Re}f\right)_{ab} D^a D^b. \tag{6.141}$$

From the explicit form of the Kähler potential, we deduce that $\mathcal{G}_i = \mathcal{G}_0 h_i$, $\mathcal{G}_{\bar{i}} = \mathcal{G}_0 h_{\bar{i}}$ and

$$\mathcal{G}^{I\bar{J}} = \frac{1}{\mathcal{G}_0} \begin{pmatrix} \frac{3}{\mathcal{G}_0} + h_k h^{k\bar{l}} h_{\bar{l}} & -h_k h^{k\bar{j}} \\ -h^{i\bar{k}} h_{\bar{k}} & h^{i\bar{j}} \end{pmatrix}, \tag{6.142}$$

where $h^{i\bar{j}} h_{\bar{j}k} = \delta^i_k$. One thus obtains

$$\mathcal{G}_I \mathcal{G}^{I\bar{J}} = \frac{3}{\mathcal{G}_0} \delta^{\bar{J}}_0 \;,\; \mathcal{G}^{I\bar{J}} \mathcal{G}_{\bar{J}} = \frac{3}{\mathcal{G}_0} \delta^I_0, \tag{6.143}$$

$$\mathcal{G}_I \mathcal{G}^{I\bar{J}} \mathcal{G}_{\bar{J}} = 3. \tag{6.144}$$

Hence the scalar potential (6.141) is reduced to its D-terms and is flat along the D-flat directions.

One may check that there is no mass splitting between the bosonic and fermionic components of the chiral supermultiplets. In order to see this, one may compute the supertrace (6.51), which reads, in this case where we have $n+1$ chiral supermultiplets [209],

$$\mathcal{S}\mathrm{Tr}M^2 = 2\, e^{\mathcal{G}}[n + \mathcal{G}_I \mathcal{G}^{I\bar{J}} R_{J\bar{K}} \mathcal{G}^{K\bar{L}} \mathcal{G}_{\bar{L}}], \quad R_{J\bar{K}} \equiv -\frac{\partial}{\partial\phi^J}\frac{\partial}{\partial\bar\phi^K}\log\det\mathcal{G}_{I\bar{J}}. \tag{6.145}$$

Using the explicit form (6.140) of the Kähler potential, one obtains (see Exercise 6) $\mathcal{S}\mathrm{Tr}M^2 = -4e^{\mathcal{G}} = -4m^2_{3/2}$. In other words, $\mathrm{STr}M^2 \equiv \mathrm{Tr}M_0^2 - 2\mathrm{Tr}M_{1/2}^2 = 0$ for this theory, which confirms the absence of splitting between the supersymmetric partners of the chiral supermultiplets.

Of course, this situation is not completely satisfactory since we wish to see some sign of supersymmetry breaking at low energy, in order to fix the gravitino mass through radiative corrections. This is done by giving a mass to gauginos through supersymmetry breaking: this communicates a nonzero mass to scalars at low energy through the renormalization group evolution.

In the case where $h(\phi^i, \bar\phi^{\bar{i}}) = \phi^i\bar\phi^{\bar{i}}$, one may easily identify the noncompact symmetry that is responsible for the F-flat direction [130]. Indeed, writing

$$z = \frac{1}{2}\frac{1 - y^0}{1 + y^0}, \quad \phi^i = \frac{y^i}{1 + y^0}, \tag{6.146}$$

the Kähler potential reads (6.140) reads

$$\mathcal{G} = -3\ln\left(1 - y^I\bar{y}^I\right) + 3\ln\left(1 + y^0\right) + 3\ln\left(1 + \bar{y}^0\right), \tag{6.147}$$

where the last two terms can be cancelled by a Kähler transformation. The corresponding Kähler metric shows that the scalar fields y^I, $I = 0, \ldots, n$ parametrize the coset space $SU(n,1)/SU(n) \times U(1)$.

Further reading

- P. van Nieuwenhuizen, *Supergravity*, Physics Reports **68** (1981) 189.
- P. Binétruy, G. Girardi and R. Grimm, *Supergravity couplings: a geometric formulation*, Physics Reports **343** (2001) 255.

Exercises

<u>Exercise 1</u> Compute the number of on-shell degrees of freedom for a massless gravitino and a massless graviton field in a D-dimensional spacetime (D even), following the counting performed in Section 6.4 for $D = 4$.

Hints: In a D-dimensional spacetime, a Majorana spinor has $2^{D/2} \times (1/2)$ degrees of freedom; a massless vector has $(D-2)$ degrees of freedom; hence for the gravitino $(D-2)2^{D/2}/2$ minus $2^{D/2}/2$ since the gauge condition $\gamma.\psi = 0$ suppresses the spin 1/2 component, *i.e.* $(D-3)2^{D/2}/2$.

For the graviton, the symmetric and traceless $h_{\mu\nu}$ has $D(D+1)/2-1$ components; the gauge conditions fix the D components $\partial^\mu h_{\mu\nu}$; the residual gauge symmetry takes care of another $D-1$ components. Hence $D(D+1)/2-1-D-(D-1) = D(D-3)/2$.

Take $D = 10$: the gravitino has 112 degrees and the graviton 35. The mismatch indicates that the gravity supermultiplet must include other dynamical fields (see Chapter 10).

<u>Exercise 2</u> One consider a single (gauge singlet) chiral superfield z coupled to supergravity, with generalized Kähler potential $\mathcal{G}$ as in (6.23) ($\kappa = 1$).

(a) Writing its scalar component $z = (A + iB)/\sqrt{2}$, show that

$$\mathrm{Tr}\, M_0^2 \equiv m_A^2 + m_B^2 = 2\mathcal{G}_{z\bar{z}}^{-1} \left.\frac{\partial^2 V}{\partial z \partial \bar{z}}\right|_{\min}$$

where V is the scalar potential.

(b) Using the explicit form (6.37) for the scalar potential and the minimization condition, show that

$$\mathrm{Tr}\, M_0^2 = 2e^{\mathcal{G}} \left.\frac{\mathcal{G}_z \mathcal{G}_{\bar{z}}}{\mathcal{G}_{z\bar{z}}} \frac{R_{z\bar{z}}}{\mathcal{G}_{z\bar{z}}}\right|_{\min} + 4e^{\mathcal{G}}\Big|_{\min}.$$

(c) Deduce the formula (6.51).

Hints: (c) Because the Goldstino does not appear in the physical spectrum, the supertrace is simply $\mathrm{Tr}\, M_0^2 - 4m_{3/2}^2$.

<u>Exercise 3</u> We wish to show that the solution to the system (6.117) is given by (6.118).

(a) Introduce the functions $f_{b,t}(t)$ defined by

$$\Delta Z_{b,t}(t) = \Delta Z_{b,t}(t_0) + Y_{b,t}(t) f_{t,b}(t).$$

Deduce from (6.117) the differential equations satisfied by $f_t(t)$ and $f_b(t)$.

(b) Solve for $f_t(t) - f_b(t)$ and then for $f_t(t)$ and $f_b(t)$ separately.

Hints: (b) $(f_t - f_b)(t) =$

$$10\left(\Delta Z_t(t_0) - \Delta Z_b(t_0)\right) e^{-2\int_0^t dt_1 \left[Y(t_1) - g_3^2(t_1)/(3\pi^2)\right]} \int_0^t dt_1\, e^{2\int_0^{t_1} dt_2 \left[Y(t_2) - g_3^2(t_2)/(3\pi^2)\right]}.$$

<u>Exercise 4</u> Using the methods of Section 6.9, prove equation (6.124).

Hints: start by solving (6.113) for Z_b and Z_t. For $Y_t(0) \gg g_3^2(0)/(16\pi^2)$,

$$Z_b(t) = Z_b(0) - \frac{16}{63} M_3^2(0) \left(\gamma_3^4 - 1\right),$$

$$Z_t(t) = -\frac{1}{6} Z_b(0) - \frac{1}{756} M_3^2(0) \left[\frac{3236(\gamma_3^{50/9} - 1)}{25(\gamma_3^{14/9} - 1)} + 32 - 192\frac{(\gamma_3^4 - 1)}{(\gamma_3^{14/9} - 1)}\right].$$

Then extract $m_{H_2}^2$ from $\widehat{m}_{H_2}^2$ in (6.116).

<u>Exercise 5</u> In the model described by the superpotential (6.128) and the Kähler potential (6.129), show that, in the limit $\kappa \to 0$, the effective scalar interactions are described by the Lagrangian (6.132).

Hints: Start from the supergravity scalar potential (6.36). Writing the full Kähler metric

$$G = \begin{pmatrix} g_{p\bar{q}} & \kappa X_{p\bar{\jmath}} \\ \kappa X_{i\bar{q}} & \delta_{i\bar{\jmath}} \end{pmatrix}, \quad X_{p\bar{\jmath}} = \sum_{\bar{\imath}} \frac{\partial \bar{\alpha}_{\bar{\imath}\bar{\jmath}}}{\partial \xi^p} \bar{\phi}^{\bar{\imath}}, \quad X_{i\bar{q}} = \sum_{j} \frac{\partial \alpha_{ij}}{\partial \bar{\xi}^{\bar{q}}} \phi^j.$$

where $g_{p\bar{q}} = g^{(0)}_{p\bar{q}} + \kappa g^{(1)}_{p\bar{q}} + \kappa^2 g^{(2)}_{p\bar{q}}$ ($g^{(i)}_{p\bar{q}}$ is the metric associated with the Kähler potential $K^{(i)}$). The inverse metric then reads:

$$G^{-1} = \begin{pmatrix} g^{\bar{q}r} + \kappa^2 \left(g^{\bar{q}p} X_{p\bar{\jmath}} \delta^{\bar{\jmath}i} X_{i\bar{s}} g^{\bar{s}r} + O(\kappa^2) \right) & \kappa \left(-g^{\bar{q}p} X_{p\bar{\jmath}} \delta^{\bar{\jmath}l} + O(\kappa^2) \right) \\ \kappa \left(-\delta^{\bar{\jmath}i} X_{i\bar{q}} g^{\bar{q}r} + O(\kappa^2) \right) & \delta^{\bar{\jmath}l} + \kappa^2 \left(\delta^{\bar{\jmath}i} X_{i\bar{q}} g^{\bar{q}p} X_{p\bar{k}} \delta^{\bar{k}l} + O(\kappa^2) \right) \end{pmatrix},$$

where $g^{\bar{q}p}$ is the inverse metric of $g_{p\bar{q}}$:

$$g^{\bar{q}p} = g^{(0)\bar{q}p} - \kappa g^{(0)\bar{q}r} g^{(1)}_{r\bar{s}} g^{(0)\bar{s}p} - \kappa^2 \left(g^{(0)\bar{q}r} g^{(2)}_{r\bar{s}} g^{(0)\bar{s}p} - g^{(0)\bar{q}r} g^{(1)}_{r\bar{s}} g^{(0)\bar{s}t} g^{(1)}_{t\bar{u}} g^{(0)\bar{u}p} \right).$$

<u>Exercise 6</u> For the no-scale model described by the Kähler potential (6.140),
(a) Show that

$$\det \mathcal{G}_{I\bar{J}} = \tfrac{1}{3} \left(\mathcal{G}_0 \right)^{n+2} \det \left[h_{i\bar{\jmath}} \right]. \tag{6.148}$$

and thus

$$R_{I\bar{J}} = -\frac{\partial}{\partial \phi^I} \frac{\partial}{\partial \bar{\phi}^{\bar{J}}} \log \det \left[h_{k\bar{l}} \right] - \frac{1}{3}(n+2)\mathcal{G}_{I\bar{J}}. \tag{6.149}$$

(b) Deduce that $\mathcal{S}TrM^2 = -4e^{\mathcal{G}}$.

Hints: (b) Use (6.143) to show that the first term in (6.149) does not contribute to $\mathcal{S}TrM^2$.

7

Phenomenology of supersymmetric models: supersymmetry at the quantum level

Since the Standard Model has been tested to a level where quantum corrections play an important rôle, it is expected that any theory beyond it will have to be tested at the same level. From this point of view, the odds are in favor of supersymmetric theories since supersymmetry has been constructed from the start to control quantum corrections. Since this chapter deals with the confrontation of supersymmetry with the real world, we first discuss in some detail this important issue of stability under quantum corrections.

Obviously, supersymmetry is broken. The way supersymmetry is broken should have some decisive impact on the search of supersymmetric particles. We thus review the main scenarios of supersymmetry breaking, mostly from a phenomenological point of view. More theoretically oriented issues are postponed till Chapter 8.

We conclude this chapter by a concise review of the search for supersymmetric particles. Since this is constantly evolving with the accumulation of new data and the progress on collider energies, I have taken the conservative approach to give the limits obtained at the LEP collider, except otherwise stated. The reader is referred to the proceedings of recent summer conferences for updated values.

7.1 Why does the MSSM survive the electroweak precision tests?

In view of the many successes of the Standard Model under precision tests (see Section A.4 of Appendix Appendix A), it might come as a surprise to realize that the MSSM is also surviving these tests. Indeed, most of the theories beyond the Standard Model fail to do so because the extra heavy fields that one introduces induce undesirable radiative corrections. Supersymmetry has been introduced in order to better control these radiative corrections. It is thus not completely surprising that it fares better in this respect. It actually does so much better than other theories that it is often stated that the more standard the theory of fundamental interactions seems to be, the more likely it is to be supersymmetric (spontaneously broken). Although we will see in Chapter 12 that such a statement has its limits, this is the reason why supersymmetry is considered as the most serious candidate for a theory beyond the Standard Model.

One of the predictions of the Standard Model that is most difficult to reproduce is the fact that the ρ parameter defined as

$$\rho = \frac{M_W^2}{M_Z^2 \cos^2 \theta_W},\tag{7.1}$$

has a value very close to 1. Indeed, this value of 1 is predicted at tree level (see equation (A.136) of Appendix Appendix A) and receives only corrections of a few per cent from radiative corrections (see Section A.4 of Appendix Appendix A). To discuss departures
from the tree level value $\rho = 1$ it proves to be useful to use a global symmetry known as the *custodial $SU(2)$ symmetry*. This concept was developed [341] in the context of technicolor theories for which this test proved to be very damaging. It can be used to discuss radiative corrections to the ρ parameter in the framework of the Standard Model as well as of any theory beyond the Standard Model.

This custodial $SU(2)$ symmetry is a *global* symmetry under which the gauge fields A_μ^a, $a = 1, 2, 3$ of the $SU(2)$ symmetry of the Standard Model transform as a triplet: as we will see in the next Section, the presence of such a symmetry ensures that the parameter ρ is 1. Deviations from 1 are associated with the breaking of this symmetry: for example, (t, b) is a doublet of custodial $SU(2)$, and the fact that the top mass is much larger than the bottom mass is a major source of breaking. The precision tests from the LEP collider have taught us that the breaking of custodial $SU(2)$ in the Standard Model is in good agreement with the one observed in the data. We develop these considerations in the next two sections.

7.1.1 Custodial symmetry

We first prove the following statement:

In any theory of electroweak interactions which conserves charge and a global $SU(2)$ symmetry under which the gauge fields A_μ^a, $a = 1, 2, 3$ of the $SU(2)$ symmetry of the Standard Model transform as a triplet, one has $\rho = 1$.

A mass term for the gauge bosons A_μ^a necessarily transforms under this custodial $SU(2)$ as[1] $(\mathbf{3} \times \mathbf{3})_s = \mathbf{1} + \mathbf{5}$. In order not to break the global $SU(2)$ the mass term must be a $SU(2)$ singlet and the corresponding mass matrix is simply of the form $M^2 \delta_{ab}$.

We then add the $U(1)_Y$ gauge field B_μ. Since charge is conserved, there is no mixed term of the form $A_\mu^a B^\mu$, $a = 1, 2$, in the mass matrix (*cf.* (A.114) of Appendix Appendix A). Thus the complete mass matrix for the electroweak gauge bosons in the basis $A_\mu^1, A_\mu^2, A_\mu^3, B_\mu$ is of the form:

$$\begin{pmatrix} M^2 & 0 & 0 & 0 \\ 0 & M^2 & 0 & 0 \\ 0 & 0 & M^2 & m'^2 \\ 0 & 0 & m'^2 & m^2 \end{pmatrix}.\tag{7.2}$$

---

[1] If we consider two triplets ϕ_1^a and ϕ_2^a of $SU(2)$, then $\sum_a \phi_1^a \phi_2^a \sim \mathbf{1}$, $\phi_1^a \phi_2^b + \phi_1^b \phi_2^a - \sum_a \phi_1^a \phi_2^a \sim \mathbf{5}$ and $\phi_1^a \phi_2^b - \phi_1^b \phi_2^a \sim \mathbf{3}$. Thus $(\mathbf{3} \times \mathbf{3})_s = \mathbf{1} + \mathbf{5}$ and $(\mathbf{3} \times \mathbf{3})_a = \mathbf{3}$.

Clearly, $M^2 = M_W^2$ the mass of the charged gauge bosons $W_\mu^\pm = (A_\mu^1 \mp A_\mu^2)/\sqrt{2}$. Since charge is assumed to be conserved, we expect one massless neutral field, the photon, the other neutral field being Z_μ^0 of mass M_Z. Thus the 2×2 neutral gauge boson mass matrix M_{neutral}^2 has vanishing determinant and a trace equal to M_Z^2. This allows us to express m^2 and m'^2 in terms of M_Z and $M = M_W$. One obtains

$$\begin{pmatrix} M_W^2 & \pm M_W \sqrt{M_Z^2 - M_W^2} \\ \pm M_W \sqrt{M_Z^2 - M_W^2} & M_Z^2 - M_W^2 \end{pmatrix}. \tag{7.3}$$

The eigenvector of vanishing mass is the photon field

$$A_\mu = \frac{1}{M_Z} \sqrt{M_Z^2 - M_W^2}\, A_\mu^3 \mp \frac{M_W}{M_Z} B_\mu. \tag{7.4}$$

Comparison with equation (A.135) of Appendix Appendix A gives

$$\cos^2 \theta_W = \frac{M_W^2}{M_Z^2} \quad \to \quad \rho = 1. \tag{7.5}$$

This proves the statement above.

This has many interesting applications:

(i) Strong interactions conserve electric charge and strong isospin. We may thus choose strong isospin as our global $SU(2)$. For it to be a symmetry in the presence of the gauge fields A_μ^a, we must transform these fields under this *global* symmetry. We conclude that the relation $\rho = 1$ remains true to all orders of strong interaction. On the other hand, weak interactions violate strong isospin and we expect violations of order $g^2 \sim G_F M_W^2$.

(ii) Let us consider the Higgs potential ((A.127) of Appendix Appendix A):

$$V\left(\Phi^\dagger \Phi\right) = -m^2 \Phi^\dagger \Phi + \lambda \left(\Phi^\dagger \Phi\right)^2. \tag{7.6}$$

We may write

$$\Phi \equiv \begin{pmatrix} \phi_2 + i\phi_1 \\ \phi_0 - i\phi_3 \end{pmatrix} \tag{7.7}$$

$$\Phi^\dagger \Phi = \phi_0^2 + \phi_1^2 + \phi_2^2 + \phi_3^2 = \vec{\phi} \cdot \vec{\phi} \quad \text{with} \quad \vec{\phi} = \begin{pmatrix} \phi_0 \\ \phi_1 \\ \phi_2 \\ \phi_3 \end{pmatrix}$$

which shows that V is invariant under $O(4) \sim SU(2) \times SU(2)$. This symmetry is spontaneously broken into $O(3) \sim SU(2)$ by $\langle \phi_a \rangle = \delta_{a0}\, v/\sqrt{2}$. Thus $\rho = 1$ to all orders of scalar field interactions. This remains true for any number of Higgs doublets (since one can redefine them in such a way that a single one acquires a vacuum expectation value). On the other hand, in presence of a triplet of Higgs, $O(3)$ is broken down to $O(2)$, which is insufficient as a custodial symmetry: $\rho \neq 1$.

(iii) We next turn to Yukawa couplings. We will consider only the third generation since it corresponds to the largest Yukawa couplings and thus to the largest source of violation of custodial symmetry. We have, following equation (A.139) of Appendix Appendix A,

$$\mathcal{L}_Y = \lambda_b \bar{\psi}_t \Phi b_R + \lambda_t \bar{\psi}_t \widetilde{\Phi} t_R + \text{ h.c.} \tag{7.8}$$

where $\psi_t = \begin{pmatrix} t_L \\ b_L \end{pmatrix}$. Writing Φ as in (7.7), in the case where $\lambda_b = \lambda_t = \lambda$, we may put (7.8) in the form

$$\mathcal{L}_Y = \lambda \left(\bar{t}_L \ \bar{b}_L \right) \left(\phi_0 + i\sigma^a \phi_a \right) \begin{pmatrix} t_R \\ b_R \end{pmatrix} + \text{ h.c.} \tag{7.9}$$

which is explicitly invariant under $SU(2) \times SU(2)$ (as in the sigma model, see equation (A.208) of Appendix Appendix A). Spontaneous breaking leaves intact the custodial $SU(2)$. Thus, in the limit $\lambda_t = \lambda_b$, we have $\rho = 1$ to all orders in the Yukawa couplings.

However, because the top quark is heavy, its Yukawa coupling λ_t is expected to be larger than λ_b, which leads to violations of $SU(2)$ custodial symmetry of order[2] $\lambda_t^2 - \lambda_b^2 \sim G_F(m_t^2 - m_b^2)$, where we have used $v^{-2} \sim G_F$ (*cf.* (A.151) of Appendix Appendix A).

To recapitulate, we expect that, in the Standard Model, the main violations of custodial $SU(2)$ symmetry are of order:

$$\rho - 1 = O\left(G_F M_W^2\right) + O\left(G_F(m_t^2 - m_b^2)\right). \tag{7.10}$$

This can be compared with the result of an exact calculation, as given in equation (A.205) of Appendix Appendix A, where we have neglected the bottom mass.

7.1.2   Electroweak precision tests in supersymmetric theories

If one considers radiative corrections to the propagators (oblique corrections), the most significant contributions come from the third generation, through the left-handed squark mass difference $m_{\tilde{t}_L}^2 - m_{\tilde{b}_L}^2 = m_t^2 - m_b^2 + M_W^2 \cos 2\beta$ and the trilinear A_t term and they are usually down by some powers of the squark masses (thus decoupling for large squark masses).

Thus if the squarks of the third generation are not too large, one may obtain predictions for $\sin^2 \theta_W$ or M_W which are sensibly different from their Standard Model values. Hence the interest of obtaining more precision on the W mass in order to start discriminating between the Standard Model and the MSSM. However if the squarks of the third generation are heavy enough (the squarks of the first two generations and the sleptons being within their experimental limits), the MSSM is difficult to distinguish from the Standard Model.

---

[2]The violations are quadratic in the Yukawa couplings because in the gauge field propagators, such as Fig. A.8 of Appendix Appendix A, the fermions appear in loops, which involve two Yukawa couplings.

To conclude, the MSSM with heavy enough squarks of the third generation, chargino and stop masses beyond the experimental limit and heavy enough pseudoscalar A^0 (that is heavy enough charged Higgs) looks very similar to the Standard Model in electroweak precision tests.

7.1.3 $b \to s\gamma$

The rare decay $b \to s\gamma$ is a good test of the FCNC structure of the Standard Model: it vanishes at tree level and appears only at one loop; see Fig. 7.1, diagram (a). It is of order $G_F^2 \alpha$ whereas most of the other FCNC processes involving leptons or photons are of order $G_F^2 \alpha^2$. It thus provides as well a key test of theories beyond the Standard Model.

The corresponding inclusive decay $B \to X_s\gamma$ has now been measured precisely: its branching ratio is found to be $\mathcal{B}[B \to X_s\gamma] = (3.55 \pm 0.32) \times 10^{-4}$ (Belle Collaboration, [261]) or $(3.27 \pm 0.18) \times 10^{-4}$ (BABAR Collaboration, [16]). The Standard Model prediction, including next to leading order QCD contributions, electroweak and power corrections is $\mathcal{B}[B \to X_s\gamma]_{\text{SM}} = (3.70 \pm 0.30) \times 10^{-4}$. This does not leave much room for new physics.

In fact, supersymmetry prevents any magnetic operator, *i.e.* any operator of the form $\bar{\Psi}\sigma_{\mu\nu}\Psi F^{\mu\nu}$ [159]. For example, in the case of supersymmetric QED [363], the one-loop contribution present in the Standard Model (see Fig. 7.2 a) is cancelled by new graphs involving sfermions and gauginos (see Fig. 7.2b,c and Exercise 1). Thus, in the supersymmetric limit $\mathcal{B}[B \to X_s\gamma]_{\text{SUSY}} = 0$. In the more realistic case of broken supersymmetry, this leaves open the possibility of partial cancellations.

In a general supersymmetric model, the one-loop contributions to the $b \to s$ transition can be distinguished according to the type of particles running in the loop (see Fig. 7.1) [32]:

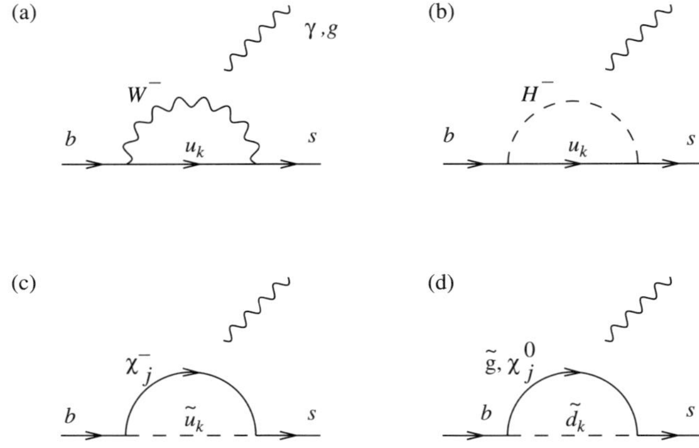

Fig. 7.1 Penguin diagrams inducing a $(b \to s)$ transition (a) in the Standard Model and (a)–(d) in supersymmetric models (the vector line, which represents a photon or a gluon, is to be attached in all possible ways).

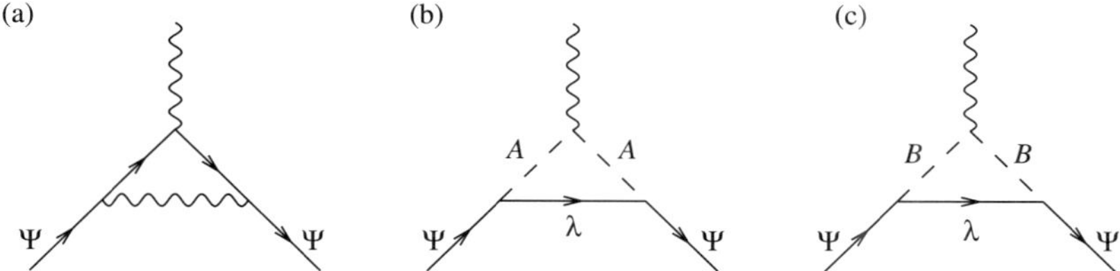

Fig. 7.2 Graphs contributing to the magnetic form factor in the case of supersymmetric QED.

(a) W^- and up-type quark;
(b) charged Higgs H^- and up-type quark;
(c) chargino χ^- and up-type squark;
(d) gluino $\tilde{g}$ (or neutralino χ^0) and down-type squark.

Before reviewing these different contributions, we provide the reader with a little background on the computation of the weak radiative B-meson decay at the leading order [60][3]. The starting point is the low energy effective Hamiltonian

$$\mathcal{H}_{\text{eff}} = -\frac{4G_F}{\sqrt{2}} V_{ts}^* V_{tb} \sum_{i=1}^{8} C_i(\mu) P_i(\mu), \tag{7.11}$$

where V_{ij} are the elements of the CKM matrix, $P_i(\mu)$ are the relevant physical operators and $C_i(\mu)$ the corresponding Wilson coefficients. We are particularly interested in the magnetic and chromomagnetic operators

$$P_7 = \frac{e}{16\pi^2} m_b \left(\bar{s}_L \sigma^{\mu\nu} b_R \right) F_{\mu\nu},$$

$$P_8 = \frac{g_3}{16\pi^2} m_b \left(\bar{s}_L \sigma^{\mu\nu} t^a b_R \right) G_{\mu\nu}^a, \tag{7.12}$$

where $G_{\mu\nu}^a$ is a gluon field and t^a the corresponding $SU(3)$ generator. The other $P_i, i = 1, \ldots, 6$, are four-fermion operators.

It proves useful [60] to turn the Wilson coefficients $C_7(\mu)$ and $C_8(\mu)$ into renormalization scheme independent coefficients $C_7^{\text{eff}}(\mu)$ and $C_8^{\text{eff}}(\mu)$, by adding to them a specific combination of the first six $C_i(\mu)$.

The decay rate for the transition $b \rightarrow s\gamma$ then reads, to leading order:

$$\Gamma[b \rightarrow s\gamma] = \frac{G_F^2 \alpha}{32\pi^4} |V_{ts}^* V_{tb}|^2 m_b^3 \bar{m}_b^2(m_b) \left| C_7^{\text{eff}}(m_b) \right|^2, \tag{7.13}$$

where m_b is the bottom quark pole mass and $\bar{m}_b$ is the running mass in the $\overline{MS}$ scheme: $\bar{m}_b(m_b) = m_b \left(1 - \frac{4}{3}\alpha_3(m_b)/\pi \right)$.

One thus has to renormalize the Wilson coefficients from the scale M_W where they can be computed (within the Standard Model or its supersymmetric extensions) down to the scale m_b. To leading order,

---

[3]For a presentation of the complete next to leading order, see [74].

$$C_7^{\text{eff}}(m_b) = \eta^{16/23} C_7(M_W) + \frac{8}{3}\left(\eta^{14/23} - \eta^{16/23}\right) C_8(M_W) + \sum_{i=1}^{8} h_i \eta_i^a, \qquad (7.14)$$

where $\eta \equiv \alpha_3(M_W)/\alpha_3(m_b) \sim 0.548$ and $\sum_{i=1}^{8} h_i \eta_i^a$ is given in [74].

The decay rate (7.13) suffers from large uncertainties because of the m_b^5 factor. One may improve this by normalizing it to the $b \to c e \bar{\nu}_e$ decay rate:

$$\Gamma\left[b \to c e \bar{\nu}_e\right] = \frac{G_F^2 m_b^5}{192\pi^3} |V_{cb}|^2 \, g(m_c/m_b) \left[1 - \frac{2}{3\pi}\alpha_3(m_b) f(m_c/m_b)\right], \qquad (7.15)$$

where $g(x) = 1 - 8x^2 + 8x^6 - x^8 - 24x^4 \ln x$ is the phase space factor, and the last factor is the next-to-leading QCD correction to the semileptonic decay ($f(m_c/m_b) \sim 2.41$) [62].

We have, to a good approximation,

$$\frac{\mathcal{B}\left[b \to X_s \gamma\right]}{\mathcal{B}\left[b \to X_c e \bar{\nu}_e\right]} \sim \frac{\Gamma\left[b \to s\gamma\right]}{\Gamma\left[b \to c e \bar{\nu}_e\right]}, \qquad (7.16)$$

where the corrections (of the order of 10 %) can be computed in the context of the Heavy Quark Effective Theory (HQET). Thus, one obtains

$$\frac{\mathcal{B}\left[b \to X_s \gamma\right]}{\mathcal{B}\left[b \to X_c e \bar{\nu}_e\right]} \sim \left|\frac{V_{ts}^* V_{tb}}{V_{cb}}\right|^2 \frac{6\alpha}{\pi g(m_c/m_b)\left[1 - \frac{2}{3\pi}\alpha_3(m_b) f(m_c/m_b)\right]} \frac{(\bar{m}_b(m_b)/m_b)^2}{} \qquad (7.17)$$

$$\times \left|\eta^{16/23} C_7(M_W) + \frac{8}{3}\left(\eta^{14/23} - \eta^{16/23}\right) C_8(M_W) + \sum_{i=1}^{8} h_i \eta_i^a\right|^2,$$

where the last factor of the first line includes some of the most important next to leading order contributions. The diagram of Fig. 7.1(a) gives

$$C_{7,8}^{(SM)}(M_W) = F_{7,8}^{(1)}\left(m_t^2(M_W)/M_W^2\right),$$

$$F_7^{(1)}(x) = \frac{x(7 - 5x - 8x^2)}{24(x-1)^3} + \frac{x^2(3x-2)}{4(x-1)^4} \ln x, \qquad (7.18)$$

$$F_8^{(1)}(x) = \frac{x(2 + 5x - x^2)}{8(x-1)^3} - \frac{3x^2}{4(x-1)^4} \ln x.$$

We now allow physics beyond the Standard Model. There is a significant contribution from charged Higgs exchange (Fig. 7.1b):

$$C_{7,8}^{(H^{\pm})}(M_W) = \frac{1}{3\tan^2\beta} F_{7,8}^{(1)}\left(m_t^2(M_W)/M_H^2\right) + F_{7,8}^{(2)}\left(m_t^2(M_W)/M_H^2\right),$$

$$F_7^{(2)}(x) = \frac{x(3 - 5x)}{12(x-1)^2} + \frac{x(3x-2)}{6(x-1)^3} \ln x, \qquad (7.19)$$

$$F_8^{(2)}(x) = \frac{x(3 - x)}{4(x-1)^2} - \frac{x}{2(x-1)^2} \ln x.$$

This, for example, puts a stringent lower bound on the charged Higgs mass in the context of the two Higgs doublet model [174]. However, in a supersymmetric context, the cancellation observed in the supersymmetric limit leads us to expect some amount of cancellation between the charged Higgs contribution and the supersymmetric particle exchange diagrams [24][4]. Moreover, as discussed in Chapter 6, Section 6.7, one expects new sources of flavor violations from the misalignment between quark and squark eigenstates.

From now on, we focus the discussion on a *minimal flavor violation* scenario where the only source of violation at the electroweak scale is the CKM matrix [24].

There are two sources that possibly enhance the chargino contribution [100]:

- Large $\tan\beta$ corrections.
  The relation between the bottom Yukawa coupling and the bottom mass receives at one loop in this limit a large correction [220]:

$$m_b = -M_W\,\sqrt{2}\frac{\lambda_b}{g}\cos\beta\,(1 + \epsilon_b\tan\beta)\,,\qquad(7.20)$$

where ϵ_b is obtained from gluino–sbottom and chargino–stop diagrams. Such an effect is taken into account by dividing the expressions for $C_{7,8}^{(\chi^{\pm})}$ by $(1+\epsilon_b\tan\beta)$. Thus, the leading chargino contribution at two loops is of order $(\tan\beta)^2$.

- Large ratio between the scale of supersymmetric particle masses M_{SUSY} and the electroweak scale M_W.
  The Higgsino–stop–bottom supersymmetric coupling $\lambda_{\tilde{t}}\,\bar{b}_L\,\tilde{t}_R\,\tilde{H}^-$ is related to the Yukawa coupling λ_t through supersymmetry. But below the scale M_{SUSY}, it becomes frozen whereas λ_t continues evolving. Large logs of the ratio M_{SUSY}/M_W are thus generated when expressing $\lambda_{\tilde{t}}$ in term of λ_t. In practice, they can be taken into account by replacing m_t in the chargino contribution by $m_t(M_{\mathrm{SUSY}})$. Similarly, the effective operators of (7.11) should be evolved from M_W to M_{SUSY}. In practice, most of the effect can be taken into account by taking directly the value of α_3 at M_{SUSY}.

7.1.4 The muon anomalous magnetic moment

It is well-known in quantum electrodynamics that the magnetic moment of the electron (or, for what concerns us here, the muon) is expressed in terms of the spin operator as $\mu = e g\mathbf{S}/(2mc)$ and that the gyromagnetic ratio g is equal to 2, up to small quantum corrections. Departure from this value induces an electromagnetic coupling

$$a\,\frac{e}{4m}\,\bar{\Psi}\sigma^{\mu\nu}\Psi F_{\mu\nu}\,,$$

where $a \equiv (g-2)/2$.

---

[4]Thus, the heavier the superpartners (squarks and charginos) are, the heavier the charged Higgs must be in order to reproduce the successes of the Standard Model. This is why a key experimental result is the limit on chargino masses.

The anomalous moment of the muon has been measured by the experiment E821 at Brookhaven [30]

$$a_\mu = 11\ 659\ 208(6) \times 10^{-10}. \tag{7.21}$$

This has to be compared with the prediction of the Standard Model, which depends mostly, at this level of precision, on the correct evaluation of the hadronic vacuum polarization and on the hadronic light-by-light contribution[5]. Depending on this one finds

$$a_\mu|_{\mathrm{exp}} - a_\mu|_{\mathrm{SM}} = (6 \text{ to } 25) \times 10^{-10}, \tag{7.22}$$

which represents a departure of 1–2.8 standard deviation.

If there is indeed a deviation, is it a sign of new physics? There are many possible candidates: anomalous couplings, lepton flavor violations, muon substructure, etc. But one should stress that, starting with [150], such a deviation has been heralded by many as a possible signature of supersymmetry.

Indeed, in the case of supersymmetry, there are new loop corrections involving neutralino–smuon or chargino–sneutrino mu. To have an idea of these new contributions, one may consider the simple situation where all superpartners have the same mass M_{SUSY}, in which case, at one loop, one has approximately

$$a_\mu|_{\mathrm{SUSY}} - a_\mu|_{\mathrm{SM}} \sim 13 \times 10^{-10} \left(\frac{100\ \mathrm{GeV}}{M_{\mathrm{SUSY}}} \right)^2 \tan\beta\ \mathrm{sign}(\mu). \tag{7.23}$$

One thus expects significant corrections for large $\tan\beta$ or a light supersymmetric spectrum. Moreover, a positive value of the supersymmetric parameter μ is strongly favored by the $(g_\mu - 2)$ data.

One should however note that the simplicity of the rule of thumb (7.23) is somewhat misleading: it does not apply, even approximately, in the case where supersymmetric partners are nondegenerate.

7.1.5 Constraints on minimal supergravity

We now illustrate on a specific model the constraints discussed in the preceding subsections. We follow the standard choice of the minimal supergravity model described in Section 6.8 of Chapter 6, not so much because it is well motivated but because it has a small number of parameters: $m_0, M_{1/2}, \tan\beta, A_0$ and the sign of μ. In fact, we choose $\mu > 0$ in order to satisfy the constraint imposed by data on the muon magnetic moment.

We give in Figures 7.3 and 7.4 the different constraints in the plane $(M = M_{1/2}, m = m_0)$ for $A_0 = 0$ and three values of $\tan\beta$: 5, 35 and 50. On each plot, the region to the left of the light grey dashed line is excluded by the lower bound on the lightest Higgs h^0 mass(se next section). The region to the left of the dotted line is excluded by the lower bound on the chargino mass $m_{\chi_1^\pm} > 103$ GeV. The region to the left of the solid line is excluded by $b \to s\gamma$. The region at the bottom (Stau LSP) is excluded because the lightest stau is the LSP.

---

[5]The latter contribution corresponds to the matrix element $\langle \mu | e j_\rho | \mu \rangle$ where $j_\rho = (2\bar{u}\gamma_\rho u - \bar{d}\gamma_\rho d - \bar{s}\gamma_\rho s)/3$ is the light quark electromagnetic current. It took some time before its sign was determined correctly [253].

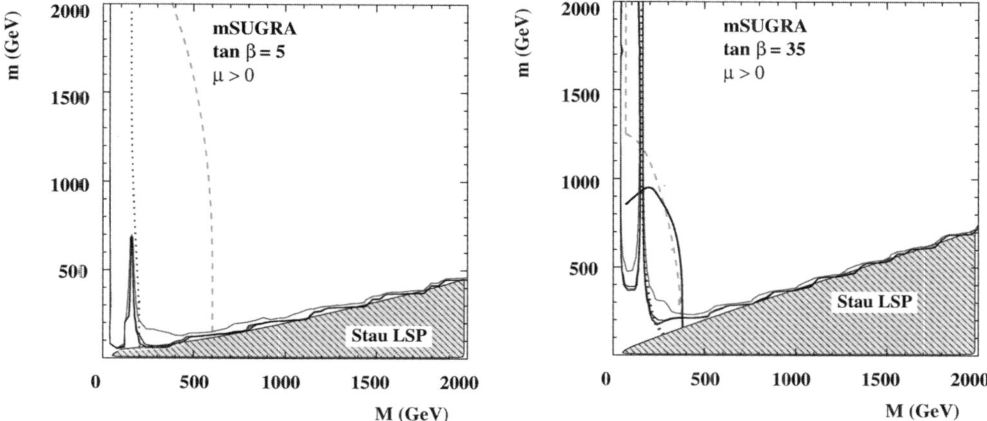

Fig. 7.3 Constraints on the mSUGRA parameter space for $\tan\beta = 5$ (left) and $\tan\beta = 35$ (right).

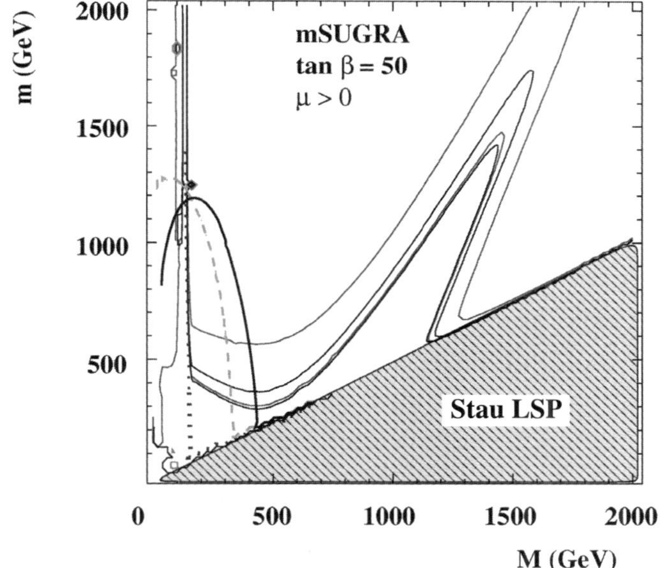

Fig. 7.4 Constraints on the mSUGRA parameter space for $\tan\beta = 50$.

The region between grey contours fulfils $0.1 \leq \Omega_\chi h^2 \leq 0.3$, whereas that between black contours indicates the WMAP range $0.094 < \omega_\chi h_0^2 \leq 0.129$ [141]. We see that the region favored by dark matter relic density is one of light scalars and gauginos, which is excluded by the negative searches of light supersymmetric particles. The bulk of the remaining parameter space yields too large a neutralino relic density. One remains with three zones where specific circumstances help to decrease this density:

- a narrow band along the "Stau LSP" region, where $\tilde{\tau}\chi$ coannihilations are at work (see the discussion in subsection 5.5.1 of Chapter 5);
- for large $\tan\beta$, a region of intermediate m_0 and $M_{1/2}$ where $\Omega_\chi h^2$ becomes smaller due to near-resonant s channel annihilation through the heavy Higgs states A or H (see equation (5.76) of Chapter 5): here m_A and m_H become smaller and their couplings to the b quark and τ lepton increase[6];
- a very thin strip almost along the vertical which corresponds to the focus points discussed in subsection 6.9.2 of Chapter 6. There are the lightest neutralino and chargino are relatively light and have a significant Higgsino content: the couplings to W and Z are large enough to increase the efficiency of the annihilation into W or Z pairs.

7.2 The Higgs sector

The prediction that there exists a light scalar is central in the search of low energy supersymmetry. It is obvious that this is not specific to supersymmetry. It is also by now clear that the (lightest) Higgs is heavier than the Z particle, and thus, in a MSSM context, that a significant contribution to its mass comes from radiative corrections. But it remains true that the vast majority of supersymmetric models predict a Higgs lighter than say 200 GeV. Given the importance of such a result, we give in what follows details on the precise determination of the Higgs scalar masses.

7.2.1 Precise estimate of the Higgs masses

We gave in Section 5.3.1 of Chapter 5 a heuristic argument to explain why one-loop corrections to the lightest Higgs h^0 may be large because they scale like m_t^4. We discuss in this section methods that allow a full determination of the one (and possibly higher) loop contribution.

The first method is based on the use of the effective potential approximation, presented in Section A.5.3 of Appendix Appendix A. The effective potential reads

$$V_{\text{eff}} \equiv V^{(0)} + \Delta V = V_{\text{eff}}^{(0)}(\mu) + V_{\text{eff}}^{(1)}(\mu) + \cdots \tag{7.24}$$

In this expression, $V_{\text{eff}}^{(0)}$ is the tree level potential $V^{(0)}$, which is given, if we restrict our attention to the neutral scalars, by equation (5.13) of Chapter 5:

$$V^{(0)} = m_1^2 |H_1^0|^2 + m_2^2 |H_2^0|^2 + B_\mu \left(H_1^0 H_2^0 + H_1^{0*} H_2^{0*} \right) + \frac{g^2 + g'^2}{8} \left(|H_1^0|^2 - |H_2^0|^2 \right)^2 . \tag{7.25}$$

It depends on the renormalization scale μ through the couplings. The μ dependence of the mass eigenvalues is softened by the inclusion of the one-loop corrections [173]:

$$V_{\text{eff}}^{(1)}(\mu) = \frac{1}{64\pi^2} \mathcal{S}\text{Tr} M^4 \left(\ln \frac{M^2}{\mu^2} - \frac{3}{2} \right) , \tag{7.26}$$

---

[6]Note that this region is very sensitive to the way radiative corrections to the bottom mass are implemented.

where we have made profit of the assumption of soft supersymmetry breaking to disregard field-independent contributions proportional to $\mathcal{S}\mathrm{Tr}M^2$.

We recall, from Appendix Appendix A, equation (A.247) that V_{eff} is the non-derivative term in the effective action $\Gamma\left[\bar{\phi}\right]$. We consider the proper Green's function $\Gamma^{(2)}(p) = \delta^2\Gamma/\delta\bar{\phi}^2$, which is the inverse propagator,

$$\Gamma^{(2)}(p) = p^2 - m^2 + \Sigma(p). \tag{7.27}$$

The physical mass is the zero of this function:

$$m^2_{\mathrm{phys}} = m^2 - \Sigma(m^2_{\mathrm{phys}}). \tag{7.28}$$

Thus we have

$$\left[\frac{\partial^2 V_{\mathrm{eff}}}{\partial\phi^2}\right]_{\phi=0} = -\Gamma^{(2)}(0) = m^2_{\mathrm{phys}} + \Sigma(m^2_{\mathrm{phys}}) - \Sigma(0). \tag{7.29}$$

If the physical mass is much smaller than the masses of particles running in the loops, one may take $\Sigma(m^2_{\mathrm{phys}}) \sim \Sigma(0)$.

Thus, at the one-loop level, masses are given by the matrix of double derivatives of the effective potential, evaluated at its minimum. Let us check that one thus recovers the preliminary results obtained in Section 5.2.1 of Chapter 5. We keep only in the supertrace (7.26) the top quark (of mass $m_t = \lambda_t H^0_2$) and the stop squarks (assumed to be degenerate in mass: $\tilde{m}^2_t = m^2_t + \tilde{m}^2$, where $\tilde{m}$ arises form the soft breaking of supersymmetry):

$$V^{(1)}_{\mathrm{eff}}(\mu) = \frac{3}{32\pi^2}\left[\tilde{m}^4_t\left(\ln\frac{\tilde{m}^2_t}{\mu^2} - \frac{3}{2}\right) - m^4_t\left(\ln\frac{m^2_t}{\mu^2} - \frac{3}{2}\right)\right]. \tag{7.30}$$

We may choose the scale $\mu = \hat{\mu}$ such that $\partial V^{(0)}_{\mathrm{eff}}/\partial\mu = 0$, in which case

$$\left.\frac{\partial V^{(1)}_{\mathrm{eff}}}{\partial\mu}\right|_{\mu=\hat{\mu}} \propto \tilde{m}^2_t\left(\ln\frac{\tilde{m}^2_t}{\hat{\mu}^2} - 1\right) - m^2_t\left(\ln\frac{m^2_t}{\hat{\mu}^2} - 1\right) = 0. \tag{7.31}$$

Then, using this condition, one finds

$$\left.\frac{\partial^2 V^{(1)}_{\mathrm{eff}}}{(\partial H^0_2)^2}(\hat{\mu})\right|_{min} = \frac{3}{4\pi^2}\lambda^4_t v^2_2 \ln\frac{\tilde{m}^2_t}{m^2_t} = \frac{3g^2}{8\pi^2}\frac{m^4_t}{M^2_W}\ln\frac{\tilde{m}^2_t}{m^2_t}, \tag{7.32}$$

where we have used $M^2_W = \frac{1}{2}g^2 v^2_2$ ($v_1 \sim 0$ in the limit that we consider here). This gives a contribution to the Higgs mass which coincides with the one obtained in (5.36) of Chapter 5).

If we take into account the mixing in the stop sector (see (5.53) of Chapter 5), then a more accurate estimate is

$$\delta m^2_h = \frac{3g^2}{8\pi^2}\frac{m^4_t}{M^2_W}\left[\ln\frac{\tilde{m}^2_{t1} + \tilde{m}^2_{t2}}{2m^2_t} + \frac{2X^2_t}{\tilde{m}^2_{t1} + \tilde{m}^2_{t2}}\left(1 - \frac{1}{6}\frac{X^2_t}{\tilde{m}^2_{t1} + \tilde{m}^2_{t2}}\right)\right], \tag{7.33}$$

where $X_t \equiv A_t - \mu\cot\beta$.

More generally, writing

$$H_i^0 \equiv v_i + \frac{S_i + iP_i}{\sqrt{2}}, \quad i = 1, 2, \tag{7.34}$$

one may obtain, in the effective potential approximation, the mass matrices from the second derivatives of the effective potential (7.24) evaluated at the minimum:

$$\left(\mathcal{M}_S^2\right)_{ij}^{\text{eff}} = \left.\frac{\partial^2 V_{\text{eff}}}{\partial S_i \partial S_j}\right|_{\text{min}}, \quad \left(\mathcal{M}_P^2\right)_{ij}^{\text{eff}} = \left.\frac{\partial^2 V_{\text{eff}}}{\partial P_i \partial P_j}\right|_{\text{min}}, \quad i, j = 1, 2. \tag{7.35}$$

As stressed earlier, V_{eff} is expressed in terms of renormalized fields and couplings. One traditionally uses the $\overline{DR}$ scheme (see Appendix E): the corresponding quantities will be written in what follows with a bar ($\bar{M}_Z,...$). Using the decomposition (7.24) with $V^{(0)}$ given by (7.25), one may write for example (see Exercise 4):

$$\left(\mathcal{M}_S^2\right)^{\text{eff}} = \left(\mathcal{M}_S^2\right)^{(0)\text{eff}} + \left(\Delta\mathcal{M}_S^2\right)^{\text{eff}}, \tag{7.36}$$

$$\left(\mathcal{M}_S^2\right)^{(0)\text{eff}} = \begin{pmatrix} \bar{M}_Z^2 \cos^2\beta + \bar{m}_A^2 \sin^2\beta & -(\bar{M}_Z^2 + \bar{m}_A^2)\sin\beta\cos\beta \\ -(\bar{M}_Z^2 + \bar{m}_A^2)\sin\beta\cos\beta & \bar{M}_Z^2 \sin^2\beta + \bar{m}_A^2 \cos^2\beta \end{pmatrix},$$

$$\left(\Delta\mathcal{M}_S^2\right)_{ij}^{\text{eff}} = \left.\frac{\partial^2 \Delta V}{\partial S_i \partial S_j}\right|_{\text{min}} - (-1)^{i+j} \left.\frac{\partial^2 \Delta V}{\partial P_i \partial P_j}\right|_{\text{min}}.$$

However, as is clear from (7.29), the effective potential describes Green's functions at vanishing momentum. The effective potential approximation thus misses some possibly important momentum corrections. This is why a purely diagrammatic approach has been developed in parallel [52]. In this approach, the physical masses m_h^2 and m_H^2 are defined as the poles of the propagator $\left(\Gamma^{(2)}(p)\right)^{-1}$. In other words, writing the proper Green's function for the scalars

$$\Gamma_S^{(2)}(p) = p^2 - \mathcal{M}_S^2(p^2), \tag{7.37}$$

the physical masses are the solutions of

$$\det\left[p^2 - \mathcal{M}_S^2(p^2)\right] = 0. \tag{7.38}$$

They can be expressed in terms of the physical masses M_Z^2 and m_A^2. To compare the two approaches, one may write [50, 52]

$$\mathcal{M}_S(p^2) = \left(\mathcal{M}_S^2\right)^{(0)} + \left(\Delta\mathcal{M}_S^2\right)^{\text{eff}} + \left(\Delta\mathcal{M}_S^2\right)^{p^2}, \tag{7.39}$$

where $\left(\mathcal{M}_S^2\right)^{(0)}$ is obtained from $\left(\mathcal{M}_S^2\right)^{(0)\text{eff}}$ by replacing the $\overline{DR}$ masses by physical masses.

Whereas, at one loop [134, 216, 299], the full corrections are known, only the leading contributions at two loops have been computed. These include corrections of order λ_t^4,

$\lambda_b^2 \alpha_3$ and $\lambda_b^2 \alpha_3$. Let us take this opportunity to answer a question which is often asked: if the one-loop corrections to the lightest Higgs are so large, should we not expect large corrections at even higher orders? The answer is no: two-loop and higher corrections are expected to be small. The reason is that, in contrary to the one-loop level where new terms with a dependence in λ_t^4 appear, we do not expect any surprise at higher orders, as can be seen for example from the list of the leading two-loop corrections just given.

7.2.2   Upper limit on the mass of the lightest Higgs

In the late 1990s, a significant amount of theoretical activity has been spent in trying to derive a precise upper limit for the lightest Higgs mass in the context of the MSSM. The main reason was that the available energy at the LEP collider allowed searches of the lightest Higgs only in a restricted range. In the present decade, that should see the first runs of the LHC collider, this question is of less importance since LHC should cover the whole range of mass for h^0 in the MSSM.

We have seen that the one-loop radiative corrections to the lightest Higgs mass depend mostly on the precise value of the top mass, on $\tan\beta$ and on the masses of the stops (and sbottoms if $\tan\beta$ is large). Obviously, if one increases the soft scalar mass parameter, one increases the squark masses and thus the corrections to the Higgs mass. Pushing to the extreme case where this mass parameter is 10 TeV, one obtains $m_{h^0} < 150$ GeV in the MSSM. Taking values of this soft mass scale more in line with a reasonable fine tuning of the parameters significantly decreases this upper limit: one finds $m_{h^0} < 110$ GeV if the stops have a common mass of 1 TeV (often referred in the literature as the "no mixing" scenario) and $m_{h^0} < 130$ GeV if there is a large splitting between the two (the "maximal mixing" scenario).

Let us note that, in the context of the MSSM model, the Higgs is light at tree level because the quartic coupling λ is of order $g^2 + g'^2$ (see 7.25): as in the Standard Model, the mass is of order λv^2. But this property is lost in extensions of the MSSM. For example, in the NMSSM model described in Section 5.6 of Chapter 5, the superpotential coupling $\lambda_S S H_2 \cdot H_1$ induces in the potential a quartic term of order λ_S^2 which is not necessarily small. Thus, in principle the lightest Higgs could be heavy.

However, if we put the low energy supersymmetric theory in the context of a more fundamental theory with a typical scale in the M_U or M_P range, then the triviality arguments presented in Section 1.2.1 of Chapter 1 apply and one does not expect the lighter scalar mass to be above 200 GeV. Let us illustrate this on the NMSSM model.

In the framework of the NMSSM model, the lightest Higgs mass satisfies the upper bound (see (5.112) of Chapter 5)

$$m_h^2 \leq M_z^2 + \left(\frac{1}{2}\lambda_S^2 v^2 - M_z^2\right)\sin^2 2\beta, \qquad (7.40)$$

where λ_S is one of the cubic couplings in the superpotential:

$$W = \frac{1}{6}\kappa_S S^3 + \lambda_S S H_2 \cdot H_1 + \lambda_t Q_3 \cdot H_2 T^c. \qquad (7.41)$$

We have assumed a discrete symmetry forbidding any quadratic term and we have kept only the top Yukawa coupling. We see that, if λ_S is allowed to be as large as possible, there is no absolute upper bound to the Higgs mass.

One can however apply triviality bounds to λ_S [45]. We will give here a simplified discussion of these bounds. The renormalization group equations describing the evolution of the superpotential couplings read [102]

$$16\pi^2 \frac{d\lambda_S}{dt} = \lambda_S \left(4\lambda_S^2 + 2\kappa_S^2 + 3\lambda_t^2 - 3g_2^2 - \frac{3}{5}g_1^2 \right),$$

$$16\pi^2 \frac{d\kappa_S}{dt} = \kappa_S \left(6\lambda_S^2 + 6\kappa_S^2 \right), \tag{7.42}$$

$$16\pi^2 \frac{d\lambda_t}{dt} = \lambda_t \left(\lambda_S^2 + 6\lambda_t^2 - \frac{16}{3}g_3^2 - 3g_2^2 - \frac{13}{15}g_1^2 \right).$$

Neglecting the gauge couplings, one obtains, for $\lambda_t \neq 0$

$$8\pi^2 \frac{d\left(\lambda_S^2/\lambda_t^2\right)}{dt} = \lambda_S^2 \left(3\frac{\lambda_S^2}{\lambda_t^2} + 2\frac{\kappa_S^2}{\lambda_t^2} - 3 \right), \tag{7.43}$$

$$8\pi^2 \frac{d\left(\kappa_S^2/\lambda_t^2\right)}{dt} = \kappa_S^2 \left(5\frac{\lambda_S^2}{\lambda_t^2} + 6\frac{\kappa_S^2}{\lambda_t^2} - 6 \right). \tag{7.44}$$

This system of differential equations has three fixed points: (a) $\lambda_S^2/\lambda_t^2 = 3/4$, $\kappa_S^2/\lambda_t^2 = 3/8$, (b) $\lambda_S^2/\lambda_t^2 = 1$, $\kappa_S = 0$, (c) $\kappa_S^2/\lambda_t^2 = 1$, $\lambda_S = 0$. Only (a) corresponds to an infrared attractive fixed point.

The presence of this attractive fixed point suggests that, if we start with random boundary conditions for the couplings λ_S, κ_S and λ_t at some superheavy scale Λ, they will tend to converge towards values which respect the ratios (a). If we thus assume that we are in a region of parameter space where the fixed point (a) is quickly reached as the scale μ decreases from Λ, then equation (7.42) for λ_S reads, using (a) and neglecting gauge couplings,

$$16\pi^2 \frac{d\lambda_S}{dt} = 9\lambda_S^3, \tag{7.45}$$

which is solved as

$$\frac{1}{\lambda_S^2(\mu)} = \frac{1}{\lambda_S^2(\Lambda)} + \frac{9}{8\pi^2} \ln(\Lambda/\mu). \tag{7.46}$$

Thus

$$\lambda_S^2(\mu) \leq \frac{8\pi^2}{9\ln(\Lambda/\mu)}. \tag{7.47}$$

This upper bound induces an upper bound on the value of m_h^2 at tree level. One must add, as in the MSSM, the radiative corrections to this mass.

7.2.3 Searching for the supersymmetric Higgs

The supersymmetric Higgs have been extensively searched for at the LEP collider. It is particularly convenient to discuss h^0 searches using the parameters $\tan\beta$ and m_A. In the case of large m_A, or for $\tan\beta$ close to one, the lighter scalar h^0 tends to be similar to the Higgs of the Standard Model. It is thus searched for using the standard

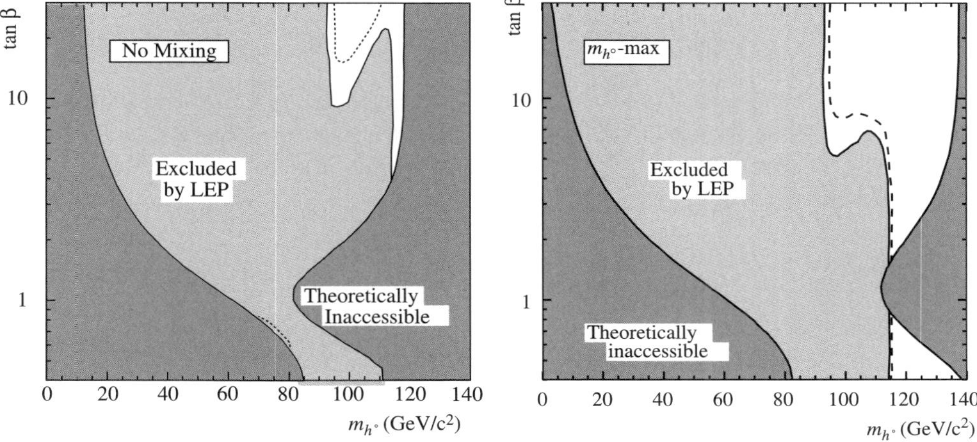

Fig. 7.5 Exclusion plots for m_h at 95% confidence level in the MSSM for the no mixing (left) and the maximal mixing (right) scenarios. Shown in light grey is the theoretically inaccessible region and in dark grey the experimentally excluded region. The dashed lines indicate the boundaries of the regions expected to be excluded on the basis of Monte Carlo simulations with no signal [271].

Higgstrahlung process $e^+e^- \rightarrow h^0 Z^0$. For lighter m_A, the modification of couplings compared to the standard case leads to a reduction of the Higgstrahlung process. This is compensated by an increase of the so-called associated production $e^+e^- \rightarrow h^0 A^0$.

Negative results from the search of both of these processes at LEP have led to the exclusion plots presented in Fig. 7.5 in the plane $(m_h, \tan\beta)$. As is traditional, one has separated the two cases of "maximal mixing" and "no mixing", the latter being the more constraining one. For low values of $\tan\beta$ above 1, the Higgstrahlung process is the decisive one: it excludes values of m_h lower than 114 GeV. For higher values of $\tan\beta$, the associated production takes over but it excludes only values of m_h lower than 91 GeV.

7.3 Avoiding instabilities in the flat directions of the scalar potential

The most remarkable feature of a supersymmetric potential is the presence of flat directions. These are valleys where the potential vanishes and thus where global supersymmetry is not broken. Since in global supersymmetry, the potential is a sum of F-terms and D-terms, one refers to them as F-flat or D-flat directions. For example, if $W(\varphi_i)$ is the superpotential, the equation $F_i = \partial W/\partial \phi_i = 0$ defines a direction in scalar field space which is a F-flat direction.

Flat directions are characteristic of supersymmetric theories and they have many interesting properties which we will discuss in the following chapters. But they represent a potential danger for phenomenology. Indeed, they are lifted once supersymmetry is broken, in particular by scalar mass terms which may lead to instabilities (if a squared mass turns negative). Such instabilities are used to spontaneously break the electroweak symmetry but they might lead also to undesirable charge or color breaking minima (we know that the quantum electrodynamics $U(1)$ or the color $SU(3)$ are good symmetries).

To simplify the discussion, we disregard the family structure and consider a single family (which could be any of the three; but since Yukawa couplings are larger for the third family, we choose it for illustration). The superpotential reads (*cf.* equations (5.1) and (5.2) of Chapter 5)

$$W = \mu H_2 \cdot H_1 + \lambda_b\, Q_3 \cdot H_1 B^c + \lambda_t\, Q_3 \cdot H_2 T^c + \lambda_\tau\, L_3 \cdot H_1 T^c \tag{7.48}$$

with obvious notation: $Q_3 = \begin{pmatrix} T \\ B \end{pmatrix}$, $L_3 = \begin{pmatrix} N_\tau \\ T \end{pmatrix}$ and as seen earlier $H_1 = \begin{pmatrix} H_1^0 \\ H_1^- \end{pmatrix}$, $H_2 = \begin{pmatrix} H_2^+ \\ H_2^0 \end{pmatrix}$. One easily extracts the F-term part of the potential: $V_F = \sum_i |\partial W/\partial \Phi_i|^2$.

The D-term part of the scalar potential reads explicitly[7]

$$\begin{aligned}
V_D &= \frac{g'^2}{8}\left[\frac{1}{3}\left(|\tilde{t}_L|^2 + |\tilde{b}_L|^2 \right) - \frac{4}{3}|\tilde{t}_R|^2 + \frac{2}{3}|\tilde{b}_R|^2 - \left(|\tilde{\nu}_{\tau L}|^2 + |\tilde{\tau}_L|^2 \right) + 2|\tilde{\tau}_R|^2 \right. \\
&\qquad \left. + |H_2^0|^2 + H_2^+ H_2^- - |H_1^0|^2 - H_1^+ H_1^- \right] \\
&\quad + \frac{g^2}{8}\left(\tilde{q}_{3L}^\dagger \vec{\tau}\, \tilde{q}_{3L} + \tilde{l}_{3L}^\dagger \vec{\tau}\, \tilde{l}_{3L} + H_1^\dagger \vec{\tau} H_1 + H_2^\dagger \vec{\tau} H_2 \right)^2 \\
&\quad + \frac{g_3^2}{8} \sum_a \left(\tilde{q}_{3L}^\dagger \lambda^a q_{3L} - \tilde{t}_R^* \lambda^a \tilde{t}_R - \tilde{b}_R^* \lambda^a \tilde{b}_R \right)^2 .
\end{aligned} \tag{7.49}$$

Finally, the soft terms read, as in equations (5.12) and (5.55) of Chapter 5,

$$\begin{aligned}
V_{soft} &= m_{H_1}^2 H_1^\dagger H_1 + m_{H_2}^2 H_2^\dagger H_2 + (B_\mu H_1 \cdot H_2 + \text{h.c.}) \\
&\quad + m_{Q_3}^2 \left(\tilde{t}_L^* \tilde{t}_L + \tilde{b}_L^* \tilde{b}_L \right) + m_T^2 \tilde{t}_R^* \tilde{t}_R + m_B^2 \tilde{b}_R^* \tilde{b}_R \\
&\quad + m_{L_3}^2 \left(\tilde{\nu}_{\tau L}^* \tilde{\nu}_{\tau L} + \tilde{\tau}_L^* \tilde{\tau}_L \right) + m_T^2 \tilde{\tau}_R^* \tilde{\tau}_R \\
&\quad + \left(A_t \lambda_t \tilde{q}_{3L} \cdot H_2 \tilde{t}_R^* + A_b \lambda_b \tilde{q}_{3L} \cdot H_1 \tilde{b}_R^* + A_\tau \lambda_\tau \tilde{l}_{3L} \cdot H_1 \tilde{\tau}_R^* + \text{h.c.} \right) .
\end{aligned} \tag{7.50}$$

The tree-level potential $V^{(0)} = V_F + V_D + V_{soft}$ must be complemented at least by the one-loop corrections. As we have seen in the preceding section, the complete one-loop potential reads, in the effective potential approximation,

$$V_1(Q) = V^{(0)}(Q) + V_{\text{eff}}^{(1)}(Q), \tag{7.51}$$

$$V_{\text{eff}}^{(1)}(Q) = \frac{1}{64\pi^2} \sum_i (-1)^{2s_i}(2s_i + 1)\, m_i^4 \left(\ln \frac{m_i^2}{Q^2} - \frac{3}{2} \right), \tag{7.52}$$

---

[7]Note that in the last term, the generators of $SU(3)$ for the charge conjugate scalars are $-\lambda^{a*}/2$ since, e.g. $\tilde{t}_R^* \to e^{-i\alpha^a \lambda^{a*}/2}\tilde{t}_R^*$. We then use the hermiticity of the Gell-Mann matrices to write: $-\left(\tilde{t}_R^*\right)^* \lambda^{a*} \tilde{t}_R^* = -\tilde{t}_R^* \lambda^a \tilde{t}_R$.

where m_i^2 is the field-dependent eigenvalue corresponding to a particle of spin s_i and Q is the renormalization scale[8].

We will identify a F or D flat direction by specifying the supermultiplets which have a nonzero vev along the direction. In other words, along the direction $(\Phi_1, \Phi_2, ..., \Phi_n)$, the scalar components $\phi_1, \phi_2, \ldots, \phi_n$ acquire a (possibly) nonvanishing vev.

Including the one-loop corrections is important in order to minimize the dependence on the renormalization scale Q : most of the dependence cancels between the running scalar masses which are present in the tree-level potential $V^{(0)}$ and the one-loop correction $V_{\text{eff}}^{(1)}$ [173]. However, instead of using the full one-loop potential, we will use the following approximation scheme [71]: we take as a scale Q the largest field-dependent mass $\hat{m}(\phi_i)$ present in (7.52) in order to minimize the value of the one-loop correction. We then use the tree-level "improved" potential $V^{(0)}\,(\hat{m}(\phi_i))$.

We note that $\hat{m}$ is in general the largest of the following two scales: a typical supersymmetric mass m_{SUSY} (usually a squark mass) or a field-dependent mass in the form of a (gauge or Yukawa) coupling times the nonzero vev along the flat direction considered. For example, in the case of a direction which involves the field H_2^0, we would have $\hat{m} = \lambda_t \langle H_2^0 \rangle$ if we consider very large H_2^0 vevs, and squark mass m_{SUSY} otherwise (*i.e.* if $\langle H_2^0 \rangle < m_{\text{SUSY}}/\lambda_t$).

Before discussing the potentially dangerous flat directions, let us review the case of the direction (H_1^0, H_2^0). This direction has already been considered in equation (5.15) of Chapter 5 for the discussion of electroweak symmetry breaking at tree level. The condition of D-flatness imposes that $\left| H_1^0 \right|^2 = \left| H_2^0 \right|^2 \equiv a^2$ and the potential then simply reads along this direction

$$V^{(0)}(Q = \hat{m}) = \left(m_1^2 + m_2^2 + 2B_\mu \cos\varphi\right) a^2, \tag{7.53}$$

as in equation (5.15) of Chapter 5 (m_1^2 and m_2^2 are defined as in (5.14) there and $\varphi = \text{Arg}\left(H_2^0/H_1^{0*}\right)$). The condition for stability in this direction is the positivity condition:

$$\mathcal{S} \equiv m_1^2 + m_2^2 - 2\left|B_\mu\right| > 0. \tag{7.54}$$

Remember that this is evaluated at a scale Q of order $\lambda_t a$ (as long as Q is larger than m_{SUSY}). If we start at the unification scale (or the scale of the underlying theory) M_U, this constraint should be valid (otherwise the breaking occurs already in the underlying theory). When we go down in scale, we will eventually encounter the scale Q_s where $\mathcal{S}$ becomes negative; we then have $\langle a \rangle \sim Q_s/h_T$.

Let us note that, if we were working solely with the tree level potential, the condition $\mathcal{S} < 0$ would be the sign of a potential unbounded from below[9]. But this is obviously not the case here: for larger values of a (hence of the renormalization scale Q) the condition (7.54) is no longer satisfied. This is just a reflection of the fact that, had we worked with the complete one-loop potential (7.51), the large logs coming from (7.52) would prevent the field vev from going to infinitely large values.

---

[8]For obvious reasons we refrain from calling it μ in this Section

[9]This is why the type of direction that we study is often called an "unbounded from below direction" and denoted UFB [71]. We will refrain from doing so here.

We will consider here only two of the better known and most constraining flat directions. We start with the direction (T, H_2^0, T^c) [165]. In order to cancel the D-terms in (7.49), we must have $|\tilde{t}_L|^2 = |\tilde{t}_R|^2 = |H_2^0|^2 = a^2$. Then the potential reads

$$V^{(0)}(a) = M^2 a^2 - 2\,|A_t \lambda_t|\,a^3 + 3\lambda_t^2 a^4, \tag{7.55}$$

where $M^2 \equiv m_{Q_3}^2 + m_T^2 + m_2^2$ and we have chosen the field phases in such a way that the coefficient of the A-term is negative. One easily checks that, when $|A_t|^2 > 8M^2/3 > 0$, a nontrivial minimum appears:

$$a_0 = \frac{3\,|A_t| + \sqrt{3(3A_t^2 - 8M^2)}}{12\,|\lambda_t|}. \tag{7.56}$$

This represents the true ground state if $V(a_0) < 0$, i.e. $|A_t|^2 > 3M^2$. We must thus impose the condition

$$|A_t|^2 < 3\left(m_{Q_3}^2 + m_T^2 + m_2^2\right). \tag{7.57}$$

Next we consider the direction[10] (B, B^c, H_2^0, N_τ) [258]. The conditions of D-flatness read:

$$\left|\tilde{b}_L\right|^2 = \left|\tilde{b}_R\right|^2,$$
$$\left|\tilde{b}_L\right|^2 + \left|\tilde{H}_2^0\right|^2 = |\tilde{\nu}_{\tau_L}|^2, \tag{7.58}$$

and the condition of F-flatness ($F_{H_1^0} = 0$):

$$\mu H_2^0 + \lambda_b \tilde{b}_L \tilde{b}_R^* = 0. \tag{7.59}$$

Up to some phases, it is solved as follows:

$$\tilde{b}_L = \tilde{b}_R = \frac{a\mu}{\lambda_b}$$
$$H_2^0 = -\frac{a^2 \mu^*}{\lambda_b}$$
$$\tilde{\nu}_{\tau_L} = \frac{|\mu|}{\lambda_b} a \sqrt{1 + a^2}. \tag{7.60}$$

We then have

$$V = \frac{a^2 |\mu|^2}{\lambda_b^2}\left[\left(m_{H_2}^2 + m_{L_3}^2\right) a^2 + m_{Q_3}^2 + m_B^2 + m_{L_3}^2\right]. \tag{7.61}$$

The potential is stabilized at large values of a if:

$$m_{H_2}^2 + m_{L_3}^2 > 0. \tag{7.62}$$

This turns out to be the most constraining relation arising from this type of constraints.

---

[10]Often referred to as UFB-3 [71].

Finally, a word about cosmological arguments. It has been argued that, even in the cases where a charge or color breaking minimum deeper than the electroweak minimum appears, the time it takes to decay into this true vacuum ground state might be longer than the age of the Universe [77]. This indeed would make most of the problems discussed in this section innocuous and the corresponding bounds nonapplicable. One should, however, be guarded against cosmological arguments in the context of this discussion. In fact, as long as the cosmological constant problem is not solved, it is unadvisable to take at face value the expression of the potential in a (global or local) minimum. Whatever is the mechanism that is flushing (most of) the vacuum energy, it is difficult to imagine that it would distinguish between the electroweak symmetry breaking minimum and any other charge or color breaking minimum. Of course, this puts the whole issue of comparing ground states on somewhat shaky grounds. But, in the preceding analysis, we did not rely on any specific value of the vacuum energy whereas estimates of tunnelling effects do. Other arguments that the system might choose the right minimum in the early Universe because of temperature corrections [265] would however be more reliable.

7.4 High-energy vs. low-energy supersymmetry breaking

7.4.1 Issues

As emphasized many times earlier, the issue of spontaneous supersymmetry breaking is central to this field: supersymmetry is not observed in the spectrum of fundamental particles. This means that, should supersymmetric particles be observed in high energy colliders, the focus of high energy physics would be to unravel the origin of supersymmetry breaking. We are not at this point yet but it is important to prepare the ground for such a task by studying general classes of supersymmetry breaking in order to understand their main experimental signatures, and how one can discriminate between them.

There are already quite a few constraints that a supersymmetry breaking scenario must obey. One may thus see how each scenario fares with respect to them. It is fair to say that not a single scenario passes all these tests *magnum cum laude*. We will return to this in more details in Chapter 12 and will only comment here on the possibly sensitive issues for each class of models.

7.4.2 Gravity mediation. The example of gaugino condensation

We have seen in Chapter 3 that it is phenomenologically undesirable to couple directly the source of supersymmetry-breaking to the quark and lepton fields. One is thus led to the general picture drawn in Fig. 7.6: an observable sector of quarks, leptons and their superpartners, a sector of supersymmetry breaking and an interaction that mediates between the two sectors.

In Chapter 6, the mediating interaction chosen was gravity. This has the advantage of really hiding the supersymmetry breaking sector: the coupling to the observable sector is the gravitational coupling $\kappa^{-1} = m_{Pl}^{-2}$. A standard example is provided by the simple Polonyi model described in Section 6.3.2 of Chapter 6. In this model, the soft supersymmetry breaking terms are universal, *i.e.* they do not depend on the flavor. This is a welcome property because this satisfies more easily the constraints on flavor-changing neutral currents discussed in Section 6.7 of Chapter 6.

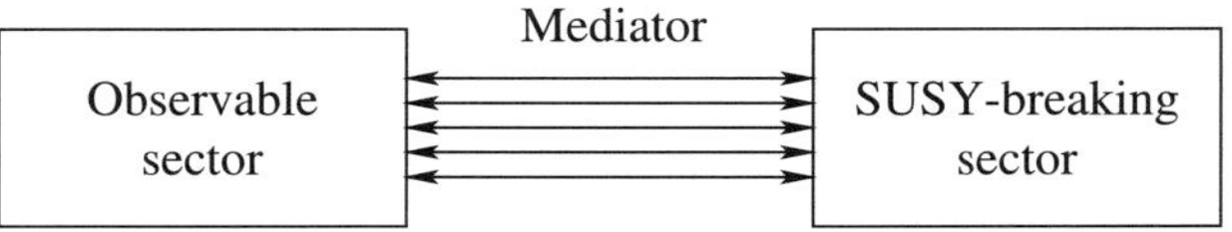

Fig. 7.6 General picture of supersymmetry breaking.

The rationale behind universality is the fact that gravity is flavor blind. In explicit theories that unify gravity with the standard gauge interactions, this is often not the case. We will return to this below. For the moment, we will describe a rather generic model of gravity mediation where supersymmetry is broken in the hidden sector by gaugino condensation.

Let us consider a theory where the gauge kinetic term f_{ab} is dynamical. To ensure universality (and gauge coupling unification), we assume that it is diagonal and expressed in terms of a single field S: $f_{ab} = S\delta_{ab}$. In other words, the gauge kinetic terms read, from equation (6.25) of Chapter 6,

$$\mathcal{L}_{\text{kin}} = -\tfrac{1}{4}\text{Re } S\; F^{a\mu\nu} F^a_{\mu\nu} + \tfrac{1}{4}\text{Im } S\; F^{a\mu\nu} \tilde{F}^a_{\mu\nu}. \tag{7.63}$$

We will encounter such a field in the context of string models: it is the famous string dilaton. As in the string case, we assume that S does not appear in the superpotential. For the time being, we just note a few facts:

- The interaction terms (7.63) are of dimension 5. This means that there is a power m_P^{-1} present which has been included in the definition of S (S is thus dimensionless). In other words, the S field has only gravitational couplings to the gauge fields, and to the other fields through radiative corrections.

- The gauge coupling is fixed by the vacuum expectation value of Re S:

$$\frac{1}{g^2} = \frac{\langle S + \bar{S}\rangle}{2}. \tag{7.64}$$

Obviously, this only provides the value of the running gauge coupling at a given scale, typically the scale of the fundamental theory where the field S appears. Because of our assumptions, this corresponds also to the scale where the gauge couplings all have the same value, *i.e.* are unified, and we will denote this scale by M_U.

- Since S does not appear in the superpotential, any minimum of the scalar potential is valid for any value of S: it corresponds to a flat direction of the potential. Hence the ground state value $\langle S\rangle$ remains undetermined: we may as well write it S, as long as some dynamics does not fix it.

We then assume that the theory considered involves a asymptotically free gauge interaction of group G_h under which the observable fields are neutral: the corresponding gauge fields A^μ_h and gauginos λ_h are part of the hidden sector.

As shown on Fig. 7.7, the associated gauge coupling explodes at a scale which is approximately given by

$$\Lambda_c = M_U e^{-8\pi^2/(b_0 g^2)} = M_U e^{-4\pi^2(S+\bar{S})/b_0}, \tag{7.65}$$

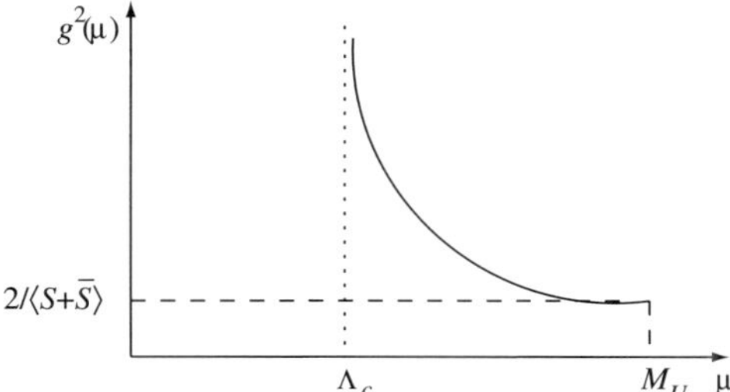

Fig. 7.7 Evolution of the hidden sector gauge coupling with energy.

where b_0 is the one-loop beta function coefficient associated with the hidden sector gauge symmetry considered and we have used (7.64). At this scale, the gauginos are strongly interacting and one expects that they will condense (we will develop in Chapter 8 more elaborate tools to study such dynamical effects). On dimensional grounds, one expects the gaugino condensates of the hidden sector to be of order

$$\left|\langle \bar{\lambda}_h \lambda_h \rangle\right|^2 \sim \Lambda_c^6 \sim M_U^6 e^{-24\pi^2 \langle S+\bar{S}\rangle / b_0}. \tag{7.66}$$

It is clear that the replacement (7.66) in the supergravity Lagrangian induces some nontrivial potential for the S field: typically the four gaugino interaction present in the supergravity Lagrangian yields an exponentially decreasing potential for S. A safe way to infer the effective theory below the condensation scale is to make use of the invariances of the complete theory [2, 353].

The symmetry that we use is Kähler invariance. We have seen in Chapter 6 that the full supergravity Lagrangian is invariant under the Kähler transformation $K \to F + \bar{F}$ if one performs a chiral $U(1)_K$ rotation (6.24) on the fermion fields. Choosing simply $F = -2i\alpha$, this simply amounts to a R-symmetry[11] where

$$\psi'_{\mu L} = e^{i\alpha}\psi_{\mu L}, \quad \lambda'_L = e^{i\alpha}\lambda_L, \quad \Psi'_L = e^{-i\alpha}\Psi_L. \tag{7.67}$$

This is, however, not a true invariance of the theory because it is anomalous. If we consider only the hidden sector, the divergence of the $U(1)_K$ current receives a contribution from the triangle anomaly associated with the gaugino fields:

$$\partial^\mu J^K_\mu = \frac{1}{3} \frac{b_0}{16\pi^2} F_{h\mu\nu} \tilde{F}_h^{\mu\nu} + \cdots \tag{7.68}$$

---

[11] R-symmetries are global $U(1)$ symmetries which do not commute with supersymmetry: supersymmetric partners transform differently under R-symmetries. We have encountered them in Section 4.1 of Chapter 4 and will discuss them in more details in the next chapter, since they play, as we already see here, a central rôle in the study of supersymmetry breaking. The transformation laws of the fields of a chiral supermultiplet are given in equation (C.43) of Appendix C (we take here $r = 0$).

where b_0 has been defined in (7.65) and the dots refer to the contributions of the observable sector. Performing a Kähler rotation thus induces an extra term in the Lagrangian

$$\delta\mathcal{L} = \frac{1}{3}\frac{b_0}{16\pi^2}\alpha\; F_{h\mu\nu}\tilde{F}_h^{\mu\nu}. \tag{7.69}$$

This may be cancelled by making a Peccei–Quinn translation on the field S

$$S \to S - i\frac{1}{3}\frac{b_0}{4\pi^2}\alpha \tag{7.70}$$

since $\mathrm{Im}S$ couples to $F_{h\mu\nu}\tilde{F}_h^{\mu\nu}$ through (7.63)

$$\delta'\mathcal{L} = -\frac{1}{3}\frac{b_0}{16\pi^2}\alpha\; F_{h\mu\nu}\tilde{F}_h^{\mu\nu}, \quad \delta\mathcal{L} + \delta'\mathcal{L} = 0. \tag{7.71}$$

Hence the full quantum theory is invariant under the combination of the R-symmetry $U(1)_K$ and the Peccei–Quinn transformation (7.70). This symmetry should remain intact through the condensation process. This allows us to determine the effective superpotential $W(S)$ below the condensation scale: just as for a standard R-symmetry, it must transform as $W \to e^{2i\alpha}W$. Hence, in the effective theory, the superpotential includes an extra term

$$W(S) = he^{-24\pi^2 S/b_0}. \tag{7.72}$$

Thus spontaneous supersymmetry breaking lifts the degeneracy associated with the field S, i.e. the flat direction of the scalar potential.

Unfortunately, this leads to a potential for S which is monotonically decreasing towards $S \to \infty$ where supersymmetry is restored ($\langle F_S \rangle \sim e^{-24\pi^2\langle S\rangle/b_0} \to 0$). We thus have to stabilize the S field since we need a determination of the gauge coupling (7.64). As we will see in Chapter 10, this problem is very general to string models, to which we borrow this example, and is known as the stabilization of moduli.

A first solution, known as the racetrack model, is to consider multiple gaugino condensation [262]. If the gauginos of two gauge groups (with respective beta function coefficients b_1 and b_2) condense, then the superpotential of the effective theory includes the terms:

$$W(S) = \alpha_1 b_1 e^{-24\pi^2 S/b_1} - \alpha_2 b_2 e^{-24\pi^2 S/b_2}, \tag{7.73}$$

with α_1 and α_2 constants of order 1. This superpotential has a stationary point at:

$$S = \frac{1}{24\pi^2}\frac{b_1 b_2}{b_1 - b_2}\ln\frac{\alpha_2}{\alpha_1}. \tag{7.74}$$

At this point, the superpotential is nonvanishing:

$$W = (b_1 - b_2)\sqrt{\alpha_1\alpha_2}\left(\frac{\alpha_1}{\alpha_2}\right)^{\frac{1}{2}\frac{b_1+b_2}{b_1-b_2}}. \tag{7.75}$$

Supersymmetry is broken through $F_S \sim K_S W \neq 0$: the scale of supersymmetry breaking is $m_{3/2} = e^{K/2}|W|$. One thus still has to specify the Kähler potential. In the case of the string models, to which we borrow this example, the S dependent part is simply $K(S) = -\ln\left(S + \bar{S}\right)$.

One sees from (7.75) that one may generate a large hierarchy between $m_{3/2}$ and M_U or M_P by having for example, if $\alpha_1 < \alpha_2$,

$$0 < b_1 - b_2 \ll b_1 + b_2. \tag{7.76}$$

For example, in the case where there is no matter in the hidden sector and the gauge groups are respectively $SU(N_1)$ and $SU(N_2)$, the beta function coefficients are simply $b_i = 3N_i$ (see equation (9.42) of Chapter 9). Writing $N_{1,2} = N \pm \Delta N$, we see from (7.73) that $m_{3/2}/M_U$ scales as $\exp(-8\pi^2 S/N)$. Since the condition (7.76) imposes to take large values of N (ΔN is an integer), we see that one generates a low energy scale if S is large, *i.e.* if the gauge coupling at unification is small. This is difficult to reconcile with indications that the coupling at unification is of the order of unity. One thus has to appeal to more elaborate models [70] (see for example Exercise 2).

Another way to stabilize the field S is to advocate the presence of a constant term in the effective superpotential [115]:

$$W = c + h e^{-24\pi^2 S/b_0}. \tag{7.77}$$

The origin of this constant could be:

- The field strength h_{MNP} of the antisymmetric tensor b_{MN} present in the massless sector of any closed string [see Chapter 10, equation (10.116) ff]. In the context of the weakly coupled heterotic string, one may associate the constant term c with the condensation of the "compact" part of h_{MNP} [103]:

$$\langle h_{ijk} \rangle = c M_P \epsilon_{ijk}, \tag{7.78}$$

  where the indices run over the three complex dimensions of the six-dimensional manifold. It should be noted that c obeys in this case a quantization condition found by topological arguments similar to the ones that lead to the quantization of a Dirac monopole [324].
- The contribution to the superpotential of a gauge singlet scalar field acquiring a large vacuum expectation value [89].

In any case, the superpotential (7.77) yields a contribution $V(S) \equiv F^S g_{S\bar{S}} \bar{F}^{\bar{S}}$ to the scalar potential which reads (using (6.31) of Chapter 6)

$$V(S) = e^K \left[c + h \left(1 + \frac{24\pi^2}{b_0}(S + \bar{S}) \right) e^{-24\pi^2 S/b_0} \right]. \tag{7.79}$$

This contribution has a supersymmetry-breaking ground state at a finite value $\langle S \rangle$: $V(\langle S \rangle) = 0$ but $m_{3/2}(\langle S \rangle) \neq 0$. The potential vanishes also for $S \to \infty$ where global supersymmetry is restored. It turns out [34, 35] that, in the case of a no-scale model where $V(S)$ is the only contribution to the scalar potential once the other fields are set to their ground state values, all soft supersymmetry-breaking terms vanish at tree level. One then has to go to the one-loop level to compute the soft terms, which thus arise through radiative corrections.

Finally, it has been proposed to use a nontrivial Kähler potential to stabilize the S field. In string theories, the simple dependance $K = -\ln\left(S + \bar{S}\right)$ receives nonperturbative corrections which may play a rôle in dilaton stabilization [20].

Let us return to a discussion of the universality of the boundary values for soft supersymmetry breaking terms. We take the example of the gaugino masses. Quite generally, the gauge kinetic terms (7.63) are related by supersymmetry to the terms

$$\mathcal{L}_m = \frac{1}{4} F_S \bar{\lambda}_R^a \lambda_L^a + \text{ h.c.} \tag{7.80}$$

which yield a universal contribution of order $\langle F_S \rangle$ to the gaugino masses. The universality of this contribution is due to the fact that there is a single field which couples in the same way to all gauge supermultiplets.

However, it is absolutely possible to have gravitational corrections through terms which involve a set of nonsinglet fields Φ_{ab} (to stress the gravitational character of this interaction term, we write the Planck scale explicitly):

$$\mathcal{L}'_{\text{kin}} = -\frac{1}{4}\frac{\text{Re}\Phi_{ab}}{M_P} \, F^{a\mu\nu} F^b_{\mu\nu} + \frac{1}{4}\frac{\text{Im}\Phi_{ab}}{M_P} \, F^{a\mu\nu} \tilde{F}^b_{\mu\nu} \tag{7.81}$$

and their supersymmetric completion:

$$\mathcal{L}'_m = \frac{1}{4}\frac{F_{\Phi_{ab}}}{M_P} \bar{\lambda}_R^a \lambda_L^b + \text{ h.c.} \tag{7.82}$$

which obviously gives a nonuniversal contribution to gaugino masses[12].

A second example is more directly related to superstring models. Indeed, in most models, the dilaton coupling (7.63) receives corrections from superheavy thresholds and takes the general form:

$$-\frac{1}{4}\left[\text{Re } S + \Delta_a\left(T, \bar{T}\right)\right] F^{a\mu\nu} F^a_{\mu\nu}, \tag{7.83}$$

where Δ_a is a function of a generic modulus field[13] T. By supersymmetric completion, one obtains

$$\frac{1}{4}\left[F_S + \frac{\partial\Delta_a}{\partial T} F_T\right] \bar{\lambda}_R^a \lambda_L^a + \text{ h.c.} \tag{7.84}$$

which thus leads to nonuniversalities.

[12] As an example, take a $SU(5)$ grand unification theory (see Chapter 9): the gauge fields belong to a **24** of $SU(5)$ and thus gauge invariance of (7.81) implies that Φ_{ab} belongs to $\mathbf{1 + 24 + 75 + 200}$. The case of a singlet Φ_{ab} (**1**) corresponds to universality whereas the others (**24, 75, 200**) generate nonuniversal masses.

[13] As we will see in Chapter 10, an example of a modulus field is a radius modulus whose value fixes the radius of a higher dimensional compact manifold. The masses of heavy Kaluza–Klein states are then radius-dependent, which explains the presence of T in formulas of the effective low energy theory such as (7.83). See Section 10.4.4 of Chapter 10 for details.

Brignole, Ibáñez and Muñoz [51] have proposed to parametrize the supersymmetry breaking effects as follows:

$$\langle F_S \rangle = C m_{3/2} \sin \theta, \quad \langle F_T \rangle = C m_{3/2} \cos \theta. \tag{7.85}$$

Thus, $\theta = \pi/2$ corresponds to gaugino mass universality (the so-called dilaton dominated scenario) whereas $\theta \ll 1$ (moduli dominated) leads to large nonuniversalities.

Universality remains in any case a preferred option from a phenomenological perspective because of the dangerous flavor-changing neutral currents (FCNC) that nonuniversalities generate. As mentioned in Section 6.8 of Chapter 6, even in the universal case, some care should be taken since renormalization to the electroweak scale induces some nonuniversalities.

7.4.3 Anomaly mediation

In the context of a locally supersymmetric theory, attention must be paid to the regularization procedure. It induces a generic contribution to the gaugino masses and the A-terms, known as the (conformal) anomaly mediated contribution. Such a contribution is dominant in models where there are no singlet fields present in the hidden sector (such as the S field of the preceding section) to generate a significant mass to gauginos. In such cases, this source of supersymmetry breaking in the observable sector is called anomaly mediation [189, 320]. But it must be stressed that this contribution is always present, although often nonleading. Let us now see more precisely how it arises.

The low energy theory may be considered as an effective theory of a deeper theory which sets boundary values at a fundamental scale Λ. Care must be taken at the cut-off scale Λ because the procedure of cutting off momenta in an integral is not consistent with *local* supersymmetry. Inspiration may be found in the case where the scale Λ is related to the expectation value of a scalar field, *i.e.* the superheavy masses are determined by a scalar vev (as in the Higgs mechanism). In a supersymmetric context, the scalar field is merged into a chiral superfield and the supersymmetry breaking effects trigger a F-term of order $m_{3/2}$. In other words, to the cut-off scale Λ must be associated an effective F-term $F_\Lambda = \Lambda m_{3/2}$.

If we now consider the observable sector gauge kinetic term

$$\mathcal{L} = -\frac{1}{4}\left[\frac{1}{g^2} + \frac{b^{(1)}}{8\pi^2}\log\frac{\Lambda}{\mu} \right] F^{\mu\nu} F_{\mu\nu}, \tag{7.86}$$

where we have included one-loop renormalization effects, the preceding considerations lead us to introduce for supersymmetric completion, according to (6.65),

$$\mathcal{L}' = \frac{1}{4}\frac{b^{(1)}}{8\pi^2}\frac{1}{\Lambda} F_\Lambda \bar{\lambda}_R \lambda_L + \text{h.c.} = \frac{1}{4}\frac{b^{(1)}}{8\pi^2} m_{3/2} \bar{\lambda}_R \lambda_L + \text{h.c.} \tag{7.87}$$

Taking into account the factor g^2 in front of the kinetic term for gauginos, this gives a universal gaugino mass:

$$M_{1/2} = -\frac{b^{(1)} g^2}{16\pi^2} m_{3/2}. \tag{7.88}$$

As discussed in Section A.5.4 of Appendix Appendix A, the type of renormalization effects that we have taken into account here are directly related to the (super)conformal anomaly. Hence the name anomaly mediation[14].

Similarly [189, 320], A-terms are generated:

$$A_{ijk} = \frac{1}{2}\left(\gamma_i + \gamma_j + \gamma_k\right)m_{3/2},\tag{7.89}$$

where the γ's are the anomalous dimensions (see Section E.3 of Appendix E) of the corresponding cubic term in the superpotential: $\lambda_{ijk}\phi^i\phi^j\phi^k$. Finally, scalar masses receive a contribution

$$m_i^2 = -\frac{1}{4}\frac{d\gamma_i}{dt}m_{3/2}^2,\tag{7.90}$$

Since γ_i is a function of gauge couplings g_j and Yukawa couplings λ_k, $d\gamma_i/dt = \sum_j(\partial\gamma_i/\partial g_j)\beta_j + \sum_k(\partial\gamma_i/\partial\lambda_k)d\lambda_k/dt$. It should be stressed that scalar masses are expected to receive other contributions than (7.90) from the superheavy fields of the fundamental theory[15].

In the situation where the anomaly mediated contribution dominates, the prediction for gaugino masses is strikingly different from gravity mediation. For example, in the case of the minimal supersymmetric model, one obtains from (7.88), using $b_1^{(1)} = -33/5, b_2^{(1)} = -1, b_3^{(1)} = 3$ (see Section E.1 of Appendix E) and $g_1^2 = 5g'^2/3$,

$$M_1 = \frac{11\alpha}{4\pi\cos^2\theta_w}m_{3/2} = 8.9\times 10^{-3}m_{3/2},$$

$$M_2 = \frac{\alpha}{4\pi\sin^2\theta_w}m_{3/2} = 2.7\times 10^{-3}m_{3/2},\tag{7.91}$$

$$M_3 = -\frac{3\alpha_3}{4\pi}m_{3/2} = -2.8\times 10^{-3}m_{3/2},$$

where the right-hand sides give the values at the electroweak scale (we have neglected here for M_1 and M_2 finite contributions coming from the Higgs sector).

This set of values implies an unexpected feature: the wino $\tilde{A}^3$ is lighter than the bino $\tilde{B}$ and becomes the LSP. Moreover, neutral ($\tilde{A}^3$) and charged ($\tilde{W}^\pm$) winos are almost degenerate. This is due to the fact that the splitting between the two arises from electroweak breaking and is of order M_w^4. For example, in the limit $\mu \gg M_{1,2}, M_w$, one finds respectively for the neutral and charged wino mass

$$m_{\chi^0} = M_2 - \frac{M_w^2}{\mu}\sin 2\beta - \frac{M_w^4\tan^2\theta_w}{(M_1 - M_2)\mu^2}\sin^2 2\beta,$$

$$m_{\chi^\pm} = M_2 - \frac{M_w^2}{\mu}\sin 2\beta + \frac{M_w^4}{\mu^3}\sin 2\beta,\tag{7.92}$$

[14]In more technical terms, the contribution given here is related to the super-Weyl anomaly. It should be stressed [19, 172] that there exists other contributions associated with the Kähler and sigma model anomalies. They depend explicitly on the form of the Kähler potential.

[15]In the case of scalar masses, the following counterterm is allowed: $\int d^4\theta\, Z^\dagger Z\Phi^\dagger\Phi/m_P^2$, where Z is a superfield of the hidden sector.

where the notation makes it clear that these are the lightest neutralino and chargino. Because wino annihilation is very efficient, the thermal relic density of such a LSP is insufficient to account for dark matter. One may appeal to nonthermal production mechanisms.

Regarding scalar masses, one obtains from (7.90) and the expressions of anomalous dimensions given in Appendix E:

$$m_Q^2 = \left(8\frac{g_3^4}{16\pi^2} - \frac{3}{2}\frac{g_2^4}{16\pi^2} - \frac{11}{50}\frac{g_1^4}{16\pi^2} + \lambda_t \frac{d\lambda_t}{dt} + \lambda_b \frac{d\lambda_b}{dt} \right) \frac{m_{3/2}^2}{16\pi^2},$$

$$m_T^2 = \left(8\frac{g_3^4}{16\pi^2} - \frac{88}{25}\frac{g_1^4}{16\pi^2} + 2\lambda_t \frac{d\lambda_t}{dt} \right) \frac{m_{3/2}^2}{16\pi^2},$$

$$m_B^2 = \left(8\frac{g_3^4}{16\pi^2} - \frac{22}{25}\frac{g_1^4}{16\pi^2} + 2\lambda_b \frac{d\lambda_b}{dt} \right) \frac{m_{3/2}^2}{16\pi^2},$$

$$m_L^2 = \left(-\frac{3}{2}\frac{g_2^4}{16\pi^2} - \frac{99}{50}\frac{g_1^4}{16\pi^2} + \lambda_\tau \frac{d\lambda_\tau}{dt} \right) \frac{m_{3/2}^2}{16\pi^2},$$

$$m_\tau^2 = \left(-\frac{198}{25}\frac{g_1^4}{16\pi^2} + 2\lambda_\tau \frac{d\lambda_\tau}{dt} \right) \frac{m_{3/2}^2}{16\pi^2},$$

$$m_{H_2}^2 = \left(-\frac{3}{2}\frac{g_2^4}{16\pi^2} - \frac{99}{50}\frac{g_1^4}{16\pi^2} + 3\lambda_t \frac{d\lambda_t}{dt} \right) \frac{m_{3/2}^2}{16\pi^2},$$

$$m_{H_1}^2 = \left(-\frac{3}{2}\frac{g_2^4}{16\pi^2} - \frac{99}{50}\frac{g_1^4}{16\pi^2} + 3\lambda_b \frac{d\lambda_b}{dt} + \lambda_\tau \frac{d\lambda_\tau}{dt} \right) \frac{m_{3/2}^2}{16\pi^2}, \tag{7.93}$$

and similarly for squarks and sleptons of the first two families. We note that, because the lepton Yukawa couplings are small, this contribution to slepton squared masses are negative. This would indicate an instability but we have noted above that that the hidden sector may give extra contributions. For phenomenological purposes, one usually introduces a universal contribution m_0 to represent these contributions.

7.4.4 Gauge mediation

The potential problems of gravity-messenger models with FCNC together with developments in dynamical supersymmetry breaking have led a certain number of authors to reconsider models with "low energy" messengers. These are the gauge-mediated models [112,113] which rely on standard gauge interactions as the mediator [109,111]: since "standard gauge interactions are flavor-blind", soft masses are universal.

One can discuss the nature of messenger fields on general grounds. These fields are charged under $SU(3) \times SU(2) \times U(1)$ and since their masses are larger than the electroweak scale they must appear in vectorlike pairs. One can parametrize this through a (renormalizable) coupling to a singlet field N, fundamental or composite: $W \ni \lambda N M \bar{M}$ with $\lambda \langle N \rangle \gg M_W, M_Z$. Since $M, \bar{M}$ must feel supersymmetry breaking, $\langle F_N \rangle \neq 0$. In other words, the Goldstino field overlaps with the supersymmetric partner

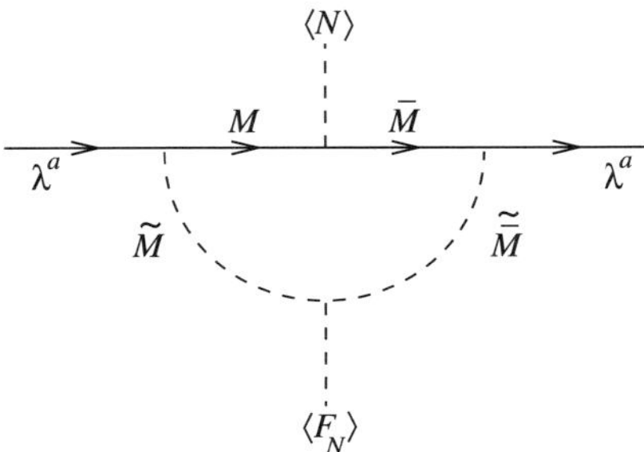

Fig. 7.8 Diagram giving rise to a gaugino mass in gauge-mediated models.

of N. In the simplest models that we consider here, it coincides with it. In order not to spoil perturbative gauge coupling unification, the messengers must appear in complete representations of $SU(5)$ (see Chapter 9). In what follows, we will use the Dynkin index n of the messenger representation: $n = 1$ for a **5** and $n = 3$ for a **10** of $SU(5)$.

Messenger fermion masses are simply $M = \lambda\langle N\rangle$ whereas the scalar mass terms receive a nondiagonal supersymmetric contribution:

$$(M^\dagger \bar{M}) \begin{pmatrix} (\lambda\langle N\rangle)^\dagger (\lambda\langle N\rangle) & (\lambda\langle F_N\rangle)^\dagger \\ (\lambda\langle F_N\rangle) & (\lambda\langle N\rangle)(\lambda\langle N\rangle)^\dagger \end{pmatrix} \begin{pmatrix} M \\ \bar{M}^\dagger \end{pmatrix}. \tag{7.94}$$

Hence scalar messenger mass eigenvalues are $\sqrt{|\lambda\langle N\rangle|^2 \pm |\lambda\langle F_N\rangle|}$. Stability requires: $|\langle F_N\rangle| < \lambda\,|\langle N\rangle|^2 = M^2$.

Gaugino masses appear at one loop from the diagram of Fig. 7.8 whereas scalar masses appear at two loops (see Fig. 7.9). In the limit where $\langle F_N\rangle \ll \lambda\langle N\rangle^2$,

$$M_a = n\frac{\alpha_a}{4\pi}\frac{\langle F_N\rangle}{\langle N\rangle} \tag{7.95}$$

$$\tilde{m}^2 = 2n\left(\frac{\langle F_N\rangle}{\langle N\rangle}\right)^2 \left[\sum_a C_i^a \left(\frac{\alpha_a}{4\pi}\right)^2\right], \tag{7.96}$$

where the index a refers to the gauge group, $C_i^{3,2}$ is the Casimir of order 2 of the sfermion representation[16], $C_i^1 = 3y_i^2/20$ and $\alpha_1 \equiv g_1^2/(4\pi) \equiv (5/3)g'^2/(4\pi)$. There are no significant contributions to A-terms. As advertized, squark masses are family independent because the $SU(3) \times SU(2) \times U(1)$ quantum numbers are so. This is welcome to suppress flavor changing neutral currents.

Since squarks and sleptons must be heavy enough, the scale $\Lambda \equiv \langle F_N\rangle/\langle N\rangle$ must be larger than 30 TeV. Since $\langle F_N\rangle < \lambda\langle N\rangle^2$, the messenger fermion mass $M = \lambda\langle N\rangle$ is larger than Λ.

In the minimal model with a $5 + \bar{5}$ of messenger fields, one typically obtains $m_{\tilde{q}}^2 \sim 8M_3^2/3$, $m_{\tilde{l}_L}^2 \sim 3M_2^2/2$, $m_{\tilde{l}_R}^2 \sim 6M_1^2/5$. Also, if the gaugino masses obey a unification

---

[16]This is $(N^2 - 1)/2N$ for a fundamental of $SU(N)$, i.e. $C_i^3 = 4/3$, $C_i^2 = 3/4$.

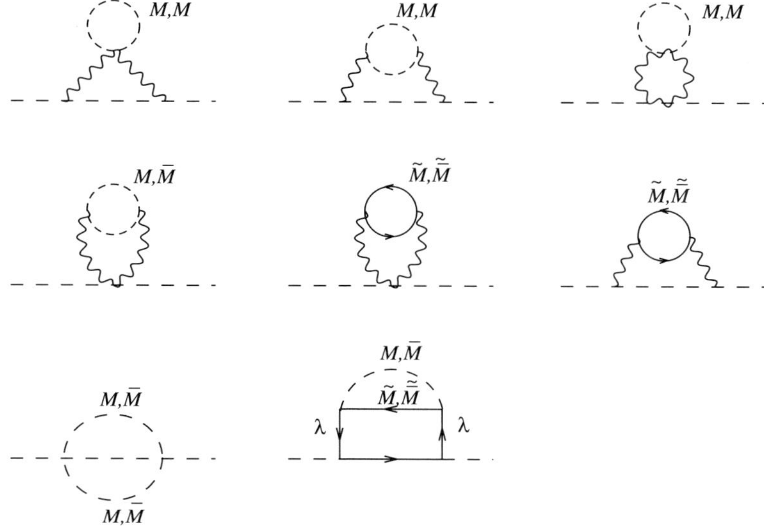

Fig. 7.9 Diagrams contributing to scalar masses in gauge-mediated models.

constraint (such as (6.90) of Chapter 6), they are in the ratios $M_1 : M_2 : M_3 = \alpha_1 : \alpha_2 : \alpha_3$ whereas $m_{\tilde{q}}^2 : m_{\tilde{l}_L}^2 : m_{\tilde{l}_R}^2 = \frac{4}{3}\alpha_3^2 : \frac{3}{4}\alpha_2^2 : \frac{3}{5}\alpha_1^2$. This leads to large squark masses, which has consequences on electroweak breaking, as we will see below. In any case, one expects strong correlations in the supersymmetric spectrum [106]. But there are possible modifications: presence of several singlet fields, nonzero D-term for $U(1)_Y$, direct couplings of messengers to the observable sector [114].

One of the reasons for the success of gauge-mediated models is that they yield a phenomenology quite different from the standard gravity-mediated approach. This has generated a lot of activity. The most notable differences come from the fact that, quite often, the gravitino is the lightest supersymmetric particle (LSP): for example $\langle F_N \rangle = (10^5 \text{ GeV})^2$ gives $m_{3/2} = \langle F_N \rangle / m_P \sim 10 \text{ eV}$. It is then important to determine which is the next lightest (NLSP). Since typically the NLSP decays into gravitino plus photon, the signatures will involve missing transverse energy and one or several photons.

On the theoretical side, gauge-mediated models face several problems. One concerns the radiative breaking of $SU(2) \times U(1)$. In standard supergravity models, as discussed in Section 6.5 of Chapter 6, it is the coupling to the top quark and large $\ln m_P / M_Z$ which are responsible for this breaking. In gauge-mediated models, it is the large stop mass: as can be inferred from the renormalization group equation (E.18) of Appendix E,

$$m_2^2(m_{\tilde{t}}) \sim m_2^2(M) - \frac{3}{8\pi^2}\lambda_t^2(m_{\tilde{t}_L}^2 + m_{\tilde{t}_R}^2)\ln M/m_{\tilde{t}}. \tag{7.97}$$

Indeed this mechanism works so well that it involves some undesirable fine-tuning [75].

The worst problem faced by gauge-mediated models is however the μ/B_μ problem: since all soft supersymmetry-breaking terms scale as Λ, it is very difficult to avoid the relation:

$$B_\mu \sim \mu\Lambda. \tag{7.98}$$

Given the value of Λ, this is obviously incompatible with a relation such as (5.25) of Chapter 5. There has been attempts to decouple the origin of μ and of B_μ [112,114]. For instance, one may introduce a new singlet N' with coupling $\lambda' N' H_2 \cdot H_1$: $\mu = \lambda'\langle N'\rangle$. But it is then extremely difficult to decouple N and N'. There does not seem to be a completely satisfactory solution to this μ/B_μ problem.

[An interesting development is the elaboration of gauge-mediated models without messengers. These models take more advantage of the developments in dynamical symmetry breaking that we will describe in the next chapter (the corresponding DSB sector was often a black box in the previous models): the rôle of the messengers is played by effective degrees of freedom of the DSB sector. This implies that the gauge symmetry of the Standard Model is a subgroup of the flavor symmetry group of the DSB sector. This group is therefore rather large and there are many effective messengers. Unless they are heavy, this spoils the perturbative unification of gauge couplings. How then to make effective messengers heavy? Remember that the messenger mass is $\lambda\langle N\rangle$ whereas $\langle F_N\rangle/\langle N\rangle$ is fixed by supersymmetry breaking; since $\langle N\rangle$ itself is not fixed, the idea is to require $\langle N\rangle \gg \langle F_N\rangle^{1/2}$. Explicit realizations involve nonrenormalisable terms [10,314] or inverted hierarchy [291].]

7.5 Limits on supersymmetric particles

Generally speaking, in the R-parity conserving scenarios that we will consider in this section, the typical supersymmetric signature is missing energy carried away by the decay LSP (which is assumed to be here the lightest neutralino).

7.5.1 Sleptons and squarks

Supersymmetric particles have been searched at the LEP collider and not been found, which allows us to put a limit on their mass. Since the limit is somewhat less model dependent for smuons than for others, we will first detail the procedure in this case.

Smuon mass limits

At LEP, the production of a pair of right-handed[17] smuons $e^+ e^- \rightarrow \tilde{\mu}_R^+ \tilde{\mu}_R^-$ goes through the exchange of a photon or a Z: thus the production cross-section depends only on the smuon mass. Then, the smuon decays into a muon and the LSP: $\tilde{\mu}_R \rightarrow \mu\chi$. Thus, besides the missing energy, one is searching for an observable final state with muons which are acoplanar with the beam (because of the momentum taken away by the LSP). As the mass of the LSP decreases, the smuons become less acoplanar and the signal becomes similar to $e^+ e^- \rightarrow W^+ W^- \rightarrow \mu\nu\mu\nu$ (although it would provide an anomalously large $W \rightarrow \mu\nu$ branching ratio). Thus the limit is expected to degrade for smaller LSP masses.

---

[17]For a given smuon mass, the $\tilde{\mu}_L$ production cross-section is larger than the $\tilde{\mu}_R$ one: the $\tilde{\mu}_R$ mass limit provides a conservative smuon mass limit.

This is what is shown in Fig. 7.10 which gives the domain excluded experimentally by the search for acoplanar muons in the plane $(m_{\tilde{\mu}_R}, m_\chi)$ assuming 100% branching ratio for the decay $\tilde{\mu}_R^- \to \mu\chi$: the dotted line gives the expected limit, *i.e.* the limit which would a *priori* be obtained if no signal was found in the data; it becomes less stringent as one goes to lower m_χ. Thanks to a welcome statistical fluctuation, the actual limit does not follow this trend however. As the mass difference $m_{\tilde{\mu}_R} - m_\chi$ becomes very small, the limit also degrades rapidly: there is not enough final state visible energy to allow a good detection efficiency.

If the branching ratio $\tilde{\mu}_R^- \to \mu\chi$ turns out not to be 100%, one must allow for decay chains, such as $\tilde{\mu}_R^- \to \mu(\chi' \to \gamma\chi)$. This reduces the efficiency of the acoplanar muon search in the region of light χ.

Other slepton limits are more model dependent. In the case of stau, mixing may be important as for the stop and the most conservative limit corresponds to a vanishing coupling of the lightest stau to Z. For the selectron, there is an extra contribution (neutralino exchange in the t channel) which leads to a higher limit as can be seen from Figure 7.10.

Stop mass limit

As emphasized in Section 5.3.3 of Chapter 5, the large mixing term in the stop mass matrix (5.53) allows for light stop mass eigenstates. Light stops have been searched for at LEP: their decay $\tilde{t} \to c\chi$ arises through a loop and the decay time may be longer than the stop hadronization time, in which case stop-hadrons are first produced. Fig. 7.11 gives the exclusion plot in the $(m_{\tilde{t}}, m_\chi)$ plane coming from LEP experiments as well as the Tevatron Run I.

Generic squarks and gluinos have also been searched for at the Tevatron where there is a strong production of $\tilde{q}\bar{\tilde{q}}$ or $\tilde{g}\tilde{g}$ pairs. In the case where squarks are lighter than gluinos, squarks decay predominantly as $\tilde{q} \to q\chi$ and the final state is a pair of acoplanar jets plus missing transverse energy. However, other decay modes such

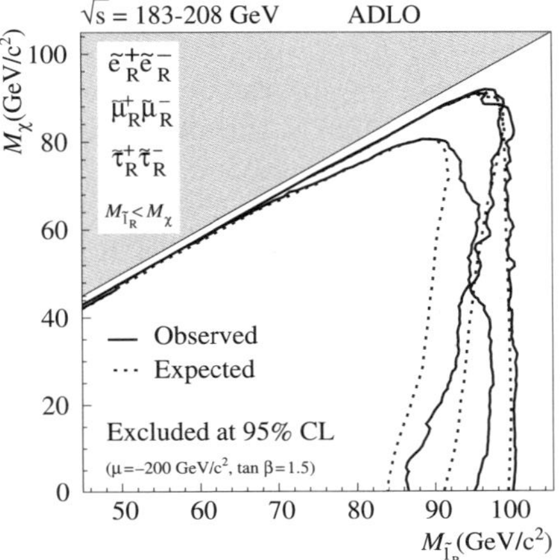

Fig. 7.10 Domain excluded at 95% CL by the four LEP experiments in the plane $(m_{\tilde{\ell}_R}, m_\chi)$ resp. from left to right for $\tilde{\tau}_R$, $\tilde{\mu}_R$ and $\tilde{e}_R$ (LEP SUSY working group).

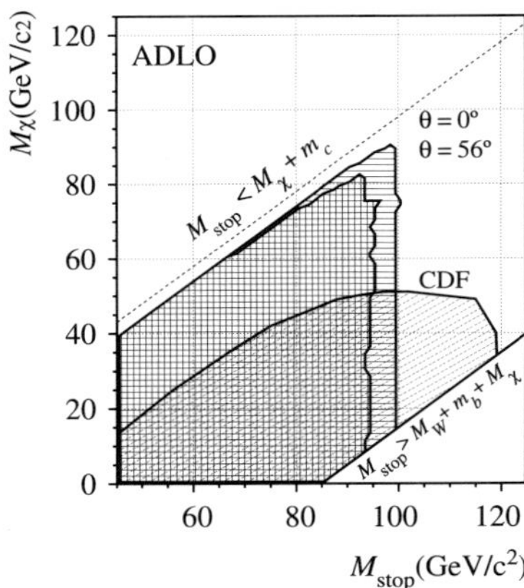

Fig. 7.11 Domain excluded by the LEP experiments in the $(m_{\tilde{t}}, m_\chi)$ plane from $\tilde{t} \to c\chi$ decays: inner (resp. outer) contour corresponds to vanishing (resp. maximal) $Z\tilde{t}\tilde{t}$ couplings. Also indicated is the domain excluded by CDF at Tevatron Run I [211].

as cascade decays ($\tilde{q} \to q\chi', q'\chi^\pm$ with subsequent decay of χ' or $\chi^\pm$) are also possible which lead to different topologies. Since the decay branching ratios are model dependent, it is difficult to obtain generic mass limits.

7.5.2 Neutralinos and charginos

Charginos (resp. neutralinos) are pair produced at LEP through s-channel γ/Z (resp. Z) exchange and t-channel sneutrino (resp. selectron) exchange: the two channels interfere destructively (resp. constructively). If sleptons are too heavy to be produced in the subsequent decays, the relevant parameters are M_2, μ and $\tan\beta$ (plus M_1 in the case of neutralinos). Typical decays are $\chi^+ \to \chi W^*$ and $\chi' \to \chi Z^*$, which make the signatures rather straightforward: the kinematic limit of 104 GeV is basically reached for the chargino at LEP [380]. For lighter sleptons, invisible final states such as $\chi' \to \tilde{\nu}\nu$ open up.

In the case where the chargino χ^+ and the neutralino χ become almost degenerate (Higgsino-like for large M_2 or wino-like in the case of anomaly mediation, see (7.92)), the above mentioned search loses its sensitivity. One has to resort to more elaborate techniques such as tagging on a low energy photon radiated from the initial state.

7.5.3 Constraints on the LSP mass

The limits on the LSP arising from LEP searches are more intricate because they result from different channels: the direct production $e^+e^- \to \tilde{\chi}^0\tilde{\chi}^0$ cannot be used since it gives an invisible final state. The mass limit in terms of $\tan\beta$ is given in Fig. 7.12. We have already seen that negative Higgs boson searches give a lower limit to the value of $\tan\beta$. For values within this bond, the main constraint comes from the chargino channel: $e^+e^- \to \tilde{\chi}^+\tilde{\chi}^-$. However, when chargino and sneutrino become almost degenerate, this channel looses its sensitivity because the decay $\tilde{\chi}^\pm \to \ell\nu$ becomes invisible. Such a configuration is called the corridor in Fig. 7.12. This is where slepton searches take over, giving an absolute mass lower limit of 45 GeV/c².

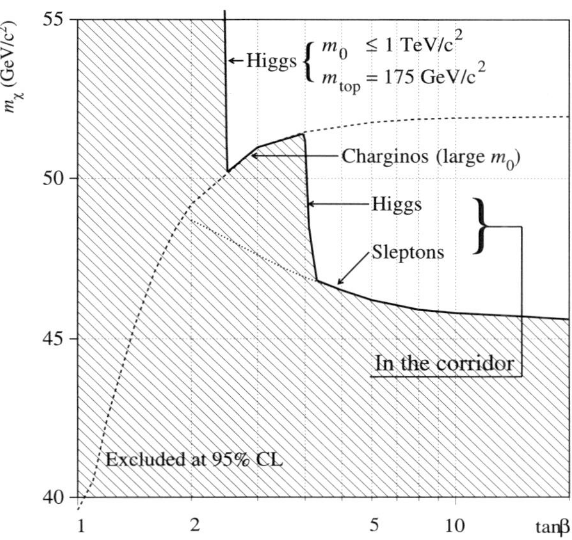

Fig. 7.12 LSP mass limit as a function of $\tan\beta$ obtained by a combination of the four LEP experiments.

In the more constrained case of the minimal SUGRA model, there are fewer pathological mass configurations, which yields a better limit of 60 GeV/c^2.

7.6 *R*-parity breaking

Violations of R-parity are often associated with violations of baryon or lepton number and are thus bounded by experimental results on baryon or lepton violating interactions. The weakest constraints are obtained when one assumes that a single R-violating coupling dominates, typically 10^{-1} to 10^{-2} times ($\tilde{m}/100$ GeV), where $\tilde{m}$ is the mass of the superpartner involved. Much more stringent constraints may arise on the product of some specific couplings.

The R-parity violating renormalizable superpotential reads (see equation (5.4) of Chapter 5)

$$W_{\not R} = \mu_i H_2 \cdot L_i + \tfrac{1}{2}\lambda_{ijk}L_i \cdot L_j E_k^c + \lambda'_{ijk}L_i \cdot Q_j D_k^c + \tfrac{1}{2}\lambda''_{ijk}U_i^c D_j^c D_k^c, \qquad (7.99)$$

where $U_i^c D_j^c D_k^c \equiv \epsilon^{\alpha\beta\gamma}U_{i\alpha}^c D_{j\beta}^c D_{k\gamma}^c$ (α, β, γ color indices). It is straightforward to show that

$$\lambda_{ijk} = -\lambda_{jik}, \qquad \lambda''_{ijk} = -\lambda''_{ikj}. \qquad (7.100)$$

At this level, it is possible to redefine the fields in order to absorb the quadratic term into the μ term: $H_1 \to H_1' \propto \mu H_1 + \mu_i L_i$. However, this is no longer equivalent once one includes soft supersymmetry breaking terms.

Let us illustrate how constraints arise on specific couplings λ_{ijk}.

We first assume that a single coupling λ_{12k} dominates. The corresponding R-parity violating operator yields an extra contribution to the muon lifetime as can be seen from Fig. 7.13.

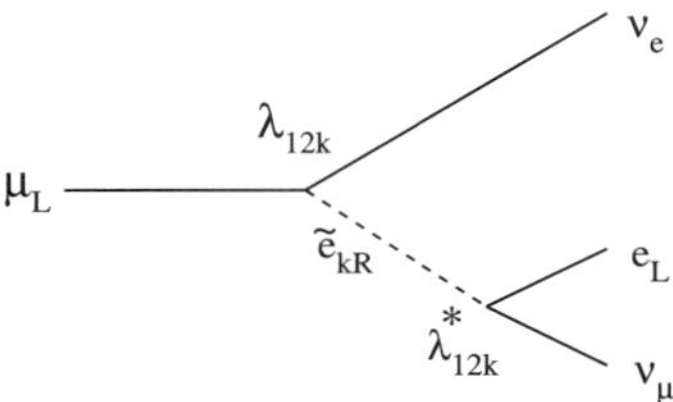

Fig. 7.13 Contribution to the muon lifetime from the R-parity violating operator $L_1 \cdot L_2 E_k^c$.

At tree level, one thus finds for the Fermi coupling measured in μ decay:

$$\frac{G_\mu}{\sqrt{2}} = \frac{g^2}{8M_W^2}\left[1 + \frac{M_W^2}{g^2 m_{\tilde{e}_{kR}}^2}|\lambda_{12k}|^2\right]. \tag{7.101}$$

This in turn induces an effect on the CKM matrix element V_{ud} determined by comparing the nuclear beta decay to muon beta decay:

$$|V_{ud}|^2 = \frac{\left|V_{ud}^{SM}\right|^2}{\left|1 + (M_W^2/g^2 m_{\tilde{e}_{kR}}^2)|\lambda_{12k}|^2\right|^2}. \tag{7.102}$$

One infers from this the 2σ limit: $\lambda_{12k} < 0.05\left(m_{\tilde{e}_{kR}}/100\text{ GeV}\right)$ $[V_{ud}]$.

Muon decay also allows for other tests, especially through its decay rate. The precise measurement of the ratio

$$R_{\tau\mu} \equiv \Gamma(\tau \to \mu\nu\bar{\nu})/\Gamma(\mu \to e\nu\bar{\nu}) \tag{7.103}$$

is sensitive to both λ_{12k} and λ_{23k} through the modification of the τ leptonic decay via $\tilde{e}_{kR}$ exchange (similar to Fig. 7.13). More precisely, we have

$$R_{\tau\mu} \simeq R_{\tau\mu}^{SM}\left[1 + \frac{2M_W^2}{g^2 m_{\tilde{e}_{kR}}^2}\left(|\lambda_{23k}|^2 - |\lambda_{12k}|^2\right)\right]. \tag{7.104}$$

Comparison with the measured value gives

$$\lambda_{12k} < 0.07\left(\frac{m_{\tilde{e}_{kR}}}{100\text{ GeV}}\right)\ [R_{\tau\mu}],\ \lambda_{23k} < 0.07\left(\frac{m_{\tilde{e}_{kR}}}{100\text{ GeV}}\right)\ [R_{\tau\mu}]. \tag{7.105}$$

Similarly, the ratio of leptonic decay widths

$$R_\tau \equiv \frac{\Gamma(\tau \to e\bar{\nu}_e\nu_\tau)}{\Gamma(\tau \to \mu\bar{\nu}_\mu\nu_\tau)} = R_\tau^{SM}\left[1 + \frac{2M_W^2}{g^2 m_{\tilde{e}_{kR}}^2}\left(|\lambda_{13k}|^2 - |\lambda_{23k}|^2\right)\right] \tag{7.106}$$

yields the following constraints

$$\lambda_{13k} < 0.07\left(\frac{m_{\tilde{e}_{kR}}}{100\text{ GeV}}\right)\ [R_\tau],\ \lambda_{23k} < 0.07\left(\frac{m_{\tilde{e}_{kR}}}{100\text{ GeV}}\right)\ [R_\tau]. \tag{7.107}$$

Table 7.1 Best limits on some of the couplings λ'_{ijk}. The bounds are obtained from experimental results on the following processes: neutrinoless double beta decay $[\beta\beta 0\nu]$, charge current universality in the quark sector $[V_{ud}]$, semileptonic decays of mesons $[B^- \to X_q \tau^- \bar{\nu}]$, atomic parity violation $[Q_W(Cs)]$.

Coupling	Constraint	Process
λ'_{111}	$3.3 \; 10^{-4}(m_{\tilde{q}}/100 \text{ GeV})^2 (m_{\tilde{g}}/100 \text{ GeV})^{1/2}$	$\beta\beta 0\nu$
$\lambda'_{11k}, \; k = 2,3$	$0.02(m_{\tilde{d}_{k_R}}/100 \text{ GeV})$	V_{ud}
λ'_{333}	$0.12(m_{\tilde{b}_{k_R}}/100 \text{ GeV})$	$B^- \to X_q \tau^- \bar{\nu}$
λ'_{1j1}	$0.04(m_{\tilde{u}_{j_L}}/100 \text{ GeV})$	$Q_W(Cs)$

We refer the reader to more extensive reviews on the subject (see for example [27]) for complete lists of constraints. It should be stressed that, whereas the bounds on the λ_{ijk} couplings scale like $m_{\tilde{e}_{k_R}}$, other bounds may have a more complicated dependence on the masses of the supersymmetric particles, as can be seen for example from Table 7.1 in the case of λ'_{ijk}.

The main phenomenological consequences of allowing for R-parity violations are as follows.

First, supersymmetric particles need not be produced in pairs, which lowers the threshold for supersymmetric particle production. This is certainly a welcome feature for experimental searches. On the other hand, the fact that, at the time of writing, no supersymmetric particle has been found thus finds a more natural explanation in the context of R-parity conservation.

Table 7.2 gives, for each class of high energy colliders, the resonant production mechanisms allowed by the different types of R-parity breaking couplings.

Second, the LSP is not stable and can decay in the detector. We note that, since the LSP is not stable, cosmological arguments about it being charge and color neutral drop. It could for example be a squark, a slepton, a chargino... We give in Table 7.3 the direct decays of supersymmetric fermions through R-parity violating couplings. Charginos and neutralinos decay into a fermion and a virtual sfermion which subsequently decays into standard fermions through R-parity violating couplings: this yields a three-fermion final state.

For example, in the case of a nonvanishing λ_{ijk} coupling, the sneutrino R-parity violating partial width reads[18]

$$\Gamma\left(\tilde{\nu}_i \to \ell_j^+ \ell_k^-\right) = \frac{1}{16\pi} \lambda_{ijk}^2 m_{\tilde{\nu}_i}. \tag{7.108}$$

The corresponding mean decay length L reads

$$\frac{L}{1 \text{ cm}} = \beta\gamma \left(\frac{100 \text{ GeV}}{m_{\tilde{\nu}_i}}\right) \left(\frac{10^{-7}}{\lambda_{ijk}}\right)^2, \tag{7.109}$$

where γ is the Lorentz boost factor.

---

[18]If the dominant coupling is λ'_{ijk}, replace in the following formulas λ_{ijk}^2 with $3\lambda'^2_{ijk}$, where 3 accounts for color summation.

Table 7.2 s-channel resonant production of sfermions at colliders in the case of R-parity violation.

Collider	Coupling	Sfermion	Process	
e^+e^-	λ_{1j1}	$\tilde{\nu}_\mu, \tilde{\nu}_\tau$	$\ell_1^+ \ell_1^- \to \tilde{\nu}_j$	$j = 2, 3$
ep	λ'_{1jk}	$\tilde{d}_R, \tilde{s}_R, \tilde{b}_R$	$\ell_1^- u_j \to \tilde{d}_{k_R}$	$j = 1, 2$
		$\tilde{u}_L, \tilde{c}_L, \tilde{t}_L$	$\ell_1^+ d_k \to \tilde{u}_{j_L}$	
$p\bar{p}$	λ'_{ijk}	$\tilde{\nu}_e, \tilde{\nu}_\mu, \tilde{\nu}_\tau$	$d_k \bar{d}_j \to \tilde{\nu}_i$	
		$\tilde{e}, \tilde{\mu}, \tilde{\tau}$	$u_j \bar{d}_k \to \tilde{\ell}_{i_L}$	$j = 1, 2$
	λ''_{ijk}	$\tilde{d}, \tilde{s}, \tilde{b}$	$\bar{u}_i \bar{d}_j \to \tilde{d}_k$	$j \neq k$
		$\tilde{u}, \tilde{c}, \tilde{t}$	$\bar{d}_j \bar{d}_k \to \tilde{u}_i$	$j \neq k$

Table 7.3 Direct decays of supersymmetric particles via R-parity violating couplings.

Supersymmetric particles	Couplings		
	λ_{ijk}	λ'_{ijk}	λ''_{ijk}
$\tilde{\nu}_{i_L}$	$\ell_{j_L}^+ \ell_{k_R}^-$	$\bar{d}_{j_L} d_{k_R}$	
$\tilde{\ell}_{i_L}^-$	$\bar{\nu}_{j_L} \ell_{k_R}^-$	$\bar{u}_{j_L} d_{k_R}$	
$\tilde{\nu}_{j_L}$	$\ell_{i_L}^+ \ell_{k_R}^-$		
$\tilde{\ell}_{j_L}^-$	$\bar{\nu}_{i_L} \ell_{k_R}^-$		
$\tilde{\ell}_{k_R}^-$	$\nu_{i_L} \ell_{j_L}^-, \ell_{i_L}^- \nu_{j_L}$		
$\tilde{u}_{i_R}$			$\bar{d}_{j_R} \bar{d}_{k_R}$
$\tilde{u}_{j_L}$		$\ell_{i_L}^+ d_{k_R}$	
$\tilde{d}_{j_L}$		$\bar{\nu}_{i_L} d_{k_R}$	
$\tilde{d}_{j_R}$			$\bar{u}_{i_R} \bar{d}_{k_R}$
$\tilde{d}_{k_R}$		$\nu_{i_L} d_{j_L}, \ell_{i_L}^- u_{j_L}$	$\bar{u}_{i_R} \bar{d}_{j_R}$
χ^0	$\ell_i^+ \bar{\nu}_j \ell_k^-, \ell_i^- \nu_j \ell_k^+,$ $\bar{\nu}_i \ell_j^+ \ell_k^-, \nu_i \ell_j^- \ell_k^+$	$\ell_i^+ \bar{u}_j d_k, \ell_i^- u_j \bar{d}_k,$ $\bar{\nu}_i \bar{d}_j d_k, \nu_i d_j \bar{d}_k$	$\bar{u}_i \bar{d}_j \bar{d}_k, u_i d_j d_k$
χ^+	$\ell_i^+ \ell_j^+ \ell_k^-, \ell_i^+ \bar{\nu}_j \nu_k,$ $\bar{\nu}_i \ell_j^+ \nu_k, \nu_i \nu_j \ell_k^+$	$\ell_i^+ \bar{d}_j d_k, \ell_i^+ \bar{u}_j u_k,$ $\bar{\nu}_i \bar{d}_j u_k, \nu_i u_j \bar{d}_k$	$u_i d_j u_k, u_i u_j d_k,$ $\bar{d}_i \bar{d}_j \bar{d}_k$

Similarly, the partial width for a pure photino neutralino decaying with λ_{ijk} is [99]

$$\Gamma = \lambda_{ijk}^2 \frac{\alpha}{128\pi^2} \frac{M_{\tilde{\gamma}}^5}{m_{\tilde{\ell}}^4}, \tag{7.110}$$

where $m_{\tilde{\ell}}$ is the mass of the virtual slepton. The corresponding decay length is then found to be

$$\frac{L}{1\text{ cm}} = 0.3\gamma \left(\frac{m_{\tilde{\ell}}}{100\text{ GeV}} \right)^4 \left(\frac{100\text{ GeV}}{M_{\tilde{\gamma}}} \right)^5 \left(\frac{10^{-5}}{\lambda_{ijk}} \right)^2. \tag{7.111}$$

Table 7.4 Charges of the fields and of the parameters under the symmetries $U(1)'$ and $U(1)'_R$ of the minimal supergravity model.

	Fields					Parameters				
	H_1	H_2	QU^c	QD^c	LE^c	λ	$M_{1/2}$	A_0	B_μ	μ
$U(1)'$	1	1	-1	-1	-1	0	0	0	-2	-2
$U(1)'_R$	1	1	1	1	1	1	-2	-2	-2	0

Thus, for couplings of order 10^{-5}, one expects displaced vertices in the detector. For much smaller couplings, the neutralino decays outside the detector whereas for larger couplings, the primary and displaced vertices cannot be separated.

7.7 The issue of phases

In our discussions above, we have not paid sufficient attention to the fact that the parameters of a supersymmetric theory are complex. The presence of nontrivial complex phases leads to possible deviations from the standard phenomenology that has been presented until now.

These phases appear already at the level of the minimal supergravity model. We may take as parameters of this model[19] $M_{1/2}$, m_0, A_0 , B_μ and μ. One can redefine the gaugino and Higgs fields in order to make $M_{1/2}$ and B_μ real. One is then left with two independent complex phases for A_0 and μ, which one defines traditionally as

$$\varphi_A \equiv \arg\left(A_0^* M_{1/2}\right), \quad \varphi_B \equiv \arg\left(\mu B_\mu^* M_{1/2}\right). \tag{7.112}$$

One may alternatively note [110] that, in the limit where both μ and the soft parameters B_μ, $M_{1/2}$ and A_0 vanish, the model acquires new abelian symmetries $U(1)'$ and $U(1)'_R$: the charges of the different superfields are given in Table 7.4. We note that the second symmetry is of a type known as R-symmetry, encountered in Chapter 4 or in Section 7.4.2: it does not commute with supersymmetry. Thus it is a symmetry of the scalar potential but not of the superpotential; similarly, gauginos transform whereas vector gauge fields are invariant. We will study in more details such symmetries in the next chapter.

If one allows the parameters of the model to transform as (spurion) fields, with charges given in Table 7.4, we may turn $U(1)'$ and $U(1)'_R$ as symmetries of the full model. We note that we have the following invariant combinations:

$$M_{1/2}\mu B_\mu^*, \quad A_0\mu B_\mu^*, \quad A_0^* M_{1/2}.$$

Two of their phases are independent: they are φ_A and φ_B.

---

[19]This list is somewhat different from the usual one given for example in Section 6.8 of Chapter 6. We have replaced $\tan\beta$ by B_μ. Moreover, in Chapter 6, we assumed implicitly B_μ to be real and $|\mu|$ was fixed by electroweak radiative breaking: we were thus left only with an ambiguity with the sign of μ. Here it is the phase of μ which remains to be fixed.

These phases are constrained by experimental data, such as the electric dipole moment of the neutron. As an example, the down quark electric dipole moment receives a one-loop gluino contribution which is computed to be [56, 309]:

$$d_N = m_d \, \frac{e\alpha_3}{18\pi} \, \frac{|m_{\tilde{g}}|}{m_{\tilde{q}}^4} \, [|A| \sin \varphi_A + |\mu| \tan \beta \sin \varphi_B] \tag{7.113}$$

$$\sim 2 \left(\frac{100 \text{ GeV}}{\tilde{m}}\right)^2 \sin \varphi_{A,B} \times 10^{-23} e.\text{cm} \le 6.3 \times 10^{-26} e.\text{cm},$$

where $\tilde{m}$ is a mass scale of the order of the gluino mass $m_{\tilde{g}}$ or a squark mass $m_{\tilde{q}}$, and A is the low energy value of the A-term (the experimental constraint is found in [144]). This shows that the $\varphi_{A,B}$ have to be small. This need not be accidental. For example, if A_0 and $M_{1/2}$ arise from a single source of supersymmetry breaking, their phases might be identical, in which case $\varphi_A = 0$.

In the case of nonminimal models, other phases appear in the squark and slepton mass matrices. We will return to this question in Chapter 12 and see that a generic supersymmetric extension of the Standard Model has 44 CP violating complex phases!

Further reading

- G.F. Giudice and R. Rattazzi, *Theories with gauge-mediated supersymmetry breaking*, Physics Reports 322 (1999) 419.
- R. Barbier et al., *R-parity violating supersymmetry*, Physics Reports 420 (2005) 1.

Exercises

<u>Exercise 1</u> *We prove the cancellation of the one loop contributions to the magnetic operator in the supersymmetric QED model of Wess and Wess and Zumino [363], studied in Exercise 5 of Chapter 3.*

Using Lorentz covariance, charge conservation and hermiticity, one may write the matrix element of the electromagnetic current in the Dirac theory in terms of

$$\bar{u}_2(p_2) M^\mu u_1(p_1) = \bar{u}_2 \left[\gamma^\mu F_1(q^2) - \sigma^{\mu\nu} \frac{q_\nu}{m} F_2(q^2) + i\gamma_5 \sigma^{\mu\nu} \frac{q_\nu}{m} F_3(q^2) \right] u_1, \tag{7.114}$$

where $q^\mu = p_2^\mu - p_1^\mu$ and $p_1^2 = p_2^2 = m^2$.

(a) Show that parity conservation implies under $F_3 = 0$. We will assume that parity is conserved from now on (hence we do not take into account weak interactions).

(b) Since $(\not{p} - m)u(p) = 0$, one may use the projectors $(\not{p} \pm m)/2m$ and write the decomposition above as

$$(\not{p}_2 + m) M^\mu (\not{p}_1 + m) = (\not{p}_2 + m) \left[\gamma^\mu F_1(q^2) - \sigma^{\mu\nu} \frac{q_\nu}{m} F_2(q^2) \right] (\not{p}_1 + m). \tag{7.115}$$

Deduce that one can write the Pauli form factor $F_2(q^2)$ as

$$F_2(q^2) = \frac{m^2}{q^2(-q^2 + 4m^2)} \text{Tr} \left[\left(\gamma_\mu - \frac{q^2 + 2m^2}{-q^2 + 4m^2} \frac{P_\mu}{m} \right) (\not{p}_2 + m) M^\mu (\not{p}_1 + m) \right], \tag{7.116}$$

where $P \equiv p_1 + p_2$.

(c) We now consider the supersymmetric QED model studied in Exercise 5 of Chapter 3. The Lagrangian is given by its equation (3.58) in terms of the two Majorana spinors Ψ_1 and Ψ_2 and their supersymmetric partners (we add mass terms to give these fields a common mass m). Show that the magnetic operator is necessary of the form $\bar{\Psi}_1 \sigma_{\mu\nu} \Psi_2 F^{\mu\nu} + $ h.c. in this model and write explicitly the indices (1 or 2) on the fermion and scalar lines of the Feynman graphs of Fig. 7.2.

(d) Compute the respective contributions of diagrams (a), (b), and (c) of Fig. 7.2 to the Pauli form factor (7.116) and show that they cancel.

Hints:

(a) Under parity $u_i(\tilde{p}_i) = \gamma^0 u_i(p_i)$ up to a phase (see equation (A.107) of Appendix A).

(b) Use (7.116) to compute $\mathrm{Tr}\left[\gamma_\mu(\not{p}_2 + m)M^\mu(\not{p}_1 + m)\right]$ and $\mathrm{Tr}\left[P_\mu(\not{p}_2 + m)M^\mu(\not{p}_1 + m)\right]$.

(c) Because of the Majorana nature (see (B.40) of Appendix B), $\bar{\Psi}_i \sigma_{\mu\nu} \Psi_i = 0$.

(d)

$$F_2(q^2)\big|_{(a)} = -2\, F_2(q^2)\big|_{(b)} = -2\, F_2(q^2)\big|_{(c)} = \cdots$$

<u>Exercise 2</u> Show that, in the case where the hidden sector gauge symmetry is broken at an intermediate scale M_I, the expression (7.66) for the gaugino condensates is replaced by

$$\left|\langle \bar{\lambda}_h \lambda_h \rangle\right|^2 \sim M^6 e^{-24\pi^2 \langle S + \bar{S} \rangle / b_0'}, \tag{7.117}$$

where one will express the scale M in terms of M_U, M_I, and the one-loop beta function coefficient b_0 (resp. b_0') of the corresponding gauge group above (resp. below) the scale M_I.

Hints: Write

$$0 = \frac{1}{g^2(M_I)} + \frac{b_0'}{8\pi^2} \ln \frac{\Lambda_c}{M_I},$$

$$\frac{1}{g^2(M_I)} = \frac{1}{g^2(M_U)} + \frac{b_0}{8\pi^2} \ln \frac{M_I}{M_U},$$

to obtain $M/M_U = (M_I/M_U)^{1-(b_0/b_0')}$.

<u>Exercise 3</u> Using Appendix C, write in components the Lagrangian

$$S = \frac{1}{4} \int d^4 y \int d^2\theta \; \mathbf{S} \, W^\alpha W_\alpha + \text{h.c.},$$

where $\mathbf{S} = S + \sqrt{2}\theta\psi_S + \theta^2 F_S$ is a chiral superfield.

Hints: Use for example Exercise 4, question (a) of Appendix C.

$$\mathcal{L} = -\frac{1}{8} S F^{\mu\nu} F_{\mu\nu} - \frac{i}{8} S F^{\mu\nu} \tilde{F}_{\mu\nu} + \frac{i}{2} S \lambda \sigma^\mu \partial_\mu \bar{\lambda} + \frac{1}{4} S D^2$$

$$+ \frac{1}{4} F_S \lambda\lambda + \frac{1}{2\sqrt{2}} \psi_S \lambda D + \frac{i}{2\sqrt{2}} (\psi_S \sigma^{\mu\nu} \lambda) F_{\mu\nu} + \text{h.c.}$$

<u>Exercise 4</u> : Consider the potential V_{eff} of (7.24) where $V^{(0)}$ is given by (5.13).

(a) Express $\partial \Delta V / \partial S_i|_{\text{min}}$, $i = 1, 2$ in terms of the parameters of $V^{(0)}$, v_1, v_2 (v_i and S_i are defined in (7.34); the minimum considered is the one of V_{eff}).

(b) Express $\left(\mathcal{M}_S^2\right)^{\text{eff}}$ and $\left(\mathcal{M}_P^2\right)^{\text{eff}}$ in (7.35) in terms of these parameters and the derivatives $\left.\frac{\partial \Delta V}{\partial S_i}\right|_{\text{min}}$, $\left.\frac{\partial^2 \Delta V}{\partial S_i \partial S_j}\right|_{\text{min}}$ and $\left.\frac{\partial^2 \Delta V}{\partial P_i \partial P_j}\right|_{\text{min}}$.

(c) Show that
$$\left(\mathcal{M}_P^2\right)^{\text{eff}} = \begin{pmatrix} \sin^2 \beta & \sin \beta \cos \beta \\ \sin \beta \cos \beta & \cos^2 \beta \end{pmatrix} \bar{m}_A^2 \,,$$

where $\bar{m}_A$ is the loop-corrected pseudoscalar mass.

(d) Deduce (7.36).

Hints:

(b)
$$\left.\frac{\partial^2 V_{\text{eff}}}{\partial P_i \partial P_j}\right|_{\text{min}} = -B_\mu \frac{v_1 v_2}{v_i v_j} - \frac{1}{\sqrt{2}} \frac{\delta_{ij}}{v_i} \left.\frac{\partial \Delta V}{\partial S_i}\right|_{\text{min}} + \left.\frac{\partial^2 \Delta V}{\partial P_i \partial P_j}\right|_{\text{min}} .$$

(c)
$$\left(\mathcal{M}_S^2\right)_{ij}^{\text{eff}} = \left.\frac{\partial^2 \Delta V}{\partial S_i \partial S_j}\right|_{\text{min}} + (-1)^{i+j} \left[\left(\mathcal{M}_P^2\right)_{ij}^{\text{eff}} - \left.\frac{\partial^2 \Delta V}{\partial P_i \partial P_j}\right|_{\text{min}} + \frac{g^2 + g'^2}{2} v_i v_j \right] .$$

8
Dynamical breaking. Duality

A motivation for supersymmetry is the stability of the electroweak breaking scale under quantum corrections arising from fundamental physics at a much higher scale (*naturalness problem*). One may however turn the argument around: in a theory with a very large fundamental scale, how does one generate a scale as low as the supersymmetry breaking scale (*hierarchy problem*)? Again, the supersymmetry breaking mechanism should be devised in order to provide a rationale for such a behavior.

We will argue in what follows that nonperturbative dynamics may be the answer: there are several examples where nonperturbative phenomena account for the generation of very large ratios of scales. This is referred to under the generic name of *dynamical breaking*. As we will see, supersymmetry provides new tools to control quantum fluctuations in the nonperturbative regime. One will uncover some duality relations between two regimes of a given theory, or two regimes of two different theories. This duality is often of a strong/weak coupling nature and is thus very promising to study strongly interacting theories.

The theories that we will consider have a dynamical scale below which some of the symmetries are broken. In the low energy regime, below this scale, the fields are usually composite bound states of the fundamental fields. This leaves the possibility that the fields of the supersymmetric theory at low energy (such as the MSSM studied in Chapter 5) are not fundamental. Supersymmetry may thus be realized at a preonic level.

8.1 Dynamical supersymmetry breaking: an overview

8.1.1 Introduction

The expression "dynamical breaking of a symmetry" refers to the spontaneous breaking of this symmetry when it is generated by a nonperturbative dynamics.

The standard example is quantum chromodynamics: keeping only the three light quark flavors u, d and s, QCD is described by the action

$$S = \int d^4x \left[-\frac{1}{4} \mathrm{Tr}\, F^{\mu\nu}\, F_{\mu\nu} + \sum_{f=1}^{3} \bar{q}_f \left(i\gamma^\mu D_\mu - m_{q_f} \right) q_f \right] \tag{8.1}$$

where $F_{\mu\nu}$ is the gluon field strength and $D_\mu q_f$ the $SU(3)$ covariant derivative of the quark field q_f. In the limit of vanishing quark masses $m_{q_f} \to 0$, this action is invariant under a global symmetry $SU(3)_L \times SU(3)_R$, known as the chiral symmetry (independent $SU(3)$ rotations respectively on the left and the right chirality quark fields). Such a symmetry is not observed in the spectrum: it would predict either a massless proton or a chiral partner for the proton. It is therefore spontaneously broken: the QCD vacuum breaks $SU(3)_L \times SU(3)_R$. It is indeed believed that the nonperturbative QCD dynamics yields nonvanishing quark condensates $U_{fg} \equiv \langle \bar{q}_{f_L} q_{g_R} \rangle$ which break spontaneously $SU(3)_L \times SU(3)_R$ into a diagonal $SU(3)_V$ (identical $SU(3)$ rotations for both quark chiralities), the symmetry which allowed Gell-Mann to classify hadrons. The octet of pseudoscalar mesons, $\pi^{\pm}$, π^0, K^0, $\bar{K}^0$, $K^{\pm}$ and η, yields the corresponding Goldstone bosons[1]. They may be interpreted as a variation in space and time of the background values and are thus described by the $SU(3)$ spacetime-dependent matrix $U_{fg}(x^{\mu})$.

The proton mass m_p is a consequence of the nonperturbative QCD dynamics:

$$m_p \propto \Lambda = M_P \, e^{-8\pi^2/(bg_s^2)} \tag{8.2}$$

where b is the one-loop beta function and g_s the value of the strong coupling at the Planck scale M_P. As noted in Chapter 1, this allows us to explain the hierarchy of scale between Λ (or m_p) and M_P. We note that the QCD Lagrangian does not contain any dimensionful parameter in the limit of vanishing quark masses. It is only at the quantum level that a low energy mass scale is generated.

In the case of supersymmetric models, we must face a similar problem, *i.e.* generate, through spontaneous supersymmetry breaking, masses for supersymmetric particles which are much smaller than the Planck scale (otherwise, the prime reason for introducing supersymmetry in the Hamiltonian – the problem of naturalness – is lost). It is thus natural to study thoroughly the possible rôle of dynamical breaking.

As we will now see, supersymmetry turns out to be an advantage, in the sense that it helps to control nonperturbative effects. This should not be seen as a complete surprise since supersymmetry was precisely devised to get a better handle on quantum fluctuations.

8.1.2   The power of holomorphy

We have seen in the preceding chapters that holomorphy is a fundamental property of the superpotential which summarizes the interactions of chiral supermultiplets (the scalar components of which are denoted here generically by ϕ): W depends on ϕ but not on ϕ^*. This is related to the chiral nature of the fermion fields associated with the scalars ϕ. One may generalize this property to account also for the dependence in the couplings.

To be explicit, let us consider the traditional Wess–Zumino model discussed in some details in Chapters 1 and 3. The superpotential reads (*cf.* (3.17) of Chapter 3)

$$W(\phi) = \tfrac{1}{2}m\phi^2 + \tfrac{1}{2}\lambda\phi^3. \tag{8.3}$$

We will extend the notion of holomorphy by considering the couplings themselves as background fields. In other words, we write $m = \langle M \rangle$ and $\lambda = \langle L \rangle$ and require W to be a holomorphic function of the scalar fields M and L as well as of ϕ.

---

[1]Restoring quark mass terms, which *explicitly* break the symmetry, generates a nonvanishing mass for these bosons.

The justification may be found in string theory where, as we will see in Chapter 10, there is a single fundamental scale, the string scale M_S and all dimensionless parameters may be expressed in terms of vacuum expectation values of scalar fields. For example, in the weakly coupled heterotic string theory, the gauge coupling g and the Planck scale M_P are expressed in terms of the vacuum expectation value of the string dilaton S:

$$\frac{1}{g^2} = \langle S \rangle, \tag{8.4}$$

$$M_P^2 = \langle S \rangle M_S^2. \tag{8.5}$$

More generally, our requirement of generalized holomorphy for the superpotential is simply a translation of the fact that, in a supersymmetric Feynman diagram which involves the interactions described by the superpotential, there is no possibility to find a coupling λ^* or a mass m^*.

Once couplings are interpreted on the same footing as fields, one may attribute them quantum numbers. For example, the superpotential (8.3) is invariant under the abelian symmetries $U(1) \times U(1)_R$, with quantum numbers given in Table 8.1.

Let us pause for a moment to discuss the difference of status of the two abelian symmetries. The $U(1)$ symmetry is of a standard nature and commutes with supersymmetry: if (ϕ_i, Ψ_{i_L}) are the bosonic and fermionic components of a chiral multiplet of charge q_i, they both transform under $U(1)$ as:

$$\phi_i \to e^{iq_i\alpha}\phi_i, \quad \Psi_{i_L} \to e^{iq_i\alpha} \Psi_{i_L}. \tag{8.6}$$

The superpotential W must be invariant under such a symmetry. On the other hand, $U(1)_R$ is known as a R-symmetry, a concept which we have encountered several times: it does not commute with supersymmetry and, as such, plays a central rôle in all discussions of supersymmetry and supersymmetry breaking. One can show [see Section 4.1 of Chapter 4] that, in the case of $N = 1$ supersymmetry, the only symmetries which may not commute with supersymmetry are abelian symmetries. Under such a symmetry, fermion and boson components of a given multiplet (ϕ_i, Ψ_{i_L}) of charge r_i transform differently:

$$\phi_i \to e^{ir_i\alpha} \phi_i \quad , \quad \phi_i^* \to e^{-ir_i\alpha} \phi_i^*$$
$$\Psi_{i_L} \to e^{i(r_i-1)\alpha} \Psi_{i_L}, \quad \Psi_{i_R}^c \to e^{-i(r_i-1)\alpha}\Psi_{i_R}^c \tag{8.7}$$

and the superpotential W has nonzero charge $r = 2$.

Table 8.1

	$U(1)$	$U(1)_R$
ϕ	1	1
m	-2	0
λ	-3	-1
W	0	2

This can be checked on the full Lagrangian (3.30) of Chapter 3. For example[2], the term

$$-\frac{1}{2} \frac{\partial^2 W}{\partial \phi_i \partial \phi_j} \overline{\Psi^c}_{iR} \Psi_{jL} \rightarrow -\frac{1}{2} \frac{e^{2i\alpha}}{e^{i(r_i+r_j)\alpha}} \frac{\partial^2 W}{\partial \phi_i \partial \phi_j} e^{i(r_i+r_j-2)\alpha} \overline{\Psi^c}_{iR} \Psi_{jL}$$

is invariant because the superpotential W has a charge $r = 2$.

The superpotential (8.3) that we started with is the tree level form. One may expect quantum corrections, in particular at the nonperturbative level, that would completely change this form. However, quantum fluctuations respect the $U(1) \times U(1)_R$ symmetry and thus $W(\phi)$ has the general form

$$W_{np}(\phi) = \sum_n c_n \frac{\lambda^n \phi^{n+2}}{m^{n-1}} \equiv m\phi^2 \sum_n c_n \left(\frac{\lambda\phi}{m}\right)^n. \tag{8.8}$$

We may take the limit $\lambda \to 0$, $m \to 0$ with λ/m fixed. Since $\lambda \to 0$, the perturbative result (8.3) holds:

$$W_{np}(\phi) = m\phi^2 \left[\frac{1}{2} + \frac{1}{3}\frac{\lambda\phi}{m}\right] \tag{8.9}$$

to all orders in λ/m. Hence $c_0 = 1/2$, $c_1 = 1/3$ and $c_n = 0$ for $n \geq 2$,

$$W_{np}(\phi) = \tfrac{1}{2}\, m\phi^2 + \tfrac{1}{3}\, \lambda\phi^2. \tag{8.10}$$

Thus, the superpotential is nonrenormalized, a well-known result which is shown to hold even at the nonperturbative level. Let us stress that holomorphy, extended to masses and couplings, has been the key assumption to prove this result, and R-symmetries are an important tool that can be used because the superpotential transforms nontrivially under them.

It is through such arguments, specific to supersymmetry, that one gets a better handle on nonperturbative dynamics. This is why dynamical breaking may be studied more thoroughly in the context of supersymmetry, as we will see in the following sections. Before doing so, let us stress that, in the context of $N = 1$ supersymmetry, arguments of holomorphy apply only to interactions (*i.e.* superpotential) and not to kinetic terms: there is still much freedom arising from wave function renormalization of the fields. One has to go to $N = 2$ supersymmetry in order to get a handle on this wave function renormalization as well, and to prove much more stringent results. This is what we will do at the end of this chapter.

8.1.3   Flat directions and moduli space

We have stressed in Chapter 7 (Section 7.3) the importance of flat directions of the potential, *i.e.* valleys where the scalar potential vanishes. Since along these directions global supersymmetry is not broken, the corresponding degeneracy is not lifted by perturbative quantum effects. Only nonperturbative effects may lift these flat directions.

---

[2][For a general treatment using superfields, see Section C.2.3 of Appendix C.]

As a first example consider a $U(1)$ supersymmetric gauge theory with two super-multiplets of charge ± 1. We denote their scalar components by $\phi_{\pm}$. In the case of a vanishing superpotential, the scalar potential is simply given by the D-term

$$V = \frac{1}{2} g^2 \left(\phi_+^\dagger \phi_+ - \phi_-^\dagger \phi_- \right)^2. \tag{8.11}$$

where g is the gauge coupling. The D-flat direction is reached for $\phi_{\pm} = \rho e^{i\theta_\pm}$. By a suitable gauge transformation, one may choose $\theta_+ = \theta_- \equiv \theta$. Thus

$$\phi_+ = \phi_- = a, \quad a = \rho e^{i\theta} \in \mathbb{C} \tag{8.12}$$

characterizes the flat direction. This flat direction consists of all the degenerate vacua; borrowing the terminology to solitons, in particular monopoles (see Chapter 4), one often refers to this set of all degenerate vacua as the classical moduli space.

For any nonvanishing value of a, the $U(1)$ gauge symmetry is spontaneously broken and the vector field becomes massive. In fact, because supersymmetry is not broken (the D-term vanishes), a whole vector supermultiplet becomes massive. As we have seen several times earlier (for example, in Section 5.1.2 of Chapter 5), such a supermultiplet has the same number of degrees of freedom as a massless vector supermultiplet plus a chiral supermultiplet (three vector, four spinor and one scalar): this may be interpreted as the supersymmetric version of the Higgs mechanism. Since we introduced in the theory one vector and two chiral supermultiplets, we are left with a single chiral superfield to describe the light degrees of freedom of the effective theory much below the scale of gauge symmetry breaking. The light scalar degree of freedom is obviously $X = \phi_+\phi_-$, since it must be gauge invariant. In the vacuum which is labelled by a, one has simply $\langle X \rangle = a^2$. The field X is called a modulus: its vacuum expectation value labels the classical moduli space. The dynamical field $X(x^\mu)$ corresponds to a continuous variation through the space of vacua as one moves in spacetime.

There is obviously no classical potential for X and no strong interaction present to generate it dynamically: $W(X) = 0$.

The Kähler potential, which fixes the normalization of the kinetic term, is obtained from the Kähler potential in the original theory. Assuming normalized kinetic terms for ϕ_+ and ϕ_-, we have

$$K\left(\phi_+, \phi_- \right) = \phi_+^\dagger \phi_+ + \phi_-^\dagger \phi_-. \tag{8.13}$$

Since $X^\dagger X = \phi_+^\dagger \phi_+ \, \phi_-^\dagger \phi_-$, along the flat direction $\phi_+ = \phi_-$, we have $\phi_+^\dagger \phi_+ = \phi_-^\dagger \phi_- = \sqrt{X^\dagger X}$ and thus

$$K(X) = 2\sqrt{X^\dagger X}. \tag{8.14}$$

The corresponding Kähler metric gives the kinetic for X

$$\frac{\partial^2 K}{\partial X \partial X^\dagger} \, \partial^\mu X^\dagger \partial_\mu X = \frac{1}{2\sqrt{X^\dagger X}} \, \partial^\mu X^\dagger \partial_\mu X. \tag{8.15}$$

We see that the theory is singular as $X \to 0$. This generally means that one is missing degrees of freedom in the effective theory. Indeed, as $a \to 0$, the symmetry remains unbroken and the low energy theory involves the two chiral supermultiplets.

The rest of this chapter lies in the "Theoretical Introduction" track.

8.2 Perturbative nonrenormalization theorems

8.2.1 Wilson effective action

Requiring holomorphy with respect to some parameters of a supersymmetric theory seems to be in contradiction with the renormalization program. Indeed, in the context of supersymmetry, all renormalization effects appear through the wave function renormalization factors Z. Since these are intrinsically nonholomorphic (recall that scalar kinetic terms are given by a real function, the Kähler potential), it seems that the relation between renormalized and tree-level couplings introduces nonholomorphicity which endangers any of the arguments presented above. The solution to this puzzle has been explained by [337, 338].

Let us illustrate this on the example of massive supersymmetric QED [338] (see Section C.3 of Appendix C or Exercise 5 of Chapter 3):

$$S = \frac{1}{4g_0^2} \int d^4x d^2\theta W^2 + \frac{1}{4\bar{g}_0^2} \int d^4x d^2\bar{\theta}\bar{W}^2 + \int d^4x d^4\theta \left(\Phi_+^\dagger e^{-V}\Phi_+ + \Phi_-^\dagger e^{V}\Phi_- \right)$$
$$+ \int d^4x d^2\theta \, m_0\Phi_+\Phi_- + \int d^4x d^2\bar{\theta} \, \bar{m}_0\Phi_+^\dagger\Phi_-^\dagger. \tag{8.16}$$

The low energy limit of this theory (*i.e.* the theory at a scale smaller to the mass scale m_0) is a theory of free massless gauge boson and gaugino. The associated gauge coupling is simply given by the exact formula [337]

$$\frac{1}{g^2} = \frac{1}{g_0^2} + \frac{1}{4\pi^2} \ln \frac{\Lambda_{UV}}{m_0}, \tag{8.17}$$

where Λ_{UV} is the ultraviolet cut-off. Since the superpotential $m_0\Phi_+\Phi_-$ is not renormalized, one obtains for the low energy physical mass $m = Zm_0$, where Z is the common field renormalization factor. One can thus express the low energy gauge coupling in terms of the physical mass as[3]:

$$\frac{1}{g^2} = \frac{1}{g_0^2} + \frac{1}{4\pi^2} \ln \frac{\Lambda_{UV}}{m} + \frac{1}{4\pi^2} \ln Z. \tag{8.19}$$

The factor Z is by essence nonholomorphic: it can be computed through the D-term renormalization and is a real function of $\ln |\Lambda_{UV}/m|$. Its presence accounts for the multiloop contribution.

It turns out that this nonholomorphic contribution is due to the presence of massless fields. The approach to effective theories that we have implicitly followed is based on the effective action formalism described in Section A.5.3 of Appendix Appendix A. The effective action Γ, which generates proper Green's functions, is expressed as a series in $\hbar$, which corresponds to an expansion in the number of loops. Because loop

---

[3]For a scale $\mu \ll m$, this becomes

$$\frac{1}{g^2} = \frac{1}{g_0^2} + \frac{1}{4\pi^2} \ln \frac{\Lambda_{UV}}{\mu} + \frac{1}{4\pi^2} \ln Z, \tag{8.18}$$

which gives the standard super-QED relation $\beta(\alpha) = \alpha^2 \left(1 - \gamma(\alpha)\right)/\pi$ with $\beta(\alpha) = \partial\alpha/\partial\ln\mu$ and $\gamma(\alpha) = \partial\ln Z/\partial\ln\mu$ (see, for a similar relation in the Wess–Zumino model, (8.39)).

integration is performed down to zero momentum, infrared effects due to the presence of massless fields are naturally included into the effective action.

An alternative approach has been pursued by [367] and [310]. If one is interested in studying the effects of a theory at a scale μ, one decomposes the quantum fields into a high frequency part ($E > \mu$) and a low frequency part ($E < \mu$). The Wilson effective action is then obtained by integrating over the high frequency modes. At a scale $\mu \gg m$, it simply reads

$$\mathcal{S}_W = \frac{1}{4g_W^2} \int d^4x d^2\theta \; W^2 + \text{h.c.} \tag{8.20}$$

with

$$\frac{1}{g_W^2} = \frac{1}{g_0^2} + \frac{1}{4\pi^2} \ln \frac{\Lambda_{UV}}{\mu}, \quad g_0^2 = g_W^2(\Lambda_{UV}), \tag{8.21}$$

which does not contain higher powers of g i.e. it is renormalized only at one loop. Thus, the relation between the gauge coupling (8.19) appearing in the effective action Γ and the one appearing in the Wilson effective action $\mathcal{S}_W$ is

$$\frac{1}{g_\Gamma^2} = \frac{1}{g_W^2} + \frac{1}{4\pi^2} \ln Z. \tag{8.22}$$

The last term may be interpreted in terms of the [259]. The parameters of the Wilson effective action are free from infrared contributions, and one-loop contributions exhaust all renormalization effects (see (8.21)), which saves holomorphicity. This is not so for the effective action Γ parameters. Infrared contributions lead to $\ln p^2$ terms at the loop level, which results in nonholomorphicity of the nonlocal action.

One may now turn to nonabelian gauge theories. In the case of a supersymmetric $SU(N)$ gauge theory, direct instanton calculations give the following result for the gluino condensate

$$\langle \lambda\lambda \rangle = C\Lambda_{UV}^3 \frac{1}{g_0^2} \exp\left(-\frac{8\pi^2}{Ng_0^2}\right), \tag{8.23}$$

where C is a constant and the prefactor $1/g_0^2$ accounts for zero modes (infrared effects). Since we are discussing holomorphy, one may allow a $\theta F_{\mu\nu}\tilde{F}^{\mu\nu}$ term which has a physical effect in an instanton background. Introducing thus the *complex* parameter

$$\tau \equiv \frac{\theta}{2\pi} + i\frac{4\pi}{g^2}, \tag{8.24}$$

the gauge action simply reads

$$\mathcal{S} = \frac{1}{16\pi}\text{Im}\left[\tau \int d^4x d^2\theta \; W^\alpha W_\alpha\right]. \tag{8.25}$$

However, if we replace all factors $1/g_0^2$ in (8.23) by $-i\tau/4\pi$, we find a dependence which is not periodic in the vacuum angle θ. Hence, the expression (8.23) does not seem to allow for a complex gauge coupling, i.e. to be consistent with holomorphy.

Again, the way out of this dilemma is to go to the Wilson effective action, whose coupling is given in terms of the coupling in the effective action as:

$$\frac{1}{g_W^2} = \frac{1}{g_0^2} - \frac{N}{8\pi^2} \ln \frac{1}{g_0^2}, \qquad (8.26)$$

when $\theta = 0$. Then, (8.23) simply reads $\langle \lambda\lambda \rangle = C\Lambda_{UV}^3 \exp\left(-8\pi^2/(Ng_W^2)\right)$, which, once the θ term is restored, reads

$$\langle \lambda\lambda \rangle = C\Lambda_{UV}^3 \exp\left(2i\pi\tau/N\right) = C\Lambda_{UV}^3 \exp\left(-\frac{8\pi^2}{Ng_W^2}\right) \exp\left(i\theta/N\right). \qquad (8.27)$$

We see that, as θ changes continuously from 0 to 2π, the N distinct vacua (with a phase dependence $e^{2i\pi k/N}$, $k = 0, \ldots, N-1$) are covered. Finally, just as in the abelian case, the evolution of the Wilson effective action gauge coupling is purely one-loop:

$$\frac{1}{g_W^2(\mu)} = \frac{1}{g_W^2} - \frac{3N}{8\pi^2} \ln \frac{\Lambda_{UV}}{\mu}, \qquad (8.28)$$

which allows us to write the θ-independent part of the condensate (8.27) as $C\mu^3 \exp\left(-8\pi^2/Ng_W^2(\mu)\right)$.

One may note that, in the Wilson approach, cut-offs play a central rôle. Just as parameters can be viewed as the vevs of chiral fields, one may consider the cut-offs as field-dependent. In order to have a holomorphic description in the effective theory, one may need to introduce different field-dependent regulators for the different sectors of the theory.

8.2.2 Flat directions

To discuss the rôle played by flat directions of the scalar potential, we have relied mostly on the tree level potential. A key property is that flat directions remain flat to all orders of perturbation theory [371]. We will prove here this statement in the context of global supersymmetry since this is the relevant framework when one discusses flat directions (see, for example, Section 6.12 of Chapter 6).

Flat directions are associated with the vanishing of auxiliary fields and are thus referred to as F-flat or D-flat directions. They are lifted if one of the auxiliary fields acquires a vacuum expectation value. The tree level potential is quadratic in these fields. In the context of renormalizable theories, a nontrivial vacuum appears if a linear term is generated at higher orders.

A term linear in F would necessarily be of the form $\int d^4x d^2\theta R(x,\theta)$ where R is a superfield which depends only on x and θ_α. The nonrenormalization theorems [210,235] discussed in Chapter 1 precisely forbid the generation of such terms at higher orders.

If a nonvanishing D-term is to lead to spontaneous supersymmetry breaking, it must be associated with a massless fermion (the Goldstino). This is only possible if the corresponding gauge symmetry is not broken. If the gauge symmetry group is semisimple (no $U(1)$ factor), a term linear in D always appears multiplied with fields which transform nontrivially under the gauge symmetry. Since the symmetry must remain unbroken, these fields have a vanishing vacuum expectation value, which forbids a nontrivial value for D. If the gauge symmetry is abelian, then a Fayet–Iliopoulos term is allowed and will be generated at higher orders, unless forbidden by symmetries (D is pseudoscalar). It has to be included from the start when discussing D-flat directions (see next section).

Thus, global supersymmetry remains unbroken to any finite order if it is unbroken at tree level. In other words, vanishing vacuum energy at tree level ensures vanishing vacuum energy at all finite orders: flat directions remain flat to all orders.

8.2.3 More on holomorphy

We emphasized the rôle of holomorphy in discussing F-flat directions. Surprisingly as it may seem, holomorphy also helps in classifying D-flat directions [1, 2, 55].

For example, when we considered in Section 8.1.3 the example of super-QED with two chiral superfields $\phi^{\pm}$ of charge ± 1, we concluded that the classical modulus space is parametrized by the holomorphic gauge invariant monomial $X \equiv \phi_+\phi_-$.

A second example may be borrowed to the lepton sector of the MSSM. The D-term restricted to the lepton superfields reads:

$$
V_D \;=\; \frac{1}{2}g^2 \left(D_1^2 + D_2^2 + D_3^2\right) + \frac{1}{2}\left(\frac{g'}{2}\right)^2 D_Y^2, \tag{8.29}
$$

$$
D_1 \;=\; \frac{1}{2}\sum_i \left(E_i^* N_i + N_i^* E_i\right), \qquad D_2 = \frac{i}{2}\sum_i \left(E_i^* N_i - N_i^* E_i\right)
$$

$$
D_3 \;=\; \frac{1}{2}\sum_i \left(|N_i|^2 - |E_i|^2\right), \qquad D_Y = \sum_i \left(2\,|E_i^c|^2 - |E_i|^2 - |N_i|^2\right).
$$

For any given set i_0, j_0, k_0 ($i_0 \neq j_0$) in $(1, 2, 3)$, the class of solutions:

$$
L_{i_0} \equiv \begin{pmatrix} N_{i_0} \\ E_{i_0} \end{pmatrix} = \begin{pmatrix} a \\ 0 \end{pmatrix}, \quad L_{j_0} = \begin{pmatrix} 0 \\ a \end{pmatrix}, \quad E_{k_0}^c = a \tag{8.30}
$$

corresponds to a flat direction of the potential V_D. It may be parametrized by the gauge invariant monomial $L_{i_0} \cdot L_{j_0} E_{k_0}^c = a^3$, which is a holomorphic function of the superfields in the MSSM. This can obviously be generalized: every class of D-flat directions (see Exercise 1 for the MSSM) may be associated with a holomorphic gauge invariant monomial. Since this combination is allowed by gauge symmetry, it should be present in the superpotential and play a rôle in the discussion of F-flat directions.

We see that the moduli space of D-flat directions is parametrized by a finite set of gauge-invariant polynomials holomorphic in the chiral superfields of the theory[4]. More precisely, the classical set of vacua can be parametrized in terms of holomorphic

---

[4]In a supersymmetric context, the gauge transformation $\Phi \to e^{2igq\Lambda}\Phi$ promotes the real gauge parameter to a complex parameter and thus the gauge symmetry group G to its complex extension G_c. In this context, the D-flatness condition may be seen as a gauge-fixing condition which breaks G_c down to G. Gauge-invariant holomorphic polynomials distinguish any two distinct G_c extended orbit [278].

monomials subject to polynomial constraints. In the case of a nonvanishing superpotential, F-flatness conditions induce additional constraints, which are also obviously holomorphic in the fields.

The presence of a Fayet–Iliopoulos term tends to lift some of the flat directions (some fields acquire a vacuum expectation value of order the Fayet–Iliopoulos coupling; see for example Exercise 2) but remains usually insufficient to lift completely the degeneracy. In any case, the corresponding vacua are still associated with holomorphic polynomials.

As an important example for the rest of this chapter, we illustrate the above considerations by considering in some details a $SU(3)$ gauge theory with two quark/antiquark flavors [2]. We denote respectively by $Q_{\alpha i} \in \mathbf{3}$ and $\bar{Q}^{\alpha i} \in \bar{\mathbf{3}}$ the quark and antiquark superfields as well as their scalar field components: $\alpha = 1, 2, 3$ is a color index and $i = 1, 2$ refers to flavor. The D-term reads[5]

$$V_D = \frac{1}{2}g^2 \sum_{a=1}^{8} \left(Q^{\dagger \alpha i} \frac{\lambda^a{}_\alpha{}^\beta}{2} Q_{\beta i} - \bar{Q}^\dagger_{\beta i} \frac{\lambda^{a*\beta}{}_\alpha}{2} \bar{Q}^{\alpha i} \right)^2$$

$$= \frac{1}{8}g^2 \sum_{a=1}^{8} \left(\lambda^a{}_\alpha{}^\beta D_\beta{}^\alpha \right)^2 \tag{8.31}$$

with

$$D_\beta{}^\alpha = Q_{\beta i} Q^{\dagger \alpha i} - \bar{Q}^\dagger_{\beta i} \bar{Q}^{\alpha i}. \tag{8.32}$$

We have used the fact that the Gell-Mann matrices are hermitian: $\lambda^a = \lambda^{a\dagger} = (\lambda^{a*})^T$.

Then it is easy to show that the flat direction corresponds to

$$V_D = 0 \leftrightarrow D_\beta{}^\alpha = \rho_0\, \delta^\alpha_\beta, \ \rho_0 \in \mathbb{R}. \tag{8.33}$$

To prove this, write $D = \sum_{b=0}^{8} \rho_b \lambda^b$ with $\lambda^0 \equiv \mathbb{1}$. Then the flat direction $V_D = 0$ corresponds to $\lambda^a{}_\alpha{}^\beta D_\beta{}^\alpha = 0$ for all $1 \le a \le 8$. Hence $\sum_{b=0}^{8} \mathrm{Tr}(\lambda^a \lambda^b)\rho_b = 2\rho_a = 0$ and $D = \rho_0 \mathbb{1}$.

We next try to find a solution to (8.33). If we define $R^\alpha{}_\beta \equiv Q^{\dagger \alpha i} Q_{\beta i}$, then

$$\det(R^\alpha{}_\beta) = Q^{\dagger 1i} Q^{\dagger 2j} Q^{\dagger 3k}\, \varepsilon_{\alpha\beta\gamma}\, Q_{\alpha i} Q_{\beta j} Q_{\gamma k} = 0,$$

since the latter part of the expression is antisymmetric under the exchange of $i, j, k \in \{1, 2\}$. Hence $R^\alpha{}_\beta$ is a positive semidefinite hermitian matrix of rank 2. Diagonalizing by a $SU(3)$ rotation yields:

$$Q^{\dagger \alpha i} Q_{\beta i} = \begin{pmatrix} v_1^2 & & \\ & v_2^2 & \\ & & 0 \end{pmatrix}. \tag{8.34}$$

---

[5] Note that, if $\phi \to e^{i\alpha^a \lambda^a} \phi$, then $\phi^* \to e^{-i\alpha^a \lambda^{a*}} \phi^*$.

A similar reasoning for $\bar{Q}^{\alpha i}\bar{Q}^{\dagger}_{\beta i}$ shows that $\rho_0 = 0$ and thus $\bar{Q}^{\alpha i}\bar{Q}^{\dagger}_{\beta i}$ is also given by (8.34). Using the symmetries to redefine the fields, one concludes that the vacuum solution is given by

$$Q_{\beta i} = \bar{Q}^{\dagger}_{\beta i} = \left| \begin{array}{cc} v_i \delta_{\beta i} & \beta = 1,2 \\ 0 & \beta = 3 \end{array} \right.$$

which we may write, using a matrix notation,

$$Q = \bar{Q} = \begin{pmatrix} v_1 & 0 & 0 \\ 0 & v_2 & 0 \end{pmatrix}. \tag{8.35}$$

The fields of the low energy effective theory should be $SU(3)$ singlets. Thus the classical moduli space is labelled by the vacuum expectation values of the meson fields described by the gauge invariant monomial $M_i{}^j \equiv Q_{\alpha i}\bar{Q}^{\alpha j}$:

$$\langle M \rangle = \begin{pmatrix} v_1^2 & 0 \\ 0 & v_2^2 \end{pmatrix}. \tag{8.36}$$

This is easily generalized to a gauge symmetry $SU(N_c)$ with N_f flavors of quarks $(Q_{\alpha i}, \alpha = 1,\ldots,N_c, i = 1,\ldots,N_f)$ and antiquarks $(\bar{Q}^{\alpha i}, \alpha = 1,\ldots,N_c, i = 1,\ldots,N_f)$. The classical moduli space is then given by:

- if $N_f < N_c$, $\qquad Q = \bar{Q} = \left(\begin{array}{c|c} \begin{matrix} v_1 & & \\ & \ddots & \\ & & v_{N_f} \end{matrix} & 0 \end{array} \right) \; N_f$ with N_c above.

The effective theory consists of meson fields $M_i{}^j = Q_{\alpha i}\bar{Q}^{\alpha j}$.

- if $N_f > N_c$, $\qquad Q = \left(\begin{array}{c} \begin{matrix} v_1 & & \\ & \ddots & \\ & & v_{N_c} \end{matrix} \\ \hline 0 \end{array} \right) \; N_f \qquad \bar{Q} = \left(\begin{array}{c} \begin{matrix} \overline{v_1} & & \\ & \ddots & \\ & & \overline{v_{N_c}} \end{matrix} \\ \hline 0 \end{array} \right) \; N_f$ with N_c above each.

with $|v_{\bar{f}}|^2 - |\bar{v}_f|^2$ independent of f.

The effective theory consists of:

- meson fields $\quad M_i{}^j = Q_{\alpha i}\bar{Q}^{\alpha j}$,
- baryon fields $\quad B_{i_1 \cdots i_{N_c}} = \varepsilon^{\alpha_1 \cdots \alpha_{N_c}} Q_{\alpha_1 i_1} \cdots Q_{\alpha_{N_c} i_{N_c}}$,
- antibaryon fields $\quad \bar{B}^{i_1 \cdots i_{N_c}} = \varepsilon_{\alpha_1 \cdots \alpha_{N_c}} \bar{Q}^{\alpha_1 i_1} \cdots \bar{Q}^{\alpha_{N_c} i_{N_c}}$.

8.3 Key issues in dynamical breaking

We continue this presentation of the main aspects of dynamical symmetry breaking by discussing the variety of phases in gauge theories. We then describe some of the key tools necessary for a detailed analysis: renormalisation group, anomalies, decoupling.

8.3.1 Phases of gauge theories

The simplicity of gauge theories is somewhat misleading. They can represent very diverse regimes with strikingly different properties (mass gap, confinement, etc.) as one probes deeper into the infrared, *i.e.* larger distances. Let us consider two test charges separated by a large distance r.

If we are in a regime where there are unconfined massless gauge fields, the interaction potential $V(r)$ is typically of the form $g^2(r)/r$, where the gauge coupling is evaluated at the distance r. The gauge theory in this regime is not asymptotically free and the gauge coupling diverges at some ultraviolet scale Λ: $g^2(r) \sim 1/\log(r\Lambda)$. Thus $V(r) \sim 1/(r\log(r\Lambda))$. This is referred to as the *free electric* phase. But, it is possible that g^2 is constant (free abelian gauge theory or renormalization group fixed point of a nonabelian gauge theory): this is the *Coulomb* phase where $V(r) \sim 1/r$. The third possibility is the presence of massless magnetic monopoles. It is now $4\pi/g^2$ which is renormalized and $V(r) \sim \log(r\Lambda)/r$. This is known as the *free magnetic* phase.

There is, however, the further possibility that the gauge symmetry be spontaneously broken, in which case the gauge bosons become massive through the Higgs mechanism. The vacuum condensate provides a constant contribution to the potential, whereas the Higgs field exchange yields a Yukawa interaction, which is exponentially decreasing at large distances. Thus $V(r)$ is constant in the *Higgs* phase.

Finally, in the case of asymptotically free nonabelian gauge theories, the interaction potential is confining at large distance: $V(r) \sim r$.

8.3.2 Renormalization group fixed points

In the renormalization group evolution, fixed points correspond to values of the coupling for which the beta function vanishes. The presence of nontrivial fixed points may enhance the symmetries of the theory. Indeed, at a fixed point, a nonsupersymmetric theory has conformal invariance. One may use this to prove general results [279]. For example, the scaling dimension d of a scalar field satisfies $d \geq 1$, with equality only for a free field.

Supersymmetry may provide useful constraints on possible fixed points. As a first example, we consider the Wess–Zumino model and show that the nonrenormalization theorems discussed above forbid the existence of nontrivial fixed points [158]. We have seen in Section 8.1.2 that the superpotential (8.3) is not renormalized. This leaves us only with wave function renormalization. Adding the appropriate counterterm to the scalar field kinetic term in the renormalized Lagrangian yields

$$\mathcal{L}_{\rm kin} = \tfrac{1}{2} Z \, \partial^\mu \phi \partial_\mu \phi, \tag{8.37}$$

where Z is the wave function renormalization constant: the bare field is $\phi_0 = Z^{1/2}\phi$. Then, expressing the superpotential (8.3) in terms of ϕ_0,

$$W(\phi) = \tfrac{1}{2} m Z^{-1} \, \phi_0^2 + \tfrac{1}{3}\lambda Z^{-3/2} \, \phi_0^3 \tag{8.38}$$

gives the renormalized coupling λ in terms of the bare coupling λ_0: $\lambda = Z^{3/2}\lambda_0$. We deduce a relation between the beta function $\beta(\lambda)$ and the anomalous dimension $\gamma(\lambda)$ of the field ϕ:

$$\beta(\lambda) = \mu\frac{d\lambda}{d\mu} = \tfrac{3}{2}\lambda\mu\frac{d\ln Z}{d\mu} = \tfrac{3}{2}\lambda\gamma(\lambda). \tag{8.39}$$

This relation is obviously a mere consequence of the nonrenormalization theorems. It forbids any nontrivial fixed point. Indeed if $\lambda^* \neq 0$ was such a point $(\beta(\lambda^*) = 0)$, it would follow that $\gamma(\lambda^*) = 0$. Since its anomalous dimension vanishes, the scalar field has its scaling dimension d coinciding with its canonical dimension: $d = 1$. Conformal symmetry implies that the field is free, in contradiction with our starting assumption that we are in a nontrivial fixed point regime.

For our next example, let us turn to nonabelian gauge theories. We suppose that the beta function reads

$$\beta(g) = \frac{1}{16\pi^2}\left(-b_1 g^3 + b_2 g^5 + O(g^7)\right) \tag{8.40}$$

with $b_1 > 0$, $b_2 > 0$ and $b_1 \ll b_2$. There is then a nontrivial infrared fixed point $g^* \sim \sqrt{b_1/b_2}$ in the perturbative regime (see Fig. 8.1).

If we take the example of nonsupersymmetric $SU(N_c)$ with N_f flavors, then $b_1 = \frac{11}{3}N_c - \frac{2}{3}N_f$ whereas b_2 is of order N_c^2 in the limit of large N_c. Thus if we choose enough flavors to almost compensate the one-loop beta function ($N_f \sim 11N_c/2$), then b_1 is of order 1 and g^* is or order $1/N_c$ at large N_c, hence in the perturbative regime [21].

In the supersymmetric case, we have [298]

$$\beta(g) = -\frac{g^3}{16\pi^2}\frac{3N_c - N_f + N_f\gamma(g^2)}{1 - N_c g^2/(8\pi^2)}, \tag{8.41}$$

$$\gamma(g) = -\frac{g^2}{8\pi^2}\frac{N_c^2 - 1}{N_c} + O(g^4),$$

where $\gamma(g)$ is the anomalous mass dimension. If again we choose a regime of large N_c where the one-loop beta coefficient almost cancels, i.e. $N_f/N_c = 3 - \epsilon$ with $\epsilon \ll 1$, then we have a zero of the beta function for $\gamma = 1 - 3N_c/N_f \sim -\epsilon/3$: we find a nontrivial infrared fixed point at

$$g^{*2}N_c = \frac{8\pi^2}{3}\epsilon + O(\epsilon^2). \tag{8.42}$$

The theory is then in a Coulomb phase, which is very different from the confining phase observed with a smaller number of flavors, e.g. QCD with three flavors.

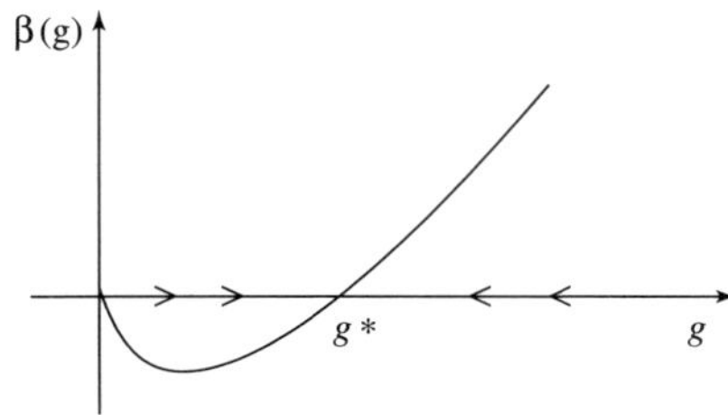

Fig. 8.1 The beta function of equation (8.40): the arrows denote how $g(\mu)$ varies as one goes into the infrared, *i.e.* for decreasing μ.

In the case of supersymmetric theories, the fixed point regime has superconformal invariance, which has some nontrivial consequences. For example, if we consider the supermultiplet of currents $(T_{\mu\nu},\ J_{r\mu},\ J_\mu^R)$ which consists of the energy–momentum tensor, the supersymmetric current and the R-symmetry current (see Section 4.2 of Chapter 4), superconformal invariance implies:

$$\Theta^\mu{}_\mu = 0, \quad \gamma^\mu_{rs} J_{s\mu} = 0, \quad \partial^\mu J_\mu^R = 0. \tag{8.43}$$

Hence, at the fixed point, the R-symmetry is conserved. Moreover scaling dimensions of superfields satisfy

$$d \geq \tfrac{3}{2}|R|, \tag{8.44}$$

where R is the R-charge, with equality for chiral or antichiral superfields. We will make use of these results below.

8.3.3 Symmetries and anomalies. The 't Hooft consistency condition

In a confining regime (such as in QCD), it is important to identify correctly the light (massless) degrees of freedom. This is not a problem for QCD since we observe them: they are for example the nucleons. When we consider other theories, identifying them might be a challenge.

't Hooft [349] has devised a general consistency condition which has proved to be a very powerful tool to identify massless fermionic bound states. It rests on a clever use of the cancellation of anomalies.

Following 't Hooft, we consider a Yang–Mills theory with gauge symmetry group G, coupled to chiral fermions in various representations of G. These massless fermions form multiplets of a global symmetry group G_F. If we now consider the low energy theory, one encounters massless composite bound states. If we now consider three currents associated with the global symmetry G_F, the corresponding triangle anomalies may be computed using the elementary fermions of the short-distance description or using the composite massless bound states. Consistency requires that the two computations coincide.

The proof is rather straightforward. Let us gauge G_F by introducing weakly-coupled "spectator" gauge bosons. In order to cancel the anomalies of the short distance theory, one must introduce massless "spectator" fermions. At low energy (large distance), one finds the gauge symmetry G_F with the massless bound states and the "spectator" fermions: anomalies must cancel *i.e.* the contribution of spectators must cancel that of bound states. Hence the latter coincides with the anomalies computed at short distance with the elementary (nonspectator) fermions.

8.3.4 Effective theories and decoupling

Another way of extracting information is based on the decoupling of heavy flavors [8]. If we make one of the fundamental fermions massive, then, in the limit of large mass, this flavor decouples: in the low energy theory, all bound states containing this fermion disappear. In the context of supersymmetric theories, this means that if we start with a theory with $N_f + 1$ massless flavors with superpotential $W(\Phi_1, \ldots, \Phi_{n+1})$ and make the $(n+1)$th flavor heavy by adding a mass term $m\Phi_{n+1}^2$, one may solve the F-flatness condition for Φ_{n+1} and thus eliminate this field from the low energy action, recovering an effective theory with N_f fundamental flavors.

We take this opportunity to note that, in order to integrate out a heavy degree of freedom in a supersymmetric way, one has to work at the level of the superpotential, and not at the level of the Lagrangian (potential).

8.4 Example of supersymmetric $SU(N_c)$ with N_f flavors. The rôle of R-symmetries

We use the methods introduced in the preceding sections to study the case of supersymmetric QCD, or more generally of a $SU(N_c)$ gauge theory with N_f flavors of quarks and antiquarks. As we will see, the discussion depends on whether the number of flavors N_c is smaller or larger than the number of colors.

8.4.1 $N_f < N_c$

If at some large scale M_0, the gauge coupling has a perturbative value g, then the running coupling $g(\mu)$ evolves, at one loop, as

$$\frac{1}{g^2(\mu)} = \frac{1}{g^2} + \frac{b}{8\pi^2} \ln \frac{\mu}{M_0} \tag{8.45}$$

where $b = 3N_c - N_f$ is the coefficient of the one-loop beta function (see Chapter 9). Due to asymptotic freedom, or rather asymptotic slavery ($b > 0$), it explodes at a scale

$$\Lambda = M_0 \, e^{-8\pi^2/(bg^2)}. \tag{8.46}$$

Above the scale Λ, we have a theory of elementary excitations, the quarks and antiquarks, whereas, below, the effective theory is a theory of mesons. We wish to study the symmetries of the original theory in order to identify the dynamical interactions of the meson fields. Under independent global rotations of left and right chiralities, $SU(N_f)_L \times SU(N_f)_R$, the quark superfields Q transform as $(\mathbf{N_f}, \mathbf{1})$ (since they include the left-handed quark field), whereas the antiquark superfields $\bar{Q}$ transform as $(\mathbf{1}, \bar{\mathbf{N}}_{\mathbf{f}})$ (they include the right-handed chirality). Similarly, baryon number conservation is associated with a global $U(1)_B$ symmetry: Q (resp. $\bar{Q}$) has charge $+1$ (resp. -1).

We now identify a R-symmetry which is nonanomalous. We have seen in (8.7) that the fermionic component of a superfield of R-charge r (the charge of its scalar component) transforms with charge $r - 1$. Thus if Q and $\bar{Q}$ transform with same R-charge r, then the quark and antiquark fields transform as:

$$\psi'_Q = e^{i(r-1)\alpha} \, \psi_Q$$
$$\psi'_{\bar{Q}} = e^{i(r-1)\alpha} \, \psi_{\bar{Q}}. \tag{8.47}$$

Also gaugino fields transform with charge $+1$ (see equation (C.77) of Appendix C):

$$\lambda' = e^{i\alpha} \, \lambda. \tag{8.48}$$

One computes the mixed $U(1) - SU(N_c) - SU(N_c)$ triangle anomalies. The goal is to choose r in order to cancel these mixed anomalies. Writing T^a (resp. t^a) the generators of $SU(N_c)$ in the adjoint (resp. fundamental) representation, we have:

$$\text{Tr } T^a T^b = C_2(G) \, \delta^{ab} \qquad C_2(G) = N_c, \tag{8.49}$$

$$\text{Tr } t^a t^b = T(R) \, \delta^{ab} \qquad T(R) = \frac{1}{2}. \tag{8.50}$$

Then the condition of anomaly cancellation reads:

$$N_c + \tfrac{1}{2}(r-1)N_f + \tfrac{1}{2}(r-1)N_f = 0.$$

Hence

$$r = \frac{N_f - N_c}{N_f}. \tag{8.51}$$

The meson superfields $M_i{}^j$ of the effective theory transform under the nonanomalous $U(1)_R$ symmetry as:

$$M_i{}^j \equiv Q_{\alpha i} \, \bar{Q}^{\alpha j} \rightarrow e^{2i \frac{N_f - N_c}{N_f} \alpha} \, M_i{}^j. \tag{8.52}$$

We now use the full symmetries in order to extract the dynamical interactions of these effective mesonic degrees of freedom. They are described by a superpotential $W_{\text{dyn}}(M_i{}^j)$. Since this superpotential cannot have a matrix structure – indeed it must be invariant under $SU(N_f)_L \times SU(N_f)_R$ – it must depend on $\det(M)$. We summarize in Table 8.2 the transformation properties of the different, fundamental or effective, fields.

Since $W_{\text{dyn}}(M)$ must have R-charge $+2$, we conclude that it must be proportional to $(\det M)^{1/(N_f - N_c)}$. Using dimensional analysis ($[W] = 3$ and $[\det M] = 2N_f$) and the fact that the only dynamical scale available is the scale Λ, we conclude that the dynamical effective potential is

$$W_{\text{dyn}}(M) = C_{N_c, N_f} \left(\frac{\Lambda^{3N_c - N_f}}{\det M} \right)^{\frac{1}{N_c - N_f}} \tag{8.53}$$

where C_{N_c, N_f} is a dimensionless constant. This result may be tested in several different regimes and the constant computed to depend on N_c and N_f as $C_{N_c, N_f} = (N_c - N_f)C^{1/(N_c - N_f)}$, with C constant (see Exercise 3).

Table 8.2

	$SU(N_f)_L$	$SU(N_f)_R$	$U(1)_B$	$U(1)_R$
Q	$\mathbf{N_f}$	$\mathbf{1}$	$+1$	$(N_f - N_c)/N_f$
$\bar{Q}$	$\mathbf{1}$	$\mathbf{\bar{N}_f}$	-1	$(N_f - N_c)/N_f$
M	$\mathbf{N_f}$	$\mathbf{\bar{N}_f}$	0	$2(N_f - N_c)/N_f$
$\det M$	$\mathbf{1}$	$\mathbf{1}$	0	$2(N_f - N_c)$

In the special case $N_f = N_c - 1$, we may then write (8.53) as

$$W_{\mathrm{dyn}}(M) = C \left(\frac{\Lambda^b}{\det M} \right) = \frac{CM_0^b}{\det M} \, e^{-8\pi^2/g^2},\tag{8.54}$$

where we have used (8.46). We recognize in the last factor the standard one-instanton contribution. Indeed, a direct instanton calculation provides the same answer and allows us to compute the remaining constant: $C = 1$ in the $\overline{DR}$ scheme [163].

Finally, taking $N_f = 0$ yields $W_{\mathrm{dyn}}(M) = \Lambda^3$. This is obviously related to gaugino condensation: the dynamical potential being a function of $\langle \lambda\lambda \rangle$ alone, dimensional analysis imposes that $W_{\mathrm{dyn}}(M) \propto \langle \lambda\lambda \rangle$. It follows from (8.27) and the remark below it that $W_{\mathrm{dyn}}(M)$ is proportional to $M_0^3 \exp\left(-8\pi^2/N_c g^2(M_0)\right)$, i.e. Λ^3.

8.4.2 $N_f = N_c$ or $N_c + 1$

Let us start with $N_f = N_c$. Then, according to Table 8.2, the quark superfields Q and $\bar{Q}$ have vanishing R-charge and it is not possible to use them only to write a superpotential of R-charge 2. However, there are now enough flavors to be able to form baryonic states:

$$B = \epsilon^{\alpha_1 \cdots \alpha_{N_c}} Q_{\alpha_1 1} \cdots Q_{\alpha_{N_c} N_c}$$

$$\bar{B} = \epsilon_{\alpha_1 \cdots \alpha_{N_c}} \bar{Q}^{\alpha_1 1} \cdots \bar{Q}^{\alpha_{N_c} N_c}.$$

Since we may write in shortened notation $B = \det[Q_{\alpha i}]$, where $[Q_{\alpha i}]$ is a $N_c \times N_f$ square matrix and similarly $\bar{B} = \det[\bar{Q}^{\alpha i}]$, we see that the low energy fields are not independent:

$$B\bar{B} = \det[Q_{\alpha i}\bar{Q}^{\alpha j}] = \det[M_i{}^j].\tag{8.55}$$

[335] has, however, argued that this relation is only valid at the classical level. At the quantum level, he proposes to include a nonperturbative contribution:

$$\det[M_i{}^j] - B\bar{B} = \Lambda^{2N_c},\tag{8.56}$$

where Λ is the dynamical scale.

One may first check that the new term Λ^{2N_c} has the right dimension $(2N_c)$ and R-charge (0). But the main argument in favour of this addition is that it yields the right theory when one flavor decouples. Let us see this in detail. As discussed above in Section 8.3.4, we make the N_fth flavor decouple by introducing the following tree level term in the superpotential:

$$W_{\mathrm{tree}} = m \, Q_{\alpha N_f} \bar{Q}^{\alpha N_f}.\tag{8.57}$$

We expect that the theory at scales much below m is supersymmetric $SU(N_c)$ with $N_c - 1$ flavors. Thus the low energy superpotential should simply read, according to (8.53),

$$W \sim \frac{\tilde{\Lambda}^{2N_c+1}}{\det_{N_c-1} M},$$
(8.58)

where $\tilde{\Lambda}$ is the dynamical scale of this effective theory with $N_c - 1$ flavors. It can be easily related to Λ by matching the two theories at scale m: since

$$0 = \frac{1}{g^2(m)} + \frac{b}{8\pi^2} \ln \frac{\Lambda}{m},$$

$$0 = \frac{1}{g^2(m)} + \frac{\tilde{b}}{8\pi^2} \ln \frac{\tilde{\Lambda}}{m},$$

where $b = 2N_c$ is the one-loop beta function coefficient of the theory with N_c flavors and $\tilde{b} = 2N_c + 1$ is the same coefficient for the theory with $N_c - 1$ flavors, we have

$$\tilde{\Lambda}^{2N_c+1} = m\Lambda^{2N_c}.$$
(8.59)

To extract the low energy theory, we must place ourselves in the F-flat direction corresponding to $Q_{\alpha N_f}$ and $\bar{Q}^{\alpha N_f}$. Since for example $dW_{\text{tree}}/d\bar{Q}^{\alpha N_f} = mQ_{\alpha N_f}$, this direction corresponds to $\langle Q_{\alpha N_f} \rangle = \langle \bar{Q}^{\alpha N_f} \rangle = 0$ (but possibly $\langle Q_{\alpha N_f} \bar{Q}^{\alpha N_f} \rangle \neq 0$). Hence $\langle B \rangle = \langle \bar{B} \rangle = \langle M_{N_f}{}^j \rangle = 0$ and $\det M = \det_{N_c-1} M \times M_{N_f}{}^{N_f}$. Then, the quantum constraint (8.56) yields $M_{N_f}{}^{N_f} = \Lambda^{2N_c} / \det_{N_c-1} M$ and we obtain from (8.57) and (8.59)

$$W_{\text{tree}} = mM_{N_f}{}^{N_f} = \frac{m\Lambda^{2N_c}}{\det_{N_c-1} M} = \frac{\tilde{\Lambda}^{2N_c+1}}{\det_{N_c-1} M},$$
(8.60)

in agreement with (8.58).

Another nontrivial check is provided by the 't Hooft anomaly condition (see Section 8.3.3 above). The complete global symmetry $SU(N_f = N_c)_L \times SU(N_f = N_c)_R \times U(1)_B \times U(1)_R$ is partially broken by the vacua which satisfy the quantum constraint (8.56). The 't Hooft consistency condition is all the more useful that the residual symmetry is richer. One may for example consider the vacuum $\langle B \rangle = \langle \bar{B} \rangle = 0$, $\langle M_i{}^j \rangle = \Lambda^2 \delta_i^j$. The symmetry is then $SU(N_f = N_c)_V \times U(1)_B \times U(1)_R$. The quantum numbers of the fundamental fermion fields (quarks ψ_Q, antiquarks $\psi_{\bar{Q}}$ and gauginos λ) as well as composite fermion fields (supersymmetric partners of the meson scalars ψ_M, baryons ψ_B and antibaryons $\psi_{\bar{B}}$) are given in Table 8.3.

We note that among the effective degrees of freedom, the mesons form an adjoint representation of dimension $N_f^2 - 1$ of $SU(N_f = N_c)_V$: one component of the matrix $M_i{}^j$ is not independent because of the quantum constraint (8.56) which fixes $\det M$.

Table 8.3

	$SU(N_f = N_c)_V$	$U(1)_B$	$U(1)_R$
ψ_Q	N_f	$+1$	-1
$\psi_{\bar{Q}}$	N_f	-1	-1
λ	1	0	$+1$
ψ_M	$N_f^2 - 1$	0	-1
ψ_B	1	N_f	-1
$\psi_{\bar{B}}$	1	$-N_f$	-1

One may then compute (see Exercise 4) the following mixed anomalies using either the fundamental degrees of freedom or the composite degrees of freedom of the low energy theory. The two results coincide:

$$
\begin{aligned}
SU(N_f) - SU(N_f) - U(1)_R & \qquad -N_f \\
U(1)_B - U(1)_B - U(1)_R & \qquad -2N_f^2 \\
U(1)_R - U(1)_R - U(1)_R & \qquad -\left(N_f^2 + 1\right).
\end{aligned}
\tag{8.61}
$$

In the case where $N_f = N_c + 1$, the (anti)baryon fields carry a flavor index:

$$
B^i = \epsilon^{ij_1 \cdots j_{N_c}} \epsilon^{\alpha_1 \cdots \alpha_{N_c}} Q_{\alpha_1 j_1} \cdots Q_{\alpha_{N_c} j_{N_c}}
$$

$$
\bar{B}_i = \epsilon_{ij_1 \cdots j_{N_c}} \epsilon_{\alpha_1 \cdots \alpha_{N_c}} \bar{Q}^{\alpha_1 j_1} \cdots \bar{Q}^{\alpha_{N_c} j_{N_c}}.
$$

The following superpotential

$$
W = \frac{1}{\Lambda^b} \left(\det M - B^i M_i{}^j \bar{B}_j \right)
\tag{8.62}
$$

allows us to recover the quantum constraint (8.56) when decoupling one flavor: it corresponds to the F-flatness condition for the decoupled flavor. Again, one may check in this case the 't Hooft consistency conditions.

8.4.3 $N_f > N_c + 1$: nonabelian electric–magnetic duality

When one reaches the value $N_f = N_c + 2$ and beyond, one encounters a growing disagreement between the computation of anomalies from the fundamental fields and from the system of baryons and mesons. This is a sign that one is misidentifying the effective theory.

We note, however, that baryon fields, having $\tilde{N}_c = N_f - N_c$ indices could be interpreted as bound states of $\tilde{N}_c$ component fields which we note q:

$$B_{i_1 \cdots i_{\tilde{N}_c}} \equiv \epsilon^{\alpha_1 \cdots \alpha_{\tilde{N}_c}} \, q_{\alpha_1 i_1} \cdots q_{\alpha_{\tilde{N}_c} i_{\tilde{N}_c}}, \tag{8.63}$$

and similarly for $\bar{B}$. This has led [336] to propose that supersymmetric QCD with N_c colors and N_f flavors could be described, for $N_f \geq N_c + 2$ as supersymmetric QCD with $\tilde{N}_c$ colors, N_f flavors of quark and antiquark supermultiplets and a field $M_i{}^j$ ($\sim Q_{\alpha i}\bar{Q}^{\alpha j}$) which is a gauge singlet coupling to the quarks through

$$W = \frac{1}{\mu}\bar{q}^{\alpha i} M_i{}^j q_{\alpha j}. \tag{8.64}$$

As we will see, the relation between these two theories is somewhat reminiscent of the electric–magnetic duality discussed in Section 4.5.2 of Chapter 4: the original $SU(N_c)$ theory provides the electric description whereas the $SU(\tilde{N}_c)$ theory is the equivalent magnetic formulation. This is why Seiberg has called this relation the nonabelian electric–magnetic duality.

One may first check that this is indeed a duality. If we start with the latter theory and perform another duality transformation, one obtains supersymmetric QCD with $N_f - \tilde{N}_c = N_c$ colors, N_f flavors $Q_{\alpha i}$ and $\bar{Q}^{\alpha i}$, and the singlet fields $M_i{}^j$ and $\widetilde{M}_i{}^j$ ($\sim q_{\alpha i}\bar{q}^{\alpha j}$) with superpotential

$$W = \frac{1}{\mu} M_i{}^j \widetilde{M}_j{}^i + \frac{1}{\tilde{\mu}}\bar{Q}^{\alpha i}\widetilde{M}_i{}^j Q_{\alpha j}. \tag{8.65}$$

The F-flatness condition for M ensures that $\widetilde{M} = 0$ and thus a vanishing superpotential. We recover the original theory.

Again, 't Hooft consistency condition provides a highly nontrivial check on the conjecture. The quantum numbers of the various superfields are given in Table 8.4[6]. Using this information, on may for example compute the $U(1)_R{}^3$ anomaly in the original picture ($N_c^2 - 1$ gauginos of charge $+1$, N_f quarks ψ_Q and antiquarks $\psi_{\bar{Q}}$ of charge $\tilde{N}_c/N_f - 1 = -N_c/N_f$) or in the dual picture ($\tilde{N}_c^2 - 1$ gauginos of charge $+1$, N_f quarks ψ_q and antiquarks $\psi_{\bar{q}}$ of charge $-\tilde{N}_c/N_f$, N_f^2 fermions ψ_M of charge $1 - 2N_c/N_f$). In both cases one finds $N_c^2 - 1 - 2N_c^4/N_f^2$.

Let us complicate somewhat the picture by noting that, for $N_f \geq 3N_c$, the original "electric" theory is no longer asymptotically free: as one goes deeper into the infrared (*i.e.* to larger distances), the coupling becomes smaller. The infrared regime is thus one of weakly coupled massless quarks and gluons: it is a free electric phase. On the other hand, we have seen in Section 8.3.2 that, for N_f just below $3N_c$, appears a nontrivial infrared fixed point: the coupling tends in the infrared to the finite value g^* given in (8.42).

---

[6] The $U(1)$ quantum numbers of q and $\bar{q}$ are obtained by noting that baryons can be made with N_c quarks Q or, in the dual picture, $\tilde{N}_c$ quarks q.

Table 8.4

	$SU(N_f)_L$	$SU(N_f)_R$	$U(1)_B$	$U(1)_R$
Q	$\mathbf{N_f}$	1	$+1$	$\tilde{N}_c/N_f$
$\bar{Q}$	1	$\mathbf{\bar{N}_f}$	-1	$\tilde{N}_c/N_f$
q	$\mathbf{N_f}$	1	$N_c/\tilde{N}_c$	N_c/N_f
$\bar{q}$	1	$\mathbf{\bar{N}_f}$	$-N_c/\tilde{N}_c$	N_c/N_f
M	$\mathbf{N_f}$	$\mathbf{\bar{N}_f}$	0	$2\tilde{N}_c/N_f$

Similarly, on the "magnetic side", for $N_f \geq 3\tilde{N}_c$, i.e. $N_f \leq 3N_c/2$, the theory becomes free in the infrared: we are in the free magnetic phase. And for N_f just above $3N_c/2$, we identify a nontrivial infrared fixed point.

One can show that nontrivial infrared fixed points may only appear for values of N_f larger than $3N_c/2$. Indeed, let us suppose that we have such a fixed point: as we have seen in Section 8.3.2, the theory has superconformal invariance. The corresponding R symmetry is nothing but the nonanomalous one that we identified in Table 8.4. Since the field $M_i{}^j$ of the magnetic theory has R-charge $2(N_f - N_c)/N_f$ and is in a chiral multiplet, we conclude from (8.44) that its scaling dimension is[7] $d = 3(N_f - N_c)/N_f$. The condition $d \geq 1$ imposes $N_f \geq 3N_c/2$.

These considerations have led Seiberg to conjecture the existence of a nontrivial infrared fixed point for $3N_c/2 < N_f < 3N_c$. The corresponding theory has superconformal invariance. Quarks and gluons are not confined but appear as interacting massless particles: this is referred to as the nonabelian Coulomb phase.

8.5   *N = 2* **supersymmetry and the Seiberg–Witten model**

We consider in this section a $N = 2$ supersymmetric $SU(2)$ gauge theory. The fundamental fields in the ultraviolet limit are the fields of a $N = 2$ vector supermultiplet. We study the infrared regime of this theory.

As we have seen in Section 4.4.1 of Chapter 4, the fundamental fields of this theory can be arranged into a $N = 1$ vector supermultiplet (A_μ, λ_α) and a $N = 1$ chiral supermultiplet (ϕ, ψ_α) in the adjoint representation. The action can then be written as a sum of the relevant $N = 1$ actions with adequate normalizations. We will slightly generalize the case studied in Chapter 4 by allowing a $\theta F_{\mu\nu}\tilde{F}^{\mu\nu}$ term. We thus write the full $N = 2$ action

$$S = \frac{1}{16\pi}\text{Im}\left[\tau \int d^4x\, d^2\theta\, W^\alpha W_\alpha\right] + \frac{1}{g^2}\int d^4x\, d^4\,\theta\Phi^\dagger e^{2V}\Phi, \tag{8.66}$$

---

[7]Note that, for $N_f = 3N_c/2$, $d = 1$. In other words, the field $M_i{}^j$ is free.

where W_α and Φ are respectively the gauge and chiral superfields and τ has been defined in (8.24) as

$$\tau = \frac{\theta}{2\pi} + i\frac{4\pi}{g^2}.$$

(8.67)

The normalization of the chiral action is different from the one chosen in equation (4.28) of Chapter 4 because the normalization of the gauge fields is different (A_μ has been rescaled by $1/g$). This can be checked on the Lagrangian written in terms of the component fields, which is obtained as in Chapter 4 using (C.81) of Appendix C:

$$\mathcal{L} = \mathrm{Tr}\left(-\frac{1}{4g^2}F_{\mu\nu}F^{\mu\nu} - \frac{\theta}{32\pi^2}F_{\mu\nu}\tilde{F}^{\mu\nu} + \frac{i}{g^2}\lambda\sigma^\mu D_\mu\bar{\lambda} + \frac{1}{2g^2}D^2\right.$$

$$\left. + \frac{1}{g^2}D^\mu\phi^\dagger D_\mu\phi + \frac{i}{g^2}\psi\sigma^\mu D_\mu\bar{\psi} + \frac{1}{g^2}F^\dagger F\right)$$

(8.68)

$$- \frac{1}{g^2}\,\mathrm{Tr}\left(D\left[\phi^\dagger,\phi\right] + \sqrt{2}\psi\left[\lambda,\phi^\dagger\right] - \sqrt{2}\bar{\psi}\left[\bar{\lambda},\phi\right]\right),$$

where the covariant derivatives are defined as:

$$D_\mu\lambda = \partial_\mu\lambda - i\left[A_\mu,\lambda\right],$$
$$D_\mu\phi = \partial_\mu\phi - i\left[A_\mu,\phi\right],$$
$$D_\mu\psi = \partial_\mu\psi - i\left[A_\mu,\psi\right],$$

(8.69)

$$F_{\mu\nu} = \partial_\mu A_\nu - \partial_\nu A_\mu - i\left[A_\mu,A_\nu\right].$$

(8.70)

Solving for the auxiliary fields gives the corresponding scalar potential, which is a D-term:

$$V(\phi) = \frac{1}{2g^2}\mathrm{Tr}\left(\left[\phi^\dagger,\phi\right]\right)^2.$$

(8.71)

The ground state is obtained for ϕ and $\phi^\dagger$ commuting. Since $\phi \equiv \phi^a\sigma^a/2$, this means that one can put it, through a gauge transformation, under the form

$$\langle\phi\rangle = a\frac{\sigma^3}{2}.$$

(8.72)

It is straightforward to compute the spectrum of the theory in such a vacuum. The fields A_μ^a, λ^a, ϕ^a, ψ^a with $a = 1,2$ all have mass $a\sqrt{2}$ whereas the corresponding $a = 3$ component has vanishing mass: in other words $A_\mu \equiv A_\mu^3$ is the $U(1)$ gauge potential and, since supersymmetry is not broken by the vacuum (8.72) ($D^a = F^a = 0$), it appears in a full $N = 2$ massless multiplet, of vanishing $U(1)$ electric charge.

Let us check these results explicitly on the bosonic fields. As usual (see Appendix Appendix A), the gauge field mass terms arise from the scalar kinetic terms once the gauge symmetry is spontaneously broken. Defining $A_\mu^{(\pm)} \equiv \left(A_\mu^1 \mp iA_\mu^2\right)/\sqrt{2}$, the mass term simply reads

$$\frac{a^2}{g^2} A^{\mu(+)} A_\mu^{(-)},$$

whereas the kinetic terms fix the normalization

$$-\frac{1}{8g^2} F_{\mu\nu}^{(+)} F^{(-)\mu\nu} + \frac{1}{2g^2} m^2 A^{\mu(+)} A_\mu^{(-)}.$$

Hence $m = a\sqrt{2}$.

The scalar fields also form two combinations $\phi^{(\pm)} \equiv \left(\phi^1 \mp i\phi^2\right)/\sqrt{2}$ of common mass $m = a\sqrt{2}$ and respective electric charge $Q_e = \pm 1$.

We conclude that the massive states can be organized in two multiplets of respective charge ± 1 but same mass $a\sqrt{2}$. Each multiplet has eight degrees of freedom. But since supersymmetry is not broken in the vacuum (8.72), these multiplets must be $N = 2$ supermultiplets. This seems in contradiction with what we have seen in Section 4.3.2 of Chapter 4: massive representations of N supersymmetry theories have a multiplicity of $2^{2N}(2j + 1)$, which gives for $N = 2$ a minimum of 16 (for $j = 0$), unless there are central charges.

This necessarily means that the theory has a central charge and that the supermultiplets that we consider are the short supermultiplets discussed in Section 4.3.3 of Chapter 4. The mass of these supermultiplets is directly related to the central charge of the theory: $m = z/2$. Let us recall that the central charge is an operator which commutes with all the generators of the supersymmetry algebra. We deduce an expression for this central charge:

$$z = 2\sqrt{2}\,|aQ_e|. \tag{8.73}$$

Let us note that this expression might be incomplete since it has been obtained in the ultraviolet regime of the theory where the degrees of freedom have only electric charge. As expected from the considerations of Chapter 4, and as seen below, we expect that, in other regimes of the theory, appear solitons which carry magnetic charge.

We now try to identify the effective low energy action. In the regime where we have easily access to it, it corresponds to a $N = 2$ supersymmetric $U(1)$ theory, which we may describe in terms of a $N = 1$ abelian gauge supermultiplet W_α and a $N = 1$ chiral supermultiplet Φ of vanishing electric charge. The action reads:

$$S = \frac{1}{16\pi} \mathrm{Im} \left[\int d^4x\, d^2\theta \; \mathcal{F}''(\Phi) W^\alpha W_\alpha + \int d^4x \int d^4\theta \; \Phi^\dagger \mathcal{F}'(\Phi) \right], \tag{8.74}$$

where the holomorphic function $\mathcal{F}(\Phi)$ is a $N = 2$ prepotential which generalizes the discussion of Section 4.4.1 of Chapter 4: it determines simultaneously the holomorphic gauge kinetic function $f(\Phi) \sim i\mathcal{F}''(\Phi)$ and the Kähler potential $K(\Phi, \Phi^\dagger) = \mathrm{Im}\mathcal{F}''(\Phi)$ which fixes the normalization of the scalar kinetic term.

We have already noted that the zeroes of the Kähler potential represent singular points in the scalar field parameter space where the parametrization is inadequate, usually because the light degrees of freedom have not been identified properly. Indeed, because $\mathcal{F}(\Phi)$ is holomorphic, $K(\Phi, \Phi^\dagger) = \text{Im} \mathcal{F}''(\Phi)$ is harmonic and has thus no minimum in the complex plane. If it is not constant, it should reach zero at some point.

We will see however that the effective action (8.74) has alternate descriptions which may thus describe other regimes of the theory. We start by performing what is known as a duality transformation.

We define

$$\Phi_D = \mathcal{F}'(\Phi) \tag{8.75}$$

$$\mathcal{F}_D(\Phi_D) = \mathcal{F}(\Phi) - \Phi\Phi_D, \tag{8.76}$$

which allows us to make a Legendre transformation since

$$\Phi = -\mathcal{F}_D'(\Phi_D). \tag{8.77}$$

We thus have

$$\text{Im} \int d^4x \int d^4\theta \; \Phi^\dagger \mathcal{F}'(\Phi) = -\text{Im} \int d^4x \int d^4\theta \; [\mathcal{F}_D'(\Phi_D)]^\dagger \, \Phi_D$$

$$= \text{Im} \int d^4x \int d^4\theta \; \Phi_D^\dagger \mathcal{F}_D'(\Phi_D). \tag{8.78}$$

Hence the second term of (8.74) allows us to identify $\mathcal{F}_D$ as the prepotential in the dual formulation. Before checking this with the first term, let us note that

$$\mathcal{F}_D''(\Phi_D) = -\frac{d\Phi}{d\Phi_D} = -\frac{1}{\mathcal{F}''(\Phi)}. \tag{8.79}$$

Hence the duality transformation yields an alternative description where the parameter τ introduced in (8.66) transforms as

$$\tau_D(a_D) = -\frac{1}{\tau(a)} \tag{8.80}$$

in moduli space where

$$\langle \phi_D \rangle = a_D \sigma^3/2 \tag{8.81}$$

as in (8.72). Moreover (8.67) tells us that this is a duality relation of the weak coupling/strong coupling type, *i.e.* $g^2 \leftrightarrow 1/g^2$, that we have encountered when discussing electric–magnetic duality.

Before dwelling more on this, let us show that the gauge part of the action (8.74) is invariant under the duality transformation. For this we recall that the gauge superfield W_α can be defined as a generic chiral superfield which satisfies the constraint (C.68) of Appendix C: $D^\alpha W_\alpha = \bar{D}_{\dot\alpha} \bar{W}^{\dot\alpha}$ which we may write $\text{Im} \; D^\alpha W_\alpha = 0$. Thus, performing a functional integration in superspace, we may write

$$\int \mathcal{D}V \exp\left[\frac{i}{16\pi} \text{Im} \int d^4x d^2\theta \mathcal{F}''(\Phi) W^\alpha W_\alpha \right]$$

$$= \int \mathcal{D}W \mathcal{D}V_D \exp\left[\frac{i}{16\pi} \text{Im} \left(\int d^4x d^2\theta \mathcal{F}''(\Phi) W^\alpha W_\alpha + 2 \int d^4x d^4\theta V_D D^\alpha W_\alpha \right)\right],$$

where V_D is a real superfield which plays the rôle of a Lagrange multiplier that ensures the constraint on W_α. We may perform an integration by part in the last term (on D^α) and use the property of Grassmann variables discussed in Section C.2.2 of Appendix C ($\int d^2\bar\theta \sim -\frac{1}{4}\bar D^2$) to write (8.81)

$$\int \mathcal{D}W \mathcal{D}V_D \exp\left[\frac{i}{16\pi}\mathrm{Im}\left(\int d^4x d^2\theta \mathcal{F}''(\Phi)W^\alpha W_\alpha - 2\int d^4x d^2\theta W_D^\alpha W_\alpha\right)\right], \quad (8.82)$$

where $W_{D\alpha} \equiv -\frac{1}{4}\bar D^2 D_\alpha V_D$. Thus the integral over the unconstrained W is a simple Gaussian integral which is easily performed. One finally obtains:

$$\int \mathcal{D}V_D \exp\left[\frac{i}{16\pi}\mathrm{Im}\int d^4x d^2\theta\left(-\frac{1}{\mathcal{F}''(\Phi)}\right)W_D^\alpha W_{D\alpha}\right], \quad (8.83)$$

which, according to (8.79) has the expected form.

One may make the duality transformation more transparent by rewriting (8.74) as

$$S = \frac{1}{16\pi}\mathrm{Im}\int d^4x d^2\theta \frac{d\Phi_D}{d\Phi}W^\alpha W_\alpha + \frac{1}{32\pi i}\int d^4x d^4\theta\left(\Phi^\dagger \Phi_D - \Phi_D^\dagger \Phi\right), \quad (8.84)$$

where we have used (8.75). The duality transformation (8.80) amounts to:

$$\begin{pmatrix}\Phi'_D \\ \Phi'\end{pmatrix} = \begin{pmatrix} 0 & -1 \\ 1 & 0\end{pmatrix}\begin{pmatrix}\Phi_D \\ \Phi\end{pmatrix}. \quad (8.85)$$

It can be supplemented by the following transformation ($\beta \in \mathbb{Z}$):

$$\begin{pmatrix}\Phi'_D \\ \Phi'\end{pmatrix} = \begin{pmatrix} 1 & \beta \\ 0 & 1\end{pmatrix}\begin{pmatrix}\Phi_D \\ \Phi\end{pmatrix}. \quad (8.86)$$

Indeed, under such a transformation, the action (8.84) receives an additional term

$$\frac{\beta}{16\pi}\int d^4x d^2\theta\, W^\alpha W_\alpha = \frac{\beta}{16\pi}\int d^4x\, F_{\mu\nu}\tilde F^{\mu\nu} = 2\pi\beta\nu, \quad (8.87)$$

where ν is an integer. Hence, S is translated by a multiple of 2π.

The two transformations (8.85) and (8.86) generate the group $SL(2,\mathbb{Z})$ of transformations

$$\begin{pmatrix}\Phi'_D \\ \Phi'\end{pmatrix} = \begin{pmatrix}\alpha & \beta \\ \gamma & \delta\end{pmatrix}\begin{pmatrix}\Phi_D \\ \Phi\end{pmatrix}, \quad \alpha\delta - \beta\gamma \in \mathbb{Z}. \quad (8.88)$$

Now we may return to (8.73) and complete it. We have introduced a duality transformation (8.85) which takes us to the magnetic regime where we expect that (8.73) is replaced by $z = 2\sqrt{2}\,|a_D Q_m|$, with a_D defined as in (8.81). But more generally, we expect the complete formula to read

$$z = 2\sqrt{2}\,|a Q_e + a_D Q_m| = 2\sqrt{2}\left|(Q_m Q_e)\begin{pmatrix} a_D \\ a \end{pmatrix}\right|. \tag{8.89}$$

Under a $SL(2,\mathbb{Z})$ transformation, the charges transform as

$$(Q'_m Q'_e) = (Q_m Q_e)\begin{pmatrix} \alpha & \beta \\ \gamma & \delta \end{pmatrix}. \tag{8.90}$$

The duality transformation (8.85) precisely exchanges electric and magnetic charges.

Seiberg and Witten [333,334] were able to determine the behavior of $a(u)$ and $a_D(u)$ where the complex variable $u \equiv \frac{1}{2}\mathrm{Tr}\phi^2$ describes the moduli space of the theory. By studying the way these functions vary as u is taken around a closed contour, they could identify the singular points which correspond to a nontrivial behavior of these functions along the closed loop. There are in fact only three such singularities: $u \to \infty$ (weak coupling) and $u = \pm u_0 \neq 0$ (strong coupling). As discussed earlier, one expects that these singularities are associated with the appearance of massless fields. The surprise here is that, in the case of the strong coupling singularities $\pm u_0$, these fields are found among the collective excitations, monopoles or dyons, of the underlying $SU(2)$ supersymmetric theory.

Further reading

- M. E. Peskin, *Duality in supersymmetric Yang–Mills theory*, Proceedings of the 1996 Theoretical Advanced Institute, Boulder, Colorado.
- A. Bilal, *Duality in $N = 2$ SUSY $SU(2)$ Yang–Mills theory*, Proceedings of the "61. Rencontre entre Physiciens Théoriciens et Mathématiciens", Strasbourg, France, December 1995.

Exercises

<u>Exercise 1</u> *We identify the D-flat directions of the MSSM using the correspondence between such flat directions and gauge invariant polynomials of chiral superfields, as discussed in Section 8.2.3.*

(a) What is the complex dimension d_S of the configuration space of the scalar fields in the MSSM? How many D-term constraints must be satisfied? Deduce the complex dimension d_D of the configuration space reduced to the D-flat directions.

   We now wish to identify a basis B of gauge-invariant monomials with the property that any gauge-invariant polynomial in the chiral superfields of the MSSM can be written as a polynomial in elements of B.

(b) First show that a basis B_3 of such monomials invariant only under the color $SU(3)$ gauge symmetry is given by E_i^c, L_i, H_1, H_2; $U^c U^c U^c$, $U^c U^c D^c$, $U^c D^c D^c$, $D^c D^c D^c$ (more precisely $U_i^{c\alpha} U_j^{c\beta} U_k^{c\gamma} \epsilon_{\alpha\beta\gamma}\epsilon^{ijk}$ with α, β, γ color indices, and so on); $Q_i U_j^c$, $Q_i D_j^c$, $Q_i^{b\alpha} Q_j^{c\beta} Q_k^{a\gamma} \epsilon_{bc}\epsilon_{\alpha\beta\gamma}$, $Q_i^{a\alpha} Q_j^{b\beta} Q_k^{c\gamma} \epsilon_{\alpha\beta\gamma}\epsilon^{ijk}$ (the latter is symmetric in the

$SU(2)$ indices a, b and c). What are their quantum numbers under $SU(2) \times U(1)$?

(c) Deduce a basis B_{32} of monomials invariant under $SU(3) \times SU(2)$.

(d) Finally deduce a basis B_{321} invariant under $SU(3) \times SU(2) \times U(1)$.

Hints:

(a) $d_S = 49$; 12 D-term constraints and 12 phases which may be gauge fixed leave $d_D = 37$ complex dimensions.

(b) see [184].

<u>Exercise 2</u> One considers an extension of the MSSM with two fields singlet under $SU(3) \times SU(2) \times U(1)$:

- a right-handed neutrino N^c;
- a Froggatt–Nielsen scalar θ (see Section 12.1.4 of Chapter 12).

These fields have respective charges x_N and x_θ under an abelian family symmetry $U(1)_X$. Because this symmetry is pseudo-anomalous, its D-term includes a Fayet–Iliopoulos term ξ:

$$D_X = x_\theta \left|\theta\right|^2 + x_N \left|N^c\right|^2 - \xi^2. \tag{8.91}$$

The scalar potential should be such that $\langle\theta\rangle \neq 0$ (to generate family hierarchies) and $\langle N^c \rangle = 0$.

(a) How should one choose the signs of x_θ and x_N? What is then a dangerous flat direction?

(b) Which terms should be present in the superpotential in order to prevent this flat direction?

Hints:

(a) $x_\theta > 0$ and $x_N < 0$; since $x_\theta x_N < 0$, one can form a holomorphic invariant using both θ and N^c: θ and N^c could be both nonvanishing;

(b) $N^c \theta^n$ would forbid $\langle\theta\rangle \neq 0$; hence $(N^c)^p \, \theta^n$, $p \geq 2$ and $n \neq 0 \bmod p$ (see [41]).

<u>Exercise 3</u> *In the case of $SU(N_c)$ with $N_f < N_c$ flavors (Section 8.4.1), we probe formula (8.53) for the dynamical superpotential in various regimes of the theory.*

1. We first assume that $\langle Q_{\alpha N_f} \rangle = v_{N_f} \delta_{\alpha N_f}$ (see end of Section 8.2.3) and study the effective theory at a scale much smaller than v_{N_f}. At scale v_{N_f}, $SU(N_c)$ is broken to $SU(N_c - 1)$. Thus the effective theory is $SU(N_c - 1)$ with $N_f - 1$ flavors.

   (a) The $SU(N_c - 1)$ gauge coupling diverges at a scale $\tilde\Lambda$. Neglecting threshold effects, express $\tilde\Lambda$ in terms of the scale Λ (dynamical scale of the original $SU(N_c)$ theory), the symmetry breaking scale v_{N_f}, and the numbers N_c and N_f.

   (b) Write $W_{\mathrm{dyn}}(M)$ in (8.53) in the limit $\langle Q_{\alpha N_f} \rangle = v_{N_f} \delta_{\alpha N_f}$ to show that $C_{N_c, N_f} = C_{N_c - 1, N_f - 1}$.

2. Alternatively, we give a large mass to the pair Q_{N_f}, $\bar{Q}^{N_f}$:

$$W_{\mathrm{tree}} = m \, Q_{\alpha N_f} \bar{Q}^{\alpha N_f}.$$

(a) What is the $U(1)_R$ charge of m? Deduce that the following combination is dimensionless and has vanishing $U(1)_R$ charge:

$$X = m \, M_{N_f}{}^{N_f} \left(\frac{\Lambda^{3N_c - N_f}}{\det M} \right)^{-1/(N_c - N_f)}.$$

(b) Write the most general superpotential compatible with the symmetries. By considering the limit of small mass m and small gauge coupling, deduce that:

$$W_{\text{dyn}} = C_{N_c, N_f} \left(\frac{\Lambda^{3N_c - N_f}}{\det M} \right)^{1/(N_c - N_f)} + m M_{N_f}{}^{N_f}. \tag{8.92}$$

(c) The effective theory at scales much smaller than m is $SU(N_c)$ with $N_f - 1$ flavors. Express the corresponding dynamical scale $\hat{\Lambda}$ in terms of Λ, m, N_c and N_f.

(d) Starting from (8.53), obtain the effective low energy superpotential and derive thus a relation between $C_{N_c, N_f - 1}$ and C_{N_c, N_f} (depending on N_c and N_f).

3. Use the preceding results to show that

$$C_{N_c, N_f} = (N_c - N_f) \, C^{1/(N_c - N_f)},$$

where C is a constant.

Hints:

1. (a) Define $b_{(N_c, N_f)} = 3N_c - N_f$. By matching the couplings of the full theory and of the effective theory at the scale $\mu = v_{N_f}$, one obtains

$$\left(\frac{\Lambda}{v_{N_f}} \right)^{b_{(N_c, N_f)}} = \left(\frac{\tilde{\Lambda}}{v_{N_f}} \right)^{b_{(N_c - 1, N_f - 1)}},$$

or

$$\frac{\Lambda^{3N_c - N_f}}{v_{N_f}^2} = \Lambda^{3(N_c - 1) - (N_f - 1)}. \tag{8.93}$$

Note that the exact scale where the matching occurs is renormalization scheme dependent. It is precisely v_{N_f} in the $\overline{DR}$ scheme (see Section E.8 of Appendix E).

(b) Following (8.53),

$$W_{\text{dyn}}(M) = C_{N_c, N_f} \left(\frac{\Lambda^{3N_c - N_f}}{v_{N_f}^2 \, \det_{N-1} M} \right)^{\frac{1}{N_c - N_f}}.$$

Use (8.93) to show that this corresponds to the formula for the $(N_c - 1, N_f - 1)$ theory.

2. (a) One obtains the $U(1)_R$ charge of m from the tree level superpotential: $2N_c/N_f$.
    (b) We have

$$W_{\text{dyn}} = F(X) \left(\frac{\Lambda^{3N_c - N_f}}{\det M} \right)^{1/(N_c - N_f)} .$$

Consider the limit $m \to 0$, $\Lambda \to 0$ but X fixed, to obtain $F(X) = C_{N_c,N_f} + X$.
    (c) $\hat{\Lambda}^{3N_c - (N_f - 1)} = m\, \Lambda^{3N_c - N_f}$.
    (d) Solve for $M_{N_f}{}^{N_f}$ by using its equation of motion derived from (8.92): $dW_{\text{dyn}}/dM_{N_f}{}^{N_f} = 0$. One obtains:

$$\left(\frac{C_{N_c,N_f-1}}{N_c - (N_f - 1)} \right)^{N_c - (N_f - 1)} = \left(\frac{C_{N_c,N_f}}{N_c - N_f} \right)^{N_c - N_f} .$$

<u>Exercise 4</u> Check the anomalies (8.61) using first the fundamental degrees of freedom, then the composite degrees of freedom of the low energy theory.

Hint: Remember that the quarks ψ_Q and antiquarks $\psi_{\bar{Q}}$ come in N_c colors and that there are $N_c^2 - 1$ gaugino fields. Thus the $U(1)_R^3$ anomaly computed with these fields reads $2N_f N_c(-1) + N_c^2 - 1 = -\left(N_f^2 + 1\right)$ since $N_f = N_c$. On the other hand, the $N_f^2 - 1$ mesons, the baryon and antibaryon fields contribute $-\left(N_f^2 - 1\right) - 2$.

9
Supersymmetric grand unification

With all its successes, the Standard Model does not really achieve complete unification. Indeed, it trades the two couplings $\alpha_{\text{e.m.}}$ and G_F describing respectively electromagnetic and weak interactions for the two gauge couplings g and g'. Moreover, even if one disregards gravitational interactions which are so much weaker, one should at some point take into account the remaining fundamental gauge interaction, namely quantum chromodynamics (QCD), which describes strong interactions. Indeed, as one goes to higher energies, the QCD running coupling becomes weaker, and thus closer in magnitude to the electroweak couplings. Earlier work on quark–lepton unification [302] helped to bridge the gap towards a real unification of all known gauge interactions and in 1973 Georgi and Glashow proposed the first such *grand unified model* based on the gauge group $SU(5)$ [181].

The most spectacular consequence of this unification is the presence of new gauge interactions which would mediate proton decay: the corresponding gauge bosons have to be very heavy in order to make the proton *almost* stable. On the more technical side, grand unified models provide a clue to one of the most intriguing aspects of the Standard Model: the cancellation of gauge anomalies. Quarks and leptons are arranged in larger representations of the grand unified symmetry which are anomaly-free [179] (as we will see, this is actually only partially true in minimal $SU(5)$ but is completely realized in the case of $SO(10)$ or larger groups).

We noted above that strong interactions become weaker as one increases the energy. Georgi, Quinn and Weinberg [183] showed that indeed strong and electroweak interactions reach the same coupling regime at a mass scale M_U of order 10^{15} GeV. The three couplings g_3 for color $SU(3)$, g and g' have approximately the same value: they are unified in the single coupling of the grand unified theory and the corresponding symmetry is effective above 10^{15} GeV. This value proved to be a blessing: it ensured that the gauge bosons which mediated proton decay were superheavy, which put (at that time) proton decay in a safe range. Also, the fundamental scale of the grand unified theory was only a few orders of magnitude away from the fundamental scale of quantum gravity, *i.e.* the Planck scale M_P: this led to possible hopes of an ultimate unification of all known fundamental unifications; it will be the subject of the next chapter.

The next step was to go supersymmetric [108] and the supersymmetric version of the minimal $SU(5)$ model was proposed by Dimopoulos and Georgi [107]. We stress that the move to supersymmetry was necessary because the fundamental scale of grand unified theories was found to be so large. In a nonsupersymmetric theory, the superheavy scale would destabilize the electroweak scale. This is the hierarchy or naturalness problem discussed in Chapter 1.

Moreover, the unification of couplings turned out to be in better agreement with the values measured for the low energy couplings [108]. Some 20 years later, with an impressive increase in the precision of measurements, this remains true and contributes significantly to the success of the grand unification scenario. In what follows, after a presentation of grand unified theories, we will thus start by discussing gauge coupling unification. We will then study some aspects specific to supersymmetric grand unification.

9.1 An overview of grand unification

9.1.1 The minimal SU(5) model

The minimal simple gauge group that incorporates $SU(3) \times SU(2) \times U(1)$ is $SU(5)$. This is why particular attention has been paid to grand unified models based on $SU(5)$.

The gauge fields transform under the adjoint of $SU(5)$ which has the same dimension 24 as the group. Under $SU(3) \times SU(2) \times U(1)$ this **24** transforms as follows:

$$\mathbf{24} = (\mathbf{8}, \mathbf{1}, 0) + (\mathbf{1}, \mathbf{3}, 0) + (\mathbf{1}, \mathbf{1}, 0) + (\mathbf{3}, \mathbf{2}, \tfrac{5}{3}) + (\bar{\mathbf{3}}, \mathbf{2}, -\tfrac{5}{3}). \tag{9.1}$$

We recognize on the right-hand side: the gluon fields $(\mathbf{8}, \mathbf{1}, 0)$, the triplet A^a_μ of $SU(2)$ gauge bosons $(\mathbf{1}, \mathbf{3}, 0)$, the hypercharge gauge boson B_μ $(\mathbf{1}, \mathbf{1}, 0)$. We note the remaining gauge fields in $(\mathbf{3}, \mathbf{2}, 5/3)$ as $X_{\alpha\mu}$, $\alpha = 1, 2, 3$ color index, for the $t_3 = 1/2$ component of charge 4/3 and $Y_{\alpha\mu}$ for the $t_3 = -1/2$ component of charge 1/3 (their antiparticles $\bar{X}_{\alpha\mu}$ and $\bar{Y}_{\alpha\mu}$ are found in $(\bar{\mathbf{3}}, \mathbf{2}, -5/3)$). They will play an important rôle in what follows since their exchange is a source of proton decay.

The $X_{\alpha\mu}$ and $Y_{\alpha\mu}$ gauge bosons acquire a mass of the order of the scale of breaking of the grand unified theory. This spontaneous breaking is realized minimally by introducing a set of scalar fields H transforming as a **24** of $SU(5)$: this breaks $SU(5)$ into $SU(3) \times SU(2) \times U(1)$.

In order to ensure the electroweak breaking, we must also incorporate the Standard Model Higgs into this framework. This is done minimally by introducing a set of scalar fields that transform as the **5** of $SU(5)$. After $SU(5)$ breaking, this splits as

$$\mathbf{5} = (\mathbf{3}, \mathbf{1}, -\tfrac{2}{3}) + (\mathbf{1}, \mathbf{2}, 1). \tag{9.2}$$

We recognize in $(\mathbf{1}, \mathbf{2})$, which we note Φ, the Standard Model $SU(2) \times U(1)$ doublet: $\langle \Phi \rangle \sim 250$ GeV. The color triplet $(\mathbf{3}, \mathbf{1})$, noted Φ_T, must have a vanishing vacuum expectation value in order not to break color: $\langle \Phi_T \rangle = 0$. However a severe problem arises in this arrangement: the mass scales of the two fields Φ and Φ_T must be vastly different. Indeed, the doublet mass is constrained to be of the order of the electroweak scale, whereas the triplet must be superheavy because it mediates proton decay. This difficulty is known as the "doublet–triplet splitting problem". In some sense, it is another disguise of the hierarchy problem discussed in Chapter 1.

The fermions of each generation are minimally grouped into two representations: (anti)fundamental $\bar{\mathbf{5}}$ and antisymmetric $\mathbf{10}$ of $SU(5)$.

The antisymmetric representation is described by a tensor χ_{ij} ($i, j = 1, \ldots, 5$) such that $\chi_{ij} = -\chi_{ji}$. It decomposes under $SU(3) \times SU(2)$ as

$$\mathbf{10} = (\mathbf{1}, \mathbf{1}) + (\bar{\mathbf{3}}, \mathbf{1}) + (\mathbf{3}, \mathbf{2}). \tag{9.3}$$

In matrix notation, it reads explicitly

$$[\chi_{ij}] = \begin{pmatrix} 0 & u_3^c & -u_2^c & -u_1 & -d_1 \\ -u_3^c & 0 & u_1^c & -u_2 & -d_2 \\ u_2^c & -u_1^c & 0 & -u_3 & -d_3 \\ u_1 & u_2 & u_3 & 0 & -e^c \\ d_1 & d_2 & d_3 & e^c & 0 \end{pmatrix}_L . \tag{9.4}$$

We are left with $d_{1_L}^c$, $d_{2_L}^c$, $d_{3_L}^c$, e_L and ν_{e_L} which have precisely the quantum numbers of the antifundamental representation $\bar{\mathbf{5}}$ (*cf.* (9.2))

$$\bar{\mathbf{5}} = (\bar{\mathbf{3}}, \mathbf{1}) + (\mathbf{1}, \mathbf{2}). \tag{9.5}$$

We denote it by η^i, $i = 1, \ldots, 5$. More precisely[1]

$$\eta^i = (d_1^c, d_2^c, d_3^c, e, -\nu_e)_L . \tag{9.6}$$

9.1.2 Simple gauge groups and Lie algebras

The idea behind grand unification is to find a unifying principle encompassing the different manifestations of gauge symmetry at low energy. The corresponding mathematical notion is the concept of a simple group. By definition a group is simple if it has no invariant subgroup (H is an invariant subgroup of G if, for every $g \in G$ and $h \in H$, $ghg^{-1} \in H$). For example, $SU(3)$ and $SU(2)$ are simple groups but $SU(3) \times SU(2)$ is not (since $SU(3)$ or $SU(2)$ are invariant subgroups). From the point of view of gauge symmetry, a single gauge coupling is associated with a simple group.

The definition follows for algebras since a group element $g \in G$ is associated with an element T_g of the associated algebra $\mathcal{G}$ through the exponential function: $g = e^{iT_g}$. An algebra $\mathcal{G}$ is simple if it has no invariant subalgebra, *i.e.* no subalgebra $\mathcal{H}$ such that $[\mathcal{G}, \mathcal{H}] \subset \mathcal{H}$.

Cartan has classified all simple Lie algebras. Their definitions are summarized in Table 9.1, together with some of their properties. The rank is defined as the maximal number of generators that commute with all the others: in a way, this is the maximal number of quantum numbers that one can fix independently. In the case of grand unification, taking into account the two generators of color $SU(3)$ (λ_3 and λ_8), the weak isospin t^3 and the hypercharge, this makes a total of four. Hence one must look for simple algebras of rank $r \geq 4$. This is why we give in Table 9.1 the explicit examples of $r = 4, 5$ and 6.

---

[1] The charge conjugates $\eta_i \equiv (d_1, d_2, d_3, e^c, -\nu^c)_R$ transform as the fundamental representation $\mathbf{5}$ of $SU(5)$. Note that $i\sigma_2 \left(\begin{smallmatrix} \nu^c \\ e^c \end{smallmatrix} \right) = \left(\begin{smallmatrix} e^c \\ -\nu^c \end{smallmatrix} \right)$ transforms as a doublet (*cf.* (A.121) of Appendix Appendix A).

Table 9.1 Cartan classification of simple Lie algebras

Cartan classification	Name	Leaves invariant the product	Definition	Dimension	Rank: 4	5	6
			$U^\dagger U = 1$				
A_n	$SU(n+1)$	$\sum_{i=1}^{n+1} y_i^* x_i$	$U = (n+1) \times (n+1)$ complex matrix	$n(n+2)$	$SU(5)$	$SU(6)$	$SU(7)$
			$O^T O = 1$				
B_n	$SO(2n+1)$	$\sum_{i=1}^{2n+1} y_i x_i$	$O = (2n+1) \times (2n+1)$ real matrix	$n(2n+1)$	$S0(9)$	$SO(11)$	$SO(13)$
C_n	$Sp(2n)$	$(y_1 \cdots y_{2n}) \begin{pmatrix} 0 & 1 & & & \\ -1 & 0 & \cdot & & \\ & & \cdot & 0 & 1 \\ & & & -1 & 0 \end{pmatrix} \begin{pmatrix} x_1 \\ \vdots \\ \vdots \\ x_{2n} \end{pmatrix}$		$n(2n+1)$	$Sp(8)$	$Sp(10)$	$Sp(12)$
			$O^T O = 1$				
D_n	$SO(2n)$	$\sum_{i=1}^{2n} y_i x_i$	$O = (2n) \times (2n)$ real matrix	$n(2n-1)$	$SO(8)$	$SO(10)$	$SO(12)$
G_2				14			
F_4	Exceptional			52	F_4		
E_6	Lie			78			E_6
E_7	algebras			133			
E_8				248			

How do we discriminate between these simple algebras? One criterion is the possibility of having complex representations. If there is no such possibility, then a mass term is always allowed by the grand unified symmetry and this mass is naturally superheavy: there is no reason why any field would survive at low energy.

Let us see in more detail how this works. If a left chirality fermion ψ_L transforms under a representation $\mathbf{f_L}$ of the gauge group, then $\psi_R = C(\bar{\psi}_L)^T$ transforms as the conjugate representation $\mathbf{\bar{f}_L}$. A kinetic term ($\bar{\psi}_L \partial\!\!\!/\psi_L$ or $\bar{\psi}_R \partial\!\!\!/\psi_R$) is always allowed by the gauge symmetry since $\mathbf{f_L} \times \mathbf{\bar{f}_L} \supset \mathbf{1}$, where $\mathbf{1}$ is the singlet representation. On the other hand, a mass term ($\bar{\psi}_L \psi_R$ or $\bar{\psi}_R \psi_L$) is only allowed if $\mathbf{f_L} \times \mathbf{f_L} \supset \mathbf{1}$. If $\mathbf{f_L}$ is equivalent to $\mathbf{\bar{f}_L}$ (meaning the generators in the two representations are related by a unitary transformation $U : \bar{T} = -UTU^\dagger$), then the two criteria coincide $\mathbf{f_L} \times \mathbf{f_L} \sim \mathbf{f_L} \times \mathbf{\bar{f}_L} \supset \mathbf{1}$ and a mass term is always allowed. The theory is said to be vectorlike. The Standard Model avoids this problem: it is a chiral (*i.e.* not vectorlike) theory and mass terms arise only through a Yukawa coupling to a nonsinglet scalar and gauge symmetry breaking.

Requiring not to have a vectorlike theory leaves us only with $SU(n)$ $(n > 2)$, $SO(4n + 2)$ and E_6.

Another criterion is the absence of gauge anomalies. In order to have a cancellation of the triangle anomalies, one must require that (see the end of Section A.6 of Appendix Appendix A)

$$A_{abc} \propto Tr\, T_L^a \left\{ T_L^b, T_L^c \right\} = 0 \tag{9.7}$$

where the T_L^a are the generators in the representation $\mathbf{f_L}$. All representations of $SO(n)$ satisfy this $(n \neq 6)$. But this is not so for the representations of $SU(n)$ $(n \geq 3)$. When we restrict the rank to be less than 6, we are left with $SO(10)$ and E_6. It must be said that the case of $SU(5)$ discussed in the previous section is atypical. Fermions fit into two representations $\mathbf{\bar{5}}$ and $\mathbf{10}$ which individually have anomalies $A(\mathbf{\bar{5}})$ and $A(\mathbf{10})$. But

$$A(\mathbf{\bar{5}}) = -A(\mathbf{10}). \tag{9.8}$$

We may finally stress that all the groups that we are considering here are compact. This implies that all the abelian subgroups that we will obtain, including the $U(1)$ gauge symmetry of quantum electrodynamics, are compact too. This is an important property if we want to account for charge quantization: if q is a multiple nq_0 of a unit charge, then the gauge transformation

$$\psi(x) \to \psi'(x) = e^{-iq\theta}\psi(x) \tag{9.9}$$

is identical for θ and $\theta + 2\pi/q_0$. In other words, the range of the abelian gauge parameter θ is compact.

9.2 Gauge coupling unification

Gauge coupling unification is one of the key predictions of grand unified theories and, as we have already emphasized in the introduction to this chapter, it is, to date, its greatest success. We will discuss how the precise measurement of the gauge couplings at low energy allows us to determine the main parameters of the theory and to check to some degree the concept of unification. Our starting point will thus be the measure of the low energy couplings. In the $\overline{\text{MS}}$ scheme, they are

$$\alpha_3(M_z) = 0.1185 \pm 0.002, \tag{9.10}$$

$$\sin^2\theta_W(M_z) = 0.23117 \pm 0.00016, \tag{9.11}$$

$$\alpha_{\text{e.m.}}^{-1}(M_z) = 127.943 \pm 0.027. \tag{9.12}$$

9.2.1 Renormalization group equations for the gauge couplings of $SU(3) \times SU(2) \times U(1)$

The diagrams which contribute to the renormalization of a gauge coupling g_i ($i = 1, 2, 3$ respectively, for $U(1)$, $SU(2)$ and $SU(3)$) are given, at one loop, by the diagrams of Fig. 9.1. These diagrams allow us to compute the beta functions to lowest order:

$$\beta(g_i) = \mu\frac{dg_i}{d\mu} = -\frac{b_i}{16\pi^2}g_i^3 + O(g^5). \tag{9.13}$$

The sclution is written in terms of $\alpha_i = g_i^2/(4\pi)$:

$$\frac{1}{\alpha_i(\mu)} = \frac{1}{\alpha_i(\mu_0)} + \frac{b_i}{2\pi}\ln\frac{\mu}{\mu_0} + O\left(\frac{\alpha_i}{\pi}\right). \tag{9.14}$$

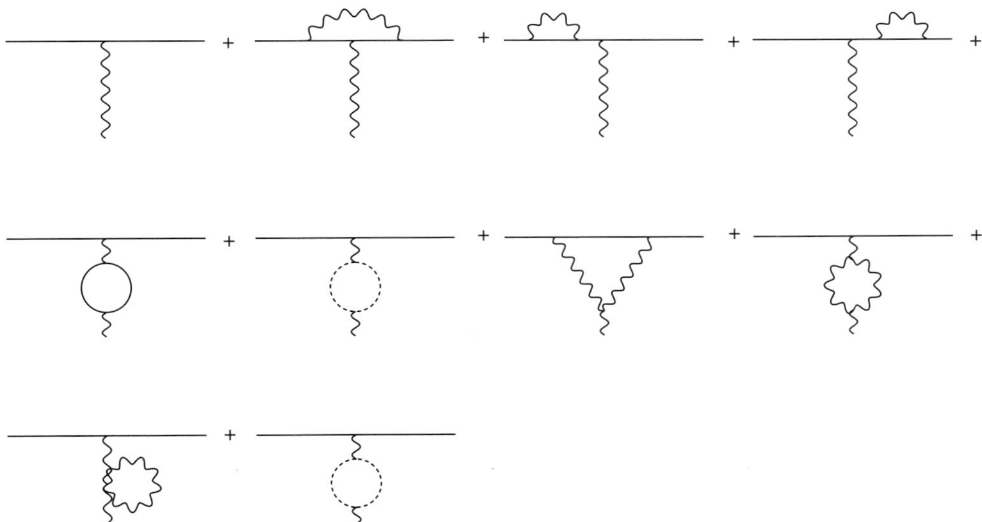

Fig. 9.1 Diagram contributing at one loop to the renormalization of the gauge coupling.

Let us consider more closely the computation of b_i, in particular concerning the group factors.

(i) Let $(t^a)_{ij}$ be the generators of the gauge group in the representation R of the fermion fields. The diagram

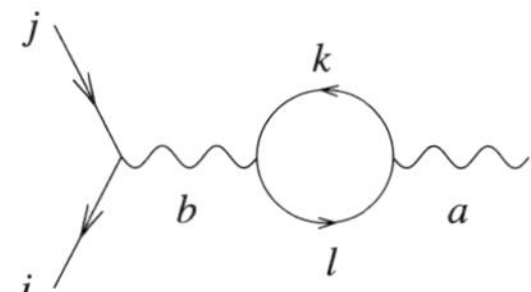

has a group factor $(t^b)_{ij} \operatorname{Tr}(t^a t^b)$. We have

$$\operatorname{Tr}(t^a t^b) = T(R)\, \delta^{ab} \tag{9.15}$$

and $T(R) = 1/2$ if R is the (anti)fundamental ($\mathbf{N}$ or $\bar{\mathbf{N}}$) of $SU(N)$. The group factor is thus $T(R)\,(t^a)_{ij}$.

(ii) Let $(T^a)_{bc}$ be the generators of the gauge group in the adjoint representation. If C_{abc} are the structure constants

$$[T^a, T^b] = iC^{abc}T^c, \tag{9.16}$$

then

$$(T^a)^{bc} = -iC^{abc} \tag{9.17}$$

and

$$\operatorname{Tr} T^a T^b = C_2(G)\delta^{ab} = C^{acd}C^{bcd}. \tag{9.18}$$

$C_2(G)$ is the Casimir of order 2 of the group G: $C_2(G) = N$ for $SU(N)$ and $C_2(G)$ vanishes for an abelian group (no gauge vector self-coupling).
The diagram

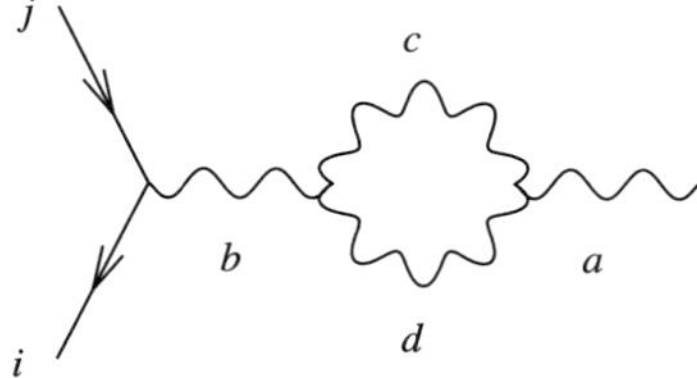

has a group factor $(t^b)_{ij} C_{acd}C_{bcd} = (t^a)_{ij}\, C_2(G)$.
In total,

$$b = \tfrac{11}{3}C_2(G) - \tfrac{2}{3}T(R)\big|_{\text{Weyl fermions}} - \tfrac{1}{3}T(R)\big|_{\text{complex scalars}}. \tag{9.19}$$

A Dirac spinor contributes with its two chirality states $(-4/3)\,T(R)$ to this beta function coefficient. A Majorana spinor contributes as a Weyl spinor.

We find, for the evolution of $SU(3)$ and $SU(2)$ couplings above M_Z, with N_F families of quarks and leptons,

$$b_3 = \tfrac{11}{3} \times 3 - \tfrac{4}{3} \times 2 \times \tfrac{1}{2}N_F = 11 - \tfrac{4}{3}N_F \tag{9.20}$$

$$b_2 = \tfrac{11}{3} \times 2 - \tfrac{2}{3} \times 4 \times \tfrac{1}{2}N_F - \tfrac{1}{3} \times \tfrac{1}{2} = \tfrac{22}{3} - \tfrac{4}{3}N_F - \tfrac{1}{6} \tag{9.21}$$

where $-1/6$ is the Higgs contribution.

The case of the abelian hypercharge symmetry $U(1)_Y$ requires special attention. Indeed, we have defined (see Section A.3.2 of Appendix Appendix A) the gauge coupling g' through the minimal coupling of a field Φ of hypercharge y:

$$D_\mu \Phi = \left(\partial_\mu - i\frac{g'}{2}yB_\mu \right) \Phi, \tag{9.22}$$

where y is related to the field charge and weak isospin through the relation:

$$q = t^3 + \frac{y}{2}. \tag{9.23}$$

In the case of an abelian group, it is not possible to use the commutation relations to normalize the generators. One may thus replace y by ky if one substitutes g'/k for g'. To make this explicit, let us write $y \equiv 2Ct^0$:

$$q = t^3 + Ct^0, \tag{9.24}$$
$$D_\mu \Phi = \left(\partial_\mu - ig'Ct^0 B_\mu \right) \Phi. \tag{9.25}$$

One then defines

$$g_1 \equiv C\, g', \quad g_2 \equiv g. \tag{9.26}$$

Since $\epsilon = g \sin\theta_W = gg'/\sqrt{g^2 + g'^2}$, we have

$$\frac{1}{e^2} = \frac{1}{g^2} + \frac{C^2}{g_1^2}, \tag{9.27}$$

or

$$\alpha_{\text{e.m.}}^{-1} = \alpha_2^{-1} + C^2\alpha_1^{-1}, \tag{9.28}$$

where $\alpha_{\text{e.m.}} \equiv e^2/(4\pi)$.

If there is an underlying theory where $U(1)_Y$ is imbedded in a nonabelian group G, then C is fixed by the normalization imposed by the commutation relations. One can obtain a simple expression for the normalization constant C under such an assumption.

Indeed, if t^3 and t^0 are the generators of a simple nonabelian group G in a representation R, then, using (9.15), we have $\text{Tr}_R \left(t^3 \right)^2 = \text{Tr}_R \left(t^0 \right)^2 = T(R)$ and $\text{Tr}_R t^3 t^0 = 0$. Then, from (9.24),

$$\text{Tr}_R q^2 = (1 + C^2)\text{Tr}_R \left(t^3 \right)^2. \tag{9.29}$$

For example, if we assume that all known fermions form one or more full representations R of the group G, then

$$\mathrm{Tr}_R q^2 = 2(1 + 1 + 1) + 3 \times 2 \left(3 \times \tfrac{4}{9} + 3 \times \tfrac{1}{9}\right) = 16,$$

where we have counted the contributions of the leptons e, μ, τ (and their antiparticles) and of the three colors of quarks u, c, t and d, s, b (and their antiparticles). Also

$$\mathrm{Tr}_R \left(t^3\right)^2 = \tfrac{1}{2}(3 + 3 \times 3) = 6$$

to account for all the fermion doublets ($T(R) = 1/2$): three leptons and three quarks in three colors. Hence,

$$C^2 = \tfrac{5}{3}. \tag{9.30}$$

This can be checked directly on the example of $SU(5)$ by identifying the explicit form of the hypercharge generator t^0 (see Exercise 1).

We then obtain the one-loop coefficient for the g_1 beta function:

$$
\begin{aligned}
b_1 &= -\frac{2}{3} \sum_f \left(t^0\right)^2{}_f - \frac{1}{3} \left(t^0\right)^2{}_\phi \\[2mm]
&= -\frac{1}{4C^2} \left(\frac{2}{3} \sum_f y_f^2 + \frac{1}{3} y_\phi^2\right) \\[2mm]
&= -\frac{4}{3} N_F - \frac{1}{10},
\end{aligned}
\tag{9.31}
$$

where $-1/10$ is the contribution from the Higgs, and we recall that N_F is the number of families.

9.2.2 The nonsupersymmetric case

We suppose the existence of an underlying gauge symmetry described by a simple gauge group G, which is broken at a superheavy scale M_U much larger than the TeV scale. This gauge invariance is only explicit above the scale M_U. This theory contains superheavy fields, such as gauge fields, with mass of order M_U. We note two important properties of the renormalization group equations for the gauge couplings α_i ($i = 1, 2, 3$) [183]:

(i) any representation R which contains only superheavy fields decouples from these equations;

(ii) any representation R which contains only light fields contributes equally to b_1, b_2 and b_3.

For example, the generators of $SU(3)$ $t^{(3)i}$ are generators of G in the representation R: $\delta^{ij} T^{(3)}(R) \equiv \mathrm{Tr}\, t^{(3)i} t^{(3)j} = \delta^{ij} T(R)$. Similarly, the generators of $SU(2)$ $t^{(2)a}$ are generators of G in R and $\delta^{ab} T^{(2)}(R) \equiv \mathrm{Tr}\, t^{(2)a} t^{(2)b} = \delta^{ab} T(R)$. Hence, they contribute equally to b_2 and b_3. This is why the coefficients of N_F in (9.20), (9.21), and (9.31) are identical: a family of quarks and leptons yields complete representations of the minimal $SU(5)$ group.

Let us thus suppose that the only multiplet which contains both light and superheavy fields is the gauge multiplet. Then, using (9.19), we have[2] $b_3 = b_1 + 11$ and $b_2 = b_1 + 22/3$. Hence

$$b_1 - 3b_2 + 2b_3 = 0, \tag{9.32}$$

and we see, using (9.14), that the combination of couplings $\alpha_1^{-1}(\mu) - 3\alpha_2^{-1}(\mu) + 2\alpha_3^{-1}(\mu)$ is not renormalized at one loop, *i.e.* is independent of the renormalization scale μ at this order.

If we suppose that these couplings are equal at the scale M_U, as expected since they unify into the single coupling g corresponding to the simple group G, then

$$\alpha_1^{-1}(\mu) - 3\alpha_2^{-1}(\mu) + 2\alpha_3^{-1}(\mu) = 0. \tag{9.33}$$

We then obtain from (9.28) and (9.33)

$$(1 + 3C^2)\alpha_2^{-1}(\mu) = \alpha_{\text{e.m.}}^{-1}(\mu) + 2C^2\alpha_3^{-1}(\mu),$$
$$(1 + 3C^2)\alpha_1^{-1}(\mu) = 3\alpha_{\text{e.m.}}^{-1}(\mu) - 2\alpha_3^{-1}(\mu), \tag{9.34}$$

from which we extract the value of $\sin^2\theta_w = g'^2/(g^2 + g'^2) = \alpha_2^{-1}/(\alpha_2^{-1} + C^2\alpha_1^{-1})$

$$\sin^2\theta_w(\mu) = \frac{1}{1 + 3C^2}\left(1 + 2C^2\frac{\alpha_{\text{e.m.}}(\mu)}{\alpha_3(\mu)}\right). \tag{9.35}$$

Thus, since C has a computable value for any given theory, (9.35) yields a testable relation between quantities measurable at $\mu = M_z$. We note the value of $\sin^2\theta_w$ at unification: $(1 + C^2)^{-1}$ that is $3/8$ for $C^2 = 5/3$.

One can also determine the scale of unification M_U and the common value α_U of the gauge couplings at unification. Indeed, since

$$5\alpha_1^{-1}(\mu) - 3\alpha_2^{-1}(\mu) - 2\alpha_3^{-1}(\mu) = \frac{44}{\pi}\ln\frac{M_U}{\mu}, \tag{9.36}$$

we deduce from (9.34)

$$\ln\frac{M_U}{M_z} = \frac{6\pi}{11(1 + 3C^2)}\left[\alpha_{\text{e.m.}}^{-1}(M_z) - (1 + C^2)\alpha_3^{-1}(M_z)\right]. \tag{9.37}$$

Let us take the example of $N_F = 3$ and $C^2 = 5/3$ as in $SU(5)$. We have, using (9.10) and (9.12),

$$M_U \sim M_z e^{30} \sim 10^{15} \text{ GeV}, \tag{9.38}$$

$$\alpha_U^{-1} = \alpha_3^{-1}(M_z) + \frac{b_3}{2\pi}\ln\frac{M_U}{M_z} \sim 42. \tag{9.39}$$

It is a remarkable coincidence that one obtains such a superheavy scale for grand unification. Indeed, any lower scale would be catastrophic for the theory since it would

---

[2]Note that this agrees with (9.20), (9.21), and (9.31), if we disregard the Higgs contribution. Indeed, the Higgs representation contains both light and heavy fields (*cf.* the doublet–triplet splitting problem discussed above for $SU(5)$).

induce rapid proton decay, the gauge bosons X^μ and Y^μ being too light. We will see that even such a large scale puts in jeopardy the minimal $SU(5)$ model.

As for the relation (9.35), it reads

$$0.23117 \pm 0.00016 = 0.2029 \pm 0.0018. \tag{9.40}$$

The order of magnitude comes out right and this is a second success of the grand unification picture. This seems to indicate that there is at least an approximate unification of couplings at a scale much larger than we might have expected. Indeed, this does not leave much room for physics in the range between M_Z and M_U: this has become known as the "grand desert hypothesis".

There is, however, still a discrepancy of some 10 standard deviations in (9.40) between the experimental result (first number) and the prediction (second number). Where could such a disagreement come from?

- Higher orders in the evolution of the renormalization group equations. One may include the two-loop contribution to the beta functions. This gives an extra contribution $\delta_{\text{h.o.}} \sim 0.0029$ on the right-hand side of (9.40).
- Threshold effects. One needs to be more quantitative to account for the decoupling of heavy particles in the evolution of the gauge couplings. One might also include possible intermediate thresholds. We call δ_{th} the corresponding contribution.
- Mixed representations. It is also possible to have representations which mix light and superheavy fields. We have seen for example that the representation for the Higgs doublet falls in such a category in the case of $SU(5)$ since the corresponding triplet must be superheavy to prevent too rapid proton decay. We call δ_{m} such a contribution.

In total, one must replace (9.35) by

$$\sin^2 \theta_w (M_Z) = \frac{1}{1 + 3C^2} \left(1 + 2C^2 \frac{\alpha_{\text{e.m.}}^{-1}(M_Z)}{\alpha_3^{-1}(M_Z)} \right) + \delta_{\text{h.o.}} + \delta_{\text{th}} + \delta_{\text{m}}. \tag{9.41}$$

The precision measurements in (9.10)–(9.12) provide useful indications on the magnitude of the last terms. In any case, they show that the minimal $SU(5)$ model is not compatible with gauge unification. In other words, given their allowed range of values, the gauge couplings of the minimal $SU(5)$ model do not intersect at a single point.

9.2.3 The supersymmetric case

We now turn to the supersymmetric case. We will describe supersymmetric models of grand unification in the next section. For the time being, it suffices to stress that, for scales μ larger than the supersymmetric thresholds, one must take into account the supersymmetric partners in the evolution of the running gauge couplings. There is no need for further computations, since equation (9.19) summarizes it all.

Indeed, for a gauge vector supermultiplet (adjoint representation), gauge bosons contribute $-\frac{11}{3} C_2(G)$ whereas (Majorana) gaugino fields contribute $\frac{2}{3} T(R_{\text{adj.}}) = \frac{2}{3} C_2(G)$; this adds up to a total of $-3C_2(G)$.

For a chiral supermultiplet in a representation R of the group G (complex) scalars contribute $\frac{1}{3} T(R)$ and the Weyl fermion $\frac{2}{3} T(R)$, adding up to $T(R)$.

Thus, (9.19) reads, in the supersymmetric case,

$$b = 3C_2(G) - T(R)|_{\text{chiral supermultiplet}} \tag{9.42}$$

which gives for N_F families and two Higgs doublets:

$$\begin{aligned}
b_1 &= -2N_F - \tfrac{3}{5} \\
b_2 &= 6 - 2N_F - 1 \\
b_3 &= 9 - 2N_F
\end{aligned} \tag{9.43}$$

where $-3/5$ and -1 are the contributions of the two doublets of Higgs supermultiplets.

If we make the hypothesis that the only multiplets which contain light and super-heavy fields are the gauge vector supermultiplets, then one recovers in the supersymmetric case, the relation (9.32) and thus (9.41) follows.

What is new is that in δ_{m}, the contribution of mixed light–superheavy representations, there are now two complete chiral supermultiplets corresponding to the representations of H_1 and H_2, instead of a single Higgs doublet in the nonsupersymmetric case.

A more complete analysis is performed by assuming that all supersymmetric particles have a common mass M_{SUSY}. Since squarks and sleptons form complete representations of $SU(5)$, they modify in the same way the evolution of all couplings. Also, the contribution of gauginos does not change the relation (9.33). In the case of Higgs, in the simplest approach, one is taken with a mass M_z whereas the other has mass M_{SUSY}. Hence the determination of δ_{m} may be turned into a determination of M_{SUSY}. It is a remarkable feature that for a value of M_{SUSY} in the TeV range, a region which was favored earlier by the naturalness criterion, there is unification of the three gauge couplings at a scale of order 10^{16} GeV (see Fig. 9.2).

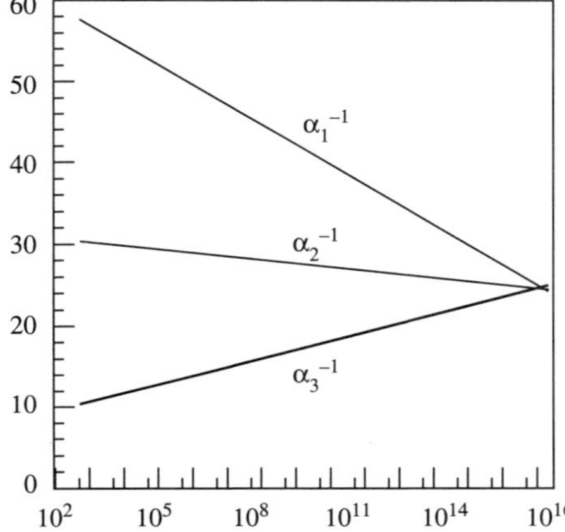

Fig. 9.2 Unification of gauge couplings in the supersymmetric case for $M_{\text{SUSY}} \sim 1\,\text{TeV}$.

9.3   The minimal supersymmetric SU(5) model

Following the procedure developed in Chapter 5 which is to introduce supersymmetric partners for every field of the nonsupersymmetric model, we introduce:

- a vector supermultiplet in the representation **24** of $SU(5)$;
- N_F chiral supermultiplets in the representations $\bar{\mathbf{5}}$ and **10** to accomodate the N_F families of quarks and leptons; as above, we denote them, respectively, η^i and χ_{ij};
- a chiral supermultiplet Σ in the representation **24** to break the grand unified symmetry $SU(5)$;
- two chiral supermultiplets, one H_1^i in the $\bar{\mathbf{5}}$ and the other H_{2i} in the **5** of $SU(5)$. Among the scalar components, one recovers the $SU(2)$ doublet H_1 in the $\bar{\mathbf{5}}$ and the $SU(2)$ doublet H_2 in the **5**: they have opposite hypercharge.

We note at this stage why we cannot identify the $\bar{\mathbf{5}}$ of Higgs with one of the $\bar{\mathbf{5}}$ of the leptons (and hence consider the Higgs doublet as the supersymmetric partner of one of the lepton doublets): the fermionic component of the corresponding $\bar{\mathbf{3}}$ would be the light d^c whereas the prevention of rapid proton decay imposes it to be superheavy (see Section 9.3.4 below).

The superpotential for the field Σ responsible for $SU(5)$ breaking is

$$W(\Sigma) = \frac{M}{2} \operatorname{Tr} \Sigma^2 + \frac{\Lambda}{3} \operatorname{Tr} \Sigma^3. \tag{9.44}$$

The corresponding scalar potential reads

$$V = \sum_{ij} \left| \frac{\partial W}{\partial \Sigma_j^i} - \frac{1}{5} \delta_i^j \operatorname{Tr} \left(\frac{\partial W}{\partial \Sigma} \right) \right|^2, \tag{9.45}$$

the unusual form being due to the implementation of the constraint $\operatorname{Tr} \Sigma = 0$ (see Exercise 2).

The minimum corresponds to

$$M\Sigma_j^i + \Lambda \left[\left(\Sigma^2 \right)_j^i - \frac{1}{5} \delta_j^i \operatorname{Tr} \Sigma^2 \right] = 0. \tag{9.46}$$

There are three solutions:

- unbroken $SU(5)$,

$$\Sigma_j^i = 0. \tag{9.47}$$

- $SU(5)$ broken into $SU(4) \times U(1)$,

$$\Sigma_j^i = \frac{M}{3\Lambda} \begin{pmatrix} 1 & & & & \\ & 1 & & & \\ & & 1 & & \\ & & & 1 & \\ & & & & -4 \end{pmatrix}. \tag{9.48}$$

- $SU(5)$ broken into $SU(3) \times SU(2) \times U(1)$,

$$\Sigma^i_j = \frac{M}{\Lambda} \begin{pmatrix} 2 & & & & \\ & 2 & & & \\ & & 2 & & \\ & & & -3 & \\ & & & & -3 \end{pmatrix}. \tag{9.49}$$

9.3.1 Gauge coupling unification

We illustrate the general discussion of supersymmetric gauge coupling unification with the case of minimal $SU(5)$. For simplicity, we assume that the scalar fields in the **24** have the same mass M_U as the superheavy vector fields, whereas the superheavy triplets have a mass denoted by M_T. We also take all supersymmetric particles and one of the two Higgs doublet to have a common mass M_{SUSY}. Finally, we limit ourselves to one-loop renormalization group equations, in the $\overline{\text{DR}}$ scheme where thresholds can be crossed with step functions (see Appendix E).

Using the results of Section 9.2, we may write the renormalization group equations for the gauge couplings:

$$\alpha_3^{-1}(M_z) = \alpha_U^{-1} + \frac{1}{2\pi}\left[\left(11 - \frac{4}{3}N_F\right)\ln\frac{M_z}{M_{\text{SUSY}}} + (9 - 2N_F)\ln\frac{M_{\text{SUSY}}}{M_U} - \ln\frac{M_T}{M_U}\right],$$

$$\alpha_2^{-1}(M_z) = \alpha_U^{-1} + \frac{1}{2\pi}\left[\left(\frac{43}{6} - \frac{4}{3}N_F\right)\ln\frac{M_z}{M_{\text{SUSY}}} + (5 - 2N_F)\ln\frac{M_{\text{SUSY}}}{M_U}\right], \tag{9.50}$$

$$\alpha_1^{-1}(M_z) = \alpha_U^{-1} + \frac{1}{2\pi}\left[\left(-\frac{1}{10} - \frac{4}{3}N_F\right)\ln\frac{M_z}{M_{\text{SUSY}}} + \left(-\frac{3}{5} - 2N_F\right)\ln\frac{M_{\text{SUSY}}}{M_U} - \frac{2}{5}\ln\frac{M_T}{M_U}\right].$$

Then, one may consider the same combinations as in (9.33) and (9.36):

$$\left[3\alpha_2^{-1} - 2\alpha_3^{-1} - \alpha_1^{-1}\right](M_z) = \frac{1}{2\pi}\left[\frac{12}{5}\ln\frac{M_T}{M_z} - 2\ln\frac{M_{\text{SUSY}}}{M_z}\right], \tag{9.51}$$

$$\left[5\alpha_1^{-1} - 3\alpha_2^{-1} - 2\alpha_3^{-1}\right](M_z) = \frac{1}{2\pi}\left[36\ln\frac{M_U}{M_z} + 8\ln\frac{M_{\text{SUSY}}}{M_z}\right]. \tag{9.52}$$

Once the supersymmetric threshold M_{SUSY} is chosen, the first relation fixes the triplet mass, or vice versa. The success of the model is associated with the fact that, for values of M_{SUSY} compatible with the requirement of naturalness, one finds values of M_T of the order of the superheavy scale M_U, thus preventing rapid proton decay (we will see however below that this is not sufficient in the case of minimal $SU(5)$). Taking for example $M_{\text{SUSY}} = 1$ TeV, and including two-loop renormalization group equations and one-loop threshold contributions, one obtains [290]:

$$3.5 \times 10^{14} \text{ GeV} \leq M_T \leq 3.6 \times 10^{15} \text{ GeV},$$
$$1.7 \times 10^{16} \text{ GeV} \leq M_U \leq 2.0 \times 10^{16} \text{ GeV}. \tag{9.53}$$

Finally, we note the value of α_U at unification: $\alpha_U \sim 1/24$ which is somewhat larger than in the non-supersymmetric case.

9.3.2   Back to the naturalness and hierarchy problem

Let us come back for a moment to the doublet–triplet splitting problem in the context of the nonsupersymmetric minimal $SU(5)$ model. The dangerous terms in the scalar potential are those which mix the **24** of Higgs (Σ) with the **5** (H):

$$V(\Sigma, H) = \alpha H^\dagger H \, \mathrm{Tr}\left(\Sigma^2\right) + \beta H^\dagger \Sigma^2 H. \qquad (9.54)$$

These terms are in any case present because they are induced by radiative corrections, as given in Fig. 9.3. They give a mass of order $\langle \Sigma \rangle$ to the triplet of Higgs within H, which prevents too fast a proton decay but a severe fine tuning is necessary in order not to give the same mass to the remaining doublet.

When we turn to the symmetric case, such terms appear already at the tree level because of the gauge interactions: the D-term of the potential necessarily involves such terms. Indeed, since the $SU(5)$ D-term is written

$$D^a = g\left(\Sigma^\dagger T^a \Sigma + H_1^\dagger t^a H_1 + H_2^\dagger t^a H_2\right), \qquad (9.55)$$

and the corresponding term $(D^a)^2$ in the potential involves the mixed term

$$g^2\left(\Sigma^\dagger T^a \Sigma\right)\left(H_i^\dagger t^a H_i\right), \quad i = 1, 2, \qquad (9.56)$$

which is a specific form of (9.54). We expect this property to be stable under radiative corrections because of the nonrenormalization theorems characteristic of supersymmetry. We have seen that the parameters of the superpotential are not renormalized: radiative corrections are not expected to produce F-terms in the potential, but only D-terms. Hence their contribution must be of the form (9.56). Indeed, there are new contributions in the form of the diagrams of Fig. 9.4: adding them to those of Fig. 9.3 yields a total contribution in the form of (9.56).

Now, with the ground state (9.49), one can easily check that $\langle \Sigma^\dagger T^a \Sigma \rangle = 0$: for generators of $SU(3) \times SU(2) \times U(1)$ this amounts to $\mathrm{Tr}\, T^a = 0$, whereas the remaining

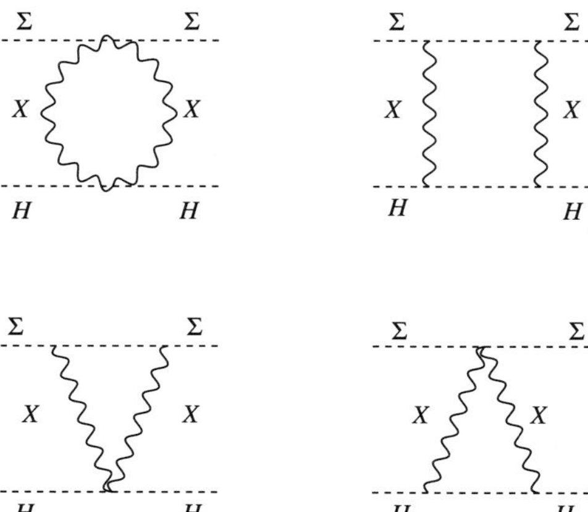

Fig. 9.3

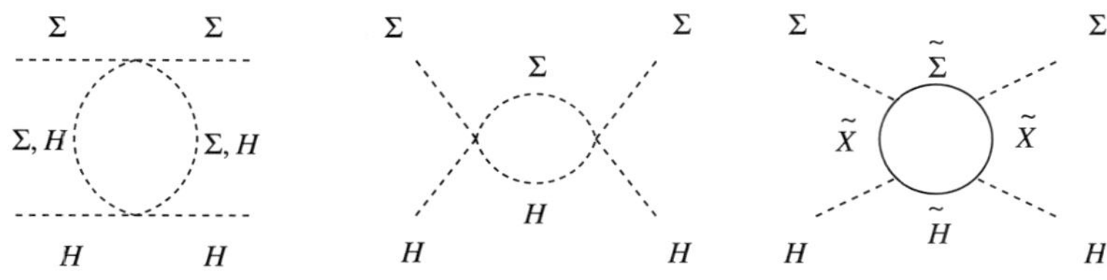

Fig. 9.4

ones have vanishing terms on the diagonal. Thus, the mixed terms (9.56) do not generate a mass term for the H_i superfields.

We still have not shown that we can generate a mass term for the triplets of Higgs while keeping the doublets light. But the preceding discussion shows that, if we succeed to do that at tree level, then we can count on supersymmetry to ensure that radiative corrections will follow the same pattern. Let us show how this may be realized in practice.

If we consider the following superpotential for H_1 and H_2:

$$W = H_1 \left(\mu + \lambda\Sigma\right) H_2, \tag{9.57}$$

the scalar potential reads

$$V = |(\mu + \lambda\Sigma)H_2|^2 + |H_1(\mu + \lambda\Sigma)|^2 + \cdots \tag{9.58}$$

where the minimization of the extra terms is supposed to fix $\langle\Sigma\rangle$ to its value (9.49). Then,

$$\mu + \lambda\Sigma = \begin{pmatrix} (\mu + 2\lambda\sigma)\mathbb{1}_3 & 0 \\ 0 & (\mu - 3\lambda\sigma)\mathbb{1}_2 \end{pmatrix},$$

with $\sigma = M/\Lambda$. Thus, the mass of the doublet is $m_2 = \mu - 3\lambda\sigma$ whereas the mass of the triplets is $M_T = \mu + 2\lambda\sigma$. If we impose that

$$\mu \sim 3\lambda\sigma \tag{9.59}$$

which is of the order of the grand unification scale, then one may realize a doublet–triplet splitting. This condition may seem at this point ad hoc but, as long as super-symmetry is unbroken, it is natural in the technical sense[3].

It remains to see which dynamics might be responsible for the condition (9.59).

In the *sliding singlet* approach [292, 372], one introduces a chiral superfield S which is a gauge singlet and one modifies the superpotential (9.57) as follows:

$$W = H_1 \left(\kappa S + \lambda\Sigma\right) H_2. \tag{9.61}$$

---

[3] Using the fact that the terms in the superpotential (9.57) are not renormalized and defining the wave function renormalization constant Z as $H_i = Z^{1/2} H_{i_R}$, we have

$$\mu_R = Z\mu, \quad \lambda_R \sigma_R = Z\lambda\sigma, \tag{9.60}$$

and thus $\mu_R \sim 3\lambda_R \sigma_R$ follows from (9.59).

The potential is now of the form (9.58) with μ replaced by κS. Its minimization leads to the conditions $(\kappa s - 3\lambda\sigma)v_i = 0$, $i = 1, 2$ where $s \equiv \langle S \rangle$. Hence, once $SU(2) \times U(1)$ is broken (v_1 or $v_2 \neq 0$), we have automatically $\kappa s - 3\lambda\sigma = 0$, which ensures massless doublets at this order. In other words, the vev of the gauge singlet slides in order to minimize the ground state energy, and correspondingly the mass of the doublets. Obviously, supersymmetry breaking will modify the analysis: it is thus important to make sure that the scale of supersymmetry breaking in this sector remains low.

Another line of attack when the scale of supersymmetry breaking is not small is to introduce a Higgs representation which contains triplets of $SU(3)$ (*i.e.* $(\mathbf{3}, \mathbf{1})$ under $SU(3) \times SU(2)$) but no doublet of $SU(2)$ (*i.e.* $(\mathbf{1}, \mathbf{2})$ under $SU(3) \times SU(2)$). Coupling this representation to H_1 and H_2, one thus gives a mass to the triplets but not to the doublets. This is the *missing doublet mechanism* [178, 208, 284].

An example is provided by the representation $\mathbf{50}$ of $SU(5)$. Its decomposition under $SU(3) \times SU(2)$ reads:

$$\mathbf{50} = (\mathbf{8}, \mathbf{2}) + (\bar{\mathbf{6}}, \mathbf{3}) + (\mathbf{6}, \mathbf{1}) + (\bar{\mathbf{3}}, \mathbf{2}) + (\mathbf{3}, \mathbf{1}) + (\mathbf{1}, \mathbf{1}).$$

One chooses to break $SU(5)$ down to $SU(3) \times SU(2) \times U(1)$ with a $\mathbf{75}$ of $SU(5)$ which we note $\tilde{\Sigma}$. Then we introduce a $\mathbf{50}$ (C) and a $\bar{\mathbf{50}}$ ($\bar{C}$); using the fact that $\mathbf{50} \times \mathbf{75} \times \bar{\mathbf{5}} \ni \mathbf{1}$, we may write the superpotential as

$$W = \rho\, C\tilde{\Sigma}H_1 + \rho'\, \bar{C}\tilde{\Sigma}H_2 + M'\bar{C}C. \tag{9.62}$$

Then, replacing $\tilde{\Sigma}$ by its vev of order M_X, we may write the part of the superpotential relevant for the triplets (T) and antitriplets ($\bar{T}$) as

$$
\begin{aligned}
W &= \rho M_X \bar{T}_{H_1} T_C + \rho' M_X \bar{T}_{\bar{C}} T_{H_2} + M' \bar{T}_{\bar{C}} T_C \\
&= M' \left(\bar{T}_{\bar{C}} + \rho \frac{M_X}{M'} \bar{T}_{H_1} \right) \left(T_C + \rho' \frac{M_X}{M'} T_{H_2} \right) - \rho\rho' \frac{M_X^2}{M'} \bar{T}_{H_1} T_{H_2}.
\end{aligned}
\tag{9.63}
$$

Hence the triplets acquire a mass of order M_X^2/M' whereas the doublets remain massless.

9.3.3   Fermion masses

Fermion masses arise from Yukawa interactions which are derived directly from the superpotential. Denoting, as above, by χ_{mn} and η^m the superfields in representations $\mathbf{10}$ and $\bar{\mathbf{5}}$, the superpotential reads:

$$W = -\lambda_d\, \eta^m \chi_{mn} H_1^n - \lambda_u \epsilon^{mnpqr} \chi_{mn} \chi_{pq} H_{2r}, \tag{9.64}$$
$$\bar{\mathbf{5}} \times \mathbf{10} \times \bar{\mathbf{5}} \qquad \mathbf{10} \times \mathbf{10} \times \mathbf{5}$$

where H_1^p and H_{2r} are respectively the Higgs superfields in representations $\bar{\mathbf{5}}$ and $\mathbf{5}$ of $SU(5)$ and ϵ is the completely antisymmetric tensor.

One easily deduces the Yukawa couplings and, setting the Higgs fields at their vacuum expectation values $\langle H_1^m \rangle = -v_1 \delta^{m5}$ and $\langle H_{2m} \rangle = v_2 \delta_{m5}$, one obtains

$$\mathcal{L}_m = -\lambda_d v_1\, \overline{\Psi}_{\eta^m} \Psi_{\chi_{m5}} + \lambda_u v_2\, \epsilon^{mnpq5} \overline{\Psi}_{\chi_{mn}} \Psi_{\chi_{pq}} + \text{h.c.} \tag{9.65}$$

with obvious notation. Reading in (9.4) and (9.6) the fermion content of the quark–lepton representations, we conclude that

$$m_d = m_e = -\lambda_d v_1, \quad m_u = \lambda_u v_2. \tag{9.66}$$

If we restore the three families, we conclude that $SU(5)$ grand unification (supersymmetric or not, as a matter of fact) makes the following extra prediction as compared to the Standard Model:

$$m_b = m_\tau, \tag{9.67}$$

$$m_s = m_\mu, \tag{9.68}$$

$$m_d = m_e. \tag{9.69}$$

These relations are valid at the scale of grand unification and must be renormalized down to low energies.

[Bottom–tau unification

We first study bottom–tau unification: the renormalization group evolution is complicated by the fact that the top Yukawa coupling is not small. In fact, we have already solved the evolution equation for the top coupling in Section 6.9.1 of Chapter 6 and we will follow the same method here. Neglecting all other Yukawa couplings than λ_t in the evolution, the renormalization group equations for the bottom and tau Yukawa couplings read:

$$\frac{\mu}{\lambda_b} \frac{d\lambda_b}{d\mu} = -\frac{1}{16\pi^2} \left(\frac{16}{3} g_3^2 + 3g_2^2 + \frac{7}{15} g_1^2 \right) + \frac{\lambda_t^2}{16\pi^2} \tag{9.70}$$

$$\frac{\mu}{\lambda_\tau} \frac{d\lambda_\tau}{d\mu} = -\frac{1}{16\pi^2} \left(3g_2^2 + \frac{9}{5} g_1^2 \right). \tag{9.71}$$

Defining as in Chapter 6, equation (6.104),

$$Y_b \equiv \frac{\lambda_b^2}{16\pi^2}, \quad Y_t \equiv \frac{\lambda_t^2}{16\pi^2}, \tag{9.72}$$

the renormalization equation for λ_b is written

$$\frac{\mu}{Y_b} \frac{dY_b}{d\mu} = -\frac{1}{16\pi^2} \left(\frac{32}{3} g_3^2 + 6g_2^2 + \frac{14}{15} g_1^2 \right) + 2Y_t, \tag{9.73}$$

where $Y_t(\mu)$ is given explicitly in equation (6.108) of Chapter 6. In the absence of the Y_t term, this would read:

$$\frac{\mu}{Y_b} \frac{dY_b}{d\mu} = \frac{32}{9} \frac{\mu}{g_3} \frac{dg_3}{d\mu} - 6 \frac{\mu}{g_2} \frac{dg_2}{d\mu} - \frac{14}{99} \frac{\mu}{g_1} \frac{dg_1}{d\mu} \tag{9.74}$$

which is readily solved as

$$Y_b(\mu) = \frac{1}{\alpha_b} E_b(\mu) \text{ with } E_b(\mu) = \left(\frac{g_3(\mu)}{g_3(\mu_0)} \right)^{32/9} \left(\frac{g_2(\mu)}{g_2(\mu_0)} \right)^{-6} \left(\frac{g_1(\mu)}{g_1(\mu_0)} \right)^{-14/99}. \tag{9.75}$$

The complete equation is solved by turning the constant α_b into a function $\alpha_b(\mu)$ which satisfies $(\mu/\alpha_b)(d\alpha_b/d\mu) = -2Y_t$. Using (6.108) of Chapter 6, this is solved as:

$$\alpha_b(\mu) = \alpha_b(\mu_0)\left[1 - 12Y_t(\mu_0)F_t(\mu)\right]^{1/6} \tag{9.76}$$

where $F_t(\mu)$ is given in equation (6.109) of the same chapter. Hence

$$Y_b(\mu) = \frac{Y_b(\mu_0)E_b(\mu)}{\left[1 - 12Y_t(\mu_0)F_t(\mu)\right]^{1/6}}. \tag{9.77}$$

Choosing $\mu_0 = M_U$, we thus have for the bottom mass

$$m_b(\mu) = m_b(M_U)\left(\frac{g_3(\mu)}{g_3(M_U)}\right)^{16/9}\left(\frac{g_2(\mu)}{g_2(M_U)}\right)^{-3}$$
$$\times \left(\frac{g_1(\mu)}{g_1(M_U)}\right)^{-7/99}\frac{1}{\left[1 - 3\lambda_t^2(M_U)F_t(\mu)/(4\pi^2)\right]^{1/12}} \tag{9.78}$$

and we would obtain along the same lines from (9.71)

$$m_\tau(\mu) = m_\tau(M_U)\left(\frac{g_2(\mu)}{g_2(M_U)}\right)^{-3}\left(\frac{g_1(\mu)}{g_1(M_U)}\right)^{-3/11}. \tag{9.79}$$

Using the unification constraint $m_b(M_U) = m_\tau(M_U)$, we deduce

$$m_b(m_b) = m_\tau(m_b)\left(\frac{g_3(m_b)}{g_3(M_U)}\right)^{16/9}\left(\frac{g_1(m_b)}{g_1(M_U)}\right)^{20/99}\frac{1}{\left[1 - 3\lambda_t^2(M_U)F_t(m_b))/(4\pi^2)\right]^{1/12}}. \tag{9.80}$$

In this equation, we can safely approximate $m_\tau(m_b)$ with the physical mass $m_\tau = 1.78$ GeV. On the other hand, QCD corrections introduce a significant difference between the running bottom quark mass and its physical mass which we will denote by M_b. More precisely, we have at one loop in the $\overline{\text{DR}}$ renormalization scheme[4]:

$$M_b = m_b(M_b)^{\overline{\text{DR}}}\left[1 + \frac{5g_3^2}{12\pi^2}\right]. \tag{9.82}$$

For a physical mass $M_b = 4.7$ GeV, this gives a current mass $m_b(M_b)^{\overline{\text{DR}}} = 4.14$ GeV or $m_b(M_b)^{\overline{\text{MS}}} = 4.24$ GeV (we use $\alpha_3(M_b) = 0.2246$).

Putting these numbers in (9.80) shows that the correction due to the top Yukawa coupling (*i.e.* the last factor) should reduce the ratios of gauge coupling constants by an approximate factor of 2. It is thus necessary to take $\lambda_t(M_U)$ rather large. As discussed in Section 6.9.1 of Chapter 6, this drives the low energy evolution of the top coupling towards the quasi-infrared fixed points.]

---

[4] As explained above, we are working in the dimensional reduction scheme $\overline{\text{DR}}$. The translation to the standard $\overline{\text{MS}}$ scheme is given by:

$$m_b(\mu)_{\overline{\text{DR}}} = m_b(\mu)_{\overline{\text{MS}}}\left[1 - \frac{g_3^2(\mu)}{3\pi}\right]. \tag{9.81}$$

First and second family unification

The same procedure can be applied to the first two families, following (9.68) and (9.69). The formulas obtained are similar to (9.80) without the correction due to the top Yukawa coupling. This gives a s quark mass of order 500 MeV, which is difficult to reconcile with estimates based on chiral symmetry breaking.

This becomes even worse when one realizes that one may get rid of most of the renormalization effects by considering the ratios:

$$\frac{m_d}{m_s} \sim \frac{m_e}{m_\mu} = \frac{1}{207}, \tag{9.83}$$

which is in gross contradiction with the current algebra value of $m_d/m_s \sim m_\pi^2/(2m_K^2 - m_\pi^2) = 1/24$ [176].

The latter problem has led to reconsider partially the problem of fermion masses in $SU(5)$. One attitude is to consider that d being a light quark, its mass is sensitive to nonrenormalizable interactions which arise from the fundamental theory behind the grand unified model (string, gravity, etc.).

Alternatively, one may envisage enlarging the scalar structure of the model. Following Georgi and Jarlskog [182], one introduces a representation **45**, Φ_{np}^m, $m, n, p = 1, \ldots, 5$, which satisfies $\Phi_{np}^m = -\Phi_{pn}^m$ and $\Phi_{mp}^m = 0$, and a $\overline{\mathbf{45}}$, $\overline{\Phi}_m^{np}$ of $SU(5)$. If the corresponding scalar fields develop a $SU(3) \times U(1)_Y$ invariant vacuum expectation value:

$$\langle \Phi_{n5}^m \rangle = a \left(\delta_n^m - 4\delta_4^m \delta_n^4 \right), \tag{9.84}$$

and similarly for $\langle \overline{\Phi}_m^{n5} \rangle$, this generates mass terms for quarks and leptons through the Yukawa couplings:

$$W' = \lambda_d' \, \eta^m \chi_{np} \overline{\Phi}_m^{np} + \lambda_u' \epsilon^{mnprs} \chi_{mn} \chi_{pq} \Phi_{rs}^q. \tag{9.85}$$
$$\mathbf{\bar{5} \times 10 \times \overline{45}} \qquad \mathbf{10 \times 10 \times 45}$$

Now, if we restrict our attention to the charged leptons and charge $-1/3$ quarks of the first two families, we may use discrete symmetries to keep only the following terms in the superpotential:

$$W|_{\text{family (1) and (2)}} = -\lambda_d \left(\eta^{(1)m} \chi_{mn}^{(2)} H_1^n + \eta^{(2)m} \chi_{mn}^{(1)} H_1^n \right) + \lambda_d' \, \eta^{(2)m} \chi_{np}^{(2)} \overline{\Phi}_m^{np} + \text{h.c.} \tag{9.86}$$

This gives a matrix structure of the form:

$$\begin{pmatrix} 0 & \lambda_d v_1 \\ \lambda_d v_1 & 3\lambda_d' a \end{pmatrix} \quad \text{for } (e, \mu),$$

$$\begin{pmatrix} 0 & \lambda_d v_1 \\ \lambda_d v_1 & \lambda_d' a \end{pmatrix} \quad \text{for } (d, s),$$

which naturally yields $m_e m_\mu \sim m_d m_s$ whereas $m_\mu + m_e = 3(m_s + m_d)$. Hence, since $m_e \ll m_\mu$ and $m_d \ll m_s$

$$\frac{m_d}{m_s} \sim 9\frac{m_e}{m_\mu} \tag{9.87}$$

which is in much better agreement with estimates.

Table 9.2 Charges under $U(1)_Y$ and $U(1)_\chi$ of the low energy fields.

	Q	L	U^c	D^c	E^c	H_1	H_2
y	$1/3$	-1	$-4/3$	$2/3$	2	-1	1
y_χ	-1	3	-1	3	-1	-2	2
$(2y - y_\chi)/5$	$1/3$	-1	$-1/3$	$-1/3$	1	0	0

Global symmetries

Before closing this section, let us note that the superpotential which consists of (9.57) and (9.64) has a *global* abelian symmetry $U(1)_\chi$ with charges:

$$y_\chi^{(\Sigma)} = 0, y_\chi^{(H_1)} = -2, y_\chi^{(H_2)} = +2, y_\chi^{(\eta)} = 3, y_\chi^{(\chi)} = -1. \tag{9.88}$$

Such a symmetry seems to lead to a disaster since it is broken by $\langle H_1 \rangle$ and $\langle H_2 \rangle$. However, the combination $(2y - y_\chi)/5$ remains unbroken: considering H_1 for example, its doublet component, which acquires a nonzero vev, has $y = -1$. Using (9.4) and (9.6), we deduce from the charges (9.88) the value of $(2y - y_\chi)/5$ for the low energy fields (see Table 9.2).

We recognize in the last line of Table 9.2 the quantum number $B - L$. Hence the $SU(5)$ model has a global $B - L$ symmetry.

9.3.4 Proton decay

Since quarks and leptons are in the same representations of the unified group, one expects quark–lepton transitions and thus the possibility of proton decay. Writing the coupling of fermions (of the first family for the time being) to the $SU(5)$ vector bosons $X_{\alpha\mu}$ or $Y_{\alpha\mu}$ ($\alpha = 1, 2, 3$ is a color index)

$$\frac{g}{\sqrt{2}} X_{\alpha\mu} \left[\epsilon_{\alpha\beta\gamma} \bar{u}_{L\gamma}^c \gamma^\mu u_{L\beta} + \bar{d}_{L\alpha} \gamma^\mu e_L^c - \bar{e}_L \gamma^\mu d_{L\alpha}^c \right]$$

$$+ \frac{g}{\sqrt{2}} Y_{\alpha\mu} \left[\epsilon_{\alpha\beta\gamma} \bar{u}_{L\gamma}^c \gamma^\mu d_{L\beta} - \bar{u}_{L\alpha} \gamma^\mu e_L^c + \bar{\nu}_L \gamma^\mu d_{L\alpha}^c \right], \tag{9.89}$$

one infers that the following decays are allowed[5]: $X \to uu, \bar{d}e^+$, $Y \to ud, \bar{u}e^+$. Thus, the exchange of X or Y, as in Fig. 9.5, provides an effective interaction of dimension 6:

$$\mathcal{H}_{\text{eff}} = \frac{g^2}{2M_X^2} \epsilon_{\alpha\beta\gamma} \left(\bar{u}_{L\gamma}^c \gamma^\mu u_{L\beta} \right) \left(\bar{d}_{L\alpha}^c \gamma_\mu e_L - \bar{e}_L^c \gamma^\mu d_{L\alpha} \right)$$

$$- \frac{g^2}{2M_Y^2} \epsilon_{\alpha\beta\gamma} \left(\bar{u}_{L\gamma}^c \gamma^\mu d_{L\beta} \right) \left(\bar{d}_{L\alpha}^c \gamma_\mu \nu_L - \bar{e}_L^c \gamma^\mu u_{L\alpha} \right), \tag{9.90}$$

which contributes to the decay channels $p \to e^+ \pi^0$ and $n \to e^+ \pi^-$.

The corresponding amplitude includes: (i) a renormalization factor due to the fact that the effective interaction (9.89) must be renormalized from M_Z down to the

---

[5]One may note that, whereas the final states have various values of B and L, they have the same value $B - L = 2/3$; thus we can define $B - L$ for X and Y.

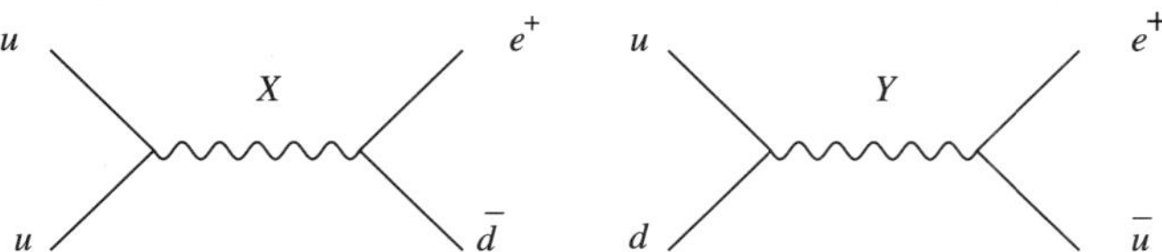

Fig. 9.5 Example of amplitude contributing to the decay $p \to e^+ \pi^0$ (left) and $n \to e^+ \pi^-$ (right).

proton mass, say 1 GeV, (ii) a flavor factor, and (ii) a hadronization factor due to the fact that the amplitude is measured between hadrons, not between quarks. One obtains:

$$
\mathcal{A} \sim \left(\frac{\alpha_3(1\ \text{GeV})}{\alpha_3(m_c)} \right)^{2/9} \left(\frac{\alpha_3(m_c)}{\alpha_3(m_b)} \right)^{6/25} \left(\frac{\alpha_3(m_b)}{\alpha_3(M_Z)} \right)^{6/23} F_{\text{flavor}} \alpha_{\text{lat}} \frac{1}{M^2_{X,Y}}, \qquad (9.91)
$$

where $\alpha_{\text{lat}} = \langle 0 | \epsilon^{\alpha\beta\gamma} d^\alpha_R u^\beta_R u^\gamma_L | p \rangle / N_L$ (N_L is is the wave function of the left-handed proton) is computed on the lattice. This yields:

$$
\tau \left(p \to e^+ \pi^0 \right) \sim 8 \times 10^{34}\ \text{years} \left(\frac{0.015\ \text{GeV}^3}{\alpha_{\text{lat}}} \right)^2 \left(\frac{M_{X,Y}}{10^{16}\ \text{GeV}} \right)^2. \qquad (9.92)
$$

This lies approximately one order of magnitude beyond the present experimental limit.

However, we have seen in Section 5.4 that, even when assuming R-parity, one remains with the possibility of dangerous B and L violating processes. In the context of grand unification, the exchange of color triplet fermions of mass M_T generates the following dimension-5 operators (see Fig. 9.6):

$$
W = -\frac{1}{M_T} \epsilon_{\alpha\beta\gamma} \left[\frac{1}{2} C_{5L}^{ijkl} Q_{i\alpha} Q_{j\beta} Q_{k\gamma} L_l + C_{5R}^{ijkl} U^c_{i\alpha} D^c_{j\beta} U^c_{k\gamma} E^c_l \right], \qquad (9.93)
$$

with α, β, γ color indices and i, j, k, l family indices.

This is why such color triplet supermultiplets must have a superheavy mass M_T. For example, the effective operator $\frac{1}{M_T} Q_1 Q_1 Q_2 L_i$ ($i = 1, 2, 3$) generates the $\Delta B = \Delta L = -1$ process $\tilde{u}_L \tilde{d}_L \to \bar{\nu}_i \bar{s}$ with coupling of order $1/M_T$, whereas $\frac{1}{M_T} U^c_1 U^c_3 D^c_1 E^c_3$ generates $ud \to \tilde{\bar{\tau}}_R \tilde{\bar{t}}_R$. Through gaugino or Higgsino exchange, such as in the amplitudes given in Fig. 9.7, this induces, respectively, the transitions $p \to K^+ \bar{\nu}_i$ (or $n \to K^0 \bar{\nu}_i$) and $p \to K^+ \bar{\nu}_\tau$.

The amplitude now reads

$$
\mathcal{A} \sim \left(\frac{\alpha_3(1\ \text{GeV})}{\alpha_3(m_c)} \right)^{2/9} \left(\frac{\alpha_3(m_c)}{\alpha_3(m_b)} \right)^{6/25} \left(\frac{\alpha_3(m_b)}{\alpha_3(M_Z)} \right)^{6/23} F_{\text{flavor}}\ F_{\text{loop}} \beta_{\text{lat}} \frac{1}{M_T}, \qquad (9.94)
$$

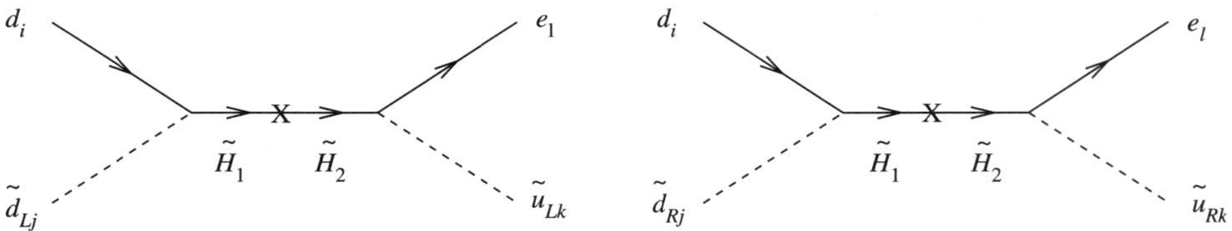

Fig. 9.6 Dimension-5 operators generated by color triplet exchange.

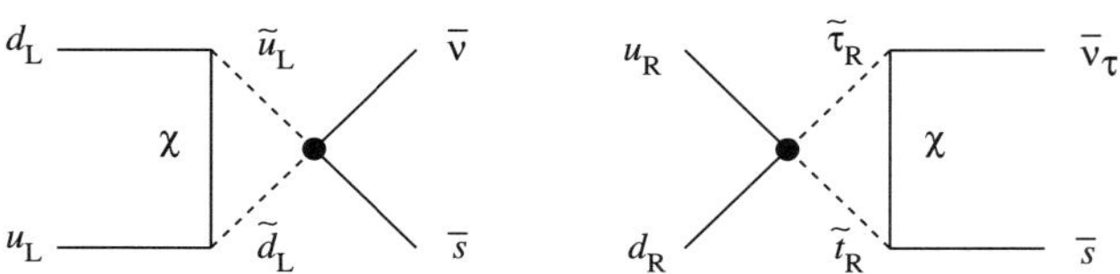

Fig. 9.7 Amplitude contributing to the decay $p \to K^+ \bar{\nu}_{e\mu\tau}$ or $n \to K^0 \bar{\nu}_{e\mu\tau}$ (left) and $p \to K^+ \bar{\nu}_\tau$ (right) through dimension-5 operators.

where the loop factor is of order $M_{1/2}/m_0^2$ for $M_{1/2} \ll m_0$ and $\beta_{\text{lat}} = \langle 0 \, | \epsilon^{\alpha\beta\gamma} d_L^\alpha u_L^\beta u_L^\gamma | \, 0 \rangle /$ N_L. This gives a lower limit on the mass of the Higgs triplet, which reads typically, for $\tan\beta \geq 5$:

$$\frac{M_T}{5.0 \times 10^{17} \text{ GeV}} \geq \left(\frac{\tau(p \to K^+ \bar{\nu})}{5.5 \times 10^{32} \text{ years}} \right)^{1/2} \left(\frac{\beta_{\text{lat}}}{0.003 \text{ GeV}^3} \right) \left(\frac{1 \text{ TeV}}{m_{\tilde{f}}} \right) \left(\frac{\tan\beta}{10} \right)^2 .$$

(9.95)

The negative search performed by the Super-Kamiokande Cherenkov detector sets a limit of 6.7×10^{32} years (90% confidence level) for the partial lifetime of the proton in the decay channel $K^+ \bar{\nu}$ [138]. This puts M_T in a mass range incompatible with the data on the gauge coupling unification (9.53). It thus rules out the minimal $SU(5)$ model [203].

This, however, does not exclude a more general $SU(5)$ unification. One may for example include more superheavy multiplets in order to push gauge unification to higher scales through threshold corrections. Alternatively, one may try to suppress the dimension-5 operators, as in the flipped $SU(5)$ model. Finally, one may enlarge the grand unification symmetry, as we will now see.

9.4 The $SO(10)$ model

The $SU(5)$ model does not truly unify the matter fields since one has to advocate two representations: $\bar{\mathbf{5}}$ and $\mathbf{10}$. It turns out that, if we view $SU(5)$ as a subgroup of $SO(10)$, these two representations make a single one of $SO(10)$: $\mathbf{16} = \bar{\mathbf{5}} + \mathbf{10} + \mathbf{1}$. The $SU(5)$ singlet that is required to make the sixteenth field is singlet under $SU(3) \times SU(2) \times U(1)$: it is interpreted as a right-handed neutrino. This provides a natural explanation for the cancellation of gauge anomalies: as explained in Section 9.1.2, anomalies naturally cancel for representations of $SO(10)$.

Since $SO(10)$ has rank 5, we expect to be able to define an extra gauge quantum number. In fact, the global $B - L$ symmetry of $SU(5)$ is promoted to the status of a gauge symmetry in the context of $SO(10)$. Noting that $\text{Tr}(B - L)|_{\bar{5}} = -3$ and $\text{Tr}(B - L)|_{10} = 2$, we see that the presence of the right-handed neutrino gives the missing contribution to ensure a vanishing total contribution for $\text{Tr}(B - L)|_{16}$, as is fit for a gauge generator. Moreover, $SO(10)$ incorporates a natural left-right symmetry and thus a spontaneous breaking of parity may be implemented. We will see below that the two issues are somewhat linked.

All these properties, compared with the mixed success of the $SU(5)$ model make $SO(10)$ unification a very interesting candidate for a grand unified theory.

9.4.1   Symmetry breaking

Among the subgroups of $SO(10)$, we first focus on $SU(5) \times U(1)$. One can easily see why, on general grounds, $SU(n) \times U(1)$ is a subgroup of $SO(2n)$. A transformation of $SU(n)$ leaves invariant the scalar product of two complex n-component vectors U and V:

$$U^\dagger \cdot V = \text{Re } U \cdot \text{Re } V + \text{Im } U \cdot \text{Im } V + i \text{ Re } U \cdot \text{Im } V - i \text{ Im } U \cdot \text{Re } V. \qquad (9.96)$$

Using instead the real $2n$-component fields $\mathcal{U} = (\text{Re } U, \text{Im } U)$ and $\mathcal{V} = (\text{Re } V, \text{Im } V)$, one sees that the following scalar product is conserved:

$$\mathcal{U} \cdot \mathcal{V} = \text{Re } U \cdot \text{Re } V + \text{Im } U \cdot \text{Im } V. \qquad (9.97)$$

Hence, a transformation of $SU(n)$ is also a transformation of $SO(2n)$. Moreover, under the $U(1)$ transformation $U \rightarrow e^{i\phi}U$, which commutes with $SU(n)$, $\mathcal{U} = (\text{Re } U, \text{Im } U)$ transforms into $(\text{Re } U \cos\phi - \text{Im } U \sin\phi, \text{Re } U \sin\phi + \text{Im } U \cos\phi)$. This transformation preserves the scalar product (9.97). One concludes that $SU(n) \times U(1)$ is indeed a subgroup of $SO(2n)$.

It is thus possible that, at some scale larger than the scale obtained earlier for $SU(5)$ unification, $SO(10)$ is broken into $SU(5) \times U(1)_\chi$. Quarks, leptons and their supersymmetric partners form N_F chiral supermultiplets in **16** of $SO(10)$, which are decomposed under $SU(5) \times U(1)_\chi$ as:

$$\mathbf{16} = \quad \bar{\mathbf{5}}_3 \quad + \quad \mathbf{10}_{-1} \quad + \quad \mathbf{1}_{-5} \qquad (9.98)$$
$$(D^c, E, N) \quad (D, U, U^c, E^c) \quad (N^c)$$

where we have indicated as subscript for each $SU(5)$ representation its charge y_χ under $U(1)_\chi$. Similarly, gauge supermultiplets transform as **45** with the decomposition:

$$\mathbf{45} = \mathbf{24}_0 + \mathbf{1}_0 + \mathbf{10}_{-4} + \overline{\mathbf{10}}_4, \qquad (9.99)$$

where we recognize in the first places the gauge bosons of $SU(5)$ and $U(1)_\chi$.

The first breaking ($SO(10)$ to $SU(5)$) may be realized through a **16** of Higgs, whereas the second breaking ($SU(5)$ to the Standard Model) uses a **45**, since **45** includes the necessary **24** of $SU(5)$. There is no constraint on the scale of $U(1)_\chi$ breaking and the corresponding gauge boson Z_χ could thus be a low energy field.

Alternatively, since the group $SO(m+n)$ contains $SO(m) \times SO(n)$, and $SO(6) \sim SU(4)$ whereas $SO(4) \sim SU(2) \times SU(2)$, we may consider the breaking of $SO(10)$ to $SU(4) \times SU(2) \times SU(2)$. One of the $SU(2)$ symmetries may be interpreted as the $SU(2)_L$ symmetry of the Standard Model and the other one is identified as its parity counterpart $SU(2)_R$. Moreover, $SU(4)$ naturally incorporates the color $SU(3)$ and the abelian group $U(1)_{B-L}$ associated with the $B-L$ quantum number.

The **16** and **45** of $SO(10)$ transform under $SU(4) \times SU(2)_L \times SU(2)_R$ as

$$\mathbf{16} = \quad (\mathbf{4,2,1}) \qquad + \qquad (\mathbf{\bar{4},1,2}), \tag{9.100}$$

$$\left[\begin{pmatrix} U \\ D \end{pmatrix}, \begin{pmatrix} N \\ E \end{pmatrix} \right] \quad \left[\begin{pmatrix} D^c \\ U^c \end{pmatrix}, \begin{pmatrix} E^c \\ N^c \end{pmatrix} \right]$$

$$\mathbf{45} = (\mathbf{15,1,1}) + (\mathbf{1,3,1}) + (\mathbf{1,1,3}) + (\mathbf{6,2,2}). \tag{9.101}$$

The breaking of $SO(10)$ is realized through a $\mathbf{54} = (\mathbf{6,2,2}) + (\mathbf{20',1,1}) + (\mathbf{1,3,3}) + (\mathbf{1,1,1})$ and the final breaking to the Standard Model uses $\mathbf{16} + \mathbf{\overline{16}}$ or $\mathbf{126} + \mathbf{\overline{126}}$. We note the following relation:

$$q = t_L^3 + t_R^3 + \frac{B-L}{2}, \tag{9.102}$$

which shows that $B-L$ is a generator of $SO(10)$ (since the others are). This provides also a nice interpretation of hypercharge, which had a somewhat mysterious origin in the context of the Standard Model:

$$y = 2t_R^3 + B - L. \tag{9.103}$$

We note also the relation with the charge y_χ introduced earlier:

$$y_\chi = 2 \left[5t_R^3 + 3(t_L^3 - q) \right] = 2y - 5(B-L). \tag{9.104}$$

9.4.2 Fermion masses

Since Yukawa couplings are trilinear, one expects to find the Higgs representations (or rather their conjugates) in the product $\mathbf{16} \times \mathbf{16} = (\mathbf{10} + \mathbf{126})_s + \mathbf{120}_a$ (the subscripts refer, respectively, to the symmetric and the antisymmetric combination). The Higgs are thus searched for in $\mathbf{10}$, $\mathbf{120}$, or $\mathbf{\overline{126}}$ (the first two are self-conjugates).

[We note that $\mathbf{16}$ is the spinor representation of $SO(10)$ constructed in Section B.2.1 of Appendix B. We thus see that matter supermultiplets are in spinor representations of $SO(10)$ whereas Higgs and gauge supermultiplets are in tensor representations (they appear in even products of spinor representations). This allows us to define a matter parity: -1 for spinors and $+1$ for tensors.]

If we take the Higgs in the representation $\mathbf{10}_H$ which decomposes under $SU(5) \times U(1)_\chi$ as $\mathbf{10} = \mathbf{5}_2 + \mathbf{\bar{5}}_{-2}$, the trilinear coupling yields the following decomposition under $SU(5)$:

$$(\mathbf{16} \times \mathbf{16}) \times \mathbf{10} \rightarrow (\mathbf{\bar{5}} \times \mathbf{10}) \times \mathbf{\bar{5}}_H + (\mathbf{10} \times \mathbf{10}) \times \mathbf{5}_H + (\mathbf{1} \times \mathbf{\bar{5}}) \times \mathbf{5}_H. \tag{9.105}$$

The first two terms were already considered in (9.64) and they yield the usual $SU(5)$ mass relations[6]. The last term yields a Dirac mass for the neutrino, equal to the up quark mass at grand unification. In other words, the $SU(5)$ prediction (9.66) is replaced by

$$m_d = m_e = -\lambda v_1, \quad m_u = m_\nu = \lambda v_2, \tag{9.107}$$

where λ is the $(\mathbf{16} \times \mathbf{16}) \times \mathbf{10}_H$ Yukawa coupling. The last prediction is obviously in contradiction with the small neutrino mass.

It is precisely in this context that Gell-Mann, Ramond and Slansky [177] and Yanagida [383] proposed the seesaw mechanism. Indeed, if one breaks $SO(10)$ down to $SU(5)$ with a $\overline{\mathbf{126}}$ (which contains a $SU(5)$ singlet $\phi^{(\overline{\mathbf{126}})}$ of charge $y_\chi = 10$ or $B - L = -2$, as seen from (9.106)), a Yukawa coupling of the form $(\mathbf{16} \times \mathbf{16}) \times \overline{\mathbf{126}}$ includes a term $(\mathbf{1}_{-5} \times \mathbf{1}_{-5}) \times \mathbf{1}_{10}$, *i.e.* $\lambda_N \bar{N}^c_L N_R \phi^{(\overline{\mathbf{126}})}$. Gauge symmetry breaking thus yields a Majorana mass for the right-handed neutrino of the order of the unification scale (*i.e.* fixed by the $B - L$ breaking scale $\langle \phi^{(\overline{\mathbf{126}})} \rangle$). This combined with the electroweak Dirac mass yields the mass matrix of the seesaw mechanism, as discussed in Section 1.1.1 of Chapter 1: $m_\nu \sim m_u^2 / \left(\lambda_N \langle \phi^{(\overline{\mathbf{126}})} \rangle \right)$. We note that R-parity remains conserved since the scalar field $\phi^{(\overline{\mathbf{126}})}$ has even $(B - L)$ value: $R = (-1)^{3(B-L)+2S}$.

In the case of theories where a $\overline{\mathbf{126}}$ representation is not available, one may also use higher-dimensional operators of the type $\mathbf{16}_i \times \mathbf{16}_j \times \overline{\mathbf{16}}_H \times \overline{\mathbf{16}}_H / M_P$ (i, j are family indices). This yields a right-handed neutrino mass proportional to $\langle \phi^{(\overline{\mathbf{16}})} \rangle^2 / M_P$, where $\phi^{(\overline{\mathbf{16}})}$ is the singlet of $\overline{\mathbf{16}}_H$ with $y_\chi = 5$ *i.e.* $B - L = -1$. For $\langle \phi^{(\overline{\mathbf{16}})} \rangle \sim M_U$, this gives a right-handed mass of order 10^{13} GeV. This time R-parity is broken since $\phi^{(\overline{\mathbf{16}})}$ has odd $(B - L)$.

It may also be necessary to introduce terms of the type $\mathbf{16}_i \times \mathbf{16}_j \times \mathbf{16}_H \times \mathbf{16}_H / M_P$ in order to generate CKM mixings.

9.4.3 [Doublet–triplet splitting

The $SO(10)$ theory allows for a nice implementation of the missing doublet mechanism (see Section 9.3.2) to generate the doublet–triplet splitting. Indeed, the $\mathbf{45}$ involves triplets $(\mathbf{3}, \mathbf{1})$ but no doublet $(\mathbf{1}, \mathbf{2})$ of $SU(3) \times SU(2)$ as can be readily checked from (9.99). In a model with two representations $\mathbf{10}_H$ and $\mathbf{10}'_H$, the coupling $\mathbf{10}_H \cdot \mathbf{45} \cdot \mathbf{10}'_H$ leaves four doublets in the light spectrum. This is too many and one has to somewhat complicate the scheme to remain with only two doublets at low energy. A superpotential that achieves this [17] involves also $\mathbf{16}_H$ and $\overline{\mathbf{16}}_H$:

$$W = \lambda \mathbf{10}_H \mathbf{45}_H \mathbf{10}'_H + \lambda' \overline{\mathbf{16}}_H \overline{\mathbf{16}}_H \mathbf{10}_H + M_{10} \mathbf{10}'^2_H + M_{16} \mathbf{16}_H \overline{\mathbf{16}}_H. \tag{9.108}$$

---

[6]Since both $\mathbf{120}$ and $\mathbf{126}$ include a $\mathbf{45}$ of $SU(5)$, one may invoke the Georgi–Jarlskog mechanism to modify these. More precisely,

$$\mathbf{120} = \mathbf{5}_2 + \bar{\mathbf{5}}_{-2} + \mathbf{10}_{-6} + \overline{\mathbf{10}}_6 + \mathbf{45}_2 + \overline{\mathbf{45}}_{-2},$$

$$\mathbf{126} = \mathbf{1}_{-10} + \bar{\mathbf{5}}_{-2} + \mathbf{10}_{-6} + \overline{\mathbf{15}}_6 + \mathbf{45}_2 + \overline{\mathbf{50}}_{-2}. \tag{9.106}$$

Table 9.3 Quantum numbers of the components $(\mathbf{3},\mathbf{3},\mathbf{1})$, $(\overline{\mathbf{3}},\mathbf{1},\overline{\mathbf{3}})$, and $(\mathbf{1},\overline{\mathbf{3}},\mathbf{3})$ of the $\mathbf{27}$ of E_6. The first column gives the decomposition under $SU(3)_c \times SU(2)_L \times SU(2)_R$: the subscripts are respectively y_L and y_R.

$SU(3)_c \times SU(2)_L \times SU(2)_R$		t_L^3	t_R^3	q_L	q_R	q	y	y_η	(y_ψ, y_χ)
$(\mathbf{3},\mathbf{2}_{1/3},\mathbf{1}_0)$	U	$+\frac{1}{2}$	0	$+\frac{2}{3}$	0	$+\frac{2}{3}$	$+\frac{1}{3}$	$-\frac{2}{3}$	$(+1,-1)$
	D	$-\frac{1}{2}$	0	$-\frac{1}{3}$	0	$-\frac{1}{3}$	$+\frac{1}{3}$	$-\frac{2}{3}$	$(+1,-1)$
$(\mathbf{3},\mathbf{1}_{-2/3},\mathbf{1}_0)$	G	0	0	$-\frac{1}{3}$	0	$-\frac{1}{3}$	$-\frac{2}{3}$	$+\frac{4}{3}$	$(-2,+2)$
$(\overline{\mathbf{3}},\mathbf{1}_0,\mathbf{2}_{-1/3})$	D^c	0	$+\frac{1}{2}$	0	$+\frac{1}{3}$	$+\frac{1}{3}$	$+\frac{2}{3}$	$+\frac{1}{3}$	$(+1,+3)$
	U^c	0	$-\frac{1}{2}$	0	$-\frac{2}{3}$	$-\frac{2}{3}$	$-\frac{4}{3}$	$-\frac{2}{3}$	$(+1,-1)$
$(\overline{\mathbf{3}},\mathbf{1}_0,\mathbf{1}_{2/3})$	G^c	0	0	0	$+\frac{1}{3}$	$+\frac{1}{3}$	$+\frac{2}{3}$	$+\frac{1}{3}$	$(-2,-2)$
$(\mathbf{1},\mathbf{2}_{-1/3},\mathbf{2}_{1/3})$	H_1^0	$+\frac{1}{2}$	$-\frac{1}{2}$	$\frac{1}{3}$	$-\frac{1}{3}$	0	-1	$+\frac{1}{3}$	$(-2,-2)$
	H_1^-	$-\frac{1}{2}$	$-\frac{1}{2}$	$-\frac{2}{3}$	$-\frac{1}{3}$	-1	-1	$+\frac{1}{3}$	$(-2,-2)$
	H_2^+	$+\frac{1}{2}$	$+\frac{1}{2}$	$+\frac{1}{3}$	$+\frac{2}{3}$	1	1	$+\frac{4}{3}$	$(-2,+2)$
	H_2^0	$-\frac{1}{2}$	$+\frac{1}{2}$	$-\frac{2}{3}$	$+\frac{2}{3}$	0	1	$+\frac{4}{3}$	$(-2,+2)$
$(\mathbf{1},\mathbf{2}_{-1/3},\mathbf{1}_{-2/3})$	N	$+\frac{1}{2}$	0	$+\frac{1}{3}$	$-\frac{1}{3}$	0	-1	$+\frac{1}{3}$	$(+1,+3)$
	E	$-\frac{1}{2}$	0	$-\frac{2}{3}$	$-\frac{1}{3}$	-1	-1	$+\frac{1}{3}$	$(+1,+3)$
$(\mathbf{1},\mathbf{1}_{2/3},\mathbf{2}_{1/3})$	E^c	0	$+\frac{1}{2}$	$+\frac{1}{3}$	$+\frac{2}{3}$	1	2	$-\frac{2}{3}$	$(+1,-1)$
	N^c	0	$-\frac{1}{2}$	$+\frac{1}{3}$	$-\frac{1}{3}$	0	0	$-\frac{5}{3}$	$(+1,-5)$
$(\mathbf{1},\mathbf{1}_{2/3},\mathbf{1}_{-2/3})$	N'^c	0	0	$+\frac{1}{3}$	$-\frac{1}{3}$	0	0	$-\frac{5}{3}$	$(+4,0)$

In $SU(5)$ notation, this yields the following mass matrix:

$$
\begin{pmatrix} \overline{\mathbf{5}}_{10_H} & \overline{\mathbf{5}}_{10'_H} & \overline{\mathbf{5}}_{16_H} \end{pmatrix}
\begin{pmatrix} 0 & \lambda\langle\mathbf{45}_H\rangle & \lambda'\langle\overline{\mathbf{16}}_H\rangle \\ -\lambda\langle\mathbf{45}_H\rangle & M_{10} & 0 \\ 0 & 0 & M_{16} \end{pmatrix}
\begin{pmatrix} \mathbf{5}_{10_H} \\ \mathbf{5}_{10'_H} \\ \mathbf{5}_{\overline{16}_H} \end{pmatrix} . \tag{9.109}
$$

As expected, the nonvanishing vacuum expectation value $\langle 45_H \rangle$ in the $(B - L)$ direction does not contribute to the doublet mass matrix. Introducing $D_1^{(10)}$ and $D_2^{(10)}$ the two doublets in 10_H (with, respectively, $y_\chi = -2$ and $y_\chi = 2$), $D^{(\overline{16})}$ the doublet and $\phi^{(\overline{16})}$ the $SU(5)$ singlet in $\overline{16}_H$ (with, respectively, $y_\chi = 5$ and $y_\chi = -3$), the doublet mass matrix simply reads:

$$\begin{pmatrix} \overline{D}_2^{(10)} & \overline{D}_1^{(10)} & \overline{D}^{(16)} \end{pmatrix} \begin{pmatrix} 0 & 0 & 0 \\ 0 & 0 & \lambda'\langle\phi^{(\overline{16})}\rangle \\ 0 & 0 & M_{16} \end{pmatrix} \begin{pmatrix} D_2^{(10)} \\ D_1^{(10)} \\ D^{(\overline{16})} \end{pmatrix}. \tag{9.110}$$

Thus the two MSSM doublets are

$$H_2 \equiv D_2^{(10)}, \quad H_1 \equiv \cos\gamma D_1^{(10)} + \sin\gamma D^{(\overline{16})}, \tag{9.111}$$

with $\tan\gamma = \lambda'\langle\phi^{(\overline{16})}\rangle/M_{16}$. We have thus $\langle D_1^{(10)}\rangle = v_1 \cos\gamma$ and $\langle D^{(\overline{16})}\rangle = v_1 \sin\gamma$.

We finally note that, in the case where nonrenormalizable terms involving 16_H and $\overline{16}_H$ are also introduced to generate right-handed Majorana masses and CKM mixings, one generates new dimension-5 operators of the form

$$\delta W \sim 16_i 16_j 16_k 16_l \langle \overline{16}_H \rangle \langle \overline{16}_H \rangle / (M_P^2 M_{16}) \tag{9.112}$$

which may lead to proton decay.]

9.5 E_6

If one further enhances the gauge symmetry, the next simple gauge group encountered is E_6, the exceptional gauge group of rank 6 with 78 generators. As we will see in more details in the next chapter, this group is naturally encountered in the context of the weakly heterotic string theory.

The fundamental representation of E_6 is complex and 27-dimensional. It decomposes under the subgroup $SO(10) \times U(1)_\psi$ as

$$27 = 16_1 + 10_{-2} + 1_4, \tag{9.113}$$

where we have indicated as subscript the charge y_ψ associated with the abelian factor, noted $U(1)_\psi$. This provides new fields: the 10 transforming under $SU(5)$ as $5 + \bar{5}$ yields new charge $-1/3$ (weak isosinglet) quarks and new (weak isodoublet) leptons; the 1 is a new neutral lepton.

The maximal subgroup $SU(3) \times SU(3) \times SU(3)$ of E_6 plays an important rôle in discussions of superstring models. One identifies the first group with color $SU(3)_c$, the second one with the $SU(3)_L$ group containing the electroweak symmetry $SU(2)_L$ and the third one with its parity counterpart $SU(3)_R$.

For example, we will consider in Section 10.4.3 of Chapter 10 the breaking of E_6 with an adjoint 78 representation which decomposes as:

$$78 = (8, 1, 1) + (1, 8, 1) + (1, 1, 8) + (3, 3, 3) + (\bar{3}, \bar{3}, \bar{3}). \tag{9.114}$$

Since $SU(3)_c$ remains unbroken, one must use $(1, 8, 1)$ to break $SU(3)_L$. But breaking $SU(3)$ down to $SU(2)$ with an octet always leaves an extra $U(1)$ factor. We thus write the residual symmetry $SU(2)_L \times U(1)_{Y_L}$. Since the corresponding abelian charge y_L cannot coincide with hypercharge (otherwise right-handed fields would have vanishing hypercharge), one must invoke a second conserved abelian charge q_R, where $U(1)_{Q_R}$ is a subgroup of $SU(3)_R$. Hypercharge is a linear combination of y_L and q_R. The orthogonal combination, noted y_η, is also conserved. The low energy symmetry is $SU(3)_c \times SU(2)_L \times U(1)_Y \times U(1)_\eta$.

For definiteness, let us consider the fundamental representation, which decomposes as:

$$\mathbf{27} = (\mathbf{3}, \mathbf{3}, \mathbf{1}) + (\bar{\mathbf{3}}, \mathbf{1}, \bar{\mathbf{3}}) + (\mathbf{1}, \bar{\mathbf{3}}, \mathbf{3}). \tag{9.115}$$

The first representation includes quark superfields, the second antiquarks and the third leptons. One may decompose $SU(3)_{L,R}$ into $SU(2)_{L,R} \times U(1)_{Y_L,Y_R}$ (both $SU(2)_{L,R}$ groups have been encountered above with $SO(10)$). Then Table 9.3 gives the full content of the **27** representation. One has

$$q_{L,R} = t^3_{L,R} + \frac{y_{L,R}}{2}, \quad q = q_L + q_R,$$
$$y = y_L + 2q_R, \quad B - L = y_L + y_R. \tag{9.116}$$

The $U(1)$ charges defined above are

$$y_\psi = 3(y_L - y_R), \quad y_\chi = 4t^3_R - 3(y_L + y_R),$$
$$y_\eta = q_R - 2y_L = t^3_R + \frac{y_R}{2} - 2y_L. \tag{9.117}$$

Obviously, the $U(1)_\eta$ charge is a combination of the other two: $y_\eta = (3y_\chi - 5y_\psi)/12$.

Exercises

<u>Exercise 1</u> *We determine explicitly in this exercise the form of the generator of $SU(5)$ associated with hypercharge in the fundamental representation* **5** *and check that its normalization agrees with the result obtained in (9.30).*

Since t^0, the generator of $SU(5)$ which corresponds to hypercharge, must commute with the generators of $SU(3)$ and $SU(2)$, we write it, in the fundamental representation of $SU(5)$, as

$$t^0 = \begin{pmatrix} \alpha & & & & \\ & \alpha & & & \\ & & \alpha & & \\ & & & \beta & \\ & & & & \beta \end{pmatrix}. \tag{9.118}$$

(a) Determine α and β.
(b) Deduce the value of the normalization constant C defined in (9.24).

Hints:

(a) Use $\mathrm{Tr}\, t^0 = 0$ and $\mathrm{Tr}\, (t^0)^2 = 1/2$ to find $\alpha = \pm 1/\sqrt{15}$ and $\beta = \mp\sqrt{3/20}$.
(b) Apply (9.24) to d_R for example (see footnote to (9.6)). One obtains $C = \mp\sqrt{5/3}$ in agreement with (9.30).

<u>Exercise 2</u> Show that, in the case where the $N \times N$ matrix field Σ^i_j is traceless, the potential that derives from the superpotential $W(\Sigma)$ takes the form:

$$V(\Sigma) = \sum_{ij} \left| \frac{\partial W}{\partial \Sigma^i_j} - \frac{1}{N} \delta^j_i \; \mathrm{Tr} \left(\frac{\partial W}{\partial \Sigma} \right) \right|^2 .$$

Hints: Introduce the superpotential $\hat{W}(\Sigma) = W(\Sigma) + \rho \, \mathrm{Tr}\Sigma$ where ρ is a Lagrange multiplier. Then

$$V = \sum_{ij} \left| \frac{\partial W}{\partial \Sigma^i_j} + \rho \delta^j_i \right|^2 + |\mathrm{Tr}\Sigma|^2 .$$

Minimization ensures that $\mathrm{Tr}\Sigma = 0$ and $N\rho = -\mathrm{Tr}\left(\partial W / \partial \Sigma \right)$.

<u>Exercise 3</u> : Writing the two Higgs supermultiplets $\bar{\mathbf{5}}$ and $\mathbf{5}$ as $H_1 = \left(\bar{T}_{H_1}, H_1^-, -H_1^0 \right)$ and $H_2 = \left(T_{H_2}, H_2^+, H_2^0 \right)$ (see (9.6) and the corresponding footnote), show that the μ parameter defined in (9.57) and the Yukawa couplings defined in (9.64) coincide with the low energy definitions given in Chapter 5, equations (5.1) and (5.2).

10

An overview of string theory and string models

It is widely believed that string theory provides the consistent framework which lies behind low energy supersymmetry. The purpose of this chapter is to provide a non-technical overview of the main ideas of string theory[1], as well as to present some of the new concepts which arise and which might lead to new developments in the study of the low energy theory. Due to lack of space, there is no room for doing justice to the richness of the complete string picture and we refer the reader to other monographs for a more thorough treatment e.g. [312]. However, string models are to be considered very seriously because they provide a complete picture of fundamental physics, including gravity. As such, they should be considered in their entirety because they have an internal coherence. Thus using them for low energy phenomenology (or cosmology) requires us at least to be aware of the general consistency of the string picture.

10.1 The general string picture

There is a well-known contradiction between the two fundamental theories which emerged in the first half of the twentieth century: quantum theory and general relativity. General relativity is a classical theory. When one tries to quantize it, one encounters divergences which prevent us from making predictions once the quantum regime of the theory is reached. As was discussed in Chapter 1, the fundamental scale of quantum gravity is obtained from Newton's constant G_N and Planck's constant $\hbar$. This gives the Planck mass[2]

$$M_P \equiv \sqrt{\frac{\hbar c}{G_N}} = 1.221 \ 10^{19} \ \text{GeV}/c^2, \tag{10.1}$$

---

[1]Some of the more technical developments are placed in boxes in this chapter. The reader interested in a nontechnical overview may skip them.

[2]A word of caution: absolute lab measurements of Newton's constant have only been conducted on distance scales in the 100 μ to m range ($G_N = 6.673 \times 10^{-11} \ \text{m}^3 \, \text{kg}^{-1} . \text{s}^{-2}$) whereas Planck's constant is by essence a microscopic scale. We are thus making the implicit assumption that gravity at the microscopic level is still described by the Newton's constant which is measured macroscopically. We will return to this in Section 10.2.1.

or the Planck length

$$\ell_P \equiv \frac{\hbar c}{M_P c^2} = \sqrt{\frac{\hbar G_N}{c^3}} = 1.616\ 10^{-20}\ \text{fm.} \tag{10.2}$$

The discussion is somewhat reminiscent of the one that we had for the low energy Fermi theory of weak interactions (see the introduction to Section A.3 of Appendix Appendix A). For a process with center of mass energy E, such as the graviton exchange in Fig. 10.1, the dimensionless coupling is $G_N E^2$. Hence for energies $E \sim G_N^{-1/2} \sim M_P$ (we now set $\hbar = c = 1$), we expect to be in the regime where quantum gravity effects are of order one (unitarity bounds lead us to expect new degrees of freedom).

The nonrenormalizability of gravity (or for that matter supergravity) is also linked to the presence of the dimensionful coupling G_N: it leads to the appearance in the computation of quantum processes of divergences which cannot be absorbed in a redefinition of the theory parameters (as in a renormalizable theory). These divergences arise from the very short distance behavior of the theory: they correspond to a divergent behavior of the loop integrals in the ultraviolet regime. These integrals should be cut off at an energy scale of order $G_N^{-1/2} \sim M_P$ (a distance scale of order ℓ_P), where we expect that a more fundamental theory takes over. But which theory?

The changes to be made are somewhat more drastic than the ones that were necessary in order to cure the high energy behavior of weak processes in Fermi theory. They involve either modifying the nature of classical spacetime or the pointlike nature of fundamental particles. String theory follows the second route and assumes that the fundamental building blocks are one-dimensional objects called strings (open if they have end points or closed if they are loops).

The quantization of extended objects is notoriously difficult because such objects have an infinite number of degrees of freedom. However, in the case of one-dimensional objects one may use the very large symmetry group of the theory in order to keep it under control. To understand where these symmetries come from, it is necessary to realize that a string covers in its motion a two-dimensional world surface known as the world-sheet (see Fig. 10.2). One may then describe the string theory as a field theory on this world-sheet. But two-dimensional surfaces have a large symmetry group which is the conformal symmetry group (well-known to anybody who has studied electromagnetic problems on two-dimensional surfaces). One may use this conformal symmetry to eliminate most of the degrees of freedom. It is thus important for the consistency of the approach to make sure that this conformal symmetry remains valid at each step.

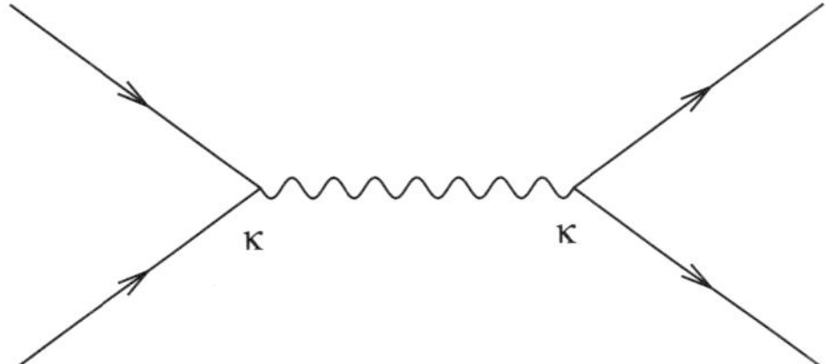

Fig. 10.1 Graviton exchange ($\kappa = (8\,\pi G_N)^{1/2} = 1/m_P$)

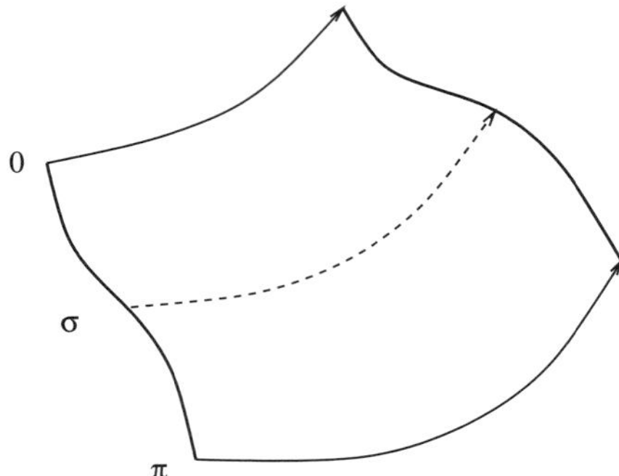

Fig. 10.2 World-sheet for an open string parametrized by the spatial coordinate σ, $0 \leq \sigma \leq \pi$, and time τ.

Let us note immediately that this type of approach does not treat spacetime in a truly quantum mechanical fashion. The string propagates in a background spacetime which is often taken to be flat and is in any case classical. In other words, the two-dimensional world-sheet is plunged into a classical background spacetime, often referred to as the target space. We will see momentarily in which sense one may still say that string theory is a quantum theory of gravity.

The "fundamental particles" appear as oscillation modes of the string. Indeed, these string vibrate and each oscillation mode is an eigenstate of the energy, and thus a particle. Just as we may reconstruct a violin string out of the sequence of its harmonics, one may reconstruct a fundamental string out of the particles which form its oscillation modes. In the low energy limit, one can only reach the fundamental mode. Similarly, our low energy world (much below M_P) has only access to the zero modes of the fundamental strings, *i.e.* the point particles that we observe.

Among these massless particles one encounters a spin 2 resonance which is interpreted as the graviton field: its long wavelength interactions are found to be in agreement with general relativity. Thus, string theory is a quantum theory of gravity (even though spacetime is treated as a classical background).

The closed string

Parametrizing the coordinate along the string as σ, $0 \leq \sigma \leq \pi$, and time along the world-line of any point of the string as τ (in a way similar to the open string of Fig. 10.2), the world-sheet is described in D-dimensional spacetime by the coordinates $X^M(\sigma, \tau)$, $M = 0, \ldots, D-1$. Making use of the symmetries of the problem, one may write the string equation of motion as a simple wave equation:

$$\left[\frac{\partial^2}{\partial \tau^2} - \frac{\partial^2}{\partial \sigma^2} \right] X^M(\sigma, \tau) = 0. \tag{10.3}$$

The standard solution is a superposition of left-moving and right-moving waves:

$$X^M = X_L^M(\tau + \sigma) + X_R^M(\tau - \sigma). \tag{10.4}$$

Introducing the notation $z \equiv \exp 2i(\tau + \sigma)$, $\bar{z} \equiv \exp 2i(\tau - \sigma)$ (complex conjugates if we perform a Wick rotation $\tau \to i\tau$), we may write them $X_L^M(z)$ and $X_R^{M}(\bar{z})$. In this language, the conformal transformations are the holomorphic and antiholomorphic reparametrizations $z \to f(z)$ and $\bar{z} \to \bar{f}(\bar{z})$: they do not mix left and right movers.

Since the string is closed, we must impose the periodicity conditions

$$X^M(\tau, \pi) = X^M(\tau, 0), \quad \partial_\sigma X^M(\tau, \pi) = \partial_\sigma X^M(\tau, 0). \tag{10.5}$$

The general solution of (10.3) is obtained as a Fourier series:

$$X^M(z, \bar{z}) = X_L^M(z) + X_R^M(\bar{z}) \tag{10.6}$$

$$X_L^M(z) = \frac{1}{2}x^M - i\frac{\alpha'}{2}p^M \ln z + i\sqrt{\frac{\alpha'}{2}} \sum_{n \neq 0} \frac{1}{n} \alpha_n^M z^{-n}$$

$$X_R^M(\bar{z}) = \frac{1}{2}x^M - i\frac{\alpha'}{2}p^M \ln \bar{z} + i\sqrt{\frac{\alpha'}{2}} \sum_{n \neq 0} \frac{1}{n} \tilde{\alpha}_n^M \bar{z}^{-n},$$

where α', which has the dimension of a length squared, can be related to the string tension $T = 1/(2\pi\alpha')$.

The usual canonical quantization yields commutation relations reminiscent of (an infinite set of) harmonic oscillators:

$$\left[x^M, p^N\right] = i\delta^{MN}, \left[\alpha_m^M, \tilde{\alpha}_n^N\right] = 0,$$
$$\left[\alpha_m^M, \alpha_n^N\right] = m\delta^{MN}\delta_{m+n,0}, \left[\tilde{\alpha}_m^M, \tilde{\alpha}_n^N\right] = m\delta^{MN}\delta_{m+n,0}. \tag{10.7}$$

In order to obtain the mass spectrum, it is easiest to work in a light-cone formalism[a] which is free from ghosts although not manifestly covariant. The mass formula reads

$$\alpha' M^2 = 4\left(\sum_{n=1}^{\infty} : \alpha_{-n}^I \alpha_n^I : -a\right), \tag{10.8}$$

where a summation over the transverse degrees of freedom $I = 1, \ldots, D - 2$ is understood and $a = (D - 2)/24$ is the zero-point energy.[b] We have the extra condition $N \equiv \sum_{n=1}^{\infty} : \alpha_{-n}^I \alpha_n^I := \sum_{n=1}^{\infty} : \tilde{\alpha}_{-n}^I \tilde{\alpha}_n^I :\equiv \tilde{N}$ which expresses the fact that the theory is invariant under translations along the string.

One constructs the Fock space of quantum states just as in the case of the harmonic oscillator. Introducing a vacuum state $|0\rangle$ annihilated by the α_n^I, we see from (10.8) that this state corresponds to a tachyon of negative squared

---

[a]One defines $X^\pm \equiv (X^0 \pm X^{D-1})/\sqrt{2}$. Residual symmetries allow us to determine $X^\pm$. One is left with the quantization of the transverse degrees of freedom X^I, $I = 1, \ldots$, $D - 2$.

[b]We use the standard notation of normal ordering ($: \ldots :$).

mass $-4a/\alpha'$. The next level appears at mass squared $4(1-a)/\alpha'$ and involves:

- $h^{IJ} = \frac{1}{2}\left(\alpha^I_{-1}\tilde{\alpha}^J_{-1} + \alpha^J_{-1}\tilde{\alpha}^I_{-1} - \frac{2}{D-2}\delta^{IJ}\sum_K \alpha^K_{-1}\tilde{\alpha}^K_{-1}\right)|0\rangle$, a symmetric traceless tensor (spin 2) field interpreted as the graviton field (*i.e.* the fluctuation of the metric: $g_{IJ} = \eta_{IJ} + h_{IJ}$);
- $b^{IJ} = \frac{1}{2}\left(\alpha^I_{-1}\tilde{\alpha}^J_{-1} - \alpha^J_{-1}\tilde{\alpha}^I_{-1}\right)|0\rangle$, an antisymmetric tensor field;
- $e^{\phi} = \sum_K \alpha^K_{-1}\tilde{\alpha}^K_{-1}|0\rangle$, a scalar field known as the string dilaton. The dilaton plays a central rôle since it determines the value of the string coupling $\lambda = \langle e^{\phi}\rangle$.

Note that when $a = 1$, *i.e.* $D = 26$, these fields, which fall into representations of the transverse Lorentz group $SO(D-2)$, are massless (see below).

Indeed, it is the presence of this spin 2 particle among the massless modes of the closed string which prompted the interpretation of string theories as a quantum theory of gravity [329]. Strings take their origin from the dual models of hadron dynamics which were devised in order to reconcile the property of scattering amplitudes known as crossing symmetry with the observation of narrow hadronic resonances with increasing spin. This was realized explicitly in the Veneziano amplitude [354] which turns out to be a string amplitude. In this context, the open string may be thought as an effective description of the (chromo-)electric flux tube which joins a quark and an antiquark when one increases their distance. The energy scale is then typically 1 GeV. It was the realization that open string theory is only consistent when one includes closed strings, and that the closed string spectrum incorporates a spin 2 particle that led to dramatically reconsider the fundamental mass scale $M_S \equiv \alpha'^{-1/2}$ and to push it all the way to the Planck scale.

This interpretation of the string leads to one possible way of introducing quantum numbers into string theory. Indeed the open string of hadrodynamics links two quarks and thus carries color quantum numbers at both ends, labelled respectively by i and j, $i,j = 1,2,3$ in Fig. 10.3. This string may emit mesons $q\bar{q}$ which fall into singlet (**1**) or octet (**8**) representations of $SU(3)_c$ (the corresponding color factor being λ^a_{ij}, where λ^0 is the identity matrix and $\lambda^a, a = 1,\ldots,8$, are the Gell-Mann matrices).

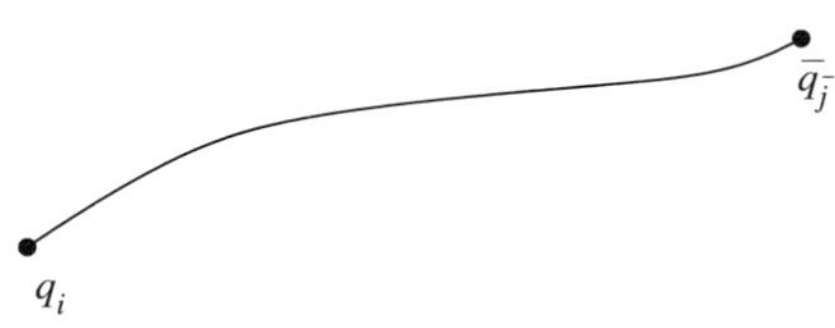

Fig. 10.3

The open string

There are two possible boundary conditions for an open string

$$\text{Dirichlet } \ X^M = \text{constant} \tag{10.9}$$

$$\text{Neumann } \ \frac{\partial X^M}{\partial \sigma} = 0. \tag{10.10}$$

For the time being we choose the latter, which reads $\partial_\sigma X^M(\tau,0) = 0 = \partial_\sigma X^M(\tau,\pi)$ for the string of Fig. 10.2. The solution of (10.3) reads:

$$X^M(z,\bar{z}) = x^M - i\frac{\alpha'}{2}p^M \ln(z\bar{z}) + i\sqrt{\frac{\alpha'}{2}} \sum_{n\neq 0} \frac{1}{n}\alpha_n^M \left(z^{-n} + \bar{z}^{-n}\right). \tag{10.11}$$

The spectrum is again obtained in the light-cone formalism from a formula similar to (10.8): $\alpha'M^2 = N - a$. The lowest-lying states are:

- $\alpha'M^2 = -a$, the ground state $|0\rangle$ which is a tachyon;

- $\alpha'M^2 = (1-a)$, a set of $D-2$ fields $\alpha_{-1}^I|0\rangle$ which form a vector representation of the transverse group $SO(D-2)$.

Again, when $a = 1$, i.e. $D = 26$, everything falls into place and the latter field is a massless vector field, with $D-2$ transverse degrees of freedom.

Generalizing the case of the color string, one may add nondynamical degrees of freedom (called Chan–Paton degrees of freedom) at both ends of the open string and describe them with indices i and j which run from 1 to N. There are then N^2 states at each mass level (tachyon, vector, etc.). One may go to a different basis (corresponding to the meson states in the case of the color string) by using the matrices λ_{ij}^a, $a = 0, \ldots, N^2 - 1$:

$$|k; a\rangle = \sum_{ij} \lambda_{ij}^a |k; ij\rangle. \tag{10.12}$$

The string diagram of Fig. 10.4 represents the tree-level interaction of two open strings. The corresponding amplitude then includes a Chan–Paton factor

$$\sum_{i,j,k,l=1}^{N} \lambda_{ij}^a \lambda_{jk}^b \lambda_{kl}^c \lambda_{li}^d = \text{Tr}\left(\lambda^a\lambda^b\lambda^c\lambda^d\right). \tag{10.13}$$

It is invariant under the $U(N)$ transformations: $\lambda \to U\lambda U^{-1}$. Since the corresponding N^2 massless vector fields transform as the adjoint representation of this $U(N)$ symmetry, this global world-sheet symmetry appears as a local gauge symmetry in spacetime.

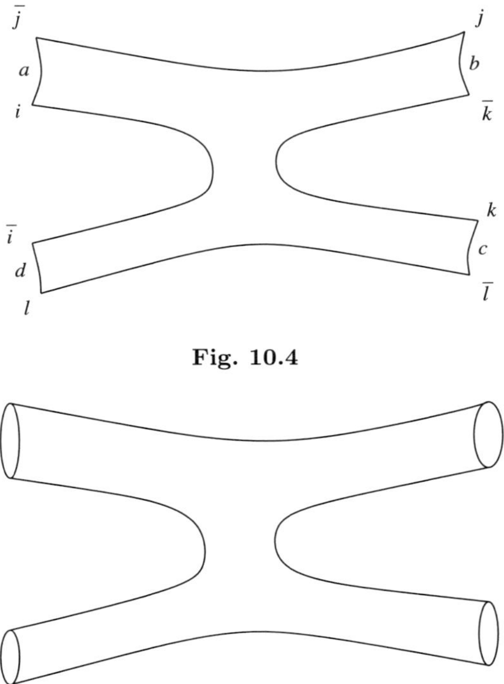

Fig. 10.4

Fig. 10.5

In the string context, the graviton exchange diagram of Fig. 10.1 is now represented by the string diagram of Fig. 10.5 which may be thought as representing the
corresponding world-sheet (in as much as the Feynman diagram of Fig. 10.1 represents
world-lines): from left to right two closed strings join to make a single closed string
and then disjoin and get apart. Conformal invariance tells us that this surface can
be deformed at will; the corresponding amplitude is not modified, as in the original
Veneziano amplitude. One may then understand qualitatively how string theory avoids
the small distance (ultraviolet) infinities: small distance singularities are smeared out
by the process which turns Feynman diagrams for point particles into surface diagrams
for extended one-dimensional objects. Indeed, string theory proves to be ultraviolet
finite.

The spectrum of string oscillation modes is very rich. One may wonder how one
can ever get oscillation modes which are spacetime fermions. The solution is to introduce internal (two-dimensional) "fermionic"[3] degrees of freedom on the world-sheet[4].
Depending on their (even or odd) boundary conditions along the strings, the corresponding oscillation modes are spacetime bosons or fermions [317]. It is then possible

---

[3]In two dimensions, the distinction between fermions and bosons is artificial. In fact, one can
"bosonize" the fermion fields.

[4]The theory on the world-sheet then has two-dimensional supersymmetry: this was in fact the first
supersymmetric theory ever written. It prompted searches for a four-dimensional supersymmetric
theory.

to write a string theory which has spacetime supersymmetry [194], *i.e.* a superstring theory.

Superstring theories are favored because they cure one problem of nonsupersymmetric theories: the requirement of supersymmetry projects out a tachyon field (*i.e.* a field of negative mass squared). The tachyon present in nonsupersymmetric string theories is presumably a sign of an instability of the theory.

Superstrings

Let us introduce D two-dimensional fermions $\psi^M = \begin{pmatrix} \psi^M_- \\ \psi^M_+ \end{pmatrix}$, $M = 0, \ldots,$ $D-1$, on the world-sheet ($\pm$ refers to the two-dimensional chirality ± 1). Their equation of motion imposes that $\psi^M_+ = \psi^M_+(\tau + \sigma)$ and $\psi^M_- = \psi^M_-(\tau - \sigma)$. We consider first the case of a closed string. A fermion $\psi^M_\pm$ may have periodic boundary conditions

$$\psi^M_\pm(\tau, \sigma + \pi) = \psi^M_\pm(\tau, \sigma). \tag{10.14}$$

It is then said to be in the [317] sector (R) and its expansion involves integer modes

$$\psi^M_\pm(\tau, \sigma) = \sqrt{2\alpha'} \sum_{n \in \mathbb{Z}} d^M_n e^{-2in(\tau \pm \sigma)}. \tag{10.15}$$

A fermion with antiperiodic boundary conditions

$$\psi^M_\pm(\tau, \sigma + \pi) = -\psi^M_\pm(\tau, \sigma) \tag{10.16}$$

is said to be in the [294] sector (NS) and has a half-integer mode expansion

$$\psi^M_\pm(\tau, \sigma) = \sqrt{2\alpha'} \sum_{r \in \mathbb{Z}+1/2} b^M_r e^{-2ir(\tau \pm \sigma)}. \tag{10.17}$$

Standard quantization yields the following anticommutation relations:

$$\{d^M_m, d^N_n\} = -\eta^{MN} \delta_{m+n,0}, \tag{10.18}$$

$$\{b^M_r, b^N_s\} = -\eta^{MN} \delta_{r+s,0}. \tag{10.19}$$

Since left- and right-moving modes are independent in the closed string, one may choose for each either boundary conditions. We thus have four possible sectors for the closed string: (R,R), (R,NS),(NS,R) and (NS,NS). In the case of an open string, the necessary boundary conditions are simply $\psi^M_+(\tau, \pi) = \pm \psi^M_-(\tau, \pi)$. Then fields in the Ramond sector (+ sign) have the following expansion:

$$\psi^M_\pm(\tau, \sigma) = \frac{1}{\sqrt{2}} \sqrt{2\alpha'} \sum_{n \in \mathbb{Z}} d^M_n e^{-in(\tau \pm \sigma)}. \tag{10.20}$$

Fields in the Neveu–Schwarz sector ($-$ sign) have the expansion

$$\psi_{\pm}^{M}(\tau, \sigma) = \frac{1}{\sqrt{2}}\sqrt{2\alpha'} \sum_{r \in \mathbb{Z}+1/2} b_r^M e^{-ir(\tau \pm \sigma)}. \tag{10.21}$$

We have, in the open string case, only the sectors (R, R) and (NS, NS). The mass spectrum of the theory is then given by[a]

$$\alpha' M^2 = N - a, \tag{10.22}$$

where (we work in the light-cone formalism and $I = 1, \ldots, D - 2$)

$$N = \left\{ \begin{array}{ll} \sum_{n=1}^{\infty} :\alpha_{-n}^{I}\alpha_{n}^{I}: + \sum_{n=1}^{\infty} n: d_{-n}^{I}d_{n}^{I}: & \text{(R)} \\ \sum_{n=1}^{\infty} :\alpha_{-n}^{I}\alpha_{n}^{I}: + \sum_{r=1/2}^{\infty} r: b_{-r}^{I}b_{r}^{I}: & \text{(NS)}. \end{array} \right. \tag{10.23}$$

The zero-point energy is $a_{\mathrm{R}} = 0$ in the R sector and $a_{\mathrm{NS}} = (D-2)/16$ in the NS sector.

The Ramond ground state $|0\rangle_{\mathrm{R}}$ is therefore massless. It is not unique: since d_0^I does not appear in (10.23), $[d_0^I, \alpha' M^2] = 0$ and $d_0^I|0\rangle$ is also a ground state. In fact, since from (10.18) the d_0^I satisfy a Clifford algebra

$$\{d_0^I, d_0^J\} = \delta^{IJ}, \tag{10.24}$$

they play the rôle of gamma matrices in the $D-2$ transverse space: their dimension is $2^{(D-2)/2}$ and they act as spinor representations of $SO(D-2)$. Thus, we see that the Ramond vacuum is in a (massless) spinor representation, and correspondingly all the quantum states built on it: states in the R sector are spacetime fermions whereas states in the NS sector are spacetime bosons. The lowest-lying states are given in Table 10.1 for the case $a_{\mathrm{NS}} = 1/2$ *i.e.* $D = 10$ in which the states fall into consistent representations of the transverse $SO(D-2)$ group.

There exists a consistent truncation of the spectrum which projects out the tachyon $|0\rangle_{\mathrm{NS}}$. This so-called GSO projection, for Gliozzi, Scherk, and Olive [194], keeps states in the NS sector with an odd two-dimensional fermion number and states in the R sector with given spacetime chirality (see Table 10.1 where these states are indicated as bold-faced). The spectrum obtained is supersymmetric and indeed spacetime supersymmetry is the rationale behind the consistency of the GSO truncation.

---

[a]The corresponding formula for the closed string is $\alpha' M^2/4 = N - a = \tilde{N} - \tilde{a}$, with obvious notation for the left- and right-moving sectors.

We have seen above the importance of ensuring that string theory has conformal invariance. This puts stringent constraints on the nature of the theory. Indeed, one has to check that conformal invariance remains valid at the quantum level: the presence of a conformal anomaly arising through quantum fluctuations would ruin the whole picture.

Table 10.1 The lowest-lying states of the Ramond–Neveu–Schwarz model for the open string. GSO projection keeps those states indicated as bold-faced.

M^2 $\downarrow$	Neveu–Schwarz	chirality +	chirality -
$1/\alpha'$	$\cdots$	$\alpha^I_{-1} d^I_{-1} |0\rangle_{R+}$	$\alpha^{\mathbf{I}}_{-1} \mathbf{d}^{\mathbf{I}}_{-1} |0\rangle_{\mathbf{R}-}$
$1/2\alpha'$	$\alpha^I_{-1}|0\rangle_{\mathrm{NS}}, b^I_{-1/2} b^J_{-1/2}|0\rangle_{\mathrm{NS}}$		
0	$\mathbf{b}^{\mathbf{I}}_{-1/2}|\mathbf{0}\rangle_{\mathrm{NS}}$	$|\mathbf{0}\rangle_{\mathbf{R}+}$	$|0\rangle_{R-}$
$-1/2\alpha'$	$|0\rangle_{\mathrm{NS}}$		
	Neveu–Schwarz	Ramond	

It turns out that the conformal anomaly depends on the number D of spacetime dimensions; more precisely it is proportional to $D - 26$ in the case of the bosonic string and to $D - 10$ in the case of the superstring. Thus the dimension is fixed to be 26 for the bosonic string and 10 for the superstring. All but four of these dimensions should be compact[5].

In fact, conformal invariance can be even more constraining. If we choose to work in a nontrivial background (*i.e.* a background with a nontrivial metric, etc.), then requiring conformal invariance induces some constraints on the background. More precisely, the violations of conformal invariance appear through the renormalization group beta functions, which depend on the background field values. Setting these beta functions to zero to restore conformal invariance imposes differential equations on the background fields which are nothing else than the equations of general relativity (or supergravity in the superstring case) with the addition of the dilaton field e^ϕ and the antisymmetric tensor field b^{IJ} present in the gravity supermultiplet [65].

10.2 Compactification

There seems to be an interesting connection between the efforts to unify fundamental interactions and the need for extra spatial dimensions. We have just seen that a quantum theory of strings seems to call for such dimensions. Already in the 1920s, T. Kaluza and O. Klein proposed to consider an extra dimension to unify the theories of electromagnetism and gravitation. This work is the basis of theories of compactification on a torus and we will start reviewing it before moving to the specific case of string theory. We will then consider a variant of compactification, known as orbifold compactification, which allows us to break more symmetries of the underlying theory.

---

[5]As we will see later, there is little difference between a compact spatial dimension and an internal dimension: in the latter case, the quantized momentum is interpreted as a quantum number. Hence, these string theories may be interpreted as four-dimensional theories with 22 (or 10) internal dimensions (*i.e.* quantum numbers).

10.2.1 Kaluza–Klein compactification

T. [245] and O. [251] proposed to unify geometrically electromagnetism with gravitation by introducing the electromagnetic field as a component of the metric of a five-dimensional spacetime. To be slightly more general, let us consider a theory of gravity in $(D \equiv d+1)$-dimensional spacetime, described by the coordinates x^M, $M = 0,\ldots,d$. The metric is g_{MN}.

We single out the spatial coordinate $x^d \equiv y$ and assume that the corresponding dimension is compact of size $L = 2\pi R$; the other d dimensions corresponding to x^μ, $\mu = 0,\ldots,d-1$ are noncompact. From the point of view of d dimensions, $g_{\mu\nu}$ is a symmetric tensor interpreted as the metric, $g_{\mu d}$ is a vector field and g_{dd} a scalar field. Correspondingly, we may write the D-dimensional metric as the following ansatz matrix:

$$g_{MN} \equiv \left(\begin{array}{cc} g_{\mu\nu}(x) & \mathcal{A}_\mu(x) \\ \mathcal{A}_\nu(x) & -e^{2\sigma(x)} \end{array} \right), \tag{10.25}$$

where we have restricted the spacetime dependence of the components to the noncompact dimensions (x^μ).

It turns out that part of the reparametrization invariance of the original D-dimensional theory may be interpreted as a gauge invariance associated with the vector field $\mathcal{A}_\mu(x) \sim g_{\mu d}$. Thus the D-dimensional gravitational theory provides a geometric unification of d-dimensional gravity and of an abelian gauge symmetry such as the one present in the theory of electrodynamics.

As for the component g_{dd} of the metric, it measures distances in the compact dimension in (higher-dimensional) Planck units. Thus the d-dimensional field $e^{\sigma(x)}$, or more precisely its vacuum value, provides a measure of the size of the compact dimension[6]. It is often called a breathing mode. From this point of view, its x-dependence reflects the fluctuations of the compact dimension (along y) in directions transverse to it (measured by x^μ). As we will see later, the purely gravitational D-dimensional action yields a vanishing potential for this scalar field: it corresponds to a flat direction of the scalar potential. In a supersymmetric set-up this real field becomes part of a complex scalar field associated with this flat direction. Such a complex field is called a modulus and traditionally written as T in four dimensions. Of course, compactification requires the size of the compact dimension to be determined: some extra dynamics must be included in order to lift the degeneracy associated with the flat direction and to determine the vacuum expectation value of the modulus field.

One may expect that, in a D-dimensional spacetime, the gravitational force between two masses m_1 and m_2 decreases as $r^{-(D-2)}$:

$$F(r) \sim G_{(D)} \frac{m_1 m_2}{r^{D-2}}. \tag{10.26}$$

---

[6]To be accurate, in our notation, the radius of the compact dimension is $R\langle e^\sigma \rangle$. One may alternatively set $R = 1$ in Planck units, in which case the metric coefficient fixes the radius; or normalize $\langle e^\sigma \rangle$ to 1: R is then the radius.

Indeed, a sphere in $(D-1)$-dimensional space (D-dimensional spacetime) has a surface which varies as its radius to the power $D-2$; we thus expect any action at distance from a point source to behave as $r^{-(D-2)}$. This should in principle be enough to discard the possibility of extra dimensions.

However, care must be taken in the case of compact dimensions when the distance r is large compared with the size L of the compact dimension(s). Let us see in more detail what happens. We model such a Universe by the infinite torus of Fig. 10.6a: the infinite dimension represents any of the d noncompact dimensions, whereas the compact dimension is visualized by the circle of length $L = 2\pi R$. As usual, this torus may be represented as in Fig. 10.6b by a series of strips with proper identification:

$$y \equiv y + 2\pi R. \tag{10.27}$$

We consider two masses m_1 and m_2 separated by a distance r on this torus. A gravitational field line may join them directly, or may make one (or more) turns round the torus. Hence the mass m_1 feels the effect of mass m_2 and of all its images (Fig. 10.6b). If r is much larger than L then these images form a continuous line. In the case of $D-d$ such compact dimensions, one obtains a $(D-d)$-dimensional continuum of masses. Then the gravitational force exerted by this continuum on m_1 reads

$$F(r) \sim G_{(D)} \frac{m_1 m_2}{r^{d-2} \, L^{D-d}} \tag{10.28}$$

which has the standard form for d noncompact dimensions. For example, if $d = 4$, Newton's constant G_N is obtained from the D-dimensional gravitational constant $G_{(D)}$:

$$G_N = \frac{G_{(D)}}{L^{D-4}}. \tag{10.29}$$

Of course, when distances become of the order or smaller than L, one recovers the law of higher-dimensional gravity in $r^{-(D-2)}$. Thus one still expects deviations from the standard law at small enough distances.

(a)

(b)

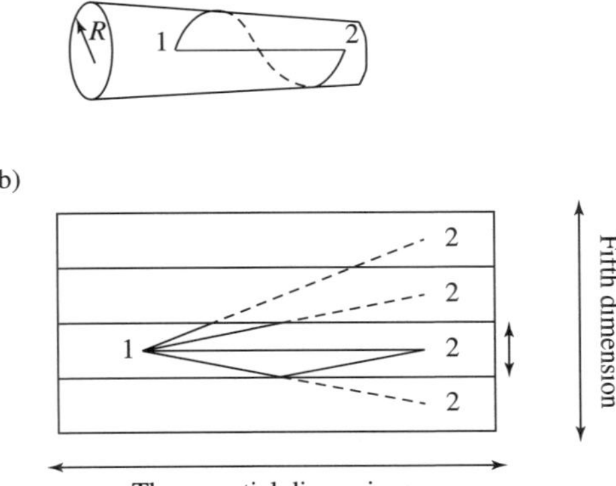

Fig. 10.6 (a) Infinite torus; (b) same torus represented by strips with opposite sides identified.

Note that we can introduce a fundamental Planck scale $M_{(D)}$ for the higher-dimensional theory, as in (10.1),

$$G_{(D)} \equiv \frac{(\hbar c)^{D-3}}{c^{2(D-4)} M_{(D)}^{D-2}}.$$

(10.30)

Then (10.29) reads

$$M_P^2 = M_{(D)}^{D-2} \left(\frac{Lc}{\hbar} \right)^{D-4}.$$

(10.31)

Thus, the larger the compact dimension is, the smaller the fundamental scale.

Effective actions

Let us consider the action of gravity in $D \equiv d+1$ dimensions (see Appendix D, Section D.1):

$$S = -\frac{1}{16\pi G_{(D)}} \int d^D x \sqrt{|g|} \, R^{(D)}$$

(10.32)

where $R^{(D)}$ is the curvature scalar associated with the D-dimensional metric. Following (10.25), we write the D-dimensional line element as ($y \equiv x^d$):

$$ds^2 = g_{MN} dx^M dx^N = g_{\mu\nu}^{(d)} dx^\mu dx^\nu - e^{2\sigma(x)} (dy + A_\mu dx^\mu)^2.$$

(10.33)

Reparametrizations of the form $y \rightarrow y + \alpha(x^\mu)$ lead to gauge transformations for the field $A_\mu(x)$: $A_\mu(x) \rightarrow A_\mu(x) - \partial_\mu \alpha$.

With the ansatz (10.33), we have (see (10.103))

$$R^{(D)} = R^{(d)} - 2D^\mu (\partial_\mu \sigma) - 2\partial^\mu \sigma \partial_\mu \sigma + \tfrac{1}{4} e^{2\sigma} F^{\mu\nu} F_{\mu\nu},$$

(10.34)

where $R^{(d)}$ is the curvature scalar built out of the metric $g_{\mu\nu}^{(d)}$. Hence, introducing

$$\frac{1}{G_{(d)}} \equiv \frac{1}{G_{(D)}} \int_0^L dy = \frac{L}{G_{(D)}},$$

(10.35)

as in (10.29), we obtain after integrating by parts

$$S_{\text{eff}} = -\frac{1}{16\pi G_{(d)}} \int d^d x \, \sqrt{|g^{(d)}|} \, e^\sigma \left[R^{(d)} + \frac{1}{4} e^{2\sigma} F^{\mu\nu} F_{\mu\nu} \right].$$

(10.36)

The absence of a kinetic term for the modulus field $\sigma(x)$ in (10.36) does not mean that it is nondynamical: the kinetic term is present in the Einstein frame[a]. The presence of the modulus flat direction may be traced back to the dilatation symmetry $y \rightarrow e^{-\lambda} y$, $A_\mu(x) \rightarrow e^{-\lambda} A_\mu(x)$, $\sigma(x) \rightarrow \sigma(x) + \lambda$, which leaves the line element (10.33) invariant.

---

[a]*i.e.* in a frame with a standard Einstein term $\sqrt{g} R$, obtained by a Weyl rescaling of the metric: $g_{\mu\nu}^{(d)} = e^{2\sigma/(2-d)} g_{\mu\nu}$.

We may also introduce matter fields in these higher-dimensional spacetimes. Since particles may be associated with waves, and extra compact dimensions with a multi-dimensional box, we expect standing waves in this higher-dimensional box. These are called Kaluza–Klein modes; they should be observed in our four-dimensional world as particles.

Let us illustrate this on the case of a particle of zero spin and mass m_0 in $D = 5$ dimensions (we use the same notation as above with $d = 4$). It is described by a scalar field $\Phi(x^M)$ with equation of motion:

$$\partial^M \partial_M \Phi = \partial^\mu \partial_\mu \Phi + \partial^y \partial_y \Phi = -m_0^2 \Phi. \tag{10.37}$$

We may decompose $\Phi(x^\mu, y)$ on the basis of plane waves in the compact dimension $e^{ip_5 y}$: because of the identification (10.27), we have $p_5 = n/R, n \in \mathbb{Z}$. Thus

$$\Phi(x^\mu, y) = \frac{1}{2\pi R} \sum_{n \in \mathbb{Z}} \phi_n(x^\mu) e^{iny/R} \tag{10.38}$$

and the equation of motion yields (we set here $g^{55} = 1$ for simplicity)

$$\partial^\mu \partial_\mu \phi_n = -\left(m_0^2 + \frac{n^2}{R^2} \right) \phi_n. \tag{10.39}$$

Thus the five-dimensional field is seen as a tower of four-dimensional particles with a spectrum characteristic of extra dimensions: particles with the same quantum numbers at regular mass-squared intervals. The spectrum has a typical mass scale M_C which is given by the inverse of the radius of the compact manifold

$$M_C \sim R^{-1}. \tag{10.40}$$

Clearly a multidimensional manifold may have several "radii" and thus several T moduli associated.

The fact that no sign of extra dimensions has been found experimentally indicates that M_C is large. In this context, it is thus important to determine the effective theory at energies much smaller than M_C. Since all massive fields ($m \sim M_C$) decouple, the task is to find out which are the "massless" fields ($m_0 = 0$ or $m_0 \ll M_C$) which are present at low energy. In the simple example that we have chosen, a massless ($m_0 = 0$) scalar field in D dimensions yields one and only one massless scalar field in four dimensions.[7]

10.2.2 String toroidal compactification

When we go from a point particle to a string, the physics of compactification becomes incomparably richer. We still find that momenta in the compact dimensions are quantized in units of $1/R$: this yields *Kaluza–Klein modes* with energies proportional to $1/R$. But compact dimensions allow the possibility of the string winding around them

---

[7] The higher-dimensional graviton also has Kaluza–Klein modes. The ansatz that we have chosen for the metric in (10.25) corresponds to neglecting the massive modes – they have a nontrivial y dependence – and restricting our attention to the zero modes.

(think of the string wrapped N times around a circle of radius R). The stable configuration thus obtained is called a *winding mode*. It has an energy proportional to R, since it would vanish for zero radius, and to the number m of wrappings: $E \sim mR$.

Hence, when R is large, the Kaluza–Klein modes are light whereas the winding modes are heavy. An effective low energy theory would include only the Kaluza–Klein states. It is the contrary when R is small[8].

It turns out that the two corresponding theories are equivalent. This can be expressed as a symmetry of the theory associated with the transformation of the modulus field $T \leftrightarrow 1/T$. Under the large/small compactification radius duality, a theory A with a large compact dimension is equivalent to a theory B with a small compact dimension. Such a duality is known as T-duality. It plays a central rôle in identifying the web of duality relations that exist between the different string theories (see next section).

Kaluza–Klein modes, winding modes, and T-duality

Let us consider a closed string on a circle $\mathcal{C}$ of radius R. As is well-known, we can describe this circle as a line with a periodic identification such as (10.27). Since a string may wind several times around the circle, we should allow for a new type of boundary condition

$$X(\sigma + \pi, \tau) = X(\sigma, \tau) + 2\pi m R, \tag{10.41}$$

and we have to change slightly the normal mode expansion (10.6):

$$X_L(z) = x_L - i\frac{\alpha'}{2}\left(p + m\frac{R}{\alpha'}\right)\ln z + i\sqrt{\frac{\alpha'}{2}}\sum_{l \neq 0}\frac{1}{l}\alpha_l z^{-l},$$

$$X_R(\bar{z}) = x_R - i\frac{\alpha'}{2}\left(p - m\frac{R}{\alpha'}\right)\ln \bar{z} + i\sqrt{\frac{\alpha'}{2}}\sum_{l \neq 0}\frac{1}{l}\tilde{\alpha}_l \bar{z}^{-l}. \tag{10.42}$$

Comparison with (10.6) reveals two new features: (i) the term linear in σ allows for a nonzero winding number m; (ii) the fact that the coordinate X obeys periodic boundary conditions – or equivalently, space is compact in this direction – imposes a quantization of the momentum $p = n/R$.
Then the string mass spectrum (10.8) is modified to[a]

$$\alpha' M^2 = 2(N + \tilde{N}) - 4 + n^2\frac{\alpha'}{R^2} + m^2\frac{R^2}{\alpha'}, \tag{10.43}$$

with the condition

$$N = \tilde{N} + nm. \tag{10.44}$$

[a]We have restored the 25 noncompact dimensions (X^M, $M = 0, \ldots, 24$) besides the compact dimension denoted $X^{25} \equiv X$ here. Then $N = \sum_{l=1}^{\infty}\left(\alpha_{-l}^M \alpha_{lM} + \alpha_{-l}\alpha_l\right)$, and similarly for $\tilde{N}$.

---

[8]One must obviously specify with respect to which scale R should be considered as large or small. If $M_S \equiv \alpha'^{-1/2}$ is the fundamental string scale, then $\ell_S \equiv \hbar c/(M_S c^2)$ is the reference length scale.

Notice that the term n^2/R^2 in (10.43) corresponds to the Kaluza–Klein modes as expected. The new term $m^2 R^2$ comes from the winding sectors.

The mass spectrum above is invariant under the following transformation:[b]

$$R \leftrightarrow \alpha'/R, \quad m \leftrightarrow n. \tag{10.46}$$

This transformation thus exchanges large and small radius compact manifolds. In the process, Kaluza–Klein modes are exchanged with winding modes: the light modes of one theory become the heavy modes of the other theory.

We note that the transformation (10.46) does not change the term in $\ln z$ of $X_L(z)$ whereas it reverses the sign of the term in $\ln \bar{z}$ of $X_R(\bar{z})$ in (10.42). We may add the transformation $\tilde{\alpha}_l \leftrightarrow -\tilde{\alpha}_l$ (note that $\tilde{N}$ and thus the spectrum formula (10.43) is invariant) to write the T-duality transformation simply as

$$X(z, \bar{z}) = X_L(z) + X_R(\bar{z}) \rightarrow X'(z, \bar{z}) = X_L(z) - X_R(\bar{z}). \tag{10.47}$$

Since we have the mass spectrum (10.43), we will take the opportunity to show that, for some special values of the compactification radius R, the massless spectrum is much richer.

First for each sector labelled by m and n, we have a ground state $|m, n\rangle$ which is annihilated by α_k^M, $\tilde{\alpha}_k^M$, α_k, $\tilde{\alpha}_k$ ($k \geq 0$). The ground state of the theory is the same as in the uncompactified theory $|m = 0, n = 0\rangle$ and corresponds to a tachyon:

$$\alpha' M^2 |m = 0, n = 0\rangle = -4. \tag{10.48}$$

At the $M^2 = 0$ level, we have as usual

$$\tilde{\alpha}_{-1}^M \, \alpha_{-1}|m = 0, n = 0\rangle, \qquad \alpha_{-1}^M \, \tilde{\alpha}_{-1}|m = 0, n = 0\rangle \tag{10.49}$$

but if we choose the radius as $R = (\alpha')^{1/2}$ we also find

$$\begin{aligned}
\alpha_{-1}^M |m = 1, n = +1\rangle, \quad &\tilde{\alpha}_{-1}^M |m = 1, n = -1\rangle \\
\alpha_{-1}^M |m = -1, n = -1\rangle, \quad &\tilde{\alpha}_{-1}^M |m = -1, n = +1\rangle.
\end{aligned} \tag{10.50}$$

The states (10.49) and (10.50) form two triplets of massless vector bosons (notice the vector index M). These are the gauge fields of $SU(2) \times SU(2)$ ($T_3 = m = n$ for the first set and $T_3 = m = -n$ for the second) which can be checked to be a symmetry of the complete spectrum. Hence for some values of the compactification radius, one finds an enhanced gauge symmetry for the compactified string theory. This is how the $E_8 \times E_8$ gauge symmetry of the heterotic string arises.

---

[b]The string coupling $\lambda = \langle e^\phi \rangle$ – hence the dilaton ϕ – is also modified. Indeed the tree level scattering amplitude for gravitons is of order $1/\lambda^2$. In the compactified theory, this amplitude is of order R/λ^2 (*cf.* (10.35)). It should be invariant (gravitons have $m = n = 0$). Hence $\lambda' = \lambda \alpha'^{1/2}/R$ and the dilaton transforms as

$$e^{\phi'} = \frac{\alpha'^{1/2}}{R} e^\phi. \tag{10.45}$$

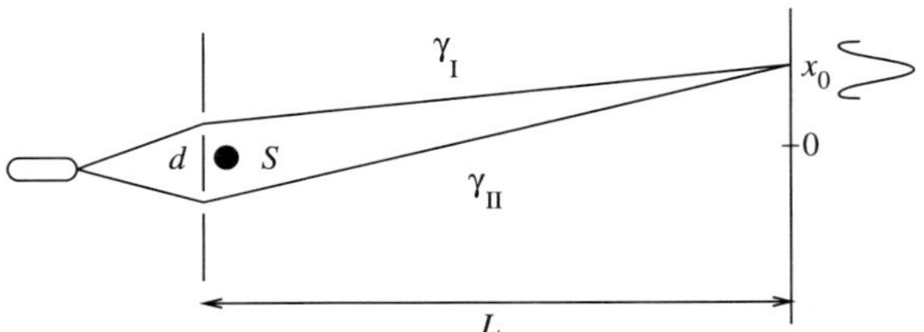

Fig. 10.7 Schematic set-up of the Aharonov–Bohm experiment.

10.2.3   Wilson lines

It is a well-known fact that nontrivial topologies may lead to interesting effects in gauge theories. One of the most striking ones is the Aharonov–Bohm effect [4] which shows that, in situations where space is not simply connected, one can have observable effects in regions where the field is zero ($\mathbf{B} = \mathbf{0}$ or $F_{\mu\nu} = 0$) if the vector potential is nonzero ($\mathbf{A} \neq \mathbf{0}$ or $A_\mu \neq 0$).

More explicitly, consider a standard interference experiment: electrons emitted by a coherent source of wavelength λ are diffracted by two slits distant by d; an interference pattern is observed on a screen at a distance L of the slits ($L \gg d$). The phase difference between the two waves, observed at a distance x from the axis of symmetry is

$$\delta = 2\pi \frac{x}{L} \frac{d}{\lambda}. \tag{10.51}$$

In a region where the vector potential A is nonzero, the phase picks up an extra contribution[9]:

$$\delta\phi = \frac{q}{\hbar} \int \mathbf{A} \cdot \mathbf{dl} \tag{10.52}$$

where q is the electric charge. Such a situation arises outside an infinitely long solenoid $S : \mathbf{B} = \mathbf{0}$ but $\mathbf{A} \neq \mathbf{0}$ ($\mathbf{B} \neq \mathbf{0}$ inside of course). The idea is therefore to put a small solenoid of infinite length between the two slits (Fig. 10.7) and see whether the presence of a nonzero vector potential perturbs the interference pattern.

A simple use of Stokes' theorem yields the new phase difference

$$\delta' = 2\pi \frac{x}{L} \frac{\delta}{\lambda} + \frac{q}{\hbar} \int_{\gamma_{\mathrm{I}} - \gamma_{\mathrm{II}}} \mathbf{A} \cdot \mathbf{dl} = 2\pi \frac{x}{L} \frac{\delta}{\lambda} + \frac{q}{\hbar} \int \mathbf{B} \cdot \mathbf{n}\, ds. \tag{10.53}$$

The contour $\gamma = \gamma_{\mathrm{I}} - \gamma_{\mathrm{II}}$ is a closed curve which can be squeezed around the solenoid ($\mathbf{B} = \mathbf{0}$ outside) but not more. Clearly, the interference pattern is shifted by a constant amount x_0.

This is generalized into the notion of a Wilson line: if the coordinate x^{25} is compactified on a circle, then one may consider the gauge configuration

$$A_{25}(x) = -\frac{\theta}{2\pi R} = -iU^{-1}\partial_{25}U, \quad U(x^{25}) = \exp\left(-\frac{i\theta x^{25}}{2\pi R}\right), \tag{10.54}$$

with θ constant. This is locally a pure gauge ($F_{MN} = 0$). But it has physical effects because, as in (10.52), the gauge potential has a nonvanishing circulation due to the

---

[9]Remember the gauge-invariant substitution $\mathbf{p} \to \mathbf{p} + q\mathbf{A}$.

nontrivial topology of the compact dimension. This is measured by the gauge-invariant Wilson line:

$$\exp\left(iq\int dx^{25}A_{25}\right) = \exp\left(-iq\theta\right).\tag{10.55}$$

In such a configuration, the canonical momentum p_{25} becomes $p_{25} + qA_{25} = p_{25} - q\theta/(2\pi R)$ and the quantification condition now reads:

$$p_{25} = \frac{2\pi n + q\theta}{2\pi R}.\tag{10.56}$$

Thus a 26-dimensional field of charge q and mass m_0 $(p^M p_M = p^\mu p_\mu - p^{25}p_{25} = m_0^2)$ appears as a tower of Kaluza–Klein states with a mass spectrum

$$m^2 = m_0^2 + \frac{(2\pi n + q\theta)^2}{4\pi^2 R^2}.\tag{10.57}$$

[If we now consider an open string with $U(N)$ Chan–Paton degrees of freedom at both ends, the constant $U(N)$ gauge potential A_{25} may be diagonalized as

$$A_{25} = -\frac{1}{2\pi R}\,\mathrm{diag}(\theta_1,\theta_2,\ldots,\theta_N).\tag{10.58}$$

The open string spectrum shows that the Wilson line breaks the $U(N)$ gauge symmetry to $U(n_1) \times \cdots \times U(n_r)$, $\sum_{i=1}^{r} n_i = N$, if the θ's appear in successive sets of n_i equal eigenvalues[10].]

Topological gauge symmetry breaking

The nontrivial nature of the topology of the Aharonov–Bohm set up or of the circle of compactification can be described using the notion of fundamental group. In both cases this group is nontrivial ($\Pi_1 = \mathbb{Z}$): space is multiply connected. In the following, we will often consider multiply-connected compact spaces K, most of the time of the form $K = K_0/G$ with K_0 simply connected and G a finite group of order n, in which case $\Pi_1(K) \cong G$. For example, we can distinguish three classes of curves in $K = K_0/\mathbb{Z}_3$; (a) [1]: curves closed in K_0; (b) $[g]$: curves closed up to a $g \in \mathbb{Z}_3$ transformation, (c) $[g^2]$: curves closed up to a g^2 transformation. These classes are the homotopy classes and they form the fundamental group of the manifold $K : \Pi_1(K) = \big([1],[g],[g^2]\big) \equiv \mathbb{Z}_3$.

[10]Let us consider a string in Chan–Paton state $|i,j\rangle$. Then, the quantification condition (10.56) yields the following mass spectrum:

$$m^2 = \frac{(2\pi n + \theta_i - \theta_j)^2}{4\pi^2 R^2} + \frac{1}{\alpha'}(N - 1).\tag{10.59}$$

Consider as above the massless vector fields ($n = 0$, $N = 1$). If all the θ's are different, out of the N^2 vectors only N are massless and the Wilson line has broken the $U(N)$ symmetry down to $U(1)^N$. If n of the θ's are equal, there correspond n^2 massless vector fields and thus a residual gauge symmetry $U(n)$.

If we consider a multiply connected compact space, then a nonvanishing Wilson line can develop along which $\langle A_k \rangle \equiv \langle A_k^a T^a \rangle \neq 0$ whereas $\langle F_{kl} \rangle = 0$. In other words, along a noncontractible loop γ ($\gamma \notin [1]$), this solution corresponds to

$$\langle U(\gamma) \rangle = \langle P \exp i \int_\gamma A_k \, dy^k \rangle \neq 1 \tag{10.60}$$

where P is a path ordering operator. Such a matrix has the following properties:

(i) $U(\gamma_1) \, U(\gamma_2) = U(\gamma_1 \gamma_2)$.

(ii) $U(\gamma)$ is identical for all γ in the same homotopy class. Take for example $\gamma_1, \gamma_2 \in [g^p]$: γ_1 and γ_2 are curves between y and $g^p y$ ($g \in G$) closed in K_0/G. Using Stokes' theorem, we have

$$\int_{\gamma_1} A_k \, dy^k - \int_{\gamma_2} A_k \, dy^k = \int_{\gamma = \gamma_1 - \gamma_2} A_k \, dy^k = \int F_{ij} \, ds^{ij} = 0. \tag{10.61}$$

Thus, if $\Pi_1(K) = G \cong \mathbb{Z}_n$, $[U(\gamma)]^n = 1$.

It is possible to set $U(\gamma)$ to 1, but at the price of modifying the boundary conditions. Take for example a curve $\gamma \in [g]$ that goes from y to gy. It is possible to perform a gauge transformation $(\exp -i \int_\gamma A_k dy^k)$ at the point gy such that it sets $U(\gamma)$ to 1. In other words, as we go along the curve γ (closed in K_0/G) we perform a gauge rotation that undoes the nonzero gauge field circulation $\int A_k dy^k$. Of course, this gauge transformation does not act only on the gauge fields and we have to gauge transform all the fields. Take for example a scalar field $\phi(y)$. Since $K = K_0/G$ is obtained by identifying y and gy, ϕ must obey the boundary condition $\phi(y) = \phi(gy)$ in order not to be multivalued. In the gauge rotated picture this becomes

$$\phi(y) = U_g^{-1} \, \phi(gy). \tag{10.62}$$

To account for that, U_g^{-1} is also called a twist in the boundary condition.

10.2.4 Orbifold compactification

Torus compactification is straightforward but it is often insufficient because it yields too many supersymmetries. Indeed, if we start with a 10-dimensional supersymmetric theory, the supersymmetric charge transforms as a Majorana spinor; hence it falls into a spinor representation **16** of the Lorentz group $SO(10)$ (see Section B.2.2 of Appendix B). When we compactify six dimensions, this yields eight four-dimensional Majorana spinors (two degrees of freedom each). Thus, one supersymmetry charge in 10 dimensions yields $N = 8$ supersymmetry charges in four dimensions. We have seen that only $N = 1$ supersymmetry is compatible with the chiral nature of the Standard Model. One must therefore look for compactifications which leave less supersymmetry charges intact.

The simplest possibility is to require further invariance under discrete symmetries of the compact manifold. For example, if we start with the circle described by the identification (10.27), we may impose a $\mathbb{Z}_2$ symmetry, often referred to as a twist,

$$y \leftrightarrow -y. \tag{10.63}$$

This then describes the segment $[0, \pi R[$. On such a compact space, strings may be closed up to the identification (10.27) as in torus compactification ($X(\sigma + \pi, \tau) = X(\sigma, \tau) + 2\pi m R$), or closed up to the identification (10.63) ($X(\sigma + \pi, \tau) = -X(\sigma, \tau)$). The latter strings are called twisted.

Such spaces obtained from manifolds by identifications based on discrete symmetries are called orbifolds. They are not as smooth as manifolds because the identifications lead to singularities of curvature. Let us illustrate this on the example of a tetrahedron which turns out ot be a two-dimensional orbifold.

We start with a torus obtained by identifying the opposite sides of a lozenge of angle $\pi/3$ (see Fig. 10.8a: $AC \sim OB$, $AO \sim CB$). This torus may thus be identified to a plane with periodic identifications of the type (10.27) in the directions corresponding to the two sides of the lozenge. Such a set-up is invariant under a rotation of π around the origin O (see Fig. 10.8b). Thanks to the periodic identifications, the points E, F, G are fixed points under this rotation.

We use this discrete symmetry to construct an orbifold out of this torus with only scissors and glue. Take the flat triangle AOB in Fig. 10.8c. The symmetry around O allows the following identifications: $EA \sim EO$, $GO \sim GB$, $FA \sim FB$. By folding along the dashed lines, one obtains a tetrahedron with vertices E, F, G, $A \sim B \sim O$. Now a tetrahedron is flat everywhere except at each apex where there is a conical singularity of deficit angle π (the sum of angles is $3 \times \pi/3 = \pi$ instead of 2π). In other words, curvature is zero everywhere except in E, F, G, O where it is infinite. For this reason, it is not a manifold and it falls into the class of orbifolds.

Obviously we have several classes of closed strings on such a space: closed strings which can shrink to a point by continuous deformations and closed strings which loop around one apex of the tetrahedron, *i.e.* one fixed point of the discrete symmetry which allowed us to construct the orbifold. The former are standard closed strings and

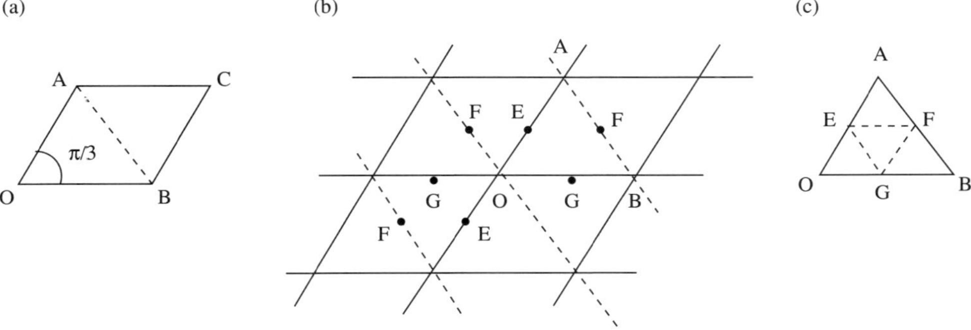

Fig. 10.8 Constructing an orbifold (a) and manifold: torus (b) same torus represented as a plane with periodic identifications (c) orbifold: tetrahedron.

are called untwisted in this context. The latter are closed only up to the identification associated with the discrete symmetry. They are the twisted strings. There is one twisted sector for each fixed point of the discrete symmetry.

Because of the curvature singularities associated with the fixed points, orbifolds were first considered only as toy models on which conjectures could be tested. It turns out that the presence of isolated singularities does not prevent from computing physical quantities in a string context, and orbifolds are now thought to be viable candidates for string compactification. Indeed, these singularities play an important rôle because they allow us to combine the simplicity of a flat metric with the richness of a curved background (curvature is located at these singularities). Moreover, it is also possible to blow up these singularities by replacing the conical singularity by some smooth surface. One obtains in this way a manifold.

Twisted sectors

Let us consider first the example of the segment $[0, \pi R[$. A closed string twisted around the origin satisfies

$$X(\sigma + \pi, \tau) = -X(\sigma, \tau) \quad \text{or} \quad X\left(ze^{2i\pi}, \bar{z}e^{-2i\pi}\right) = -X(z, \bar{z}). \tag{10.64}$$

Its expansion thus reads

$$X(z, \bar{z}) = i\sqrt{\frac{\alpha'}{2}} \sum_{n=-\infty}^{+\infty} \frac{1}{n + 1/2} \left(\frac{\alpha_{n+1/2}}{z^{n+1/2}} + \frac{\tilde{\alpha}_{n+1/2}}{\bar{z}^{n+1/2}} \right). \tag{10.65}$$

Comparison with the untwisted case (10.42) shows the absence of terms relative to the center of mass: it is fixed at the fixed point considered here[a] *i.e.* $x = 0$. The oscillators satisfy the commutation relations:

$$\left[\alpha_{m+1/2}, \alpha_{n-1/2}\right] = (m + 1/2)\delta_{m+n,0}, \tag{10.66}$$

and the mass formula reads:

$$\alpha' M^2 = 4\left(N - \frac{15}{16} \right) = 4\left(\tilde{N} - \frac{15}{16} \right). \tag{10.67}$$

The lowest levels are tachyonic[b]: the vacuum state $|0\rangle_0$ of mass squared $-15/(4\alpha')$ and the first excited state $\alpha_{-1/2}\tilde{\alpha}_{-1/2}|0\rangle_0$ of mass squared $-7/(4\alpha')$. The same procedure can be followed for the tetrahedron of Fig. 10.8 where the discrete symmetry used for orbifold identification is $\mathbb{Z}_2$. We may note that the torus of Fig. 10.8b is also invariant under rotations of angle $2\pi/3$ around the origin. The corresponding identifications lead to a $\mathbb{Z}_3$ orbifold. In this case, the oscillators of any given twisted sector are of the form $\alpha_{n\pm1/3}$ (see Exercise 3).

---

[a]A closed string twisted around the other fixed point would have the same expansion with an extra additive term πR (the coordinate of the fixed point).

[b]The indices 0 refer to the fixed point considered. We have similar states with index πR for the other fixed point.

10.3   String dualities and branes

The ultimate goal of the unification program is to find eventually a single string theory. There are actually five distinct superstring theories in 10 dimensions. But very detailed relations – known as duality relations – exist between these theories, which may be compatible with having finally a single mother theory, the elusive M-theory.

Let us start by identifying the five superstring theories. We have already stressed that, for the closed string, the oscillation modes in one direction along the string decouple from the modes in the other direction. These are respectively the left-movers and the right-movers. Once one introduces fermions, the left-moving fermions may have an opposite or the same chirality as the right-moving fermions. A closed super-string theory on which left and right-moving fermions have opposite (resp. the same) chirality, is called a type IIA (resp. IIB) superstring. Type IIA or IIB strings have $N = 2$ supersymmetry as can be checked on the massless mode spectrum: it contains two gravitinos.

One may also consider open strings together with closed strings (for consistency and in order to obtain a graviton among the massless modes). In this case only one supersymmetry charge is allowed and the corresponding theory is referred to as type I. As we have seen, open strings may carry gauge charges at their ends, which allows them to describe gauge theories.

Orientifolds

One way to obtain $N = 1$ supersymmetry from type IIB theories is to correlate left and right movers by requiring invariance under a world-sheet symmetry. Let us introduce the world-sheet parity $\Omega : \sigma \leftrightarrow \pi - \sigma$ or $z \leftrightarrow \bar{z}$, then we obtain from (10.6)

$$\Omega \alpha_n^I \Omega^{-1} = \tilde{\alpha}_n^I, \quad \Omega \tilde{\alpha}_n^I \Omega^{-1} = \alpha_n^I. \tag{10.68}$$

If we gauge this discrete symmetry, only states invariant under the symmetry remain in the spectrum of this by now unoriented string. This means for example that we must discard the antisymmetric tensor field among the massless modes of the closed string. This procedure is somewhat reminiscent of the orbifold twist except that we have used a $\mathbb{Z}_2$ *world-sheet* symmetry instead of a *spacetime* symmetry. The corresponding string theory is called a type IIB orientifold. As in the orbifold case, consistency of the theory requires the addition of twisted strings with respect to Ω. These are nothing else but the type I open strings.

Finally, since left-movers and right-movers may be quantized independently, it has been realized that one can describe simultaneously the right-movers by a superstring theory (and thus obtain $N = 1$ spacetime supersymmetry) in 10 dimensions and the left-movers by a standard bosonic string theory in 26 dimensions. Spacetime obviously has only the standard 10 dimensions. The extra 16 compact dimensions found in the right-movers are considered as internal: the corresponding momenta are quantized (because the dimensions are compact) and can thus be interpreted as quantum numbers. Indeed, it was shown that they can describe a nonabelian gauge symmetry with a gauge group of rank 16. These form the so-called heterotic string theories. The cancellation of quantum anomalies imposes that the gauge groups are either $SO(32)$ or $E_8 \times E_8$.

The heterotic string theory

We have seen in Section 10.2.2 that it is possible to generate gauge symmetry by compactifying some coordinates on the torus. More precisely, we found that, by requiring the coordinates $X^{25} = X_L^{25} + X_R^{25}$ to lie on a circle of radius $R = (\alpha')^{1/2}$, one generates a gauge symmetry $SU(2) \times SU(2)$. Here we have separated the left- and right-moving coordinate because a closer look at (10.50) shows that each $SU(2)$ is associated with one sector (α_{-1}^M are the creation operators for the left-moving sector, and $\tilde{\alpha}_{-1}^M$ for the right-moving sector). The rank of the group (the rank of $SU(2)$ is one: t_3 is the only diagonal generator) is equal to the number of compactified coordinates in each sector.

In the context of the heterotic string, one thus expects a gauge symmetry group of rank 16 (*i.e.* 16 gauge quantum numbers) from the available degrees of freedom in the left-moving sector. One finds in fact $SO(32)$ or $E_8 \times E_8$. We describe here briefly the $E_8 \times E_8$ heterotic string construction of [212, 213].

We start with the right-moving sector. It is constructed along the lines of the superstring theory described in Section 10.1. We work in the light-cone formalism. Besides the bosonic degrees of freedom $X_R^I(\bar{z})$, $I = 1, \ldots, 8$, we have fermions $\psi^I(\bar{z})$ in the Ramond sector or in the Neveu–Schwarz sector.

The mass spectrum is given by

$$\frac{\alpha'}{4} M^2 = N - a = \tilde{N} - \tilde{a}, \tag{10.69}$$

where $\tilde{a}_R = 0$ in the R case, $\tilde{a}_{NS} = 1/2$ in the NS case. The Ramond vacuum is a spinor which we write $|A\rangle$, where $A = 1, \ldots, 16$ are indices in the spinor representation $\mathbf{8}_+ + \mathbf{8}_-$ of the transverse Lorentz group $SO(8)$. In the NS sector, the vacuum $|0\rangle_{NS}$ is a tachyon ($M^2 = -2/\alpha'$) whereas the massless state is $|I\rangle = \tilde{b}_{-1/2}^I |0\rangle_{NS}$, *i.e.* a state in the vector representation of $SO(8)$. We conclude that we have spacetime vectors $|I\rangle \otimes \cdots$ and spacetime spinors $|A\rangle \otimes \cdots$, where $\cdots$ represents the left-moving state that we now proceed to determine.

The left-moving sector consists of 24 transverse coordinates, 16 of which are internal degrees of freedom[a]. For these 16 coordinates, we use the following property: in two dimensions, one boson is equivalent to two Majorana fermions[b]. We therefore replace the 16 internal coordinates by 2×16 fermions λ^i. In the same way that the coordinates X^M, $M = 1, \ldots, 2n$, are in the vector representation of $SO(2n)$, these fermions form a vector representation of a $SO(16) \times SO(16)'$ group. They can have periodic or antiperiodic boundary conditions. However modular invariance imposes restrictions on the possible choices. If all 32 fermions have the same boundary conditions, one obtains

---

[a]The remaining eight spacetime coordinates $X^I(z)$ provide the $N = 1$ supergravity multiplet which consists of $|I\rangle \otimes \alpha_{-1}^J|0\rangle$ (graviton [35 degrees of freedom], antisymmetric tensor [28] and dilaton [1]) and $|A\rangle \otimes \alpha_{-1}^J|0\rangle$ (gravitino [56] and a spinor field called the dilatino [8]).

[b]Schematically, the correspondence reads at the level of currents: $\partial_z X \equiv \psi_1 \psi_2$.

$SO(32)$ gauge symmetry. If one divides them into two sets of 16 fermions, and allow for each set R or NS boundary conditions, one finds, as we will now see, $E_8 \times E_8$.

We note in the left-moving sector: $X_L^I(z)$, $I = 1, \ldots, 8$; $\lambda^i(z)$, $i = 1, \ldots, 16$, in the vector representation of $SO(16)$; $\lambda^i(z)$, $i = 17, \ldots, 32$, in the vector representation of $SO(16)'$. For each possible choice of boundary conditions on the two sets of fermions, we have the following zero-point energy: $a_{(R,R)} = -1$, $a_{(R,NS)} = a_{(NS,R)} = 0$, $a_{(NS,NS)} = +1$. We obtain the mass spectrum from (10.69). The lowest-lying state in the (R, R) sector is at level $M^2 = 4/\alpha'$. For the mixed R, NS sectors, it is massless. Take for example (R, NS), which corresponds to

$$\lambda^i(z) = \sqrt{2\alpha'} \sum_{n \in \mathbb{Z}} d_n^i z^{-n}, \quad i = 1, \ldots, 16,$$

$$\lambda^i(z) = \sqrt{2\alpha'} \sum_{r \in \mathbb{Z}+1/2} b_r^i z^{-r}, \quad i = 17, \ldots, 32. \tag{10.70}$$

The R vacuum is annihilated by the d_0^i, $i = 1, \ldots, 16$, whose commutation relations show that they are 2^8-dimensional gamma matrices. Hence the R vacuum is in the spinorial representation $\mathbf{128_+} + \mathbf{128_-}$ of $SO(16)$ (the NS vacuum is a singlet). Similarly, in the (NS, R) case, the ground state is in $\mathbf{128_+} + \mathbf{128_-}$ of $SO(16)'$, corresponding to the indices $i = 17, \ldots, 32$. Finally, in the (NS, NS) sector, the ground state (a singlet) is at level $M^2 = -4/\alpha'$. The massless states are $b_{-1/2}^i b_{-1/2}^j |0>$, $i, j = 1, \ldots, 16$, and $b_{-1/2}^i b_{-1/2}^j |0>$, $i, j = 17, \ldots, 32$, which are respectively in representations[a] $(\mathbf{120}, \mathbf{1})$ and $(\mathbf{1}, \mathbf{120})$ of $SO(16) \times SO(16)'$. The states with $M^2 \leq 0$ are represented in Table 10.2.

All that is left to do is to perform the GSO projection. In the NS sectors, it removes the half-integer $|\alpha' M^2/4|$ levels and in the R sectors, it projects onto one definite chirality. The (NS, NS) ground state at mass level $-4/\alpha'$ must be removed because it has no corresponding state in the left-moving sector (we must satisfy (10.69)). Hence all tachyons are projected out of the spectrum and the massless fields are

$$[|I\rangle + |A\rangle] \otimes [(\mathbf{120} + \mathbf{128_+}, \mathbf{1}) + (\mathbf{1}, \mathbf{120} + \mathbf{128_+})]. \tag{10.71}$$

We thus obtain a 10-dimensional gauge supermultiplet, *i.e.* states in the vector representation $\mathbf{8_v}$ and in the spinor representation $\mathbf{8_s}$ of $SO(8)$. It turns out that the adjoint representation $\mathbf{248}$ of the exceptional group E_8 decomposes under $SO(16)$ exactly as $\mathbf{120} + \mathbf{128}$. We therefore find the gauge fields of $E_8 \times E_8'$ gauge symmetry.

---

[a]Note that the $b_{-1/2}^i$ anticommute; there are therefore $(16 \times 15)/2$ different states in each case.

Table 10.2 The lowest-lying states of the $E_8 \times E_8$ heterotic string theory.

M^2 $\downarrow$					
0	$\lvert A \rangle$ $\lvert I \rangle$	$(120, 1) + (1, 120)$	$(1, 128_+ + 128_-)$	$(128_+ + 128_-, 1)$	
$-2/\alpha'$	$\lvert 0 \rangle$	$(16, 1) + (1, 16)$			
$-4/\alpha'$		$(1, 1)$			
	R NS	(NS, NS)	(R, NS)	(NS, R)	
	R-moving sector	L-moving sector			

Type IIA, IIB, I, heterotic $SO(32)$ and $E_8 \times E_8$ form the five known types of superstring theories. There are some unexpected equivalences between them. These equivalences are basically of two types:

- Large/small compactification radius duality (the T-duality discussed in Section 10.2.2). Under the large/small compactification radius duality, a theory $\mathcal{T}_1$ with a large compact dimension is equivalent to a theory $\mathcal{T}_2$ with a small compact dimension. For example, type IIA superstring theory compactified on a circle of radius R is equivalent to type IIB compactified on a circle of radius ℓ_s^2/R. Similarly, the two heterotic string theories are related by this duality.

- Strong/weak coupling duality (or S-duality). The five string theories discussed above have been defined in their perturbative regimes. In other words, if λ_s is the string coupling for one of these theories, the theory is defined by its perturbative expansion: corresponding amplitudes are expressed as power series in λ_s. Nonperturbative effects appear to vanish at small coupling, as for example an instanton contribution of order e^{-1/g^2} in a gauge theory of coupling g. Strong/weak coupling duality relates a theory $\mathcal{T}_1$ in its strong coupling regime to a theory $\mathcal{T}_2$ in its weak coupling regime: an amplitude $\mathcal{M}^{(1)}(\lambda_s^{(1)})$ in theory $\mathcal{T}_1$ can be understood as amplitude $\mathcal{M}^{(2)}(\lambda_s^{(2)} = 1/\lambda_s^{(1)})$. [This is somewhat reminiscent of the electric–magnetic duality that we have discussed in Chapter 4. As discussed there, supersymmetry is a key ingredient to prove powerful results.] Such a type of duality relates for example type I superstring to the $SO(32)$ heterotic string theory whereas type IIB string theory is self-dual.

  The name S-duality refers to a four-dimensional scalar field which is present in string theory. The field S is the four-dimensional massless mode of the string dilaton encountered among the massless states of the closed string. Its value determines the value of the string coupling (which in turn fixes the value of the gauge couplings): $\langle S \rangle / m_P = 1/g^2$. Hence strong/weak duality corresponds to the duality $S \leftrightarrow 1/S$.

As long as supersymmetry is not broken, S and T correspond to flat directions of the scalar potential. Their value thus specifies a given fundamental state among a

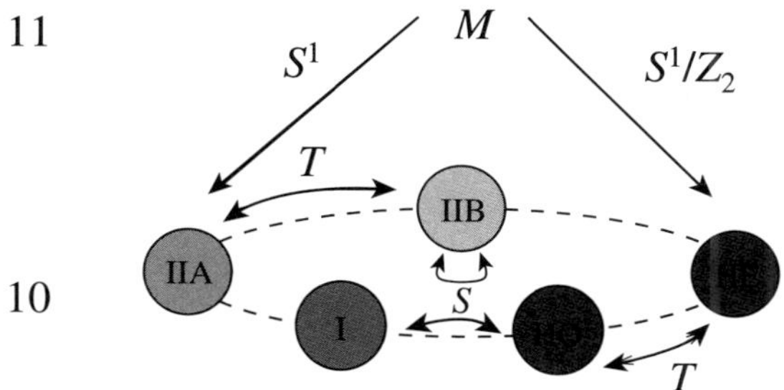

Fig. 10.9 M-theory and the five known string theories.

continuum. As such, they can be considered as moduli fields[11]. The low energy physics obviously depends on the explicit values of these fields (through the gauge coupling or through the value of the compactification radii). It is thus important to devise supersymmetry breaking schemes which completely lift the degeneracy associated with these moduli.

Duality relations allow us to have access to nonperturbative effects in a given superstring theory by studying perturbatively its dual theory. The most striking discovery in this respect was the realization that there exists an eleventh dimension. More precisely, the spectrum of type IIA superstring theory was found to include states with mass M_s/λ_s at weak coupling. Supersymmetry helped to solve the problem of the bound state of N such particles which was found to have precisely a mass of NM_s/λ_s. This is strongly reminiscent of the mass spectrum of Kaluza–Klein modes with a radius of compactification $R_{(11)} \equiv \lambda_s/M_s$.

The reason why this eleventh dimension was not found in the perturbative string approach is that perturbation precisely means an expansion around $\lambda_s = 0$ and thus, at fixed string scale M_s, around $R_{(11)} = 0$. This is why a perturbative expansion does not "see" this eleventh dimension. In all generality, one expects the ultimate M-theory to be eleven dimensional (at least). It is connected with the five known superstring theories as shown in Fig. 10.9 (HO and HE denote the two heterotic string theories).

There is therefore another limit of the superstring ultimate theory which is simply 11-dimensional supergravity. This solution does not have string excitations. Instead, it has branes. Branes are extended mem*branes* with one time dimension and p spatial dimensions: one then talks of a p-brane. They appear in string theory at the nonperturbative level. For example, they describe the locus of the extremities of open strings. We have seen above that open string ends carry nontrivial charges. Hence gauge degrees of freedom can be localized on branes. To be precise, the branes encountered in the context of strings are Dirichlet p-branes or Dp-branes: Dirichlet refers to the type of boundary condition imposed to open strings which end on the brane.

It is believed [227,228] that the strong coupling limit of the 10-dimensional $E_8 \times E_8$ heterotic string theory is 11-dimensional M-theory compactified on a line segment $[0, \pi R_{(11)}]$ (*i.e.* a S^1/Z_2 orbifold, to use the notions introduced above). The gauge

---

[11]Sometimes, in the literature, the term "modulus" is reserved in this context for the T fields.

fields belong to 10-dimensional supermultiplets that propagate on the boundaries of spacetime, that is on the 10-dimensional surfaces at 0 and $\pi R_{(11)}$. The effective field theory is a version of 11-dimensional supergravity on a manifold with boundaries, known as Hořava–Witten supergravity. Eleven-dimensional supergravity also contains D2-branes and D5-branes.

Open strings, T-duality, and branes

Branes were encountered [311] by studying the behavior of open strings under T-duality. Let us see this in more details.

There does not seem to be any winding mode associated with an open string since, in principle, an open string can always unwind itself. On the other hand, under a duality transformation (10.47), the Neumann boundary condition is turned into a Dirichlet condition: since $z = e^{2i(\tau+\sigma)}$, $\bar z = e^{2i(\tau-\sigma)}$ we have

$$\partial_\sigma X(z, \bar z) = \partial_\tau X'(z, \bar z). \tag{10.72}$$

Hence in the dual theory, the end of the open string is fixed (in the associated compact dimension). One can go one step further and show that all endpoints lie on the same hyperplane. For example we have

$$X'(\pi) - X'(0) = \int_0^\pi d\sigma \partial_\sigma X'(z, \bar z) = \int_0^\pi d\sigma \partial_\tau X(z, \bar z)$$

$$= \frac{2\pi \alpha' n}{R} = 2\pi n R'. \tag{10.73}$$

In other words, the corresponding open string winds n times around the dual dimension. This corresponds to a stable configuration because the ends of the string are attached to the hyperplane (the open string cannot unwind itself).

It is reasonable to expect that this hyperplane is a dynamical object which can fluctuate in shape and position. Indeed, one finds among the massless string states a scalar field which describes the position of this plane. This dynamical object is called a Dirichlet brane or D-brane.

We have seen that open strings allow us to introduce a gauge symmetry through the Chan–Paton degrees of freedom. Let us consider a Wilson line with a constant gauge potential (10.58). Following the discussion of Section 10.2.3, in particular (10.56), we see that a string state $|ij\rangle$ has momentum $p_{25} = (2\pi n + q\theta_i - q\theta_j)/(2\pi R)$. Thus (10.73) reads in this case

$$X'(\pi) - X'(0) = (2\pi n + \theta_i - \theta_j)\, R'. \tag{10.74}$$

Thus the end points of the open string $|ij\rangle$ lie at branes located at $\theta_i R'$ and $\theta_j R'$ respectively (up to a constant). If n values of θ are equal, that is if n branes are coincident, we have seen in Section 10.2.3 that there is an enhanced gauge symmetry $U(n)$.

We have discussed compactification from the point of view of duality relations between string theories. Of course, we need to compactify six out of the 10 dimensions of the perturbative superstring (or seven out of 11 in the strongly interacting case). This is unfortunately where we lose the uniqueness of the string picture (if indeed M-theory provides a unique framework): the vast number of compactification schemes leads to an equally vast number of low energy models. It is only a complete understanding of the nonperturbative dynamics of compactification which would determine which four-dimensional model corresponds to the true ground state. Meanwhile, however, the string picture gets richer.

We have already encountered orbifolds: they provide standard examples of six or seven-dimensional spaces on which to compactify string theories. For instance, one may take the product of three two-dimensional orbifolds such as the ones constructed in Section 10.2.4 (a classic example uses the $\mathbb{Z}_3$ orbifold studied in Exercise 3). Another type of compact space which is widely used are Calabi–Yau spaces. They are devised in such a way that compactification on these six-dimensional spaces yields $N = 1$ supersymmetry in four dimensions, as well as scalar fields charged under the gauge symmetry.

New gauge symmetries find their origin in the symmetries of the six or seven-dimensional compact manifold. As stressed before, symmetries, including supersymmetry, can also be broken by the boundary conditions that one chooses to impose on the fields for each of the compact coordinates. In the case of brane models, we have just seen that nonabelian gauge symmetries arise at the place where several branes coincide.

Some low energy observables which remain unexplained by the Standard Model find a new interpretation in the context of compactified string models. For example, in the context of compactification of the heterotic string model on a Calabi–Yau manifold, the number of families is interpreted as a topological number which can be computed for each Calabi–Yau manifold. Models with three families have been found, which resemble closely the Standard Model at low energy. Similar determination of the number of families exists also in the context of brane models.

In the context of the Kaluza–Klein compactification, the size of the compact dimensions is microscopic: the nonobservation of Kaluza–Klein modes at high energy colliders implies a bound of approximately $(1 \text{ TeV})^{-1}$. The constraint is different in the context of brane models. There, if the standard matter and gauge interactions are localized on the brane (or coincident branes), one must distinguish compact dimensions along the brane and orthogonal to the brane. The previous limit holds for dimensions along the brane. Compact dimensions orthogonal to the brane only support excitations with gravitational-type forces. The experimental limits then only arise from Cavendish-type experiments and give a bound of a few fractions of a millimeter for R. We will see in what follows that such macroscopic values for the radius of compactification allow us to decrease dramatically the string scale M_S, or the Planck scale of the higher-dimensional theory, down to a few TeV. This opens the remarkable possibility that the next high energy colliders probe directly the quantum gravity regime.

Calabi–Yau manifolds

Our starting point is the heterotic string theory which has been discussed above. Our discussion here is purely field theoretical and for this purpose we consider that, to a first approximation, the heterotic string theory yields a 10-dimensional supersymmetric Yang–Mills theory with gauge group $E_8 \times E_8$ coupled to $N = 1$ supergravity.

More precisely, this means that, in the Yang–Mills sector, the massless fields are the gauge fields and their supersymmetric partners, the gauginos, in the adjoint representation of $E_8 \times E_8$, $(\mathbf{1}, \mathbf{248}) + (\mathbf{248}, \mathbf{1})$. These are the only massless fields with gauge degrees of freedom. The algebra of $N = 1$ supersymmetry in 10 dimensions reads:

$$\{Q_A, Q_B\} = 2\Gamma^M_{AB}\, P_M \qquad (10.75)$$

where A, B are $SO(1,9)$ spinor indices. Q is a Majorana–Weyl spinor. Indeed, it is possible to impose both conditions in $2n = 2$ mod 8 dimensions, hence in 10 dimensions (see Section B.2.2 of Appendix B). As such, Q represents $2^{5-2} = 8$ degrees of freedom.

We now need to compactify the theory on $M_4 \times K$ where K is a compact manifold. We narrow down the possible choices by making the following requirements:

(i) $N = 1$ supergravity in four dimensions;

(ii) the low energy four-dimensional theory includes some scalar fields with nonzero gauge quantum numbers (typically Higgs fields).

We now follow the general strategy of Candelas, Horowitz, Strominger and Witten [66].We want to find a vacuum solution $|\Omega\rangle$ which is the solution of the equations of motion and which respects one four-dimensional supersymmetry charge Q. Clearly, the vacuum must be Lorentz invariant which implies, for any fermion F,

$$\langle\Omega|F|\Omega\rangle = 0. \qquad (10.76)$$

To determine $\langle\Omega|B|\Omega\rangle$ for the boson fields, we use the supersymmetry transformations in the following way: denoting by δF the transformation of F under supersymmetry, it follows from the fact that the vacuum is supersymmetric $(Q|\Omega\rangle = 0)$ that

$$\langle\Omega|\delta F|\Omega\rangle \;=\; \langle\Omega|\{Q, F\}|\Omega\rangle = 0. \qquad (10.77)$$

Explicitly, δF can be expressed as a function of the boson fields B. Hence the system of constraints (10.77) can be used to determine the bosonic expectation values $\langle\Omega|B|\Omega\rangle$ [66].

We can now write the supersymmetry transformations of the fermion fields[a]:

$$\delta\psi_M = \frac{1}{\kappa_{10}} D_M \eta$$

$$\delta\lambda^a = -\frac{1}{4g\sqrt{\phi}}\, \Gamma^M\, \Gamma^N\, F^a_{MN}\eta \tag{10.78}$$

where ϕ is the dilaton field, η is the 10-dimensional supersymmetry transformation parameter (a Majorana–Weyl spinor), $D_M\eta$ its covariant derivative and κ_{10} is the 10-dimensional gravitational constant. We have made in (10.78) the simplifying assumptions[b] that ϕ is constant and the antisymmetric tensor field b_{MN} has vanishing field strength h_{MNP}. The latter hypothesis imposes ($dh = \mathrm{tr}\,R \wedge R - \mathrm{tr}\,F \wedge F$)

$$F^a_{[MN}F^a_{PQ]} = R^{KL}{}_{[MN}R_{PQ]KL}. \tag{10.79}$$

Then (10.77) reads

$$D_k\eta = 0, \tag{10.80}$$

$$\Gamma^k\,\Gamma^l\,F^a_{kl}\eta = 0. \tag{10.81}$$

where $k, l = 4, \ldots, 9$ ($F^a_{\mu\nu} = 0$ in the vacuum $|\Omega\rangle$ to ensure Lorentz invariance). First look at (10.80) which expresses the fact that there exists a covariantly constant spinor. One can show that this implies that the manifold K is a complex Kähler manifold (*cf.* equation (C.55) of Appendix C). We recall in Section D.1 of Appendix D the notion of holonomy group and mention that for a general six-dimensional complex manifold, the holonomy group H is $SO(6)$. But (10.80) expresses the fact that there is one direction in tangent space which remains untouched by the action of H

$$(\forall\, h_\gamma \in H)\ \ h_\gamma\eta = \eta. \tag{10.82}$$

The spinor η is Majorana–Weyl and is therefore in the **4** of $SO(6)$. The holonomy group H is the subgroup of $SO(6) \cong SU(4)$ that leaves invariant a **4**; hence $H = SU(3)$. This is where Calabi–Yau manifolds enter the game.

A **Calabi–Yau manifold** is a complex Kähler manifold with a metric of $SU(3)$ holonomy (we will denote the holonomy group by $SU(3)_H$).

We may now write the decomposition of Q_A under the subgroup $SO(2) \times SU(3)_H$ of the transverse Lorentz group $SO(2) \times SO(6)$ (the quantum number associated with the four-dimensional $SO(2)$ is the helicity):

$$Q_A = (h = +1/2, \mathbf{1}) + (h = -1/2, \mathbf{1}) + (h = +1/2, \mathbf{3}) + (h = -1/2, \mathbf{\bar{3}}). \tag{10.83}$$

---

[a]Since we only consider values of the fields in the vacuum $|\Omega\rangle$, we will write from now on F or B instead of $\langle\Omega|F|\Omega\rangle$ and $\langle\Omega|B|\Omega\rangle$ and we use (10.76) to set F to zero.

[b]The dilatino variation $\delta\lambda$ is identically zero under these assumptions.

The residual charges at low energy are those which leave invariant the background fields, which are in $SU(3)_H$. As expected, the only supersymmetry charge that survives is the singlet under $SU(3)_H$ which gives two degrees of freedom, *i.e.* $N = 1$ supergravity in four dimensions[a].

We also want to fulfill condition (ii) and must therefore take a closer look at the only nonsinglet bosons available, the gauge fields. Their decomposition under $SO(2) \times [SO(6) \equiv SU(4)]$ is given by

$$\mathbf{8}_v = (h = +1, \mathbf{1}) + (h = -1, \mathbf{1}) + (h = 0, \mathbf{6}). \tag{10.84}$$

Since $\mathbf{6} \equiv \mathbf{3} + \overline{\mathbf{3}}$ under $SU(3)_H$, the decomposition of a 10-dimensional gauge field under $SO(2) \times SU(3)_H$ reads:

$$\mathbf{8}_v = (h \pm 1, \mathbf{1}) + (h = 0, \mathbf{3}) + (h = 0, \overline{\mathbf{3}}). \tag{10.85}$$

Hence the only fields invariant under $SU(3)_H$ are gauge fields ($h = \pm 1$) and there does not seem to be any four-dimensional scalar with gauge interactions. However proper care must be taken of the condition (10.79). The easiest way to satisfy this condition is to identify the spin connection (Christoffel symbol, see (D.5) of Appendix D) with some of the Yang–Mills fields (gauge connection). We therefore choose one of the two E_8 (refer to the other one as E_8'), decompose it into $E_6 \times SU(3)_{\mathrm{YM}}$ and set

$$A_M = A_M^a \, T^a = \begin{pmatrix} 0 & 0 \\ 0 & \Gamma^i_{Mj} \end{pmatrix} \begin{matrix} \}E_6 \\ \}SU(3)_{\mathrm{YM}} \end{matrix} \tag{10.86}$$

(remember that A_M stands for $\langle \Omega | A_M | \Omega \rangle$). For $M = 4, \ldots, 9$, the $(A_M)^i{}_j$ are (four-dimensional) scalar fields in the adjoint of $SU(3)_{\mathrm{YM}}$, with a nonzero vacuum expectation value through (10.86) (if K is not flat, the spin connection Γ^i_{Mj} cannot be set to zero globally). They therefore break the E_8 gauge symmetry to E_6.

A Yang–Mills field A_M^a $a = 1, \ldots, \dim E_8 = 248$ transforms under $SO(8)$ and E_8 as $(\mathbf{8}_v, \mathbf{248})$, with the following decomposition:

$$
\begin{array}{ccccccc}
& E_6 & SU(3)_{\mathrm{YM}} & & & & \\
& \uparrow & \uparrow & & & & \\
\mathbf{248} = & (\mathbf{78}, & \mathbf{1}) & + & (\mathbf{1}, \mathbf{8}) & + (\mathbf{27}, \mathbf{3}) & + (\overline{\mathbf{27}}, \overline{\mathbf{3}}) \\
& & & & & & \updownarrow \\
\mathbf{8}_v = & (h = \pm 1, & \mathbf{1}) & & + & (h = 0, \mathbf{3}) & + (h = 0, \overline{\mathbf{3}}) \\
& & \downarrow & & & & \\
& & SU(3)_H & & & &
\end{array}
$$

[a]One may alternatively count the number of (Weyl–Majorana) supersymmetry charges on the two-dimensional world-sheet. Just by counting the number of degrees of freedom, one concludes that two charges are needed in the world-sheet to make $N = 1$ spacetime supersymmetry. In the context of the heterotic string, they come from the right-moving sector and the minimal model has $(0, 2)$ supersymmetry, *i.e.* 0 left-moving charges and two right-moving ones. In the case of Calabi–Yau supersymmetry, the identifications made enhance the system of two-dimensional supersymmetry charges to $(2, 2)$ world-sheet supersymmetry.

The arrow $\updownarrow$ connects representations for which a $SU(3)_H$ rotation can be undone by a $SU(3)_{\mathrm{YM}}$ rotation. Hence, the fields which commute with the nontrivial background are in the representations:

$$(h = \pm 1, \mathbf{78}) + (h = 0, \mathbf{27}) + (h = 0, \overline{\mathbf{27}}), \tag{10.87}$$

where the second entry is the E_6 representation content. We find gauge fields $(h = \pm 1)$ in the adjoint of E_6 (**78**) and scalar fields in **27** and $\overline{\mathbf{27}}$. We also have gauge fields in the adjoint representation **248** of E_8'.

For completeness we give the four-dimensional decomposition of the 10-dimensional Yang–Mills supermultiplet in Table 10.3 and of the 10-dimensional supergravity multiplet in Table 10.4. The numbers $h_{p,q}$ are topological numbers called Hodge numbers. They count the number of (p,q)-forms (that is differential forms with p holomorphic indices and q antiholomorphic indices; remember that a Calabi–Yau manifold is a complex manifold) that are annihilated by the Laplacian. For a manifold of complex dimension 3 (hence $p, q \leq 3$), the property of $SU(3)$ holonomy implies $h_{0,0} = h_{3,3} = h_{0,3} = 1$ and $h_{0,1} = h_{0,2} = h_{1,3} = h_{2,3} = 0$. Since $h_{p,q} = h_{q,p}$, the only Hodge numbers to determine for a specific Calabi–Yau manifold are $h_{1,1}$ and $h_{2,1}$. We note that the Euler characteristics (see Section 10.4.1) of the compact manifold is given in terms of the Hodge numbers by the relation

$$\chi(K) = \sum_{p,q}(-)^{p+q}h_{p,q} = 2(h_{1,1} - h_{2,1}). \tag{10.88}$$

A closer look at Table 10.3 shows that the combination $h_{2,1} - h_{1,1}$ is the number of fermions in **27** of E_6 minus the number of fermions in $\overline{\mathbf{27}}$, that is the net number of families (as seen in Chapter 9, a representation **27** of E_6 encompasses a full family of the Standard Model). Thus, in the context of Calabi–Yau models, the number of families is directly related to the Euler characteristics of the manifold. Manifolds with $\chi = -6$ have been actively searched for.

Looking at Table 10.4, one notes the presence of real scalar fields. In order to fall into chiral supermultiplets, they should pair up to form complex scalar fields. Indeed, the spin 0 component of the chiral supermultiplets reads:

$$S = \frac{1}{(2\pi)^7}e^{-2\phi+6\sigma} + ia \ , \ T = \frac{1}{(2\pi)^7}e^{2\sigma} - i\sqrt{2}\beta^{(h_{1,1})},$$

$$T^{(j)} = \alpha^{(j)} + i\beta^{(j)} \ , \quad j = 1, \ldots, h_{1,1} - 1, \tag{10.89}$$

where a is the pseudoscalar field obtained by duality transformation from the antisymmetric tensor field $b_{\mu\nu}$ (see Section 10.4.2 below).

Table 10.3 Decomposition of the 10-dimensional Yang–Mills supermultiplet.

$D = 10$			$D = 4$	
(E_8, E_8') representation			(E_6, E_8') representation	
		$A_\mu^a(x, y)$	$(\mathbf{78}, \mathbf{1})$	$h_{0,0} = 1$ gauge fields
$(\mathbf{248}, \mathbf{1})$	$A_M^a(x, y)$	$A_K^a(x, y)$	$\begin{cases} (\mathbf{27}, \mathbf{1}) & h_{2,1} \text{ complex scalar fields } A_\alpha^{(i)}(x) \\ (\overline{\mathbf{27}}, \mathbf{1}) & h_{1,1} \text{ complex scalar fields } A_{\bar\alpha}^{(j)}(x) \end{cases}$	
			$(\mathbf{1}, \mathbf{1})$	n_E (not topological number) complex scalar fields
			$(\mathbf{78}, \mathbf{1})$	$h_{0,0} = 1$ Majorana spinor
$(\mathbf{248}, \mathbf{1})$	$\lambda_A^a(x, y)$		$\begin{cases} (\mathbf{27}, \mathbf{1}) & h_{2,1} \text{ L-handed spinors } \psi_\alpha^{(i)}(x) \\ (\overline{\mathbf{27}}, \mathbf{1}) & h_{1,1} \text{ L-handed spinors } \psi_{\bar\alpha}^{(j)}(x) \end{cases}$	
			$(\mathbf{1}, \mathbf{1})$	n_E L-handed spinors
$(\mathbf{1}, \mathbf{248})$	$A_M^{a'}(x, y)$		$(\mathbf{1}, \mathbf{248})$	1 gauge field
$(\mathbf{1}, \mathbf{248})$	$\lambda_A^{a'}(x, y)$		$(\mathbf{1}, \mathbf{248})$	1 Majorana spinor

10.4 Phenomenological aspects of superstring models

We discuss in this section the most salient features of string theory applied to the description of fundamental interactions. We have seen in preceding sections that string models generically have a gauge symmetry large enough to include the Standard Model interactions, and thus provide a scheme unifying all known interactions, including gravity. However, the gravitational sector is richer than in standard Einstein gravity: it involves moduli fields, the value of which determines the low energy physics. The determination of these values thus puts, once again, the issue of supersymmetry breaking center stage.

Before we address these questions, it is important to discuss the energy scales present in these models, and their possible range, since this determines the possible ways to put these issues to the experimental test.

10.4.1 Scales

The different string theories that we have discussed in Section 10.3 involve various relations between the mass scales and couplings [378]. This is best seen by looking at some key terms in the effective field theory action. The different regimes are partly due to the fact that, depending on the string theory considered, the relevant terms appear

Table 10.4 Decomposition of the 10-dimensional supergravity supermultiplet.

	$D = 10$		$D = 4$
	$g_{\mu\nu}$		$g_{\mu\nu}(x)$ graviton
$g_{MN}(x,y)$	$g_{\mu k}$ g_{kl}	$g_{a\bar{b}}$	$h_{1,1}$ real scalar fields e^σ, $\alpha^{(j)}$ $j = 1,\dots,h_{1,1} - 1$
		$g_{\bar{a}\bar{b}}, g_{ab}$	$h_{2,1}$ complex scalar fields (moduli) $U^{(i)}$ $i = 1,\dots,h_{2,1}$
	$b_{\mu\nu}$		$b_{\mu\nu}(x) \leftrightarrow 1$ real scalar field a
	$b_{\mu k}$		$h_{0,1} = 0$ spin 1 field
$b_{MN}(x,y)$	b_{kl}	$b_{a\bar{b}}$	$h_{1,1}$ real scalar fields $\beta^{(j)}$ $j = 1,\dots,h_{1,1}$
		$b_{ab}, b_{\bar{a}\bar{b}},$	$h_{0,2}$ complex scalar field
$\phi(x,y)$			1 real scalar field ϕ
	ψ_μ		gravitino
$\psi_M(x,y)$			$h_{2,1}$ Majorana spinors $\psi^{(i)}$ $i = 1,\dots,h_{2,1}$
	ψ_k		$h_{1,1}$ Majorana spinors $\chi^{(j)}$ $j = 1,\dots,h_{1,1}$
$\psi(x,y)$			1 Majorana spinor $\psi^{(0)}$

at different orders of string perturbation theory. We recall that the string coupling is given by the vacuum expectation value of the string dilaton

$$\lambda = \langle e^\phi \rangle. \tag{10.90}$$

It is well-known that, in the case of the Yang–Mills action $\int d^4x F^{\mu\nu} F_{\mu\nu}$, one can redefine the gauge fields in order to have an overall factor g^{-2} as only gauge coupling dependence. Then propagators appear with a factor g^2 and vertices with a factor g^{-2}. Thus a diagram with L loops[12] appears with an overall factor $g^{2(L-1)}$.

The same property is valid in string theory if we work with the fields which appear on the world-sheet: the corresponding metric is called the string frame metric and the only mass scale appearing in the action is the string scale $M_s = (\alpha')^{1/2}$. If we consider a closed string, a tree-level diagram ($L = 0$) has the topology of a sphere

---

[12]We have the standard relation $L - 1 = I - V$ between L, the number I of internal lines, and the number V of vertices.

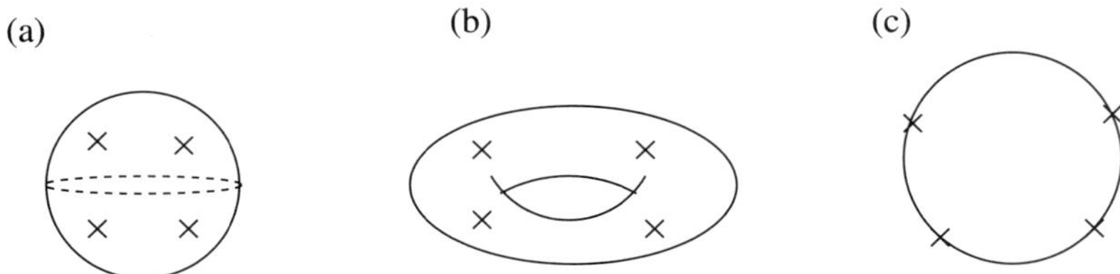

Fig. 10.10 (a) Topology of a tree-level closed string diagram such as in Fig. 10.5 (the crosses indicate vertices of particle emission corresponding to the incoming and outgoing strings of Fig. 10.5); (b) topology of a one-loop closed string diagram; (c) topology of a tree-level open string diagram such as in Fig. 10.4.

(Euler characteristics $\chi = 2$), a one-loop diagram the topology of a torus ($\chi = 0$), and so on (see Fig. 10.10). Thus a general diagram has an overall coupling factor $\lambda^{-\chi}$, hence an overall dilaton dependence $e^{-\chi\phi}$. This result extends to the open string case: a tree-level diagram has the topology of a disk ($\chi = 1$) and thus a dependence $e^{-\phi}$.

Euler characteristics

The Euler characteristics is easily computed in the case of simple compact surfaces. A standard method uses the triangulation of the compact manifold. In the case of a two-dimensional surface, it consists in dividing the surface into triangles, making sure that any two distinct triangles either are disjoint, have a single vertex or an entire edge in common. Examples of triangulations are given in Fig. 10.11.
The Euler characteristic is then given by

$$\chi = b_0 - b_1 + b_2 \tag{10.91}$$

where b_0, b_1, b_2 are respectively the number of vertices, edges, and triangles. From looking at Fig. 10.11a,c, one obtains $\chi(S_1) = 3 - 3$ and $\chi(S_2) = 5 - 9 + 6 = 2$. This generalizes to $\chi(S_n) = 1 + (-1)^n$ for a sphere S_n in $(n+1)$ dimensions. We also obtain for a disk (Fig. 10.11b), $\chi(D_2) = 4 - 6 + 3 = 1$. From this last result, one can infer another very convenient formula for the Euler characteristics of a two-dimensional surface:

$$\chi = 2 - 2h - b \tag{10.92}$$

where h is the number of handles and b the number of boundaries. Indeed, a surface with neither handles nor boundaries is the sphere S_2. Adding a boundary amounts to puncturing a hole in this sphere, that is removing a disk ($\delta\chi = -1$). Adding a handle amounts to puncturing two holes ($\delta\chi = -2$) and gluing their boundaries. We deduce that $\chi = 0$ for the torus: by deforming a torus, one easily obtains a sphere with one handle.

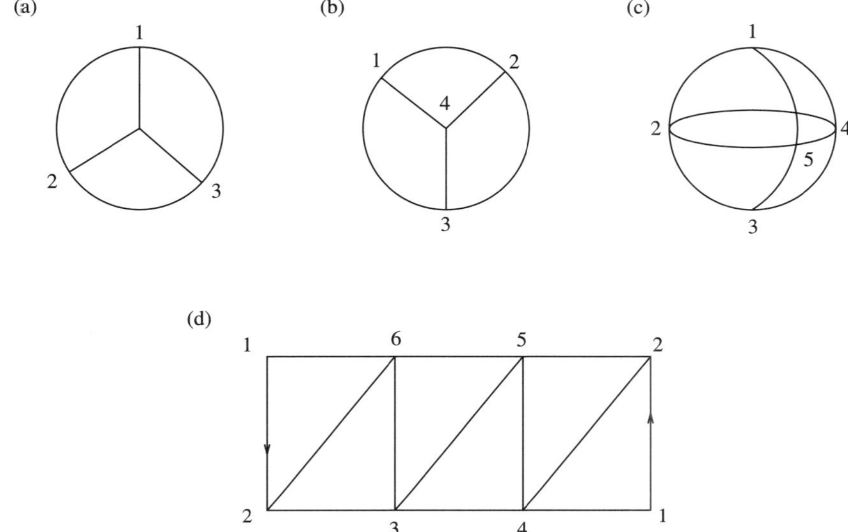

Fig. 10.11 Examples of triangulation: (a) circle S_1; (b) disk D_2; (c) sphere S_2; (d) Möbius band.

We start with the weakly coupled heterotic string. The 10-dimensional effective supergravity action includes the following terms which all appear at closed string tree level (hence the overall dependence in $e^{-2\phi}$ in the string frame):

$$\mathcal{S} = -\int \frac{d^{10}x}{(2\pi)^7} \sqrt{|g|}\, e^{-2\phi} \left[\frac{1}{(\alpha')^4} \left(R^{(10)} + 4\partial^\mu \phi \partial_\mu \phi \right) + \frac{1}{(\alpha')^3} \frac{1}{4} \mathrm{Tr} F^2 + \cdots \right]. \quad (10.93)$$

Once one compactifies on a six-dimensional manifold of volume V_6, one obtains

$$\mathcal{S} = -\int \frac{d^4x}{(2\pi)^7} \sqrt{|g|} \left(\frac{V_6}{(\alpha')^4} e^{-2\phi} R^{(4)} + \frac{V_6}{(\alpha')^3} e^{-2\phi} \frac{1}{4} \mathrm{Tr} F^2 + \cdots \right),$$

$$= -\int d^4x \sqrt{|g|} \left(\frac{1}{2} m_P^2 R^{(4)} + \frac{1}{16\pi \alpha_U} \mathrm{Tr} F^2 + \cdots \right), \quad (10.94)$$

from which we read the four-dimensional Planck scale as well as the value of the gauge coupling α_U at the string scale. Introducing the string scale $M_H \equiv \alpha'^{-1/2}$ (the subscript H for heterotic) and the compactification scale $M_C \equiv V_6^{-1/6}$, we obtain an expression for the string scale and the string coupling λ_H:

$$M_H^2 = 2\pi \alpha_U m_P^2, \quad \lambda_H = \langle e^\phi \rangle = \frac{\alpha_U^2}{2\pi^{3/2}} \frac{m_P^3}{M_C^3}. \quad (10.95)$$

This shows that the string scale is of the order of the Planck scale (taking for α_U the value at unification: $1/24$). Moreover, if the compact manifold is isotropic, M_C represents, to a first approximation, the scale where the theory becomes truly unified and is thus interpreted as the gauge coupling unification scale M_U. It is clear in this context that (10.95) implies a large string scale M_H.

We next turn to the strongly coupled $SO(32)$ heterotic string which is equivalent to the weakly coupled type I open string. We now note the string scale $M_I = (\alpha')^{1/2}$ and the string coupling λ_I. The effective supergravity action reads

$$
S = - \int \frac{d^{10}x}{(2\pi)^7} \sqrt{|g|} e^{-2\phi} \frac{1}{(\alpha')^4} R^{(10)} - \int \frac{d^{p+1}x}{(2\pi)^{p-2}} \sqrt{|g|} e^{-\phi} \frac{1}{(\alpha')^{(p-3)/2}} \frac{1}{4} \mathrm{Tr} F^2_{(p)} + \cdots,
$$

$$(10.96)$$

where the gauge symmetry arises from p-branes ($p = 3, 5, 7$ or 9) [312]: the dilaton dependence is $e^{-\phi}$ because the gauge term corresponds to an open string tree level amplitude (disk of Fig. 10.10c). One obtains after compactification

$$
S = - \int \frac{d^4x}{(2\pi)^7} \sqrt{|g|} \left(e^{-2\phi} \frac{V_6}{(\alpha')^4} R^{(4)} + e^{-\phi}(2\pi)^{9-p} \frac{V_{p-3}}{(\alpha')^{(p-3)/2}} \frac{1}{4} \mathrm{Tr} F^2 + \cdots \right).
$$

$$(10.97)$$

Writing $V_6 = M_C^{-6}$ and $V_{p-3} = M_C^{-(p-3)}$, we obtain

$$
M_I = 2\pi \left(\frac{\sqrt{\pi}}{\alpha_U} \frac{M_C^{p-6}}{m_P} \right)^{1/(p-7)}, \qquad \lambda_I = \sqrt{4\pi} \left[\frac{\pi^2}{\alpha_U^4} \left(\frac{M_C}{m_P} \right)^{p-3} \right]^{1/(p-7)}. \qquad (10.98)
$$

We infer

$$
\frac{M_I}{m_P} = \sqrt{\pi} \left(8\alpha_U^3 \lambda_I^{p-6} \right)^{1/(p-3)}. \qquad (10.99)
$$

Thus, at least for $p = 9$ or 7, the string scale can be as low as the type I string coupling λ_I can be taken small. In the case of 9-branes, whose world-volume fills the 10-dimensional spacetime, we simply have $M_I = m_P \left(2\pi\alpha_U \lambda_I \right)^{1/2}$.

Finally, we may write the effective 11-dimensional Hořava–Witten [227, 228] supergravity action for M-theory compactified on the orbifold S^1/Z_2 or line segment $[0, \pi R_{(11)}]$

$$
S = - \int_{\mathcal{M}^{11}} \frac{d^{11}x}{(2\pi)^8} \sqrt{|g|} M_M^9 R^{(11)} - \sum_{i=1}^{2} \left(\frac{3}{2} \right)^{1/3} \int_{\mathcal{M}_i^{10}} \frac{d^{10}x}{(2\pi)^7} \sqrt{|g|} M_M^6 \mathrm{Tr} F_i^2 + \cdots
$$

$$(10.100)$$

where $\mathcal{M}_i^{10}$, $i = 1, 2$, are the two boundaries of spacetime located at the two ends of the line segment, 0 and $\pi R_{(11)}$, and we have introduced the fundamental scale M_M $(2\kappa_{11}^2 \equiv (2\pi)^8/M_M^9)$. We obtain, after compactification on a six-dimensional manifold,

$$M_M = \pi\sqrt{2}\ \left(\frac{2}{3}\right)^{1/18} \frac{M_C}{\alpha_U^{1/6}}, \qquad \pi R_{(11)} = \frac{4\sqrt{3}}{\pi}\alpha_U^{3/2}\frac{m_P^2}{M_C^3}. \tag{10.101}$$

Clearly then, the M-theory scale is of the order of the gauge coupling unification scale $M_U \sim M_C$.

10.4.2 Dilaton and moduli fields

Our discussion of mass scales shows the pre-eminent rôle played by fundamental scalar fields in string theory: the dilaton fixes the string coupling, other fields determine the radii and shape of the compact manifold (hence its volume V). More generally, because there is only one fundamental scale, all other scales are fixed in terms of it by vacuum expectation values of scalar fields. In many instances such as the ones listed above, the scalar field corresponds to a flat direction of the scalar potential. We have already stressed the importance of flat directions in supersymmetric theories. We have seen that such fields are called moduli: contrary to the case of Goldstone bosons, different values of the moduli fields lead to different physical situations. For example, different values of the dilaton lead to different values of the gauge coupling, hence possibly different regimes of the gauge interaction.

Before we explain why dilaton and radii correspond to flat directions, we have to show how these *real* scalar fields fit into supersymmetric multiplets. The antisymmetric tensor b_{MN} which is present among the massless modes of the closed string plays a crucial rôle to provide the missing bosonic degrees of freedom (remember for example that the scalar component of a chiral supermultiplet is complex).

For example, to form the complex modulus field T, the radius-squared R^2 of the compact manifold is paired up with an imaginary part which is related to the antisymmetric tensor field b_{kl} (with k and l six-dimensional *compact* indices; hence the corresponding components are four-dimensional scalars). Similar interpretations apply to the other radii moduli, known as Kähler moduli. The gauge invariance of the antisymmetric tensor $(\delta b_{MN} = \partial_M \Lambda_N - \partial_N \Lambda_M)$ induces a Peccei–Quinn symmetry for $\mathrm{Im}\ T$ ($\mathrm{Im}\ T \to \mathrm{Im}\ T +$ constant) which has only derivative couplings, just like the axion. Hence the superpotential cannot depend on $\mathrm{Im}\ T$, and being analytic in the fields, cannot depend on T as a whole [377].

Through supersymmetry, the string dilaton ϕ is related to the antisymmetric tensor $b_{\mu\nu}$ (this time with four-dimensional indices). Together with a Majorana fermion, the dilatino, they form what is known as a linear supermultiplet L, which is real. The superpotential, being analytic in the fields, cannot depend on L. This is related again to the gauge invariance associated with the antisymmetric tensor. This in turn ensures that the superpotential cannot depend on ϕ.

The latter result may be interpreted from the point of view of standard nonrenormalization theorems [117, 283]. Indeed, since e^ϕ is the string coupling, it ensures that the superpotential is not renormalized, to all orders of string perturbation theory.

Before we proceed, let us be more explicit on the way the string dilaton and the T modulus appear in four dimensions. We work here in the string frame and the corresponding fundamental mass scale is the string scale M_s. In order to introduce the degree of freedom associated with the overall size of the compact manifold, we introduce the "breathing mode" e^σ through the compact space part of the metric:

$$g_{kl}(x,y) = e^{2\sigma(x)}g_{kl}^{(0)}(y), \quad k,l = 4,\ldots,9, \quad \int d^6 y \sqrt{|g^{(0)}|} = M_s^{-6}. \tag{10.102}$$

Thus the volume of the 6-dimensional compact manifold is $V_6 = \int d^6 y \sqrt{|g|} = M_s^{-6}\langle e^{6\sigma}\rangle$ and $\langle e^{2\sigma}\rangle$ measures R^2 in string units.

If we consider specifically the weakly coupled heterotic string, then the terms in (10.93) give, after compactification[13],

$$S = \int d^4 x \sqrt{|g^{(4)}|}\; \frac{1}{(2\pi)^7}\; e^{-2\phi+6\sigma} \left[M_s^2 \left(-R^{(4)} + 12 D^\mu \partial_\mu\sigma + 42\partial^\mu\sigma\partial_\mu\sigma - 4\partial^\mu\sigma\partial_\mu\phi\right) \right.$$
$$\left. -\frac{1}{4}\mathrm{Tr}F^{\mu\nu}F_{\mu\nu} \right]. \tag{10.104}$$

In terms of the real fields

$$s = \frac{1}{(2\pi)^7}e^{-2\phi+6\sigma}, \quad t = \frac{1}{(2\pi)^7}e^{2\sigma}, \tag{10.105}$$

the action reads, after integrating by parts,

$$S = \int d^4 x \sqrt{|g^{(4)}|}\; s\left[M_s^2 \left(-R^{(4)} + \frac{3}{2}\frac{\partial^\mu t\partial_\mu t}{t^2} - \frac{\partial^\mu s\partial_\mu s}{s^2}\right) - \frac{1}{4}\mathrm{Tr}F^{\mu\nu}F_{\mu\nu} \right]. \tag{10.106}$$

We conclude that

$$m_P^2 = 2\langle s\rangle M_s^2, \quad g^{-2} = \langle s\rangle. \tag{10.107}$$

The couplings of the s field are reminiscent of the Wess–Zumino terms which restore scale invariance through a dilaton field (see (A.268) of Appendix Appendix A).

---

[13]We use the following useful formula which is a generalization of (10.34): we start with a $(D = d+N)$-dimensional theory with metric $g_{MN}^{(D)}$ and scalar curvature $R^{(D)}$ and compactify N coordinates. Then writing

$$g_{\mu\nu}^{(D)} = e^{a\sigma(x^\mu)}g_{\mu\nu}^{(d)}, \quad \mu,\nu = 1,\ldots,d,$$
$$g_{kl}^{(D)} = -e^{b\sigma(x^\mu)}\delta_{kl}, \quad k,l = d+1,\ldots,d+N = D,$$

which defines the d-dimensional metric $g_{\mu\nu}^{(d)}$ (scalar curvature $R^{(d)}$), we have

$$e^{a\sigma}R^{(D)} = R^{(d)} - [a(d-1)+bN]\,D^\mu\partial_\mu\sigma - \left[\frac{a^2}{4}(d-1)(d-2) + \frac{Nab}{2}(d-2) + \frac{b^2}{4}N(N+1)\right]\partial^\mu\sigma\partial_\mu\sigma. \tag{10.103}$$

The sign of the kinetic term for s is not a problem because s is coupled to the spacetime curvature in the string frame metric $g_{\mu\nu}^{(4)}$. If we go to the Einstein frame by performing a Weyl transformation on the four-dimensional metric:

$$g_{\mu\nu}^{(4)} \equiv \frac{1}{2s}\frac{m_P^2}{M_S^2}g_{\mu\nu}, \tag{10.108}$$

the action takes the standard form

$$\mathcal{S} = \int d^4x \sqrt{|g|} \left[-\frac{1}{2}m_P^2 R + \frac{1}{4}\frac{\partial^\mu s \partial_\mu s}{s^2} + \frac{3}{4}\frac{\partial^\mu t \partial_\mu t}{t^2} - \frac{1}{4}\, s\, \mathrm{Tr} F^{\mu\nu} F_{\mu\nu} \right]. \tag{10.109}$$

The fields s and t appear to be the real parts of complex scalar fields S and T (see (10.89)) with Kähler potential

$$K(S,T) = -\ln\left(S + S^\dagger\right) - 3\ln\left(T + T^\dagger\right). \tag{10.110}$$

One readily checks that the corresponding kinetic terms (see (6.19) of Chapter 6) yield the kinetic terms for s and t just found.

The Lagrangian (10.109) is invariant under the group $SL(2, Z)$ of modular transformations

$$T \to \frac{aT - ib}{icT + d}, \quad ad - bc = 1, \quad a, b, c, d \in Z, \tag{10.111}$$

among which we recognize T-duality $(T \to 1/T)$[14]. Indeed, such a transformation corresponds to a Kähler transformation for K:

$$K \to K + F + \bar{F}, \quad F = 3\ln\left(icT + d\right). \tag{10.112}$$

Chiral supermultiplets describing matter may be added: as we have seen earlier, they arise from the components of the 10-dimensional gauge supermultiplet with compact space indices. In the simplest compactification scheme [375], the Kähler potential is then generalized into

$$K(S,T) = -\ln\left(S + S^\dagger\right) - 3\ln\left(T + T^\dagger - \sum_i \Phi^i \Phi^{i\dagger}\right), \tag{10.113}$$

where Φ^i are the matter fields. The modular transformations (10.111) must thus be complemented by

$$\Phi^i \to \frac{\Phi^i}{icT + d}. \tag{10.114}$$

Their interactions are described through the cubic superpotential

$$W = d_{ijk}\Phi^i \Phi^j \Phi^k. \tag{10.115}$$

[We recognize in the structure thus unravelled the no-scale supergravity discussed in Section 6.12 of Chapter 6.]

---

[14][The S field does not transform under T-duality because of the transformation law of the string dilaton (*cf.* (10.45)).]

The antisymmetric tensor field

We have seen in Section 10.1 that an antisymmetric tensor $b_{MN} = -b_{NM}$ is present among the massless modes of the closed string. There is a gauge invariance associated with such a tensor, namely

$$\delta b_{MN} = \partial_M \Lambda_N - \partial_N \Lambda_M. \tag{10.116}$$

The gauge invariant field strength correspondingly reads

$$h_{MNP} = \partial_M b_{NP} + \partial_N b_{PM} + \partial_P b_{MN}, \tag{10.117}$$

and the Lagrangian is simply (compare with a Yang–Mills field)

$$\mathcal{L} = \frac{1}{4} h^{MNP} h_{MNP}. \tag{10.118}$$

An antisymmetric tensor in D dimensions corresponds to $(D-2)(D-3)/2$ degrees of freedom[a]. If we restrict our attention to four dimensions, this gives a single degree of freedom. Indeed, an antisymmetric tensor field is equivalent on-shell to a pseudoscalar field.

In order to prove this equivalence, we start with the generalized action

$$S = \int d^4 x \left[\frac{1}{4} h^{\mu\nu\rho} h_{\mu\nu\rho} - \frac{1}{\sqrt{12}} \theta \epsilon^{\mu\nu\rho\sigma} \partial_\mu h_{\nu\rho\sigma} \right], \tag{10.119}$$

where $h_{\mu\nu\rho}$ is a general 3-index antisymmetric tensor and $\theta(x)$ a real scalar. This field θ plays the rôle of a Lagrange multiplier: its equation of motion simply yields $\epsilon^{\mu\nu\rho\sigma} \partial_\mu h_{\nu\rho\sigma} = 0$ which is the Bianchi identity, *i.e.* the necessary condition for $h_{\mu\nu\rho}$ to be considered as the field strength of a 2-index antisymmetric tensor (*cf.* (10.117)).

Alternatively, we may minimize with respect to $h_{\mu\nu\rho}$. This is easier to do after having performed an integration by parts on the second term. One obtains the Hodge duality relation

$$h^{\mu\nu\rho} = \frac{1}{\sqrt{3}} \epsilon^{\mu\nu\rho\sigma} \partial_\sigma \theta, \tag{10.120}$$

which establishes the equivalence. After replacement, the action (10.119) is simply the action of a free real scalar field.

---

[a]We first recall the counting for a vector field. Out of the D components A_M, 1 is fixed by the gauge condition $\partial^M A_M = 0$; we are left with the residual symmetry $A_M \to A_M - \partial_M \Lambda$ with $\Box \Lambda = 0$ which corresponds to one degree of freedom (just as a massless scalar field). Hence we find $D - 1 - 1 = D - 2$ degrees of freedom.

For the tensor b_{MN}, which has $D(D-1)/2$ components, we fix $D-1$ of them by the gauge condition $\partial^M b_{MN} = 0$ (note that the vector $t_N \equiv \partial^M b_{MN}$ is transverse: $\partial^N t_N = 0$); we are left with a residual symmetry (10.116) which satisfies $\partial^M (\partial_M \Lambda_N - \partial_N \Lambda_M) = 0$ *i.e.* the equation of motion of a massless vector field. Hence $D(D-1)/2 - (D-1) - (D-2) = (D-2)(D-3)/2$.

This equivalence is generalized to the corresponding supermultiplets: a linear supermultiplet is equivalent on-shell to a chiral supermultiplet. For example the linear supermultiplet with bosonic components s and $b_{\mu\nu}$ is equivalent to a chiral supermultiplet with scalar field S (see Section C.4 of Appendix C). In the case of the heterotic string, it turns out that the antisymmetric tensor transforms nontrivially under Yang–Mills gauge transformations:

$$\delta b_{MN} = -\frac{\kappa}{\sqrt{2}} \mathrm{Tr}\left(\alpha F_{MN}\right),\tag{10.121}$$

if $\delta A_M = -\partial_M \alpha$. We may introduce the 3-index antisymmetric tensor, called the Chern–Simons 3-form[a]

$$\omega_{MNP} = \mathrm{Tr}\left(A_{[M}\partial_N A_{P]} + \frac{2}{3}g A_{[M} A_N A_{P]}\right)\tag{10.122}$$

which satisfies

$$\partial_{[M}\omega_{NPQ]} = \mathrm{Tr}\left(F_{[MN}F_{PQ]}\right),\quad \delta\omega_{MNP} = \partial_{[M}\mathrm{Tr}\left(\alpha\partial_N A_{P]}\right).\tag{10.123}$$

The gauge invariant field strength of the antisymmetric tensor is then

$$\hat{h}_{MNP} \equiv 3\partial_{[M}b_{NP]} - \frac{\kappa}{\sqrt{2}}\omega_{MNP}.\tag{10.124}$$

This extra term plays a central rôle in the cancellation of gauge anomalies in 10 dimensions. Indeed, the hexagonal diagram of Fig. 10.12a, which is responsible of the anomaly in 10 dimensions (just as a triangular diagram is responsible of the anomaly in four dimensions, see Section A.6 of Appendix Appendix A), is cancelled by the diagram of Fig. 10.12b which represents the tree level exchange of a b_{MN} field. The vertex on the left-hand side originates from the kinetic term $\hat{h}^{MNP}\hat{h}_{MNP}$ which includes a term $\partial^{[M}b^{NP]}A_{[M}\partial_N A_{P]}$. The vertex on the right-hand side is associated with the counterterm introduced by [205]

$$\mathcal{S}_{\mathrm{GS}} = \int d^{10}x\, \epsilon^{M_1\cdots M_{10}} b_{M_1 M_2} \mathrm{Tr}\left(F_{M_3 M_4} F_{M_5 M_6} F_{M_7 M_8} F_{M_9 M_{10}}\right).\tag{10.125}$$

We note that the gauge completion of the 3-index field strength in (10.124) induces, after a Hodge duality transformation, a coupling of the pseudoscalar field θ to the gauge fields. More precisely, the kinetic term $\hat{h}^{MNP}\hat{h}_{MNP}/4$ induces in four dimensions

$$\int -\frac{\kappa}{2\sqrt{2}}\partial_{[\mu}b_{\nu\rho]}\,\mathrm{Tr}\left(A_{[\mu}\partial_\nu A_{\rho]} - \frac{2}{3}g A_{[\mu} A_\nu A_{\rho]}\right)$$

$$= \int -\frac{\kappa}{2\sqrt{6}}\epsilon^{\mu\nu\rho\sigma}\partial_\sigma\theta\,\mathrm{Tr}\left(A_{[\mu}\partial_\nu A_{\rho]} - \frac{2}{3}g A_{[\mu} A_\nu A_{\rho]}\right)$$

---

[c]For N indices between brackets, we average (factor $1/N!$) the sum of the terms obtained by permutations of the N indices, with ± 1 for even or odd perturbations.

$$= \int \frac{\kappa}{2\sqrt{6}} \, \theta \, \mathrm{Tr} F^{\mu\nu} \tilde{F}_{\mu\nu}. \tag{10.126}$$

This is a Wess–Zumino term of the type encountered in anomaly cancellation mechanisms. Since this is reminiscent of the way the axion field couples to the gauge fields, the field $a \equiv \sqrt{2/3}\theta$ is called the string axion. The S field introduced earlier is $S = s + ia$ and the action term (10.126) is simply a supersymmetric completion of the last term in (10.109).

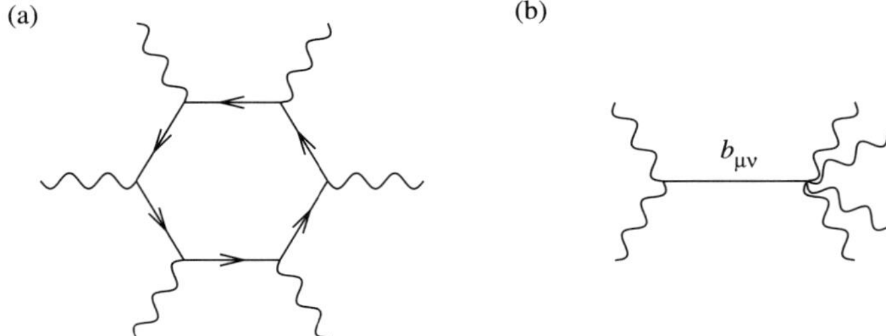

Fig. 10.12 Hexagonal anomaly diagram (a) and antisymmetric tensor field exchange tree diagram (b).

In realistic compactifications, the situation is of course more involved than the one that we have presented. Let us illustrate this on the example of Calabi–Yau compactification. Matter fields appear in representations **27** and $\overline{\mathbf{27}}$ of gauge group E_6. As can be checked from Tables 10.3 and 10.4, there are as many T fields, *i.e.* radius or Kähler moduli, as there are $\overline{\mathbf{27}}$ representations (their number is given by the Hodge number $h_{1,1}$).

There exists another class of moduli (noted $U^{(i)}$ in Table 10.4) which describes the complex structure of the compact manifold. Let us illustrate the difference on the example of a torus. We may represent a torus in the complex plane by the square of unit length $(0, 1, i, 1 + i)$ with opposite sides identified. Changing the radius of this torus amounts to multiply all its dimensions by a factor λ. On the other hand, the case where only the imaginary direction is dilated cannot be described in the complex plane by a holomorphic transformation $z \to \lambda z$: it corresponds to a change in the complex structure of the torus. In Calabi–Yau compactification, there are as many complex structure moduli as there are **27** ($h_{2,1}$).

[One may show in this context that the same holomorphic function determines the superpotential of the matter fields and the Kähler potential of the moduli[15]. More

---

[15]Such a relation between superpotential and Kähler potential is reminiscent of a $N = 2$ supersymmetry which lies in the background of this type of compactification.

explicitly, both the Kähler potential for the T fields and the superpotential W for the matter fields $\bar{\Phi}$ in $\overline{\mathbf{27}}$ are fixed by the same function $\mathcal{F}_1(T)$:

$$K_1(T) = -\ln Y_1, \quad W = \frac{1}{3}\partial_\ell \partial_m \partial_n \mathcal{F}_1 \bar{\Phi}^\ell \bar{\Phi}^m \bar{\Phi}^n,$$

$$Y_1 = \sum_{\ell=1}^{h_{1,1}} \left(\partial_\ell \mathcal{F}_1 + \partial_\ell \bar{\mathcal{F}}_1\right)\left(T^\ell + T^{\ell\dagger}\right) - 2\left(\mathcal{F}_1 + \bar{\mathcal{F}}_1\right). \tag{10.127}$$

Note that for $h_{1,1} = 1$ and $\mathcal{F}_1 = \lambda T^3$, one recovers the expressions found above in (10.110) and (10.115)[16].

A similar expression exists between the Kähler potential $K_2(U)$ for the U moduli and the superpotential for the matter fields in $\mathbf{27}$. Finally, the normalization of the kinetic terms for the matter fields can be expressed in terms of $K_1(T)$ and $K_2(U)$.]

10.4.3 Symmetries

One remarkable property of string theories is the absence of continuous global symmetries: any continuous symmetry must be a gauge symmetry. This is certainly a welcome property for a theory of gravity since it is believed that quantum gravity effects (wormholes) tend to break any kind of global symmetry (continuous or discrete) [263]. The underlying reason for this property is that there exists a deep connection between global symmetries on the world-sheet and local symmetries in space-time[17].

Discrete symmetries may also be viewed in string theory as local, in the sense that, at certain points of moduli space, they give rise to full blown local gauge theories. In other words they may be seen as resulting from the breakdown of continuous local theories as the field measuring the departure from the point of enhanced symmetry becomes nonvanishing. Such symmetries play an important rôle in taking care of the dangerous baryon and lepton violating interactions discussed in Section 5.4 of Chapter 5. They may be of the general matter parity type, *i.e.* Z_N, $N > 2$, and could be R-symmetries [232].

Returning to continuous gauge symmetries, we may find a grand unified gauge group but the use of Wilson lines could also break it directly to a product of simple groups, in which case there is no grand unified symmetry but just partial unification. As we will see in the next section, there remains usually a unification of the gauge couplings.

Let us illustrate this on the example of the E_6 gauge symmetry obtained by Calabi–Yau compactification of the heterotic string theory. We assume, as in Section 10.2.3 (you may have interest to browse through the box "Topological gauge symmetry breaking"), that the compact manifold is of the form $K = K_0/G$ with K_0 simply connected and G a finite group of order n. Then, a Wilson line corresponds to a nontrivial pure gauge configuration such that $U(\gamma) \cong \exp i \int_\gamma A_k^a T^a dy^k \neq 1$ for a noncontractible loop γ. Since $A_k = A_k^a T^a$ is in the adjoint representation of the algebra E_6, $U(\gamma)$ is in the adjoint representation of the group E_6. The situation is

---

[16]In this case, the index runs over the component of the single $\overline{\mathbf{27}}$.

[17]Using complex variables z and $\bar{z}$ to parametrize the world-sheet, we see that a world-sheet Noether current j^z allows us to construct a spacetime operator $j^z \partial_z X^\mu$ which may be associated with a massless gauge field.

therefore very similar to the case of a Higgs field ϕ in the adjoint representation of the gauge group ($U \cong e^{i\phi}$) with a nonzero vacuum expectation value[18]. Indeed, in much the same way, E_6 is broken to the gauge group that commutes with $\langle U \rangle$. This is sometimes called the Hosotani mechanism [226].

Gauge symmetry is broken to the gauge group that commutes with U. Since we want the residual symmetry to contain at least $SU(3)_c \times SU(2)_L \times U(1)_Y$, this imposes constraints on the form of U.

To derive them, it is easier to consider the $SU(3)_c \times SU(3)_L \times SU(3)_{\mathrm{R}}$ subgroup of E_6 considered in Section 9.5 of Chapter 9 and to write U as

$$U = U_c \otimes U_L \otimes U_{\mathrm{R}} \tag{10.128}$$

where $U_{c,L,R}$ are 3×3 matrices, elements of the gauge *groups* $SU(3)_{c,L,R}$. To fully fix our notation, we have to say which $SU(2)$ subgroup of $SU(3)$ is $SU(2)_L$. We will therefore suppose that the t_L^3 element of the *algebra* $SU(2)_L$ is

$$t_L^3 = \frac{1}{2} \begin{pmatrix} 1 & & \\ & -1 & \\ & & 0 \end{pmatrix}. \tag{10.129}$$

Now, from the fact that all the generators of $SU(3)_c \times SU(2)_L$ commute with U, one easily infers that

$$U_c = \begin{pmatrix} \eta & & \\ & \eta & \\ & & \eta \end{pmatrix}, \quad U_L = \begin{pmatrix} \beta & & \\ & \beta & \\ & & \beta^{-2} \end{pmatrix} \tag{10.130}$$

where $\det U_c = \eta^3 = 1$. For simplicity we will take in the following $\eta = 1$.

The case of $U(1)_Y$ is a little less straightforward. Remember that $y = 1/3$, $4/3$, $-2/3$ for $(u, d)_L$, u_{R} and d_{R} respectively. This means that the (algebra) generator corresponding to y reads

$$[Y] = (0)_c \oplus \begin{pmatrix} 1/3 & & \\ & 1/3 & \\ & & -2/3 \end{pmatrix}_L \oplus \begin{pmatrix} 4/3 & & \\ & -2/3 & \\ & & -2/3 \end{pmatrix}_{\mathrm{R}}. \tag{10.131}$$

But any matrix that commutes with $[Y]$ will commute with the two matrices (notation follows from Section 9.5 of Chapter 9):

$$[Y_L] = (0)_c \oplus \begin{pmatrix} 1/3 & & \\ & 1/3 & \\ & & -2/3 \end{pmatrix}_L \oplus (0)_{\mathrm{R}},$$

$$[Q_{\mathrm{R}}] = (0)_c \oplus (0)_L \oplus \begin{pmatrix} 2/3 & & \\ & -1/3 & \\ & & -1/3 \end{pmatrix}_{\mathrm{R}}. \tag{10.132}$$

Therefore the gauge symmetry is broken by topological breaking at most to $SU(3)_c \times SU(2)_L \times U(1)_{Y_L} \times U(1)_{Q_{\mathrm{R}}}$. One combination of the two $U(1)$ charges yields the weak hypercharge $y = y_L + 2q_{\mathrm{R}}$; the orthogonal combination is $y_\eta = q_{\mathrm{R}} - 2y_L$.

---

[18]Except that, since $[U(\gamma)]^n = 1$, the eigenvalues of U are quantized.

Thus, an extra $U(1)_\eta$ appears in the gauge symmetry pattern at the scale of compactification. It remains to be seen whether this extra $U(1)$ remains unbroken down to the low energies where it could be observed.

Since U must commute with $[Y_L]$ and $[Y_R]$, we find

$$U_\mathrm{R} = \begin{pmatrix} \gamma & \\ & V \end{pmatrix} \tag{10.133}$$

where V is 2×2 matrix and $\gamma \det V = 1$. If $\Pi_1(K) = G$ is commutative, then all matrices U_g, $g \in G$ must commute and V is diagonal. In this case

$$U = (1)_c \otimes \begin{pmatrix} \beta & & \\ & \beta & \\ & & \beta^{-2} \end{pmatrix} \otimes \begin{pmatrix} \gamma & & \\ & \delta & \\ & & \epsilon \end{pmatrix}_\mathrm{R} , \qquad \gamma\delta\epsilon = 1. \tag{10.134}$$

For example if $G \cong \mathbb{Z}_n$, $\beta^n = \gamma^n = \delta^n = \epsilon^n = 1$.

Let us turn to a specific example where $G \cong \mathbb{Z}_3$. To build U_R, it is easy to convince oneself that only two cases arise: (I) all its entries are equal, (II) all its entries are different.

In case (I), we have

$$U = (1)_c \otimes \begin{pmatrix} \alpha_2 & & \\ & \alpha_2 & \\ & & \alpha_2 \end{pmatrix}_L \otimes \begin{pmatrix} \widetilde{\alpha}_2 & & \\ & \widetilde{\alpha}_2 & \\ & & \widetilde{\alpha}_2 \end{pmatrix}_\mathrm{R} \qquad \alpha_2, \widetilde{\alpha}_2 \in \mathbb{Z}_3. \tag{10.135}$$

To see which gauge symmetry remains unbroken, we consider the generators in the adjoint representation of E_6. Their decomposition under $SU(3)_c \times SU(3)_L \times SU(3)_\mathrm{R}$ was given in Chapter 9, equation (9.114):

$$\mathbf{78} = (\mathbf{8},\mathbf{1},\mathbf{1}) + (\mathbf{1},\mathbf{8},\mathbf{1}) + (\mathbf{1},\mathbf{1},\mathbf{8}) + (\mathbf{3},\mathbf{3},\mathbf{3}) + (\overline{\mathbf{3}},\overline{\mathbf{3}},\overline{\mathbf{3}}). \tag{10.136}$$

Clearly, all the generators in the adjoint representation of $SU(3)_c \times SU(3)_L \times SU(3)_\mathrm{R}$ (the first three terms) are invariant ($[T, U] = 0$). For the remaining ones, we have

$$U(\mathbf{3},\mathbf{3},\mathbf{3}) = 1 \times \alpha_2 \times \widetilde{\alpha}_2 \ (\mathbf{3},\mathbf{3},\mathbf{3})$$

$$U(\overline{\mathbf{3}},\overline{\mathbf{3}},\overline{\mathbf{3}}) = 1 \times \overline{\alpha_2} \times \overline{\widetilde{\alpha}_2} \ (\overline{\mathbf{3}},\overline{\mathbf{3}},\overline{\mathbf{3}}).$$

Therefore if $\widetilde{\alpha}_2 \neq \alpha_2^{-1}$, E_6 is broken to $SU(3)_c \times SU(3)_L \times SU(3)_\mathrm{R}$ and if $\widetilde{\alpha}_2 = \alpha_2^{-1}$, E_6 remains unbroken.

In case (II), all entries are different in U_R; for example

$$U = (1)_c \otimes \begin{pmatrix} \alpha & & \\ & \alpha & \\ & & \alpha \end{pmatrix}_L \otimes \begin{pmatrix} \alpha & & \\ & \alpha^{-1} & \\ & & 1 \end{pmatrix}_\mathrm{R} . \tag{10.137}$$

Among the generators of $SU(3)_c \times SU(3)_L \times SU(3)_R$ only $(\mathbf{8}, \mathbf{1}, \mathbf{1})$, $(\mathbf{1}, \mathbf{8}, \mathbf{1})$ and the two diagonal generators λ_3, λ_8 in $(\mathbf{1}, \mathbf{1}, \mathbf{8})$ are invariant. Among the rest, only $(\mathbf{3}_c, \mathbf{3}_L, t_R^3 = -1/2)$ and $(\overline{\mathbf{3}}_c, \overline{\mathbf{3}}_L, t_R^3 = +1/2)$ are invariant; for example $U(\mathbf{3}_c, \mathbf{3}_L, t_R^3 = -1/2) = 1 \times \alpha \times \alpha^{-1}(\mathbf{3}_c, \mathbf{3}_L, t_R^3 = -1/2)$. One can check that these $8 + 8 + 2 + 9 + 9 = 36$ generators are the generators of $SU(6) \times U(1)$. This last example is interesting because it shows that although we chose to represent U under the maximal subgroup $SU(3)^3$ of E_6, it does not mean that the residual symmetry has to be a subgroup of $SU(3)^3$.

We have seen that in the case of a compactification on a Calabi–Yau manifold matter fields fall into full representations $\mathbf{27}$ and $\overline{\mathbf{27}}$ of E_6, the net number of families being fixed by the Euler characteristics of the manifold. This is modified when E_6 is broken by topological gauge symmetry breaking: matter fields fall in representations of the residual gauge symmetry group – the subgroup of E_6 that commutes with $\langle U \rangle$. Indeed we saw that matter fields survive compactification if they obey the twisted boundary conditions (10.62).

10.4.4 Gauge coupling unification

There is no guaranteed unification of the gauge couplings in string theory. For example in type I models the couplings of gauge groups which correspond to different stacks of D-branes have no a priori reason to be equal. Because of the apparent success of the gauge coupling unification one might therefore be inclined to search for string models with gauge coupling unification.

One should stress at this point that string theories yield more possibilities of gauge coupling unification than standard grand unified theories. In fact the relation (10.107) may be generalized to

$$\frac{1}{g_i^2} = k_i \langle s \rangle = k_i \frac{m_P^2}{2M_S^2} \equiv \frac{k_i}{g_S^2}, \qquad (10.138)$$

where k_i is known as a Kač–Moody level. Crudely speaking, in the case of a nonabelian symmetry, it represents the relative strength between the gravitational coupling and the trilinear self-coupling of the gauge bosons[19]; it is an integer. In the case of an abelian symmetry, k_i is simply a real number necessary to account for the normalization fixed by (10.138). Thus the unification condition reads

$$k_1 g_1^2(M) = k_2 g_2^2(M) = k_3 g_3^2(M). \qquad (10.140)$$

The successes of coupling unification do impose to take $k_2 = k_3$.

---

[19][This may be expressed more quantitatively through the operator product expansion between two world-sheet currents associated with the gauge symmetry:

$$j_i^a(z) j_i^b(w) \sim k_i \frac{\vec{\alpha}_i}{2} \frac{\delta^{ab}}{(z-w)^2} + \frac{iC^{abc}}{z-w} j_i^c(w) + \cdots \qquad (10.139)$$

where $\vec{\alpha}_i$ is the longest root. The double pole term corresponds to a gravitational coupling whereas the single pole term is associated with the trilinear self-coupling.]

There are reasons to go beyond $k = 1$. Indeed, the choice $k = 1$ puts some restrictions on the possible representations. Such restrictions are welcome in the case of the Standard Model since they yield as only possible representations $\mathbf{1}$ and $\mathbf{2}$ for $SU(2)$ and $\mathbf{1}$, $\mathbf{3}$ and $\bar{\mathbf{3}}$ for $SU(3)$. But, in the case of grand unified groups, they limit the representations (e.g. $\mathbf{1}$, $\mathbf{5}$ and $\bar{\mathbf{5}}$, $\mathbf{10}$ and $\overline{\mathbf{10}}$ for $SU(5)$, $\mathbf{1}$, $\mathbf{27}$ and $\overline{\mathbf{27}}$ for E_6) in such a way as to prevent further breaking. In this case one should go to higher Kač–Moody level or resort to partial unification.

Regarding the running of couplings at scales below M, it is important to note the following: since string theory amplitudes are finite, there are no ultraviolet divergences and thus no renormalization group type of running. The couplings that we have defined above are couplings of the effective field theory which describes the interactions of the *massless* string modes; this effective theory has ultraviolet divergences and thus running couplings. Thus, in a sense, the running of these couplings is associated with the *infrared* properties of the underlying string theory.

The effective theory is obtained by integrating out the massive string modes. One thus expects some corresponding threshold contributions at the string scale:

$$\frac{16\pi^2}{g_i^2(\mu)} = k_i \frac{16\pi^2}{g_S^2} + b_i \ln \frac{M_S^2}{\mu^2} + \Delta_i, \tag{10.141}$$

where the second (resp. third) term on the right-hand side is due to massless (resp. massive) string modes running in loops. The massive string threshold corrections Δ_i depend on the precise distribution of the massive string spectrum; they are generally moduli-dependent.

They have been computed in some explicit cases. We will take the example of four-dimensional string models which have $N = 2$ spacetime supersymmetry [121,247]: they are obtained through the toroidal compactification of six-dimensional string models with $N = 1$ spacetime supersymmetry. One obtains

$$\Delta_i(T, U) = -b_i \log\left[(\mathrm{Re}T) \, |\eta\,(iT)|^4 \, (\mathrm{Re}U) \, |\eta\,(iU)|^4\right] + b_i X, \tag{10.142}$$

where X is a numerical constant, T, U are the complex moduli that parametrize the two-dimensional compactification from six dimensions, and the Dedekind η function is defined as: $\eta(\tau) = e^{i\pi\tau/12} \prod_{n=1}^{\infty} \left(1 - e^{2i\pi n\tau}\right)$. Under the modular transformation (10.111), $\eta(iT)$ transforms into $(icT + d)^{1/2}\, \eta(iT)$; threshold corrections are thus invariant under modular transformations for the T (and U) field.

We note that, in the limit of large moduli ($T \to \infty$), we have $\log\left[(\mathrm{Re}T) \, |\eta\,(iT)|^4\right] \sim -\frac{\pi}{3}\mathrm{Re}T$ and

$$\Delta_i \sim \frac{\pi}{3} b_i \left(\mathrm{Re}T + \mathrm{Re}U\right), \tag{10.143}$$

which thus behaves like the radius squared. Hence, in the decompactification limit, such threshold corrections may become large (in the $N = 2$ case considered here, they may in fact be absorbed into the definition of the unification scale[20]). However, in realistic cases, moduli are found at or near the self-dual point $T = 1$ where threshold corrections are small. In the context of heterotic string models for example, it is thus difficult to reconcile through string threshold corrections the factor of 20 between the string scale (10.95) and the grand unification scale M_U obtained in Chapter 9.

10.4.5   Axions and pseudo-anomalous $U(1)$ symmetries

The coupling of the string axion to the gauge fields allows in many string models the presence of a seemingly anomalous abelian symmetry, which we note here $U(1)_X$. We have seen in (10.126) that the introduction of the Green-Schwarz [205] counterterm leads in four dimensions to a coupling of the string axion to the gauge fields:

$$\mathcal{L} = -\frac{1}{4}s(x)\sum_i k_i F_{i\mu\nu}F^i_{\mu\nu} + \frac{1}{4}a(x)\sum_i k_i F_{i\mu\nu}\tilde{F}^i_{\mu\nu} + \cdots \tag{10.144}$$

where $s(x)$ and $a(x)$ are the dilaton and the axion fields, and k_i is the Kač–Moody level of the corresponding gauge group G_i (taken to be 1 in (10.126)).

The anomaly cancellation mechanism is based on this coupling. Performing a $U(1)_X$ gauge transformation: $A^X_\mu(x) \to A^X_\mu(x) - \partial_\mu\theta(x)$ yields

$$\delta\mathcal{L} = \frac{1}{8}\sum_i C_i\theta(x)F^{i\mu\nu}\tilde{F}^i_{\mu\nu}, \tag{10.145}$$

where C_i is the mixed $U(1)_X G_i G_i$ anomaly coefficient. We can complement this with a Peccei–Quinn transformation of the axion: $a(x) \to a(x) - \theta(x)\delta_{\mathrm{GS}}/2$ where δ_{GS} is a number,

$$\delta'\mathcal{L} = -\frac{1}{8}\delta_{\mathrm{GS}}\theta\sum_i k_i F^{i\mu\nu}\tilde{F}^i_{\mu\nu}. \tag{10.146}$$

The total transformation is an invariance of the Lagrangian if $C_i = \delta_{\mathrm{GS}}k_i$. Hence the necessary condition for the cancellation of anomalies with the Green–Schwarz counterterm is

$$\frac{C_1}{k_1} = \frac{C_2}{k_2} = \frac{C_3}{k_3} = \frac{C_X}{k_X} = \delta_{\mathrm{GS}}. \tag{10.147}$$

Such a symmetry is often present in string models. As we will see in the next chapters it may play an important phenomenological rôle.

---

[20]This is not so in a more general $N = 1$ case because b_i in the formulas above should then be replaced by some $b_{ia}^{(N=2)}$ which no longer coincides with the beta function coefficient: $b_i = b_i^{(N=1)} + \sum_a b_{ia}^{(N=2)}$.

Couplings of the pseudo-anomalous $U(1)$

We use the superfield formalism described in Appendix C to obtain the bosonic couplings of a pseudo-anomalous $U(1)_X$ gauge supermultiplet [116]. We note V_X the gauge vector superfield and $S = s + ia$ the dilaton chiral superfield. The usual sigma model term

$$\mathcal{L}_S = \int d^4\theta K, \tag{10.148}$$

with $K = -\ln\left(S + S^\dagger\right)$ is not invariant under the nonanomalous symmetry discussed above: $V_X \to V_X + i\left(\Lambda - \Lambda^\dagger\right)$, $S \to S + i\delta_{\rm GS}\Lambda$. The obvious modification is to take

$$K = -\ln\left(S + S^\dagger - \delta_{\rm GS}V_X\right). \tag{10.149}$$

Adding the standard kinetic term for the gauge fields

$$\mathcal{L}_V = \int d^2\theta \left[\frac{1}{4}k_X S W^\alpha W_\alpha + {\rm h.c.}\right], \tag{10.150}$$

gives explicitly the following terms:

$$\begin{aligned}
&\mathcal{L}_S + \mathcal{L}_V \\
&= \frac{1}{4s^2}\left(\partial^\mu s\partial_\mu s + \partial^\mu a\partial_\mu a\right) - \frac{\delta_{\rm GS}}{4s^2}A_X^\mu \partial_\mu a + \frac{\delta_{\rm GS}}{4s}D_X \\
&\quad -\frac{\delta_{\rm GS}}{16s^2}A_X^\mu A_{X\mu} - \frac{1}{4}k_X s F_X^{\mu\nu}F_{X\mu\nu} + \frac{1}{4}k_X a F_X^{\mu\nu}\tilde{F}_{X\mu\nu} + \frac{1}{2}k_X s D_X^2.
\end{aligned} \tag{10.151}$$

We note the presence of a term which is a remnant of the Green–Schwarz counterterm (10.125) in four dimensions (by integration by parts $\epsilon_{\mu\nu\rho\sigma}b^{\mu\nu}F_X^{\rho\sigma} \sim \epsilon_{\mu\nu\rho\sigma}h^{\mu\nu\rho}A_X^\sigma \sim \partial_\sigma a A_X^\sigma$) as well as a mass term for the gauge fields: the pseudo-anomalous gauge symmetry is broken by the gauge anomalies. We note that the corresponding mass, being of order $1/s = g^2$, is a loop effect.

Adding a standard D-term coupling $D_X\left(\sum_i x_i\Phi_i^\dagger\Phi_i\right)$, the full D-term Lagrangian reads

$$\begin{aligned}
\mathcal{L}_D = \frac{1}{2}k_X s &\left[D_X + \frac{1}{k_X s}\left(\sum_i x_i\Phi_i^\dagger\Phi_i + \frac{\delta_{\rm GS}}{4s}\right)\right]^2 \\
&- \frac{1}{2k_X s}\left(\sum_i x_i\Phi_i^\dagger\Phi_i + \frac{\delta_{\rm GS}}{4s}\right)^2.
\end{aligned} \tag{10.152}$$

Once we solve for D_X, we obtain a scalar potential in the form of a D-term:

$$V = \frac{1}{2}g_X^2\left(\sum_i x_i\Phi_i^\dagger\Phi_i + \xi_X\right)^2, \quad \xi_X \equiv \frac{1}{4}k_X g_X^2\delta_{\rm GS}m_P^2 = \frac{1}{2}\delta_{\rm GS}M_s^2, \tag{10.153}$$

> where we have have restored the Planck scale and used (10.107) and (10.138):
> $1/s = k_X g_X^2$. The presence of the field-dependent Fayet–Iliopoulos term usually induces a nonzero vacuum expectation value for one or more field of x charge opposite to $\delta_{\rm GS}$. The scale $\sqrt{|\xi_X|}$ determines the energy scale at which the $U(1)_X$ symmetry is broken. A string computation gives in the context of the heterotic string [13]
>
> $$\delta_{\rm GS} = \frac{1}{192\pi^2}\, {\rm Tr} X, \tag{10.154}$$
>
> which is of the form C_g/k_g as in (10.147), C_g being the mixed gravitational anomaly proportional to ${\rm Tr} X$. In this case, the scale $\sqrt{|\xi_X|}$ is at most one order of magnitude smaller than M_s.
>
> We finally note that the dilaton supermultiplet consists of the dilaton, dilatino, and antisymmetric tensor and, strictly speaking, fits into a linear supermultiplet (dual to the chiral supermultiplet S used above). The corresponding real linear superfield L is introduced in Section C.4 of Appendix C^a. The Green–Schwarz counterterm then takes in this formulation the following form
>
> $$\mathcal{L}_{\rm GS} = \frac{1}{2}\delta_{\rm GS} \int d^4\theta L V_X = \frac{1}{4}\delta_{\rm GS}\ell D + \frac{1}{24}\epsilon_{\mu\nu\rho\sigma} A_X^\mu \hat{h}^{\nu\rho\sigma} \tag{10.155}$$
>
> where $\hat{h}^{\nu\rho\sigma}$ is defined in (10.124).
>
> ---
>
> ^{a}More precisely, the linear multiplet $\hat{L}$ that we must consider here includes Chern–Simons forms, as discussed in Section 10.4.2 and satisfies the generalized constraints $D^2\hat{L} = -{\rm Tr}\bar{W}_{\dot\alpha}\bar{W}^{\dot\alpha}$ and $\bar{D}^2\hat{L} = -{\rm Tr}W^\alpha W_\alpha$.

The set of relations (10.147) allowed L. [234] to relate the value of the Weinberg angle to the mixed anomaly coefficients of the anomalous $U(1)$. Indeed, using (10.140), one finds:

$$\tan^2\theta_W(M) = \frac{g_1^2(M)}{g_2^2(M)} = \frac{k_2}{k_1} = \frac{C_2}{C_1}. \tag{10.156}$$

We finally note that, in the context of type I or type IIB strings, pseudo-anomalous $U(1)$ symmetries also appear (the corresponding axion field is not the string axion but may be provided by the imaginary part of moduli fields). The scale is no longer restricted to be of the order of the string scale and may be much smaller. Moreover, there may be several such symmetries in a given string model.

10.4.6 Supersymmetry breaking

As for any supersymmetric scenario, supersymmetry breaking is a key issue in string models. A favoured mechanism is gaugino condensation in a hidden sector which we have discussed in some details in subsection 7.4.2. But gaugino condensation is formulated in the context of the effective field theory and is not strictly speaking of a stringy nature (although it is largely motivated by string models).

Other mechanisms are more directly relevant to the string context, in the sense that identifying them within the experimental spectrum of supersymmetric particles would be a clear sign of some of the more typical aspects of string models. We will discuss briefly here the Scherk–Schwarz mechanism, the role of the pseudo-anomalous $U(1)$ and fluxes.

Scherk–Schwarz mechanism

This mechanism, proposed by Scherk and Schwarz [330], makes use of the symmetries of the higher dimensional theory to break supersymmetry through the boundary conditions imposed on fields in the compact dimensions. These symmetries are symmetries that do not commute with supersymmetry, e.g. R-symmetries or fermion number $(-1)^F$.

Let us illustrate this on the example of a single compact dimension of radius R: $0 \le y \le 2\pi R$. We consider bosonic and fermionic fields $\Phi_i(x^\mu, y)$ which obey the following boundary conditions:

$$\Phi_i(x^\mu, 2\pi R) = U_{ij}(\omega)\Phi_j(x^\mu, 0) \; , \tag{10.157}$$

where the matrix U is associated with a symmetry that does not commute with supersymmetry (and is thus different for bosons and fermions). The corresponding expansion is thus (compare with (10.38))

$$\Phi_i(x^\mu, y) = U_{ij}(\omega, y/R) \sum_{n \in Z} \Phi_{jn}(x^\mu) e^{iny/R} \; . \tag{10.158}$$

Scherk and Schwarz take $U = \exp(\omega M y/R)$ where M is an anti-Hermitian matrix. Kinetic terms in the compact direction generate fermion-boson splittings of order ω/R.

Let us apply this to the case of one massless hypermultiplet, introduced in Section 4.4.2 of Chapter 4 [151, 152]. As can be seen from the Lagrangian (4.43)

$$\mathcal{L} = \partial^\mu \phi^{i*} \partial_\mu \phi^i + i\bar{\Psi}\gamma^\mu \partial_\mu \Psi \; , \tag{10.159}$$

there is a R-symmetry that leaves the fermion field Ψ invariant and rotates the two complex scalar fields. We thus choose $M = i\sigma^2$. The decomposition (10.158) then gives the following four-dimensional Lagrangian

$$\mathcal{L} = \sum_n \left[i\psi_n^1 \sigma^\mu \partial_\mu \bar{\psi}_n^1 + i\psi_n^2 \sigma^\mu \partial_\mu \bar{\psi}_n^2 + \partial^\mu \phi_n^{1*} \partial_\mu \phi_n^1 + \partial^\mu \phi_n^{2*} \partial_\mu \phi_n^2 \right.$$
$$\left. -\frac{n}{R} \left(\psi_n^1 \psi_n^2 + \bar{\psi}_n^1 \bar{\psi}_n^2\right) - \frac{n^2 + \omega^2}{R^2} \left(|\phi_n^1|^2 + |\phi_n^1|^2\right) \right] \; , \tag{10.160}$$

where we have introduced two-component spinors $\Psi_n \equiv \begin{pmatrix} \psi_n^1 \\ \bar{\psi}_n^2 \end{pmatrix}$. We see that this mechanism generates soft masses of order ω/R.

Supergravity versions of this mechanism may be worked out using for example the string dilaton and Kähler moduli (see for example [126]). They show that, although in the strict sense supersymmetry is broken explicitly, Scherk–Schwarz breaking is in many ways similar to an F-type spontaneous breaking.

Pseudo-anomalous U(1)

Pseudo-anomalous $U(1)_X$ symmetries may play a significant rôle in supersymmetry breaking. Since they have mixed anomalies with the other gauge symmetries –those of the Standard Model as well as of the hidden sector–, it is not surprising that the whole issue of supersymmetry breaking through gaugino condensation is modifed in such models. Moreover, because in the Green–Schwarz mechanism all the mixed anomalies are non-vanishing and proportional to one another, there must exist fields charged under $U(1)_X$ in the observable as well as in the hidden sector. The $U(1)_X$ gauge symmetry thus serves as a messenger interaction competitive with the gravitational interaction.

We will give an explicit example to stress the modifications that the presence of such an anomalous $U(1)_X$ symmetry is bringing to the scenario of a dynamical supersymmetry breaking through gaugino condensation.

The model that we consider [33] is an extension of the model of Section 8.4.1: supersymmetric $SU(N_c)$ with $N_f < N_c$ flavors. Quarks Q_i in the fundamental of $SU(N_c)$ have $U(1)_X$ charge q and antiquarks $\bar{Q}^i$ in the antifundamental of $SU(N_c)$ have charge $\tilde{q}$.

Since we want to avoid $SU(N_c)$ breaking in the $U(1)_X$ flat direction, we require that the charges q and $\tilde{q}$ are positive. We then need at least one field of negative charge in order to cancel the D-term. For simplicity we introduce a single field ϕ of $U(1)_X$ charge normalized to -1.

We write the classical Lagrangian compatible with the symmetries in the superfield language of Appendix C. The reader interested only in the phenomenological results may go directly to equation (10.171).

[We write $\mathcal{L} = \mathcal{L}_{\text{kin}} + \mathcal{L}_{\text{couplings}}$, where we assume a flat Kähler potential for the matter fields and (10.149) for the dilaton field S:

$$\mathcal{L}_{\text{kin}} = \int d^4\theta \left[Q^+ e^{2qV_X + V_N} Q + \bar{Q} e^{2\tilde{q}V_X - V_N} \tilde{Q}^+ + \phi^+ e^{-2V_X} \phi \right]$$
$$+ \int d^4\theta\, K + \int d^2\theta \frac{1}{4} S \left[k_X W^\alpha W_\alpha + k_N \text{Tr} \mathcal{W}^\alpha \mathcal{W}_\alpha \right] \qquad (10.161)$$

$$\mathcal{L}_{\text{couplings}} = \int d^2\theta \left(\frac{\phi}{M_P} \right)^{q+\tilde{q}} m^i{}_j Q_i \bar{Q}^j + h.c. \,, \qquad (10.162)$$

where k_X and k_N are the respective Kač–Moody levels of $U(1)_X$ and $SU(N_c)$.

The mixed anomaly $U(1)_X [SU(N_c)]^2$ which fixes, through (10.147), all the mixed anomalies in the model is given by

$$C_N = \frac{1}{4\pi^2} N_f (q + \tilde{q}) = k_N \delta_{GS}. \qquad (10.163)$$

We thus require $q + \tilde{q} > 0$, which in turn justifies the presence of the superpotential term (10.162).

The two scales present in the problem are:
- the scale at which the anomalous $U(1)_X$ symmetry is broken which is set by

$$\sqrt{\xi} = \frac{1}{2} k_X^{1/2} g_X \delta_{GS}^{1/2} M_P \,. \qquad (10.164)$$

- the scale at which the gauge group $SU(N_c)$ enters in a strong coupling regime:

$$\Lambda = M_p e^{-8\pi^2 k_N S/(3N_c - N_f)} \,, \tag{10.165}$$

where we have used (8.46) with $b = (3N_c - N_f)$.

We suppose that $\Lambda \ll \sqrt{\xi}$. Below the scale Λ, the appropriate degrees of freedom are the field ϕ and the mesons $M_i{}^j = Q_i \bar{Q}^j$. The effective superpotential is fixed uniquely by the global symmetries as in (8.53):

$$W = (N_c - N_f) \left(\frac{\Lambda^{3N_c - N_f}}{\det M} \right)^{\frac{1}{N_c - N_f}} + \left(\frac{\phi}{M_P} \right)^{q + \tilde{q}} m^i{}_j M_i{}^j \tag{10.166}$$

and is seen to be automatically $U(1)_X$ invariant. Similarly for the gaugino condensation scale

$$\langle \lambda \lambda \rangle = \left(\Lambda^{3N_c - N_f} / \det M \right)^{\frac{1}{N_c - N_f}} \,. \tag{10.167}$$

The gauge contributions to the scalar potential can be computed along the $SU(N_c)$ classical flat directions. The result is

$$V_D = \frac{g_X^2}{2} \left[(q + \tilde{q}) \mathrm{Tr}(M^\dagger M)^{1/2} - \phi^\dagger \phi + \xi \right]^2 \,. \tag{10.168}$$

Auxiliary fields F_S, F_M, F_ϕ and D_X may easily be computed from (10.166) and (10.168).

In order to obtain analytic solutions, one may make a few simplifying asumptions [36]. First, since we are only interested in orders of magnitude, we make the assumption that $m^i{}_j = m\delta^i_j$ and search for solutions $M_i{}^j = M\delta^j_i$ of the equations of motion. Secondly, we linearize the minimization procedure by looking for a minimum in the vicinity of:

a) $\phi_0 = \sqrt{\xi}$, the field value which minimizes V_D in the absence of condensates;

b) M_0, the solution of $F_M = 0$:

$$M_0 = m^{\frac{N_f - N_c}{N_c}} \Lambda^{\frac{3N_c - N_f}{N_c}} \left(\frac{\xi}{M_P^2} \right)^{\frac{N_f - N_c}{2N_c}(q + \tilde{q})} \,. \tag{10.169}$$

The minimum is obtained by making around the field configuration M_0 an expansion in the parameter

$$\epsilon \equiv \frac{M_0}{\xi} = \left(\frac{\Lambda}{\sqrt{\xi}} \right)^{\frac{3N_c - N_f}{N_c}} \left[\frac{m}{M_P} \left(\frac{\sqrt{\xi}}{M_P} \right)^{q + \tilde{q} - 1} \right]^{\frac{N_f - N_c}{N_c}} \,. \tag{10.170}$$

One finds that all auxiliary fields are all of the same order:

$$\langle D_X^{1/2} \rangle \sim \left\langle \frac{F_\phi}{\phi} \right\rangle \sim \left\langle \frac{F_M}{M} \right\rangle \sim \epsilon m \left(\frac{\sqrt{\xi}}{M_P} \right)^{q + \tilde{q}} \,. \tag{10.171}$$

The magnitude of the soft terms in the observable sector is fixed by the values of the auxiliary fields F_ϕ, F_M and D_X. At the tree level of the Lagrangian of our model, we find soft scalar masses $\tilde{m}_i^2$ and trilinear soft terms A_{ijk} given by the expressions[21]:

$$\tilde{m}_i^2 = X_i \langle D_X \rangle, \quad A_{ijk} = (X_i + X_j + X_k) \left\langle \frac{F_\phi}{\phi} \right\rangle, \tag{10.172}$$

where X_i is the $U(1)_X$ charge of the corresponding field Φ^i. Gaugino masses in the hidden sector are also induced: $M_\lambda \sim N_f \langle F_M / M \rangle$. The gaugino masses in the observable sector are absent at tree level and are induced by standard gauge loops.

From (10.171) one obtains

$$\tilde{m} \sim N_f(q + \tilde{q}) \frac{\Lambda^3}{\xi} \left[\frac{m}{\Lambda} \left(\frac{\sqrt{\xi}}{M_P} \right)^{q+\tilde{q}} \right]^{N_f/N_c} = N_f(q + \tilde{q}) \frac{\langle \lambda\lambda \rangle}{\xi}, \tag{10.173}$$

where $\tilde{m}$ generically denotes a soft-breaking term (10.172) and we have used (10.167) in order to derive the last relation. This relation is indeed central to the kind of models described here and stresses the connected role of the relevant scales: ξ as the scale of messenger interaction and the gaugino condensate as the seed of supersymmetry breaking (although, as stressed earlier, the chiral nature of the $U(1)_X$ plays an important role: $q \neq -\tilde{q}$).

Fluxes

We have insisted several times on the fact that supersymmetry breaking scenarios in the context of string models need to address the question of the stabilization of moduli. It has been realized that fluxes may play an important rôle in this stabilisation.

These fluxes are generalizations of the familiar electromagnetic fluxes. More precisely, we have encountered above fully antisymmetric tensor fields $A_{\mu_1\mu_2\cdots\mu_{q-1}}$. Their field strengths

$$F_{\mu_1\mu_2\cdots\mu_q} = \partial_{\mu_1} A_{\mu_2\cdots\mu_q} \pm \text{permutations of } (\mu_1, \cdots, \mu_q) . \tag{10.174}$$

are antisymmetric in their q indices, and are therefore associated with q-forms, which turn out to be generated by D-branes. These fluxes obey quantization conditions. If they encompass d compact dimensions described by a generic parameter R, their energy $E_q = \int F_q^2$ scales like $R^d.R^{-2q}$, where the first factor is a volume element and the second finds its origin in the quantization condition. If p-branes are wrapped around some of these dimensions (as well as extend over the three infinite dimensions), their energy E_p scales like R^{p-3}. These two contributions lead, in the four dimensional effective theory, to a potential $V(R) \propto E_q(R) + E_p(R)$, the minimization of which may lead to a dynamical determination of the corresponding modulus, not necessarily a Kähler modulus. Indeed, in the simplest case, the dilaton and the complex structure moduli [185] are thus stabilized whereas the stabilisation of the Kähler modulus requires non-perturbative effects as well as explicit supersymmetry breaking [243].

In any case, it is not surprising that field strength fluxes play a rôle in supersymmetry breaking. In fact, a generic choice of fluxes often leads to a nonsupersymmetric model. This is presently a very active field, still under development, and we refer the reader to the rapidly expanding literature on the subject.

---

[21]We assume the presence in the superpotential of terms of the form $(\phi/M_P)^{X_i+X_j+X_k} \Phi_i \Phi_j \Phi_k$, as allowed by the $U(1)_X$ symmetry.

Further reading

- J. Pclchinski, *String theory*, volume 1, Cambridge University Press.
- M.B. Green, J.H. Schwarz and E. Witten, *Superstring theory*, volume 2, Cambridge monographs in mathematical physics, Cambridge University Press.

Exercises

Exercise 1 Compute, in a D-dimensional spacetime with $d - 1$ infinite spatial dimensions ard $D - d$ spatial dimensions of finite size L, the gravitational force between two masses m_1 and m_2 placed at a distance r much larger than L (*cf.* (10.28)).

Exercise 2 *We generalize the analysis of the Kaluza–Klein modes of a scalar field performed in Section 10.2.1 to the case of several compact dimensions. We analyze the problem using this time the action (instead of the equation of motion).*

Let us consider a massive real scalar field Φ in $D = 4 + N$ dimensions. Its action in flat $(4 + N)$-dimensional spacetime reads

$$
S = \int d^{4+N}x \left[\frac{1}{2} \partial^M \Phi \partial_M \Phi - \frac{1}{2} m_0^2 \, \Phi^2 \right]
$$

$$
= \int d^4 x \prod_{k=1}^{N} dy^k \left[\frac{1}{2} \partial^\mu \Phi \partial_\mu \Phi + \frac{1}{2} \partial^k \Phi \partial_k \Phi - \frac{1}{2} m_0^2 \, \Phi^2 \right] \tag{10.175}
$$

where henceforth $\mu = 0, \ldots, 3$, $M = 0 \cdots (3 + N)$, $k = 1, \ldots, N$. The N dimensions parametrized by y^k are compactified on circles of radius R; we thus have the identification:

$$
y^k = y^k + 2\pi R.
$$

(a) Show that the field Φ can be decomposed as:

$$
\Phi \left(x^\mu, y^k \right) = \frac{1}{(2\pi R)^{N/2}} \sum_{n_k \in Z} \phi_{\{n_k\}} \left(x^\mu \right) e^{i \sum_k n_k y^k / R} \tag{10.176}
$$

where $\{n_k\}$ is a set of N integers (one for each compact coordinate).

(b) How does the reality condition on Φ translate on the components $\phi_{\{n_k\}}$?

(c) Using the decomposition (10.176), show that the action (10.175) can be written as:

$$
S = \int d^4 x \sum_{n_k} \frac{1}{2} \partial^\mu \phi^*_{\{n_k\}} \partial_\mu \phi_{\{n_k\}} - \frac{1}{2} m^2_{\{n_k\}} \phi^*_{\{n_k\}} \phi_{\{n_k\}}
$$

and determine $m^2_{\{n_k\}}$.

(d) Show that all mass levels except the lowest one are degenerate.

Hints:

(b) $\Phi_{\{-n_k\}} = \Phi^*_{\{n_k\}}$.

(c) Use $\int dy\, e^{i(n+\bar{n})y/R} = (2\pi R)^{1/2}\,\delta_{n+\bar{n},0}$, to prove

$$\int \left(\prod_k dy^k\right)\partial^k\Phi\partial_k\Phi = \sum_{n_k}\phi_{\{n_k\}}\phi_{\{-n_k\}}\frac{n_k^2}{R^2}.$$

$m^2_{\{n_k\}} = m_0^2 + \sum_{k=1}^{N} n_k^2/R^2$.

(d) There is only one state of mass $m_0(\{n_k\} = \{0,\ldots,0\})$; all other mass levels are degenerate (for example $\{n_k\} = \{1,0,\ldots,0\}$, $\{0,1,0,\ldots,0\}$, $\{0,0,1,\ldots,0\}$,... correspond to the first such level, with degeneracy N).

Exercise 3 Let us consider the torus T_2 represented in Fig. 10.8b. We construct in this exercise the orbifold $T_2/\mathbb{Z}_3$ which is obtained by identifying points which are transformed into one another by a rotation of $2\pi/3$ around the origin.

(a) Draw the fixed points of the transformation.

(b) It is more convenient to use complex notations and to represent the torus of Fig. 10.8b as the complex plane $w = x^1 + ix^2$ with identification $w \equiv w + n \equiv w + me^{i\pi/3}$ $(m, n \in \mathbb{Z})$. The orbifold group $\mathbb{Z}_3$ is then generated by the transformation:

$$g : w \to we^{2i\pi/3}.$$

What are the complex coordinates of the fixed points?

(c) Explain why there are two twisted sectors for each fixed point.

(d) One defines the complex string coordinates $X(z,\bar{z}) \equiv X^1(z,\bar{z}) + iX^2(z,\bar{z})$, $\bar{X}(z,\bar{z}) \equiv X^1(z,\bar{z}) - iX^2(z,\bar{z})$. Give the oscillator expansion and the corresponding oscillator commutation relations.

Hints:

(b) $w = 0$, $w = \frac{1}{\sqrt{3}}e^{i\pi/6}$, $w = \frac{2}{\sqrt{3}}e^{i\pi/6}$.

(c) Strings may be twisted by g or g^2.

(d) For a string twisted by g for example, we have

$$X(z,\bar{z}) = x_0 + i\sqrt{\frac{\alpha'}{2}}\sum_{n\in Z}\left[\frac{1}{n+1/3}\frac{\alpha_{n+1/3}}{z^{n+1/3}} + \frac{1}{n-1/3}\frac{\tilde{\alpha}_{n-1/3}}{\bar{z}^{n-1/3}}\right],$$

$$\bar{X}(z,\bar{z}) = \bar{x}_0 + i\sqrt{\frac{\alpha'}{2}}\sum_{n\in Z}\left[\frac{1}{n-1/3}\frac{\alpha^\dagger_{-n+1/3}}{z^{n-1/3}} + \frac{1}{n+1/3}\frac{\tilde{\alpha}^\dagger_{-n-1/3}}{\bar{z}^{n+1/3}}\right],$$

where x_0 is the coordinate of the fixed point considered. The commutation relations are of the form:

$$\left[\alpha_{m+1/3}, \alpha^\dagger_{-n+1/3}\right] = (m+1/3)\delta_{m+n,0}. \tag{10.177}$$

<u>Exercise 4</u> In D dimensions, how many degrees of freedom are associated with a symmetric traceless tensor h_{MN}, such as the metric fluctuation tensor[22] with gauge transformation $\delta h_{MN} = \partial_M \Lambda_N + \partial_N \Lambda_M$? Compare with the computation for an antisymmetric tensor (see footnote following (10.118)).

Hints: $D(D+1)/2 - 1$ independent components; gauge condition $\partial^M h_{MN} = 0$ eliminates D. Residual gauge transformations satisfy $\partial^M \Lambda_M = 0$ and $\Box \Lambda_M = 0$. Hence $D(D+1)/2 - 1 - D - (D-1) = D(D-3)/2$. Note that $(D \leq 3)$-dimensional gravities are very different from four-dimensional gravity.

---

[22]We write the metric tensor $g_{MN} = \eta_{MN} + h_{MN}$, with η_{MN} the Minkowski metric. The gauge transformation is then obtained by linearizing the general transformation law, equation (D.3) of Appendix D: $x'^M = x^M - \Lambda^M(x)$.

11

Supersymmetry and the early Universe

11.1   The ultimate laboratory

The introduction of supersymmetry allows us to introduce new physics with a fundamental mass scale which is very large. A certain number of observations plead for such a possibility, among which the value of gauge couplings compatible with unification and the scale of neutrino masses. More generally the fact that fundamental physics is described by gauge theories, which are known to dominate in the infrared (*i.e.* low energy) limit, tends to accommodate the idea that the scale of underlying physics is much higher.

There are indirect ways to test at low energy a theory with a large fundamental scale, through effective interactions which scale like inverse powers of the fundamental scale. The best-known example is proton decay: an excessively rare event because the corresponding amplitude scales like the inverse of the square of the grand unification scale.

On the other hand, there is no hope to construct accelerators that would reach the desired mass scales to test these theories directly. One has to resort to natural cosmic accelerators or to the study of the early Universe , in which, at least in the hot Big-Bang scenario, temperatures have reached high enough values to excite superheavy degrees of freedom.

From this point of view, the recent successes of observational cosmology, which have turned it into a quantitative science, are a strong encouragement. It is fair however to say that we have, until now, tested quantitatively the evolution of our Universe up to nucleosynthesis (see Appendix D and Table D.1), which corresponds to an energy scale which is a fraction of MeV. The physics at higher scales thus merely provides boundary conditions for the observationally testable cosmological evolution. But this is sufficient to test indirectly the evolution in the very early Universe. The most famous example is inflation: inflationary expansion takes place at a very early stage of the expansion of the Universe but provides fluctuations that develop in later stages of the evolution to show up in the cosmic microwave background.

Supersymmetric models provide a wealth of new fields and new mechanisms that may play a significant rôle in cosmology. In particular, fundamental scalar fields, which are one of the building blocks of supersymmetry, have played an increasing rôle in cosmology. They are nowadays called upon to solve all kinds of problems from the time variation of constants to dark energy or inflation. If the associated energy scale is much larger than the TeV, they should be definitely considered in a supersymmetric context.

Of particular interest for cosmology are light scalar fields. There are two types of fields which are naturally light: pseudo-Goldstone bosons and moduli. In the latter case, different values of the fields lead to different physics. This makes them especially valuable in a cosmological context. As we have seen throughout this book, they are typical of supersymmetric theories. For example, superstring theories lead to a large number of them with properties which can be precisely evaluated in the context of a specific string model (see Section 10.4.2 of Chapter 10). We will start this chapter by discussing what has been dubbed as "modular cosmology".

Since flat directions are a natural property for a modulus and a desired feature for inflation, this will lead us naturally to discussing inflationary scenarios in a supersymmetric context. Other topics considered here are topological defects and baryogenesis.

11.2 Cosmological relevance of moduli fields

11.2.1 Dilaton and scalar-tensor theories

We have encountered in the preceding chapter moduli fields which couple to matter with gravitational strength. Examples are, in the context of Kaluza–Klein compactification, the radius modulus e^σ (see equation (10.36) of Chapter 10) or, in string theory, the dilaton e^ϕ or its four-dimensional counterpart, the S field (see equation (10.106) of Chapter 10). If these fields remain light they induce a long rang force similar to gravity which might lead to difficulties when confronting observation.

One of the stringent constraints on gravitational-type interactions comes from the high accuracy at which the equivalence principle has been tested[1]. In its weak form, the equivalence principle states the universality of free fall: two test bodies at the same location and at rest with respect to each other, fall in the same way in an external gravitational field, independently of their mass and composition (hence inertial and gravitational masses are identical). In the Einstein formulation, at every point of an arbitrary gravitational field, it is possible to define locally a coordinate system such that the laws of nature take the same form as in special relativity (see the book by C. Will [366] for a more detailed formulation).

Let us consider for example the string dilaton coupling to gauge fields, as obtained in Chapter 10 (equation (10.106)),

$$S = -\frac{1}{4} \int d^4x \sqrt{|g^{(4)}|} \; s \; F^{\mu\nu} F_{\mu\nu}. \tag{11.1}$$

As long as the dilaton s is not stabilized, the gauge coupling constants depend on space and time ($1/g^2 = s$). Since the mass of hadrons is mostly gluon field energy, it follows that these masses also depend on space and time and we lose the universality of free fall.

---

[1]To give an idea of the orders of magnitude involved, the relative difference in acceleration $|\Delta\mathbf{a}|/|\mathbf{a}|$ between two bodies of different composition in the Earth's gravitational field is presently measured to be smaller than 10^{-12}.

It should be noted that the scalar field dependence in (11.1) cannot be absorbed in a Weyl transformation of the metric,

$$g^{(4)}_{\mu\nu} = A^2(\phi)g_{\mu\nu},\tag{11.2}$$

because $\sqrt{|g^{(4)}|}g^{(4)\mu\rho}g^{(4)\nu\sigma}$ is Weyl invariant. *A contrario* the easiest way to satisfy the stringent constraints imposed by the apparent absence of violations of the equivalence principle is to consider a scalar-tensor theory for which the matter fields couple to a universal metric of the form (11.2) where ϕ stands for one or several (ϕ^a, $a = 1,\ldots,n$) scalar fields: lengths and times are measured by rods and clocks in the frame defined by this unique metric.

Let us thus consider the following action

$$S = S_{\text{gravity}} + S_{\text{matter}}\left(g^{(4)}_{\mu\nu}, \Psi_m\right),\tag{11.3}$$

$$S_{\text{gravity}} = \frac{1}{8\pi G}\int d^4x\sqrt{|g^{(4)}|}A^{-2}(\phi)\left[-\frac{1}{2}R^{(4)} + \frac{1}{2}\gamma^{(4)}_{ab}(\phi)g^{(4)\mu\nu}\partial_\mu\phi^a\partial_\nu\phi^b - v(\phi)\right].$$

The Weyl transformation (11.2) gives the following action in the Einstein frame[2]:

$$S_{\text{gravity}} = \frac{1}{8\pi G}\int d^4x\sqrt{|g|}\left[-\frac{1}{2}R + \frac{1}{2}\gamma_{ab}(\phi)g^{\mu\nu}\partial_\mu\phi^a\partial_\nu\phi^b - V(\phi)\right],\tag{11.4}$$

$$\gamma_{ab}(\phi) \equiv \gamma^{(4)}_{ab}(\phi) + 6\alpha_a(\phi)\alpha_b(\phi), \quad \alpha_a(\phi) \equiv \frac{\partial\ln A}{\partial\phi^a}, \quad V(\phi) = A(\phi)^4 v(\phi).$$

The functions $\alpha_a(\phi)$ just defined play an important rôle since they measure in the equations of motion the strength of the coupling of the fields ϕ^a to the energy–momentum of matter. Indeed,

$$\frac{1}{\sqrt{|g|}}\frac{\delta S_{\text{matter}}\left(A^2(\phi)g_{\mu\nu}, \Psi_m\right)}{\delta\phi^a} = \frac{1}{\sqrt{|g|}}g_{\mu\nu}\frac{\delta S_{\text{matter}}}{\delta g_{\mu\nu}}\frac{\partial\ln A^2(\phi)}{\partial\phi^a} = \alpha_a(\phi)g_{\mu\nu}T^{\mu\nu},\tag{11.5}$$

where the matter energy–momentum tensor is defined as

$$T^{\mu\nu} = \frac{2}{\sqrt{|g|}}\frac{\delta S_{\text{matter}}}{\delta g_{\mu\nu}}.\tag{11.6}$$

Actually, if one considers this theory in the Newtonian limit (which corresponds to the limit where velocities are much smaller than c), one finds for the interaction between two pointlike bodies of mass m_1 and m_2 distant by r

$$F(r) = G\frac{m_1 m_2}{r^2}\left[1 + \sum_{a=1}^{n}\alpha_a(\phi_0)\gamma^{ab}(\phi_0)\alpha_b(\phi_0)e^{-m_a r}\right],\tag{11.7}$$

where ϕ_0^a is the minimum of the potential (one assumes $V(\phi_0) = 0$ and $A(\phi_0) = 1$), m_a the mass of ϕ^a and γ^{ab} the inverse of γ_{ab}.

---

[2]One may use (10.103) of Chapter 10, with $N = 0$ to perform the Weyl transformation on the metric.

For the case of a single scalar field, this gives limits on $|\alpha(\phi_0)|^2$ which depend on the range $\lambda = 1/m$ of the interaction, from 10^{-3} in the (10 m, 10 km) range to 10^{-8} in the $(10^4, 10^5)$ km range. As we have stressed in Chapter 10, the law of gravity is poorly determined below the mm region.

In the case of massless fields, the expression (11.7) does not modify the $1/r^2$ law; it simply leads to a rescaling of Newton's constant. One has then to resort to the post-Newtonian limit (terms smaller by a factor v^2/c^2 or $Gm/(rc^2)$) to put constraints on such theories). The limit obtained on $|\alpha|^2$ is typically 10^{-3}.

It is also possible to appeal to the cosmological evolution to account for the smallness of such coefficients in scalar-tensor theories. For example, [95] have found an attractor mechanism towards general relativity.

This mechanism exploits the stabilization of the dilaton-type scalar through its conformal coupling to matter. Indeed, assuming that this scalar field ϕ couples to matter with equation of state parameter w_B through the action (11.3)–(11.4) (with $\gamma(\phi) = 1$), then its equation of motion takes the form:

$$\frac{2}{3 - \phi'^2}\phi'' + (1 - w_B)\phi' = -(1 - 3w_B)\alpha(\phi), \tag{11.8}$$

where $\phi' = d\phi/d\ln a$. This equation can be interpreted as the motion of a particle of velocity-dependent mass $2/(3 - \phi'^2)$ subject to a damping force $(1 - w_B)\phi'$ in an external force deriving from a potential $v_{\text{eff}}(\phi) = (1 - 3w_B)\ln A^2(\phi)$. If this effective potential has a minimum, the field quickly settles there.

An efficient mechanism of this type has been devised for the string dilaton by Damour and Polyakov [96] under the assumption that the dilaton dependent coupling functions terms entering the effective string action have a common extremum (*cf.* the tree level action (10.93) of Chapter 10 where these functions are a universal $e^{-2\phi}$).

11.2.2 Time variation of fundamental constants

Since couplings and scales are often given in terms of moduli fields, it is tempting to consider that, since some moduli may not have been stabilized, some of these quantities are still presently varying with time or have been doing so in the course of the cosmological evolution. This leads to the fascinating possibility that some of the fundamental constants of nature are time-dependent.

Such an idea was put forward by Dirac [119, 120]. According to him, a fundamental theory should not involve fundamental dimensionless parameters (*i.e.* dimensionless ratios of fundamental parameters) which are very large numbers. Such numbers should instead be considered as resulting from the evolution of the Universe and the corresponding dimensionless parameters be variables characterizing the evolving state of the Universe. Obviously this leads to some time-dependent fundamental parameters.

To illustrate Dirac's approach, one may consider, besides dimensionless parameters such as the fine structure constant $\alpha = e^2/(4\pi\hbar c)$ or the strong gauge coupling α_3:

- the ratio of the electromagnetic to gravitational force between a proton and an electron $e^2/(G_N m_p m_e) \sim 10^{39}$;

- the age of the Universe (of the order of H_0^{-1}) in microscopic time units (a typical atomic time scale is $\hbar^3/(m_e e^4)$, as can be checked on Bohr's model of the hydrogen atom) $me^4/(\hbar^3 H_0) \sim 10^{34} h_0^{-1}$.

The second ratio obviously evolves linearly with time (just replace the Hubble constant H_0 by the Hubble parameter $H \propto t^{-1}$). The dependence of the first one with time would involve for example the time dependence of Newton's constant.

In the context of supersymmetric theories where many of these dimensionless ratios are fixed by the values of moduli fields, one may expect some time dependence. For example, in heterotic string theory we have seen that the Planck scale (hence Newton's constant) is given in terms of the string scale by the vacuum expectation value of the string dilaton. Similarly for the four-dimensional coupling, evaluated at the string scale (close to unification scale). If the dilaton is not properly stabilized at low energy, that is if the flat direction is not lifted or if its minimum remains too shallow, one thus expects a possible time dependence of the dimensionless ratio M_P/M_S or of the fine structure constant.

There are, however, some stringent bounds on the possible time evolution of fundamental constants [350]. For example, present limits on $\left| \dot{G}_N/G_N \right|$ are in the 10^{-12} yr^{-1} region whereas the presence in the Oklo uranium in Gabon of a natural fission reactor which operated some 10^9 yr ago puts a limit [94] on $|\dot{\alpha}/\alpha|$ in the 10^{-17} yr^{-1} region.

11.2.3 Moduli and gravitino problems

Because moduli are light and have gravitational interactions, they are long lived. There are then two potential dangers. If their lifetime is smaller than the age of our Universe, their decay might have released a very large amount of entropy in the Universe and diluted its content. If their lifetime is larger than the age of our Universe, they might presently still be oscillating around their minimum and the energy stored in these oscillations may overclose the Universe. One refers to these problems as the moduli problem or sometimes as the Polonyi problem since they were first discussed in the context of the Polonyi model described in Section 6.3.2 of Chapter 6 [90,133,199]. Taken at their face values, such constraints forbid any modulus field which is not superlight or very heavy. We now proceed to make these statements quantitative.

We first define two quantities which play a central rôle in this discussion. A modulus field ϕ has typically gravitational interactions and thus its decay constant Γ_ϕ scales like $\kappa^2 \sim m_P^{-2}$. Since the only available scale is the scalar field mass m_ϕ, one infers from simple dimensional analysis that

$$\Gamma_\phi = \frac{m_\phi^3}{m_P^2}. \tag{11.9}$$

One thus deduces that the modulus will decay at present times if $\Gamma_\phi \sim H_0$, that is if its mass m_ϕ is of order $\left(H_0 m_P^2\right)^{1/3} \sim 20$ MeV.

The other relevant quantity is the initial value f_ϕ of the scalar field with respect to its ground state value ϕ_0. Presumably at very high energy (that is above the phase transition associated with dynamical supersymmetry breaking) where the flat direction is restored, one expects generically that $f_\phi \sim m_P$ since this is the only scale available.

Let us first consider the case where $m_\phi < 20$ MeV, that is a field which has not yet decayed at present time. The equation of evolution for the field ϕ reads (compare with (D.103) of Appendix D)

$$\ddot{\phi} + 3H\dot{\phi} + V'(\phi) = -\Gamma_\phi \dot{\phi}, \tag{11.10}$$

where the last term accounts from particle creation due to the time variation of ϕ.

If we assume that the Universe is initially radiation dominated, then $H \sim T^2/m_P$ (see (D.62) of Appendix D). As long as $H > m_\phi$, the friction term $3H\dot{\phi}$ dominates in the equation of motion and the field ϕ remains frozen at its initial value f_ϕ. When $H \sim m_\phi$, i.e. for $T_{\rm I} \sim (m_\phi m_P)^{1/2}$, the field ϕ starts oscillating around the minimum ϕ_0 of its potential which we approximate as:

$$V(\phi) = \tfrac{1}{2}m_\phi^2(\phi - \phi_0)^2 + O\left[(\phi - \phi_0)^3\right]. \tag{11.11}$$

Thus one looks for a solution of the form

$$\phi = \phi_0 + A(t)\cos\left(m_\phi t\right), \tag{11.12}$$

with $|\dot{A}/A| \ll m_\phi$ and $A(t_{\rm I}) \sim f_\phi$. Since $\Gamma_\phi < H_0 < H$, one may neglect the right-hand side term in equation (11.10) which simply reads, within our approximations, $\dot{A} = -3HA/2$. Since the energy density stored in the ϕ field is $\rho_\phi = \dot{\phi}^2/2 + V(\phi) \sim A^2 m_\phi^2$, one can write this equation as:

$$\dot{\rho}_\phi = -3H\rho_\phi. \tag{11.13}$$

We recover the standard result that coherent oscillations behave like nonrelativistic matter *i.e.*

$$\rho_\phi(T) = \rho_\phi(T_{\rm I})\left(\frac{T}{T_{\rm I}}\right)^3 \sim m_\phi^2 f_\phi^2 \left(\frac{T}{T_{\rm I}}\right)^3. \tag{11.14}$$

Since the radiation energy density $\rho_R(T)$ behaves as T^4, ρ_ϕ/ρ_R increases as the temperature of the Universe decreases and one reaches a time where the energy of the scalar field oscillations dominates the energy density of the Universe. One should then make sure that $\rho_\phi(T_0) < \rho_c$. Using (11.14) and $T_{\rm I} = (m_\phi m_P)^{1/2}$, one may write this condition as

$$m_\phi < m_P\left(\frac{\rho_c m_P}{f_\phi^2 T_0^3}\right)^2 \sim 10^{-26} \text{ eV}, \tag{11.15}$$

where we have set $f_\phi \sim m_P$. Thus, if 10^{-26} eV $< m_\phi < 20$ MeV, there is too much energy stored in the ϕ field (which has not yet decayed in present times).

We next consider the case where $m_\phi > 20$ MeV, that is the scalar field has already decayed at present times. Decay occurs at a temperature T_D when $H(T_D) \sim \Gamma_\phi$, i.e.

$$\Gamma_\phi^2 \sim \frac{\rho_\phi(T_D)}{m_P^2} = \frac{\rho_\phi(T_I)}{m_P^2} \left(\frac{T_D}{T_I}\right)^3, \tag{11.16}$$

where we have used (11.14), assuming that, at T_D, the scalar field energy density dominates over radiation; $\rho_\phi(T_I) \sim m_\phi^2 f_\phi^2$. At decay, all energy density is transferred into radiation. Thus, the reheating temperature T_{RH}, that is the temperature of radiation issued from the decay, is given by the condition

$$\rho_\phi(T_D) \sim g_* T_{RH}^4 \tag{11.17}$$

where we have used (D.61) of Appendix D. Using (11.16) to express $\rho_\phi(T_D)$, we obtain

$$T_{RH} \sim m_P^{1/2} \Gamma_\phi^{1/2} \sim \frac{m_\phi^{3/2}}{m_P^{1/2}}. \tag{11.18}$$

The entropy release is, according to (D.64) of Appendix D,

$$\sigma \equiv \frac{S_{RH}}{S_D} = \left(\frac{T_{RH}}{T_D}\right)^3 \sim \frac{1}{m_P^{1/2} \Gamma_\phi^{1/2}} \frac{\rho_\phi(T_I)}{T_I^3}. \tag{11.19}$$

This gives, using $T_I \sim m_\phi^{1/2} m_P^{1/2}$ and $\rho_\phi(T_I) \sim m_\phi^2 f_\phi^2$, $\sigma \sim f_\phi^2/(m_\phi m_P)$. With $f_\phi \sim m_P$, this gives a very large entropy release as long as the modulus mass remains much smaller than the Planck scale.

This entropy release must necessarily precede nucleosynthesis since otherwise it would dilute away its effects. This condition, namely $T_{RH} > 1$ MeV, gives $m_\phi > 10$ TeV. Thus for 20 MeV $< m_\phi < 10$ TeV, the entropy release following the decay of the modulus field is too large to be consistent with present observations.

In the absence of other effects, we are left with only superlight moduli fields ($m_\phi < 10^{-26}$ eV) or heavy ones ($m_\phi > 10$ TeV).

This moduli problem may be discussed in parallel with a similar problem associated with gravitinos [360]. Indeed gravitinos are also fields with gravitational interactions and thus a decay constant given by (11.9) with m_ϕ replaced by $m_{3/2}$. The formula (11.19) for entropy release remains valid with $\rho_\phi(T_I)$ replaced by $\rho_{3/2}(T_I) = m_{3/2} n_{3/2}(T_I)$. Using (D.74) and (D.77) of Appendix D which give the number density of a particle species which is relativistic at freezing (as is the gravitino), we obtain

$$\sigma \sim \frac{m_{3/2}}{m_P^{1/2} \Gamma_\phi^{1/2}} \sim \frac{m_P^{1/2}}{m_{3/2}^{1/2}}. \tag{11.20}$$

Again, if 20 MeV $< m_{3/2} < 10$ TeV, this gives too large an entropy release which dilutes away the products of nucleosynthesis.

The preferred value of the order of a TeV thus seems ruled out by such a constraint. This was also a theoretically preferred value for the moduli mass since $m_{3/2}$ is the characteristic mass scale describing the effects of supersymmetry breaking at low energy (and modulus mass arises through supersymmetry breaking). It is, however, possible to ease the bound by allowing for a late period of inflation. This requires a rather low reheat temperature, estimated to be smaller than 10^8 or 10^9 GeV [248] (or even lower if the gravitino has hadronic decay modes).

11.3   Inflation scenarios

Since the central prediction of inflation, namely that the total energy density of the Universe is very close to the critical energy density ρ_c for which space is flat (*i.e.* $\Omega \equiv \rho/\rho_c \sim 1$), seems to be in good agreement with observation, any theory of fundamental interactions should provide an inflation scenario.

Such a scenario was first imagined by Guth [214] in the context of the phase transition associated with grand unification. There is thus an obvious connection with supersymmetry. Let us see indeed why supersymmetry provides a natural setting for inflation scenarios.

We recall in Section D.4 of Appendix D that a standard scenario for inflation involves a scalar field ϕ evolving slowly in its potential $V(\phi)$. This is obtained by requiring two conditions on the form of the potential:

$$\varepsilon \equiv \frac{1}{2}\left(\frac{m_P V'}{V}\right)^2 \ll 1, \tag{11.21}$$

$$\eta \equiv \frac{m_P^2 V''}{V} \ll 1. \tag{11.22}$$

In the de Sitter Phase, *i.e.* in the phase of exponential growth of the cosmic scale factor, quantum fluctuations of the scalar field value are transmitted to the metric. Because the size of the horizon is fixed (to H^{-1}) in this phase, the comoving scale a/k associated with these fluctuations eventually outgrows the horizon, at which time the fluctuations become frozen. It is only much later when the Universe has recovered a radiation or matter dominated regime that these scales reenter the horizon and evolve again. They have thus been protected from any type of evolution throughout most of the evolution of the Universe (this is in particular the case for the fluctuations on a scale which reenters the horizon *now*). Fluctuations in the cosmic microwave background provide detailed information on the fluctuations of the metric. In particular, the observation by the COBE satellite of the largest scales puts a important constraint on inflationary models. Specifically, in terms of the scalar potential, this constraint known as COBE normalization, reads:

$$\frac{1}{m_P^3}\frac{V^{3/2}}{V'} = 5.3 \times 10^{-4}. \tag{11.23}$$

Using the slow-roll parameter introduced above, this can be written as

$$V^{1/4} \sim \varepsilon^{1/4}\, 6.7 \times 10^{16} \text{ GeV.} \tag{11.24}$$

In most of the models that we will be discussing, ε is very small. However as long as $\varepsilon \gg 10^{-52}$, we have $V^{1/4} \gg 1$ TeV. In other words, the typical scale associated with inflation is then much larger than the TeV, in which case it makes little sense to work outside a supersymmetric context.

From the point of view of supersymmetry, one might expect that the presence of numerous flat directions may ease the search for an inflating potential[3]. One possible difficulty arises from the condition (11.22) which may be written as a condition on the mass of the inflation field

$$m^2 \ll H^2. \tag{11.25}$$

Since supersymmetry breaking is expected to set the scale that characterizes departures from flatness, it should control both m and $V^{1/4}$. For example, in the case of gravity mediation, we expect both m^2 and $H^2 \sim V/m_P^2$ to be of the order of $m_{3/2}^2$. If one does not want to be playing with numbers of order one to explain the $N = 50$ e-foldings of exponential evolution necessary to a satisfactory inflation scenario, one should be ready to introduce a second scale into the theory.

Since supersymmetric scalar potentials consist of F-terms and D-terms, the discussion of suitable potentials for inflation naturally follows this classification. As we will see in the following, they naturally provide models for what is known as hybrid inflation (see Appendix D) which involves two directions in field space: one is slow-rolling whereas the other ensures the exit from inflation (and is fixed during slow-roll).

11.3.1 F term inflation

Let us start with a simple illustrative model [86]. We consider two chiral supermultiplets of respective scalar components σ and χ with superpotential

$$W(\sigma, \chi) = \sigma \left(\lambda \psi^2 - \mu^2 \right). \tag{11.26}$$

Writing $|\sigma| \equiv \phi/\sqrt{2}$, one obtains for the scalar potential

$$V = 2\lambda^2 \phi^2 |\psi|^2 + \left| \lambda \psi^2 - \mu^2 \right|^2. \tag{11.27}$$

The global supersymmetric minimum is found for $\psi^2 = \mu^2/\lambda$ and $\phi = 0$ but, for fixed ϕ, we may write the potential as ($\psi \equiv A + iB$)

$$V = \mu^4 + 2\lambda(\lambda\phi^2 - \mu^2)A^2 + 2\lambda(\lambda\phi^2 + \mu^2)B^2 + \lambda^2(A^2 + B^2)^2. \tag{11.28}$$

We conclude that, for $\phi^2 > \phi_c^2 \equiv \mu^2/\lambda$, there is a local minimum at $A = B = 0$ for which $V = \mu^4$. In other words, the ϕ direction is flat for $\phi > \phi_c$ with a nonvanishing potential energy. This may lead to inflation if one is trapped there. Since global supersymmetry is broken along this direction, one expects that loop corrections yield some slope which allows slow-roll . Once ϕ reaches ϕ_c, ψ starts picking up a vacuum expectation value and one quickly falls into the global minimum.

This simple example of F-term hybrid inflation may easily be generalized. However F-term inflation suffers from a major drawback when one tries to consider it in the

---

[3]although we have seen that this leads to new problems, the solution of which may require late inflation.

context of supergravity [86, 345]. We recall the form of the scalar potential in supergravity, as obtained in Section 6.2 of Chapter 6,

$$V = e^{K/m_P^2} \left[D_i W g^{i\bar{j}} D_{\bar{j}} \bar{W} - 3 \frac{|W|^2}{m_P^2} \right] + \frac{g^2}{2} \mathrm{Re} f_{ab}^{-1} D^a D^b \tag{11.29}$$

where

$$D_i W = \frac{\partial W}{\partial \phi^i} + \frac{1}{m_P^2} \frac{\partial K}{\partial \phi^i} W \tag{11.30}$$

and

$$D^a = -\frac{\partial K}{\partial \phi^i} (t^a)^i{}_j \phi^j + \xi^a. \tag{11.31}$$

Here ξ^c is a Fayet–Iliopoulos term which is present only in the case of a $U(1)$ symmetry.

In what follows the crucial rôle is played by the exponential factor e^{K/m_P^2} in front of the F-terms. Inflation necessarily breaks supersymmetry. Let us assume for a moment that the inflation is dominated by some of the F-terms and that the D-terms are vanishing or negligible. Then the slow-roll conditions (11.21) and (11.22) can be written as

$$\varepsilon = \frac{1}{2} \left(\frac{K_I}{m_P} + \cdots \right)^2 \ll 1, \quad \eta = K_{I\bar{I}} + \cdots \ll 1. \tag{11.32}$$

Here the subscript I denotes a derivative with respect to the inflation denoted ϕ^I. The extra terms denoted $\cdots$ in these expressions are typically of the same order as the ones written explicitly. Their precise value is model dependent. They might lead to cancellations but generically, this requires a fine tuning.

During inflation, unless a very special form is chosen, K_I is typically of the order of ϕ^I. Thus, in principle, one can satisfy the ε constraint if the inflation scenario is of the small field type ($\phi^I \ll 1$, see Section D.4 of Appendix D). But the η condition is more severe. The quantity $K_{I\bar{I}}$ stands in front of the kinetic term and therefore in the true vacuum it should be normalized to one. Then it is very unlikely to expect it to be much smaller during inflation. Indeed, this condition can be written as (11.25) since the mass of the inflaton m^2 receives a contribution $K_{I\bar{I}} V/m_P^2 \sim K_{I\bar{I}} H^2$.

These arguments indicate that it is not easy to implement F-type inflation in supergravity theories. All the solutions proposed involve specific nonminimal forms of the Kähler potential [345].

11.3.2 *D* term inflation

What is interesting about inflation supported by D-terms is that the problems discussed above can be automatically avoided because of the absence of a factor e^{K/m_P^2} in front of them. Indeed for inflation dominated by some of the D-terms the slow-roll conditions can be easily satisfied.

Let us show how such a scenario can naturally emerge in a theory with a $U(1)$ gauge symmetry [37, 221][4]. We first consider an example with global supersymmetry and a nonanomalous $U(1)$ symmetry. We introduce three chiral superfields

---

[4]An earlier version was proposed in [345] which uses F-terms to let the fields roll down the flat direction.

Φ_0, Φ_+ and Φ_- with charges equal to 0, $+1$ and -1, respectively. The superpotential has the form

$$W = \lambda \Phi_0 \Phi_+ \Phi_-$$ (11.33)

which can be justified by several choices of discrete or continuous symmetries and in particular by R-symmetry. The scalar potential in the global supersymmetry limit reads:

$$V = \lambda^2 |\phi_0|^2 \left(|\phi_-|^2 + |\phi_+|^2 \right) + \lambda^2 |\phi_+ \phi_-|^2 + \frac{g^2}{2} \left(|\phi_+|^2 - |\phi_-|^2 - \xi \right)^2$$ (11.34)

where g is the gauge coupling and ξ is a Fayet–Iliopoulos D-term (which we choose to be positive). This system has a unique supersymmetric vacuum with broken gauge symmetry

$$\phi_0 = \phi_- = 0, \quad |\phi_+| = \sqrt{\xi}.$$ (11.35)

Minimizing the potential, for fixed values of ϕ_0, with respect to other fields, we find that for $|\phi_0| > \phi_c \equiv g\sqrt{\xi}/\lambda$, the minimum is at $\phi_+ = \phi_- = 0$. Thus, for $|\phi_0| > \phi_c$ and $\phi_+ = \phi_- = 0$ the tree level potential has a vanishing curvature in the ϕ_0 direction and large positive curvature in the remaining two directions ($m_\pm^2 = \lambda^2 |\phi_0|^2 \mp g^2 \xi$). Along the ϕ_0 direction ($|\phi_0| > \phi_c$, $\phi_+ = \phi_- = 0$), the tree level value of the potential remains constant: $V = g^2 \xi^2/2 \equiv V_0$. Thus ϕ_0 provides a natural candidate for the inflaton field.

Along the inflationary trajectory all the F-terms vanish and the universe is dominated by the D-term which splits the masses of the Fermi–Bose components in the ϕ_+ and ϕ_- superfields. Such splitting results in a one-loop effective potential (see Section A.5.3 of Appendix Appendix A). In the present case this potential can be easily evaluated and for large ϕ_0 it behaves as

$$V_{eff} = \frac{g^2}{2} \xi^2 \left(1 + \frac{g^2}{16\pi^2} \ln \frac{\lambda^2 |\phi_0|^2}{\Lambda^2} \right) \equiv V_0 \left(1 + \frac{Cg^2}{8\pi^2} \ln \frac{\lambda \varphi}{\Lambda} \right),$$ (11.36)

where $\varphi \equiv |\phi_0|$ and $C \sim 1$.

Along this potential, the value of φ that leads to the right number $N \sim 50$ of e-foldings can be directly obtained from (D.112) of Appendix D:

$$\frac{\varphi}{m_P} = \sqrt{\frac{NCg^2}{4\pi^2}}.$$ (11.37)

This is safely of order g in the model that we consider but might be dangerously close to 1 for models which yield a larger value for C (see below). The values of the slow-roll parameters (11.21) and (11.22) are correspondingly

$$\varepsilon = \frac{Cg^2}{32N\pi^2}, \quad \eta = -\frac{1}{2N},$$ (11.38)

which yields a spectral index ((D.116) of Appendix D)

$$1 - n_S = \frac{1}{N} \left(1 + \frac{3Cg^2}{32\pi^2} \right).$$ (11.39)

Finally, the COBE normalization (11.24) fixes the overall scale:

$$\xi^{1/2} \sim \left(\frac{C}{N}\right)^{1/4} \times 1.9 \ 10^{16} \mathrm{GeV}. \tag{11.40}$$

Let us now consider the supergravity extension of our model. For definiteness we will assume canonical normalization for the gauge kinetic function f and the Kähler potential (*i.e.* $K = |\phi_-|^2 + |\phi_+|^2 + |\phi_0|^2$; note that this form maximizes the problems for the F-type inflation). The scalar potential reads

$$V = e^{(|\phi_-|^2 + |\phi_+|^2 + |\phi_0|^2)/m_P^2} \lambda^2 \left[|\phi_+ \phi_-|^2 \left(1 + \frac{|\phi_0|^4}{m_P^4}\right) \right.$$

$$\left. + |\phi_+ \phi_0|^2 \left(1 + \frac{|\phi_-|^4}{m_P^4}\right) + |\phi_- \phi_0|^2 \left(1 + \frac{|\phi_+|^4}{m_P^4}\right) - 3\frac{|\phi_+ \phi_- \phi_0|^2}{M^2} \right]$$

$$+ \frac{g^2}{2} \left(|\phi_+|^2 - |\phi_-|^2 - \xi \right)^2 \tag{11.41}$$

Again for values of $|\phi_0| > \phi_c$, other fields than ϕ_0 vanish and the behavior is much similar to the global supersymmetry case. The zero tree level curvature of the inflaton potential is not affected by the exponential factor in front of the first term since this term is vanishing during inflation. This solves the problems of the F-type inflation.

Let us now consider the case of a pseudo-anomalous $U(1)$ symmetry. Such symmetries usually appear in the context of string theories and have been discussed in Section 10.4.5 of Chapter 10: the anomaly is cancelled by the Green–Schwarz mechanism [205]. To see how D-term inflation is realized in this case, let us consider the simple example of such a $U(1)$ symmetry under which n_+ chiral superfields Φ_+^i and n_- superfields Φ_-^A carry one unit of positive and negative charges, respectively. For definiteness let us assume that $n_- > n_+$, so that the symmetry is anomalous and Tr $Q \neq 0$. We assume that some of the fields transform under other gauge symmetries, since the Green–Schwarz mechanism requires nonzero mixed anomalies. Let us introduce a single gauge-singlet superfield Φ_0. Then the most general trilinear coupling of Φ_0 with the charged superfields can be put in the form:

$$W = \sum_{A=1}^{n_+} \lambda_A \Phi_0 \Phi_+^A \Phi_-^A \tag{11.42}$$

(for simplicity we assume additional symmetries that forbid direct mass terms). Thus there are $n_- - n_+$ superfields Φ_-^i with negative charge that are left out of the superpotential. The potential has the form

$$V = \lambda_A^2 |\phi_0|^2 \left(|\phi_-^A|^2 + |\phi_+^A|^2 \right) + \lambda_A^2 |\phi_+^A \phi_-^A|^2 + \frac{g^2}{2} \left(|\phi_+^A|^2 - |\phi_-^A|^2 - |\phi_-^i|^2 - \xi_{\mathrm{GS}} \right)^2 \tag{11.43}$$

where summation over $A = 1, 2, \ldots, n_+$ and $i = 1, 2, \ldots, (n_- - n_+)$ is assumed. Again, minimizing this potential for fixed values of ϕ_0 we find that for

$|\phi_0| > \phi_c = \xi_{\rm GS}^{1/2} {\rm max}_A (g/\lambda_A)$, the minimum for all ϕ_+ and ϕ_- fields is at zero. Thus, the tree level curvature in the ϕ_0 direction is zero and inflation can occur. During inflation masses of $2n_+$ scalars are $m_{A\pm}^2 = \lambda_A^2 |\phi_0|^2 \mp g^2 \xi_{\rm GS}$ and the remaining $n_- - n_+$ negatively charged scalars have masses squared equal to $g^2 \xi_{\rm GS}$. We see that inflation proceeds much in the same way as for the nonanomalous $U(1)$ example discussed above. The interesting difference is that in the latter case the scale of inflation is an arbitrary input parameter (although in concrete cases it can be determined by the grand unified scale), whereas in the anomalous case it is predicted by the Green–Schwarz mechanism.

11.4   Cosmic strings

In the context of supersymmetry, cosmic strings have a remarkable property: they carry currents. This tends to stabilize them, which turns out to be a mixed blessing, since their relic density might overclose the Universe. As stressed in Section D.5 of Appendix D, which presents a short introduction to cosmic strings, an advantage of strings over other types of defects is that they interact and may thus disappear with time. This is lost in the context of supersymmetry and one must take into account the corresponding constraints on parameters, for each phase transition that may lead to the formation of strings.

Just as for inflation, the form of the scalar potential as a sum of F-terms and D-terms leads to the two main types of supersymmetric cosmic strings: F-strings and D-strings. We review their construction and some of their properties.

11.4.1   *F*-term string

We consider a $U(1)$ supersymmetric gauge theory with two charged superfields $\Phi_\pm$ (of respective charge ± 1) and a neutral superfield Φ_0. The superpotential is chosen to be [98]

$$W = \rho \Phi_0 \left(\Phi_+ \Phi_- - \eta^2 \right), \tag{11.44}$$

with η and ρ real. The scalar potential (F and D terms) vanishes for $\phi_0 = 0$, $\phi_\pm = \eta e^{\pm i\alpha}$.

We look for a solution of the equations of motion corresponding to a string extending over the z axis. Following Section D.5 of Appendix D, we thus consider the following ansatz in cylindrical coordinates (r, θ, z)

$$\phi_0 = 0, \quad \phi_\pm = \eta e^{\pm in\theta} f(r), \quad A_\mu = \frac{n}{g} \frac{a(r)}{r} \delta_\mu^\theta, \tag{11.45}$$

with the boundary conditions:

$$f(0) = a(0) = 0, \quad \lim_{r \to \infty} f(r) = \lim_{r \to \infty} a(r) = 1. \tag{11.46}$$

We note that, whereas the ground state of the theory (which coincides here with the state far away from the string) conserves supersymmetry and breaks gauge symmetry, supersymmetry is broken in the core of the string ($F_0 = \rho\eta^2(1 - f^2)$) and gauge symmetry is restored ($\phi_\pm = 0$).

The equations of motion are then written

$$f'' + \frac{f'}{r} - n^2 \frac{(1-a)^2}{r^2} = \rho^2 \eta^2 (f^2 - 1) f,$$

$$a'' - \frac{a'}{r} = -4g^2 \eta^2 (1-a) f^2, \tag{11.47}$$

which allows us to determine the profile functions $f(r)$ and $a(r)$.

We then turn our attention to the fermionic degrees of freedom: zero-energy solutions $\Psi_i(r,\theta)$ $(i = 0, \pm)$ and $\lambda(r,\theta)$ exist in the string core. According to standard index theorems [97, 237, 358], there are $2n$ of them. As in the nonsupersymmetric superconducting string [376], such solutions may be turned into more general solutions propagating along the string. Indeed, let us write the ansatz

$$\widetilde{\Psi}_i(r,\theta,z,t) = \Psi_{i_L}(r,\theta)\alpha(t,z) \tag{11.48}$$

(and similarly for λ). The Dirac operator can be written as $\slashed{D} = \gamma^0 D_0 + \gamma^3 D_3 + \slashed{D}_T$. In the interior of the strings, the Yukawa couplings vanish since all scalar fields are zero and $\slashed{D}_T \Psi_{i_L}(r,\theta) = 0$. One infers from the Dirac equation for $\widetilde{\Psi}_i$

$$\left[\gamma^0 \frac{\partial}{\partial t} + \gamma^3 \frac{\partial}{\partial z} \right] \alpha(t,z) = 0. \tag{11.49}$$

If Ψ_{i_L} satisfies (see below)

$$i\gamma^1\gamma^2 \Psi_{i_L} = \pm\Psi_{i_L}, \quad i\gamma^1\gamma^2 = \begin{pmatrix} \sigma^3 & 0 \\ 0 & \sigma^3 \end{pmatrix}, \tag{11.50}$$

then, using $i\gamma^0\gamma^1\gamma^2\gamma^3\widetilde{\Psi}_i = -\widetilde{\Psi}_i$, we have $\gamma^0\gamma^3\alpha = \mp\alpha$ and

$$\left[\frac{\partial}{\partial t} \mp \frac{\partial}{\partial z} \right] \alpha(t,z) = 0, \tag{11.51}$$

i.e. $\alpha(t,z) = f(t \pm z)$.

Thus fermions which are trapped in the transverse zero modes (satisfying (11.50)) travel at the speed of light along the string (in the $\mp z$ direction). The presence of these excitations is responsible for turning the string into a superconducting wire.

[For the sake of completeness, we derive here the explicit form of the fermion zero modes. We use two-component spinors (see Appendix B) and write

$$\psi_i(r,\theta) = \begin{pmatrix} \psi_{i1}(r,\theta) \\ \psi_{i2}(r,\theta) \end{pmatrix}, \quad \lambda(r,\theta) = \begin{pmatrix} \lambda_1(r,\theta) \\ \lambda_2(r,\theta) \end{pmatrix}. \tag{11.52}$$

We note that setting the second (resp. first) components to zero, Ψ_{i_L} satisfies (11.50) with a plus (resp. minus) sign. We proceed with the first choice. Equations of motion read

$$e^{-i\theta}\left(\partial_r - \frac{i}{r}\partial_\theta\right)\bar\lambda_1 + g\sqrt{2}\eta f\left(e^{in\theta}\psi_{-1} - e^{-in\theta}\psi_{+1}\right) = 0,$$

$$e^{-i\theta}\left(\partial_r - \frac{i}{r}\partial_\theta\right)\bar\psi_{01} + i\rho\eta f\left(e^{in\theta}\psi_{-1} + e^{-in\theta}\psi_{+1}\right) = 0, \tag{11.53}$$

$$e^{-i\theta}\left(\partial_r - \frac{i}{r}\partial_\theta \pm n\frac{a}{r}\right)\bar\psi_{\pm1} + \eta f e^{\mp in\theta}\left(i\rho\psi_{01} \mp g\sqrt{2}\lambda_1\right) = 0.$$

A shortcut that allows us to find two of the zero modes (*i.e.* all of them if $n = 1$) makes direct use of the supersymmetry transformations: one acts through supersymmetry on the bosonic zero modes to obtain fermionic zero modes. Using equations (C.29) and (C.70) of Appendix C, one obtains in the bosonic background considered:

$$\delta\psi_{\pm\alpha} = -i\sqrt{2}\sigma^{\mu\alpha\dot\alpha}\bar\xi^{\dot\alpha}D_\mu\phi_\pm = -i\eta\sqrt{2}e^{\pm in\theta}\left[f'\sigma^r \pm in\frac{(1-a)f}{r}\sigma^\theta\right]_{\alpha\dot\alpha}\bar\xi^{\dot\alpha},$$

$$\delta\psi_{0\alpha} = \sqrt{2}\xi_\alpha F_0 = \sqrt{2}\rho\eta^2(1 - f^2)\xi^\alpha, \tag{11.54}$$

$$\delta\lambda_\alpha = -\frac{i}{2}(\sigma^\mu\bar\sigma^\nu)_\alpha{}^\beta\xi_\beta F_{\mu\nu} = -\frac{n}{g}\frac{a'(r)}{r}(\sigma^3)_\alpha{}^\beta\xi^\beta,$$

where we have used $F_{12} = (n/g)a'(r)/r$ and defined

$$\sigma^r \equiv \sigma^1\cos\theta + \sigma^2\sin\theta = \begin{pmatrix} 0 & e^{-i\theta} \\ e^{i\theta} & 0 \end{pmatrix},$$

$$\sigma^\theta \equiv -\sigma^1\sin\theta + \sigma^2\cos\theta = \begin{pmatrix} 0 & -ie^{-i\theta} \\ ie^{i\theta} & 0 \end{pmatrix}. \tag{11.55}$$

We thus obtain (we set the supersymmetry transformation parameter ξ_2 to zero and note ξ_1 as the complex number α)

$$\lambda_1 = -\alpha\frac{n}{g}\frac{a'}{r},$$

$$(\psi_\pm)_1 = i\eta\sqrt{2}\alpha^*\left[f' \pm n\frac{(1-a)f}{r}\right]e^{\pm i(n\mp1)\theta},$$

$$(\psi_0)_1 = \alpha\rho\eta^2\sqrt{2}(1 - f^2). \tag{11.56}$$

We note that theses modes remain confined in the core of the string: f', a', $1 - a$ and $1 - f^2$ vanish away from the string. Similar expressions can be obtained by setting $\xi_1 = 0$: as discussed above, they correspond to modes propagating along the string in the other direction.]

11.4.2 *D*-term string

We now turn to the case of a phase transition where the gauge symmetry breaking occurs through a *D*-term:

$$V = \frac{g^2}{2}\left(|\phi|^2 - \xi\right)^2.$$

(11.57)

For convenience, we disregard other scalar fields than the one responsible for gauge symmetry breaking and we write $\xi \equiv \eta^2$. The local string ansatz

$$\phi = \eta e^{in\theta} f(r), \quad A_\mu = \frac{n}{g}\frac{a(r)}{r}\delta^\theta_\mu,$$

(11.58)

with the same boundary conditions as in (11.46), leads to the equations of motion:

$$f' = n\frac{(1-a)}{r}f, \quad n\frac{a'}{r} = g^2\eta^2(1-f^2).$$

(11.59)

Again, supersymmetry is broken $(D = g\eta^2(1 - f^2))$ and gauge symmetry restored in the string core.

The surprise comes from the fermionic zero modes: they only travel in one direction, which expresses the fact that supersymmetry is only half broken. [As above, we find the fermionic zero modes by performing a supersymmetry transformation on the bosonic background (11.58). Using (11.58), we find

$$\delta\psi_\alpha = -i\sqrt{2}\sigma^{\mu\alpha\dot{\alpha}}\bar{\xi}^{\dot{\alpha}}D_\mu\phi$$
$$= -i\sqrt{2}\eta n e^{in\theta}\frac{f(1-a)}{r}\left(\sigma^r + i\sigma^\theta\right)_{\alpha\dot{\alpha}}\bar{\xi}^{\dot{\alpha}},$$
$$\delta\lambda_\alpha = -\xi^\alpha D - \frac{i}{2}\left(\sigma^\mu\bar{\sigma}^\nu\right)_\alpha{}^\beta\xi_\beta F_{\mu\nu}$$
$$= -g\eta^2(1 - f^2)\left[1 + \sigma^3\right]_\alpha{}^\beta\xi^\beta.$$

(11.60)

We see that, if we set $\xi_1 = 0$, the two expressions vanish. We thus only obtain zero modes by considering the case $\xi_1 = \alpha$ and $\xi_2 = 0$:

$$\lambda_1 = -2g\alpha\eta^2(1 - f^2),$$
$$\psi_1 = 2i\sqrt{2}n\eta\alpha^*\frac{(1-a)f}{r}e^{i(n-1)\theta}.$$

(11.61)

According to the discussion in the preceding section, only modes moving in one direction are present.]

11.5   Baryogenesis

Any complete cosmological scenario should give a satisfactory explanation as to why matter dominates over antimatter in the observable Universe. This is summarized in the following observational constraint (see Section D.3.5 of Appendix D):

$$\eta_B \equiv \frac{n_B - n_{\bar{B}}}{n_\gamma} = \left(6.1^{+0.3}_{-0.2}\right) \times 10^{-10}. \tag{11.62}$$

We recall in Appendix D that, in the context of the Standard Model, the electroweak phase transition has some difficulty to fulfill the Sakharov requirements necessary to generate a baryon asymmetry: it is not sufficiently first order and CP violation is not strong enough to generate enough baryons.

In the supersymmetric framework of the MSSM, the source of CP violation comes from the chargino sector. If we consider the chargino mass matrix given in equation (5.44) of Chapter 5, the complex phase ϕ_μ of the μ parameter[5] leads to the dominant contribution to the baryon asymmetry: the scattering of charginos against the expanding bubble wall creates an asymmetry between Higgsinos which is converted into a chiral quark asymmetry through Higgsino scattering off gluinos and stops; this chiral quark asymmetry is then translated into a baryon asymmetry through sphaleron processes [67, 229]. Regarding the order of the transition, it has been recognized that the lighter the right-handed stop, the more first order the phase transition is. It turns out that this is becoming marginally consistent with the present limit on the stop mass (a little room is left because the experimental limit is somewhat model dependent).

Grand unification has also been the framework for one of the first models of baryon generation proposed in the context of gauge theories. In this case, the departure from equilibrium is achieved through the decay of superheavy particles. Such particles are coupled to the superheavy gauge fields which ensures that they have B and CP violating interactions. Thus all three Sakharov conditions are fulfilled. This scenario is, however, difficult to realize in practice because any baryon asymmetry produced at the grand unification transition is washed away by a subsequent inflation period and limits on the reheat temperature (see the end of Section 11.2.3) prevent from producing the superheavy particles after reheating.

We thus present below the two scenarios which seem to be favored in the context of supersymmetric theories.

11.5.1   Leptogenesis

One may combine some of the ingredients used above (decay of heavy particles, B and L violating sphaleron process) to overcome the difficulties encountered so far.

Because a sphaleron interaction (see (D.93) in Appendix D) violates both B and L but not $B - L$, a nonzero value of $B + L$ is not washed out in the high temperature symmetric phase (*i.e.* between 100 GeV and 10^{12} GeV) if we start with

---

[5] See also the discussion of Section 7.7 of Chapter 7.

nonvanishing $B - L$. A study of the chemical potential of all particle species present in this phase yields the following relation between B, L and $B - L$:

$$Y_B = aY_{B-L} = \frac{a}{a-1}Y_L, \tag{11.63}$$

where a is a number of order one depending on the other processes at equilibrium and $Y_B = (n_B - n_{\bar{B}})/s$ (s is the entropy density as defined in (D.64) of Appendix D), with similar definitions for Y_L and Y_{B-L}. Thus a baryon asymmetry can be generated from a lepton asymmetry: lepton violating interactions, such as the ones encountered in grand unified theories, may lead to the generation of the baryon asymmetry. This is the scenario of leptogenesis [169].

In thermal leptogenesis, one obtains the departure from equilibrium from the decay of heavy weakly interacting particles. The favored model considers Majorana neutrino[6] decays into a lepton–Higgs pair: $N \rightarrow \ell\phi, \bar{\ell}\bar{\phi}$. If the corresponding couplings are CP violating, a lepton asymmetry arises through the unbalance between the decay rates:

$$\Gamma(N \rightarrow \ell\phi) = \frac{1}{2}(1 + \epsilon)\Gamma, \quad \Gamma(N \rightarrow \bar{\ell}\bar{\phi}) = \frac{1}{2}(1 - \epsilon)\Gamma. \tag{11.64}$$

If we consider the lightest of the heavy Majorana neutrinos N_1 of mass M_1, then the CP asymmetry parameter ϵ, which arises through interference between tree level and one loop diagrams, reads explicitly:

$$\epsilon = \frac{3}{16\pi} \frac{M_1}{\left(\lambda_\nu^\dagger \lambda_\nu\right)_{11} \langle\phi\rangle^2} \mathrm{Im}\left(\lambda_\nu^\dagger m_\nu \lambda_\nu^*\right)_{11}. \tag{11.65}$$

In the case of a hierarchical spectrum for neutrinos (with m_{ν_3} the largest neutrino mass), we have typically

$$\epsilon \sim \frac{3}{16\pi} \frac{M_1 m_{\nu_3}}{\langle\phi\rangle^2} \sim 0.1 \frac{M_1}{M_3}. \tag{11.66}$$

Hence, leptogenesis is a direct window over neutrino Majorana masses [57]. Sphaleron processes then transform the lepton asymmetry into a baryon asymmetry. Typically, one finds

$$\eta_B = \frac{\kappa}{f} c_S \epsilon \sim 10^{-4} \frac{M_1}{M_3}, \tag{11.67}$$

where c_S is the sphaleron conversion factor of order 1, $f \sim 10^2$ is a dilution factor to account for the increase of photons in a comoving volume between baryogenesis and today, and κ is a washout factor of order 0.1 which depends on neutrino masses. Thus $M_1/M_3 \sim 10^{-5}$ gives the right order of magnitude.

---

[6] We recall (see Chapter 1, Section 1.1.1) that Majorana neutrinos appear in the seesaw mechanism: if the right-handed Majorana neutrino 3×3 mass matrix is M, then the light neutrino mass matrix is $m_\nu = -\lambda_\nu \frac{1}{M} \lambda_\nu^T \langle\phi\rangle^2$, where λ_ν is the 3×3 matrix of Yukawa couplings.

We note that the baryogenesis temperature is typically

$$T_B \sim M_1 \sim 10^{10} \text{ GeV}. \tag{11.68}$$

Just as in the grand unification scenario, it remains necessary to produce these heavy Majorana particles after the reheating phase of inflation, which sets an upper limit on their mass of the order of 10^8 to 10^9 GeV. This might be difficult to reconcile with standard neutrino mass scales.

Ways to circumvent this difficulty have been proposed. Interestingly enough, they depend on the scenario for supersymmetry breaking and on the nature of the LSP. For example in gauge mediation, we have seen that the gravitino is the LSP: in the case where $m_{3/2} \sim 10$ eV, there is no gravitino problem[7]. In anomaly mediation, a gravitino with mass greater than 100 TeV has no cosmological problem. Since gravitino decay products include a LSP, there is a nonthermal production of LSP through gravitino decay which must be taken into account.

One may also consider nonthermal leptogenesis. For example, inflaton decay may be the source of Majorana particles [12, 191, 264]. In this type of scenario, the inflaton decays into a pair of N_1 Majorana neutrinos, which subsequently decay into $\ell\phi$ and $\bar{\ell}\bar{\phi}$. The condition on the reheating temperature $T_R < M_1$ ensures that one is out of equilibrium, as required by Sakharov conditions.

11.5.2 Affleck–Dine mechanism

Another mechanism for generating the baryon asymmetry has been proposed by [3]. It makes use of the presence of numerous flat directions in the scalar potential of supersymmetric theories.

Indeed, let us consider one of these flat directions, labelled by the field ϕ. We assume that the fundamental high energy theory, characterized by a scale M, violates baryon number (as for example grand unified theories). At high energies, that is at an early stage of the evolution of the Universe, the field ϕ sits along the flat direction at an arbitrary value $\phi_0 \sim M$. As temperature lowers, one reaches the energy scale associated with supersymmetry breaking. The degeneracy associated with the flat direction is lifted and the field direction acquires a nontrivial structure $V(\phi)$. There is a priori no reason to have ϕ_0 as a minimum of $V(\phi)$. Hence the field ϕ starts oscillating around the minimum of $V(\phi)$, with a frequency of the order of its mass m_ϕ.

In this way, one fulfills the three Sakharov requirements:

- baryon number violation from the fundamental theory;
- CP violation through the CP-violating phases φ of the soft terms (as discussed in Section 7.7 of Chapter 7),
- departure from equilibrium because of the oscillations.

It is thus not surprising that one generates some net baryon number. One obtains [3]:

$$n_B \sim \varphi m_\phi |A(t)|^2 \left(\frac{\phi_0^2}{M^2} \right), \tag{11.69}$$

---

[7] Axions may provide the dark matter candidate.

where $A(t)$ is the amplitude of oscillations at time t. The energy of oscillations is thus $m_\phi^2 |A(t)|^2$: it can be viewed as a coherent state of particles of mass m_ϕ and number density $m_\phi^2 |A(t)|^2$. The baryon number per particle is thus of order $\varphi \, \phi_0^2/M^2$. More precisely one finds

$$\frac{n_B}{n_\gamma} \sim 10^2 \varphi \left(\frac{\phi_0^2}{M^2} \right) . \tag{11.70}$$

This shows that the Affleck–Dine mechanism is very efficient to generate a baryon asymmetry.

Further reading

- T. Damour, *Gravitation, experiment and cosmology*, Proceedings of the 5th Hellenic School of Elementary Particle Physics (arXiv:gr-qc/9606079).
- D. Lyth and A. Riotto, *Particle physics models of inflation and the cosmological density perturbation*, Phys. Rep. **314** (1999) 1.

12

The challenges of supersymmetry

Although supersymmetry provides a satisfactory, and probably necessary, framework to address the questions left aside by the Standard Model, it is facing some difficult challenges. We take the opportunity in this last chapter to discuss two of the most difficult questions: the flavor problem and the cosmological constant problem. The lack of a clear solution to either of these problems would be (is?) probably a sign that we are missing an important piece of the jigsaw puzzle and that, if there is supersymmetry, it is probably not realized in the exact way that we presently think it is.

12.1 The flavor problem

One of the motivations for going beyond the Standard Model is to try to explain the variety of masses, *i.e.* the variety of Yukawa couplings. Any theory of mass, in fact any theory beyond the Standard Model, introduces new degrees of freedom. The quantum corrections associated with these new degrees of freedom tend to spoil the delicate balance achieved by the Standard Model in the flavor sector: mainly the strong suppression of Flavor Changing Neutral Currents (FCNC) and the presence of a single CP-violating parameter, the phase δ_{CKM} (besides the poorly understood θ_{QCD} parameter).

The latter property has been successfully verified in recent years. A useful tool to describe these experimental results is the unitary triangle. We refer the reader to Section A.3.4 of Appendix Appendix A for the definition of this triangle in the context of the Standard Model and we give here in Fig. 12.1 the experimental status in 2005. The parameters ρ and η appear in the Wolfenstein parametrization [379] of the Cabibbo–Kobayashi–Maskawa matrix:

$$
V_{\mathrm{CKM}} \sim \begin{pmatrix} 1 - \frac{1}{2}\lambda^2 & \lambda & A\lambda^3(\rho - i\eta) \\ -\lambda & 1 - \frac{1}{2}\lambda^2 & A\lambda^2 \\ A\lambda^3(1 - \rho - i\eta) & -A\lambda^2 & 1 \end{pmatrix}
\tag{12.1}
$$

The fact that all experimental data available are consistent with a small region with nonvanishing η shows that CP is violated and that its violation is consistent with a single origin, the phase of the CKM matrix.

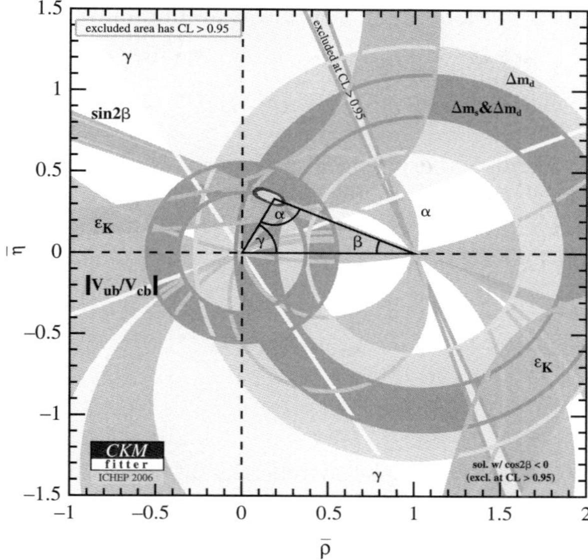

Fig. 12.1 Unitarity triangle and the experimental limits obtained in 2005 [76].

We note that the question of CP violation is intimately connected with the question of mass. For example, the condition for CP violation in the Standard Model may be summarized as (see (A.171) of Appendix Appendix A)

$$(m_t^2 - m_c^2)(m_t^2 - m_u^2)(m_c^2 - m_u^2)(m_b^2 - m_s^2)(m_b^2 - m_d^2)(m_s^2 - m_d^2)J_{\text{CKM}} \neq 0,$$

$$(12.2)$$

which clearly involves the fermion mass spectrum besides the complex phases encapsulated in the quantity J_{CKM}. CP violation has been a recurring theme in the last chapters. We know that all experimental results are in agreement with the CKM framework of the Standard Model. On the other hand, we expect other sources of CP violation to explain how baryon number was generated in the Universe. This is one of the basis for expecting physics beyond the Standard Model.

The natural procedure to account for the diversity of fermion masses is to introduce family symmetries. Such symmetries are not observed in the spectrum and thus must be broken at some large scale. A possible hint about this scale may be found in the seesaw mechanism for neutrino masses. In any case, fine tuning arguments require us to discuss such symmetries in a supersymmetric context.

But supersymmetry introduces many new fields, in particular sfermions, and thus many new sources of FCNC and CP violation. As discussed below [217], a generic extension has 44 CP-violating phases! This poses a severe problem and ways to control these sources have to be devised. There is thus a necessary connection between the discussion of supersymmetry breaking and the resolution of the (s)fermion mass problem.

In a supersymmetric framework, the Yukawa coupling matrices appear in the superpotential

$$W^{(3)} = \Lambda_{d_{ij}} Q_i \cdot H_1 D_j^c + \Lambda_{u_{ij}} Q_i \cdot H_2 U_j^c + \Lambda_{e_{ij}} L_i \cdot H_1 E_j^c, \qquad (12.3)$$

generalizing the one-family superpotential considered in (5.2) of Chapter 5.

If we organize quark and lepton fields in columns associated with family and lines associated with Standard Model quantum numbers

$$
\begin{array}{ccc}
u & c & t \\
d & s & b \\
e & \mu & \tau \\
\nu_e & \nu_\mu & \nu_\tau
\end{array}
$$

we can distinguish

- vertical symmetries, such as the symmetries of the Standard Model or grand unification; such symmetries may be advocated to explain the intragenerational hierarchies of masses. Infrared fixed points or grand unified relations provide examples of how such vertical symmetries can be used.
- horizontal (or family) symmetries; such symmetries address the question of intergenerational hierarchies, in particular the question of why the first family is lighter than the second which is lighter than the third.

We will review the possible uses of both types of symmetries in what follows. Before doing this, we will recall where we stand experimentally to try to evaluate the seriousness of the mass problem in the context of supersymmetric extensions of the Standard Model.

We conclude this section by computing, as promised, the number of independent phases in a generic supersymmetric Standard Model. The three Yukawa 3×3 complex matrices in (12.3) involve 3×9 complex parameters. Similarly for the A-terms. The five 3×3 hermitian mass-squared sfermion masses involve 5×3 real and 5×3 complex parameters. Finally in the gauge sector, we count four real parameters (g_1, g_2, g_3 and θ_{QCD}) and three complex (M_1, M_2, M_3); in the Higgs sector, two real ($m_{H_1}^2, m_{H_2}^2$) and two complex (μ, B_μ) parameters. In total, 95 real parameters and 74 imaginary phases.

We have not yet taken into account the possible field redefinitions. We follow the method already used in Section 7.7 of Chapter 7: the various couplings are spurion fields that break the global symmetry $U(3)^5 \times U(1)' \times U(1)'_R$ (the charges under the latter two symmetries are given in Table 7.4 of Chapter 7) into $U(1)_B \times U(1)_L$. We can thus remove 15 complex parameters and 15 phases. We are left with 80 real parameters and 44 imaginary phases.

12.1.1 The supersymmetric CP problem

When we consider supersymmetric extensions of the Standard Model, new sources of CP violation appear both in flavor diagonal and flavor violating interactions.

We have encountered the first case when we discussed the neutron electric dipole moment in Section 7.7 of Chapter 7. Even in a universal model (with two residual

CP violating phases $\varphi_{A,B}$), one computes a neutron electric dipole moment of the order of

$$d_N \sim 2 \left(\frac{100 \text{ GeV}}{\tilde{m}} \right)^2 \sin \varphi_{A,B} \times 10^{-23} ecm, \tag{12.4}$$

which misses the experimental bound [144] by two orders of magnitude for $\tilde{m} \sim 100$ GeV and $\varphi_{A,B} \sim 1$.

As for flavor violating interactions, sfermion mixing leads to new sources of FCNC processes both CP conserving and CP violating. It has become customary to parametrize these flavor violations in the so-called mass insertion approximation [219]. The idea is to work in the basis for fermion and sfermion states where all couplings to neutral gauginos are diagonal: flavor violating effects appear through the nondiagonality of mass matrices for sfermions of the same electric charge. Denoting by $\tilde{m}^2 \delta$ such nondiagonal terms ($\tilde{m}$ is an average sfermion mass), one may parametrize the main CP conserving and CP violating flavor violations by considering only the first term in an expansion in the δ matrix elements.

Thus, writing the squark mass matrix obtained in Section 5.3.3 of Chapter 5 as

$$\widetilde{M}_q^2 = \begin{pmatrix} \widetilde{M}_{q,LL}^2 & \widetilde{M}_{q,LR}^2 \\ \widetilde{M}_{q,RL}^2 & \widetilde{M}_{q,RR}^2 \end{pmatrix} \tag{12.5}$$

where each entry is a 3×3 matrix in flavor space, we introduce

$$(\delta_{MN}^q)_{ij} = \left(V_M^q \, \widetilde{M}_{q,MN}^2 \, V_N^{q\dagger} \right)_{ij} / \tilde{m}^2 \tag{12.6}$$

where $M, N \in \{L, R\}$. In this parametrization, gluino–quark–squark couplings are diagonal but squark mass matrices are not. If V^q was equal to $\widetilde{V}_{L,R}^q$, δ_{MN}^q would be diagonal[1].

We illustrate how these parameters are constrained by the experimental data on $\epsilon_K : \epsilon_K = (2.28 \pm 0.02) \times 10^{-3} e^{i\pi/4}$.

The supersymmetric contribution to ϵ_K is dominated by box diagrams involving $\tilde{s}_{L,R}$ and $\tilde{d}_{L,R}$ squarks in the loop. Assuming $m_{\tilde{g}} = m_{\tilde{q}} \equiv \tilde{m}$, we obtain for the contributions of the first two families [170]:

$$\epsilon_K \Delta m_K = \frac{1}{\sqrt{2}} \, \text{Im} \, \langle K^0 | \mathcal{H}_{\text{eff}}^{\Delta S=2} | \bar{K}^0 \rangle \tag{12.7}$$

$$= \frac{5\alpha_3^2}{162\sqrt{2}} \frac{f_K^2 m_K}{\tilde{m}^2} \left[\left(\frac{m_K}{m_s + m_d} \right)^2 + \frac{3}{25} \right] \text{Im} \left[(\delta_{LL}^d)_{12} \, (\delta_{RR}^d)_{12} \right].$$

Thus

$$\frac{(\epsilon_K \Delta m_K)^{\text{SUSY}}}{(\epsilon_K \Delta m_K)^{\text{EXP}}} = 4 \times 10^6 \left(\frac{500 \text{ GeV}}{\tilde{m}} \right)^2 \text{Im} \left[(\delta_{LL}^d)_{12} \, (\delta_{RR}^d)_{12} \right]. \tag{12.8}$$

---

[1]In Chapter 6, Section 6.7, we had passed over the fact that there are different rotation matrices $\widetilde{V}_L^q$ and $\widetilde{V}_R^q$ for L and R squarks.

More precise limits are given in Table 12.1. One sees that the experimental constraint imposes fine tunings on the imaginary part of $\left(\delta^d_{LL}\right)_{12}\left(\delta^d_{RR}\right)_{12}$ of a few parts to 10^8. The constraint on the real part arising from $\Delta m_K = 2\,\mathrm{Re}\,\langle K^0\,|\mathcal{H}^{\Delta S=2}_{\mathrm{eff}}|\,\bar{K}^0\rangle$ is less stringent.

As for the $\Delta F = 1$ processes, experimental data on ϵ' and $\mathcal{B}(B \to X_s\gamma)$ put more stringent limits on δ^d_{LR} than on δ^d_{LL}. In the case of ϵ', this is because of a cancellation between the contributions of the box and the penguin diagrams to δ^d_{LL}. In the case of $b \to s\gamma$, the helicity flip needed with a δ^d_{LR} mass insertion is found in the internal gluino line, which leads to an enhancement factor of $m_{\tilde{g}}/m_b$ over the amplitude with a δ^d_{LL} insertion.

We give also in Table 12.1 limits on the flavor-conserving CP-violating mass insertions $\left(\delta^{u,d}_{LR}\right)_{11}$ which arise from the limit on the electric dipole moment of the neutron already discussed at the end of Chapter 7: $d_N \le 6.3 \times 10^{-26}$ e cm. Indeed, the corresponding contribution is [170] $d_N = (4d_d - d_u)/3$ with

$$d_u = \frac{e}{27\pi}\frac{\alpha_s}{\tilde{m}}\,\mathrm{Im}\,(\delta^u_{LR})_{11}\,, \qquad d_d = -\frac{e}{54\pi}\frac{\alpha_s}{\tilde{m}}\,\mathrm{Im}\,(\delta^d_{LR})_{11}\,, \tag{12.9}$$

in the limit $m_{\tilde{g}} = m_{\tilde{q}} \equiv \tilde{m}$.

12.1.2   Supersymmetry breaking versus flavor dynamics

Obviously the spectrum of masses and mixing angles in the supersymmetric sector plays a key rôle in the solution to the supersymmetric FCNC and CP problem.

If we go to the basis of squark mass eigenstates, we have for example

$$\left(\delta^d_{LL}\right)_{12} = \frac{\delta m^2_Q}{m^2_Q}\left|K^{dL}_{12}\right|\,, \qquad \left(\delta^d_{RR}\right)_{12} = \frac{\delta m^2_D}{m^2_D}\left|K^{dR}_{12}\right|\,, \tag{12.10}$$

where $\delta m_{Q(D)}$ is the mass difference between the two left (right) down squarks and $K^{dL(R)}_{12}$ the gluino couplings to left (right) handed down quarks and squarks.

We have seen in Section 6.7 of Chapter 6 that there are three possible solutions to the supersymmetric flavor problem which requires small mass insertions:

- universality (e.g. $\delta m^2_Q \ll m^2_Q$);
- effective supersymmetry (e.g. $m^2_Q \gg M^2_W$);
- quark-squark alignment (e.g. $\left|K^d_{12}\right| \ll 1$);

Regarding CP violations specifically, an approximate CP symmetry which would ensure all CP-violating phases to be small is no longer a valid possibility since the phase measured in $B \to \psi K_S$ is of order one [136, 137].

A general discussion of these issues relies on two scales: the scale Λ_S of supersymmetry breaking and Λ_F the scale of flavor dynamics. If $\Lambda_F \gg \Lambda_S$, one expects a sparticle spectrum which is approximately flavor-free. Universality is thus favored. On the other hand, if $\Lambda_F \le \Lambda_S$, supersymmetry breaking and flavor dynamics are intimately connected: one has to resort to other solutions such as heavy supersymmetric partners or alignment.

We now discuss models illustrating the different possibilities.

Table 12.1 Experimental constraints on mass insertions parameters coming from experimental data on flavor-changing CP-conserving and CP-violating interactions [143, 170]. One assumes $m_{\tilde{g}} = m_{\tilde{q}} = 500$ GeV.

Δm_K	$\sqrt{\left\|\mathrm{Re}\left(\delta^d_{LL}\right)^2_{12}\right\|}$	4.6×10^{-2}
	$\sqrt{\left\|\mathrm{Re}\left(\delta^d_{LR}\right)^2_{12}\right\|}$ for $\left\|(\delta^d_{LR})_{12}\right\| \gg \left\|(\delta^d_{RL})_{12}\right\|$	2.8×10^{-3}
	$\sqrt{\left\|\mathrm{Re}\left(\delta^d_{LR}\right)^2_{12}\right\|}$ for $\left\|(\delta^d_{LR})_{12}\right\| = \left\|(\delta^d_{RL})_{12}\right\|$	2.8×10^{-2}
	$\sqrt{\left\|\mathrm{Re}\left(\delta^d_{LL}\right)_{12}\left(\delta^d_{RR}\right)_{12}\right\|}$	9.6×10^{-4}
ϵ_K	$\sqrt{\left\|\mathrm{Im}\left(\delta^d_{LL}\right)^2_{12}\right\|}$	6.1×10^{-3}
	$\sqrt{\left\|\mathrm{Im}\left(\delta^d_{LR}\right)^2_{12}\right\|}$ for $\left\|(\delta^d_{LR})_{12}\right\| \gg \left\|(\delta^d_{RL})_{12}\right\|$	3.7×10^{-4}
	$\sqrt{\left\|\mathrm{Im}\left(\delta^d_{LR}\right)^2_{12}\right\|}$ for $\left\|(\delta^d_{LR})_{12}\right\| = \left\|(\delta^d_{RL})_{12}\right\|$	3.7×10^{-3}
	$\sqrt{\left\|\mathrm{Im}\left(\delta^d_{LL}\right)_{12}\left(\delta^d_{RR}\right)_{12}\right\|}$	1.3×10^{-4}
$B^0 - \bar{B}^0$	$\sqrt{\left\|\mathrm{Re}\left(\delta^d_{LL}\right)^2_{13}\right\|}$	9.8×10^{-2}
	$\sqrt{\left\|\mathrm{Re}\left(\delta^d_{LR}\right)^2_{13}\right\|}$ for $\left\|(\delta^d_{LR})_{13}\right\| = \left\|(\delta^d_{RL})_{13}\right\|$	3.3×10^{-2}
	$\sqrt{\left\|\mathrm{Re}\left(\delta^d_{LL}\right)_{13}\left(\delta^d_{RR}\right)_{13}\right\|}$	1.8×10^{-2}
ϵ'/ϵ	$\left\|\mathrm{Im}\left(\delta^d_{LL}\right)_{12}\right\|$	4.8×10^{-1}
	$\left\|\mathrm{Im}\left(\delta^d_{LR}\right)_{12}\right\|$ for $\left\|(\delta^d_{LR})_{12}\right\| = \left\|(\delta^d_{RL})_{12}\right\|$	2.0×10^{-5}
$b \to s\gamma$	$\left\|(\delta^d_{LL})_{23}\right\|$	8.2
	$\left\|(\delta^d_{LR})_{23}\right\|$ for $\left\|(\delta^d_{LR})_{23}\right\| = \left\|(\delta^d_{RL})_{23}\right\|$	1.6×10^{-2}
d_N	$\left\|\mathrm{Im}\left(\delta^d_{LR}\right)_{11}\right\|$	1.8×10^{-6}
	$\left\|\mathrm{Im}\left(\delta^u_{LR}\right)_{11}\right\|$	3.6×10^{-6}

12.1.3 Minimal flavor violation

In the case where $\Lambda_F \gg \Lambda_S$, one may expect that the Yukawa couplings are the only source of flavor and CP violation. This is the so-called minimal flavor violation hypothesis [93].

Let us consider again the gauge group $U(3)^5$ of unitary transformations that commutes with the gauge symmetry group. One may write it as

$$
SU(3)_{Q_L} \times SU(3)_{U_R} \times SU(3)_{D_R} \times SU(3)_{L_L} \times SU(3)_{E_R}
$$
$$
\times\, U(1)_B \times U(1)_L \times U(1)_Y \times U(1)_{DE} \times U(1)_E
$$

where the horizontal nonabelian $SU(3)$ symmetries refer to the different types of fermions (*i.e.* different sets of gauge quantum numbers) and $U(1)_{DE}$ transforms only D_R and E_R (in the same way) whereas $U(1)_E$ transforms only E_R. This symmetry is broken by the Yukawa couplings (12.3). As usual, one may restore it by treating the Yukawa couplings as spurion fields transforming under $SU(3)_{Q_L} \times SU(3)_{U_R} \times SU(3)_{D_R} \times SU(3)_{L_L} \times SU(3)_{E_R}$ as

$$
\Lambda_u \sim (\mathbf{3}, \bar{\mathbf{3}}, \mathbf{1}; \mathbf{1}, \mathbf{1}), \quad \Lambda_d \sim (\mathbf{3}, \mathbf{1}, \bar{\mathbf{3}}; \mathbf{1}, \mathbf{1}), \quad \Lambda_\ell \sim (\mathbf{1}, \mathbf{1}, \mathbf{1}; \mathbf{3}, \bar{\mathbf{3}}). \tag{12.11}
$$

One may note that the leading flavor-changing corrections arise from $\Lambda_u \Lambda_u^\dagger \sim (\mathbf{8}, \mathbf{1}, \mathbf{1}; \mathbf{1}, \mathbf{1})$ since $\left(\Lambda_u \Lambda_u^\dagger\right)_{ij} \sim \lambda_t^2 V^u_{Li3} V^{u*}_{Lj3}$ (in the case of large $\tan\beta$, one should also include $\Lambda_d \Lambda_d^\dagger$).

One may, in this context, discuss corrections to the soft supersymmetry breaking parameters. We generalize the squark mass matrix formulas (5.53) and (5.54) of Chapter 5 to include flavor nonconservation

$$
\widetilde{M}_u^2 = \begin{pmatrix} \widetilde{M}_{u,LL}^2 & \widetilde{M}_{u,LR}^2 \\[2mm] \widetilde{M}_{u,RL}^2 & \widetilde{M}_{u,RR}^2 \end{pmatrix}
$$

$$
= \begin{pmatrix} M_Q^2 + \Lambda_u \Lambda_u^\dagger v_2^2 + \frac{1}{6}(4M_W^2 - M_Z^2)\cos 2\beta & v_2\left(A_u - \mu\Lambda_u \cot\beta\right) \\[3mm] v_2\left(A_u - \mu\Lambda_u \cot\beta\right)^\dagger & M_U^2 + \Lambda_u^\dagger \Lambda_u v_2^2 + \frac{2}{3}(-M_W^2 + M_Z^2)\cos 2\beta \end{pmatrix},
$$

$$
\widetilde{M}_d^2 = \begin{pmatrix} \widetilde{M}_{d,LL}^2 & \widetilde{M}_{d,LR}^2 \\[2mm] \widetilde{M}_{d,RL}^2 & \widetilde{M}_{d,RR}^2 \end{pmatrix}
$$

$$
= \begin{pmatrix} M_Q^2 + \Lambda_d \Lambda_d^\dagger v_1^2 - \frac{1}{6}(2M_W^2 + M_Z^2)\cos 2\beta & v_1\left(A_d - \mu\Lambda_d \tan\beta\right) \\[3mm] v_1\left(A_d - \mu\Lambda_d \tan\beta\right)^\dagger & M_D^2 + \Lambda_d^\dagger \Lambda_d v_1^2 + \frac{1}{3}(M_W^2 - M_Z^2)\cos 2\beta \end{pmatrix}.
$$

The soft terms are defined by (*cf.* (5.55) of Chapter 5)

$$
\mathcal{L}_{\text{soft}} = - \left(\widetilde{u}^{*}_{i_L} \left[M^2_Q \right]_{ij} \widetilde{u}_{j_L} + \widetilde{d}^{*}_{i_L} \left[M^2_Q \right]_{ij} \widetilde{d}_{j_L} \right) - \widetilde{u}^{*}_{i_R} \left[M^2_U \right]_{ij} \widetilde{u}_{j_R} - \widetilde{d}^{*}_{i_R} \left[M^2_D \right]_{ij} \widetilde{d}_{j_R}
$$
$$
- \left(\widetilde{q}_{i_L} \cdot H_1 \left[\mathcal{A}_d \right]_{ij} \widetilde{d}^{*}_{j_R} + \widetilde{q}_{i_L} \cdot H_2 \left[\mathcal{A}_u \right]_{ij} \widetilde{u}^{*}_{j_R} + \text{h.c.} \right). \tag{12.12}
$$

Now, using the transformation properties of the spurions, one may easily infer that the leading corrections to these soft terms are as follows [93]

$$
M^2_Q = \tilde{m}^2 \left(a_1 + b_1 \Lambda_u \Lambda^{\dagger}_u + b_2 \Lambda_d \Lambda^{\dagger}_d + b_3 \Lambda_d \Lambda^{\dagger}_d \Lambda_u \Lambda^{\dagger}_u + b_4 \Lambda_u \Lambda^{\dagger}_u \Lambda_d \Lambda^{\dagger}_d \right),
$$
$$
M^2_U = \tilde{m}^2 \left(a_2 + b_5 \Lambda^{\dagger}_u \Lambda_u \right),
$$
$$
M^2_D = \tilde{m}^2 \left(a_3 + b_6 \Lambda^{\dagger}_d \Lambda_d \right), \tag{12.13}
$$
$$
\mathcal{A}_u = A \left(a_4 + b_7 \Lambda_d \Lambda^{\dagger}_d \right) \Lambda_u,
$$
$$
\mathcal{A}_d = A \left(a_5 + b_8 \Lambda_u \Lambda^{\dagger}_u \right) \Lambda_d.
$$

At tree level (all b_i vanish), one recovers universality. Other terms arise through quantum corrections (see, for example, Fig. 12.2 for b_1 or b_2). We note that the condition $\Lambda_F \gg \Lambda_S$ is rather constraining, especially in the case of gravity mediation: it imposes that the high energy dynamics up to scale Λ_S satisfies minimal flavor violation.

In the case of minimal flavor violation, the unitarity triangle may be determined on the basis of $|V_{ub}/V_{cb}|$, γ, $A_{\text{CP}} \left(B \to J/\Psi K^{(*)} \right)$, β (from $BB \to D^0 h^0$), α and $\Delta m_s / \Delta m_d$ [58, 139]. We see from Fig. 12.3 that basically the same region is obtained as in the Standard Model. One may then extract upper bounds on new physics effects.

12.1.4　Alignment: family symmetries

An alternative way to solve the supersymmetric flavor problem is to make the case for small mixing angles. Indeed, a remarkable feature of the CKM mixing matrix (12.1) is its hierarchical structure: mixing angles appear in increasing powers of the small coupling $\lambda \equiv \sin\theta_c \sim 0.22$. One may go a step further since a hierarchical pattern is also observed among the masses of the quarks and the charged leptons. For example,

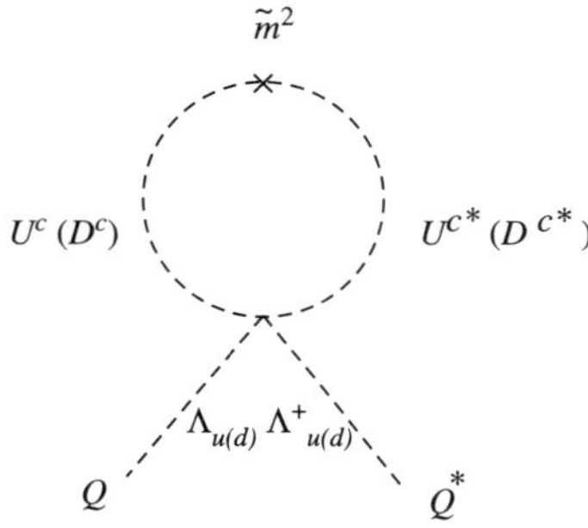

Fig. 12.2 Radiative contribution to M^2_Q through squark field loops.

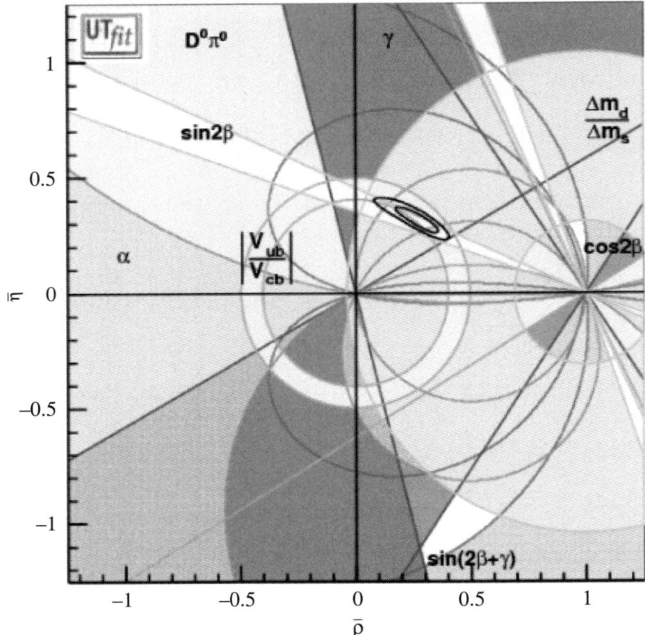

Fig. 12.3 Selected region in $(\bar{\rho}, \bar{\eta})$ plane in the case of the minimal flavor violation extension of the Standard Model [139].

when one renormalizes the quark and charged lepton masses up to the scale of grand unification, one observes the following hierarchical structure:

$$m_u : m_c : m_t \sim \lambda^8 : \lambda^4 : 1$$
$$m_d : m_s : m_b \sim \lambda^4 : \lambda^2 : 1 \tag{12.14}$$
$$m_e : m_\mu : m_\tau \sim \lambda^4 : \lambda^2 : 1$$

where only orders of magnitude are given[2]. One may extract for future use from (12.14) the following constraint

$$\frac{m_d m_s m_b}{m_e m_\mu m_\tau} \sim O(1). \tag{12.15}$$

The obvious question is whether one can explain such structures with the help of a family or horizontal symmetry.

How such a symmetry would work was explained by C. Froggatt and H. Nielsen [167, 168] more than 20 years ago, when they proposed an illustrative example which remains the prototype of such models. They assume the existence of a symmetry which requires some quark and lepton masses to be zero: a finite mass is generated at some order in a symmetry breaking interaction.

---

[2]In other words, constants of order one are not written explicitly; because the value of λ is not very small compared to one, the exponents in (12.14) are to be understood up to one unit.

Let us illustrate here this line of reasoning on an example [44]. We consider an abelian gauge symmetry $U(1)_X$ which forbids any renormalizable Yukawa coupling except the top quark coupling. Hence the Yukawa coupling matrix Λ_u in (12.3) has the form:

$$\Lambda_u = \begin{pmatrix} 0 & 0 & 0 \\ 0 & 0 & 0 \\ 0 & 0 & 1 \end{pmatrix} \tag{12.16}$$

where 1 in the last entry means a matrix element of order one. The presence of such a nonzero entry means that the charges under $U(1)_X$ obey the relation:

$$x_{Q_3} + x_{U_3^c} + x_{H_2} = 0 \tag{12.17}$$

whereas similar combinations for the other field are nonvanishing and prevent the presence of a nonzero entry elsewhere in the matrix Λ_u.

We assume that this symmetry is spontaneously broken through the vacuum expectation value of a field θ of charge x_θ normalized to -1: $\langle\theta\rangle \neq 0$. The presence of nonrenormalizable terms of the form $Q_i U_j^c H_2 (\theta/M)^{n_{ij}}$ induces in the effective theory below the scale of $U(1)_X$ breaking an effective Yukawa matrix of the form:

$$\Lambda_u = \begin{pmatrix} \lambda^{n_{11}} & \lambda^{n_{12}} & \lambda^{n_{13}} \\ \lambda^{n_{21}} & \lambda^{n_{22}} & \lambda^{n_{23}} \\ \lambda^{n_{31}} & \lambda^{n_{32}} & 1 \end{pmatrix} \tag{12.18}$$

where $\lambda = \langle\theta\rangle/M$ and

$$n_{ij} = x_{Q_i} + x_{U_j^c} + x_{H_2} \tag{12.19}$$

($n_{33}=0$). Such nonrenormalizable interactions may arise through integrating out heavy fermions of mass M as in the Froggatt–Nielsen model or appear if the underlying theory incorporates gravity, e.g. in string theories, in which case the scale M is the Planck scale m_P.

We note however that the form (12.18) is valid only if all n_{ij} are positive. If one n_{ij} is strictly negative, the corresponding entry vanishes: barring nonperturbative effects, fields appear only in the superpotential with positive exponents and holomorphicity prevents us from using θ^* in the superpotential (and thus from writing a term of the form $Q_i U_j^c H_2 (\theta^*/M)^{-n_{ij}}$). Such zero entries are thus called holomorphic zeros[3].

---

[3] Alternatively, one may introduce a vectorlike couple of scalars θ and $\bar\theta$ with opposite $U(1)_X$ charges ± 1 [233]. They acquire equal vacuum expectation values along a D-flat direction and we may write $\lambda \equiv \langle\theta\rangle/M = \langle\bar\theta\rangle/M$. In this case, there are no holomorphic zeros and $(\Lambda_u)_{ij} \sim \lambda^{|n_{ij}|}$.

Let us denote for example the X-charges of the standard supermultiplets as given in Table 12.2 (we assume that $3a_8 + b_8 > a_3 + b_3 > 0$ and $3a_8 + b_8 > a_3 + b_3 > 0$ and similarly for $b_{3,8}$ replaced by $c_{3,8}$).

Then the CKM matrix reads [43]

$$V = \begin{pmatrix} 1 & \lambda^{2a_3} & \lambda^{3a_8+a_3} \\ \\ \lambda^{2a_3} & 1 & \lambda^{3a_8-a_3} \\ \\ \lambda^{3a_8+a_3} & \lambda^{3a_8-a_3} & 1 \end{pmatrix}. \tag{12.20}$$

One has in particular $V_{us}V_{cb} = V_{ub}$. As for the mass ratios, one finds (use the result proven in Exercise 1)

$$\frac{m_u}{m_t} \sim \lambda^{3(a_8+b_8)+a_3+b_3}, \quad \frac{m_c}{m_t} \sim \lambda^{3(a_8+b_8)-a_3-b_3}$$

$$\frac{m_d}{m_b} \sim \lambda^{3(a_8+c_8)+a_3+c_3}, \quad \frac{m_s}{m_b} \sim \lambda^{3(a_8+c_8)-a_3-c_3}. \tag{12.21}$$

For example, if $a_3 = c_3$, one obtains

$$V_{us} \sim \lambda^{2a_3} \sim \sqrt{\frac{m_d}{m_s}}, \tag{12.22}$$

which is a classical relation [175].

It might seem on this example that, by choosing the charges of the different fields, one may accommodate any observed pattern of masses. There are however constraints on the symmetry: in particular those coming from the cancellation of anomalies. This is indeed one of the reasons to choose a local gauge symmetry. We will see that this gives very interesting constraints on the model.

Table 12.2 $U(1)_X$ charges for the standard supermultiplets (e.g. the charge of Q_3 is a_0-2a_8).

	Q_1	Q_2	Q_3
$x = a_0+$	$a_8 + a_3$	$a_8 - a_3$	$-2a_8$
	U^c	C^c	T^c
$x = b_0+$	$b_8 + b_3$	$b_8 - b_3$	$-2b_8$
	D^c	S^c	B^c
$x = c_0+$	$c_8 + c_3$	$c_8 - c_3$	$-2c_8$
	L_1	L_2	L_3
$x = d_0+$	$d_8 + d_3$	$d_8 - d_3$	$-2d_8$
	E^c	M^c	T^c
$x = e_0+$	$e_8 + e_3$	$e_8 - e_3$	$-2e_8$

Anomalies

In order to make sense of the horizontal symmetry, one must make sure that the anomalies are cancelled: not only the anomaly C_X corresponding to the triangle diagram with three $U(1)_X$ gauge bosons, but also the mixed anomalies C_i corresponding with triangle diagrams with one $U(1)_X$ gauge boson and two gauge bosons of the Standard Model gauge symmetry: $U(1)_Y$, $SU(2)$ and $SU(3)$ for $i \in \{1, 2, 3\}$, respectively.

Since they are linear in the $U(1)_X$ charges the coefficients C_i depend only on the family independent part of the quantum numbers, respectively a_0, b_0, c_0, d_0 and e_0 for Q_i, U_i^c, D_i^c, L_i and E_i^c. They read explicitly

$$C_1 = a_0 + 8b_0 + 2c_0 + 3d_0 + 6e_0 + h_1 + h_2$$

$$C_2 = 3(3a_0 + d_0) + h_1 + h_2 \tag{12.23}$$

$$C_3 = 3(2a_0 + b_0 + c_0)$$

where h_1 and h_2 are the x-charges of H_1 and H_2.

On the other hand, mass ratios are also given by the X-charges. In the example given above, which is fairly general, one finds:

$$m_u m_c m_t = v_2^3 \det \Lambda_u \sim \lambda^{3(a_0+b_0+h_2)} \tag{12.24}$$

$$m_d m_s m_b = v_1^3 \det \Lambda_d \sim \lambda^{3(a_0+c_0+h_1)} \tag{12.25}$$

$$m_e m_\mu m_\tau = v_1^3 \det \Lambda_e \sim \lambda^{3(d_0+e_0+h_1)}. \tag{12.26}$$

Anomaly cancellation would require $C_1 = C_2 = C_3 = \cdots = 0$ which gives, after a redefinition of the x charge (see Exercise 2), $a_0 + b_0 = a_0 + c_0 = 0$ and $3(d_0 + e_0) = -(h_1 + h_2)$. Then comparing (12.24) and (12.25) with the data (12.14) yields $h_1 = 2$ and $h_2 = 4$. The last equation (12.26) is then incompatible with the same data. This shows that the observed pattern of masses is incompatible with a nonanomalous family symmetry: the $U(1)_X$ symmetry must be anomalous [33, 44, 233]. Is this the end of the story?

Before we address this question, let us derive from the mass spectrum (12.14) a relation among the anomaly coefficients. We have

$$\frac{m_d m_s m_b}{m_e m_\mu m_\tau} = \lambda^{3(a_0+c_0-d_0-e_0)} = \lambda^{h_1+h_2-(C_1+C_2-\frac{8}{3}C_3)/2}. \tag{12.27}$$

Assuming the presence of a mu-term $\mu H_2 \cdot H_1$ imposes that $h_1 + h_2 = 0$. Since the data (12.15) imposes this ratio of masses to be of order one, one then finds:

$$C_1 + C_2 - \frac{8}{3}C_3 = 0. \tag{12.28}$$

We have encountered in Section 10.4.5 of Chapter 10 a seemingly anomalous symmetry: in superstring models, there is a $U(1)$ symmetry whose anomaly is compensated by the four-dimensional version [116] of the Green–Schwarz mechanism [205]. This is possible through the couplings of the gauge fields to a dilaton–axion–dilatino supermultiplet: the anomalous terms that arise when we perform a gauge transformation are cancelled by a Peccei–Quinn transformation of the axion. The necessary condition for the cancellation of anomalies à *la* Green–Schwarz is

$$\frac{C_1}{k_1} = \frac{C_2}{k_2} = \frac{C_3}{k_3} = \frac{C_X}{k_X} = \delta_{GS}, \tag{12.29}$$

where the k_i are the Kač–Moody levels. Combined with the gauge unification condition $k_1 g_1^2(M) = k_2 g_2^2(M) = k_3 g_3^2(M) = k_X g_X^2(M)$, this gives

$$\tan^2\theta_w(M) = \frac{g_1^2(M)}{g_2^2(M)} = \frac{k_2}{k_1} = \frac{C_2}{C_1}. \tag{12.30}$$

The relation (12.28) can now be discussed in this context. For example, in the standard case where all the nonabelian symmetries appear at the same Kac–Moody level, $k_2 = k_3$ and thus $C_2 = C_3$, we find:

$$C_1 = \frac{5}{3}C_2 \tag{12.31}$$

and

$$\sin^2\theta_w(M) = \frac{3}{8}. \tag{12.32}$$

Thus, if the horizontal abelian symmetry is precisely the pseudo-anomalous $U(1)_X$, observed hierarchies of fermion masses are compatible with the standard value of $\sin^2\theta_w$ at gauge coupling unification (see (9.35)). And this result is obtained without ever making reference to a grand unified gauge group (which is rarely present in superstring models).

There is another advantage of working in the context of string models. Indeed, in this case, the properties of the anomalous $U(1)_X$ are constrained. For example in the case of the weakly coupled heterotic string model, the absolute normalization is fixed [13] and

$$\lambda^2 = \frac{\langle\theta\rangle^2}{M^2} = \frac{g^2}{192\pi^2}\,\mathrm{Tr}\,X \sim 10^{-2}\ \text{to}\ 10^{-1}. \tag{12.33}$$

Hence, one naturally obtains the small parameter (the Cabibbo angle) that was necessary for the whole picture to make sense.

Let us stress, however, the main drawback of this approach. In all the preceding formulas we have neglected factors of order one and discussed only the orders of magnitude as powers of the small parameter $\lambda \sim \sin\theta_c \sim 1/5$. But since this is not such a small parameter, the actual value of the factor of order one introduces some uncertainties: for example $\lambda^n/2 \sim 3\lambda^{n+1}$.

Now, returning to our original motivation: does this help to align sfermion and fermion mass eigenstates? Obviously, sfermion masses are also constrained by the symmetry $U(1)_X$ ([272, 273, 297]).

Let us consider the LL squark mass matrix. If $x_{Q_i} > x_{Q_j}$, the following mass term is allowed:

$$\tilde{m}^2 \tilde{q}_i \tilde{q}_j^* \left(\frac{\theta}{M}\right)^{x_{Q_i} - x_{Q_j}} + \text{h.c.} \tag{12.34}$$

where $\tilde{m}$ is an overall supersymmetry-breaking scale, and if $x_{Q_j} > x_{Q_i}$,

$$\tilde{m}^2 \tilde{q}_i \tilde{q}_j^* \left(\frac{\theta^\dagger}{M}\right)^{x_{Q_j} - x_{Q_i}} + \text{h.c.} \tag{12.35}$$

(the use of hermitian conjugates of fields is allowed since this is a supersymmetry-breaking contribution).

Thus, after $U(1)_X$ breaking, this yields $\tilde{m}^2 \tilde{q}_i \tilde{q}_j^* \lambda^{|x_{Q_i} - x_{Q_j}|} + \text{h.c.}$, where $\lambda = \langle\theta\rangle/M = \langle\theta^\dagger\rangle/M$ and the scalar squared mass matrix reads:

$$\tilde{M}^2_{q,LL} \sim \tilde{m}^2 \begin{pmatrix} 1 & \lambda^{|x_{Q_1} - x_{Q_2}|} & \lambda^{|x_{Q_1} - x_{Q_3}|} \\ \lambda^{|x_{Q_2} - x_{Q_1}|} & 1 & \lambda^{|x_{Q_2} - x_{Q_3}|} \\ \lambda^{|x_{Q_3} - x_{Q_1}|} & \lambda^{|x_{Q_3} - x_{Q_2}|} & 1 \end{pmatrix}. \tag{12.36}$$

Thus the squark mass matrices are approximately diagonal: the family symmetry induces a partial alignment [297].

One finds for the parameters introduced in (12.6)

$$(\delta^q_{LR})_{ij} \sim \frac{\sqrt{m_{q_i} m_{q_j}}}{\tilde{m}} \ll 1, \tag{12.37}$$

and more constraining conditions for δ_{LL} and δ_{RR}. Typically [125],

$$\left(\delta^d_{12}\right)^2 = (x_{Q_1} - x_{Q_2})(x_{D_1^c} - x_{D_2^c}) \left(V_L^Q\right)_{12} \left(V_L^{D^c}\right)_{12} \frac{m_d}{m_s} \tag{12.38}$$

where the matrix elements on the right-hand side are to be renormalized down to low energies where they tend to decrease.

This still gives some severe constraints on the models that do not seem to be satisfied without the presence of holomorphic zeros in the down quark mass matrix [296].

Neutrino masses

The neutrino sector is specific since both Dirac and Majorana mass terms are allowed for neutral leptons (see Appendix Appendix A, Section A.3). Introducing a right-handed neutrino N_R, we write the Dirac mass term as

$$\mathcal{L}_{\text{Dirac}} = -m_D \bar{\nu}_L N_R + \text{h.c.} \tag{12.39}$$

where m_D arises through $SU(2) \times U(1)$ breaking and is of the order of the electroweak scale. On the other hand, a Majorana mass term involves a single chirality. Standard

Model gauge symmetry forbids such a term for the left-handed neutrino but not for the right-handed neutrino:

$$\mathcal{L}_{\text{Majorana}} = -\tfrac{1}{2}M\overline{(N^c)_L}N_R + \text{h.c.} \tag{12.40}$$

where M is not constrained by electroweak symmetry. The seesaw model [177, 383], which represents the prototype of neutrino mass models in all theories which involve large scales such as grand unified or superstring theories, includes both Dirac and Majorana mass terms:

$$\mathcal{L} = -\tfrac{1}{2}(\overline{\nu}_L \ \overline{(N^c)_L})\begin{pmatrix} 0 & m_D \\ m_D & M \end{pmatrix}\begin{pmatrix} (\nu^c)_R \\ N_R \end{pmatrix} + \text{h.c.} \tag{12.41}$$

In the case where $M \gg m_D$, the eigenvalues are respectively:

$$m_1 \sim \frac{m_D^2}{M}, \quad m_2 \sim M \tag{12.42}$$

and, due to the presence of a zero in the matrix, the mixing angle is given in terms of mass ratios:

$$\tan\theta \sim \sqrt{\frac{m_1}{m_2}} \sim \frac{m_D}{M}. \tag{12.43}$$

We have discussed the case of one family but the discussion easily generalizes to the three-family case with a mass matrix of the form:

$$\mathcal{M} = \begin{pmatrix} 0 & \mathcal{M}_D \\ \mathcal{M}_D & \mathcal{M}_M \end{pmatrix}, \tag{12.44}$$

where the Dirac and Majorana mass matrices, respectively $\mathcal{M}_D$ and $\mathcal{M}_M$, are 3×3 matrices. Then the light neutrino mass matrix reads:

$$\mathcal{M}_\nu = -\mathcal{M}_D\mathcal{M}_M^{-1}\mathcal{M}_D^T. \tag{12.45}$$

Of course, given the freedom we have on each of the specific entries in $\mathcal{M}_D$ and $\mathcal{M}_M$, seesaw models really form a class of models and one has to go to specifics in order to discuss their phenomenology. This is precisely what a family symmetry provides us with.

In order to discuss neutrino masses in this context, we introduce one right-handed neutrino for each family: N_i^c, $i \in \{1, 2, 3\}$. The neutrino Dirac mass term is generated from the nonrenormalizable couplings:

$$L_i \cdot H_2 N_j^c \left(\frac{\theta}{M}\right)^{p_{ij}} \tag{12.46}$$

where L_i is the left-handed lepton doublet and $p_{ij} = x_{L_i} + x_{N_j^c} + x_{H_2}$ is assumed to be positive (otherwise, this coupling is absent, which leads to a supersymmetric zero in the mass matrix). This yields a Dirac mass matrix:

$$(\mathcal{M}_D)_{ij} \sim \langle H_2 \rangle \left(\frac{\langle \theta \rangle}{M} \right)^{p_{ij}} . \tag{12.47}$$

The entries of the Majorana matrix $\mathcal{M}_M$ are generated in the same way, with nonrenormalizable interactions of the form:

$$M N_i^c N_j^c \left(\frac{\theta}{M} \right)^{q_{ij}} \tag{12.48}$$

giving rise to effective Majorana masses

$$(\mathcal{M}_M)_{ij} \sim M \left(\frac{\langle \theta \rangle}{M} \right)^{q_{ij}} \tag{12.49}$$

provided that $q_{ij} = x_{N_i^c} + x_{N_j^c}$ is a positive integer (see above).

The order of magnitude of the entries of the light neutrino mass matrix $\mathcal{M}_\nu = -\mathcal{M}_D \mathcal{M}_M^{-1} \mathcal{M}_D^T$ are therefore fixed by the $U(1)_X$ symmetry.

Let us suppose that all entries of $\mathcal{M}_D$ and $\mathcal{M}_M$ are nonzero (*i.e.* $p_{ij}, q_{ij} \geq 0$); one obtains [43]:

$$(\mathcal{M}_\nu)_{ij} \sim \frac{\langle H_2 \rangle^2}{M} \lambda^{x_{L_i} + x_{L_j} + 2x_{H_2}} \tag{12.50}$$

which leads to the following light neutrino masses and lepton mixing matrix:

$$m_{\nu_i} \sim \frac{\langle H_2 \rangle^2}{M} \lambda^{2x_{L_i} + 2x_{H_2}}, \quad U_{ij} \sim \lambda^{|x_{L_i} - x_{L_j}|}. \tag{12.51}$$

One therefore finds that the neutrino spectrum is hierarchical. Moreover, the structure of the lepton mixing matrix is very similar to the CKM matrix (12.1) with generically small mixing angles: $U_{ij}^2 \sim m_{\nu_i}/m_{\nu_j}$ for $m_{\nu_i} < m_{\nu_j}$. This is in disagreement with the experimental evidence for large angles both for $\mu - \tau$ mixing (atmospheric neutrinos) and $\mu - e$ mixing (solar neutrinos).

It would be wrong, however, to consider that a hierarchical spectrum is a generic feature of this type of models. Indeed, one can work out models which allow for degeneracies in the light neutrino spectrum [42]. It is easy to see that such degeneracies are associated in these models with large mixings.

One may note that the situation in the neutrino sector might be very different from the quark and charged lepton sector where one Yukawa coupling (the top) dominates over all the rest, thus providing a clear starting point for the $U(1)_X$ symmetric situation, summarized in (12.16). In the case of neutrinos, the $U(1)_X$ symmetry, even though it is abelian, may induce some degeneracies. Consider for example the following matrix:

$$\mathcal{M}_\nu = \begin{pmatrix} 0 & 0 & 0 \\ 0 & 0 & a \\ 0 & a & 0 \end{pmatrix} \tag{12.52}$$

where a is a number of order one. This pattern corresponds to the conservation of a combination of lepton number à la Zeldovich-Konopinsky–Mahmoud ([260, 386]): indeed $\mathbf{L_e}$ and $\mathbf{L_\mu} + \mathbf{L_\tau}$ are separately conserved. It has two degenerate eigenvalues and the corresponding diagonalizing matrix R_ν has one large mixing angle:

$$D_\nu = \begin{pmatrix} 0 & 0 & 0 \\ 0 & -a & 0 \\ 0 & 0 & a \end{pmatrix} \qquad R_\nu = \begin{pmatrix} 1 & 0 & 0 \\ 0 & \frac{1}{\sqrt{2}} & \frac{1}{\sqrt{2}} \\ 0 & -\frac{1}{\sqrt{2}} & \frac{1}{\sqrt{2}} \end{pmatrix} \tag{12.53}$$

where $D_\nu = R_\nu^T \mathcal{M}_\nu R_\nu$.

Of course, as soon as the $U(1)_X$ symmetry is broken, the vanishing entries in (12.52) are filled by powers of the small parameter λ. This lifts the degeneracy at a level which is fixed by the charges under the $U(1)_X$ symmetry.

The large mixing angles observed in the neutrino sector may also point towards a nonabelian nature of the family symmetry, for example $SU(3)$. This goes beyond the scope of this book and we refer the reader to reviews on the subject such as [325].

12.1.5 Split supersymmetry

If one is ready to raise the mass of the supersymmetric particles to alleviate the flavor and CP problem, one may wonder how far one can go. Obviously, the fine tuning problem becomes more acute. However, we will see in the next section that supersymmetric theories have to deal with an even more severe problem of fine tuning associated with the vacuum energy. This has recently led several groups to propose to set aside the naturalness problem that was one of the bases of low energy supersymmetry: if the supersymmetric spectrum is heavy enough, one may expect to ease problems with flavor and CP violation, fast proton decay through dimension-5 operators, tight Higgs mass limits etc.

The next issue is obviously the status of gauge coupling unification. Arkani-Hamed and Dimopoulos [9] have shown that unification can still be achieved in a supersymmetric model, now referred to as split supersymmetry [192], where all scalars but one Higgs doublet are much heavier than the electroweak scale. Next comes the question of a dark matter candidate: if the presence of such a "light" state is imposed on the theory, this leaves hope to find some direct or indirect signal of this type of models at high energy colliders.

12.2 Cosmological constant

As is well known, the cosmological constant appears as a constant in Einstein's equations:

$$R_{\mu\nu} - \tfrac{1}{2}g_{\mu\nu}R = 8\pi G_N T_{\mu\nu} + \lambda g_{\mu\nu}, \tag{12.54}$$

where G_N is Newton's constant, $T_{\mu\nu}$ is the energy–momentum tensor and $R_{\mu\nu}$ the Ricci tensor, which is obtained from the Riemann tensor measuring the curvature of spacetime (see Appendix D). The cosmological constant λ is thus of the dimension of an inverse length squared. It was introduced by [127] in order to build a static Universe model, its repulsive effect compensating the gravitational attraction, but, as we will now see, constraints on the expansion of the Universe impose for it a very small upper limit.

It is more convenient to work in the specific context of a homogeneous and isotropic Friedmann–Lemaître universe, with a Robertson–Walker metric:

$$ds^2 = dt^2 - a^2(t)\left[\frac{dr^2}{1 - kr^2} + r^2\left(d\theta^2 + \sin^2\theta d\phi^2\right)\right], \tag{12.55}$$

where $a(t)$ is the cosmic scale factor, which is time dependent in an expanding or contracting Universe . Implementing energy conservation into the Einstein equations then leads to the Friedmann equation, which gives an expression for the Hubble parameter H measuring the rate of the expansion of the Universe:

$$H^2 \equiv \frac{\dot{a}^2(t)}{a^2(t)} = \frac{1}{3}\left(\lambda + 8\pi G_N \rho\right) - \frac{k}{a^2}. \tag{12.56}$$

In this equation, we use standard notation: $\dot{a}$ is the time derivative of the cosmic scale factor, $\rho = T^0{}_0$ is the energy density and the term proportional to k is a spatial curvature term (see (12.55)). Note that the cosmological constant appears as a constant contribution to the Hubble parameter.

Evaluating each term of the Friedmann equation at present time t_0 allows for a rapid estimation of an upper limit on λ (for a more precise determination see Appendix D). Indeed, we have for the Hubble constant H_0 i.e. the present value of the Hubble parameter, $H_0 = h_0 \times 100$ km.s^{-1}Mpc^{-1} with h_0 of order one, whereas the present energy density ρ_0 is certainly within one order of magnitude of the critical energy density $\rho_c = 3H_0^2/(8\pi G_N) = h_0^2\,2.10^{-26}$ kg m^{-3}; moreover the spatial curvature term certainly does not represent presently a dominant contribution to the expansion of the Universe. Thus, (12.56) considered at present time implies the following constraint on λ:

$$|\lambda| \leq H_0^2. \tag{12.57}$$

In other words, the length scale $\ell_\Lambda \equiv |\lambda|^{-1/2}$ associated with the cosmological constant must be larger than the Hubble length $\ell_{H_0} \equiv cH_0^{-1} = h_0^{-1}\, 10^{26}$ m, and thus be a cosmological distance.

This is not a problem as long as one remains classical: ℓ_{H_0} provides a natural cosmological scale for our present Universe. The problem arises when one tries to combine gravity with quantum theory. Indeed, from Newton's constant *and* the Planck constant $\hbar$, we have seen that we can construct the Planck mass scale $m_P = \sqrt{\hbar c/(8\pi G_N)} = 2.4 \times 10^{18}$ GeV/c^2. The corresponding length scale is the Planck length

$$\ell_P = \frac{\hbar}{m_P c} = 8.1 \times 10^{-35} \text{ m.} \tag{12.58}$$

The above constraint now reads:

$$\ell_\Lambda \equiv |\lambda|^{-1/2} \geq \ell_{H_0} = \frac{c}{H_0} \sim 10^{60}\, \ell_P. \tag{12.59}$$

In other words, there are more than 60 orders of magnitude between the scale associated with the cosmological constant and the scale of quantum gravity.

A rather obvious solution is to take $\lambda = 0$. This is as valid a choice as any other in a pure gravity theory. Unfortunately, it is an unnatural one when one introduces any kind of matter. Indeed, set λ to zero but assume that there is a nonvanishing vacuum (*i.e.* ground state) energy: $\langle T_{\mu\nu}\rangle = \rho_{\text{vac}} g_{\mu\nu}$; then the Einstein equations (12.54) read

$$R_{\mu\nu} - \tfrac{1}{2} g_{\mu\nu} R = 8\pi G_N T_{\mu\nu} + 8\pi G_N \rho_{\text{vac}} g_{\mu\nu}. \tag{12.60}$$

The last term is interpreted as an effective cosmological constant (from now on, we set $\hbar = c = 1$):

$$\lambda_{\text{eff}} = 8\pi G_N \rho_{\text{vac}} \equiv \frac{\Lambda^4}{m_P^2}. \tag{12.61}$$

Generically, ρ_{vac} receives a nonzero contribution from symmetry breaking: for instance, the scale Λ would be typically of the order of 100 GeV in the case of the electroweak gauge symmetry breaking or 1 TeV in the case of supersymmetry breaking. But the constraint (12.59) now reads:

$$\Lambda \leq 10^{-30}\, m_P \sim 10^{-3} \text{ eV.} \tag{12.62}$$

It is this very unnatural fine tuning of parameters (in explicit cases ρ_{vac} and thus Λ are functions of the parameters of the theory) that is referred to as the *cosmological constant problem*, or more accurately the vacuum energy problem.

The most natural reason why vacuum energy would be vanishing is a symmetry argument. We have seen in Section 2.3 of Chapter 2 that global supersymmetry indeed provides such a rationale. The problem is that, at the same time, supersymmetry predicts equal boson and fermion masses and therefore needs to be broken. The amount of breaking necessary to push the supersymmetric partners high enough not to have been observed yet, is incompatible with the limit (12.62).

Moreover, in the context of cosmology, we should consider supersymmetry in a gravity context and thus work with its local version, supergravity. We have seen in Section 6.3.1 of Chapter 6 that, in this context, the criterion of vanishing vacuum energy is traded for one of vanishing gravitino mass. Local supersymmetry is then absolutely compatible with a nonvanishing vacuum energy, preferably a negative one (although possibly also a positive one). This is both a blessing and a problem: supersymmetry may be broken while the cosmological constant remains small, but we have lost our rationale for a vanishing, or very small, cosmological constant and fine tuning raises again its ugly head.

In some supergravity theories, however, one may recover the vanishing vacuum energy criterion.

Over the last few years, there has been an increasing number of indications that the Universe is presently undergoing accelerated expansion. This appears to be a strong departure from the standard picture of a matter-dominated universe. Indeed, the standard equation for the conservation of energy,

$$\dot{\rho} = -3(p + \rho)H, \tag{12.63}$$

allows us to derive from the Friedmann equation (12.56), written in the case of a Universe dominated by a component with energy density ρ and pressure p:

$$\frac{\ddot{a}}{a} = -\frac{4\pi G_N}{3}(\rho + 3p). \tag{12.64}$$

Obviously, a matter-dominated $(p \sim 0)$ Universe is decelerating. One needs instead a component with a negative pressure.

A cosmological constant is associated with a contribution to the energy–momentum tensor as in (12.60), (12.61):

$$T^\mu_\nu = \Lambda^4 \delta^\mu_\nu = (\rho, -p, -p, -p). \tag{12.65}$$

The associated equation of motion is therefore

$$p = -\rho. \tag{12.66}$$

It follows from (12.64) that a cosmological constant tends to accelerate expansion.

As explained in Appendix D, recent cosmological observation expressed in the plane $(\Omega_M, \Omega_\Lambda)$ leads to the constraint given in the plot of Fig. D.2. It is certainly remarkable that a very diverse set of data singles out the same region in this parameter space: $\Omega_M \sim 0.2$ to 0.3 and $\Omega_\Lambda \sim 0.7$ to 0.8.

This raises a new problem. Since matter and a cosmological constant evolve very differently, why should they be of the same order at present times? Indeed, for a component of equation of state $p = w\rho$, we may rewrite (12.63) as

$$\frac{\dot{\rho}}{\rho} = -3\frac{\dot{a}}{a}(1 + w). \tag{12.67}$$

Thus matter $(p \sim 0)$ energy density evolves as a^{-3} whereas a cosmological constant stays constant, as expected. Why should they be presently of similar magnitude (see

Fig. 12.4)? This is known as the *cosmic coincidence problem*. In order to avoid any reference to us (and hence any anthropic interpretation, see below), we may rephrase the problem as follows: why does the vacuum energy starts to dominate at a time t_Λ (redshift $z_\Lambda \sim 1$) which almost coincides with the epoch t_G (redshift $z_G \sim 3$ to 5) of galaxy formation?

12.2.1 Relaxation mechanisms

From the point of view of high energy physics, it is however difficult to imagine a rationale for a pure cosmological constant, especially if it is nonzero but small compared to the typical fundamental scales (electroweak, strong, grand unified, or Planck scale). There should be some dynamics associated with this form of energy.

For example, we have seen in Chapter 10 that, in the context of string models, any dimensionful parameter is expressed in terms of the fundamental string scale M_S and of vacuum expectation values of scalar fields. The physics of the cosmological constant would then be the physics of the corresponding scalar fields. Indeed, it is difficult to envisage string theory in the context of a true cosmological constant. The corresponding spacetime is known as de Sitter spacetime and has an event horizon (see Appendix D). This is difficult to reconcile with the S-matrix approach of string theory in the context of conformal invariance. More precisely, in the S-matrix approach, states are asymptotically (*i.e.* at times $t \to \pm\infty$) free and interact only at finite times: the S-matrix element between an incoming set of free states and an outgoing set yields the probability associated with such a transition. In string theory, the states are strings and a diagram such as the one given in Fig. 10.5 of Chapter 10 gives a contribution to the S-matrix element. But conformal invariance, which we noted in Chapter 10 to be a key element, imposes that the string world-sheet can be deformed at will: this is difficult to reconcile with the presence of a horizon and the requirement of asymptotically free states. Recent advances have been obtained in this domain by the use of fluxes.

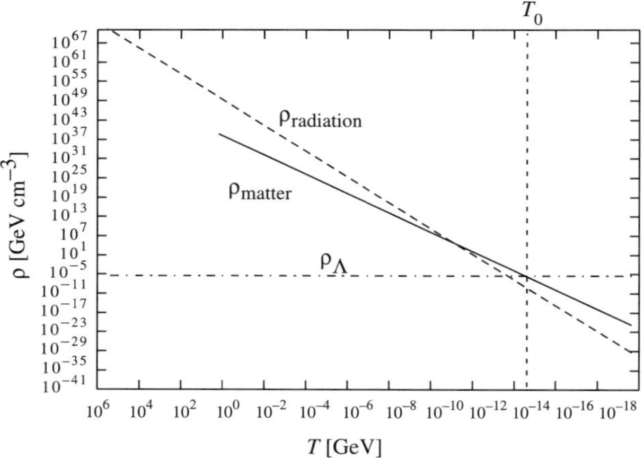

Fig. 12.4 Evolution of radiation, matter and vacuum energy densities with temperature.

Steven Weinberg [361] has constrained the possible mechanisms for the relaxation of the cosmological constant by proving the following "no-go" theorem: *it is not possible to obtain a vanishing cosmological constant as a consequence of the equations of motion of a finite number of fields.*

Indeed, let us consider N such fields φ_n, $n = 1, \ldots, N$. In the equilibrium configuration these fields are constant and their equations of motion simply read

$$\frac{\delta \mathcal{L}}{\delta \varphi_n} = 0. \tag{12.68}$$

Remembering that $\lambda_{\text{eff}} \sim \langle T^\mu{}_\mu \rangle$ where the energy-momentum tensor may be obtained from the metric[4], we see that the vanishing of the cosmological constant is a consequence of the equations (12.68) if we can find N functions $f_n(\varphi)$ such that

$$2g_{\mu\nu} \frac{\delta \mathcal{L}}{\delta g_{\mu\nu}} = \sum_n \frac{\delta \mathcal{L}}{\delta \varphi_n} f_n(\varphi). \tag{12.69}$$

This amounts to a symmetry condition, the invariance of the Lagrangian $\mathcal{L}$ under

$$\delta g_{\mu\nu} = 2\alpha g_{\mu\nu}, \quad \delta \varphi_n = -\alpha f_n(\varphi). \tag{12.70}$$

However, one can redefine the fields φ_n, $n = 1, \ldots, N$, into σ_a, $a = 1, \ldots, N - 1$, and φ in such a way that the invariance reads

$$\delta g_{\mu\nu} = 2\alpha g_{\mu\nu}, \quad \delta \sigma_a = 0, \quad \delta \varphi = -\alpha. \tag{12.71}$$

The Lagrangian which satisfies this invariance is written

$$\mathcal{L} = \sqrt{\text{Det } (e^{2\varphi} g_{\mu\nu})} \mathcal{L}_0(\sigma) = e^{4\varphi} \sqrt{|g|} \, \mathcal{L}_0(\sigma) \tag{12.72}$$

which does not provide a solution to the relaxation of the cosmological constant, as can be seen by redefining the metric: $\hat{g}_{\mu\nu} = e^{2\varphi} g_{\mu\nu}$ (in the new metric, the field φ has only derivative couplings).

Obviously, Weinberg's no-go theorem relies on a series of assumptions: Lorentz invariance, *finite* number of *constant* fields, possibility of globally redefining these fields, etc.

An example of a relaxation mechanism is provided by the Brown–Teitelboim mechanism [53, 54] where the quantum creation of closed membranes leads to a reduction of the vacuum energy inside. This is easier to understand on a toy model with a single spatial dimension.

Let us thus consider a line and establish along it a constant electric field $E_0 > 0$: the corresponding (vacuum) energy is $E_0^2/2$. Quantum creation of a pair of $\pm q$-charged particles ($q > 0$) leads to the formation of a region (between the two charges) where the electric field is partially screened to the value $E_0 - q$ and thus the vacuum energy is decreased to the value $(E_0 - q)^2/2$. Quantum creation of pairs in the new region will

---

[4]See equation (D.15).

subsequently decrease the value of the vacuum energy. The process ends in flat space when the electric field reaches the value $E \leq q/2$ because it then becomes insufficient to separate the pairs created.

In a truly three-dimensional universe, the quantum creation of pairs is replaced by the quantum creation of membranes and the one-dimensional electric field is replaced by a tensor field $A_{\mu\nu\rho}$. There are two potential problems with such a relaxation of the cosmological constant.

First, since the region of small cosmological constant originates from regions with large vacuum energies, hence exponential expansion, it is virtually empty: matter has to be produced through some mechanism yet to be specified. The second problem has to do with the multiplicity of regions with different vacuum energies: why should we be in the region with the smallest value? Such questions are crying for an anthropic type of answer: some regions of spacetime are preferred because they allow the existence of observers.

More generally, the anthropic principle approach can be sketched as follows. We consider regions of spacetime with different values of t_G (time of galaxy formation) and t_Λ, the time when the cosmological constant starts to dominate, *i.e.* when the Universe enters a de Sitter phase of exponential expansion. Clearly galaxy formation must precede this phase otherwise no observer (similar to us) would be able to witness it. Thus $t_G \leq t_\Lambda$. On the other hand, regions with $t_\Lambda \gg t_G$ have not yet undergone any de Sitter phase of reacceleration and are thus "phase-space suppressed" compared with regions with $t_\Lambda \sim t_G$. Hence the regions favored have $t_\Lambda \overset{>}{\sim} t_G$ and thus $\rho_\Lambda \sim \rho_M$.

12.2.2   Dark energy

In this section, we will take a slightly different route. We assume that some unknown mechanism relaxes the vacuum energy to zero or to a very small value. We then introduce some new dynamical component which accounts for the present observations. Introducing dynamics generally modifies the equation of state (12.66) of this new component of the energy density to the more general form with negative pressure[5]:

$$p = w\rho, \quad w < 0. \tag{12.73}$$

For example, a network of light, nonintercommuting topological defects [343, 355] gives $w = -n/3$ where n is the dimension of the defect, *i.e.* 1 for a string and 2 for a domain wall. The equation of state for a minimally coupled scalar field necessarily satisfies the condition $w \geq -1$.

Experimental data may constrain such a dynamical component, referred to in the literature as dark energy, just as it did with the cosmological constant. For example, in a spatially flat Universe with only matter and an unknown component X with equation of state $p_X = w_X \rho_X$, one obtains from (12.64) with $\rho = \rho_M + \rho_X$, $p = w_X \rho_X$ the following form for the deceleration parameter (see Section D.2.1 of Appendix D)

$$q_0 = -\frac{\ddot{a}_0 a_0}{\dot{a}_0^2} = \frac{\Omega_M}{2} + (1 + 3w_X)\frac{\Omega_X}{2}, \tag{12.74}$$

where $\Omega_X = \rho_X/\rho_c$. Supernovae results give a constraint on the parameter w_X.

---

[5] We recall that nonrelativistic matter (dust) has an equation of state $p \sim 0$ whereas $p = \rho/3$ corresponds to radiation.

Another important property of dark energy is that it does not appear to be clustered (just as a cosmological constant). Otherwise, its effects would have been detected locally, as for the case of dark matter.

A particularly interesting candidate in the context of fundamental theories is a scalar[6] field ϕ slowly evolving in its potential $V(\phi)$. Indeed, since the corresponding energy density and pressure are, for a minimally coupled scalar field,

$$\rho_\phi = \tfrac{1}{2}\dot\phi^2 + V(\phi) \tag{12.75}$$

$$p_\phi = \tfrac{1}{2}\dot\phi^2 - V(\phi), \tag{12.76}$$

we have

$$w_\phi = \frac{\tfrac{1}{2}\dot\phi^2 - V(\phi)}{\tfrac{1}{2}\dot\phi^2 + V(\phi)}. \tag{12.77}$$

If the kinetic energy is subdominant ($\dot\phi^2/2 \ll V(\phi)$), we clearly obtain $-1 \le w_\phi \le 0$.

A consequence is that the corresponding speed of sound $c_s = \delta p/\delta\rho$ is of the order of the speed of light: the scalar field pressure resists gravitational clustering.

We will see below that the scalar field must be extremely light. Two situations have been considered:

- A scalar potential slowly decreasing to zero as ϕ goes to infinity ([63, 322, 365]). This is often referred to as *quintessence* or runaway quintessence.

- A very light field (pseudo-Goldstone boson) which is presently relaxing to its vacuum state [166].

In both cases one is relaxing to a position where the vacuum energy is zero. This is associated with our assumption that some unknown mechanism wipes the cosmological constant out. We discuss the two cases in turn.

Runaway quintessence

A runaway potential is frequently present in models where supersymmetry is dynamically broken. We have seen that supersymmetric theories are characterized by a scalar potential with many flat directions, *i.e.* directions ϕ in field space for which the potential vanishes. The corresponding degeneracy is lifted through dynamical supersymmetry breaking. In some instances (dilaton or compactification radius), the field expectation value $\langle\phi\rangle$ actually provides the value of the strong interaction coupling. Then at infinite ϕ value, the coupling effectively goes to zero together with the supersymmetry breaking effects and the flat direction is restored: the potential decreases monotonically to zero as ϕ goes to infinity.

Let us take the example of supersymmetry breaking by gaugino condensation in effective superstring theories. We recall briefly the discussion of Section 7.4.2. The value g_0 of the gauge coupling at the string scale M_s is provided by the vacuum expectation value of the dilaton field s (taken to be dimensionless by dividing by m_P)

---

[6]A vector field or any field which is not a Lorentz scalar must have settled down to a vanishing value, otherwise, Lorentz invariance would be spontaneously broken.

present among the massless string modes: $g_0^2 = \langle s \rangle^{-1}$. If the gauge group has a one-loop beta function coefficient $b > 0$, then the running gauge coupling becomes strong at the scale

$$\Lambda \sim M_S e^{-8\pi^2/(bg_0^2)} = M_S e^{-8\pi^2 s/b}. \tag{12.78}$$

At this scale, the gaugino fields are expected to condense. Through dimensional analysis, the gaugino condensate $\langle \bar{\lambda}\lambda \rangle$ is expected to be of order Λ^3. Terms quadratic in the gaugino fields thus yield in the effective theory below condensation scale a potential for the dilaton:

$$V \sim \left| \langle \bar{\lambda}\lambda \rangle \right|^2 \propto e^{-48\pi^2 s/b}. \tag{12.79}$$

The s-dependence of the potential is of course more complicated and one usually looks for stable minima with vanishing cosmological constant. But the behavior (12.78) is characteristic of the large s region and provides a potential slopping down to zero at infinity as required in the quintessence approach. Similar behavior is observed for moduli fields whose vev describes the radius of the compact manifolds which appear from the compactification from 10 or 11 dimensions to four in superstring theories.

Let us take therefore the example of an exponentially decreasing potential. More explicitly, we consider the following action

$$S = \int d^4x \sqrt{g} \left[-\frac{m_P^2}{2} R + \frac{1}{2}\partial^\mu \phi \partial_\mu \phi - V(\phi) \right], \tag{12.80}$$

which describes a real scalar field ϕ minimally coupled with gravity and the self-interactions of which are described by the potential:

$$V(\phi) = V_0 e^{-\lambda \phi/m_P}, \tag{12.81}$$

where V_0 is a positive constant.

The energy density and pressure stored in the scalar field are given by (12.75) and (12.76). We will assume that the background (matter and radiation) energy density ρ_B and pressure p_B obey a standard equation of state

$$p_B = w_B \rho_B. \tag{12.82}$$

If one neglects the spatial curvature ($k \sim 0$), the equation of motion for ϕ simply reads

$$\ddot{\phi} + 3H\dot{\phi} = -\frac{dV}{d\phi}, \tag{12.83}$$

with

$$H^2 = \frac{1}{3m_P^2}(\rho_B + \rho_\phi). \tag{12.84}$$

This can be rewritten as

$$\dot{\rho}_\phi = -3H\dot{\phi}^2. \tag{12.85}$$

We are looking for *scaling solutions*, i.e. solutions where the ϕ energy density scales as a power of the cosmic scale factor: $\rho_\phi \propto a^{-n_\phi}$ or $\dot{\rho}_\phi/\rho_\phi = -n_\phi H$ (n_ϕ being constant).

In this case, one easily obtains from (12.75), (12.76), and (12.85) that the ϕ field obeys a standard equation of state

$$p_\phi = w_\phi \rho_\phi, \tag{12.86}$$

with

$$w_\phi = \frac{n_\phi}{3} - 1. \tag{12.87}$$

Hence

$$\rho_\phi \propto a^{-3(1+w_\phi)}. \tag{12.88}$$

If one can neglect the background energy ρ_B, then (12.84) yields a simple differential equation for $a(t)$ which is solved as:

$$a \propto t^{2/[3(1+w_\phi)]}. \tag{12.89}$$

Since $\dot{\phi}^2 = (1 + w_\phi)\rho_\phi \sim t^{-2}$, one deduces that ϕ varies logarithmically with time. One then easily obtains from (12.83,12.84) that

$$\phi = \phi_0 + \frac{2}{\lambda} m_P \ln(t/t_0). \tag{12.90}$$

and[7]

$$w_\phi = \frac{\lambda^2}{3} - 1, \tag{12.91}$$

It is clear from (12.91) that, for λ sufficiently small, the field ϕ can play the rôle of quintessence. We note that, even if we started with a small value ϕ_0, ϕ reaches a value of order m_P.

But the successes of the standard Big Bang scenario indicate that clearly ρ_ϕ cannot have always dominated: it must have emerged from the background energy density ρ_B. Let us thus now consider the case where ρ_B dominates. It turns out that the solution just discussed with $\rho_\phi \gg \rho_B$ and (12.91) is a late time attractor [218] only if $\lambda^2 < 3(1 + w_B)$. If $\lambda^2 > 3(1 + w_B)$, the global attractor turns out to be a scaling solution [87, 160, 365] with the following properties:

$$\Omega_\phi \equiv \frac{\rho_\phi}{\rho_\phi + \rho_B} = \frac{3}{\lambda^2}(1 + w_B) \tag{12.92}$$

$$w_\phi = w_B. \tag{12.93}$$

The second equation (12.93) clearly indicates that this does not correspond to a dark energy solution (12.73).

The semirealistic models discussed earlier tend to give large values of λ and thus the latter scaling solution as an attractor. For example, in the case (12.79) where the scalar field is the dilaton, $\lambda = 48\pi^2/b$ with $b = 90$ for a E_8 gauge symmetry down to $b = 9$ for $SU(3)$. Moreover [161], on the observational side, the condition that

---

[7] Under the condition $\lambda^2 \leq 6$ ($w_\phi \leq 1$ since $V(\phi) \geq 0$).

ρ_ϕ should be subdominant during nucleosynthesis (in the radiation-dominated era) imposes to take rather large values of λ. Typically requiring $\rho_\phi/(\rho_\phi + \rho_B)$ to be then smaller than 0.2 imposes $\lambda^2 > 20$.

Ways to obtain a quintessence component have been proposed however. Let us sketch some of them in turn.

One is the notion of *tracker field*[8] [387]. This idea also rests on the existence of scaling solutions of the equations of motion which play the rôle of late time attractors, as illustrated above. An example is provided by a scalar field described by the action (12.80) with a potential

$$V(\phi) = \lambda \, \frac{\Lambda^{4+\alpha}}{\phi^\alpha} \tag{12.94}$$

with $\alpha > 0$. In the case where the background density dominates, one finds an attractor scaling solution [304, 322] $\phi \propto a^{3(1+w_B)/(2+\alpha)}$, $\rho_\phi \propto a^{-3\alpha(1+w_B)/(2+\alpha)}$ (see Exercise 1). Thus ρ_ϕ decreases at a slower rate than the background density ($\rho_B \propto a^{-3(1+w_B)}$) and tracks it until it becomes of the same order at a given value a_Q. We thus have:

$$\frac{\phi}{m_P} \sim \left(\frac{a}{a_Q} \right)^{3(1+w_B)/(2+\alpha)} , \tag{12.95}$$

$$\frac{\rho_\phi}{\rho_B} \sim \left(\frac{a}{a_Q} \right)^{6(1+w_B)/(2+\alpha)} . \tag{12.96}$$

One finds

$$w_\phi = -1 + \frac{\alpha(1 + w_B)}{2 + \alpha}. \tag{12.97}$$

Shortly after ϕ has reached for $a = a_Q$ a value of order m_P, it satisfies the standard slow-roll conditions (equations (D.109) and (D.110) of Appendix D) and therefore (12.97) provides a good approximation to the present value of w_ϕ. Thus, at the end of the matter-dominated era, this field may provide the quintessence component that we are looking for.

Two features are interesting in this respect. One is that this scaling solution is reached for rather general initial conditions, *i.e.* whether ρ_ϕ starts of the same order or much smaller than the background energy density [387]. Regarding the cosmic coincidence problem, it can be rephrased here as follows (since ϕ is of order m_P in this scenario): why is $V(m_P)$ of the order of the critical energy density ρ_c? It is thus the scale Λ which determines the time when the scalar field starts to emerge and the Universe expansion reaccelerates. Indeed, using (12.94), the constraint reads:

$$\Lambda \sim \left(H_0^2 m_P^{2+\alpha} \right)^{1/(4+\alpha)} . \tag{12.98}$$

We may note that this gives for $\alpha = 2$, $\Lambda \sim 10$ MeV, not such an atypical scale for high energy physics.

---

[8]Somewhat of a misnomer since in this solution, as we see below, the field ϕ energy density tracks the radiation-matter energy density before overcoming it, in contradistinction with (12.92). One should rather describe it as a *transient tracker field*.

A model [11] has been proposed which goes one step further: the dynamical component, a scalar field, is called k-essence and the model is based on the property observed in string models that scalar kinetic terms may have a nontrivial structure. Tracking occurs only in the radiation-dominated era; a new attractor solution where quintessence acts as a cosmological constant is activated by the onset of matter domination.

Models of dynamical supersymmetry breaking easily provide a model of the tracker field type just discussed [46]. Let us consider supersymmetric QCD with gauge group $SU(N_c)$ and $N_f < N_c$ flavors, *i.e.* N_f quarks Q_g (resp. antiquarks $\bar{Q}^g$), $g = 1, \ldots, N_f$, in the fundamental $\mathbf{N_c}$ (resp. antifundamental $\bar{\mathbf{N}}_{\mathbf{c}}$) of $SU(N_c)$ (as studied in Section 8.4.1 of Chapter 8). At the scale of dynamical symmetry breaking Λ where the gauge coupling becomes strong, bound states of the meson type form: $M_f{}^g = Q_f \bar{Q}^g$. The dynamics is described by a superpotential which can be computed nonperturbatively using standard methods, see equation (8.53) of Chapter 8:

$$W = (N_c - N_f) \frac{\Lambda^{(3N_c - N_f)/(N_c - N_f)}}{(\det M)^{1/(N_c - N_f)}}.$$

(12.99)

Such a superpotential has been used in the past but with the addition of a mass or interaction term (*i.e.* a positive power of M) in order to stabilize the condensate. One does not wish to do that here if M is to be interpreted as a runaway quintessence component. For illustration purpose, let us consider a condensate diagonal in flavor space: $M_f{}^g \equiv \phi^2 \delta_f^g$. Then the potential for ϕ has the form (12.94), with $\alpha = 2(N_c + N_f)/(N_c - N_f)$. Thus,

$$w_\phi = -1 + \frac{N_c + N_f}{2N_c}(1 + w_B),$$

(12.100)

which clearly indicates that the meson condensate is a potential candidate for a quintessence component.

One may note that, in the tracker model, when ϕ reaches values of order m_P, it satisfies the slow-roll conditions of an inflation model. The last possibility that I will discuss goes in this direction one step further. It is known under several names: deflation [344], kination [240], quintessential inflation [305]. It is based on the remark that, if a field ϕ is to provide a dynamical cosmological constant under the form of quintessence, it is a good candidate to account for an inflationary era where the evolution is dominated by the vacuum energy. In other words, are the quintessence component and the inflaton the same unique field?

In this kind of scenario, inflation (where the energy density of the Universe is dominated by the ϕ field potential energy) is followed by reheating where matter-radiation is created by gravitational coupling during an era where the evolution is driven by the ϕ field kinetic energy (which decreases as a^{-6}). Since matter-radiation energy density is decreasing more slowly, this turns into a radiation-dominated era until the ϕ energy density eventually emerges as in the quintessence scenarios described above.

Quintessential problems

However appealing, the quintessence idea is difficult to implement in the context of realistic models [68, 256]. The main problem lies in the fact that the quintessence field must be extremely weakly coupled to ordinary matter. This problem can take several forms:

- We have assumed until now that the quintessence potential monotonically decreases to zero at infinity. In realistic cases, this is difficult to achieve because the couplings of the field to ordinary matter generate higher order corrections that are increasing with larger field values, unless forbidden by a symmetry argument. For example, in the case of the potential (12.94), the generation of a correction term $\lambda_d\, m_P^{4-d}\phi^d$ puts in jeopardy the slow-roll constraints on the quintessence field, unless very stringent constraints are imposed on the coupling λ_d. But one typically expects from supersymmetry breaking $\lambda_d \sim M_{\rm SB}^4/m_P^4$ where $M_{\rm SB}$ is the supersymmetry breaking scale.

  Similarly, because the vev of ϕ is of order m_P, one must take into account the full supergravity corrections. One may then argue [49] that this could put in jeopardy the positive definiteness of the scalar potential, a key property of the quintessence potential. This may point towards models where $\langle W \rangle = 0$ (but not its derivatives, see (6.36) of Chapter 6) or to no-scale type models: in the latter case, the presence of three moduli fields T^i with Kähler potential $K = -\sum_i \ln(T^i + \bar{T}^i)$ cancels the negative contribution $-3|W|^2$ in the scalar potential (see (6.36) of Chapter 6).

- The quintessence field must be very light. If we return to our example of supersymmetric QCD in (12.94), $V''(m_P)$ provides an order of magnitude for the mass-squared of the quintessence component:

$$
m_\phi \sim \Lambda \left(\frac{\Lambda}{m_P}\right)^{1+\alpha/2} \sim H_0 \sim 10^{-33} \text{ eV} \tag{12.101}
$$

using (12.98). This might argue for a pseudo-Goldstone boson nature of the scalar field that plays the rôle of quintessence. This field must in any case be very weakly coupled to matter; otherwise its exchange would generate observable long range forces. Eötvös-type experiments put very severe constraints on such couplings. Again, for the case of supersymmetric QCD, higher order corrections to the Kähler potential of the type

$$
\kappa(\phi_i, \phi_j^\dagger) \left[\beta_{ij}\left(\frac{Q^\dagger Q}{m_P^2}\right) + \bar{\beta}_{ij}\left(\frac{\bar{Q}\bar{Q}^\dagger}{m_P^2}\right)\right] \tag{12.102}
$$

will generate couplings of order 1 to the standard matter fields ϕ_i, $\phi_j^\dagger$ since $\langle Q \rangle$ is of order m_P.

- It is difficult to find a symmetry that would prevent any coupling of the form $\beta(\phi/m_P)^n F^{\mu\nu}F_{\mu\nu}$ to the gauge field kinetic term. Since the quintessence behavior is associated with time-dependent values of the field of order m_P, this would generate, in the absence of fine tuning, corrections of order one to the gauge coupling. But we have seen in Chapter 11 that the time dependence of the fine structure constant for example is very strongly constrained: $|\dot{\alpha}/\alpha| < 5 \times 10^{-17} \mathrm{yr}^{-1}$. This yields a limit [68]:

$$|\beta| \leq 10^{-6} \frac{m_P H_0}{\langle \dot{\phi} \rangle},$$

(12.103)

where $\langle \dot{\phi} \rangle$ is the average over the last 2×10^9 years.

Pseudo-Goldstone boson

There exists a class of models [166] very close in spirit to the case of runaway quintessence: they correspond to a situation where a scalar field has not yet reached its stable groundstate and is still evolving in its potential.

More specifically, let us consider a potential of the form:

$$V(\phi) = M^4 v \left(\frac{\phi}{f} \right),$$

(12.104)

where M is the overall scale, f is the vacuum expectation value $\langle \phi \rangle$ and the function v is expected to have coefficients of order one. If we want the potential energy of the field (assumed to be close to its vev f) to give a substantial fraction of the energy density at present time, we must set

$$M^4 \sim \rho_c \sim H_0^2 m_P^2.$$

(12.105)

However, requiring that the evolution of the field ϕ around its minimum has been overdamped by the expansion of the Universe until recently imposes

$$m_\phi^2 = \frac{1}{2} V''(f) \sim \frac{M^4}{f^2} \leq H_0^2.$$

(12.106)

Let us note that this is again one of the slow-roll conditions familiar to the inflation scenarios.

From (12.105) and (12.106), we conclude that f is of order m_P (as the value of the field ϕ in runaway quintessence) and that $M \sim 10^{-3}$ eV (not surprisingly, this is the scale Λ typical of the cosmological constant, see (12.62)). As we have seen, the field ϕ must be very light: $m_\phi \sim h_0 \times 10^{-60} m_P \sim h_0 \times 10^{-33}$ eV. Such a small value is only natural in the context of an approximate symmetry: the field ϕ is then a pseudo-Goldstone boson. A typical example of such a field is provided by the string axion field. In this case, the potential simply reads:

$$V(\phi) = M^4 \left[1 + \cos(\phi/f) \right].$$

(12.107)

All the preceding shows that there is extreme fine tuning in the couplings of the quintessence field to matter, unless they are forbidden by some symmetry. This is somewhat reminiscent of the fine tuning associated with the cosmological constant. In fact, *the quintessence solution does not claim to solve the cosmological constant (vacuum energy) problem* described above. If we take the example of a supersymmetric theory, the dynamical cosmological constant provided by the quintessence component clearly does not provide enough amount of supersymmetry breaking to account for the mass difference between scalars (sfermions) and fermions (quarks and leptons): at least 100 GeV. There must be other sources of supersymmetry breaking and one must fine tune the parameters of the theory in order not to generate a vacuum energy that would completely drown ρ_ϕ.

However, the quintessence solution shows that, once this fundamental problem is solved, one can find explicit fundamental models that effectively provide the small amount of cosmological constant that seems required by experimental data.

Further reading

- Y. Nir, *B Physics and CP violation*, Proceedings of the 2005 Les Houches summer school, ed. by D. Kazakov and S. Lavignac.
- S. Weinberg, *The cosmological constant*, Rev. Mod. Phys. **61** (1989) 1.

Exercises

<u>Exercise 1</u> Consider a 3×3 matrix Λ where the entries Λ_{ij} are of order $\lambda^{n_{ij}}$, λ being a small parameter and $n_{ij} \geq 0$. Defining

$$p \equiv \min\,(n_{11}, n_{22}, n_{12}, n_{21}), \quad q \equiv \min\,(n_{11} + n_{22}, n_{12} + n_{21}), \qquad (12.108)$$

show that we have the following eigenvalue patterns:

$$p \geq \frac{q}{2} : O(1), \pm O(\lambda^p)$$

$$p \leq \frac{q}{2} : O(1), O(\lambda^p), O(\lambda^{q-p}).$$

Hints: Consider the characteristic equation of the hermitian combination $\Lambda^\dagger \Lambda$ ([43, 44]).

<u>Exercise 2</u> Introducing the mixed $U(1)_X$ gravitational anomaly $C_g = \mathrm{Tr}\,x$, express the charges a_0, b_0, c_0, d_0 and e_0 defined in Table 12.2, in terms of the anomaly coefficients C_1, C_2, C_3 in (12.23), C_g and $a_0 + b_0$. Show that one can redefine the x charge by combining it to the y charge in order to obtain $a_0 + b_0 = a_0 + c_0 = 0$ and $3(d_0 + e_0) = -(h_1 + h_2)$ in the absence of anomalies.

Hints: $C_g = 3(6a_0 + 3b_0 + 3c_0 + 2d_0 + e_0)$

$$a_0 = +\tfrac{1}{3}(a_0 + c_0) + \tfrac{1}{3}\mathcal{C}_D$$

$$b_0 = -\tfrac{4}{3}(a_0 + c_0) - \tfrac{1}{3}\mathcal{C}_D + \frac{1}{3}C_3$$

$$c_0 = +\frac{2}{3}(a_0 + c_0) - \frac{1}{3}\mathcal{C}_D$$

$$d_0 = -1(a_0 + c_0) - 1\mathcal{C}_D + \frac{1}{3}C_2 - \frac{1}{3}(h_1 + h_2)$$

$$e_0 = +2(a_0 + c_0) + 1\mathcal{C}_D - \frac{1}{3}C_2 + \frac{1}{6}(C_1 + C_2 - \frac{8}{3}C_3).$$

with $\mathcal{C}_D = -C_g/3 + C_1/6 + C_2/2 + 5C_3/9$.

<u>Exercise 3</u> *Tracker field solution.*
In the case where the evolution of the Universe is driven by the background (matter or radiation) energy, we search a scaling solution to the set of equations (12.83), (12.84) and (12.75), for the potential (12.94), *i.e.* $V(\phi) = \lambda\Lambda^{4+\alpha}/\phi^\alpha$, with $\alpha > 0$. In other words, we are looking for a solution such that

$$\rho_\phi = \rho_0 a^x, \quad \phi = \phi_0 a^y. \tag{12.109}$$

1. One assumes that the background has equation of state $p_B = w_B\rho_B$. If this background energy determines the evolution of the Universe, how does $a(t)$ evolves with time?
2. By plugging the scaling solution (12.109) into (12.75) and (12.83), express x and y in terms of α and w_B.
3. Show that ρ_ϕ/ρ_B increases in time until it reaches the value a_Q where it becomes of order 1 (and where our starting assumptions are no longer valid).
4. Compute ϕ/m_P and $\rho_\phi/(\lambda\Lambda^{4+\alpha}/m_P^\alpha)$ in terms of a/a_Q, w_B and α.
5. Compute the pressure p_ϕ and deduce (12.97).
6. Show that, as long as $\phi \gg \phi(a_Q)$, one reaches the slow-roll regime $\varepsilon, \eta \ll 1$ (*cf.* equations (D.109) and (D.110) of Appendix D). Solve the equations of motion in this context (note that $\rho_\phi \gg \rho_B$ then).

Hints:

1. $a(t) \propto t^{2/[3(1+w_B)]}$ (see the beginning of Section D.3 of Appendix D)

2. All terms in a given equation should have the same time dependence:

$$y = \frac{3}{2+\alpha}(1 + w_B), \quad x = -\frac{3\alpha}{2+\alpha}(1 + w_B).$$

4. One obtains:

$$\phi = m_P\sqrt{\frac{\alpha(2+\alpha)}{3(1+w_B)}}\left(\frac{a}{a_Q}\right)^{3(1+w_B)/(2+\alpha)},$$

$$\rho_\phi = \frac{2+\alpha}{2+\alpha(1-w_B)/2}\left(\frac{3}{\alpha}\frac{1+w_B}{2+\alpha}\right)^{\alpha/2}\lambda\frac{\Lambda^{4+\alpha}}{m_P^\alpha}\left(\frac{a}{a_Q}\right)^{-3\alpha(1+w_B)/(2+\alpha)}.$$

6. $\varepsilon = (\alpha/2)(m_P/\phi)^2, \quad \eta = \alpha(\alpha+1)(m_P/\phi)^2.$

Appendix A

A review of the Standard Model and of various notions of quantum field theory

We review in this appendix the basics of the Standard Model of electroweak interactions. We will take this opportunity to recall some basic notions of quantum field theory, especially in connection with the concepts of symmetry, spontaneous symmetry breaking, and quantum anomalies.

A.1 Symmetries

A.1.1 Currents and charges

We consider a general action

$$\mathcal{S} = \int d^4x \; \mathcal{L}\left(\Phi(x), \partial_\mu \Phi(x)\right) \tag{A.1}$$

which depends on a generic field $\Phi(x)$ and its spacetime derivatives. The equations of motion for the field $\Phi(x)$ are expressed by the Euler–Lagrange equation:

$$\partial_\mu \left(\frac{\delta \mathcal{L}}{\delta(\partial_\mu \Phi)}\right) - \frac{\delta \mathcal{L}}{\delta \Phi} = 0. \tag{A.2}$$

We suppose that the action is invariant under a transformation $\Phi(x) \to \Phi'(x')$ which depends on a continuous parameter α (or a collection of such parameters). If this parameter has an infinitesimal value $\delta\alpha$, one may write the transformation as:

$$\Phi(x) \to \Phi'(x') = \Phi(x) + \delta\Phi(x). \tag{A.3}$$

For each given transformation, $\delta\Phi(x)$ will be expressed explicitly in terms of $\Phi(x)$ and $\partial_\mu \Phi(x)$. Let us illustrate this on two examples, a spacetime symmetry $(x'^\mu \neq x^\mu)$ and an internal symmetry $(x'^\mu = x^\mu)$.

If the transformation is a translation of infinitesimal vector δa^μ $(x^\mu \to x^\mu + \delta x^\mu = x^\mu + \delta a^\mu)$, then

$$\Phi\left(x^\mu\right) \to \Phi'\left(x'^\mu\right) = \Phi\left(x^\mu + \delta a^\mu\right) = \Phi\left(x^\mu\right) + \delta a^\mu \, \partial_\mu \Phi(x). \tag{A.4}$$

We see that in the case of spacetime transformations, $\delta\Phi(x)$ depends on the spacetime derivatives of the field: $\delta\Phi(x) = \delta a^\mu \partial_\mu \Phi(x)$.

Next, we consider a (global) phase transformation:

$$\Phi\left(x^{\mu}\right) \to \Phi'\left(x'^{\mu}\right) = \Phi'\left(x^{\mu}\right) = e^{-i\theta}\Phi\left(x^{\mu}\right). \tag{A.5}$$

Infinitesimally,

$$\Phi'\left(x^{\mu}\right) = (1 - i\delta\theta)\Phi\left(x^{\mu}\right) \tag{A.6}$$

and $\delta\Phi(x) = -i\delta\theta\Phi(x)$: for an internal symmetry, $\delta\Phi(x)$ only depends on the field $\Phi(x)$ itself.

Whenever the action has a continuous symmetry, there is a procedure, known as the Niether procedure (see for example [318]), which allows us to construct a conserved current J^{μ}:

$$\partial_{\mu}J^{\mu} = 0. \tag{A.7}$$

Explicitly, the current reads

$$J^{\mu}(x) = \left[\mathcal{L}g_{\rho}^{\mu} - \frac{\delta\mathcal{L}}{\delta\left(\partial_{\mu}\Phi(x)\right)}\partial_{\rho}\Phi\right]\frac{\delta x^{\rho}}{\delta\alpha} + \frac{\delta\mathcal{L}}{\delta\left(\partial_{\mu}\Phi(x)\right)}\frac{\delta\Phi(x)}{\delta\alpha}. \tag{A.8}$$

Note that, in the case where the transformation is internal, *i.e.* leaves spacetime unchanged ($\delta x^{\mu} = 0$),

$$J^{\mu}(x) = \frac{\delta\mathcal{L}}{\delta\left(\partial_{\mu}\Phi(x)\right)}\frac{\delta\Phi(x)}{\delta\alpha}. \tag{A.9}$$

In this case, if the transformation is not a symmetry of the Lagrangian, the current is no longer conserved but one may still write a simple equation:

$$\partial_{\mu}J^{\mu} = \frac{\delta\mathcal{L}}{\delta\alpha}. \tag{A.10}$$

If the current is conserved, then the associated charge

$$Q \equiv \int d^{3}x\, J^{0}(x) \tag{A.11}$$

is a constant of motion. Indeed

$$\frac{dQ}{dt} = \int d^{3}x\, \partial_{0}J^{0} = \int d^{3}x\left(\partial_{0}J^{0} + \partial_{i}J^{i}\right) = \int d^{3}x\, \partial_{\mu}J^{\mu} = 0 \tag{A.12}$$

where we have assumed that the fields vanish at infinity (and thus the integral of a total spatial divergence vanishes). Since in the quantum theory $dQ/dt = i[H, Q]$, we have

$$[H, Q] = 0. \tag{A.13}$$

The transformation is said to be a symmetry of the Hamiltonian. We recall the standard equal-time commutation relations (assuming from now on that the field $\Phi(x)$ is a scalar field $\phi(x)$):

$$[\phi(t, \mathbf{x}), \phi(t, \mathbf{y})] = 0 \tag{A.14}$$

$$[\phi(t, \mathbf{x}), \pi(t, \mathbf{y})] = i\,\delta^{3}(\mathbf{x} - \mathbf{y}) \tag{A.15}$$

where π is the canonical conjugate momentum

$$\pi(x) = \frac{\delta\mathcal{L}}{\delta\partial_0\phi(x)}.$$ (A.16)

We note therefore that the time component of the Noether current (A.9) is simply expressed in terms of the conjugate momentum as $J^0 = \pi\delta\phi/\delta\alpha$. In the case of an internal symmetry, where $\delta\phi$ is a function of ϕ only (and not of its spacetime derivatives), one may deduce an interesting commutation relation. Indeed, from (A.14), (A.15)

$$[J_0(y), \phi(x)]_{x^0=y^0} = [\pi(y), \phi(x)]_{x^0=y^0}\frac{\delta\phi}{\delta\alpha}(y)$$
$$= -i\delta^3(\mathbf{x}-\mathbf{y})\frac{\delta\phi}{\delta\alpha}(x).$$

Hence, using the fact that the charge is a constant of motion (*i.e.* time independent)

$$[Q, \phi(x)] = [Q(x^0), \phi(x^0,\mathbf{x})]$$
$$= \int d^3y\,[J^0(x^0,\mathbf{y}), \phi(x^0,\mathbf{x})]$$
$$= -i\frac{\delta\phi}{\delta\alpha}(x)$$

and we have

$$\delta\phi(x) = i\delta\alpha\,[Q, \phi(x)].$$ (A.17)

In other words, once we have the charge operator, we can reconstruct the infinitesimal transformation, and even the finite transformation. Indeed the unitary transformation $U \equiv e^{-i\alpha Q}$ acts infinitesimally on ϕ as follows:

$$U^{-1}\phi(x)U = e^{i\delta\alpha Q}\phi(x)e^{-i\delta\alpha Q} = \phi(x) + i\delta\alpha[Q, \phi(x)] = \phi(x) + \delta\phi(x).$$ (A.18)

It thus corresponds to the finite transformation.

We now illustrate these notions on several examples.

A.1.2 Isospin symmetry

The Lagrangian describing a Dirac spinor field

$$\mathcal{L} = \overline{\psi}\,(i\gamma^\mu\partial_\mu - m)\,\psi$$ (A.19)

has the well-known global phase invariance

$$T_0 : \psi(x) \to e^{-i\theta}\,\psi(x).$$ (A.20)

These transformations form an abelian group: $T_{\theta_1}T_{\theta_2} = T_{\theta_1+\theta_2} = T_{\theta_2}T_{\theta_1}$. The associated charge just counts the number of ψ fields (if ψ is the quark, it can be interpreted as the baryon number).

Such abelian global transformations have a nonabelian generalization. The simplest case is provided by the formalism of isospin which we now describe. From the point of view of strong interactions, protons and neutrons have identical properties. This can be formalized by saying that physics should remain the same if one makes a unitary rotation on these states. Let us thus write a doublet

$$\psi_N = \begin{pmatrix} \psi_p \\ \psi_n \end{pmatrix} \tag{A.21}$$

which we call the nucleon. Physics should be invariant under the $SU(2)$ transformation

$$\psi_N \to \psi_N' = \begin{pmatrix} \psi_{p'} \\ \psi_{n'} \end{pmatrix} = U\psi_N = U\begin{pmatrix} \psi_p \\ \psi_n \end{pmatrix}. \tag{A.22}$$

where U is a unitary 2×2 matrix of determinant unity.

This is obviously reminiscent of the spin formalism. It is why such a transformation is called an isospin transformation. The two components of the doublet (A.21) have respectively $I_3 = +1/2$ for the proton and $I_3 = -1/2$ for the neutron. We may write the $SU(2)$ matrix U as:

$$\begin{aligned} U(\alpha^a) &= e^{-i\alpha^a \sigma^a / 2} \\ &= \cos\frac{|\alpha|}{2}\mathbb{1} - i\sin\frac{|\alpha|}{2}\frac{\alpha^a}{|\alpha|}\sigma^a \end{aligned} \tag{A.23}$$

where $|\alpha| \equiv \sqrt{(\alpha^1)^2 + (\alpha^2)^2 + (\alpha^3)^2}$. The Pauli matrices σ^a, $a = 1, 2, 3$, are explicitly

$$\sigma^1 = \begin{pmatrix} 0 & 1 \\ 1 & 0 \end{pmatrix}, \quad \sigma^2 = \begin{pmatrix} 0 & -i \\ i & 0 \end{pmatrix}, \quad \sigma^3 = \begin{pmatrix} 1 & 0 \\ 0 & -1 \end{pmatrix}. \tag{A.24}$$

For future reference, we note that they satisfy the trace formula

$$\mathrm{Tr}\left(\frac{\sigma^a}{2}\frac{\sigma^b}{2}\right) = \frac{1}{2}\delta^{ab}, \tag{A.25}$$

and the (anti)commutation relations

$$\left[\frac{\sigma^a}{2}, \frac{\sigma^b}{2}\right] = i\epsilon^{abc}\frac{\sigma^c}{2}, \quad \left\{\frac{\sigma^a}{2}, \frac{\sigma^b}{2}\right\} = \frac{1}{2}\delta^{ab}, \tag{A.26}$$

where ϵ^{abc} is the completely antisymmetric tensor ($\epsilon^{123} = 1$).

The following Lagrangian is invariant under the isospin transformation:

$$\mathcal{L} = \overline{\psi}_N\left(i\gamma^\mu\partial_\mu - m\right)\psi_N \tag{A.27}$$

because the transformation $\psi_N \to U\psi_N$ is global and the unitary matrix commutes with the spacetime derivatives ∂_μ. One may develop (A.27)

$$\begin{aligned} \mathcal{L} &= \left(\overline{\psi}_{ps}\overline{\psi}_{ns}\right)\left(i\gamma^\mu_{st}\partial_\mu - m\delta_{st}\right)\begin{pmatrix} \psi_{pt} \\ \psi_{nt} \end{pmatrix} \\ &= \overline{\psi}_p\left(i\gamma^\mu\partial_\mu - m\right)\psi_p + \overline{\psi}_n\left(i\gamma^\mu\partial_\mu - m\right)\psi_n \end{aligned} \tag{A.28}$$

where we have written explicitly spinor indices (s, t) to distinguish them from the indices of the internal symmetry (p, n).

Thus, proton and neutron must have the same mass if isospin is to be a symmetry (see Exercise 1): this is the standard way of identifying a symmetry (such as isospin) in the spectrum of a theory.

A.1.3   Local abelian gauge transformation. Quantum electrodynamics

It is well known that, in the case of quantum electrodynamics, the phase invariance (A.20) is promoted to the level of a local transformation, *i.e.* the phase θ depends on the spacetime point considered

$$\psi(x) \to \psi'(x) = e^{-iq\theta(x)}\,\psi(x) \tag{A.29}$$

(we introduce a charge coupling q for future use).

Obviously the Dirac Lagrangian (A.19) is not invariant since

$$\partial_\mu \psi'(x) = e^{-iq\theta(x)}\left[\partial_\mu\psi(x) - iq\partial_\mu\theta(x)\psi(x)\right]. \tag{A.30}$$

However, introducing a vector field $A_\mu(x)$ (the photon field) which transforms simultaneously as

$$A_\mu(x) \to A'_\mu(x) = A_\mu(x) - \frac{1}{g}\partial_\mu\theta(x) \tag{A.31}$$

i.e. as under a gauge transformation, we may build a new derivative

$$D_\mu\psi(x) = \partial_\mu\psi(x) - igqA_\mu(x)\,\psi(x) \tag{A.32}$$

which transforms under (A.29) and (A.31) as

$$(D_\mu\psi)' = \partial_\mu\psi'(x) - igqA'_\mu(x)\psi'(x) = e^{-iq\theta(x)}D_\mu\psi. \tag{A.33}$$

It is called a covariant derivative to express the fact that it transforms as the field itself. Then obviously the generalized Lagrangian

$$\mathcal{L} = \overline{\psi}(x)\left(i\gamma^\mu D_\mu - m\right)\psi(x) \tag{A.34}$$

is invariant under the local gauge transformation, *i.e.* (A.29) and (A.31).

In the case of quantum electrodynamics (QED), one sets $g = e$ and $q = -1$ (the electron charge) and one introduces a dynamics for the photon field

$$\mathcal{L}_\gamma = -\frac{1}{4}F_{\mu\nu}\,F^{\mu\nu} \tag{A.35}$$

where $F_{\mu\nu} = \partial_\mu A_\nu - \partial_\nu A_\mu$ is the field strength *invariant* under the gauge transformation (A.31). The complete Lagrangian is the QED Lagrangian

$$\mathcal{L} = -\tfrac{1}{4}F_{\mu\nu}\,F^{\mu\nu} + \overline{\psi}(x)\left(i\gamma^\mu\partial_\mu - m\right)\psi(x) - e\overline{\psi}(x)\gamma_\mu\psi(x)A^\mu(x). \tag{A.36}$$

The electron–photon coupling is called minimal coupling since it has been obtained by a purely geometric argument: completing the spacetime derivative into the covariant derivative to ensure invariance under local gauge transformations. We note that this coupling is of the form $J_\mu(x)A^\mu(x)$ where J_μ is the Noether current (*cf.* (A.9)).

Note that a mass term for the vector field is forbidden by the gauge symmetry (A.31).

A.1.4 Nonabelian local gauge transformations: Yang–Mills theory

Following Yang and Mills [384], we now try to make the $SU(2)$ isospin symmetry local. For future reference, we will adopt slightly more general notations than in Section A.1.2 and consider a general doublet $\Psi(x) = \begin{pmatrix} \psi_1(x) \\ \psi_2(x) \end{pmatrix}$ which transforms under $SU(2)$ as

$$\Psi(x) \rightarrow \Psi'(x) = e^{-i\alpha^a(x)t^a}\,\Psi(x) = U\left(\alpha^a(x)\right)\Psi(x) \tag{A.37}$$

where $t^a = \sigma^a/2$ and the three parameters $\alpha^a(x)$ $a = 1, 2, 3$, are now position dependent. Everything which follows in this section will be equally valid for a general symmetry group G: the t^a are the generators of the group ($a = 1, \ldots, n = \dim G$) in an N-dimensional matrix representation; $\Psi(x)$ is then an N-dimensional vector. In the case of $SU(2)$ specifically considered here, $n = 3$ and $N = 2$.

Since $\partial_\mu \Psi(x) \rightarrow U(\alpha)\partial_\mu \Psi(x) + [\partial_\mu U(\alpha)]\psi(x)$, we must complete the spacetime derivative to make it covariant. By analogy with the abelian case, we introduce n vector fields $A_\mu^a(x)$ (because there are n independent transformations associated with the n parameters $\alpha^a(x)$) and we define the derivative as

$$D_\mu\Psi(x) = \partial_\mu\Psi(x) - igA_\mu^a(x)t^a\Psi(x). \tag{A.38}$$

It is a covariant derivative if it transforms like Ψ itself, *i.e.* if

$$(D_\mu\Psi(x))' = U(\alpha)\,D_\mu\Psi(x). \tag{A.39}$$

It is straightforward to check that this corresponds to transforming the gauge fields as follows (beware of noncommuting matrices!)

$$A_\mu'^a t^a = U(\alpha)A_\mu^a t^a U(\alpha)^{-1} - \frac{i}{g}\left[\partial_\mu U(\alpha)\right]U(\alpha)^{-1}. \tag{A.40}$$

From now on, we will introduce the $N \times N$ matrix:

$$A_\mu \equiv A_\mu^a t^a. \tag{A.41}$$

More specifically, in our $SU(2)$ example,

$$A_\mu \equiv A_\mu^a\frac{\sigma^a}{2} = \begin{pmatrix} \dfrac{A_\mu^3}{2} & \dfrac{A_\mu^1 - iA_\mu^2}{2} \\[2mm] \dfrac{A_\mu^1 + iA_\mu^2}{2} & -\dfrac{A_\mu^3}{2} \end{pmatrix}. \tag{A.42}$$

Then the transformation law (A.40) reads

$$A_\mu' = UA_\mu U^{-1} - \frac{i}{g}\,\partial_\mu U\,U^{-1}. \tag{A.43}$$

Comparison with the abelian case is more transparent in the case of infinitesimal transformations:

$$U(\alpha) \simeq 1 - i\alpha^a(x)t^a + \cdots \tag{A.44}$$

In order to expand (A.43) we need the commutator of two generators. In the general case, we have

$$\left[t^a, t^b\right] = iC^{abc}t^c, \tag{A.45}$$

where the C^{abc} are the structure constants of the group G. Comparison with (A.26) shows that, in the case of $SU(2)$, $C^{abc} = \epsilon^{abc}$. One then easily obtains from expanding (A.43) to linear order in α^a,

$$A_\mu^{'a} = A_\mu^a + C^{abc}\,\alpha^b A_\mu^c - \frac{1}{g}\,\partial_\mu\alpha^a + \cdots \tag{A.46}$$

The last term is similar to the abelian case in (A.31). The second term is specific to the nonabelian situation: it involves the nonvanishing structure constants C^{abc} of the nonabelian gauge group. Introducing the generators T^a in the adjoint representation $((T^a)^{bc} = -iC^{abc})$, one may write the infinitesimal transformation (A.46) as

$$\delta A_\mu^a = -\frac{1}{g}\left[\partial_\mu\alpha^a - igA_\mu^c\,(T^c)^{ab}\,\alpha^b\right] \equiv -\frac{1}{g}D_\mu\alpha^a. \tag{A.47}$$

It remains to write a dynamics for the vector fields, and thus to define a field strength. By analogy with the abelian case, we may consider $\partial_\mu A_\nu - \partial_\nu A_\mu$. Using (see exercise 2)

$$\partial_\mu U^{-1} = -U^{-1}\left[\partial_\mu U\right]U^{-1}, \tag{A.48}$$

we obtain from (A.43)

$$\begin{aligned}
\partial_\mu A_\nu' - \partial_\nu A_\mu' = {}& U\left(\partial_\mu A_\nu - \partial_\nu A_\mu\right)U^{-1} \\
&+ \Big[\partial_\mu U A_\nu U^{-1} - U A_\nu U^{-1}\partial_\mu U U^{-1} \\
&+ \frac{i}{g}\partial_\nu U U^{-1}\partial_\mu U U^{-1} - (\mu \leftrightarrow \nu)\Big].
\end{aligned}$$

From the form of the first term on the right-hand side, it is clearly hopeless to expect a gauge invariant field strength as in the abelian case. We may at best expect that $F_{\mu\nu}$ transforms as $UF_{\mu\nu}U^{-1}$: a covariant field strength. Indeed, because

$$\begin{aligned}
-ig\left[A_\mu' A_\nu'\right] = {}& -ig\,U\left[A_\mu A_\nu\right]U^{-1} \\
&+ \left\{-\left[\partial_\mu U U^{-1}, U A_\nu U^{-1}\right] + \frac{i}{g}\partial_\mu U U^{-1}\partial_\nu U U^{-1} - (\mu \leftrightarrow \nu)\right\},
\end{aligned}$$

we check that

$$F_{\mu\nu} = \partial_\mu A_\nu - \partial_\nu A_\mu - ig\left[A_\mu A_\nu\right] \tag{A.49}$$

transforms covariantly:

$$F_{\mu\nu}' = U\,F_{\mu\nu}\,U^{-1}. \tag{A.50}$$

We could have obtained directly the field strength from the commutation of two co-variant derivatives (as in the abelian case):

$$[D_\mu, D_\nu] = -igF_{\mu\nu}. \tag{A.51}$$

We also note that $Tr\ F^{\mu\nu}F_{\mu\nu}$ is invariant. Since $F_{\mu\nu}$ is a $N \times N$ matrix, we may write it $F_{\mu\nu} = F_{\mu\nu}^a t^a$ where, from (A.45) and (A.49)

$$F_{\mu\nu}^a = \partial_\mu A_\nu^a - \partial_\nu A_\mu^a + gC^{abc}A_\mu^b A_\nu^c. \tag{A.52}$$

The complete Lagrangian reads

$$\mathcal{L} = -\tfrac{1}{2}\mathrm{Tr}\left(F^{\mu\nu}\ F_{\mu\nu}\right) + \overline{\psi}i\gamma^\mu D_\mu\psi - m\overline{\psi}\psi. \tag{A.53}$$

Normalizing the generators t^a as

$$\mathrm{Tr}\left(t^a t^b\right) = \tfrac{1}{2}\delta^{ab} \tag{A.54}$$

as in (A.25), we have $\tfrac{1}{2}\mathrm{Tr}(F^{\mu\nu}\ F_{\mu\nu}) = \tfrac{1}{4}F^{a\mu\nu}\ F_{\mu\nu}^a$. As in the abelian case, the fermion–gauge field coupling is of the form

$$\mathcal{L}_{\mathrm{int}} = gJ^{a\mu}A_\mu^a, \tag{A.55}$$

where $J_\mu^a = \bar{\Psi}\gamma_\mu t^a\Psi$ is the fermionic part of the Noether current.

The corresponding equation of motion is obtained by varying with respect to A_μ^a. It reads

$$D_\mu F^{a\mu\nu} = \partial_\mu F^{a\mu\nu} - gC^{bac}A_\mu^b F^{c\mu\nu} = -gJ^{a\nu}, \tag{A.56}$$

where we have introduced the covariant derivative in the adjoint representation:

$$D_\rho F_{\mu\nu}^b = \partial_\rho F_{\mu\nu}^b - igA_\rho^a \left(T^a\right)^{bc} F_{\mu\nu}^c \quad , \quad \left(T^a\right)^{bc} = -iC^{abc}. \tag{A.57}$$

One important difference between abelian and nonabelian gauge theories is the notion of charge or quantum number. In (A.29) and (A.32), we have included a real number q whose value is not fixed by the symmetry. If we try to introduce such a number in the nonabelian case and replace (A.37) and (A.38), respectively, by

$$\Psi'(x) = e^{-iX\alpha^a(x)t^a}\ \Psi(x) \tag{A.58}$$

and

$$D_\mu\Psi(x) = \partial_\mu\Psi(x) - ig\ X\ A_\mu^a(x)t^a\Psi(x), \tag{A.59}$$

then it is easy to show that $D_\mu\Psi$ is covariant under an infinitesimal transformation only if $X = 0$ or 1. This shows that, in a nonabelian gauge theory, charges (*i.e.* quantum numbers) can only take quantized values fixed by group theory (*i.e.* matrix elements of generators t^a). This will be one of the motivations for embedding the quantum electrodynamics $U(1)$ symmetry into a nonabelian grand unified gauge symmetry group.

A.2   Spontaneous breaking of symmetry

A.2.1   Example of a global symmetry

We start with the simplest example of a global abelian symmetry and consider a complex scalar field $\phi(x)$ with a Lagrangian density

$$\mathcal{L} = \partial^\mu \phi^\dagger \partial_\mu \phi - V(\phi^\dagger \phi). \tag{A.60}$$

This has a global phase invariance $\phi(x) \rightarrow \phi'(x) = e^{-i\theta}\phi(x)$; and the associated conserved Noether current is $J_\mu = -i\phi\partial_\mu\phi^\dagger + i\phi^\dagger\partial_\mu\phi$. Requirement of renormalizability imposes to limit the interaction potential V to quartic terms:

$$V(\phi^\dagger \phi) = a\,\phi^\dagger\phi + \frac{\lambda}{2}(\phi^\dagger\phi)^2, \tag{A.61}$$

where $\lambda > 0$ to avoid instability at large values of the field. The corresponding equation of motion reads:

$$(\Box + a)\phi = -\lambda\phi(\phi^\dagger\phi). \tag{A.62}$$

If $a > 0$, we may write $a \equiv m^2$ and the equation of motion

$$(\Box + m^2)\phi = j(\phi) \tag{A.63}$$

where $j(\phi) \equiv -\lambda\phi(\phi^\dagger\phi)$ describes the self-interaction of the field ϕ. It is then possible to quantize the theory: first quantize the free field theory corresponding to vanishing self-interaction (the field obeys the Klein–Gordon equation $(\Box + m^2)\phi = 0$) and then use perturbation theory in λ.

We are interested in the vacuum structure of the theory, *i.e.* in identifying its ground state. Using Noether formalism, one may compute the energy of the system

$$P_0 = \int d^3x\,\mathcal{H}(x)$$

$$\mathcal{H}(x) = \pi\partial_0\phi(x) + \pi^\dagger\partial_0\phi^\dagger(x) - \mathcal{L} \tag{A.64}$$

where π (resp. $\pi^\dagger$) is the canonical conjugate momentum of $\phi(x)$ (resp. $\phi^\dagger(x)$) as in (A.16). This gives in the present case:

$$\mathcal{H}(x) = \partial_0\phi^\dagger(x)\partial_0\phi(x) + \partial_i\phi^\dagger(x)\partial_i\phi(x) + V(\phi^\dagger\phi). \tag{A.65}$$

Since the first two terms are positive definite, minimization of the energy density is obtained by setting them to zero, *i.e.* by taking a constant value for the field ϕ, and by minimizing the potential energy. In other words, we minimize the energy density by choosing a constant value ϕ_0 which corresponds to the minimum of V. If $a \geq 0$, the minimum is at $\phi = 0$ (see Fig. A.1) *i.e.* for vanishing field. We are familiar with this situation: for example, the electromagnetic energy density $\frac{1}{8\pi}(\mathbf{E}^2 + \mathbf{B}^2)$ is minimized for $\mathbf{E} = \mathbf{B} = \mathbf{0}$. In this latter case, a nonvanishing electric or magnetic field in the vacuum (*i.e.* the ground state) would obviously break the isotropy of space. There is no such risk with a scalar field.

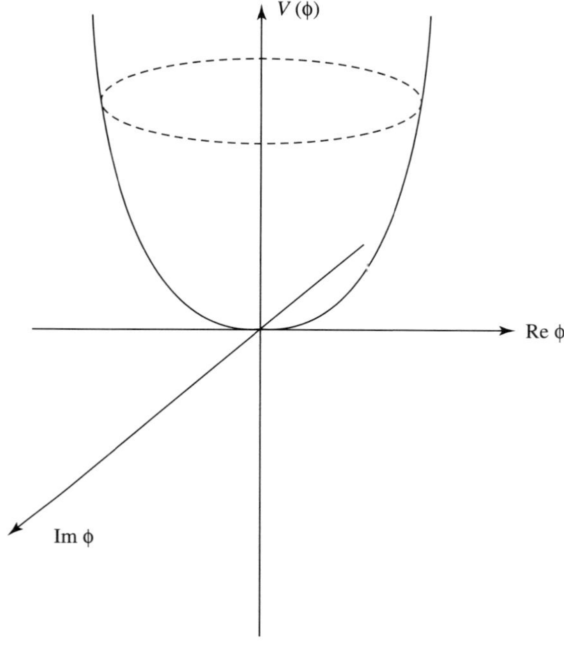

Fig. A.1

Indeed, when $a < 0$, the ground state corresponds to a nonvanishing value for the scalar field: $\phi_0^\dagger \phi_0 = -a/\lambda$. If we set $a \equiv -m^2$, the equation of motion reads

$$(\Box - m^2)\phi = j(\phi). \tag{A.66}$$

In the limit of vanishing self-interaction ($j \to 0$), the equation is not the standard Klein–Gordon equation because of the sign of m^2 and we thus do not know how to quantize the theory. Writing $-m^2 = (im)^2$, we may identify the origin of this sign as the instability of the system for small values of the field, *i.e.* around $\phi = 0$ (see Fig. A.2). In other words, the problem might be that we evaluate the scalar field ϕ around an unstable value. It is then preferable to evaluate it around one of its stable values defined by $\phi_0^\dagger \phi_0 = m^2/\lambda$.

We note that the potential has a rotation symmetry corresponding to $\phi \to e^{-i\theta}\phi$. On the other hand, the ground state of the system (which corresponds to the choice of one point on the set of degenerate minima) necessarily breaks this symmetry. We are in a situation of spontaneous symmetry breaking:

- the Lagrangian (the Hamiltonian) is invariant under the symmetry;
- the ground state (the vacuum) is not left invariant by the symmetry.

Such a situation is often encountered in classical physics with macroscopic objects (a ball at the top of a Mexican hat falls into the bottom of the hat, a vertical force down a column will shatter it into pieces). In ordinary quantum mechanics, one can always

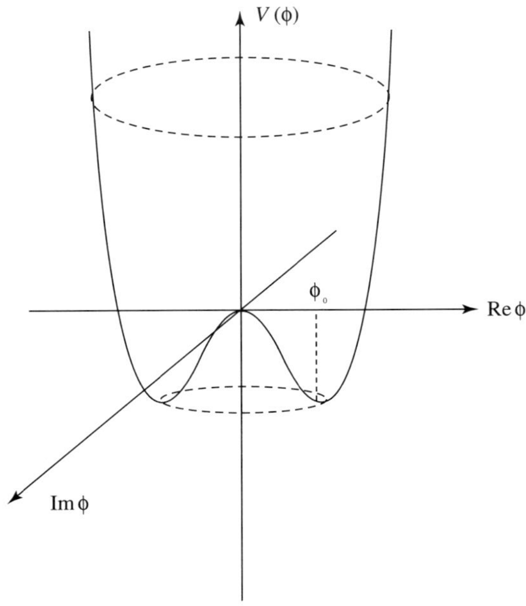

Fig. A.2

form linear combinations of ground states which are invariant under the symmetry. This is not possible in the case of quantum field theory because of the infinite number of degrees of freedom (infinite volume): the mixed Hamiltonian matrix elements between two different ground states vanish.

Now, following our remark above, let us return to our example and express the scalar field around its ground state value, say $e^{i\theta_0}\sqrt{m^2/\lambda}$. We write

$$\phi = e^{i\theta_0}\left(\sqrt{\frac{m^2}{\lambda}} + \frac{\varphi_1 + i\varphi_2}{\sqrt{2}}\right). \tag{A.67}$$

Then

$$V(\phi^\dagger\phi) = -\frac{m^4}{2\lambda} + m^2\varphi_1^2 + \text{ cubic and quartic terms.} \tag{A.68}$$

We conclude that the field φ_1 has mass squared $2m^2$ (beware of the normalization of the kinetic term $\frac{1}{2}\partial^\mu\varphi_1\partial_\mu\varphi_1$) whereas φ_2 is massless. Thus the mass spectrum does not reflect the symmetry: even though φ_1 and φ_2 are interchanged in the symmetry transformation, they are not degenerate in mass. This is a consequence of the spontaneous breaking of the symmetry. Moreover, the fact that φ_2 is massless is not a coincidence. Indeed, it is a general theorem due to [198] that *to every continuous global symmetry spontaneously broken, there corresponds a massless particle.* In the case of internal symmetries such as considered here, it is a boson called the Goldstone boson.

Before proceeding to prove and discuss Goldstone's theorem, let us stress one important property of a Goldstone boson. If we write the infinitesimal transformation $\delta\phi = -i\theta\phi$ in the parametrization (A.67), we obtain

$$\delta\varphi_1 = \theta\varphi_2, \quad \delta\varphi_2 = -\theta\sqrt{\frac{2m^2}{\lambda}} - \theta\varphi_1. \tag{A.69}$$

The Goldstone transformation is nonhomogeneous and is characterized by a constant term proportional to the field vacuum expectation value.

This is even better seen by using the following more appropriate parametrization of the scalar field ϕ:

$$\phi(x) = e^{i(\theta_0 + \gamma(x))}\left[\sqrt{\frac{m^2}{\lambda}} + \rho(x)\right]. \tag{A.70}$$

Since $V(\phi^\dagger\phi) = V([\rho + \sqrt{m^2/\lambda}]^2)$, the field $\gamma(x)$ does not appear in the potential and is thus massless: it is the Goldstone boson, *i.e.* the massless excitation along the continuum of vacua. Under a gauge transformation we simply have $\gamma(x) \to \gamma(x) - \theta$. We will see that, when the symmetry is local ($\theta(x)$), this allows us to gauge away the corresponding degree of freedom.

A.2.2 Goldstone theorem

We have defined two distinct notions of a symmetry:

- A symmetry of the Hamiltonian: the corresponding charge commutes with the Hamiltonian

$$[Q, H] = 0 \tag{A.71}$$

which is a statement of the conservation of the associated current

$$\partial^\mu j_\mu = 0. \tag{A.72}$$

It also implies that the finite transformation $U \equiv e^{-i\alpha Q}$ leaves the Hamiltonian invariant:

$$UHU^\dagger = H = U^\dagger HU. \tag{A.73}$$

- A symmetry of the vacuum, *i.e.* a symmetry which leaves the ground state $|0\rangle$ invariant:

$$U|0\rangle = |0\rangle \tag{A.74}$$

or

$$Q\,|0\rangle = 0. \tag{A.75}$$

[83] has shown that a symmetry of the vacuum is a symmetry of the Hamiltonian[1]. But the converse is not true. If $[Q, H] = 0$ or $\partial^\mu j_\mu = 0$, then one may distinguish two cases:

(a) Wigner realization $Q|0\rangle = 0$

This has well-known properties. Two states which transform into one another under the symmetry are degenerate in energy. Indeed, if $\phi' = U^\dagger \phi U$, then

$$\langle 0|\phi'^\dagger H \phi'|0\rangle \;=\; \langle 0|U^\dagger \phi^\dagger U H U^\dagger \phi\, U|0\rangle \;=\; \langle 0|\phi^\dagger H \phi|0\rangle,$$

using *both* (A.73) and (A.74).

Also, Green's functions (*i.e.* the vacuum values of chronological products of fields $\langle 0|T\phi(x_1)\cdots\phi(x_n)|0\rangle$) are invariant under the symmetry: this leads to Ward identities which translate the symmetry at the level of Green's functions. Indeed, under an infinitesimal transformation, using (A.17) (which is derived from (A.13) or (A.71)), we have, for $x_1^0 > x_2^0 > \cdots > x_n^0$,

$$\begin{aligned}
\delta\langle 0|T\phi(x_1)\cdots\phi(x_n)|0\rangle &= \langle 0|\delta\phi(x_1)\phi(x_2)\cdots\phi(x_n)|0\rangle \\
&\quad +\langle 0|\phi(x_1)\delta\phi(x_2)\cdots\phi(x_n)|0\rangle + \cdots \\
&= i\delta\alpha\langle 0|(Q\phi(x_1) - \phi(x_1)Q)\phi(x_2)\cdots\phi(x_n)|0\rangle \\
&\quad +i\delta\alpha\langle 0|\phi(x_1)(Q\phi(x_2) - \phi(x_2)Q)\cdots\phi(x_n)|0\rangle + \cdots \\
&= 0.
\end{aligned}$$

(b) Goldstone realization $Q|0\rangle \neq 0$

Two states which are exchanged under the symmetry are not necessarily degenerate. Hence the symmetry is not manifest in the energy (mass) spectrum. We have seen an explicit example of this in the previous Section. The symmetry is said[2] to be spontaneously broken.

According to case (a), it suffices to prove that at least one Green's function is not invariant: $\delta\langle 0|T\phi(x_1)\cdots\phi(x_n)|0\rangle \neq 0$. In most cases, it is the one-field Green's function $\langle 0|\phi(x)|0\rangle$ which is not invariant: in other words, the vacuum expectation value of a field is not invariant. This was precisely the case of the explicit example presented in the previous Section. We now prove Goldstone's theorem in this case.

Defining

$$\eta \equiv \langle 0|[Q(t), \phi(0)]|0\rangle = \int d^3x\, \langle 0|\,[J_0(t, \mathbf{x}), \phi(0)]\,|0\rangle \tag{A.76}$$

we see that $\eta \neq 0$ because the symmetry is not an invariance of the vacuum (expressed as $0 \neq \langle 0|\delta\phi(x)|0\rangle = i\langle 0|[Q, \phi(x)]|0\rangle$ using (A.17)) whereas $d\eta/dt = 0$ because the symmetry is an invariance of the Hamiltonian (*i.e.* the charge is a constant of motion: $dQ/dt = i[H, Q] = 0$).

---

[1] In terms of the current j_μ, (A.75) reads $\int d^3x j_0(t, \mathbf{x})|0\rangle = 0$. If $|n\rangle$ is a state of vanishing 3-momentum, one may apply the bra $\langle n|$ to the previous relation. Thus, $\langle n|\int d^3x e^{-i\mathbf{k}\cdot\mathbf{x}} j_0(t, \mathbf{x})|0\rangle = 0$ and $\langle n|j_0(x)|0\rangle = 0$ from which follows $\langle n|\partial^\mu j_\mu(x)|0\rangle = 0$. Lorentz invariance implies that this is true for any state $|n\rangle$ and thus $\partial^\mu j_\mu(x)|0\rangle = 0$ from which (A.72) follows.

[2] Somewhat improperly since the symmetry remains an invariance of the Hamiltonian.

We introduce a complete basis of eigenstates $|n\rangle$ of energy ($H|n\rangle = E_n|n\rangle$) and 3-momentum ($\mathbf{P}|n\rangle = \mathbf{p_n}|n\rangle$). The zero of energy is taken at the ground state: $H|0\rangle = E_0|0\rangle = 0$, $\mathbf{P}|0\rangle = \mathbf{p_0}|0\rangle = \mathbf{0}$. Then using the closure relation $\sum_n |n\rangle\langle n| = \mathbb{1}$, and translation operators:

$$J_0(t,\mathbf{x}) = e^{-i\mathbf{P}\cdot\mathbf{x}}e^{iHt}J_0(0)e^{-iHt}e^{i\mathbf{p}\cdot\mathbf{x}},$$

we may write η as:

$$\eta = \sum_n \int d^3x \left[\langle 0|J_0(t,\mathbf{x})|n\rangle\langle n|\phi(0)|0\rangle - \langle 0|\phi(0)|n\rangle\langle n|J_0(t,\mathbf{x})|0\rangle\right]$$

$$= \sum_n \int d^3x \left[e^{-iE_n t}e^{i\mathbf{p_n}\cdot\mathbf{x}}\langle 0|J_0|n\rangle\langle n|\phi(0)|0\rangle\right.$$

$$\left. - e^{iE_n t}e^{-i\mathbf{p_n}\cdot\mathbf{x}}\langle 0|\phi(0)|n\rangle\langle n|J_0)|0\rangle\right]$$

$$= \sum_n (2\pi)^3\delta^3(\mathbf{p_n})\left[e^{-iE_n t}\langle 0|J_0|n\rangle\langle n|\phi(0)|0\rangle\right.$$

$$\left. - e^{iE_n t}\langle 0|\phi(0)|n\rangle\langle n|J_0(0)|0\rangle\right]. \tag{A.77}$$

Obviously,

$$\frac{d\eta}{dt} = -i\sum_n (2\pi)^3 E_n\delta^3(\mathbf{p_n})\left[e^{-iE_n t}\langle 0|J_0(0)|n\rangle\langle n|\phi(0)|0\rangle\right.$$

$$\left. + e^{iE_n t}\langle 0|\phi(0)|n\rangle\langle n|J_0(0)|n\rangle\right]. \tag{A.78}$$

For this to be identically vanishing, whereas $\eta \neq 0$, there must be a state $|n_0\rangle$ such that $E_{n_0}\delta^3(\mathbf{p_{n_0}}) = 0$ (*i.e.* the state $|n_0\rangle$ is a massless excitation of the system) and

$$\langle 0|J_0(0)|n_0\rangle \neq 0 \quad \text{and} \quad \langle n_0|\phi(0)|0\rangle \neq 0. \tag{A.79}$$

This massless excitation is the Goldstone boson.

A.2.3 General discussion in the case of global symmetries

We now consider the general case of a group G of continuous global transformations. The scalar fields are taken to be real and transform under G as:

$$\phi'(x) = e^{-i\alpha^a t^a}\,\phi(x) \tag{A.80}$$

where the t^a are the generators of the group, in a n-dimensional representation[3]. Infinitesimally

$$\phi'_m(x) = \phi_m(x) - i\alpha^a t^a_{mn}\phi_n(x) \tag{A.81}$$

where $m = 1, \ldots, n$.

---

[3]The t^a are taken to be hermitian, $t^{a\dagger} = t^a$. Since ϕ is real, $(it^a)^* = it^a$. Hence $t^{a*} = -t^a$ and $t^{aT} = -t^a$: the generators are antisymmetric.

Since the potential $V(\phi)$ is invariant, we have

$$\frac{\partial V}{\partial \phi_m} \, \alpha^a \, t^a_{mn} \, \phi_n = 0 \tag{A.82}$$

from which we obtain, by differentiating with respect to ϕ_ℓ,

$$\frac{\partial^2 V}{\partial \phi_m \partial \phi_\ell} \, \alpha^a \, t^a_{mn} \, \phi_n + \frac{\partial V}{\partial \phi_m} \, \alpha^a \, t^a_{m\ell} = 0. \tag{A.83}$$

At the minimum ϕ_0 of the potential, we thus have

$$M^2_{\ell m} \, \alpha^a \, t^a_{mn} \, \phi_{0n} = 0 \tag{A.84}$$

where

$$M^2_{\ell m} = \left. \frac{\partial^2 V}{\partial \phi_\ell \partial \phi_m} \right|_{\phi_0} \tag{A.85}$$

is the scalar squared mass matrix. In the case of spontaneous breaking, the vacuum ϕ_0 is not invariant and thus $\alpha^a t^a_{mn} \phi_{0n}$ is a nonzero eigenvector of the squared mass matrix with vanishing eigenvalue: it corresponds to a Goldstone boson.

It is possible to separate the generators t^a $(a = 1 \cdots \dim G)$ into:

- Generators t^i which leave the vacuum invariant: $t^i \phi_0 = 0$. These generators correspond to the residual symmetry of gauge group H and $i = 1, \ldots, \dim H$.
- Generators θ^α which do not leave the vacuum invariant: $\theta^\alpha \phi_0 \neq 0$ $(\alpha = 1, \ldots, \dim G - \dim H)$. Let us show that the vectors $\theta^\alpha \phi_0$ are independent and thus generate a vector space of dimension $\dim G - \dim H$. We introduce a scalar product $(x, y) = \sum_{i=1}^n x_i^* y_i$ and consider the matrix $X^{\alpha\beta} = (\theta^\alpha \phi_0, \theta^\beta \phi_0)$. Since X is symmetric, it can be diagonalized: $D = OXO^T$ with

$$D^{\alpha\alpha} = \left(O^{\alpha\gamma} \, \theta^\gamma \, \phi_0, O^{\alpha\delta} \, \theta^\delta \, \phi_0 \right). \tag{A.86}$$

But $O^{\alpha\gamma} \theta^\gamma \phi_0 \neq 0$. Hence all diagonal elements are positive definite and the vector space generated by the $O^{\alpha\gamma} \theta^\gamma$ or the θ^γ is $(\dim G - \dim H)$ dimensional.

[We may parametrize nonlinearly the scalar degrees of freedom as in (A.70) to make explicit the rôle of the Goldstone bosons. Indeed, let us consider the function

$$F(g, x) = (\phi(x), g\phi_0), \tag{A.87}$$

where $g = e^{i\alpha^a t^a}$ is an element of G. Since, for every element h of the residual symmetry group H $(h\phi_0 = \phi_0)$, we have $F(g, x) = F(gh, x)$, F is in fact a function on right cosets of G with respect to H, i.e. on G/H.[4] Since G is compact, the continuous function

---

[4] One defines the right coset with respect to H of any element in G as the equivalence class $\dot{g} = \{\gamma \in G; \exists h \in H, \gamma = gh\} = gH$. If H is an invariant subgroup of G, the set of these right cosets has a group structure, the quotient group G/H.

F is bounded and reaches a maximum for some element $\hat{g}(x)$. Using $\delta g = i\alpha^a g t^a$, we have

$$0 = \delta F(\hat{g}, x) = i\alpha^a \left(\phi(x), \hat{g}(x)t^a\phi_0\right) = i\alpha^a \left(\hat{g}^{-1}(x)\phi(x), t^a\phi^0\right). \tag{A.88}$$

Hence $\hat{\phi}(x) \equiv \hat{g}^{-1}(x)\phi(x)$ is orthogonal to the directions $\theta^\alpha\phi_0$ in field space: it parametrizes the (in general) massive excitations.

Since any element of G can be written as $e^{i\gamma^\alpha\theta^\alpha}e^{i\beta^i t^i}$ and since $\hat{g}(x)$ is in fact a right coset, we may choose to represent it by

$$\hat{g}(x) = e^{i\gamma^\alpha(x)\,\theta^\alpha}. \tag{A.89}$$

Correspondingly, the field $\phi(x)$ is parametrized as

$$\phi(x) = \hat{g}(x)\hat{\phi}(x) = e^{i\gamma^\alpha(x)\,\theta^\alpha}\,\hat{\phi}(x), \tag{A.90}$$

which generalizes (A.70): the (dim G − dim H) fields $\gamma^\alpha(x)$ are the Goldstone bosons. We note [82] that they transform nonlinearly under G: $\gamma^\alpha \to \gamma^{\alpha\prime}(x)$ defined by

$$g\,e^{i\gamma^\alpha(x)\,\theta^\alpha} = e^{i\gamma^{\alpha\prime}(x)\,\theta^\alpha}e^{i\beta^i(\gamma^\alpha,g)\,t^i}.] \tag{A.91}$$

A.2.4 Spontaneous breaking of a local symmetry. Higgs mechanism

The conclusions that we have reached in the preceding section, especially the Goldstone theorem, are drastically changed when we turn to local transformations. Let us illustrate this on the case of a complex scalar field minimally coupled to an abelian gauge symmetry, a theory known as scalar electrodynamics. The Lagrangian reads

$$\mathcal{L} = -\tfrac{1}{4}F^{\mu\nu}F_{\mu\nu} + D^\mu\phi^\dagger D_\mu\phi - V(\phi^\dagger\phi) \tag{A.92}$$

with $D_\mu\phi = \partial_\mu\phi - igqA_\mu\phi$ and $V(\phi^\dagger\phi)$ given by (A.61) with $a = -m^2$. The scalar field kinetic term thus yields a term quadratic in the vector field:

$$D^\mu\phi^\dagger D_\mu\phi = \partial^\mu\phi^\dagger\partial_\mu\phi - igq\left(\phi\partial^\mu\phi^\dagger - \phi^\dagger\partial^\mu\phi\right)A_\mu + g^2q^2\phi^\dagger\phi A^\mu A_\mu. \tag{A.93}$$

Expressing the scalar field around its vacuum expectation value $\phi_0 = e^{i\theta_0}\sqrt{m^2/\lambda}$ as in (A.67) thus yields what seems to be a mass term for the vector field: $g^2q^2\phi_0^\dagger\phi_0 A^\mu A_\mu$.

Let us pause for a while to discuss massive vector bosons. The free field action is

$$S = \int d^4x\left[-\frac{1}{4}F^{\mu\nu}F_{\mu\nu} + \frac{1}{2}M^2 A^\mu A_\mu\right]. \tag{A.94}$$

Gauge invariance is only recovered in the limit of vanishing mass M. The corresponding equation of motion is the Proca equation:

$$\partial_\mu F^{\mu\nu} + M^2 A^\nu = 0. \tag{A.95}$$

Acting with a derivative ∂_ν and using the antisymmetry of $F^{\mu\nu}$, we infer from the Proca equation the Lorentz condition $\partial_\nu A^\nu = 0$. Thus the massive vector field has $4 - 1 = 3$ degrees of freedom: more precisely two transverse (as the massless photon) and one longitudinal. Using the Lorentz condition, we may finally write the Proca equation

$$(\Box + M^2)A^\nu = 0, \tag{A.96}$$

which shows that, indeed, M is the mass of the gauge field.

Returning to our case of interest, we deduce that, once the gauge symmetry is spontaneously broken ($\phi_0 \neq 0$), the gauge field acquires a mass

$$M_A^2 = 2g^2 q^2 \phi_0^\dagger \phi_0 = 2g^2 q^2 \frac{m^2}{\lambda}. \tag{A.97}$$

There seems, however, to be a discrepancy when we count the number of degrees of freedom. Using the gauge invariance built in the Lagrangian, we count two scalar degrees of freedom and two vector (transverse) degrees of freedom. However, once one translates the scalar field around its vev $\phi_0 \neq 0$, one seems to identify:

- a massless Goldstone boson φ_2 and a real scalar φ_1 of mass $2m^2$ (as in Section A.2.1);
- a vector field of mass M_A, with three degrees of freedom (two transverse and one longitudinal).

It turns out that making the symmetry local has changed one of the scalar fields into a spurious degree of freedom. This is most easily seen by using the nonlinear parametrization (A.70) of the scalar degrees of freedom. We see that, in the local case, one can gauge away the field $\gamma(x)$ by performing a *local* gauge transformation:

$$\phi(x) \to \phi'(x) = e^{-iq\theta(x)}\phi(x) = \left[\rho(x) + \sqrt{\frac{m^2}{\lambda}}\right] e^{i\theta_0} \tag{A.98}$$

with the choice $\theta(x) = \gamma(x)/q$. In other words, because the symmetry is local, the would be Goldstone boson is now a gauge artifact. It reappears in the theory as the longitudinal component of the massive vector particle. This is the famous Higgs mechanism.

The choice (A.98) corresponds to a gauge choice known as the unitary gauge: with this choice of gauge, all fields are physical and the S-matrix is unitary, but gauge symmetry is no longer apparent. One often prefers covariant gauges which retain some of the gauge invariance (through the dependence on some gauge parameters, which should drop from physical results) but include the spurious boson.

We may generalize this analysis to nonabelian gauge symmetries and will illustrate it on the example of $SU(2)$ which provides the basis for the Standard Model. We introduce a complex scalar field $\Phi(x) = \left(\begin{smallmatrix}\phi_1(x)\\\phi_2(x)\end{smallmatrix}\right)$ which transforms as a doublet under $SU(2)$ (*cf.* (A.37))

$$\Phi(x) \to \Phi'(x) = e^{-i\alpha^a(x)t^a}\,\Phi(x). \tag{A.99}$$

Then the full Lagrangian reads:

$$\mathcal{L} = D^\mu \Phi^\dagger \, D_\mu \Phi - V(\Phi^\dagger \Phi) - \tfrac{1}{4} F^{a\mu\nu} F^a_{\mu\nu} \tag{A.100}$$

where $F^a_{\mu\nu}$ is given in (A.52), and the potential is given by

$$V(\Phi^\dagger \Phi) = -m^2 \Phi^\dagger \Phi + \lambda(\Phi^\dagger \Phi)^2. \tag{A.101}$$

The ground state is reached for $\Phi_0^\dagger \Phi_0 = m^2/(2\lambda)$. We choose to orient it along the second component of the doublet:

$$\Phi_0 = \begin{pmatrix} 0 \\ \dfrac{v}{\sqrt{2}} \end{pmatrix}, \quad v = \sqrt{\dfrac{m^2}{\lambda}}. \tag{A.102}$$

Then, studying as before the quadratic terms $A^a_\mu A^{a\mu}$ coming from the term $D^\mu \Phi^\dagger D_\mu \Phi$, one concludes that all three gauge fields A^a_μ, $a = 1, 2, 3$, acquire a mass

$$M_A = \frac{gv}{2} \tag{A.103}$$

through the Higgs mechanism. This requires three longitudinal degrees of freedom. Reexpressing the scalar field around its vev as $\Phi = \widetilde{\Phi} + \Phi_0$, and inspecting the scalar potential, one finds three would be Goldstone bosons: $(\widetilde{\phi}_1 + \widetilde{\phi}_1^*)/\sqrt{2}$, $(\widetilde{\phi}_j - \widetilde{\phi}_j^*)/(i\sqrt{2})$ with $j = 1, 2$. A more transparent parametrization is the nonlinear one:

$$\Phi(x) = e^{i\gamma^a(x)t^a} \begin{pmatrix} 0 \\ \dfrac{v + \eta(x)}{\sqrt{2}} \end{pmatrix}. \tag{A.104}$$

Since $V(\Phi^\dagger \Phi) = V([v + \eta]^2/2)$, the three fields $\gamma^a(x)$ are the would-be Goldstone bosons. But a gauge transformation (A.99) with $\alpha^a(x) = \gamma^a(x)$ shows that they are spurious degrees of freedom.

A.3 The Standard Model of electroweak interactions

From the very beginning, electrodynamics has inspired the theory of weak interactions. In electrodynamics, a vector current of matter fields couples to the photon. Electromagnetic interactions through photon exchange is thus described by a current–current interaction with a photon propagator to account for the action at distance. In nuclear beta decay, which we write here at the level of nucleons $n \rightarrow p + e^- + \overline{\nu}_e$, the range seemed to be zero or very small and [156, 157] proposed to keep a current–current structure but to make it a local one (in order to account for the zero range):

$$\mathcal{H} = \frac{G_F}{\sqrt{2}} \, [\overline{p}(x)\gamma^\mu n(x)] \, [\overline{e}(x)\gamma_\mu \nu_e(x)] + \text{h.c.} \tag{A.105}$$

where the letters $p, n, \ldots$ denote the corresponding spinor fields (proton, neutron, $\ldots$) and G_F is the Fermi constant:

$$G_F \simeq \frac{10^{-5}}{m_p^2}, \tag{A.106}$$

where m_p is the proton mass.

It was soon realized that the interaction (A.105) does not account for all nuclear beta decays (in particular, the so-called Gamow–Teller transitions) and that other processes are of the same type: in particular, the constant involved in the muon decay $\mu^+ \to e^+ + \nu_e + \overline{\nu}_\mu$ is basically the Fermi constant. Weak interactions were discovered to be universal.

It was during the 1950s and 1960s that the full structure of low energy weak interactions was unravelled. A key stage was the discovery of parity violation [269, 382] which showed that one had to include an axial vector into the current–current interaction. This was complemented by the measurement of the neutrino helicity by [197].

We take this opportunity to recall that under parity a spinor transforms as:

$$\mathcal{P}\psi(x)\mathcal{P}^\dagger = \eta_p \gamma^0 \psi(\widetilde{x}) \tag{A.107}$$

where η_p is a phase and $\widetilde{x}^\mu \equiv x_\mu$. One deduces

$$\begin{aligned}
\text{scalar (S)} \quad & \mathcal{P}\overline{\psi}(x)\psi(x)\mathcal{P}^\dagger = \overline{\psi}(\widetilde{x})\psi(\widetilde{x}) \\
\text{pseudoscalar (P)} \quad & \mathcal{P}\overline{\psi}(x)\gamma^5\psi(x)\mathcal{P}^\dagger = -\overline{\psi}(\widetilde{x})\gamma^5\psi(\widetilde{x}) \\
\text{vector (V)} \quad & \mathcal{P}\overline{\psi}(x)\gamma^\mu\psi(x)\mathcal{P}^\dagger = \overline{\psi}(\widetilde{x})\gamma_\mu\psi(\widetilde{x}) \\
\text{axial (A)} \quad & \mathcal{P}\overline{\psi}(x)\gamma^\mu\gamma^5\psi(x)\mathcal{P}^\dagger = -\overline{\psi}(\widetilde{x})\gamma_\mu\gamma^5\psi(\widetilde{x}).
\end{aligned} \tag{A.108}$$

Finally, the V-A theory was proposed by [346] and by [162]. It reads for the beta decay

$$\mathcal{H} = \frac{G_F}{\sqrt{2}} \left[\overline{p}(x)\gamma^\mu \left(g_V + g_A \gamma_5\right) n(x)\right] \left[\overline{e}(x)\gamma_\mu(1 - \gamma_5)\nu_e(x)\right] + \text{ h.c.} \tag{A.109}$$

where $g_V \simeq 1$ (the conservation of the vector current – the so-called CVC hypothesis – ensures that strong interaction do not renormalize this coupling) whereas $g_A/g_V \simeq -1.262 \pm 0.005$ (an effect of strong interactions). Taking into account other weak processes such as muon decay and turning off strong interactions, one may write the pure weak interaction Hamiltonian density (this time at the quark level) as:

$$\mathcal{H} = \frac{G_F}{\sqrt{2}} J_\mu^\dagger \, J^\mu + \text{ h.c.} \tag{A.110}$$

where

$$J_\mu(x) = J_\mu^{(h)}(x) + J_\mu^{(\ell)}(x) \tag{A.111}$$
$$J_\mu^{(h)}(x) = \overline{u}(x)\gamma_\mu(1 - \gamma_5)d(x) + \overline{c}(x)\gamma_\mu(1 - \gamma_5)s(x) + \cdots$$
$$J_\mu^{(\ell)}(x) = \overline{\nu}_e(x)\gamma_\mu(1 - \gamma_5)e(x) + \overline{\nu}_\mu(x)\gamma_\mu(1 - \gamma_5)\mu(x) + \cdots$$

The main problem with the theory of weak interactions described by (A.111) is that it is nonrenormalizable: the Fermi coupling G_F has negative mass dimension. This has some undesirable consequences at high energies: cross-sections increase monotonically with the energy available in the center of mass. There is thus an energy where the cross

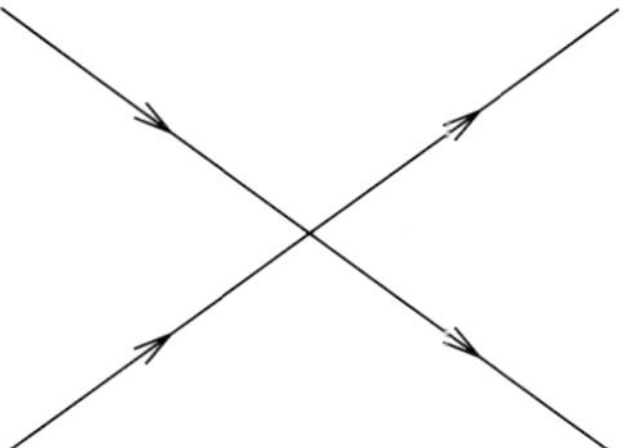

Fig. A.3 Tree-level Fermi interaction for nuclear beta decay.

sections have become so large that they are no longer compatible with the unitarity of the S-matrix (which expresses the conservation of probability). At such an energy the fundamental theory must be modified.

Since this type of reasoning will be repeated for gravity (another nonrenormalizable theory where the coupling, Newton's constant G_N, has also mass dimension -2), let us sketch the argument more precisely.

If we consider the tree-level process depicted in Fig. A.3, then, on purely dimensional grounds, we may deduce that the corresponding cross-section σ (given by the square of the amplitude, and thus proportional to G_F^2) behaves as $G_F^2 E^2$, where E is the energy available in the center of mass. But conservation of probability (expressed through the unitarity of the S-matrix) imposes that $\sigma \sim E^{-2}$. Thus at energies $E \sim G_F^{-1/2}$, the theory is no longer compatible with first principles and must be replaced by a more complete theory (which involves new fields whose exchange modifies the high energy behavior of the cross-section). This is the so-called unitarity limit that we consider in more details in Section 1.2.1 of Chapter 1.

This behavior of the cross-sections for tree-level processes may be related here to the nonrenormalizable character of the theory. If we compute the one-loop (second order) amplitude given in Fig. A.4, we obtain a contribution of order $G_F^2 \int^\infty E dE$ (to be compared with the tree-level G_F as in Fig. A.3) which is infinite if we assume that the theory remains valid up to arbitrarily large energies, *i.e.* to arbitrarily small distances. The renormalization procedure may take care of a finite number of such infinities but, in the case considered, new infinities appear at each order of perturbation theory and thus perturbative renormalization cannot help.

These considerations show that, because the coupling G_F of the weak theory has mass dimension -2, the theory at energy E is characterized by the dimensionless combination[5] $G_F E^2$. When E reaches a scale of order $G_F^{-1/2} \sim 300$ GeV, the dimensionless combination is of order one and the low energy weakly coupled theory is no longer valid: it must be replaced by a more complete fundamental theory[6].

Since the problem occurs at short distance (*i.e.* when the two vertices of the graph of Fig. A.4 become arbitrarily close), an obvious way of curing the problem is to replace

---

[5]For example, one may write the tree-level cross-section as $\left(G_F E^2\right)^2 / E^2$ whereas the ratio of the one-loop amplitude to the tree-level amplitude reads $G_F E_{\max}^2$ where $E_{\max}$ is the cut-off in the integral over the energy.

[6]The precise value of scale may be found by imposing the unitarity bound on varied processes. The best limit thus obtained is 630 GeV.

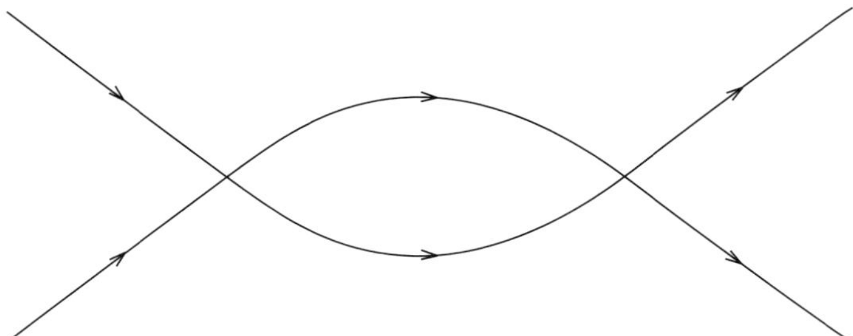

Fig. A.4 One-loop contribution to beta decay.

the contact interaction between the four fermions in the Fermi interaction by a vector particle exchange as in Fig. A.5, much in the spirit of quantum electrodynamics. This (charged) vector particle was introduced long before the days of the Standard Model and called the intermediate vector boson. It had to be be massive in order to account for the short range interaction. The above arguments showed that its mass was expected to be smaller than a few hundred GeVs. Once it was realized that a nonvanishing mass for a vector boson was compatible with gauge symmetry (although spontaneously broken), it was natural to try to fit weak interactions into a gauge theory. We will now see that the minimal nontrivial possibility leads to the Standard Model.

A.3.1 Identifying the gauge structure and quantum numbers

We will try here to fit the current–current theory of weak interactions just described with the structure of a gauge theory. In other words, pursuing the analogy with QED which motivated Fermi, we try to identify the intermediate vector boson $W^\pm$ as a gauge field. Our guiding principle will be here minimality: there is no a priori reason that the theory of weak interactions should be the minimal one but, as we will see, it turns out to be the case.

From the graph of Fig. A.5 (and a similar one for muon beta decay), we infer that the couplings of the intermediate vector boson to the quarks u and d and the leptons e, ν_e, μ, ν_μ must be of the form

$$
\begin{aligned}
\mathcal{L} \sim\ & g\,\bar{u}\,\gamma^\mu \frac{1-\gamma_5}{2} d\ W_\mu^+ + g\,\bar{d}\,\gamma^\mu \frac{1-\gamma_5}{2} u\ W_\mu^- \\
& + g\,\bar{\nu}_e\,\gamma^\mu \frac{1-\gamma_5}{2} e\ W_\mu^+ + g\,\bar{e}\,\gamma^\mu \frac{1-\gamma_5}{2} \nu_e\ W_\mu^- \\
& + g\,\bar{\nu}_\mu\,\gamma^\mu \frac{1-\gamma_5}{2} \mu\ W_\mu^+ + g\,\bar{\mu}\,\gamma^\mu \frac{1-\gamma_5}{2} \nu_\mu\ W_\mu^- .
\end{aligned}
\tag{A.112}
$$

Since a charged field is complex, we need to introduce two real gauge fields; minimality leads to consider an abelian symmetry $U(1) \times U(1)$, with a global symmetry between the two $U(1)$, or a nonabelian gauge symmetry $SU(2)$. The first choice is very close to QED. Since the physics of weak interactions seems quite different from electrodynamics, we pursue the second possibility.

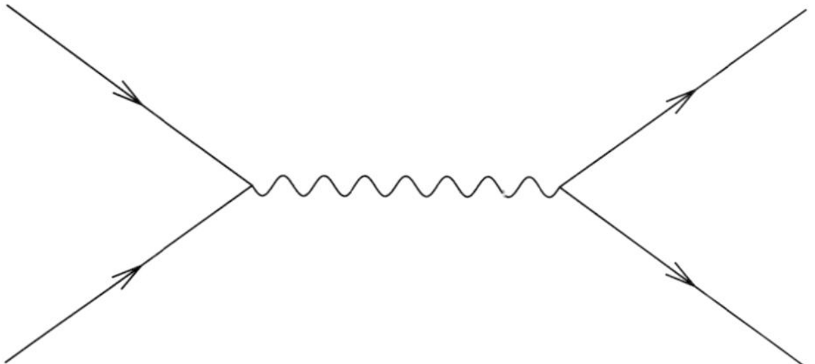

Fig. A.5 Intermediate vector boson exchange.

Following (A.55), we may rewrite the interaction term between a $SU(2)$ gauge field $A_\mu^a(x)$ and a doublet fermion $\Psi(x) = \left(\begin{smallmatrix} \psi_1(x) \\ \psi_2(x) \end{smallmatrix}\right)$ as

$$
\begin{aligned}
\mathcal{L}_{\text{int}} &= g\ \bar{\Psi}\gamma^\mu A_\mu^a \frac{\sigma^a}{2}\Psi = g\ J^{a\mu}\ A_\mu^a \\
&= \frac{g}{2}\left(J^{1\mu} + iJ^{2\mu}\right)\left(A_\mu^1 - iA_\mu^2\right) + \frac{g}{2}\left(J^{1\mu} - iJ^{2\mu}\right)\left(A_\mu^1 + iA_\mu^2\right) + g\ J^{3\mu}A_\mu^3
\end{aligned}
$$

$$(A.113)$$

where we have introduced the complex gauge fields:

$$
\frac{A_\mu^1 \mp iA_\mu^2}{\sqrt{2}} \equiv W_\mu^\pm.
$$

$$(A.114)$$

The associated currents are $J_\pm^\mu \equiv J^{1\mu} \pm iJ^{2\mu}$ and we have

$$
\mathcal{L}_{\text{int}} = \frac{g}{\sqrt{2}}\left(J_+^\mu W_\mu^+ + J_-^\mu W_\mu^-\right) + g\ J^{3\mu}A_\mu^3.
$$

$$(A.115)$$

Since $J_+^\mu = \bar{\psi}_1\gamma^\mu\psi_2$ and $J_-^\mu = \bar{\psi}_2\gamma^\mu\psi_1$, comparison with (A.112) tells us how the quarks and leptons must be grouped into $SU(2)$ doublets. Indeed since, for example

$$
\bar{u}\ \gamma^\mu\frac{1-\gamma_5}{2}\ d = \bar{u}\ \frac{1+\gamma_5}{2}\gamma^\mu\frac{1-\gamma_5}{2}\ d = \bar{u}_L\ \gamma^\mu\ d_L,
$$

we conclude that only left-handed components of quarks and leptons behave as $SU(2)$ doublets:

$$
\Psi(x) = \begin{pmatrix} u_L(x) \\ d_L(x) \end{pmatrix},\ \begin{pmatrix} \nu_{e_L}(x) \\ e_L(x) \end{pmatrix},\ \begin{pmatrix} \nu_{u_L}(x) \\ \mu_L(x) \end{pmatrix}.
$$

The right-handed chiralities do not transform under $SU(2)$: they are singlets under $SU(2)$.

We note immediately that this local $SU(2)$ symmetry has nothing to do with the global $SU(2)$ isospin symmetry of strong interactions that Yang and Mills attempted to make local. Strong isospin is a *global* symmetry of strong interactions and it does not apply to leptons for example. It is true however that $\left(\begin{smallmatrix} u(x) \\ d(x) \end{smallmatrix}\right)$ form a strong interaction doublet (note the absence of chiral indices though!). This has let to the regrettable use of the expression "weak isospin" for the $SU(2)$ symmetry that we consider here : weak isospin is a *local* symmetry of (electro)weak interactions and leptons transform under it.

It remains to identify the nature of the real field A_μ^3 which completes the gauge multiplet. Since it cannot be paired to another real gauge field, it is obviously neutral. The natural candidate would be the photon but the corresponding charge operator can be written now that we have identified the content of the doublets ($J_\mu^3 = \frac{1}{2}\bar\psi_1\gamma_\mu\psi_1 - \frac{1}{2}\bar\psi_2\gamma_\mu\psi_2$):

$$Q^3 \equiv \int d^3x\, J_0^3 = \int d^3x \frac{1}{2} \left[u_L^\dagger u_L - d_L^\dagger d_L + \nu_{eL}^\dagger \nu_{eL} - e_L^\dagger e_L + \nu_{\mu L}^\dagger \nu_{\mu L} - \mu_L^\dagger \mu_L \right]$$

$$(A.116)$$

which has nothing to do with the electric charge operator[7] ($J_\mu^Q = \sum_i q_i \bar\psi_i \gamma_\mu \psi_i$, with q_i the charge of fermion ψ_i):

$$Q \equiv \int d^3x\, J_0^Q = \int d^3x \left[\frac{2}{3} u^\dagger u - \frac{1}{3} d^\dagger d - e^\dagger e - \mu^\dagger \mu \right]. \qquad (A.117)$$

However, we may note that

$$Y \equiv 2(Q - Q^3) = \int d^3x \left[\frac{1}{3} \left(u_L^\dagger u_L + d_L^\dagger d_L \right) + \frac{4}{3} u_R^\dagger u_R - \frac{2}{3} d_R^\dagger d_R \right.$$

$$- \left(\nu_{eL}^\dagger \nu_{eL} + e_L^\dagger e_L \right) - 2 e_R^\dagger e_R$$

$$\left. - \left(\nu_{\mu L}^\dagger \nu_{\mu L} + \mu_L^\dagger \mu_L \right) - 2 \mu_R^\dagger \mu_R \right]. \qquad (A.118)$$

In other words, all the elements of a given representation of $SU(2)$ – whether a doublet of two left-handed fields, say u_L and d_L, or a singlet right-handed field, say u_R – appear in the sum with the same coefficient. This means that out of the $SU(2)$ gauge symmetry associated with the generator t^3 and the $U(1)$ QED gauge symmetry, one may form an abelian $U(1)$ symmetry which commutes with the weak isospin $SU(2)$ symmetry. The associated quantum number is noted y and called the weak hypercharge ; the gauge group is noted $U(1)_Y$ to distinguish it from quantum electrodynamics. From (A.118), we infer the following relation between the charge q, the weak isospin t^3 and the weak hypercharge y of a fermion

$$q = t^3 + \frac{y}{2}. \qquad (A.119)$$

Table A.1 gives the quantum numbers of the low energy quarks and leptons.

We have introduced a right-handed neutrino which did not appear in the earliest formulations of the Standard Model, since neutrino were assumed to be massless. As noted in Chapter 1, this field is a gauge singlet under $SU(2) \times U(1)_Y$.

---

[7]Indeed, if the photon field A_μ is one of the gauge fields of $SU(2)$, then the corresponding generator is traceless and the electric charges in a given representation must add to zero. One then needs to introduce extra leptons (and quarks) as in the Georgi and Glashow model [180] (see Exercise 9).

Table A.1

Field	q	t^3	y
u_L	$+2/3$	$+1/2$	$+1/3$
d_L	$-1/3$	$-1/2$	$+1/3$
u_R	$+2/3$	0	$+4/3$
d_R	$-1/3$	0	$-2/3$
ν_{eL}	0	$+1/2$	-1
e_L	-1	$-1/2$	-1
e_R	-1	0	-2
N_R	0	0	0

The total gauge symmetry has now four gauge bosons: A^a_μ, $a = 1, 2, 3$, for $SU(2)$ and the abelian gauge field of $U(1)_Y$ which we will note B_μ. It follows from our construction that the photon field A_μ (associated with the charge operator Q) is a combination of A^3_μ and B_μ. The orthogonal combination is a real, thus neutral, vector field denoted by Z^0_μ. Its exchange leads to new low energy weak interactions known as neutral currents. It is their discovery in 1973 at CERN which led to the first experimental verification of the Standard Model.

In order to give a mass to the intermediate vector boson $W^\pm_\mu$ (and to the new Z^0_μ since there does not seem to exist a long range force associated with it), we must break spontaneously the gauge symmetry, while keeping the photon A_μ massless. In other words, we must break $SU(2) \times U(1)_Y$ down to the $U(1)$ symmetry of QED. We need at least three real scalar fields to provide for the three longitudinal degrees of freedom $W^\pm_L$, Z^0_L. The most economical choice is provided by a *complex* doublet $\Phi = \binom{\phi^+}{\phi^0}$ with quantum numbers given in Table A.2.

After spontaneous breaking, there remains $4 - 3 = 1$ degree of freedom: the neutral scalar field known as Higgs and actively searched at high energy colliders.

For future reference, we note that, since Φ transforms under $SU(2)$ as a doublet

$$\Phi(x) \to e^{-i\alpha^a(x)t^a} \, \Phi(x). \tag{A.120}$$

Φ^* which describes the charge conjugate does not transform as a doublet. It is the field

$$\widetilde{\Phi} = i\sigma_2 \, \Phi^* = \begin{pmatrix} \phi^{0*} \\ -\phi^- \end{pmatrix} \tag{A.121}$$

which transforms as (A.120), as can be seen by using

$$\sigma_2 \, \frac{\sigma^{a*}}{2} \, \sigma_2 = -\frac{\sigma^a}{2}. \tag{A.122}$$

This shows that, under $SU(2)$, the representation **2** is equivalent to the conjugate representation $\bar{\mathbf{2}}$. We note also that $\widetilde{\Phi}$ has hypercharge $y = -1$.

We are now in a position to write the Lagrangian of the Standard Model and extract its consequences.

Table A.2

Field	q	t^3	y
ϕ^+	$+1$	$+1/2$	$+1$
ϕ^0	0	$-1/2$	$+1$

A.3.2 The Glashow–Weinberg–Salam or SU(2) × U(1) model

We start with the gauge and scalar sector of the theory. The Lagrangian simply reads

$$\mathcal{L} = -\tfrac{1}{4} F^{a\mu\nu} F^a_{\mu\nu} - \tfrac{1}{4} B^{\mu\nu} B_{\mu\nu} + D^\mu \Phi^\dagger D_\mu \Phi - V(\Phi^\dagger \Phi) \tag{A.123}$$

where $F^a_{\mu\nu}$ ($a = 1, 2, 3$) is the $SU(2)$ covariant field strength given in (A.52), $B_{\mu\nu} = \partial_\mu B_\nu - \partial_\nu B_\mu$ is the $U(1)_Y$ invariant field strength and Φ is the Higgs doublet. Its derivative $D_\mu \Phi$ is covariant under both $SU(2)$ and $U(1)_Y$:

$$D_\mu \Phi = \partial_\mu \Phi - ig\, A^a_\mu \frac{\sigma^a}{2} \Phi - i\frac{g'}{2} y_\phi\, B_\mu \Phi, \tag{A.124}$$

where g is the $SU(2)$ gauge coupling, $g'/2$ is the $U(1)_Y$ gauge coupling and $y_\phi = +1$ is the Higgs hypercharge. This Lagrangian is invariant under local transformations of $SU(2) \times U(1)_Y$, *i.e.* infinitesimally, following (A.46), (A.31), (A.98) and (A.99),

$$\delta A^a_\mu = -\frac{1}{g}\, \partial_\mu \alpha^a + \varepsilon^{abc}\, \alpha^b\, A^c_\mu$$

$$\delta B_\mu = -\frac{2}{g'} \partial_\mu \beta$$

$$\delta \Phi = -i\, \alpha^a \frac{\sigma^a}{2}\, \Phi - iy_\phi \beta \Phi \tag{A.125}$$

where α^a, $a = 1, 2, 3$, are the parameters of the $SU(2)$ transformation and β of the abelian $U(1)_Y$. We note that the choice $\alpha^1 = \alpha^2 = 0$ and $\alpha^3 = 2\beta = \theta$ corresponds to the $U(1)$ QED gauge transformation (A.29):

$$\delta \phi^+ = -i\theta \phi^+, \quad \delta \phi^0 = 0. \tag{A.126}$$

We take as usual a scalar potential

$$V(\Phi^\dagger \Phi) = -m^2 \Phi^\dagger \Phi + \lambda (\Phi^\dagger \Phi)^2, \tag{A.127}$$

with the ground state[8]

$$\langle \Phi \rangle \equiv \Phi_0 = \begin{pmatrix} 0 \\ \frac{v}{\sqrt{2}} \end{pmatrix}, \quad v = \sqrt{\frac{m^2}{\lambda}}. \tag{A.128}$$

---

[8]We note that it is this choice of orientation which determines the exact nature of the residual symmetry, $U(1)_{\mathrm{QED}}$, and thus the electric charge of the fields. In order to remain simple in the presentation, we have worked backwards here.

As usual, the gauge field mass matrix is obtained from the covariant completion of the scalar kinetic term. Setting ϕ at its ground state value, we obtain from (A.124):

$$\langle D_\mu \Phi \rangle = \begin{pmatrix} -ig\dfrac{v}{\sqrt{2}} \dfrac{A_\mu^1 - iA_\mu^2}{2} \\[2ex] +i\dfrac{v}{\sqrt{2}} \dfrac{gA_\mu^3 - g'B_\mu}{2} \end{pmatrix} \tag{A.129}$$

and we read the mass terms directly from $\langle D^\mu \Phi^\dagger D_\mu \Phi \rangle$, following (A.93). The mass eigenstates are:

$$W_\mu^\pm = \frac{A_\mu^1 \mp iA_\mu^2}{\sqrt{2}}, \quad M_W = \frac{1}{2}gv, \tag{A.130}$$

$$Z_\mu^0 = \frac{gA_\mu^3 - g'B_\mu}{\sqrt{g^2 + g'^2}}, \quad M_Z = \frac{1}{2}\sqrt{g^2 + g'^2}\, v, \tag{A.131}$$

$$A_\mu = \frac{g'A_\mu^3 + gB_\mu}{\sqrt{g^2 + g'^2}}, \quad M_A = 0. \tag{A.132}$$

We thus check that the $SU(2) \times U(1)_Y$ symmetry is spontaneously broken down to $U(1)_{\text{QED}}$: only the photon field A_μ remains massless. Correspondingly, the Higgs doublet can be parametrized as in (A.104):

$$\Phi(x) = e^{i\gamma^a(x)t^a} \begin{pmatrix} 0 \\ \dfrac{v+h(x)}{\sqrt{2}} \end{pmatrix}. \tag{A.133}$$

The three fields $\gamma^a(x)$ can be set to zero through a gauge transformation (unitary gauge): they provide the longitudinal degrees of freedom of $W^\pm$ and Z^0.

One defines the mixing angle, called the Weinberg angle, between the neutral gauge bosons as:

$$\sin\theta_W = \frac{g'}{\sqrt{g^2 + g'^2}}, \quad \cos\theta_W = \frac{g}{\sqrt{g^2 + g'^2}}, \quad \text{tg}\,\theta_W = \frac{g'}{g}. \tag{A.134}$$

Then

$$\begin{aligned} Z_\mu &= \cos\theta_W\, A_\mu^3 - \sin\theta_W\, B_\mu, \\ A_\mu &= \sin\theta_W\, A_\mu^3 + \cos\theta_W\, B_\mu. \end{aligned} \tag{A.135}$$

We also note the important relation between the gauge boson masses and the mixing angle:

$$\rho \equiv \frac{M_W^2}{M_Z^2 \cos^2\theta_W} = 1. \tag{A.136}$$

We will see that this tree-level prediction of the Standard Model is a key test of the theory, and a major source of problems for theories beyond the Standard Model.

We now turn to the fermion sector. We will only consider here a single family of quarks and leptons: e, ν_e, u, and d. We have identified in the previous section the quantum numbers of these fields. Restoring the color $SU(3)$ degrees of freedom, we thus define

$$
\begin{array}{ccccccc}
& & & SU(3) & SU(2) & U(1)_Y & \\
\psi_\ell = & \begin{pmatrix} \nu_L \\ e_L \end{pmatrix} & \in & (3, & 2, & y = -1) & \\
& N_R & \in & (1, & 1, & y = 0) & \\
& e_R & \in & (1, & 1, & y = -2) & \\
\psi_q = & \begin{pmatrix} u_L \\ d_L \end{pmatrix} & \in & (3, & 2, & y = 1/3) & \\
& u_R & \in & (3, & 1, & y = 4/3) & \\
& d_R & \in & (3, & 1, & y = -2/3). &
\end{array}
$$

The minimal coupling to gauge fields is obtained from a standard action of the type (A.53), *i.e.* $\bar\psi i \gamma^\mu D_\mu \psi$ with the covariant derivatives:

$$
D_\mu \psi_\ell = \left(\partial_\mu - ig\, A_\mu^a \frac{\sigma^a}{2} + i\frac{g'}{2} B_\mu \right) \psi_\ell,
$$

$$
D_\mu N_R = \partial_\mu N_R, \quad D_\mu e_R = \left(\partial_\mu + ig'\, B_\mu \right) e_R,
$$

$$
D_\mu \psi_q = \left(\partial_\mu - ig\, A_\mu^a \frac{\sigma^a}{2} - i\frac{g'}{6} B_\mu \right) \psi_q, \tag{A.137}
$$

$$
D_\mu u_R = \left(\partial_\mu - \frac{2ig'}{3} B_\mu \right) u_R, \quad D_\mu d_R = \left(\partial_\mu + i\frac{g'}{3} B_\mu \right) d_R.
$$

On the other hand, a mass term $-m(\bar\psi_L \psi_R + \bar\psi_R \psi_L)$ is not allowed by the gauge symmetry because the left and right chiralities transform differently.

This chiral nature of the Standard Model is one of its key properties: fermion masses as well as gauge boson masses only appear once the symmetry is spontaneously broken. This is especially important when one considers the Standard Model as a low energy effective theory of an underlying fundamental theory with a very high mass scale Λ. If the Standard Model was vectorlike, *i.e.* if left and right chiralities were transforming the same way, then masses of order Λ would be allowed. Because the Standard Model is chiral, only masses of order v are allowed.

We note also that the only dimensionful parameter (mass scale) in the Standard Model is the scale m in the scalar potential (A.127): it is this scale which fixes the scale of electroweak symmetry breaking v in (A.128). It is also this scale which fixes, at tree-level, the Higgs mass. This can easily be seen by considering the scalar potential in the unitary gauge $V[(v + h(x))^2/2]$:

$$
m_h^2 = 2m^2 = 2\lambda v^2. \tag{A.138}
$$

This establishes the key rôle of the Higgs mass parameter in the Standard Model. As discussed in Chapter 1, this is one of the most sensitive issues when one tries to go beyond the Standard Model.

Fermion mass terms arise from the coupling of fermions to the scalar sector. One may check that the following renormalizable couplings, known as Yukawa couplings, are the most general couplings allowed by the $SU(2) \times U(1)_Y$ local gauge symmetry:

$$\mathcal{L}_Y = \lambda_e \, \bar{\psi}_\ell \, \Phi \, e_R + \lambda_\nu \, \bar{\psi}_\ell \, \widetilde{\Phi} \, N_R + \lambda_d \, \bar{\psi}_q \, \Phi \, d_R + \lambda_u \, \bar{\psi}_q \, \widetilde{\Phi} \, u_R + \text{h.c.} \qquad (A.139)$$

where $\widetilde{\Phi} \equiv i\tau_2 \Phi^*$ has been defined in (A.121). We note that such couplings automatically satisfy baryon (B) and lepton (L) number conservation.

If we set the scalar field at its ground state value (A.128), we obtain the following mass terms:

$$\mathcal{L}_m = \langle \mathcal{L}_Y \rangle = \lambda_e \frac{v}{\sqrt{2}} \bar{e}_L e_R + \lambda_\nu \frac{v}{\sqrt{2}} \bar{\nu}_L N_R + \lambda_d \frac{v}{\sqrt{2}} \bar{d}_L d_R + \lambda_u \frac{v}{\sqrt{2}} \bar{u}_L u_R + \text{h.c.} \quad (A.140)$$

Thus

$$m_e = -\lambda_e \frac{v}{\sqrt{2}}, \quad m_d = -\lambda_d \frac{v}{\sqrt{2}}, \quad m_u = -\lambda_u \frac{v}{\sqrt{2}}. \qquad (A.141)$$

The case of the neutrino mass is complicated by the possible presence of a Majorana mass term which is not constrained by the $SU(2) \times U(1)_Y$ symmetry since N_R is an electroweak singlet:

$$\mathcal{L}_M = -\tfrac{1}{2} M \, \overline{N_L^c} \, N_R + \text{h.c.} \qquad (A.142)$$

As discussed in Chapter 1, this leads in a fairly straightforward way to the seesaw mechanism for neutrino masses[9].

It remains for us to check that the low energy limit of the Standard Model is indeed the Fermi theory described by the current–current interaction (A.110). This will provide some important information on the mass scales involved.

A.3.3 Low energy effective theory

Let us identify the currents coupled to the different gauge bosons of the Standard Model. We have already done this in (A.115): it suffices to add the coupling to the $U(1)_Y$ gauge boson B_μ. Thus

$$\mathcal{L}_{\text{int}} = \frac{g}{\sqrt{2}} \left(J_+^\mu \, W_\mu^+ + J_-^\mu \, W_\mu^- \right) + g \, J^{3\mu} \, A_\mu^3 + \frac{g'}{2} J^{B\mu} B_\mu \qquad (A.144)$$

with

$$J_+^\mu = \bar{\nu}_e \, \gamma^\mu \frac{1-\gamma_5}{2} e + \bar{u} \, \gamma^\mu \frac{1-\gamma_5}{2} d, \quad J_-^\mu = (J_+^\mu)^\dagger \qquad (A.145)$$

$$J_\mu^3 = \sum_i t_i^3 \, \bar{\psi}_i \, \gamma_\mu \, \psi_i, \quad J_\mu^B = \sum_i y_i \, \bar{\psi}_i \, \gamma_\mu \, \psi_i, \qquad (A.146)$$

---

[9] Alternatively, if one does not introduce N_R at all, one obtains a nonvanishing Majorana neutrino mass by allowing the following nonrenormalizable coupling:

$$\mathcal{L}_M = \frac{\hat{\lambda}_\nu}{M} \, \overline{\psi_\ell^c} \psi_\ell \phi \phi + \text{h.c.} \qquad (A.143)$$

This term is interpreted as arising in the context of an effective theory valid at energies smaller than the scale M of a more fundamental theory.

where $\psi_i = (\psi_\ell, e_R, N_R, \psi_q, u_R, d_R)$. One may express A_μ^3 and B_μ in terms of the mass eigenstates A_μ and Z_μ^0 using (A.135). Using further (A.119) and (A.134), we obtain

$$g J^{3\mu} A_\mu^3 + \frac{g'}{2} J^{B\mu} B_\mu = g \sin\theta_w \sum_i q_i \,\bar{\psi}_i \gamma^\mu \psi_i \, A_\mu \tag{A.147}$$

$$+ \frac{g}{\cos\theta_w} \sum_i \left[t_i^3 - q_i \sin^2\theta_W \right] \bar{\psi}_i \gamma^\mu \psi_i Z_\mu^0.$$

We conclude that the QED coupling e is simply given by

$$e = g \sin\theta_w = \frac{g g'}{\sqrt{g^2 + g'^2}}. \tag{A.148}$$

On the weak interaction side, the W exchange in the diagram of figure A.5 leads at low energy to the current–current interaction

$$\mathcal{L}_{\text{eff}}^W = -\frac{g^2}{2 M_W^2} \, J_{+\mu} \, J^{-\mu} \tag{A.149}$$

since each vertex contributes a factor $(-ig/\sqrt{2})$ and the W propagator $-ig_{\mu\nu}/(k^2 - M_W^2) \sim ig_{\mu\nu}/M_W^2$ at low momentum transfer $(k^2 \ll M_W^2)$. Comparison with (A.110) thus gives

$$\frac{G_F}{\sqrt{2}} = \frac{g^2}{8 M_W^2}. \tag{A.150}$$

Expressing the W mass in terms of the Higgs vacuum expectation value as in (A.130), we obtain the important result:

$$v = \left(\frac{1}{G_F \sqrt{2}} \right)^{1/2} \simeq 246 \text{ GeV}. \tag{A.151}$$

Thus the electroweak symmetry breaking scale is completely fixed by the low energy Fermi theory. We have seen earlier that this scale is related to the mass parameter in the Higgs potential, and thus to the Higgs mass. We conclude from (A.138) that, if the Higgs is light compared to the scale v, the symmetry breaking (Higgs) sector is weakly coupled $(\lambda < 1)$, whereas it becomes strongly coupled for Higgs masses much larger than v, *i.e.* in the TeV range.

The Z^0 exchange also leads to low energy effective four-fermion interaction. One sees from (A.147) that one can write the fermionic current J_μ^Z coupled with Z^0 as

$$J_\mu^Z = J_\mu^3 - \sin^2\theta_w \, J_\mu^Q, \quad J_\mu^Q = \sum_i q_i \,\bar{\psi}_i \gamma_\mu \psi_i, \tag{A.152}$$

with coupling constant $g/\cos\theta_w$. The exchange of Z^0 at low energy thus leads to the effective "neutral current" interaction:

$$\mathcal{L}_{\text{eff}}^Z = -\frac{1}{2} \frac{g^2}{\cos^2\theta_w M_Z^2} J^{Z\mu} J_\mu^Z = -\frac{g^2}{2 M_W^2} J^{Z\mu} J_\mu^Z, \tag{A.153}$$

where we have included a symmetry factor of $1/2$ (absent in the charged current effective interaction because $J_\mu^+ \neq J_\mu^-$). Such an interaction scales like the Fermi constant G_F but was much more difficult to detect because it does not correspond to a change in the nature of the quarks (as in the beta decay). It was discovered at CERN in 1973, following the prediction of the Standard Model.

A.3.4 Flavor structure

The Standard Model has a very simple family structure which leads to specific predictions, such as the quasi-absence of flavor changing neutral currents, which are difficult to reproduce in the context of its extensions. We will thus first review the main results.

In the Standard Model, we are in presence of three families which are replicas (from the point of view of gauge quantum numbers) of the set (ν_e, e, u, d). We will thus denote the fermions of each three family as

$$(\nu_i, e_i, u_i, d_i)$$
$$i = 1 \quad (\nu_e, e, u, d)$$
$$i = 2 \quad (\nu_\mu, \mu, c, s)$$
$$i = 3 \quad (\nu_\tau, \tau, t, b)$$

where color indices have been suppressed on the quark fields. We call the index i a family index. If we include the three families, then the Yukawa couplings have a matrix structure in family space: all couplings in

$$\mathcal{L}_Y = \Lambda_{\ell_{ij}} \bar{\psi}_{\ell_i} \Phi e_{Rj} + \Lambda_{\nu_{ij}} \bar{\psi}_{\ell_i} \widetilde{\Phi} N_{Rj} + \Lambda_{d_{ij}} \bar{\psi}_{q_i} \Phi d_{Rj} + \Lambda_{u_{ij}} \bar{\psi}_{q_i} \widetilde{\Phi} u_{Rj} + \text{h.c.} \quad \text{(A.154)}$$

are allowed by the gauge symmetries of the Standard Model (since these are *horizontal*, *i.e.* do not depend on the family of the quark or lepton). In (A.154), $\psi_{\ell_i} = \binom{\nu_{Li}}{e_{Li}}$ and $\psi_{q_i} = \binom{u_{Li}}{d_{Li}}$.

The Yukawa matrices Λ_ℓ, Λ_d, and Λ_u are generic complex 3×3 matrices. In all generality, a complex matrix has a polar decomposition, *i.e.* it can be written as a product of a hermitian matrix H and a unitary matrix U (see Exercise 3)

$$\Lambda = HU. \quad \text{(A.155)}$$

The matrix H, being hermitian , can be diagonalized with a unitary matrix V_L: $H = V_L D V_L^\dagger$. Then the general complex matrix Λ is diagonalized with the help of two unitary matrices V_L and $V_R \equiv U^\dagger V_L$

$$V_L^\dagger \Lambda V_R = D = \begin{pmatrix} \lambda_1 & & \\ & \lambda_2 & \\ & & \lambda_3 \end{pmatrix}. \quad \text{(A.156)}$$

Using this general result, one may then write for example the electron mass term which arises from (A.154) as

$$\mathcal{L}_m = \bar{e}_L \Lambda_\ell \frac{v}{\sqrt{2}} e_R = \bar{e}_L V_L^\ell D_\ell V_R^{\ell\dagger} \frac{v}{\sqrt{2}} e_R$$

$$= \overline{\left(V_L^{\ell\dagger} e_L\right)}_i \lambda_i \frac{v}{\sqrt{2}} \left(V_R^{\ell\dagger} e_R\right)_i, \quad \text{(A.157)}$$

where we have suppressed some of the summed family indices. Thus, charge eigenstates, *i.e.* fields belonging to definite representations of the gauge group such as (u, d)

or (ν_τ, τ), should not be confused with mass eigenstates *i.e.* fields which are observed in the detectors:

$$
\begin{aligned}
\widehat{e}_L &= V_L^{\ell\dagger} e_L, & \widehat{e}_R &= V_R^{\ell\dagger} e_R, \\
\widehat{\nu}_L &= V_L^{\nu\dagger} \nu_L, & \widehat{\nu}_R &= V_R^{\nu\dagger} \nu_R, \\
\widehat{u}_L &= V_L^{u\dagger} u_L, & \widehat{u}_R &= V_R^{u\dagger} u_R, \\
\widehat{d}_L &= V_L^{d\dagger} d_L, & \widehat{d}_R &= V_R^{d\dagger} d_R.
\end{aligned}
\tag{A.158}
$$

Charged and neutral currents should be expressed in terms of the observable fields, that is in terms of the mass eigenstates. This induces some new structure. For example, the charged current reads

$$
\begin{aligned}
J_+^\mu &= \sum_i \bar{u}_{Li} \gamma^\mu d_{Li} + \bar{\nu}_{Li} \gamma^\mu e_{Li} \\
&= \sum_{kl} \overline{\widehat{u}}_{Lk} \left(V_L^{u\dagger} V_L^d \right)_{kl} \gamma^\mu \widehat{d}_{Ll} + \overline{\widehat{\nu}}_{Lk} \left(V_L^{\nu\dagger} V_L^\ell \right)_{kl} \gamma^\mu \widehat{e}_{Ll}.
\end{aligned}
\tag{A.159}
$$

Hence the quark charged current coupled to the W^+ is not diagonal. It involves the mixing matrix

$$
V_{\text{CKM}} \equiv V_L^{u\dagger} V_L^d
\tag{A.160}
$$

known as the Cabibbo–Kobayashi–Maskawa matrix. And similarly for the leptons

$$
V_{\text{MNS}} \equiv V_L^{\nu\dagger} V_L^\ell
\tag{A.161}
$$

where V_{MNS} is the Maki–Nakagawa–Sakata [280] matrix.

On the other hand, neutral currents, because they are flavor diagonal, remain diagonal when expressed in terms of mass eigenstates. For example, J_μ^3 which appears in $J_\mu^Z = J_\mu^3 - \sin^2\theta_W J_\mu^Q$ (see equation (A.152)), is written:

$$
\begin{aligned}
J_\mu^3 &= \sum_i \frac{1}{2} \bar{u}_{Li} \gamma_\mu u_{Li} - \frac{1}{2} \bar{d}_{Li} \gamma_\mu d_{Li} + \cdots \\
&= \sum_i \frac{1}{2} \overline{\widehat{u}}_{Li} \gamma_\mu \widehat{u}_{Li} - \frac{1}{2} \overline{\widehat{d}}_{Li} \gamma_\mu \widehat{d}_{Li} + \cdots
\end{aligned}
\tag{A.162}
$$

since the diagonalization matrices cancel ($V_L^{u\dagger} V_L^u = 1, \ldots$). Thus neutral current interactions proceed without any change of flavor (*i.e.* type of quarks). This is why neutral currents were so much more difficult to detect experimentally than charged currents (nuclear beta decay), although they are of the same strength (see (A.181) below).

The absence of flavor changing neutral currents extends to the one-loop level through a cancellation known as the GIM mechanism [193]. We illustrate it here on the simpler case of two families where the quark mixing matrix can be written (see below) as the 2×2 orthogonal Cabibbo matrix

$$
V_C = \begin{pmatrix} \cos\theta_C & \sin\theta_C \\ -\sin\theta_C & \cos\theta_C \end{pmatrix},
\tag{A.163}
$$

where θ_C is the Cabibbo angle ($\sin\theta_C \sim 0.2$). Since charged currents are flavor changing, we should expect that the simultaneous exchange of a W^+ and a W^- in a one-loop

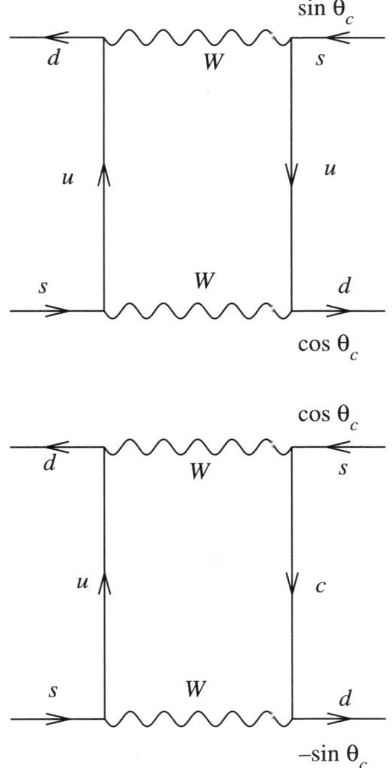

Fig. A.6 Two contributions to $K^0(s\bar{d}) - \bar{K}^0(d\bar{s})$ oscillations.

process (box diagram) yields a flavor-changing neutral current. Take for example two of the diagrams which contribute to the $K^0 - \bar{K}^0$ oscillation, as shown in Fig. A.6 (there are two other diagrams with the u quark internal line on the left replaced by a c quark).

In the limit of vanishing masses or equal masses, it is clear that the two diagrams have the same contribution apart from the couplings of the internal quark line on the right, which is $\cos\theta_c \sin\theta_c$ for the first diagram and $-\sin\theta_c \cos\theta_c$ for the second diagram. Thus they cancel and this flavor-changing process vanishes in the limit of equal masses: it is at most of the order of the quark mass differences and thus small. A careful study of the size of the effect allowed [171] to estimate the charm quark mass.

We close this presentation of the flavor sector of the Standard Model by a discussion of CP violation: it turns out that the only source of CP violation in the Standard Model is a single phase in the fermion sector.

Generally speaking, invariance under CP of a fermion mass term

$$-\bar{\psi}_L M \psi_R + \text{ h.c.} = -\tfrac{1}{2}\bar{\psi}\left[\left(M + M^\dagger\right) + \left(M - M^\dagger\right)\gamma_5\right]\psi \qquad (A.164)$$

requires that the mass matrix is real: $M = M^*$. Indeed, invariance under parity P (under which $\bar{\psi}_1\psi_2$ is invariant whereas $\bar{\psi}_1\gamma_5\psi_2$ changes sign, *cf.* (A.108)) imposes $M = M^\dagger$; invariance under charge conjugation C (under which $\bar{\psi}_1\psi_2 \to \bar{\psi}_2\psi_1$ and $\bar{\psi}_1\gamma_5\psi_2 \to \bar{\psi}_2\gamma_5\psi_1$) imposes $M = M^T$.

Thus, we will have a natural violation of the CP symmetry if there exists at least one physical phase in the fermion mass matrix or, if we are working with the mass eigenstates, in the Cabibbo–Kobayashi–Maskawa matrix [254]. By "physical" we mean a phase which cannot be set to zero by a redefinition of the fermion fields. Let us count the number of physical phases of V_{CKM} in the general case of N_F families of quarks: V_{CKM} being a complex $N_F \times N_F$ matrix which satisfies the unitarity condition, it has N_F^2 real parameters. Out of these, $2N_F - 1$ can be absorbed by redefining the phases of the quark fields[10] which leaves $(N_F - 1)^2$ physical parameters. If V_{CKM} was real, it would be a $N_F \times N_F$ orthogonal matrix and would have $N_F(N_F-1)/2$ real parameters. The remaining $(N_F - 1)(N_F - 2)/2$ parameters are complex phases.

Thus, with $N_F = 2$ families of quarks, there is no source of CP violation and the Cabibbo matrix (A.163) can be made real by redefining the phases of the quark fields. And $N_F = 3$ corresponds to the minimum number of families necessary to have a CP violating phase in the Standard Model, as noted first by Kobayashi and Maskawa. This single phase is generically referred to as δ_{CKM}.

We may for example write the Cabibbo–Kobayashi–Maskawa matrix in the Wolfenstein parametrization [379][11]:

$$
V_{\mathrm{CKM}} \equiv V = \begin{pmatrix} 1 - \frac{1}{2}\lambda^2 & \lambda & A\lambda^3(\rho - i\eta) \\ -\lambda & 1 - \frac{1}{2}\lambda^2 & A\lambda^2 \\ A\lambda^3(1 - \rho - i\eta) & -A\lambda^2 & 1 \end{pmatrix},
\qquad \text{(A.165)}
$$

where the CP-violating phase is now parametrized by η.

The unitarity relation

$$
V_{ud}V_{ub}^* + V_{cd}V_{cb}^* + V_{td}V_{tb}^* = 0
\qquad \text{(A.166)}
$$

may be represented geometrically in the complex plane and is called a unitarity triangle. It has become customary to rescale the length of one side, *i.e.* $|V_{cd}V_{cb}^*|$ which is well-known, to 1 and to align it along the real axis. The unitarity triangle thus looks like Fig. A.7.

The angles α, β, γ are defined as:

$$
\alpha \equiv \arg\left(-\frac{V_{td}V_{tb}^*}{V_{ud}V_{ub}^*}\right), \quad \beta \equiv \arg\left(-\frac{V_{cd}V_{cb}^*}{V_{td}V_{tb}^*}\right), \quad \gamma \equiv \arg\left(-\frac{V_{ud}V_{ub}^*}{V_{cd}V_{cb}^*}\right),
\qquad \text{(A.167)}
$$

and the lengths

$$
R_t \equiv \left|\frac{V_{td}V_{tb}^*}{V_{cd}V_{cb}^*}\right|, \quad R_u \equiv \left|\frac{V_{ud}V_{ub}^*}{V_{cd}V_{cb}^*}\right|.
\qquad \text{(A.168)}
$$

Fig. 12.1 of Chapter 12 shows how the experimental limits obtained in 2005 constrain the unitarity triangle. The fact that all data is consistent with a small region with nonvanishing η shows that CP is violated and that its violation is consistent with a single origin: the phase of the CKM matrix.

---

[10]A global phase redefinition would have no effect; hence the -1.

[11]We neglect terms of order λ^4 and higher [59]; in the same spirit, we use in what follows the variables $\bar{\rho}$ and $\bar{\eta}$ which are precisely defined as $\bar{\rho} \equiv \rho(1 - \lambda^2/2)$ and $\bar{\eta} \equiv \eta(1 - \lambda^2/2)$.

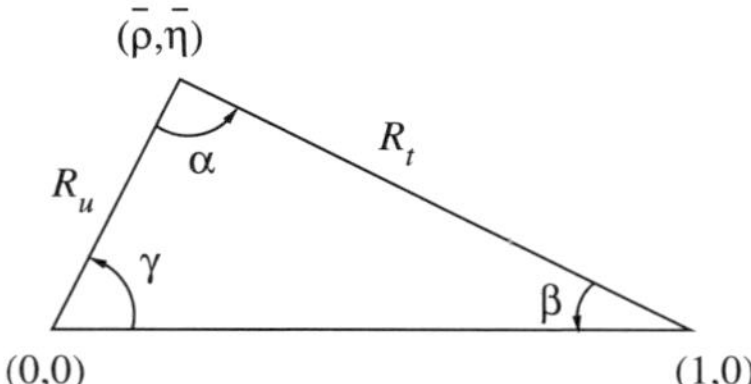

Fig. A.7 Unitarity triangle

More quantitatively, one may show that CP is violated if and only if [238]

$$\mathrm{Im}\left(\det\left[\Lambda_d\Lambda_d^\dagger, \Lambda_u\Lambda_u^\dagger\right]\right) \neq 0. \tag{A.169}$$

Introducing the quantity

$$\mathrm{Im}\left(V_{ij}V_{kl}V_{il}^*V_{kj}^*\right) = J_{\mathrm{CKM}}\sum_{m,n=1}^{3}\epsilon_{ikm}\epsilon_{jln} \tag{A.170}$$

($J_{\mathrm{CKM}} \sim \lambda^6 A^2\eta$ in the Wolfenstein parametrization), one may write the condition (A.169) in the mass eigenstate basis as

$$(m_t^2 - m_c^2)(m_t^2 - m_u^2)(m_c^2 - m_u^2)(m_b^2 - m_s^2)(m_b^2 - m_d^2)(m_s^2 - m_d^2)J_{\mathrm{CKM}} \neq 0. \tag{A.171}$$

A.4 Electroweak precision tests

The Standard Model has been tested at the CERN electron–positron collider LEP to a precision of the order of the per mil (‰). Such a precision allows us to go beyond the tree-level which has been described above and to test the theory at the quantum level. Indeed, if quantum fluctuations were not included, the Standard Model would be in disagreement with experimental data. We review in this section how these quantum corrections are included in order to confront the theory with experiment.

A.4.1 Principle of the analysis of the radiative corrections in the Standard Model

As soon as renormalization is involved, it is important to rest the discussion on observables: only observable quantities are free of the arbitrariness inherent to the renormalization procedure.

Let us consider for example the three parameters which, besides the Higgs mass, describe the gauge sector of the Standard Model: the $SU(2)$ gauge coupling g, the hypercharge $U(1)$ gauge coupling $g'/2$, and the Higgs vacuum expectation value $v/\sqrt{2}$. Since they are not directly measurable, one prefers to replace them by observable quantities:

- α, the fine structure constant;
- G_μ, the Fermi constant measured in μ decay;
- M_z, the Z boson mass.

They are presently measured to be:

$$\alpha^{-1} = 137.0359895(61) \tag{A.172}$$

$$G_\mu = 1.16637(2) \times 10^{-5} \ \mathrm{GeV}^{-2} \tag{A.173}$$

$$M_z = 91.1872 \pm 0.0021 \ \mathrm{GeV}. \tag{A.174}$$

The relation between the theory parameters g, g', v, and these observable quantities is, at tree-level of the theory, given by (A.148), (A.151) and (A.131):

$$\alpha = \frac{e^2}{4\pi} = \frac{1}{4\pi} \frac{g^2 g'^2}{g^2 + g'^2}, \tag{A.175}$$

$$G_\mu = \frac{1}{v^2 \sqrt{2}}, \tag{A.176}$$

$$M_z = \frac{1}{2} \sqrt{g^2 + g'^2} \ v. \tag{A.177}$$

Given this set of data, it is interesting to ask how to define the central parameter of the electroweak theory, $\sin^2 \theta_W$. There are several equivalent ways to do it *at tree-level*:

- Using the relation $M_W^2 = M_z^2 \cos^2 \theta_W$, we may write

$$\sin^2 \theta_W = 1 - \frac{M_W^2}{M_z^2} \equiv s_W^2. \tag{A.178}$$

- We could use α and G_μ instead of M_z. Indeed, using (A.134) we have

$$\frac{\alpha}{G_\mu} = g^2 v^2 \sin^2 \theta_W \frac{\sqrt{2}}{4\pi}, \tag{A.179}$$

and, since $M_W^2 = g^2 v^2/4$,

$$\sin^2 \theta_W = \frac{A}{M_W^2}, \qquad A \equiv \frac{\pi \alpha}{\sqrt{2} G_\mu}. \tag{A.180}$$

- One may also use the ratio between the charged and neutral current effective interactions. According to (A.149) and (A.153), they read, in the limit of vanishing transfer momentum q:

$$\left. \frac{G_{\mathrm{NC}}}{G_{\mathrm{CC}}} \right|_{q^2=0} = \frac{M_W^2}{M_z^2 \cos^2 \theta_W} = \rho. \tag{A.181}$$

In other words, we could define

$$\sin^2 \theta_W = 1 - \frac{M_W^2}{M_z^2} \frac{G_{\mathrm{CC}}}{G_{\mathrm{NC}}}. \tag{A.182}$$

These three definitions are obviously equivalent at tree-level but are affected by different corrections at higher orders. Thus, if one wants to define a consistent procedure, one must decide on a single definition. We will choose here (A.178): in other words, we define $\sin^2\theta_W$, and note s_W^2, the function $1 - M_W^2/M_Z^2$, *to all orders of perturbation theory*. When one uses this definition, then (A.180) reads

$$M_W^2 s_W^2 = \frac{A}{1 - \delta r} \tag{A.183}$$

and (A.181)

$$\rho \equiv \left.\frac{G_{\mathrm{NC}}}{G_{\mathrm{CC}}}\right|_{q^2=0} = 1 + \delta\rho, \tag{A.184}$$

where δr and $\delta\rho$ account for radiative corrections.

It is easy to obtain first order expressions for these quantities. Denoting bare quantities with a superscript 0, we may write for example (A.180) as

$$M_W^{0\,2} = \frac{\pi\alpha^0}{\sqrt{2}G_\mu^0 \sin^2\theta_W^0}. \tag{A.185}$$

We then replace bared quantities by renormalized ones:

$$\alpha^0 = \alpha\left(1 - \frac{\delta\alpha}{\alpha}\right),$$
$$G_\mu^0 = G_\mu\left(1 - \frac{\delta G_\mu}{G_\mu}\right), \dots \tag{A.186}$$

Then (A.185) reads, in terms of renormalized quantities, and to first order,

$$M_W^2 = \frac{\pi\alpha}{\sqrt{2}G_\mu \sin^2\theta_W}\left[1 - \frac{\delta\alpha}{\alpha} + \frac{\delta G_\mu}{G_\mu} + \frac{\delta\sin^2\theta_W}{\sin^2\theta_W} + \frac{\delta M_W^2}{M_W^2}\right] \tag{A.187}$$

and, since we have defined $\sin^2\theta_W$ through (A.178),

$$\delta\sin^2\theta_W = -\frac{\delta M_W^2}{M_Z^2} + \frac{M_W^2}{M_Z^4}\delta M_Z^2. \tag{A.188}$$

Hence

$$\delta r = -\frac{\delta\alpha}{\alpha} + \frac{\delta G_\mu}{G_\mu} + \frac{c_W^2}{s_W^2}\frac{\delta M_Z^2}{M_Z^2} + \left(1 - \frac{c_W^2}{s_W^2}\right)\frac{\delta M_W^2}{M_W^2} \tag{A.189}$$

with $c_W^2 = 1 - s_W^2$. Similarly, from (A.184),

$$\delta\rho = \frac{\delta G_{\mathrm{NC}}}{G_{\mathrm{NC}}} - \frac{\delta G_\mu}{G_\mu}, \tag{A.190}$$

where we have used $\rho = 1$ at zeroth order ($G_{\mathrm{CC}} \equiv G_\mu$).

One may distinguish three types of radiative corrections in the Standard Model:

(i) α and G_μ are measured at low energy and must be renormalized up to the scale at which present measurements are performed (typically M_z). For example, summing the large $(\ln M_z/m_e)^n$ contributions into a running coupling $\alpha(M_z)$, one obtains from the boundary value (A.172) at a scale m_e

$$\alpha^{-1}(M_z) = 127.934 \pm 0.027, \qquad (A.191)$$

where the error quoted is mainly due to the uncertainty on the low energy hadronic contribution to vacuum polarization.

On the other hand, G_μ does not receive large $\ln(M_z/m_\mu)$ corrections because of a nonrenormalization theorem[12].

(ii) Large corrections due to the top or the Higgs may appear through vacuum polarization diagrams. These are traditionally called *oblique corrections*. There is a well-defined procedure to take them into account.

One may introduce in general the two point-function:

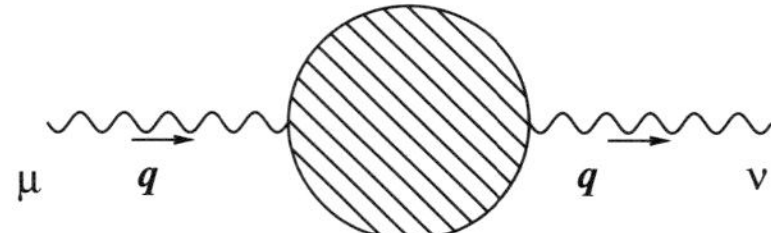

$$\Pi_{\mu\nu}\left(q^2\right) = -ig_{\mu\nu}\left[A + Fq^2 + Gq^4 + \cdots\right] + O\left(q_\mu q_\nu\right). \qquad (A.192)$$

Since this is usually contracted with external fermionic currents, the terms $O\left(q_\mu q_\nu\right)$ give contributions of the order of the light fermion masses, which we neglect (see however below). If we denote by M the mass of the heavy field (Z, top or Higgs), then dimensional analysis tells us that $A \sim M^2$, $F \sim M^0$ and $G \sim M^{-2}$ and thus the term in q^4 is negligible for $q^2 \ll M^2$.

We thus have to consider

$$\Pi_{\mu\nu}^{WW}\left(q^2\right) = -ig_{\mu\nu}\left[A_{WW} + q^2 F_{WW}\right],$$

$$\Pi_{\mu\nu}^{ZZ}\left(q^2\right) = -ig_{\mu\nu}\left[A_{ZZ} + q^2 F_{ZZ}\right],$$

$$\Pi_{\mu\nu}^{\gamma\gamma}\left(q^2\right) = -ig_{\mu\nu}\left[0 + q^2 F_{\gamma\gamma}\right],$$

$$\Pi_{\mu\nu}^{\gamma Z}\left(q^2\right) = -ig_{\mu\nu}\left[0 + q^2 F_{\gamma Z}\right], \qquad (A.193)$$

where $A_{\gamma\gamma} = A_{\gamma Z} = 0$ because the electromagnetic current is conserved: $q^\mu \Pi_{\mu\nu}^{\gamma\gamma} = 0 = q^\mu \Pi_{\mu\nu}^{\gamma Z}$ for $q^2 = 0$.

---

[12]Through Fierz reordering, one may write the effective current–current interaction responsible for muon decay as

$$\mathcal{L}_{\text{eff}} = \frac{G_\mu}{\sqrt{2}}\ [\bar{e}\gamma^\mu(1 - \gamma_5)\mu]\ [\bar{\nu}_\mu\gamma_\mu(1 - \gamma_5)\nu_e].$$

The lepton-changing $(e \to \mu)$ vector and axial currents that appear satisfy a nonrenormalization theorem that allows only finite corrections and hence forbid any dependence in the effective cut-off M_z.

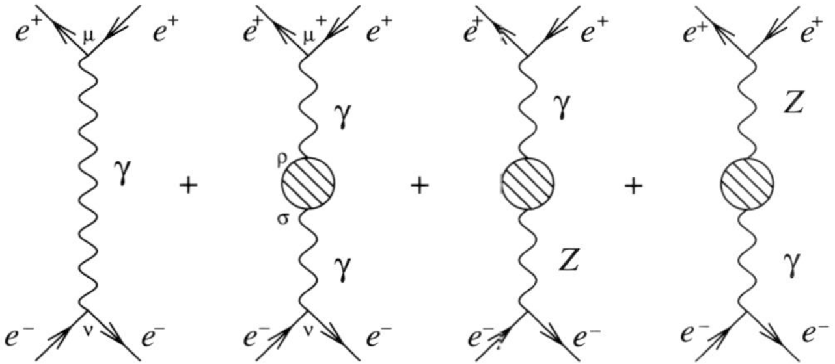

Fig. A.8

Now, if we consider the diagrams of Fig. A.8, only the tree-level and $\gamma\gamma$ propagator contribute to the pole in q^2. Thus the pole part of the amplitude reads

$$\mathcal{M}|_{\text{pole}} = -\frac{i}{q^2}g^{\mu\nu}e_0^2\left(1 - F_{\gamma\gamma}\right) = -\frac{i}{q^2}g^{\mu\nu}\left(e^2 - \delta e^2\right)\left(1 - F_{\gamma\gamma}\right)$$

$$\equiv -\frac{i}{q^2}g^{\mu\nu}e^2, \tag{A.194}$$

where we have introduced, as in (A.186), the bare e_0 and renormalized e couplings. Thus, to first order, $\delta e^2/e^2 = -F_{\gamma\gamma}$.

We obtain, with similar analyses for G_μ and M_z,

$$\frac{\delta\alpha}{\alpha} = -F_{\gamma\gamma}, \tag{A.195}$$

$$\frac{\delta G_\mu}{G_\mu} = \frac{A_{WW}}{M_W^2}, \tag{A.196}$$

$$\frac{\delta M_z^2}{M_z^2} = -\frac{A_{ZZ}}{M_z^2} - F_{ZZ}. \tag{A.197}$$

Thus, three out of six of the independent quantities that we have introduced in (A.193) are used to renormalize the variables α, G_μ and M_z. We are left with three variables which fully describe the oblique corrections [6, 307].

One prefers to work in the original $SU(2) \times U(1)$ basis (A_μ^3, B_μ) for the gauge fields. One then introduces[13]:

$$\epsilon_1 = \frac{A_{33} - A_{WW}}{M_W^2}, \tag{A.200}$$

---

[13] Alternatively, in the language of [307, 308], one introduces the three variables S, T and U which, to lowest order, are related to the variables of [6] by

$$\epsilon_1 = \alpha\, T,$$

$$\epsilon_2 = -\frac{\alpha}{4s_W^2}\, U, \tag{A.198}$$

$$\epsilon_3 = \frac{\alpha}{4s_W^2}\, S. \tag{A.199}$$

$$\epsilon_2 = F_{WW} - F_{33}, \tag{A.201}$$

$$\epsilon_3 = \frac{c_W}{s_W} F_{3B}. \tag{A.202}$$

Using (A.196) and its neutral current equivalent ($\delta G_{\mathrm{NC}}/G_{\mathrm{NC}} = A_{ZZ}/M_Z^2$), we obtain from (A.190)

$$\delta\rho = \frac{A_{ZZ}}{M_Z^2} - \frac{A_{WW}}{M_W^2} = \epsilon_1, \tag{A.203}$$

where we have used (A.285). Hence ϵ_1 represents the departure of the ρ parameter from the value 1. Similarly (see Exercise 4),

$$\delta r = -\frac{c_W^2}{s_W^2}\epsilon_1 - \left(1 - \frac{c_W^2}{s_W^2}\right)\epsilon_2 + 2\epsilon_3. \tag{A.204}$$

It is rather easy to guess the order of magnitude of the dimensionless parameters ϵ_i. At one loop, which is the first order at which they are nonvanishing, the corresponding diagrams involve two gauge couplings, and thus are of order $G_F M_W^2$. Take first the example of ϵ_3. Clearly $F_{3B} \neq 0$ requires the breaking of $SU(2) \times U(1)$ which confirms that $\epsilon_3 \sim M_W^2$. Moreover, the presence of F_{3B} requires to factor q out (as in (A.192)): we are left with a logarithmic dependence on the high scales, typically $\ln(m_t/M_Z)$ or $\ln(m_H/M_Z)$. The case is somewhat different with ϵ_1. The same argument as just given tells us that $A_{33} - A_{WW} \sim G_F M_W^2 m^2$ where dimensional analysis tells us that we are still missing a squared mass factor m^2 which turns out to be the top mass-squared. Hence $\epsilon_1 \sim G_F m_t^2$.

A complete calculation gives:

$$\epsilon_1 = \frac{3G_F m_t^2}{8\pi^2\sqrt{2}} - \frac{3G_F M_W^2}{4\pi^2\sqrt{2}}\tan^2\theta_W \ln\frac{m_H}{M_Z} + \cdots \tag{A.205}$$

$$\epsilon_2 = -\frac{G_F M_W^2}{2\pi^2\sqrt{2}}\ln\frac{m_t}{M_Z} + \cdots \tag{A.206}$$

$$\epsilon_3 = \frac{G_F M_W^2}{12\pi^2\sqrt{2}}\ln\frac{m_H}{M_Z} - \frac{G_F M_W^2}{6\pi^2\sqrt{2}}\ln\frac{m_t}{M_Z} + \cdots \tag{A.207}$$

Hence, $\epsilon_1 \gg \epsilon_2, \epsilon_3$ and, for example, $\delta r \sim -\left(c_W^2/s_W^2\right)\delta\rho$.

We also note that there is no term of order $G_F m_H^2$. The absence of such terms may be attributed to the symmetries of the theory in the limit of large Higgs mass. In this limit, the Higgs sector has the structure of a nonlinear sigma-model coupled to gauge fields. Indeed, setting for a moment the gauge fields to zero, one may write the scalar Lagrangian as (We write $\phi^+ \equiv \phi_2 + i\phi_1, \phi^0 \equiv h - i\phi_3$ and $M \equiv h + i\sigma^a\phi_a$)

$$\mathcal{L} = \frac{1}{4}\,\mathrm{Tr}\,\partial^\mu M^\dagger \partial_\mu M - V(M, M^\dagger), \tag{A.208}$$

$$V(M, M^\dagger) = \frac{\lambda}{4}\left[\mathrm{Tr}\,M^\dagger M - v^2\right]^2,$$

where $M \equiv h + i\tau \cdot \xi$. This Lagrangian has an invariance $SU(2)_L \times SU(2)_R$: $M \to U_L M U_R^\dagger$, where U_L, U_R are 2×2 unitary matrices (to understand the meaning of the L, R subscripts, see Equation (7.9) of Chapter 7). When we restore the gauge fields, there remains a $SU(2)_R$ invariance in the limit $\theta_W \to 0$ ($g' \to 0$). When $m_H \to \infty$, the model becomes nonlinear with the constraint $\mathrm{Tr}\, M^\dagger M = v^2$. The $SU(2)_R$ symmetry still constrains the structure of the counterterms and forbids terms of the form $G_F m_H^2$ (m_H is the physical cut-off).

This result is known as the screening theorem [351]. The sensitivity of low energy physics in the Higgs mass is only logarithmic. But tests at the LEP collider have achieved such a precision that they now allow us to put limits on the Higgs mass within the Standard Model.

(iii) In the case of processes involving the third generation of quarks, one has a possible large dependence in m_t through vertex corrections (e.g. $Z \to b\bar{b}$).

A.5 Dilatations and renormalization group

Scale invariance and its violations as described by the renormalization group approach play an important rôle in the study of supersymmetry. We present the basic notions in this section. We also review the notion of effective potential which is used in the main text to discuss quantum corrections.

A.5.1 Dilatations and conformal transformations

A dilatation or scaling transformation is a spacetime transformation of the form

$$x \to x' = e^{-\alpha} x. \tag{A.209}$$

It acts linearly on the fields:

$$\Phi(x) \to \Phi'(x') = e^{\alpha d}\Phi(x), \tag{A.210}$$

which we may write by keeping spacetime fixed

$$\Phi(x) \to \Phi'(x) = e^{\alpha d}\Phi(e^\alpha x). \tag{A.211}$$

The number d is characteristic of the field Φ and is called its scaling dimension. Infinitesimally,

$$\delta\Phi(x) = \alpha\left(d + x^\mu \partial_\mu\right)\Phi(x). \tag{A.212}$$

At the classical level, the scaling dimension coincides with the canonical dimension: $d = 1$ for the scalar fields, $\frac{3}{2}$ for spin $\frac{1}{2}$ or $\frac{3}{2}$ fermions.

We may consider as an example the following action involving a scalar field $\phi(x)$ and a Dirac spinor field $\Psi(x)$:

$$S = S_0 + S_1 = \int d^4x \mathcal{L}_0(x) + \int d^4x \mathcal{L}_1(x), \tag{A.213}$$

$$\mathcal{L}_0 = \frac{1}{2}\partial^\mu\phi\partial_\mu\phi + \frac{i}{2}\bar{\Psi}\gamma^\mu\partial_\mu\Psi + \lambda_Y \phi\bar{\Psi}\Psi - \frac{\lambda_0}{4!}\phi^4, \tag{A.214}$$

$$\mathcal{L}_1 = -\frac{1}{2}m_0^2\phi^2. \tag{A.215}$$

Under the dilatation transformation (A.212), we have

$$\delta \mathcal{L}_0 = \alpha \left(4 + x^\mu \partial_\mu\right) \mathcal{L}_0. \tag{A.216}$$

and thus, by integration by parts,

$$\delta \mathcal{S}_0 = \alpha \int d^4x \, \partial_\mu \left(x^\mu \mathcal{L}_0\right). \tag{A.217}$$

Thus $\mathcal{S}_0$ is invariant by dilatation. On the other hand,

$$\frac{1}{\alpha}\delta \mathcal{S}_1 = \int d^4x \, (-m_0^2)\phi \left(1 + x^\mu \partial_\mu\right)\phi = \int d^4x \, (-m_0^2)\left(1 + \frac{1}{2}x^\mu \partial_\mu\right)\phi^2$$
$$= \int d^4x \, (-m_0^2)\left(1 - \frac{1}{2}\partial_\mu x^\mu\right)\phi^2 = \int d^4x \, m_0^2 \phi^2.$$

Hence

$$\delta \mathcal{S} = \alpha \int d^4x \, \Delta \quad \text{with} \quad \Delta = m_0^2 \phi^2. \tag{A.218}$$

It follows that the dilatation current, or scaling current, D_μ is not conserved: from (A.10) we obtain

$$\partial^\mu D_\mu = \Delta. \tag{A.219}$$

Using (A.8), one obtains from (A.209) and (A.210) the general explicit form for the dilatation current

$$D_\mu = -x^\rho \left[\mathcal{L}g_{\mu\rho} - \frac{\delta \mathcal{L}}{\delta\left(\partial^\mu \Phi(x)\right)}\partial_\rho \Phi\right] + \frac{\delta \mathcal{L}}{\delta\left(\partial^\mu \Phi(x)\right)} d\Phi(x)$$
$$= x^\rho T_{\rho\mu} + \frac{\delta \mathcal{L}}{\delta\left(\partial^\mu \Phi(x)\right)} d\Phi(x), \tag{A.220}$$

where $T_{\mu\nu}$ is the canonical energy–momentum tensor:

$$T_{\mu\nu} = \frac{\delta \mathcal{L}}{\delta\left(\partial^\mu \Phi(x)\right)}\partial_\nu \Phi - g_{\mu\nu}\mathcal{L}. \tag{A.221}$$

It is, however, possible to define [64], without affecting the construction of the Lorentz generators P_μ and $M_{\mu\nu}$, a symmetric energy–momentum tensor $\Theta_{\mu\nu}$ such that

$$D_\mu = x^\rho \Theta_{\rho\mu}. \tag{A.222}$$

Explicitly, in our example (A.213) above (see Exercise 7),

$$\Theta_{\mu\nu} = T_{\mu\nu} - \frac{1}{6}\left(\partial_\mu \partial_\nu - g_{\mu\nu}\partial^\rho \partial_\rho\right)\phi^2. \tag{A.223}$$

However, the construction of $\Theta_{\mu\nu}$ is only possible under some conditions which are easily identified in the limit $\Delta \to 0$. Indeed, in this case, the conservation of the dilatation current (A.222) is expressed by the condition of tracelessness for the new energy–momentum tensor:

$$\partial^\mu D_\mu = \Theta^\mu{}_\mu = 0. \tag{A.224}$$

It then follows that the following four ($\lambda = 0, 1, 2, 3$) currents are conserved:

$$\mathcal{K}_{\lambda\mu} \equiv 2x_\lambda x^\nu \,\Theta_{\nu\mu} - x^2 \,\Theta_{\lambda\mu}. \tag{A.225}$$

These are actually the four currents associated with the four generators of conformal transformations $K_\lambda \equiv \int d^4 x \mathcal{K}_{\lambda 0}$. The corresponding transformations are obtained by combining the four translations $x^\mu \to x^\mu - a^\mu$ with the inversion $x^\mu \to -x^\mu/x^2$:

$$x^\mu \to x'^\mu = e^{-\alpha(x)} \left(x^\mu + c^\mu x^2\right), \quad e^{\alpha(x)} \equiv 1 + 2c \cdot x + c^2 x^2. \tag{A.226}$$

We note that $x'^\mu/x'^2 = x^\mu/x^2 + c^\mu$. By construction, an invariant line element transforms as

$$g_{\mu\nu} dx^\mu dx^\nu \to (g_{\mu\nu} dx^\mu dx^\nu)' = e^{-2\alpha(x)} g_{\mu\nu} dx^\mu dx^\nu. \tag{A.227}$$

The possibility of defining a tensor $\Theta_{\mu\nu}$ is associated with the presence of conformal invariance in the limit $\Delta \to 0$. One may note here the close link between conformal and dilatation invariance. Thus, any Lorentz invariant theory which is also invariant under conformal transformations has dilatation invariance. This can be seen from the general relation

$$\partial^\mu \mathcal{K}_{\lambda\mu} = 2x_\lambda \partial^\mu D_\mu. \tag{A.228}$$

This can also be obtained from the algebra. Indeed, defining the charges

$$D = \int d^3 x \, D_0, \quad K_\mu = \int d^3 x \mathcal{K}_{\mu 0}, \tag{A.229}$$

we have the following algebra

$$\begin{aligned}
&[D, M_{\rho\sigma}] = 0 \ , \ [K_\mu, M_{\rho\sigma}] = i \left(\eta_{\mu\rho} K_\sigma - \eta_{\mu\sigma} K_\rho\right), \\
&[P_\mu, D] = iP_\mu \ , \ [K_\mu, D] = -iK_\mu, \\
&[K_\mu, K_\nu] = 0 \ , \ [P_\mu, K_\nu] = 2i \left(\eta_{\mu\nu} D - M_{\mu\nu}\right),
\end{aligned} \tag{A.230}$$

which complements the Poincaré algebra (given in (2.10) of Chapter 2) to form the algebra of the conformal group. According to the last relation, invariance under P_μ, $M_{\mu\nu}$, and K_μ implies dilatation invariance.

We end this section with a few words on the conformal group in D dimensions. As we just saw in (A.227) for the case $D = 4$, the conformal group is the subgroup of coordinate transformations or reparametrizations such that the metric transforms as[14]

$$g_{\mu\nu}(x) \to g'_{\mu\nu}(x) = e^{-2\alpha(x)} g_{\mu\nu}(x). \tag{A.231}$$

This group obviously includes the Poincaré group since it leaves the metric invariant.

---

[14]This transformation preserves the angle between two vectors a^μ and b^μ: $g_{\mu\nu} a^\mu b^\nu / \sqrt{g_{\mu\nu} a^\mu a^\nu g_{\rho\sigma} b^\rho b^\sigma}$.

Writing infinitesimally $x'^\mu = x^\mu - \xi^\mu(x)$, this becomes, following (D.3) of Appendix D, $\nabla_\mu \xi_\nu + \nabla_\nu \xi_\mu = -2\alpha g_{\mu\nu}$. Contraction with $g^{\mu\nu}$ yields $\nabla_\mu \xi^\mu = -4\alpha$, and it follows

$$\nabla_\mu \xi_\nu + \nabla_\nu \xi_\mu = \tfrac{1}{2} g_{\mu\nu} \nabla_\rho \xi^\rho. \tag{A.232}$$

A vector ξ^μ satisfying this equation is called a conformal Killing vector.

We now restrict ourselves to flat spacetime. One obtains from this equation that, for $D > 2$, ξ^μ is at most quadratic in x (see Exercise 8). In increasing order in x, one finds:

- translations $\xi^\mu = a^\mu$;
- rotations $\xi^\mu = \omega^\mu{}_\nu x^\nu$;
- dilatations $\xi^\mu = \alpha x^\mu$;
- special conformal transformations $\xi^\mu = c^\mu x^2 - 2x^\mu c \cdot x$.

The corresponding algebra has dimension $D + \tfrac{1}{2}D(D-1) + 1 + D = \tfrac{1}{2}(D+1)(D+2)$ and is locally isomorphic to $SO(2, D)$. Conformal invariance imposes severe constraints on the N-point functions of a quantum theory. In particular, it determines completely, up to a constant, the two and three point functions [187]. The special case $D = 2$ is considered in Section 10.1 of Chapter 10. We only note here that it is very specific since in $D = 2$, the conformal group is infinitely dimensional.

A.5.2 The dilaton

It is possible to modify the Lagrangian in order to break dilatation symmetry spontaneously rather than explicitly, as with a mass term. One must make the Lagrangian invariant: only the fundamental state breaks the invariance. According to Goldstone's theorem, a massless boson appears in the spectrum. This Goldstone boson associated with dilatations is called dilaton. We denote it by $\sigma(x)$.

In order to construct an invariant Lagrangian describing the interactions of the dilaton with other fields, one may use a method which is inspired from the construction of chiral Lagrangians describing the pion interactions (the pion is the Goldstone boson associated with chiral symmetry breaking).

We consider a scalar field $\Phi(x)$ which we write:

$$\Phi(x) = f \, e^{\sigma(x)/f}. \tag{A.233}$$

The field $\sigma(x)$ corresponds to fluctuations of $\Phi(x)$ around its vacuum value f, which determines the scale of spontaneous breaking of dilatation symmetry. Since under dilatations, the scalar field $\Phi(x)$ transforms as (A.211), the field $\sigma(x)$ transforms as:

$$\sigma(x) \to \sigma'(x) = \sigma\left(e^\alpha x\right) + \alpha f, \tag{A.234}$$

or infinitesimally

$$\delta\sigma(x) = \alpha\left(f + x^\mu \partial_\mu \sigma\right). \tag{A.235}$$

We note that the symmetry is realized nonlinearly: as we have seen in Section A.2.1, the constant term is a sign that $\sigma(x)$ is a Goldstone boson.

One may use factors $e^{\sigma(x)/f}$ to make dilatation-breaking terms invariant. Thus $m_0^2\phi^2$ is replaced by $m_0^2\phi^2 e^{2\sigma/f}$ since $\delta\left(\phi^2 e^{2\sigma/f}\right) = (4 + x^\mu\partial_\mu)\left(\phi^2 e^{2\sigma/f}\right)$. One also introduces a kinetic term for $\sigma(x)$:

$$\mathcal{L}_{\text{kin}}^{(\sigma)} = \tfrac{1}{2}\partial^\mu\Phi\partial_\mu\Phi = \tfrac{1}{2}e^{2\sigma/f}\partial^\mu\sigma\partial_\mu\sigma. \tag{A.236}$$

Then the action corresponding to the Lagrangian

$$\mathcal{L} = \mathcal{L}_0 + \tfrac{1}{2}e^{2\sigma/f}\partial^\mu\sigma\partial_\mu\sigma - \tfrac{1}{2}m_0^2\phi^2 e^{2\sigma/f}, \tag{A.237}$$

with $\mathcal{L}_0$, given in (A.213), is dilatation invariant. Under these conditions, the dilatation current is conserved:

$$\partial^\mu D_\mu = 0. \tag{A.238}$$

A.5.3 Effective potential

We first recall how the effective action is introduced. We consider for simplicity the theory of a scalar field in interaction. One defines the functional of the current J

$$Z\left[J\right] = \langle\Omega\left|Te^{i\int d^4x J(x)\phi(x)}\right|\Omega\rangle, \tag{A.239}$$

where $|\Omega\rangle$ is the vacuum of the interacting theory, $\phi(x)$ the scalar field (in Heisenberg representation) and the symbol T denotes as usual the chronological product. $Z\left[J\right]$ is the generating functional of Green's functions, in the sense that a general Green's function is given by

$$g(x_1,\ldots,x_n) \equiv \langle\Omega\left|T\phi(x_1)\cdots\phi(x_n)\right|\Omega\rangle = (-i)^n \left.\frac{\delta^n Z\left[J\right]}{\delta J(x_1)\cdots\delta J(x_n)}\right|_{J=0}. \tag{A.240}$$

The functional $W\left[J\right] = \ln Z\left[J\right]$ is the generating functional of the connected Green's functions:

$$g_c(x_1,\ldots,x_n) = (-i)^n \left.\frac{\delta^n W\left[J\right]}{\delta J(x_1)\cdots\delta J(x_n)}\right|_{J=0}. \tag{A.241}$$

One then defines

$$\overline{\phi}(x) \equiv -i\frac{\delta W}{\delta J(x)}. \tag{A.242}$$

$\overline{\phi}(x)$ is itself a functional of $J(x)$ but its value at $J=0$ gives the value of the field $\phi(x)$ in the ground state (vacuum):

$$-i\frac{\delta W}{\delta J(x)} = \overline{\phi}(x)\big|_{J=0} = g_c(x) = \langle\Omega|\phi(x)|\Omega\rangle\big|_{\text{connected}}. \tag{A.243}$$

One is then looking for a functional $\Gamma\left[\overline{\phi}\right]$ such that

$$J(x) = -\frac{\delta\Gamma\left[\overline{\phi}\right]}{\delta\overline{\phi}(x)}. \tag{A.244}$$

$\Gamma\left[\overline{\phi}\right]$ is obtained by a Legendre transformation:

$$\Gamma\left[\overline{\phi}\right] + iW\left[J\right] = -\int d^4y\,\overline{\phi}(y)J(y), \tag{A.245}$$

as can be verified explicitly by functionally differentiating this relation with respect to $\overline{\phi}(x)$.

The functional $\Gamma\left[\overline{\phi}\right]$ is called the effective action. Two of its properties will be of special interest to us here:

(i) The analog of the relation (A.243) for Γ is written

$$\left.\frac{\delta\Gamma}{\delta\overline{\phi}(x)}\right|_{\overline{\phi}(x)=\langle\Omega|\phi|\Omega\rangle} = 0. \tag{A.246}$$

In other words, the extrema of the effective action allow us to determine the values of the fields in the fundamental state of the interacting theory (the vacuum $|\Omega\rangle$). This turns out to be useful when studying the spontaneous breaking of a symmetry. If this aspect alone is of interest to us, one may restrict one's attention to the nonderivative terms in the effective action which define the effective potential. More precisely, on may write $\Gamma\left[\overline{\phi}\right]$ as a series with an increasing number of derivatives of $\overline{\phi}$:

$$\Gamma\left[\overline{\phi}\right] = \int d^4x\left[-V_{\text{eff}}\left(\overline{\phi}\right) + \frac{1}{2}\,\partial^\mu\overline{\phi}\partial_\mu\overline{\phi}\,Z\left(\overline{\phi}\right) + \cdots\right]. \tag{A.247}$$

(ii) The effective action is also the generating functional of proper Green's functions (*i.e.* one-particle irreducible and truncated Green's functions). In momentum space,

$$\Gamma^{(n)}(p_1,\ldots,p_n) = \frac{\delta^n\Gamma\left[\overline{\phi}\right]}{\delta\overline{\phi}(p_1)\cdots\delta\overline{\phi}(p_n)} \tag{A.248}$$

yields proper Green's functions. This allows us to compute the effective action from truncated Green's functions. In particular, the effective potential corresponds to vanishing momentum on external legs:

$$-V_{\text{eff}}\left[\overline{\phi}(x)\right] = \sum_n \frac{1}{n!}\,\overline{\phi}(x)^n\,\Gamma^{(n)}(p_1 = 0,\ldots,p_n = 0). \tag{A.249}$$

The effective action is usually computed in powers of $\hbar$, which corresponds to an expansion in the number L of loops. Let us illustrate, in the example of a theory in $\lambda\phi^4$, how one determines the effective potential. We consider the theory of a real massless scalar field with action:

$$\mathcal{L} = \frac{1}{2}\partial^\mu\phi\partial_\mu\phi - \frac{\lambda}{4!}\phi^4 + \frac{1}{2}Z\partial^\mu\phi\partial_\mu\phi + \frac{1}{2}\delta m^2\,\phi^2 + \frac{1}{4!}\delta\lambda\,\phi^4, \tag{A.250}$$

where we have included counterterms in the second line.

At tree level $(L = 0)$, the one-particle irreducible Green's function is simply the one given in Fig. A.9(a) $i\Gamma^{(4)}(0,0,0,0) = -i\lambda$ and

$$V_{\text{eff}}^{(0)}\left(\overline{\phi}\right) = \frac{\lambda}{4!}\overline{\phi}^4 - \frac{1}{2}\delta m^2\,\overline{\phi}^2 - \frac{1}{4!}\delta\lambda\,\overline{\phi}^4. \tag{A.251}$$

The one-loop diagrams of Fig. A.9(b) contribute

$$i\Gamma^{(2n)}(0,\ldots,0) = \left(\frac{-i\lambda}{2}\right)^n \frac{n!}{2n} \int \frac{d^4k}{(2\pi)^4}\left(\frac{i}{k^2}\right)^n,$$

where we have included a factor $n!$ for the exchange of the n vertices and a symmetry factor $2n$ (left-right symmetry and rotation of an angle which is a multiple of $2\pi/n$). Then, after a Wick rotation,

$$\begin{aligned}
V_{\text{eff}}^{(1)}\left(\overline{\phi}\right) &= -\int \frac{d^4k}{(2\pi)^4}\sum_{n=1}^{\infty}\frac{(-)^n}{2n}\left(\frac{\lambda\overline{\phi}^2/2}{k^2}\right)^n \\
&= \frac{1}{2}\int \frac{d^4k}{(2\pi)^4}\ln\left(1 + \frac{\lambda\overline{\phi}^2}{2k^2}\right).
\end{aligned} \tag{A.252}$$

Cutting off the integral at momentum Λ, we obtain from (A.251) and (A.252) the effective potential to one-loop order:

$$\begin{aligned}
V_{\text{eff}}\left(\overline{\phi}\right) &= \frac{\lambda}{4!}\overline{\phi}^4 - \frac{1}{2}\delta m^2\overline{\phi}^2 - \frac{1}{4!}\delta\lambda\,\overline{\phi}^4 \\
&\quad + \frac{\lambda\Lambda^2}{64\pi^2}\overline{\phi}^2 + \frac{\lambda^2\overline{\phi}^4}{256\pi^2}\left(\ln\frac{\lambda\overline{\phi}^2}{2\Lambda^2} - \frac{1}{2}\right) + O\left(\frac{\overline{\phi}^2}{\Lambda^2}\right).
\end{aligned} \tag{A.253}$$

We choose renormalization conditions for the mass and the coupling:

$$\left.\frac{d^2V_{\text{eff}}}{d\overline{\phi}^2}\right|_{\overline{\phi}=0} = 0, \qquad \left.\frac{d^4V_{\text{eff}}}{d\overline{\phi}^4}\right|_{\overline{\phi}=\mu} = \lambda, \tag{A.254}$$

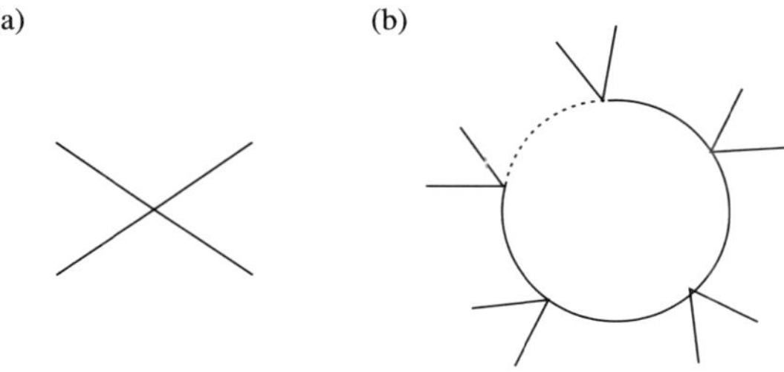

Fig. A.9 Zero and one-loop contributions to the effective potential.

the first corresponding to the requirement of the scalar field being massless, the second introducing a renormalization scale μ. These determine δm^2 and $\delta\lambda$. It then follows that

$$V_{\text{eff}}\left(\overline{\phi}\right) = \frac{\lambda}{4!}\overline{\phi}^4 + + \frac{\lambda\Lambda^2}{64\pi^2}\overline{\phi}^2 + \frac{\lambda^2\overline{\phi}^4}{256\pi^2}\left(\ln\frac{\lambda\overline{\phi}^2}{2\Lambda^2} - \frac{25}{6}\right). \tag{A.255}$$

We note that the bare coupling is, to this order,

$$\lambda_0 = \lambda - \delta\lambda = \lambda\left[1 - \frac{3\lambda^2}{32\pi^2}\left(\ln\frac{\lambda\mu^2}{2\Lambda^2} + \frac{11}{3}\right)\right] \tag{A.256}$$

or alternatively, the renormalized coupling is expressed in terms of the bare coupling as

$$\lambda = \lambda_0\left[1 + \frac{3\lambda_0^2}{32\pi^2}\left(\ln\frac{\lambda_0\mu^2}{2\Lambda^2} + \frac{11}{3}\right)\right]. \tag{A.257}$$

Thus, the beta function, which describes the dependence of the renormalized coupling with respect to the renormalization scale, reads

$$\beta(\lambda) = \mu\frac{d\lambda}{d\mu} = \frac{3\lambda^2}{16\pi^2} + O(\lambda^3). \tag{A.258}$$

More generally, in a theory involving fields of various spins, the effective potential reads at one loop $V_{\text{eff}}\left(\overline{\phi}\right) = V_{\text{eff}}^{(0)}\left(\overline{\phi}\right) + V_{\text{eff}}^{(1)}\left(\overline{\phi}\right)$ where $V_{\text{eff}}^{(0)}$ is the tree level potential and

$$V_{\text{eff}}^{(1)}\left(\overline{\phi}\right) = \frac{1}{32\pi^2}\Lambda^2 \mathcal{S}\text{Tr}M^2\left(\overline{\phi}\right) + \frac{1}{64\pi^2}\mathcal{S}\text{Tr}M^4\left(\overline{\phi}\right)\left[\ln\frac{M^2\left(\overline{\phi}\right)}{\Lambda^2} - \frac{3}{2}\right]. \tag{A.259}$$

In this expression, $M\left(\overline{\phi}\right)$ are field-dependent masses and the supertrace is defined as:

$$\mathcal{S}\text{Tr}F\left[M^2\left(\overline{\phi}\right)\right] \equiv \sum_J(-1)^{2J}(2J+1)F\left[M_J^2\left(\overline{\phi}\right)\right]. \tag{A.260}$$

A.5.4 Conformal anomaly

As we have seen in the previous example, renormalization introduces a scale μ. The question of scale invariance must therefore be revisited at the quantum level.

For example, we see from (A.255) that the logarithmic correction breaks dilatation invariance: indeed, under $\overline{\phi} \to \overline{\phi}' = e^\alpha\overline{\phi}$, the effective potential is modified as

$$\delta V_{\text{eff}}(\overline{\phi}) = \alpha\frac{3\lambda}{16\pi^2}\frac{\lambda}{4!}\overline{\phi}^4, \tag{A.261}$$

or

$$\delta\Gamma = \alpha\frac{3\lambda}{16\pi^2}\int d^4x\left(-\frac{\lambda}{4!}\overline{\phi}(x)^4\right). \tag{A.262}$$

Thus, the dilatation current is no longer conserved:

$$\partial^\mu D_\mu = \frac{3\lambda}{16\pi^2}\left(-\frac{\lambda}{4!}\bar\phi(x)^4\right).$$

(A.263)

This is interpreted as an anomaly, called the conformal anomaly: quantum corrections perturb the naive scaling laws of the classical theory. We note that we may rewrite the latter equation as

$$\partial^\mu D_\mu = \frac{\beta(\lambda)}{\lambda}\left(-\frac{\lambda}{4!}\bar\phi(x)^4\right),$$

(A.264)

where the beta function is defined in (A.258). The beta function is thus interpreted as the coefficient of the conformal anomaly.

More generally, when one computes the quantum corrections to the action through the effective action Γ, the necessity of introducing a renormalization scale μ induces a fundamental breaking of scale invariance. Since the dilatation transformation (A.209) corresponds to a renormalization scale (momentum) transformation $\mu' = e^\alpha\mu$, we may write that, under dilatations, the variation of the effective action due to renormalization is

$$\delta\Gamma|_{\rm ren} = \mu\frac{\partial\Gamma}{\partial\mu} \equiv \int d^4x\,\mathcal{A}(x),$$

(A.265)

where $\mathcal{A}(x)$ is called the conformal anomaly. The divergence of the dilatation current then reads:

$$\partial^\mu D_\mu = \mathcal{A}(x) + \cdots$$

(A.266)

where the extra terms would come from dilatation breaking terms in the classical action. The general form of the conformal anomaly involves the beta functions associated with the different couplings:

$$\mathcal{A}(x) = -\frac{\lambda}{4!}\,\phi^4(x)\frac{\beta(\lambda)}{\lambda} - \frac{1}{4}\,F^{\mu\nu}F_{\mu\nu}\,\frac{2\beta(g)}{g} + \cdots$$

(A.267)

This equation may indeed serve as a definition for beta functions.

If we have introduced a dilaton $\sigma(x)$ in the theory, it is possible to restore the dilatation invariance by adding the following Wess–Zumino term in the action:

$$\mathcal{S}_{\rm wz} = -\int d^4x\,\mathcal{A}(x)\,\frac{\sigma(x)}{f}.$$

(A.268)

Indeed, using the nonlinear transformation of the dilaton field (A.235), we have

$$\delta\left(\mathcal{S} + \mathcal{S}_{\rm wz}\right) = 0.$$

(A.269)

A.6   Axial anomaly

We have seen that a key property of the Standard Model is that it is chiral: left and right chirality fermions transform independently. In fact, in the limit of vanishing mass or Yukawa couplings, left and right chiralities completely decouple from one another. This can be interpreted at the level of symmetries by considering the axial transformation:

$$\psi(x) \to e^{-i\alpha\gamma_5}\psi(x) \quad \text{or} \quad \begin{cases} \psi_L(x) \to e^{i\alpha}\psi_L(x) \\ \psi_R(x) \to e^{-i\alpha}\psi_R(x). \end{cases} \tag{A.270}$$

The corresponding axial current

$$J_A^\mu(x) = \bar\psi(x)\gamma^\mu\gamma_5\psi(x) \tag{A.271}$$

is conserved.

However, this axial symmetry is, generally speaking, not compatible at the quantum level, with a vector symmetry. Such a symmetry may arise from a continuous invariance of the action. The Noether procedure then allows us to construct a conserved vector current:

$$J_V^\mu(x) = \bar\psi(x)\gamma^\mu\psi(x). \tag{A.272}$$

This may for example be the electromagnetic current.

If we insist on conserving the vector current, then there generically appears a source of nonconservation for the axial current. In the case of quantum electrodynamics, we have, in the limit of massless fermions,

$$\partial_\mu J_A^\mu(x) = \frac{\alpha}{2\pi} F^{\mu\nu}\tilde F_{\mu\nu} \tag{A.273}$$

where α is the fine structure constant and $F_{\mu\nu}$ is the photon field strength. Thus the axial current is not conserved and axial symmetry is not a good symmetry at the quantum level. The physical reason is that, even in the limit of vanishing fermion mass, left and right chirality remain coupled through the emission of virtual photons.

Before considering the case of the Standard Model, where this issue is a key one because chiral symmetry is built into the gauge symmetry, we will compute the anomalous term and discuss in more details the origin of this axial anomaly on the simple theory of a single massless fermion $\psi(x)$ with a conserved vector current $J_V^\mu(x)$ given in (A.272). For illustration purpose, we will couple this vector current to the photon. We define the Green's function

$$T^{\mu\nu\lambda}(p_1,p_2) = i\int d^4x_1 d^4x_2\, e^{i(p_1 x_1 + p_2 x_2)} \langle 0|TJ_V^\mu(x_1)J_V^\nu(x_2)J_A^\lambda(0)|0\rangle. \tag{A.274}$$

Then the conservation of currents $\partial_\mu J_V^\mu = 0$ and $\partial_\mu J_A^\mu = 0$ translate, respectively, into the Ward identities[15]

$$p_1^\mu T_{\mu\nu\lambda} = p_2^\nu T_{\mu\nu\lambda} = 0, \tag{A.275}$$
$$(p_1 + p_2)^\lambda T_{\mu\nu\lambda} = 0. \tag{A.276}$$

At one loop, $T_{\mu\nu\lambda}$ is given by the diagrams of Fig. A.10.

---

[15] Changing variables ($x_i' = x + x_i$) and using translation invariance of the Green's function, one may rewrite (A.274) as

$$T^{\mu\nu\lambda}(p_1,p_2) = i\int d^4x_1' d^4x_2'\, e^{i[p_1 x_1' + p_2 x_2' - (p_1+p_2)x)]} \langle 0|TJ_V^\mu(x_1')J_V^\nu(x_2')J_A^\lambda(x)|0\rangle.$$

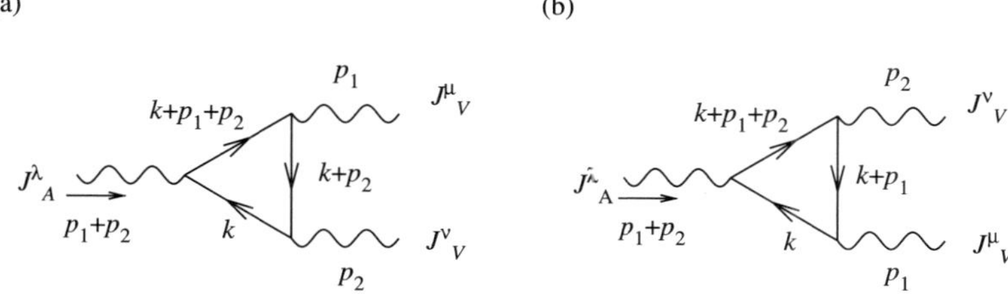

Fig. A.10 Triangle diagrams contributing to the axial anomaly.

Their contribution reads

$$T^{\mu\nu\lambda}_{(1)} = (-)i^3 \int \frac{d^4k}{(2\pi)^4} \, \mathrm{Tr}\left[\gamma^\mu \frac{i}{\not{k}+\not{p}_1+\not{p}_2} \gamma^\lambda \gamma_5 \frac{i}{\not{k}} \gamma^\nu \frac{i}{\not{k}+\not{p}_2}\right] + (\mu \leftrightarrow \nu, p_1 \leftrightarrow p_2).$$

$$\text{(A.277)}$$

When computing $p_{1\mu}T^{\mu\nu\lambda}$, one writes in the numerator $\not{p}_1 = (\not{k}+\not{p}_1+\not{p}_2) - (\not{k}+\not{p}_2)$ to obtain

$$p_{1\mu}T^{\mu\nu\lambda}_{(1)} = \int \frac{d^4k}{(2\pi)^4} \, \mathrm{Tr}\left[\gamma^\lambda \gamma_5 \frac{1}{\not{k}} \gamma^\nu \frac{1}{\not{k}+\not{p}_2}\right] - \mathrm{Tr}\left[\frac{1}{\not{k}+\not{p}_1+\not{p}_2} \gamma^\lambda \gamma_5 \frac{1}{\not{k}} \gamma^\nu\right] + \cdots \quad \text{(A.278)}$$

Each of the terms depends on a single momentum (p_2^μ or $(p_1+p_2)^\mu$). It is easy to convince oneself, given the properties under parity ($p_{1\mu}T^{\mu\nu\lambda} \to -p_1^\mu T_{\mu\nu\lambda}$ according to (A.108)), that each term vanishes after integration[16].

For the axial current Ward identity, one writes $(\not{p}_1 + \not{p}_2)\gamma_5 = (\not{k}+\not{p}_1+\not{p}_2)\gamma_5 + \gamma_5\not{k}$:

$$(p_1+p_2)_\lambda T^{\mu\nu\lambda}_{(1)} = \int \frac{d^4k}{(2\pi)^4} \, \mathrm{Tr}\left[\gamma^\mu \gamma_5 \frac{1}{\not{k}} \gamma^\nu \frac{1}{\not{k}+\not{p}_2}\right] + \mathrm{Tr}\left[\gamma^\mu \frac{1}{\not{k}+\not{p}_1+\not{p}_2} \gamma_5 \gamma^\nu \frac{1}{\not{k}+\not{p}_2}\right] + \cdots$$

$$\text{(A.279)}$$

One would be tempted to change the integration variable in the second term to $k'_\mu = k_\mu + p_{2\mu}$ and conclude that this vanishes. However the integrals are divergent and one must be extremely careful with changes of variables. Indeed, if we consider a linearly divergent integral, such as $T^{\mu\nu\lambda}_{(1)}$, $I = \int d^4k F(k)$ with $F(k) \sim k^{-3}$ for $k \to \infty$, then under the change of variables $k' = k + a$

$$I = \int d^4k' F(k' - a) = \int d^4k' \left[F(k') - a^\mu \frac{\partial F}{\partial k'^\mu} + \cdots\right]$$

$$= \int d^4k' F(k') - a^\mu \int_{\Sigma_\infty} d^3\sigma_\mu F(k), \quad \text{(A.280)}$$

---

[16]For example, the only possible form for the first term is $\epsilon_{\nu\lambda\rho\sigma}p_2^\rho p_2^\sigma$ which vanishes identically.

where the last term is nonzero because on the surface at infinity Σ_∞, $d^3\sigma_\mu \sim k^3$ and $F(k) \sim k^{-3}$ (on the other hand, the terms denoted $\cdots$ are of order k^{-4} and do not contribute)[17].

If we apply this to the divergent integral in $T^{\mu\nu\lambda}_{(1)}$

$$D^{\mu\nu\lambda} = \int \frac{d^4k}{(2\pi)^4} \frac{\mathrm{Tr}\left[\gamma^\mu \slashed{k} \gamma^\lambda \gamma_5 \slashed{k} \gamma^\nu \slashed{k}\right]}{k^6} = 4i\epsilon^{\mu\nu\lambda\rho} \int \frac{d^4k}{(2\pi)^4} \frac{k_\rho}{k^4}, \qquad (A.281)$$

we obtain, after a change of variable $k'^\mu = k^\mu + a^\mu$,

$$D^{\mu\nu\lambda} = D^{(a)\mu\nu\lambda} - a_\sigma 4i\epsilon^{\mu\nu\lambda\rho} \int \frac{d^4k}{(2\pi)^4} \frac{\partial}{\partial k_\sigma} \left(\frac{k_\rho}{k^4}\right) = D^{(a)\mu\nu\lambda} + \frac{1}{8\pi^2}\epsilon^{\mu\nu\lambda\sigma} a_\sigma. \qquad (A.282)$$

If we change the variable k^μ into $k'^\mu = k^\mu + p_2^\mu$, we may prove that the contribution of Fig. A.10(a) to $(p_1 + p_2)_\lambda T^{\mu\nu\lambda}$ vanishes. Similarly for Fig. A.10(b) with $k'^\mu = k^\mu + p_1^\mu$ (and exchange of μ and ν). Hence

$$(p_1 + p_2)^\lambda T_{\mu\nu\lambda} = -\frac{1}{8\pi^2}\epsilon_{\mu\nu\lambda\sigma}(p_1 + p_2)^\lambda(p_1 - p_2)^\sigma = \frac{1}{4\pi^2}\epsilon_{\mu\nu\lambda\sigma}p_1^\lambda p_2^\sigma.$$

This corresponds to an anomalous conservation law for the axial current[18]:

$$\partial_\lambda J_A^\lambda = \frac{1}{8\pi^2}\mathcal{F}_{\mu\nu}\tilde{\mathcal{F}}^{\mu\nu}, \qquad (A.283)$$

where $\mathcal{F}_{\mu\nu} = \partial_\mu J_{V\nu} - \partial_\nu J_{V\mu}$ and $\tilde{\mathcal{F}}^{\mu\nu} = \frac{1}{2}\epsilon^{\mu\nu\rho\sigma}\mathcal{F}_{\rho\sigma}$. If one couples the electromagnetic current to the vector current, one may write

$$\partial_\lambda J_A^\lambda = \frac{1}{8\pi^2}F_{\mu\nu}\tilde{F}^{\mu\nu}, \qquad (A.284)$$

which corresponds to triangle diagrams similar to Fig. A.10 with one axial current and two photons ($F_{\mu\nu}$ is the photon field strength).

We may now return to a discussion of axial anomalies in the Standard Model. Since $SU(2)$ and $U(1)_Y$ have an axial character, there is a danger to have an anomaly that would destroy the symmetry at the quantum level. For example, the diagrams of Fig. A.11 may lead respectively to a (a) $SU(2)$ (b) mixed $U(1)_Y - SU(2)$ and (c) $U(1)_Y$ anomaly. However, the $SU(2)$ anomaly (a) has an overall group factor

$$\mathrm{Tr}\left(t^a\{t^b, t^c\}\right) = \frac{1}{2}\delta^{bc}\mathrm{Tr}(t^a) = 0,$$

[17]More precisely, $\int d^3\sigma_\mu F(k) = \lim_{|K| \to +\infty} \frac{K_\mu}{K} iS_3(|K|)F(|K|)$ where $S_3(K) = 2\pi^2 K^3$.

[18]We note in this computation the important rôle played by chirality and the matrix γ_5. If we were working in dimensional regularization where there is no linear divergence, we would encounter a difficulty defining γ_5 in $4 + \epsilon$ dimensions. Proper treatment [84] leads to the same result.

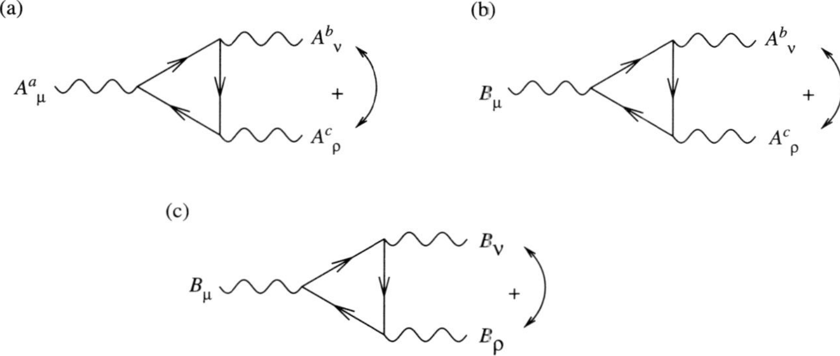

Fig. A.11 Triangle anomalies for the Standard Model.

where we have used (A.26). Similarly, the mixed $U(1)_Y - SU(2)$ anomaly is proportional to $\mathrm{Tr}\left(Y\{t^b, t^c\}\right) \sim \delta^{bc}\mathrm{Tr}(Y)$ and $\mathrm{Tr}(Y) = \sum_j y_j$ summed over the fermion doublets of the Standard Model vanishes between quarks and leptons of a given family. Finally, the $U(1)$ anomaly is proportional to $\mathrm{Tr}(Y^3) = \sum_i y_i^3$ summed over all fermions, which is also computed to vanish: $2/9 - 2 - 64/9 + 8/9 + 8$ (resp. for (u_L, d_L), (ν_L, e_L), u_L^c, d_L^c, e_L^c; note that the axial charges are opposite for the two chiralities). Thus, the Standard Model is free of axial anomalies and its symmetries remain good symmetries at the quantum level. It remains to understand the origin of this somewhat miraculous cancellation (see Chapter 9).

Further reading

- P. Ramond, *Field theory: a modern primer*, 2nd edition, Frontiers in physics, Addison-Wesley publishing company 1990.
- S. Coleman, *Secret symmetry* in *Laws of hadronic matter*, Proceedings of the 1973 International School of Subnuclear Physics, Erice, ed. A. Zichichi, pp. 138-223.

Exercises

<u>Exercise 1</u> Experimentally, the neutron and proton are not degenerate in mass: $m_n c^2 - m_p c^2 \sim 1.29$ MeV. We have seen that isospin, which is a symmetry of strong interactions, imposes the same mass. Is the observed mass difference an effect of electromagnetic or weak interactions?

Hint: If the main effect was electromagnetic, one would have $m_p > m_n$ since the proton, being charged, has a larger electromagnetic mass. Thus the effect comes mainly from weak interactions which are at the origin of quark masses: the d quark is heavier than the u and thus the neutron (ddu) is heavier than the proton (uud). Let us note that most of the nucleon mass comes from the strong interaction that confines the quarks. Quark masses only account for a small fraction of the nucleon mass.

<u>Exercise 2</u> Prove (A.48).

Hint: Expand $\partial_\mu \left(UU^{-1}\right) = \partial_\mu \mathbb{1} = 0$.

<u>Exercise 3</u> *Polar decomposition: any complex matrix may be written as the product of a hermitian matrix and a unitary matrix.* Let Λ be a general complex matrix. Since $\Lambda\Lambda^\dagger$ is hermitian, we can introduce a unitary matrix V such that $V^\dagger\left(\Lambda\Lambda^\dagger\right)V = D^2$, a real diagonal matrix. We note $H \equiv VDV^\dagger$, a hermitian matrix, and $U \equiv H^{-1}\Lambda$.

(a) Prove that $U = H\Lambda^{\dagger-1}$.
(b) Show that U is unitary.

<u>Exercise 4</u> We define two-point functions $\Pi^{33}_{\mu\nu}$, $\Pi^{3B}_{\mu\nu}$, and $\Pi^{BB}_{\mu\nu}$ for the $SU(2) \times U(1)$ gauge fields A^3_μ and B_μ as in (A.192).

(a) Express the corresponding quantities A and F in terms of the original quantities in (A.193). One will show in particular that

$$A_{3B} = -\frac{c_w}{s_w}A_{33}, \qquad A_{BB} = \frac{s_w^2}{c_w^2}A_{33}, \qquad A_{ZZ} = \frac{A_{33}}{c_w^2}. \qquad (A.285)$$

(b) Use these relations to obtain from (A.189) the value of δr, as given in (A.204).

Hints:

(a) To prove (A.285), use $A_{\gamma\gamma} = A_{\gamma Z} = 0$ to obtain two relations between A_{3B}, A_{BB} and A_{33}.

<u>Exercise 5</u> Compute explicitly the leading contribution to the electroweak parameter ϵ_1 in the Standard Model (*i.e.* the first term in (A.205)).

<u>Exercise 6</u>
(a) Prove that $\partial_\mu\Phi$ has scaling dimension $d + 1$ if Φ is a field of scaling dimension d.
(b) Prove (A.216).

Hints:

(a) $\partial'_\mu\Phi'(x') = \frac{\partial}{\partial(e^{-\alpha}x^\mu)}e^{\alpha d}\Phi(x) = e^{\alpha(d+1)}\partial_\mu\Phi(x)$.
(b) We have from (a) the infinitesimal transformation: $\delta\partial_\lambda\phi = \alpha(2+x^\mu\partial_\mu)\partial_\lambda\phi$. Hence, for example,

$$\frac{1}{\alpha}\delta\left(\partial^\lambda\phi\partial_\lambda\phi\right) = 2\partial^\lambda\phi(2 + x^\mu\partial_\mu)\partial_\lambda\phi = 2\left(2 + \frac{1}{2}x^\mu\partial_\mu\right)\left(\partial^\lambda\phi\partial_\lambda\phi\right)$$

$$= (4 + x^\mu\partial_\mu)\left(\partial^\lambda\phi\partial_\lambda\phi\right).$$

<u>Exercise 7</u> We consider the theory described by the action (A.214) setting for simplicity the fermion field to zero and we derive some of the properties of the energy–momentum $\Theta^{\mu\nu}$.

(a) Check that $\Theta^{\mu\nu}$ is conserved if and only if $T^{\mu\nu}$ is. Prove that the generators P^μ and $M^{\mu\nu}$ are identical, whether constructed from $T^{\mu\nu}$ or $\Theta^{\mu\nu}$.
(b) We write

$$\Theta^{\mu\nu} = T^{\mu\nu} - \tfrac{1}{6}\partial_\lambda\partial_\tau X^{\lambda\tau\mu\nu}, \qquad X^{\lambda\tau\mu\nu} = \left(g^{\lambda\mu}g^{\tau\nu} - g^{\lambda\tau}g^{\mu\nu}\right)\phi^2.$$

Infer from (A.220) that

$$D^\mu = x_\nu \Theta^{\mu\nu} + \tfrac{1}{6}\partial_\lambda \partial_\tau \left(x_\nu X^{\lambda\tau\mu\nu} \right).$$

(c) The second term on the right-hand side of the last equation reads $\partial_\tau A^{\tau\mu}$ with $A^{\tau\mu} = -A^{\mu\tau}$. Show that such a term does not modify the conservation of current or the charge and that it can thus be ignored. One then recovers (A.224).

Hints:

(a) For example, $\Theta^{00} = T^{00} - \tfrac{1}{6}\partial_i\partial_i\phi^2$ and $\Theta^{i0} = T^{i0} - \tfrac{1}{6}\partial_i\partial_0\phi^2$. The generators P^0 and P^i are obtained by taking the spatial integrals: the second terms in each expression do not contribute, being total divergences.

(b) $D^\mu = x_\nu T^{\mu\nu} + \tfrac{1}{2}\partial^\mu\phi^2$
$\quad = x_\nu \Theta^{\mu\nu} + \tfrac{1}{6}\partial_\lambda\partial_\tau \left(x_\nu X^{\lambda\tau\mu\nu} \right) - \tfrac{1}{6}\left(g_{\lambda\nu}\partial_\tau + g_{\tau\nu}\partial_\lambda \right) X^{\lambda\tau\mu\nu} + \tfrac{1}{2}\partial^\mu\phi^2.$
The last two terms cancel.

(c) $\partial_\mu\partial_\tau A^{\tau\mu} = 0$ and $\partial_\tau A^{\tau 0} = \partial_i A^{i0}$ does not contribute to an integral over space.

<u>Exercise 8</u> We wish to show that infinitesimal coordinate transformations $x'^\mu = x^\mu - \xi^\mu(x)$ which belong to the conformal group of a d-dimensional spacetime $(d > 2)$ are such that ξ^μ is at most quadratic in x. We assume that spacetime is flat with metric $\eta_{\mu\nu}$.

(a) Deduce from (A.232) that

$$[\eta_{\mu\nu}\Box + (d-2)\partial_\mu\partial_\nu]\,(\partial^\rho\xi_\rho) = 0.$$

(b) Conclude that ξ is at most quadratic in x.

Hints: For example, if ξ was cubic in x, it would be of the form $\xi^\mu = \lambda x^2 x^\mu + \cdots$ where λ is infinitesimal.

<u>Exercise 9</u> *We study in this exercise the model of unification of [180]. This model provides a unification of electroweak interactions within the simple gauge group $SO(3)$. It does not require the introduction of neutral currents but of extra leptons, and was not confirmed by experiment. But it provides the simplest example where the electromagnetic $U(1)$ symmetry is part of a larger nonabelian symmetry which allows us to consider the magnetic charge as a topological charge (see Section 4.5.3 of Chapter 4).*

We consider a gauge theory with gauge group $SO(3) \sim SU(2)$. In the standard basis for the generators of the adjoint representation $(T^a)^{bc} = -i\epsilon^{abc}$, the gauge fields are A^a_μ, $a = 1,2,3$. We want to interpret these fields as the intermediate vector boson $W^\pm_\mu$ and the photon A_μ:

$$\begin{pmatrix} W^+_\mu \\ A_\mu \\ W^-_\mu \end{pmatrix} = \begin{pmatrix} 1/\sqrt{2} & -i/\sqrt{2} & 0 \\ 0 & 0 & 1 \\ 1/\sqrt{2} & i/\sqrt{2} & 0 \end{pmatrix} \begin{pmatrix} A^1_\mu \\ A^2_\mu \\ A^3_\mu \end{pmatrix} \equiv U \begin{pmatrix} A^1_\mu \\ A^2_\mu \\ A^3_\mu \end{pmatrix}. \tag{A.286}$$

1. Write the adjoint (vector) representation **3** in the charge eigenstate basis (A.286), which we will denote $\hat{T}^a$.

2. We consider a fermion field (ψ^+, ψ^0, ψ^-) in the representation $\hat{T}^a$ (the superscript gives the electric charge). Write the minimal coupling to the intermediate vector boson and to the photon. Show that the gauge coupling g must be equal to the electromagnetic coupling e.

3. In the Georgi–Glashow model (1972b), one introduces charged E^+ and neutral E^0 leptons besides the electron and the neutrino in the following representations of $SO(3)$:
   - $(E_L^+, \nu_L \sin\theta_{GG} + N_L^0 \cos\theta_{GG}, e_L^-) \in \mathbf{3}$
   - $(E_R^+, N_R^0, e_R^-) \in \mathbf{3}$
   - $N_L^0 \sin\theta_{GG} - \nu_L \cos\theta_{GG} \in \mathbf{1}$
   - $\nu_R \in \mathbf{1}$.

   Explain why by writing explicitly the couplings to the gauge bosons. Show that the coupling constant to the intermediate vector boson is $g_W = e \sin\theta_{GG}$.

4. One introduces a real scalar field $\phi^a, a = 1, 2, 3$, in the vector representation T^a with the following action (with summation over repeated indices):

$$\mathcal{L} = \frac{1}{2} D^\mu \phi^a D_\mu \phi^a - V(\phi^a \phi^a) \tag{A.287}$$

   where the potential V is assumed to have a nontrivial ground state for $\langle \phi^a \phi^a \rangle = \phi_0^2$ (see Section 4.5.3 of Chapter 4). Aligning the vev of ϕ along ϕ^3 and defining

$$\phi^\pm = \frac{\phi^1 \mp \phi^2}{\sqrt{2}}, \quad \phi^0 = \phi^3$$

   compute $D_\mu \phi^{\pm,0}$ and show that $SU(2)$ is broken down to $U(1)$. What is the physical spectrum of the theory?

Hints:

1. $\hat{T}^a = U T^a U^\dagger$. For example, $\hat{T}^3 = \operatorname{diag}(1, 0, -1)$.

2. $\mathcal{L} = g(-\bar{\psi}^+ \gamma^\mu \psi^0 + \bar{\psi}^0 \gamma^\mu \psi^-) W_\mu^+ + \text{h.c.} + g(\bar{\psi}^+ \gamma^\mu \psi^+ - \bar{\psi}^- \gamma^\mu \psi^-) A_\mu.$

3. $\mathcal{L} = -g\, \bar{e}\gamma^\mu e\, A_\mu + g\sin\theta_{GG} \left(\bar{e}_L \gamma^\mu \nu_L W_\mu^- + \bar{\nu}_L \gamma^\mu e_L W_\mu^+\right) + \cdots.$
   Hence $g = e$ and $g_W = e \sin\theta_{GG}$.

4. $D_\mu \phi^\pm = \partial_\mu \phi^\pm \mp ie A_\mu \phi^\pm \pm ie W_\mu^\pm \phi^0,$

   $D_\mu \phi^0 = \partial_\mu \phi^0 - ie W_\mu^+ \phi^- + ie W_\mu^- \phi^+.$

   Since $\frac{1}{2} D^\mu \phi^a D_\mu \phi^a \ni e^2 \phi_0^2\, W_\mu^+ W^{-\mu}$, $SO(3)$ is broken down to $U(1)$ and the photon field remains massless. Out of the scalar degrees of freedom, $\phi^\pm$ provide $W_L^\pm$ and ϕ^0 is the physical Higgs.

Appendix B
Spinors

B.1 Spinors in four dimensions

In this appendix, we first discuss spinors in four dimensions and introduce two-component spinors, a notation that we heavily use in the superspace formulation of supersymmetry. We then review spinors in higher dimensional spacetime, which are in spinor representations of a general $SO(1, D - 1)$ Lorentz group. This is relevant when we discuss higher dimensions, for example in the context of string theories (chapter 10).

B.1.1 Van der Waerden notation

The metric used is $\eta_{\mu\nu} = (1, -1, -1, -1)$. Let us consider the spinor representation of the Lorentz group $SO(1, 3)$ which decomposes into irreducible representations of given chiralities: $4 = 2_L + 2_R$. The corresponding spinor fields are called Weyl spinors. We will denote the representation $2_L \equiv (\xi^\alpha, \alpha = 1, 2)$ by $(1/2, 0)$. Under a proper Lorentz transformation[1]

$$\xi^{\alpha'} = M^\alpha{}_\beta\, \xi^\beta, \quad \det M = 1. \tag{B.1}$$

One can then check that, introducing a second such spinor η^α, $\xi^2\eta^1 - \xi^1\eta^2$ is invariant under M.

Let us note

$$\xi_\alpha = \varepsilon_{\alpha\beta}\, \xi^\beta \tag{B.2}$$

where $\varepsilon_{\alpha\beta}$ is the antisymmetric tensor $(\varepsilon_{\alpha\beta} = -\varepsilon_{\beta\alpha}$, we choose $\varepsilon_{21} = 1)$, invariant under a Lorentz transformation:

$$\varepsilon_{\alpha\beta} M^\alpha{}_\gamma\, M^\beta{}_\delta = \varepsilon_{\gamma\delta}\, \det M = \varepsilon_{\gamma\delta}. \tag{B.3}$$

One may write also

$$\xi^\alpha = \varepsilon^{\alpha\beta}\, \xi_\beta \tag{B.4}$$

where $\varepsilon^{\alpha\beta}$ is the invariant antisymmetric tensor which is the inverse of $\varepsilon_{\alpha\beta}$:

$$\varepsilon^{\alpha\beta}\, \varepsilon_{\beta\gamma} = \delta^\alpha_\gamma$$

$(\varepsilon^{12} = 1)$.

---

[1]Since the dimension of space time is different from 2 mod 8, ξ^α being of definite chirality cannot be chosen real (otherwise it would be a Majorana–Weyl spinor) and $M \in SL(2, C)$.

Then using $\eta^1 = \eta_2$ and $\eta^2 = -\eta_1$, one may write, using the Einstein convention

$$\xi^2\eta^1 - \xi^1\eta^2 = \xi^1\eta_1 + \xi^2\eta_2 \equiv \xi^\alpha\eta_\alpha.$$

Let us note that the spinor ξ_α introduced in (B.2) transforms as:

$$\begin{aligned}
\xi'_\alpha &= \varepsilon_{\alpha\beta}\,\xi^{\beta'} = \varepsilon_{\alpha\beta}\,M^\beta{}_\delta\,\xi^\delta = \varepsilon_{\gamma\delta}(M^{-1})^\gamma{}_\alpha\,\xi^\delta \\
&= \left(M^{-1T}\right)_\alpha{}^\gamma\,\xi_\gamma
\end{aligned} \tag{B.5}$$

where we have used (B.3).

One may then check that $\xi^\alpha\eta_\alpha$ is invariant under a Lorentz transformation:

$$\xi'^\alpha\eta'_\alpha = M^\alpha{}_\beta\,\xi^\beta\left(M^{-1T}\right)_\alpha{}^\gamma\,\eta_\gamma = \left(M^{-1}M\right)^\gamma{}_\beta\,\xi^\beta\,\eta_\gamma = \xi^\beta\eta_\beta.$$

We will use from now on the following notation:

$$\xi^\alpha\eta_\alpha \equiv \xi\eta, \qquad \xi\xi \equiv \xi^2. \tag{B.6}$$

Our convention is that spinors always anticommute. It follows that

$$\xi\eta = \eta\xi. \tag{B.7}$$

The complex conjugate $(\xi^\alpha)^*$ transforms with the matrix M^*. To make the difference explicit, one introduces a different notation

$$\bar{\xi}^{\dot\alpha'} = (M^*)^{\dot\alpha}{}_{\dot\beta}\,\bar{\xi}^{\dot\beta}. \tag{B.8}$$

The $\bar{\xi}^{\dot\alpha}$ form an inequivalent spinor representation: $\left(\bar{\xi}^{\dot\alpha}, \dot\alpha = 1, 2\right) = 2_R = (0, 1/2)$. We use the antisymmetric tensors $\varepsilon_{\dot\alpha\dot\beta}$ and $\varepsilon^{\dot\alpha\dot\beta}$ to raise and lower dotted indices $(\varepsilon_{\dot2\dot1} = 1 = \varepsilon^{\dot1\dot2})$, and we note:

$$\bar{\xi}_{\dot\alpha}\,\bar{\eta}^{\dot\alpha} \equiv \bar{\xi}\,\bar{\eta}, \qquad \bar{\xi}\bar{\xi} \equiv \bar{\xi}^2. \tag{B.9}$$

The convention is the following: *when indices are not written, undotted indices are descending and dotted indices are ascending.*

We will also take as a convention to reverse the order of spinors when performing complex conjugation:

$$(\xi\eta)^\dagger = (\xi^\alpha\eta_\alpha)^\dagger = \bar{\eta}_{\dot\alpha}\bar{\xi}^{\dot\alpha} = \bar{\eta}\bar{\xi} = \bar{\xi}\bar{\eta}. \tag{B.10}$$

B.1.2 Relation of spinors with vectors

As we have just seen, one may form a spin-zero object $\xi\eta$ out of two spinors. One may also form a spin-one object. In order to do so, let us introduce the Pauli matrices σ^μ, $\mu = 0, \ldots, 3$,

$$\sigma^0 = \begin{pmatrix} 1 & 0 \\ 0 & 1 \end{pmatrix} \qquad \sigma^1 = \begin{pmatrix} 0 & 1 \\ 1 & 0 \end{pmatrix}$$

$$\sigma^2 = \begin{pmatrix} 0 & -i \\ i & 0 \end{pmatrix} \qquad \sigma^3 = \begin{pmatrix} 1 & 0 \\ 0 & -1 \end{pmatrix} \tag{B.11}$$

and form the vector $V^\mu \equiv \xi^\alpha\,\sigma^\mu_{\alpha\dot\alpha}\,\bar{\xi}^{\dot\alpha}$.

The matrices σ^μ, being hermitian, satisfy:

$$(\sigma^\mu{}_{\alpha\dot\beta})^* = (\sigma^{\mu*})_{\dot\alpha\beta} = (\sigma^{\mu\dagger})_{\beta\dot\alpha} = (\sigma^\mu)_{\beta\dot\alpha} \; . \tag{B.12}$$

One introduces also the matrices $\bar\sigma^\mu$ defined by:

$$\bar\sigma^{\mu\,\dot\alpha\alpha} = \varepsilon^{\dot\alpha\dot\beta}\,\varepsilon^{\alpha\beta}\,\sigma^\mu{}_{\beta\dot\beta} \tag{B.13}$$

$(\bar\sigma^0 = \sigma^0,\ \bar\sigma^i = -\sigma^i)$. They have the property

$$\sigma^\mu{}_{\alpha\dot\alpha}\bar\sigma^{\nu\dot\alpha\beta} + \sigma^\nu{}_{\alpha\dot\alpha}\bar\sigma^{\mu\dot\alpha\beta} = 2\eta^{\mu\nu}\delta_\alpha^\beta$$

$$\bar\sigma^{\mu\dot\alpha\alpha}\sigma^\nu{}_{\alpha\dot\beta} + \bar\sigma^{\nu\dot\alpha\alpha}\sigma^\mu{}_{\alpha\dot\beta} = 2\eta^{\mu\nu}\delta_{\dot\beta}^{\dot\alpha}, \tag{B.14}$$

from which one deduces

$$\mathrm{Tr}\,(\sigma^\mu\bar\sigma^\nu) = 2\,\eta^{\mu\nu}. \tag{B.15}$$

One may then show using (B.61) and (B.15) that

$$V^\mu V_\mu = 2\,\xi^2\bar\xi^2.$$

Hence $V^\mu V_\mu$ is invariant under the transformations (B.1) and (B.8) and $V^\mu \to V'^\mu = \xi'^\alpha\,\sigma^\mu{}_{\alpha\dot\alpha}\,\bar\xi'^{\dot\alpha}$ is a Lorentz transformation. The vector representation may thus be referred to as $(1/2, 1/2)$.

B.1.3 Dirac spinors

A four-component Dirac spinor describes two Weyl spinors:

$$\Psi = \begin{pmatrix} \chi_\alpha \\ \bar\xi^{\dot\alpha} \end{pmatrix}. \tag{B.16}$$

One may introduce the following basis for gamma matrices, called the Weyl basis:

$$\gamma^\mu = \begin{pmatrix} 0 & \sigma^\mu{}_{\alpha\dot\beta} \\ \bar\sigma^{\mu\dot\alpha\beta} & 0 \end{pmatrix}, \quad \gamma_5 = \gamma^5 = i\gamma^0\gamma^1\gamma^2\gamma^3 = -\frac{i}{4!}\epsilon^{\mu\nu\rho\sigma}\gamma_\mu\gamma_\nu\gamma_\rho\gamma_\sigma = \begin{pmatrix} -I & 0 \\ 0 & I \end{pmatrix} \tag{B.17}$$

where $\epsilon^{\mu\nu\rho\sigma}$ is the fully antisymmetric tensor with the convention:

$$\epsilon^{0123} = 1 = -\epsilon_{0123}. \tag{B.18}$$

Indeed one checks, using (B.14), that

$$\{\gamma^\mu, \gamma^\nu\} = 2\eta^{\mu\nu}. \tag{B.19}$$

If we define

$$\gamma^{[\mu}\gamma^\nu\gamma^{\rho]} = \frac{1}{3!}\left(\gamma^\mu\gamma^\nu\gamma^\rho + \gamma^\rho\gamma^\mu\gamma^\nu + \gamma^\nu\gamma^\rho\gamma^\mu - \gamma^\nu\gamma^\mu\gamma^\rho - \gamma^\mu\gamma^\rho\gamma^\nu - \gamma^\rho\gamma^\nu\gamma^\mu\right), \tag{B.20}$$

antisymmetric in $\mu\nu\rho$, it must necessarily be of the form $\epsilon^{\mu\nu\rho\sigma}A_\sigma$ where A_σ can be determined by taking specific values of $\mu,\ \nu,\ \rho$:

$$\gamma^{[\mu}\gamma^\nu\gamma^{\rho]} = -i\,\epsilon^{\mu\nu\rho\sigma}\gamma_5\gamma_\sigma. \tag{B.21}$$

Similarly, with obvious notation,

$$\gamma^{[\mu}\gamma^{\nu}\gamma^{\rho}\gamma^{\sigma]} = -i\epsilon^{\mu\nu\rho\sigma}\gamma_5. \tag{B.22}$$

We will also use

$$\sigma^{\mu\nu} = \frac{1}{4}[\gamma^{\mu},\gamma^{\nu}] = \begin{pmatrix} \sigma^{\mu\nu}{}_{\alpha}{}^{\beta} & 0 \\ 0 & \bar{\sigma}^{\mu\nu\dot{\alpha}}{}_{\dot{\beta}} \end{pmatrix}, \tag{B.23}$$

where

$$\sigma^{\mu\nu}{}_{\alpha}{}^{\beta} = \frac{1}{4}\left(\sigma^{\mu}_{\alpha\dot{\alpha}}\bar{\sigma}^{\nu\dot{\alpha}\beta} - \sigma^{\nu}_{\alpha\dot{\alpha}}\bar{\sigma}^{\mu\dot{\alpha}\beta}\right)$$

$$\bar{\sigma}^{\mu\nu\dot{\alpha}}{}_{\dot{\beta}} = \frac{1}{4}\left(\bar{\sigma}^{\mu\dot{\alpha}\alpha}\sigma^{\nu}_{\alpha\dot{\beta}} - \bar{\sigma}^{\nu\dot{\alpha}\alpha}\sigma^{\mu}_{\alpha\dot{\beta}}\right). \tag{B.24}$$

We have

$$\sigma^{\mu\nu} = -\frac{i}{4}\epsilon^{\mu\nu\rho\sigma}\gamma_5\gamma_{\rho}\gamma_{\sigma}. \tag{B.25}$$

The chirality eigenstates are:

$$\Psi_L = \frac{1-\gamma^5}{2}\Psi = \begin{pmatrix} \chi_{\alpha} \\ 0 \end{pmatrix} \quad \text{and} \quad \Psi_R = \frac{1+\gamma_5}{2}\Psi = \begin{pmatrix} 0 \\ \bar{\xi}^{\dot{\alpha}} \end{pmatrix}.$$

Since we have $\bar{\Psi} = \Psi^{\dagger}A$ where[2] $A = \begin{pmatrix} 0 & \delta^{\dot{\alpha}}_{\dot{\beta}} \\ \delta^{\beta}_{\alpha} & 0 \end{pmatrix}$,

$$\bar{\Psi} = (\bar{\chi}_{\dot{\alpha}} \ \xi^{\alpha})\begin{pmatrix} 0 & \delta^{\dot{\alpha}}_{\dot{\beta}} \\ \delta^{\beta}_{\alpha} & 0 \end{pmatrix} = (\xi^{\beta} \ \bar{\chi}_{\dot{\beta}}). \tag{B.27}$$

One then easily proves that, for two Dirac spinors Ψ_1 and Ψ_2:

$$\begin{aligned}
\bar{\Psi}_1 \Psi_2 &= \xi_1 \chi_2 + \bar{\chi}_1 \bar{\xi}_2 \\
\bar{\Psi}_1 \gamma_5 \Psi_2 &= -\xi_1 \chi_2 + \bar{\chi}_1 \bar{\xi}_2 \\
\bar{\Psi}_1 \gamma^{\mu} \Psi_2 &= \xi_1 \sigma^{\mu} \bar{\xi}_2 - \chi_2 \sigma^{\mu} \bar{\chi}_1 \\
\bar{\Psi}_1 \gamma^{\mu}\gamma^5 \Psi_2 &= \xi_1\sigma^{\mu} \bar{\xi}_2 + \chi_2 \sigma^{\mu} \bar{\chi}_1.
\end{aligned} \tag{B.28}$$

The completeness of the 16 Dirac matrices can be used to derive the Fierz rearrangement formula for two anticommuting spinors

$$\Psi_2\bar{\Psi}_1 = -\frac{1}{4}\mathbf{1}\left(\bar{\Psi}_1\Psi_2\right) - \frac{1}{4}\gamma_{\mu}\left(\bar{\Psi}_1\gamma^{\mu}\Psi_2\right) - \frac{1}{8}\sigma_{\mu\nu}\left(\bar{\Psi}_1\sigma^{\mu\nu}\Psi_2\right)$$

$$+\frac{1}{4}\gamma_{\mu}\gamma_5\left(\bar{\Psi}_1\gamma^{\mu}\gamma^5\Psi_2\right) - \frac{1}{4}\gamma_5\left(\bar{\Psi}_1\gamma^5\Psi_2\right). \tag{B.29}$$

[2] A is the matrix that intertwines the representation γ_{μ} with the equivalent representation $\gamma_{\mu}^{\dagger}$:

$$A\gamma_{\mu}A^{-1} = \gamma_{\mu}^{\dagger} \tag{B.26}$$

(C plays a similar rôle between γ_{μ} and $-\gamma_{\mu}^{T}$, as seen from (B.31)). As is well known, A coincides numerically with γ^0. We refrain from using the same notation because the two-component indices do not match as can be seen by comparing the explicit form of A with (B.17).

The charge conjugation matrix C reads, in this notation,

$$C = \begin{pmatrix} \varepsilon_{\alpha\beta} & 0 \\ 0 & \varepsilon^{\dot\alpha\dot\beta} \end{pmatrix}, \qquad C^{-1} = \begin{pmatrix} \varepsilon^{\alpha\beta} & 0 \\ 0 & \varepsilon_{\dot\alpha\dot\beta} \end{pmatrix} \tag{B.30}$$

and satisfies

$$C\,\gamma_\mu^T\,C^{-1} = -\gamma_\mu. \tag{B.31}$$

The charge conjugate spinor Ψ^c is thus simply:

$$\Psi^c = C\bar\Psi^T = \begin{pmatrix} \xi_\alpha \\ \bar\chi^{\dot\alpha} \end{pmatrix}. \tag{B.32}$$

Thus a Majorana mass term reads:

$$\bar\Psi^c\Psi = \chi^\alpha\chi_\alpha + \bar\xi_{\dot\alpha}\bar\xi^{\dot\alpha} = \overline{\Psi}^c_{\,R}\Psi_L + \overline{\Psi}^c_{\,L}\Psi_R. \tag{B.33}$$

The following properties may be checked from (B.28) and (B.64):

$$\begin{aligned}
\bar\Psi^c_1\Psi^c_2 &= \bar\Psi_2\Psi_1, \\
\bar\Psi^c_1\gamma_5\Psi^c_2 &= \bar\Psi_2\gamma_5\Psi_1, \\
\bar\Psi^c_1\gamma_\mu\Psi^c_2 &= -\bar\Psi_2\gamma_\mu\Psi_1, \\
\bar\Psi^c_1\gamma_\mu\gamma_5\Psi^c_2 &= \bar\Psi_2\gamma_\mu\gamma_5\Psi_1,
\end{aligned} \tag{B.34}$$

and more generally

$$\begin{aligned}
\bar\Psi^c_1\gamma_{\mu_1}\cdots\gamma_{\mu_n}\Psi^c_2 &= (-)^n\bar\Psi_2\gamma_{\mu_n}\cdots\gamma_{\mu_1}\Psi_1, \\
\bar\Psi^c_1\gamma_{\mu_1}\cdots\gamma_{\mu_n}\gamma_5\Psi^c_2 &= \bar\Psi_2\gamma_{\mu_n}\cdots\gamma_{\mu_1}\gamma_5\Psi_1,
\end{aligned} \tag{B.35}$$

and

$$\begin{aligned}
\left(\bar\Psi_1\gamma_{\mu_1}\cdots\gamma_{\mu_n}\Psi_2\right)^* &= (-)^n\bar\Psi^c_1\gamma_{\mu_1}\cdots\gamma_{\mu_n}\Psi^c_2, \\
\left(\bar\Psi_1\gamma_{\mu_1}\cdots\gamma_{\mu_n}\gamma_5\Psi_2\right)^* &= (-)^{n+1}\bar\Psi^c_1\gamma_{\mu_1}\cdots\gamma_{\mu_n}\gamma_5\Psi^c_2.
\end{aligned} \tag{B.36}$$

The Majorana condition

$$\Psi = C\bar\Psi^T \tag{B.37}$$

reads simply

$$\chi_\alpha = \xi_\alpha. \tag{B.38}$$

Thus a Majorana spinor Ψ_M is expressed in terms of a single two-dimensional spinor as

$$\Psi_M = \begin{pmatrix} \chi_\alpha \\ \bar\chi^{\dot\alpha} \end{pmatrix}. \tag{B.39}$$

Majorana spinors have the following properties, as can be checked from (B.35) and (B.36):

$$\begin{aligned}
\bar\Psi_1\gamma_{\mu_1}\cdots\gamma_{\mu_n}\Psi_2 &= (-)^n\bar\Psi_2\gamma_{\mu_n}\cdots\gamma_{\mu_1}\Psi_1, \\
\bar\Psi_1\gamma_{\mu_1}\cdots\gamma_{\mu_n}\gamma_5\Psi_2 &= \bar\Psi_2\gamma_{\mu_n}\cdots\gamma_{\mu_1}\gamma_5\Psi_1,
\end{aligned} \tag{B.40}$$

and

$$\begin{aligned}
\left(\bar\Psi_1\gamma_{\mu_1}\cdots\gamma_{\mu_n}\Psi_2\right)^* &= (-)^n\bar\Psi_1\gamma_{\mu_1}\cdots\gamma_{\mu_n}\Psi_2, \\
\left(\bar\Psi_1\gamma_{\mu_1}\cdots\gamma_{\mu_n}\gamma_5\Psi_2\right)^* &= (-)^{n+1}\bar\Psi_1\gamma_{\mu_1}\cdots\gamma_{\mu_n}\gamma_5\Psi_2.
\end{aligned} \tag{B.41}$$

B.2 Spinors in higher dimensions

We first review the representations of the group $SO(2n)$ before defining spinors in D dimensions.

B.2.1 Representations of $SO(2n)$

$SO(2n)$ is defined as the group of $2n$-dimensional orthogonal matrices Λ of determinant unity, *i.e.*

$$\Lambda^T \Lambda = 1, \quad \text{Det } \Lambda = 1. \tag{B.42}$$

Turning to the algebra, $\Lambda = e^J \sim 1 + J + \cdots$, these two conditions become

$$J^T = -J. \tag{B.43}$$

There are $2n(2n-1)/2$ independent $2n$-dimensional matrices satisfying (B.43). Hence the dimension of the $SO(2n)$ algebra is $n(2n-1)$. A basis of generators is ($1 \leq K, L \leq 2n$)

$$J_{KL} = \begin{pmatrix} & & & \cdot & & & \cdot & \\ \cdots & \cdots & \cdots & 1 & \cdots & \\ & & & \cdot & & & \cdot \\ \cdots & 1 & \cdots & \cdots & \cdots & \cdots & \\ & \cdot & & & & & \cdot \\ & K & & & L & & \end{pmatrix} \begin{matrix} \\ K \\ \\ L \\ \\ \end{matrix} . \tag{B.44}$$

From this one obtains the commutation relations

$$[J_{IJ}, J_{KL}] = \delta_{JK} J_{IL} - \delta_{JL} J_{IK} + \delta_{IL} J_{JK} - \delta_{IK} J_{JL} , \tag{B.45}$$

which define the $SO(2n)$ algebra.

A maximal set of commuting generators (known as the Cartan subalgebra) is given by the n generators[3] $N_k \equiv J_{2k-1,2k}$, $k = 1, \ldots, n$. The number of such generators is the rank of the algebra, hence the rank of $SO(2n)$ is n.

The representation (B.44) is the vector representation of dimension $2n$. The dimension of the algebra gives the dimension of the adjoint representation: $n(2n-1)$. In the following, we build the spinor representations of dimension 2^{n-1}.

The standard procedure is to consider a set of n fermion creation and annihilation operators $a_k^\dagger, a_k$, $k = 1, \ldots, n$:

$$\left[a_k, a_l^\dagger\right] = \delta_{kl}, \quad [a_k, a_l] = 0 = \left[a_k^\dagger, a_l^\dagger\right] . \tag{B.46}$$

One then constructs a Fock space by defining a vacuum $|0\rangle$ and acting on it with creation operators:

$$|0\rangle, \quad a_k^\dagger |0\rangle, \quad a_k^\dagger a_l^\dagger |0\rangle, \quad \ldots, \quad a_1^\dagger \cdots a_n^\dagger |0\rangle \tag{B.47}$$

form a set of 2^n independent states.

---

[3]Each of them is a block diagonal matrix with only nonzero block $\begin{pmatrix} 0 & 1 \\ -1 & 0 \end{pmatrix}$ in the $(2k-1, 2k)$ entry.

The following $2n$ matrices

$$\Gamma_K \equiv a_K + a_K^\dagger, \quad K = 1, \ldots, n$$

$$\Gamma_K \equiv \frac{a_{K-n} - a_{K-n}^\dagger}{\imath}, \quad K = n+1, \ldots, 2n \tag{B.48}$$

act on this 2^n-dimensional Fock space. They satisfy the Clifford algebra

$$\{\Gamma_K, \Gamma_L\} = 2\delta_{KL} , \tag{B.49}$$

and are therefore generalizations of the gamma matrices to dimensions higher than four. The point of interest to us is that the Σ matrices built out of them have precisely the commutation relations (B.45) of $SO(2n)$:

$$\Sigma_{KL} \equiv \frac{1}{4}[\Gamma_K, \Gamma_L]$$

$$[\Sigma_{IJ}, \Sigma_{KL}] = \delta_{JK}\Sigma_{IL} - \delta_{JL}\Sigma_{IK} + \delta_{IL}\Sigma_{JK} - \delta_{IK}\Sigma_{JL} . \tag{B.50}$$

Since the Σ_{KL} matrices act on the 2^n-dimensional Fock space, we have constructed a (2^n)-dimensional representation known as the spinor representation.

This representation is not irreducible, however. Indeed, define

$$\Gamma^{(2n+1)} \equiv \imath^{n(2n-1)}\Gamma_1 \cdots \Gamma_{2n}, \quad \left(\Gamma^{(2n+1)}\right)^2 = 1 . \tag{B.51}$$

Let us choose a Fock vacuum of definite chirality, *i.e.* $\Gamma^{(2n+1)}|0\rangle = +|0\rangle$. Then, since $\left\{\Gamma^{2n+1}, a_k^\dagger\right\} = 0$,

$$\Gamma^{(2n+1)}a_{k_1}^\dagger \cdots a_{k_p}^\dagger |0\rangle = (-1)^p a_{k_1}^\dagger \cdots a_{k_p}^\dagger |0\rangle , \tag{B.52}$$

and we can divide our Fock space (B.47) into states of opposite chirality. We have therefore built two (2^{n-1})-dimensional irreducible representations, the spinor representations of $SO(2n)$ of opposite chirality.

B.2.2 Spinors in D-dimensional spacetimes

Consider a D-dimensional Riemannian[4] manifold. We need to introduce the tangent space T_p, the set of tangent vectors at point p which is generated by the infinitesimal displacements dx^M, $M = 1, \ldots, D$, at this point. The discussion will rest on the notion of tangent space group, the set of orthogonal transformations on T_p. Writing such a transformation as $dx^M = \Lambda^M{}_N dx^N$, the property of orthogonality is expressed as usual by (g_{MN} is the metric of the manifold)

$$g_{MN}\, dx^M\, dx^N = g_{MN}\, \Lambda^M{}_P\, \Lambda^N{}_S\, dx^P\, dx^S = g_{PS}\, dx^P\, dx^S$$

that is

$$g_{MN}\, \Lambda^M{}_P\, \Lambda^N{}_S = g_{PS} . \tag{B.53}$$

---

[4]A Riemannian manifold is a smooth manifold with a symmetric metric tensor g_{MN} such that the bilinear form $g_{MN}V^M W^N$ ($V, W \in T_p$) is nondegenerate ($g_{MN}V^M W^N = 0$ for all V iff $W = 0$).

In the case of Minkowski spacetime M_4, with $g_{\mu\nu} = (+1, -1, -1, -1)$, this tangent space group is the proper Lorentz group $SO(1,3)$ (if we restrict to $\det \Lambda = +1$). For a D-dimensional manifold, with metric $g_{MN} = (+1, -1, -1, \ldots, -1)$, the tangent space group is $SO(1, D-1)$.

The corresponding Clifford algebra reads

$$\{\Gamma_M, \Gamma_N\} = 2g_{MN} \,, \tag{B.54}$$

where the gamma matrices have dimension $2^{[D/2]}$, with $[D/2]$ the integral part of $D/2$. As in the previous section, the generators of $SO(1,9)$ are the $\Sigma_{MN} = 1/4[\Gamma_M, \Gamma_N]$.

Just as before, one may introduce for *even* spacetime dimension D, the matrix $\Gamma^{(D+1)}$ which anticommutes with all gamma matrices[5]:

$$\Gamma^{(D+1)} \equiv i^{(D-2)/2}\Gamma^0\Gamma^1 \cdots \Gamma^{D-1} \,. \tag{B.55}$$

We note that γ^5 defined in (B.17) is precisely $\Gamma^{(5)}$ for $D = 4$. One then defines the chirality eigenstates by the condition:

$$\Gamma^{(D+1)}\Psi_{L,R} = \mp\Psi_{L,R} \,. \tag{B.56}$$

Thus, a Dirac spinor has $2^{D/2}$ degrees of freedom whereas a Weyl[6] fermion has $2^{(D-2)/2}$.

Can we always define a Majorana spinor? In order to answer this question, let us define the matrix B which intertwines the representation Γ^M with the equivalent representation $-\Gamma^{M*}$:

$$B\Gamma^M B^{-1} = -\Gamma^{M*} \,. \tag{B.57}$$

We have $BB^* = \eta I$ with $\eta = \pm 1$. Clearly, $B = C\Gamma^0$ where C is defined as in (B.31) by

$$C^{-1}\Gamma^M C = -\Gamma^{MT} \,. \tag{B.58}$$

If a spinor field Ψ of charge q satisfies the Dirac equation coupled with the electromagnetic field, one easily checks that $B^{-1}\Psi^*$ satisfies it with charge $-q$: it is associated with the antiparticle. One defines a Majorana spinor by the condition that particle and antiparticle are identical:

$$\Psi^* = B\Psi \,. \tag{B.59}$$

This Majorana condition is easily shown (see Exercise 6) to be equivalent to the condition (B.32) given above. It obviously implies $\eta = 1$.

It turns out that η can be computed as a function of the dimension D (see [331] for a proof):

$$\eta = -\sqrt{2}\cos\left[\frac{\pi}{4}(D+1)\right] \,. \tag{B.60}$$

---

[5] The difference in the overall factor between (B.51) and (B.55) is due to the change of metric between Euclidean and Minkowski. In both cases, it is chosen in such a way that $\left(\Gamma^{(D+1)}\right)^2 = 1$.

[6] In the case of odd dimension D', there are no Weyl spinors. If we write $D' = D + 1$ where D is even, then $\Gamma^0, \Gamma^1, \ldots, \Gamma^{D-1}$ and $i\Gamma^{(D+1)}$ given in (B.55) yield a set of gamma matrices. In what follows, we only consider the case of even D.

Thus $\eta = 1$ for $D = 2, 4$ mod 8. One can then show that there exists a representation of the gamma matrices which is purely imaginary, in which case $B = 1$ and $C = \Gamma^0$: the Majorana condition reads $\Psi^* = \Psi$. A Weyl condition (B.56) is then only possible if $\Gamma^{(D+1)}$ in (B.42) is real *i.e.* for $D/2$ odd. Hence one can define a Majorana–Weyl spinor only in $D = 2$ mod. 8 dimensions.

Further reading

- J. Wess and J. Bagger, *Supersymmetry and Supergravity*, Princeton series in physics, Appendix A (note the opposite metric signature in this reference).
- J. Scherk, *Extended supersymmetry and supergravity theories*, in *Advanced Study Institute on Gravitation: Recent Developments*, Cargèse, 1978.

Exercises

<u>Exercise 1</u> Prove the following relations which you will find to be very useful

$$\xi^\alpha \, \xi^\beta = -\frac{1}{2} \, \varepsilon^{\alpha\beta} \, \xi^2, \qquad \xi_\alpha \, \xi_\beta = \frac{1}{2} \, \varepsilon_{\alpha\beta} \, \xi^2,$$

$$\bar{\xi}^{\dot\alpha} \, \bar{\xi}^{\dot\beta} = \frac{1}{2} \, \varepsilon^{\dot\alpha\dot\beta} \, \bar{\xi}^2, \qquad \bar{\xi}_{\dot\alpha} \, \bar{\xi}_{\dot\beta} = -\frac{1}{2} \, \varepsilon_{\dot\alpha\dot\beta} \, \bar{\xi}^2. \tag{B.61}$$

Deduce that

$$\left(\theta\sigma^\mu\bar\theta\right) \left(\theta\sigma^\nu\bar\theta\right) = \frac{1}{2}\eta^{\mu\nu}\theta^2\bar\theta^2. \tag{B.62}$$

<u>Exercise 2</u> Prove that

$$\chi \, \sigma^\mu \, \bar\psi = -\bar\psi \, \bar\sigma^\mu \, \chi, \tag{B.63}$$

$$\left(\chi \, \sigma^\mu \, \bar\psi\right)^* = \psi \, \sigma^\mu \, \bar\chi. \tag{B.64}$$

<u>Exercise 3</u> Use

$$\sigma^\mu_{\alpha\dot\alpha}\sigma_{\mu\beta\dot\beta} = 2\varepsilon_{\alpha\beta}\varepsilon_{\dot\alpha\dot\beta} \tag{B.65}$$

to show the following Fierz reordering formula:

$$\left(\psi_1\psi_2\right) \left(\bar\chi_1\bar\chi_2\right) = \frac{1}{2} \left(\psi_1\sigma^\mu\bar\chi_1\right) \left(\psi_2\sigma_\mu\bar\chi_2\right) . \tag{B.66}$$

<u>Exercise 4</u> Show the following useful relations:

$$\left(\sigma^{\mu\nu}\right)_\alpha{}^\beta = \frac{i}{2} \, \epsilon^{\mu\nu\rho\sigma} \left(\sigma_{\rho\sigma}\right)_\alpha{}^\beta ,$$

$$\left(\bar\sigma^{\mu\nu}\right)^{\dot\alpha}{}_{\dot\beta} = -\frac{i}{2} \, \epsilon^{\mu\nu\rho\sigma} \left(\bar\sigma_{\rho\sigma}\right)^{\dot\alpha}{}_{\dot\beta} , \tag{B.67}$$

$$\left(\sigma^{\mu\nu}\sigma^{\rho}\right)_{\alpha\dot\beta} = \frac{1}{2}\left(g^{\nu\rho}g^{\mu\lambda} - g^{\mu\rho}g^{\nu\lambda} - i\epsilon^{\mu\nu\rho\lambda}\right)\sigma_{\lambda\alpha\dot\beta}\ ,$$

$$\left(\sigma^{\mu}\bar\sigma^{\nu\rho}\right)_{\alpha\dot\beta} = \frac{1}{2}\left(g^{\mu\nu}g^{\rho\lambda} - g^{\mu\rho}g^{\nu\lambda} - i\epsilon^{\mu\nu\rho\lambda}\right)\sigma_{\lambda\alpha\dot\beta}\ ,$$

$$\left(\bar\sigma^{\mu\nu}\bar\sigma^{\rho}\right)^{\dot\alpha\beta} = \frac{1}{2}\left(g^{\nu\rho}g^{\mu\lambda} - g^{\mu\rho}g^{\nu\lambda} + i\epsilon^{\mu\nu\rho\lambda}\right)\bar\sigma_\lambda{}^{\dot\alpha\beta}\ ,$$

$$\left(\bar\sigma^{\mu}\sigma^{\nu\rho}\right)^{\dot\alpha\beta} = \frac{1}{2}\left(g^{\mu\nu}g^{\rho\lambda} - g^{\mu\rho}g^{\nu\lambda} + i\epsilon^{\mu\nu\rho\lambda}\right)\bar\sigma_\lambda{}^{\dot\alpha\beta}\ . \tag{B.68}$$

Hints: To derive (B.67), use (B.25).

<u>Exercise 5</u> *We construct the generators of the Lorentz group in the spinor representation.*

Let P_μ be a covariant vector. Define $P_{\alpha\dot\alpha} \equiv P_\mu \sigma^\mu_{\alpha\dot\alpha}$ or alternatively

$$P_\mu = \frac{1}{2}\,\bar\sigma_\mu^{\dot\alpha\alpha}P_{\alpha\dot\alpha}. \tag{B.69}$$

(a) Using (B.69), express the Lorentz transformation $L_\mu{}^\nu$ in the vector representation in terms of $\left(M^{-1T}\right)_\alpha{}^\beta$, $(M^*)^{\dot\alpha}{}_{\dot\beta}$ and σ matrices.

(b) Making these transformations infinitesimal :

$$L_\mu{}^\nu = \delta_\mu^\nu + \omega_\mu{}^\nu$$

$$\left(M^{-1T}\right)_\alpha{}^\beta = \delta_\alpha^\beta + \eta_\alpha{}^\beta$$

$$(M^*)^{\dot\alpha}{}_{\dot\beta} = \delta_{\dot\beta}^{\dot\alpha} + \bar\eta^{\dot\alpha}{}_{\dot\beta}$$

show that

$$\left(M^{-1T}\right)_\alpha{}^\beta = \delta_\alpha^\beta + \omega_{\mu\nu}\left(\sigma^{\nu\mu}\right)_\alpha{}^\beta. \tag{B.70}$$

<u>Exercise 6</u> Show that the Majorana condition (B.59) is equivalent to the condition (*cf.* (B.32) and (B.37) in four dimensions)

$$\Psi = \Psi^c \equiv C\bar\Psi^T\ , \tag{B.71}$$

where $\bar\Psi = \Psi^\dagger\Gamma^0$.

Hints: Using (B.59) and $C^2 = -1$, we have

$$C\bar\Psi^T = C\Gamma^{0T}\Psi^* = C\Gamma^{0T}B\Psi = -\Gamma^0 CB\Psi = -\Gamma^0 C^2\Gamma^0\Psi = \Psi\ .$$

Appendix C
Superfields

In this appendix, we construct representations of the supersymmetry algebra by using a trick that allows us to put under the rug the difficulty that arises from having anticommutation relations. Supersymmetry transformations are understood as translations in a generalized space, known as superspace, where one adds to the standard spacetime coordinates anticommuting (Grassmann) variables. Fields in this superspace, known as superfields, then describe general supermultiplets. Specific covariant constraints on these superfields yield irreducible representations of the supersymmetry algebra.

Since this appendix lies on the theorist's track, we use throughout two-component spinor notation, as introduced in appendix B.

C.1   Superspace and superfields

Let us first consider translations. A translation of vector y^μ on a scalar field $\phi(x)$ reads:

$$\phi(x) \rightarrow e^{iy^\mu P_\mu}\phi(x)e^{-iy^\mu P_\mu}, \tag{C.1}$$

where P_μ is the generator of translations. We can write $\phi(x)$ as the value of ϕ translated from the origin:

$$\phi(x) \rightarrow e^{ix^\mu P_\mu}\phi(0)e^{-ix^\mu P_\mu}. \tag{C.2}$$

Thus, under a translation,

$$\phi(x) \rightarrow e^{i(x^\mu+y^\mu)P_\mu}\phi(0)e^{-i(x^\mu+y^\mu)P_\mu}. \tag{C.3}$$

If the translation is infinitesimal, then according to (C.2) and (C.3), we have

$$\delta\phi = -iy^\mu[\phi, P_\mu] = y^\mu\frac{\partial}{\partial x^\mu}\phi(x), \tag{C.4}$$

which we can write

$$\delta\phi = -iy^\mu \mathcal{P}_\mu\phi, \tag{C.5}$$

with

$$\mathcal{P}_\mu = i\frac{\partial}{\partial x^\mu}. \tag{C.6}$$

In other words, $\mathcal{P}_\mu$ is the representation of the translation generator P_μ as a differential operator. For the simplicity of notation, we will denote it simply by P_μ in what follows.

To obtain a similar representation of supersymmetry transformations in terms of differential operators, we have to introduce Grassmann variables. Let us recall that a Grassmann variable θ is an anticommuting variable

$$\{\theta, \theta\} = 0 \iff \theta^2 = 0.$$

A general function $F(\theta)$ can be expanded in series

$$F(\theta) = \sum_{n=0}^{\infty} a_n\, \theta^n = a_0 + a_1\, \theta.$$

The derivative is simply

$$\frac{dF}{d\theta} = a_1.$$

One may introduce two Grassmann variables θ^α, $\alpha = 1, 2$:

$$
\begin{aligned}
F(\theta^\alpha) &= a_0 + \theta^1\, a_1 + \theta^2\, a_2 + \theta^2\theta^1\, a_3 \\
&= a_0 + \theta^\alpha\, a_\alpha + \tfrac{1}{2}\, \theta\theta\, a_3.
\end{aligned}
$$

where we used the notation introduced in Appendix B (equation (B.6)) since α is to be interpreted as a spinor index. In fact, we will need two sets θ^α, $\alpha = 1, 2$, and $\bar{\theta}_{\dot\alpha}$, $\dot\alpha = 1, 2$, of Grassmann variables:

$$\{\theta^\alpha, \theta^\beta\} = \{\bar{\theta}_{\dot\alpha}, \bar{\theta}_{\dot\beta}\} = \{\theta^\alpha, \bar{\theta}_{\dot\alpha}\} = 0,$$

corresponding to the supersymmetry charges Q_α and $\bar{Q}^{\dot\alpha}$. One may then multiply the supersymmetry algebra

$$\{Q_\alpha, \bar{Q}_{\dot\alpha}\} = 2\sigma^\mu_{\alpha\dot\alpha}\, P_\mu, \quad \{Q_\alpha, Q_\beta\} = 0 = \{\bar{Q}_{\dot\alpha}, \bar{Q}_{\dot\beta}\} \tag{C.7}$$

by $\theta^\alpha\bar{\theta}^{\dot\alpha}$ and sum over α , $\dot\alpha$ to obtain

$$[\theta Q, \bar{\theta}\bar{Q}] = 2\left(\theta\sigma^\mu\bar{\theta}\right) P_\mu, \quad [\theta Q, \theta'Q] = 0 = [\bar{\theta}\bar{Q}, \bar{\theta}'\bar{Q}] \tag{C.8}$$

thus recovering commutation relations (because the θ's anticommute with the supersymmetry charges).

Using the standard procedure described above, we now realize the algebra (C.8) through differential operators acting on a "superspace" described by the spacetime variable x^μ and the Grassmann variables θ^α, $\bar{\theta}_{\dot\alpha}$. An object in this superspace is a superfield $\mathcal{F}(x, \theta, \bar{\theta})$.

We consider the "superspace translation" generalizing (C.1):

$$\mathcal{F}(x,\theta,\bar\theta) \to S\left(y^\mu,\eta,\bar\eta\right) \mathcal{F}(x,\theta,\bar\theta) S^{-1}\left(y^\mu,\eta,\bar\eta\right), \tag{C.9}$$

with

$$S\left(y^\mu,\eta,\bar\eta\right) = e^{i\left(y^\mu P_\mu + \eta Q + \bar\eta\bar Q\right)}. \tag{C.10}$$

Using the Hausdorff formula

$$e^A\, e^B = e^{A+B+\frac{1}{2}[A,B]}$$

if $[A,B]$ commutes with A and B, we obtain from the supersymmetry algebra (C.8):

$$S\left(y^\mu,\eta,\bar\eta\right)\, S\left(x^\mu,\theta,\bar\theta\right) = S\left(x^\mu + y^\mu + i\eta\sigma^\mu\bar\theta - i\theta\sigma^\mu\bar\eta,\, \theta+\eta,\, \bar\theta+\bar\eta\right). \tag{C.11}$$

Since any point can be obtained from the origin by translation, one may always write

$$\mathcal{F}(x,\theta,\bar\theta) = S(x,\theta,\bar\theta)\, \mathcal{F}(0,0,0) S^{-1}(x,\theta,\bar\theta). \tag{C.12}$$

Thus using (C.11), a superspace translation of parameters y^μ, η, and $\bar\eta$ acts as follows on the superfield:

$$\mathcal{F}(x^\mu,\theta,\bar\theta) \to \mathcal{F}\left(x^\mu + y^\mu + i\eta\sigma^\mu\bar\theta - i\theta\sigma^\mu\bar\eta,\, \theta+\eta,\, \bar\theta+\bar\eta\right). \tag{C.13}$$

Infinitesimally,

$$\delta_s\mathcal{F}\left(x^\mu,\theta,\bar\theta\right) = \left[\mathcal{F}\left(x^\mu,\theta,\bar\theta\right),\, -i\left(y^\mu P_\mu + \eta Q + \bar\eta\bar Q\right)\right] \tag{C.14}$$
$$= \left[\eta^\alpha\frac{\partial}{\partial\theta^\alpha} + \bar\eta_{\dot\alpha}\frac{\partial}{\partial\bar\theta_{\dot\alpha}} + \left(y^\mu + i\eta\sigma^\mu\bar\theta + i\bar\eta\bar\sigma^\mu\theta\right)\frac{\partial}{\partial x^\mu}\right]\mathcal{F}.$$

This explicitly shows the action of the different generators on the superfield $\mathcal{F}$:

$$\delta_s\,\mathcal{F}\left(x^\mu,\theta,\bar\theta\right) = -iy^\mu P_\mu\mathcal{F} - i\eta Q\mathcal{F} - i\bar\eta\bar Q\mathcal{F} \tag{C.15}$$

with

$$P_\mu = i\,\partial_\mu,$$
$$Q_\alpha = i\left[\frac{\partial}{\partial\theta^\alpha} + i\,\sigma^\mu_{\alpha\dot\alpha}\,\bar\theta^{\dot\alpha}\,\partial_\mu\right],$$
$$\bar Q_{\dot\alpha} = -i\left[\frac{\partial}{\partial\bar\theta^{\dot\alpha}} + i\,\theta^\alpha\sigma^\mu_{\alpha\dot\alpha}\,\partial_\mu\right], \tag{C.16}$$
$$\bar Q^{\dot\alpha} = i\left[\frac{\partial}{\partial\bar\theta_{\dot\alpha}} + i\,\bar\sigma^{\mu\dot\alpha\alpha}\theta_\alpha\,\partial_\mu\right].$$

(C.16) gives a representation of the supersymmetry algebra in terms of differential operators. One may check directly that

$$\{Q_\alpha,\bar Q_{\dot\alpha}\} = 2i\,\sigma^\mu_{\alpha\dot\alpha}\,\partial_\mu$$
$$\{Q_\alpha,Q_\beta\} = 0 = \{\bar Q_{\dot\alpha},\bar Q_{\dot\beta}\}. \tag{C.17}$$

One may expand the superfield $\mathcal{F}(x^\mu, \theta, \bar\theta)$ in terms of the Grassmann variables. The series stops at the term

$$\theta^2\, \theta^1\, \bar\theta^{\dot2}\, \bar\theta^{\dot1}\; \propto\; (\theta^\alpha\, \theta_\alpha)\,(\bar\theta_{\dot\alpha}\, \bar\theta^{\dot\alpha})\,.$$

One has

$$\begin{aligned}
\mathcal{F}\left(x,\theta,\bar\theta\right) &= f(x) + \theta\, \rho(x) + \bar\theta\, \bar\chi(x) + \theta^2\, m(x) + \bar\theta^2\, n(x)\\
&\quad + \left(\theta\sigma^\mu\bar\theta\right) v_\mu(x) + \theta^2\, \bar\theta\bar\lambda(x) + \bar\theta^2\, \theta\psi(x) + \theta^2\bar\theta^2 d(x)
\end{aligned} \tag{C.18}$$

with f, m, n, d complex scalar fields, ρ_α, $\bar\chi^{\dot\alpha}$, $\bar\lambda^{\dot\alpha}$, and ψ_α spinor fields and v_μ complex vector field. These fields form a representation of the supersymmetry algebra since one can write the explicit transformations using (C.15):

$$\begin{aligned}
\delta_s\, \mathcal{F}(x,\theta,\bar\theta) &= \left[-i\eta^\alpha Q_\alpha - i\bar\eta_{\dot\alpha}\bar Q^{\dot\alpha}\right] \mathcal{F}(x,\theta,\bar\theta)\\
&= \delta_s f(x) + \theta\, \delta_s\rho(x) + \bar\theta\, \delta_s\bar\chi(x) + \cdots
\end{aligned} \tag{C.19}$$

But this representation is obviously reducible: there is more to it than simply a chiral supermultiplet, or a vector supermultiplet. The problem will be to impose constraints that reduce the representation and make it irreducible.

For that matter, we will need to introduce covariant spinor derivatives on superfields, that is spinor derivatives such that

$$D_\alpha[\delta_s\, \mathcal{F}] = \delta_s[D_\alpha\, \mathcal{F}]. \tag{C.20}$$

In other words, covariant derivatives are spinor derivatives which commute with supersymmetry transformations.

It is easy to show that

$$D_\alpha = \frac{\partial}{\partial\theta^\alpha} - i\,\sigma^\mu_{\alpha\dot\alpha}\,\bar\theta^{\dot\alpha}\,\partial_\mu \tag{C.21}$$

$$\bar D_{\dot\alpha} = \frac{\partial}{\partial\bar\theta^{\dot\alpha}} - i\,\theta^\alpha\sigma^\mu_{\alpha\dot\alpha}\,\partial_\mu$$

$$\bar D^{\dot\alpha} = -\frac{\partial}{\partial\bar\theta_{\dot\alpha}} + i\bar\sigma^{\mu\dot\alpha\alpha}\theta_\alpha\,\partial_\mu \tag{C.22}$$

satisfy

$$\{D_\alpha, \bar D_{\dot\alpha}\} = -2i\,\sigma^\mu_{\alpha\dot\alpha}\,\partial_\mu \qquad \{D_\alpha, D_\beta\} = 0 = \{\bar D_{\dot\alpha}, \bar D_{\dot\beta}\}$$

$$\{D_\alpha, Q_\beta\} = \{D_\alpha, \bar Q_{\dot\beta}\} = \{\bar D_{\dot\alpha}, Q_\beta\} = \{\bar D_{\dot\alpha}, \bar Q_{\dot\beta}\} = 0. \tag{C.23}$$

C.2 The chiral superfield

C.2.1 Definition

We want to reduce the field content of a general superfield to the content of a chiral supermultiplet $(\phi,\ \psi_\alpha,\ F)$. Let us take this opportunity to note that we are dealing with the off-shell formulation here because the supersymmetry transformations are obviously linear. Hence some of the fields in the general decomposition (C.18) of the superfield $\mathcal{F}$ are auxiliary fields.

There is unfortunately no straightforward procedure for, given a specific super-multiplet (of the type constructed in Chapter 4), identifying the relevant constraint. One often adopts a trial and error process which sometimes leaves the nonexpert bewildered.

One possibility is to use the covariant derivative which has just been introduced and impose the constraint

$$\bar{D}^{\dot{\alpha}}\, \Phi\left(x^{\mu}, \theta, \bar{\theta}\right) = 0. \tag{C.24}$$

Such a superfield Φ is called chiral and, as we will show below, describes the fields of a chiral supermultiplet. The constraint (C.24) is chosen because it is invariant by supersymmetry. Indeed, using (C.20),

$$\bar{D}^{\dot{\alpha}}\left(\delta_S\, \Phi\right) = \delta_S\, \bar{D}^{\dot{\alpha}}\, \Phi = 0.$$

It is easy to see how the constraint (C.24) reduces the representation of the supersymmetry algebra. One notes that

$$\bar{D}^{\dot{\alpha}}\, \theta_{\alpha} = 0$$
$$\bar{D}^{\dot{\alpha}}\left(x^{\mu} - i\, \theta\sigma^{\mu}\bar{\theta}\right) = \bar{D}^{\dot{\alpha}}\left(x^{\mu} + i\, \bar{\theta}\bar{\sigma}^{\mu}\theta\right)$$
$$= \left(-\frac{\partial}{\partial\bar{\theta}_{\dot{\alpha}}} + i\, \bar{\sigma}^{\nu\dot{\alpha}\alpha}\, \theta_{\alpha}\, \partial_{\nu}\right)\left(x^{\mu} + i\, \bar{\theta}_{\dot{\beta}}\, \bar{\sigma}^{\mu\dot{\beta}\beta}\, \theta_{\beta}\right)$$
$$= i\, \bar{\sigma}^{\mu\dot{\alpha}\alpha}\, \theta_{\alpha} - i\, \bar{\sigma}^{\mu\dot{\alpha}\beta}\, \theta_{\beta} = 0.$$

Hence any function of θ^{α} and $y^{\mu} = x^{\mu} - i\theta\sigma^{\mu}\bar{\theta}$ automatically satisfies the constraint[1]. This can also be seen trivially by expressing everything in terms of θ, $\bar{\theta}$ and y^{μ}:

$$D_{\alpha} = \frac{\partial}{\partial\theta^{\alpha}} - 2i\, \sigma^{\mu}_{\alpha\dot{\alpha}}\, \bar{\theta}^{\dot{\alpha}}\, \frac{\partial}{\partial y^{\mu}}$$
$$\bar{D}^{\dot{\alpha}} = -\frac{\partial}{\partial\bar{\theta}_{\dot{\alpha}}} \tag{C.25}$$

and therefore the solution of $\bar{D}^{\dot{\alpha}}\Phi = 0$ is $\Phi = \Phi(y^{\mu}, \theta)$. A chiral superfield is thus written as:

$$\Phi(y, \theta) = \phi(y) + \sqrt{2}\, \theta^{\alpha}\, \psi_{\alpha}(y) + \theta^2\, F(y). \tag{C.26}$$

We are left with three independent fields. Counting dimensions may suggest that F is an auxiliary field: taking ϕ as a scalar field of canonical dimension 1, we find a dimension $-1/2$ for θ^{α} (so that ψ_{α} be of dimension $3/2$) and a dimension $+2$ for F; the latter dimension suggests that F gives a total divergence under a supersymmetry transformation and thus, as explained in Chapter 3, that it is an auxiliary field.

---

[1] Note that, strictly speaking, the variable y^{μ} is complex (use (B.64)). This is one of the many illustrations that supersymmetry "complexifies" spacetime.

One may rewrite (C.26) in terms of the variables θ, $\bar\theta$, and x^μ, using (B.15) and (B.61),

$$\Phi\left(x^\mu,\theta,\bar\theta\right) = \phi(x) - i\,\theta\sigma^\mu\bar\theta\,\partial_\mu\phi(x) - \frac{1}{4}\theta^2\,\bar\theta^2\,\Box\phi(x)$$

$$+\sqrt{2}\,\theta^\alpha\psi_\alpha(x) + \frac{i}{\sqrt{2}}\,\theta^2\,\partial_\mu\psi(x)\sigma^\mu\bar\theta + \theta^2 F(x) \qquad (C.27)$$

which clearly shows how the condition (C.24) constrains the component fields of a general superfield (C.18).

Note that we can define the component fields of the chiral superfield as:

$$\phi = \Phi\big|_0\,,\quad \chi_\alpha = \frac{1}{\sqrt{2}}D_\alpha\Phi\bigg|_0\,,\quad F = -\frac{1}{4}D^2\Phi\bigg|_0\,, \qquad (C.28)$$

where the index 0 means that we take $\theta = \bar\theta = 0$.

The supersymmetry transformations are easily obtained by expressing Q_α and $\bar Q^{\dot\alpha}$ in terms of θ, $\bar\theta$ and y^μ

$$Q_\alpha = i\frac{\partial}{\partial\theta^\alpha}\,,$$

$$\bar Q^{\dot\alpha} = i\left[\frac{\partial}{\partial\bar\theta_{\dot\alpha}} + 2i\,\bar\sigma^{\mu\dot\alpha\alpha}\,\theta_\alpha\,\frac{\partial}{\partial y^\mu}\right],$$

and using (C.19). Indeed

$$-i\eta^\alpha\,Q_\alpha\,\Phi = \sqrt{2}\,\eta\,\psi(y) + 2\,\eta\theta\,F(y)$$
$$-i\eta_{\dot\alpha}\,\bar Q^{\dot\alpha}\,\Phi = 2i\,\bar\eta\bar\sigma^\mu\theta\,\partial_\mu\phi + 2i\sqrt{2}\,(\bar\eta\bar\sigma^\mu\theta)\,\theta^\beta\,\partial_\mu\psi_\beta$$
$$= -2i\,\theta\sigma^\mu\bar\eta\,\partial_\mu\phi - 2i\sqrt{2}\,\theta^\alpha\sigma^\mu_{\alpha\dot\alpha}\theta^\beta\,\partial_\mu\psi_\beta\,\bar\eta^{\dot\alpha}$$
$$= -2i\,\theta\sigma^\mu\bar\eta\,\partial_\mu\phi + i\sqrt{2}\,\theta^2\,\partial_\mu\psi\sigma^\mu\bar\eta$$

where we have used $\theta^\alpha\theta^\beta = -\frac{1}{2}\,\varepsilon^{\alpha\beta}\,\theta^2$ (see (B.61) of Appendix B). Hence

$$\delta_s\phi = \sqrt{2}\,\eta\,\psi$$
$$\delta_s\psi_\alpha = \sqrt{2}\,\eta_\alpha\,F - i\sqrt{2}\,\sigma^\mu_{\alpha\dot\alpha}\bar\eta^{\dot\alpha}\,\partial_\mu\phi$$
$$\delta_s\bar\psi^{\dot\alpha} = \sqrt{2}\,\bar\eta^{\dot\alpha}\,F^* - i\sqrt{2}\,\bar\sigma^{\mu\dot\alpha\alpha}\eta_\alpha\,\partial_\mu\phi^*$$
$$\delta_s F = i\sqrt{2}\,\partial_\mu\psi\sigma^\mu\bar\eta = -i\sqrt{2}\,\bar\eta\bar\sigma^\mu\partial_\mu\psi \qquad (C.29)$$

which are the familiar supersymmetry transformations of a chiral supermultiplet. We check that F is indeed the auxiliary field: it transforms into a total derivative.

C.2.2 Actions

Just as in Minkowski spacetime where we obtain actions invariant under translations by integrating over all spacetime, we will obtain supersymmetric actions by integrating over all superspace. In order to fulfill this program, we first have to discuss the integration of a Grassmann variable.

We start with one variable θ. We want to define an integration law for a general function $f(\theta) = a_0 + a_1 \theta$ such that one has the standard property of linearity and one recovers the rule

$$\int \frac{df}{d\theta} \, d\theta = 0.$$

Clearly this requires $\int 1 \cdot d\theta = 0$ and, if we want the integral to be nonvanishing, $\int \theta \, d\theta \neq 0$. Choosing the overall normalization, we thus set

$$\int d\theta = 0 \qquad \int \theta \, d\theta = 1. \tag{C.30}$$

Hence

$$\frac{df}{d\theta} = a_1 = \int f(\theta) \, d\theta. \tag{C.31}$$

For two variables θ^α, $\alpha = 1, 2$, we define

$$\int d^2\theta \equiv -\frac{1}{4} \int \varepsilon_{\alpha\beta} \, d\theta^\alpha \, d\theta^\beta = \frac{1}{2} \int d\theta^1 \, d\theta^2.$$

Hence

$$\int d^2\theta \cdot 1 = 0 \qquad \int d^2\theta \, \theta^\alpha = 0$$

$$\int d^2\theta \, \theta\theta = \frac{1}{2} \int d\theta^1 \, d\theta^2 \, 2\theta^2\theta^1 = 1. \tag{C.32}$$

It is then straightforward to write a supersymmetric action using a chiral superfield Φ:

$$S = \int d^4y \, F = \int d^4y \int d^2\theta \, \Phi = \int d^4y \, \Phi \Big|_{\theta^2} = -\frac{1}{4} \int d^4y \, D^2\Phi \Big|_0 \tag{C.33}$$

where $\Phi|_{\theta^2}$ is the θ^2 component of Φ and $D^2\Phi|_0$ is the scalar component of $D^2\Phi$ (we have $D^\alpha D_\alpha \Phi|_0 = \varepsilon^{\alpha\beta} D_\beta D_\alpha \Phi|_0 = \varepsilon^{\alpha\beta} \frac{\partial}{\partial\theta^\beta} \frac{\partial}{\partial\theta^\alpha} \Phi|_0 = -4F$ using the results of Exercise 1).

Indeed since $\delta_s F$ is a total derivative

$$\delta_s S = \int d^4y \, \delta_s F = 0.$$

Hence any F component of a chiral superfield provides a good supersymmetric action.

Let us note, as an aside, that, as in (C.31), an integration over a Grassmann variable is equivalent to a differentiation. In supersymmetric covariant notation, $d^2\theta \sim -\frac{1}{4}D^2$.

We may now use the previous remark in order to derive supersymmetric Lagrangians. Let us consider a set of n chiral superfields Φ_i, $i = 1, \ldots, n$. Any product of these (and thus any analytic function $W(\Phi_i)$) is also a chiral superfield. For example,

$$\Phi_i = \phi_i + \sqrt{2} \, \theta \, \psi_i + \theta^2 \, F_i$$

$$\Phi_i \, \Phi_j = \phi_i \, \phi_j + \sqrt{2} \, \theta \, (\psi_i \, \phi_j + \psi_j \, \phi_i) + \theta^2 \, (\phi_i \, F_j + \phi_j \, F_i - \psi_i\psi_j)$$

$$\Phi_i \, \Phi_j \, \Phi_k = \phi_i \, \phi_j \, \phi_k + \sqrt{2} \, \theta \, (\psi_i \, \phi_j \, \phi_k + \text{perm.}) + \theta^2 \, (F_i \, \phi_j \, \phi_k - \psi_i\psi_j \, \phi_k + \text{perm.})$$

and more generally for a general analytic function $W(\Phi_i)$

$$W(\Phi_i)|_{\theta^2} = \Sigma_i \frac{\partial W}{\partial \Phi^i}(\phi)\, F_i - \frac{1}{2}\sum_{ij} \frac{\partial^2 W}{\partial \Phi_i \partial \Phi_j}(\phi)\, \psi_i \psi_j. \tag{C.34}$$

We thus recover the terms of a supersymmetric Lagrangian which arise from the superpotential. Let us note that the analyticity of the superpotential is intrinsically connected with the requirement of supersymmetry: a product of Φ_i and $\Phi_i^\dagger$ fields is not a chiral superfield; thus in this construction we must restrict ourselves to sums of products of Φ_i's and thus to analytic functions.

This, however, does not provide us with kinetic terms for the component fields. Let us therefore consider more closely $\Phi^\dagger$: $\Phi^\dagger$ is not chiral since it satisfies

$$D^\alpha\, \Phi^\dagger = 0. \tag{C.35}$$

This is known as an antichiral superfield. In any case, $\Phi^\dagger$ is not a function of y and θ only (it is in fact a function of y^* and $\bar{\theta}$) and it is better to revert to the x variable.

Using (C.27) (and (B.64) of Appendix B)

$$\Phi^\dagger(x) = \phi^*(x) + i\theta\sigma^\mu\bar{\theta}\, \partial_\mu\phi^*(x) - \frac{1}{4}\, \theta^2\, \bar{\theta}^2\, \Box\, \phi^*(x)$$

$$+\sqrt{2}\, \bar{\theta}_{\dot\alpha}\bar{\psi}^{\dot\alpha}(x) - \frac{i}{\sqrt{2}}\, \bar{\theta}^2\, \theta\sigma^\mu\partial_\mu\bar{\psi}(x) + \bar{\theta}^2\, F^*(x)\, . \tag{C.36}$$

We may now consider the real superfield $\mathcal{F} = \Phi^\dagger\Phi$. The highest component of such a real superfield (known as a vector superfield) transforms under supersymmetry into a total derivative. This may be guessed since this component has the highest mass dimension and must therefore transform into derivatives of the other fields (of lower dimensions). Thus $\mathcal{F}|_{\theta^2\bar{\theta}^2}$ provides us with a good supersymmetric action

$$S = \int d^4x\, \mathcal{F}\,\Big|_{\theta^2\bar{\theta}^2} = \int d^4x\, d^2\theta d^2\bar{\theta}\, \mathcal{F} = \frac{1}{16}\int d^4x\, D^2\bar{D}^2\mathcal{F}\,\Big|_0 \tag{C.37}$$

with obvious notation for $d^2\bar{\theta}$ ($\int d^2\bar{\theta}\, \bar{\theta}^2 = 1$). Indeed

$$\delta_s S = -i\int d^4x\, d^2\theta\, d^2\bar{\theta}\, (\eta Q + \bar{\eta}\bar{Q})\, \mathcal{F} = 0$$

since Q_α and $\bar{Q}^{\dot\alpha}$ involve only derivatives $\frac{\partial}{\partial x^\mu}$, $\frac{\partial}{\partial \theta^\alpha}$ or $\frac{\partial}{\partial \theta_{\dot\alpha}}$. Again, integrating over all superspace provides an action invariant under superspace translations, *i.e.* supersymmetry transformations.

But using (C.27) and (C.36)

$$\mathcal{F}|_{\theta^2\bar{\theta}^2} = -\frac{1}{4}\, \phi\Box\phi^* - \frac{1}{4}\, \phi^*\Box\phi + \frac{1}{2}\, \partial^\mu\phi^*\partial_\mu\phi + F^*F$$

$$+\frac{i}{2}\psi\sigma^\mu\partial_\mu\bar{\psi} - \frac{i}{2}\, \partial_\mu\psi\sigma^\mu\bar{\psi}$$

$$= \partial^\mu\phi^*\partial_\mu\phi + i\psi\sigma^\mu\partial_\mu\bar{\psi} + F^*F +\ \text{total derivatives.} \tag{C.38}$$

To recapitulate, we may write the following supersymmetric action in the case of a single chiral superfield Φ

$$
\begin{aligned}
S &= \int d^4x \, d^2\theta \, d^2\bar{\theta} \; \Phi^\dagger(x,\theta,\bar{\theta}) \; \Phi(x,\theta,\bar{\theta}) + \left[\int d^4y \, d^2\theta \; W\left(\Phi(y,\theta)\right) + \text{h.c.} \right] \\
&= \int d^4x \left\{ \partial^\mu \phi^* \partial_\mu \phi + i\,\psi\sigma^\mu \partial_\mu \bar{\psi} + F^* F + \frac{\partial W}{\partial \Phi}(\phi) F + \left[\frac{\partial W}{\partial \Phi}(\phi) \right]^* F^* \right. \\
&\qquad \left. - \frac{1}{2} \frac{\partial^2 W}{\partial \Phi^2}(\phi)\psi\psi - \frac{1}{2} \left[\frac{\partial^2 W}{\partial \Phi^2}(\phi) \right]^* \bar{\psi}\bar{\psi} \right\}.
\end{aligned}
\tag{C.39}
$$

This is by construction invariant under the supersymmetry transformations (C.29). The field F has no kinetic term: we may solve for this auxiliary field:

$$
F = - \left[\frac{\partial W}{\partial \Phi}(\phi) \right]^*
$$

to obtain

$$
S = \int d^4x \left\{ \partial^\mu \phi^* \partial_\mu \phi + i\psi\sigma^\mu \partial_\mu \bar{\psi} - \left| \frac{\partial W}{\partial \Phi}(\phi) \right|^2 - \left(\frac{1}{2} \frac{\partial^2 W}{\partial \Phi^2}(\phi)\,\psi\psi + \text{h.c.} \right) \right\}.
\tag{C.40}
$$

C.2.3 R-symmetry

We saw in Section 4.1 of Chapter 4 that $N = 1$ supersymmetry allows for a global $U(1)$ symmetry which does not commute with supersymmetry: the R-symmetry whose generator satisfies

$$
\begin{aligned}
[Q_\alpha, R] &= Q_\alpha \\
[\bar{Q}_{\dot\alpha}, R] &= -\bar{Q}_{\dot\alpha}.
\end{aligned}
\tag{C.41}
$$

Let us illustrate on the example of a single chiral superfield how this R-symmetry can be realized. Because the R-symmetry does not commute with supersymmetry, it must act differently on the different components of the superfield and thus cannot leave the Grassmann variable θ invariant. The R-symmetry thus acts generically on the chiral superfield $\Phi(y,\theta)$ as

$$
\begin{aligned}
R\,\Phi(y,\theta) &= e^{ir\alpha}\,\Phi\left(y, e^{-i\alpha}\theta\right) \\
R\,\Phi^\dagger(y^*,\bar{\theta}) &= e^{-ir\alpha}\,\Phi^\dagger\left(y^*, e^{i\alpha}\bar{\theta}\right)
\end{aligned}
\tag{C.42}
$$

where r is the R-charge of the supermultiplet.

In terms of the component fields, this reads

$$
\begin{aligned}
R\,\phi(x) &= e^{ir\alpha}\,\phi(x) \\
R\,\psi(x) &= e^{i(r-1)\alpha}\,\psi(x) \\
R\,F(x) &= e^{i(r-2)\alpha}\,F(x).
\end{aligned}
\tag{C.43}
$$

Let us note that, because of the Grassmann nature of the θ variable, if $\theta' = e^{-i\alpha}\theta$, $d^2\theta' = e^{2i\alpha}d^2\theta$ (*cf.* (C.32)). Obviously, the kinetic term $\int d^2\theta\, d^2\bar{\theta}\, \Phi^\dagger\Phi$ is R-invariant; the superpotential term $\int d^2\theta\, W(\Phi)$ is invariant if

$$R\, W\left(\Phi(y,\theta)\right) = e^{2i\alpha}\, W\left(\Phi(y, e^{-i\alpha}\theta)\right) \tag{C.44}$$

in which case

$$\int d^2\theta R\, W(\Phi(y,\theta)) = \int d^2\theta\, e^{2i\alpha}\, W\left(\Phi(y, e^{-i\alpha}\theta)\right)$$

$$= \int d^2\theta'\, W\left(\Phi(y,\theta')\right). \tag{C.45}$$

This is the case if $W(\Phi)$ is a monomial in Φ and the R-charges add up to 2: for example $W(\Phi) = \lambda\Phi^3$ with Φ of R-charge $r = \frac{2}{3}$. This is not the case if $W(\Phi)$ is a general polynomial in Φ.

C.2.4   Supersymmetric nonlinear sigma models and Kähler invariance

If one considers several chiral superfields Φ_i, the kinetic term for each of them is provided by $\Phi_i^\dagger\Phi_i$. But one may try to be more general and consider a generic real function K of the Φ_i and $\Phi_i^\dagger$ [389]:

$$S = \int d^4x\, d^2\theta\, d^2\bar{\theta}\, K(\Phi_i, \Phi_i^\dagger). \tag{C.46}$$

Obviously, this leads to nonnormalized kinetic terms for the scalar fields. Indeed, following the same steps as before, one obtains

$$S = \int d^4x\, \frac{\partial^2 K}{\partial\Phi_i\partial\Phi_j^\dagger}(\phi, \phi^*)\, \partial^\mu\phi_i\partial_\mu\phi_j^* + \cdots \tag{C.47}$$

A field redefinition $\phi_i \to \widehat{\phi}_i(\phi_j)$ may lead in some cases to a normalized kinetic term $\partial^\mu\widehat{\phi}_i^*\partial_\mu\widehat{\phi}_i$ but this is not general. This is reminiscent of the situation encountered in the nonlinear sigma model. Let us recall that this model is obtained as a limit of the linear sigma model whose Lagrangian is described in terms of a triplet of 'pion' fields $\pi^i, i = 1, 2, 3$, and a σ field by:

$$\mathcal{L} = \tfrac{1}{2}\left(\partial^\mu\sigma\partial_\mu\sigma + \partial^\mu\pi^i\partial_\mu\pi^i\right) - \tfrac{1}{4}\lambda\left(\sigma^2 + \pi^i\pi^i - v^2\right)^2. \tag{C.48}$$

In the limit $\lambda \to +\infty$, we must impose the constraint $\sigma^2 + \pi^i\pi^i = v^2$ to keep the energy of the configuration finite. Differentiation of this constraint yields $\sigma\partial_\mu\sigma = -\pi^i\partial_\mu\pi^i$ and (C.48) can be rewritten as

$$\mathcal{L} = \frac{1}{2}\left[\delta^{ij} + \frac{\pi^i\pi^j}{v^2 - \pi^k\pi^k}\right]\partial^\mu\pi^i\partial_\mu\pi^j, \tag{C.49}$$

which is the Lagrangian of the nonlinear sigma model: the fields π^i cannot be redefined to give a normalized kinetic term.

Indeed, the Lagrangian (C.46) provides the supersymmetric generalization of a generic nonlinear sigma model. We now show that supersymmetry provides a very nice geometrical description. Let us first rewrite the complete component form of (C.46):

$$
\begin{aligned}
S &= \frac{1}{16}\int d^4x \ D^2\bar{D}^2 K\big|_0 \\
&= \int d^4x \ K_{i\bar{j}}(\phi,\phi^*)\left(\partial^\mu\phi_i\partial_\mu\phi_j^* - \frac{i}{2}\psi_i\sigma_\mu\partial^\mu\bar{\psi}_j + \frac{i}{2}\partial^\mu\psi_i\sigma_\mu\bar{\psi}_j + F_iF_j^*\right) \\
&\quad - \tfrac{1}{2}K_{i\bar{j}k}(\phi,\phi^*)\left(\psi_i\psi_k F_j^* - i\psi_i\sigma^\mu\bar{\psi}_j\partial_\mu\phi_k\right) \\
&\quad - \tfrac{1}{2}K_{i\bar{j}\bar{k}}(\phi,\phi^*)\left(\bar{\psi}_j\bar{\psi}_k F_i + i\psi_i\sigma^\mu\bar{\psi}_j\partial_\mu\phi_k^*\right) \\
&\quad + \tfrac{1}{4}K_{i\bar{j}k\bar{l}}(\phi,\phi^*)\left(\psi_i\psi_k\right)\left(\bar{\psi}_j\bar{\psi}_l\right),
\end{aligned}
\tag{C.50}
$$

where we have used the notation

$$
K_{i\bar{j}} = \frac{\partial^2 K}{\partial\Phi_i\partial\Phi_j^\dagger}, \quad
K_{i\bar{j}k} = \frac{\partial^2 K}{\partial\Phi_i\partial\Phi_j^\dagger\partial\Phi_k^\dagger}, \quad \cdots
\tag{C.51}
$$

All these seemingly complicated terms have a beautiful geometric interpretation which we now present. Indeed, one may note that the transformation

$$
K(\Phi,\Phi^\dagger) \to K(\Phi,\Phi^\dagger) + F(\Phi) + \bar{F}(\Phi^\dagger),
\tag{C.52}
$$

where $F(\Phi)$ is an analytic function of the Φ fields ($\bar{F}(\Phi^\dagger)$ is the complex conjugate in order that the total expression remains real), is an invariance of the action (C.46):

$$
\int d^4x\, d^2\theta\, d^2\bar{\theta}\ F\left(\Phi(x,\theta,\bar{\theta})\right) = \int d^4y\int d^2\vartheta\int d^2\bar{\theta}\ F\left(\Phi(y,\theta)\right) = 0.
\tag{C.53}
$$

An invariance such as (C.52) is known as a Kähler invariance and $K(\Phi,\Phi^\dagger)$ is called a Kähler potential. Let us note that it is a transformation on functions of fields and it may not be realized as a transformation on the fields themselves[2].

The space parametrized by the complex scalar fields $\phi^i \equiv \phi_i$, $\bar{\phi}^{\bar{j}} \equiv \phi_j^*$ is a complex manifold. The kinetic term (C.47) may be rewritten

$$
S = \int d^4x\ g_{i\bar{j}}\,\partial^\mu\phi^i\partial_\mu\bar{\phi}^{\bar{j}} + \cdots
\tag{C.54}
$$

where

$$
g_{i\bar{j}} = g_{\bar{j}i} \equiv \frac{\partial K}{\partial\Phi_i\partial\Phi_j^\dagger}(\phi,\phi^*), \quad g_{ij} = 0, \quad g_{\bar{i}\bar{j}} = 0
\tag{C.55}
$$

---

[2]although it does in some well-known cases (see Exercise 9).

is interpreted as a metric on the complex manifold. It is said that the metric derives from the Kähler potential K and a complex manifold with such a metric is called a Kähler manifold[3].

Kähler manifolds have very specific properties; for example, the only nonzero components of the Christoffel symbol (see Section D.1 of Appendix D) are

$$\Gamma^i{}_{jk} = g^{i\bar{l}}\partial_j g_{k\bar{l}},$$
$$\Gamma^{\bar{i}}{}_{\bar{j}k} = g^{\bar{i}l}\partial_{\bar{j}}g_{k\bar{l}}, \tag{C.56}$$

where $g^{i\bar{j}}$ is the inverse metric ($g^{i\bar{j}}g_{\bar{j}k} = \delta^i_k$) and $\partial_i \equiv \partial/\partial\Phi_i$, $\partial_{\bar{j}} \equiv \partial/\partial\Phi^\dagger_j$. We have used the fact that, since the metric derives from a Kähler potential, $\partial_j g_{k\bar{l}} = K_{jk\bar{l}} = \partial_k g_{j\bar{l}}$. Similarly, the Ricci tensor is simply

$$R_{i\bar{j}} = -\partial_i\partial_{\bar{j}}\log\left[\det g\right]. \tag{C.57}$$

Finally, the action (C.50) may be written as:

$$S = \int d^4x\; g_{i\bar{j}}\left[\partial^\mu\phi^i\partial_\mu\bar{\phi}^{\bar{j}} - \frac{i}{2}\psi^i\sigma_\mu\mathcal{D}^\mu\bar{\psi}^{\bar{j}} + \frac{i}{2}\mathcal{D}^\mu\psi^i\sigma_\mu\bar{\psi}^{\bar{j}} + \hat{F}^i\hat{\bar{F}}^{\bar{j}}\right]$$
$$+\frac{1}{4}R_{i\bar{i}j\bar{j}}\left(\psi^i\psi^j\right)\left(\bar{\psi}^{\bar{i}}\bar{\psi}^{\bar{j}}\right), \tag{C.58}$$

where we have defined the Kähler covariant derivatives:

$$\mathcal{D}_\mu\psi^i_\alpha \equiv \left(\partial_\mu\delta^i_l + \Gamma^i{}_{kl}\partial_\mu\phi^k\right)\psi^l_\alpha,$$
$$\mathcal{D}_\mu\bar{\psi}^{\bar{j}\dot\alpha} \equiv \left(\partial_\mu\delta^{\bar{j}}_{\bar{l}} + \Gamma^{\bar{j}}{}_{\bar{k}\bar{l}}\partial_\mu\bar{\phi}^{\bar{k}}\right)\bar{\psi}^{\bar{l}\dot\alpha}, \tag{C.59}$$

which expresses the fact that the fermion fields transform as contravariant vectors under a reparametrization of the Kähler manifold (see Exercise 8); the Riemann tensor

$$R_{i\bar{i}j\bar{j}} = \partial_j\partial_{\bar{j}}g_{i\bar{i}} - g_{k\bar{k}}\Gamma^k{}_{ij}\Gamma^{\bar{k}}{}_{\bar{i}\bar{j}}, \tag{C.60}$$

and we have redefined the auxiliary fields

$$\hat{F}^i = F^i - \tfrac{1}{2}\Gamma^i{}_{jk}\psi^i\psi^j, \quad \hat{\bar{F}}^{\bar{j}} = \bar{F}^{\bar{j}} - \tfrac{1}{2}\Gamma^{\bar{j}}{}_{\bar{k}\bar{l}}\bar{\psi}^{\bar{k}}\bar{\psi}^{\bar{l}} \tag{C.61}$$

in order that the solution of the corresponding equation of motion is simply $\hat{F}^i = 0 = \hat{\bar{F}}^{\bar{j}}$.

The action (C.58) is invariant under the supersymmetry transformations:

$$\delta_S\phi^i = \sqrt{2}\,\eta\,\psi^i$$
$$\delta_S\psi^i_\alpha = \sqrt{2}\,\eta_\alpha\,\hat{F}^i - i\sqrt{2}\sigma^\mu_{\alpha\dot\alpha}\bar{\eta}^{\dot\alpha}\,\mathcal{D}_\mu\phi^i \tag{C.62}$$
$$\delta_S\hat{F}^i = i\sqrt{2}\,\mathcal{D}_\mu\psi^i\sigma^\mu\bar{\eta}$$

---

[3]For a mathematical introduction to Kähler manifolds, see Section 5.3 of the book by [196].

which are basically (C.29) up to the replacement of standard derivatives with Kähler covariant ones and a redefinition of the auxiliary fields. This shows that the structure of superspace may be accommodated with the Kähler invariance: this leads to the notion of Kähler superspace where Kähler invariance is built in the superspace geometry [38, 39].

The presence of nonrenormalizable terms in (C.50) or (C.58) makes it necessary to consider nonlinear sigma models in the context of supergravity (if we consider that the underlying physics responsible for such nonrenormalizable terms appears at a scale close to the Planck scale).

C.3   The vector superfield

Let us identify the superfield which describes the vector supermultiplet $(A_\mu(x), \lambda(x), D(x))$ introduced in Chapter 3. Since $A_\mu(x)$ is a real field, we must impose a reality condition on the general superfield $\mathcal{F}$, whose decomposition was written in (C.18) (so that its vector component is real). From now on, we will denote it by V:

$$V(x, \theta, \bar\theta) = V^\dagger(x, \theta, \bar\theta). \tag{C.63}$$

We will then write the field decomposition in a slightly different way (and justify it immediately):

$$\begin{aligned}
V(x, \theta, \bar\theta) = {} & C(x) + i\theta\chi(x) - i\bar\theta\bar\chi(x) + \frac{i}{2}\theta^2[M(x) + iN(x)] \\[4pt]
& - \frac{i}{2}\,\bar\theta^2[M(x) - iN(x)] + \theta\sigma^\mu\bar\theta A_\mu(x) \\[4pt]
& + \theta^2\bar\theta_{\dot\alpha}\left[\bar\lambda^{\dot\alpha}(x) + \frac{1}{2}\left(\bar\sigma^\mu\partial_\mu\chi(x)\right)^{\dot\alpha}\right] \\[4pt]
& + \bar\theta^2\theta^\alpha\left[\lambda_\alpha(x) - \frac{1}{2}\left(\sigma^\mu\partial_\mu\bar\chi(x)\right)_\alpha\right] \\[4pt]
& - \frac{1}{2}\theta^2\bar\theta^2\left[D(x) + \frac{1}{2}\Box C(x)\right].
\end{aligned} \tag{C.64}$$

A vector field is associated with a gauge transformation: $A_\mu(x) \rightarrow A_\mu(x) + \partial_\mu\alpha(x)$ in the abelian case. In order to write it in a supersymmetric way, let us introduce a chiral superfield

$$\Lambda(y, \theta) = a(y) + \sqrt{2}\theta\psi(y) + \theta^2 F(y) \tag{C.65}$$

and recall (C.27) and (C.36):

$$\begin{aligned}
\Lambda(x, \theta, \bar\theta) - \Lambda^\dagger(x, \theta, \bar\theta) = {} & [a(x) - a^*(x)] + \sqrt{2}\left[\theta\psi(x) - \bar\theta\bar\psi(x)\right] \\[4pt]
& - i\theta\sigma^\mu\bar\theta\,\partial_\mu[a(x) + a^*(x)] + \theta^2 F(x) - \bar\theta^2 F^*(x) \\[4pt]
& - \frac{i}{\sqrt{2}}\,\theta^2\left[\bar\theta\bar\sigma^\mu\partial_\mu\psi(x)\right] + \frac{i}{\sqrt{2}}\,\bar\theta^2\left[\theta\sigma^\mu\partial_\mu\bar\psi(x)\right] \\[4pt]
& - \frac{1}{4}\,\theta^2\bar\theta^2\Box[a(x) - a^*(x)].
\end{aligned}$$

Note therefore that the transformation

$$V \to V + i(\Lambda - \Lambda^\dagger) \qquad (C.66)$$

corresponds to a gauge transformation with parameter $\alpha(x) \equiv -g(a(x) + a^*(x))$ for the vector component (*cf.* (A.31) of Appendix Appendix A). The complete gauge transformation reads

$$C(x) \to C(x) + i[a(x) - a^*(x)]$$

$$\chi(x) \to \chi(x) + \sqrt{2}\psi(x)$$

$$M(x) + iN(x) \to M(x) + iN(x) + 2F(x)$$

$$A_\mu(x) \to A_\mu(x) - \frac{1}{g}\partial_\mu\alpha(x)$$

$$\lambda(x) \to \lambda(x)$$

$$D(x) \to D(x).$$

We see that, through a choice of the parameters $i[a(x) - a^*(x)]$, $\psi_\alpha(x)$ and $F(x)$, we may set the components $C(x)$, $\chi_\alpha(x)$, $M(x)$ and $N(x)$ to zero. This is the so-called Wess–Zumino gauge, which is obviously not consistent with supersymmetry transformations. The gauge invariant degrees of freedom on the other hand are: the gauge field strength $F_{\mu\nu} = \partial_\mu A_\nu - \partial_\nu A_\mu$, the gaugino field $\lambda(x)$, and the auxiliary field $D(x)$. They turn out to be the component fields of the chiral superfield:

$$W_\alpha = -\tfrac{1}{4}\,\bar{D}_{\dot\alpha}\bar{D}^{\dot\alpha}D_\alpha V. \qquad (C.67)$$

It is obvious that W_α is chiral[4]

$$\bar{D}^{\dot\beta}W_\alpha = 0$$

and gauge invariant since:

$$\bar{D}_{\dot\alpha}\bar{D}^{\dot\alpha}D_\alpha(\Lambda - \Lambda^\dagger) = \bar{D}_{\dot\alpha}\bar{D}^{\dot\alpha}D_\alpha\Lambda = -\bar{D}^{\dot\alpha}\{\bar{D}_{\dot\alpha}, D_\alpha\}\Lambda$$
$$= 2i\,\sigma^\mu_{\alpha\dot\alpha}\bar{D}^{\dot\alpha}\partial_\mu\Lambda = 2i\,\sigma^\mu_{\alpha\dot\alpha}\partial_\mu\bar{D}^{\dot\alpha}\Lambda = 0$$

where we used the fact that Λ is chiral and $\Lambda^\dagger$ antichiral. We note that W_α satisfies the condition

$$D^\alpha W_\alpha = \bar{D}_{\dot\alpha}\bar{W}^{\dot\alpha}. \qquad (C.68)$$

In fact, W^α can be defined as the chiral superfield which satisfies this constraint. It is reminiscent of a similar property in nonsupersymmetric gauge theories: $F_{\mu\nu}$ can be defined from the gauge potential as $\partial_\mu A_\nu - \partial_\nu A_\mu$ or as a generic antisymmetric tensor field which satisfies the Bianchi identity $\epsilon^{\mu\nu\rho\sigma}\partial_\nu F_{\rho\sigma} = 0$ (see Exercise 6).

---

[4]Since the covariant derivatives $\bar{D}_{\dot\alpha}$ anticommute between themselves and $\dot\alpha$, $\dot\beta$ can only take two values $\dot{1}$ or $\dot{2}$: $\bar{D}^{\dot\beta}\bar{D}_{\dot\alpha}\bar{D}^{\dot\alpha} = 0$.

One can check that

$$W_\alpha = \lambda_\alpha(y) - \left[\delta_\alpha^\beta D(y) + \frac{i}{2}(\sigma^\mu\bar{\sigma}^\nu)_\alpha{}^\beta F_{\mu\nu}(y)\right]\theta_\beta + i\theta^2\sigma^\mu_{\alpha\dot\alpha}\partial_\mu\bar{\lambda}^{\dot\alpha}(y). \tag{C.69}$$

Note that one may avoid altogether to work in the Wess–Zumino gauge and take (C.69) as defining the gauge invariant degrees of freedom λ_α, $\bar{\lambda}^{\dot\alpha}$, D and $F_{\mu\nu}$.

One infers the supersymmetry transformations from (C.29)

$$\delta_s F_{\mu\nu} = \left[\eta\sigma_\nu\partial_\mu\bar{\lambda} - \bar{\eta}\bar{\sigma}_\nu\partial_\mu\lambda\right] - (\mu \leftrightarrow \nu)$$

$$\delta_s \lambda_\alpha = -\eta_\alpha D - \frac{i}{2}(\sigma^\mu\bar{\sigma}^\nu)_\alpha{}^\beta \eta_\beta F_{\mu\nu} \tag{C.70}$$

$$\delta_s D = i\eta\sigma^\mu\partial_\mu\bar{\lambda} + i\bar{\eta}\bar{\sigma}^\mu\partial_\mu\lambda.$$

A supersymmetric action must involve a term $F^{\mu\nu}F_{\mu\nu}$ and therefore be quadratic in W^α. An obvious candidate is

$$S = \frac{1}{4}\int d^4y\, d^2\theta\; W^\alpha W_\alpha + \text{h.c} \tag{C.71}$$

which reads in terms of component fields after some partial integration

$$S = \int d^4x\left[-\frac{1}{4}F^{\mu\nu}F_{\mu\nu} + i\lambda\sigma^\mu\partial_\mu\bar{\lambda} + \frac{1}{2}D^2\right]. \tag{C.72}$$

Thus one recovers the Lagrangian described in Chapter 3.

One may construct the supersymmetry current associated with this system (as we saw in Section 3.1.5 of Chapter 3, this should not be confused with the Noether current associated with the transformation (C.70)):

$$j_{\mu\alpha} = F^{\rho\sigma}\left(\sigma_{\rho\sigma}\sigma_\mu\right)_{\alpha\dot\beta}\bar{\lambda}^{\dot\beta},$$

$$\bar{j}_\mu^{\dot\alpha} = -F^{\rho\sigma}\left(\bar{\sigma}_{\rho\sigma}\bar{\sigma}_\mu\right)^{\dot\alpha\beta}\lambda_\beta. \tag{C.73}$$

If we want to couple this to matter described by a chiral supermultiplet ϕ, we must implement a gauge transformation at the superfield level. Since Λ in (C.65) is a chiral superfield and its scalar component gives the gauge parameter $(\alpha(x) = -2g\text{Re }a(x))$, an obvious candidate is

$$\Phi \to \Phi' = e^{2igq\Lambda}\,\Phi$$

which preserves the chirality of the superfield.

Then a term such as $\Phi^\dagger\Phi$ is not gauge invariant and must be replaced by $\Phi^\dagger e^{-2gqV}\Phi$ since

$$\Phi^\dagger\, e^{-2gqV}\,\Phi \to \Phi^\dagger\, e^{-2igq\Lambda^\dagger}\, e^{-2gq[V+i(\Lambda-\Lambda^\dagger)]}\, e^{2igq\Lambda}\,\Phi = \Phi^\dagger\, e^{-2gqV}\,\Phi.$$

Hence the action (see (C.39)) is replaced by

$$S = \int d^4x\, d^2\theta\, d^2\bar{\theta}\; \Phi^\dagger\, e^{-2gqV}\,\Phi \tag{C.74}$$

which reads in terms of component fields

$$S = \int d^4x \left[D^\mu \phi^* D_\mu \phi + i\psi\sigma^\mu D_\mu \bar\psi + F^* F \right.$$
$$\left. + gq \left(D\phi^* \phi + \sqrt{2}\,\lambda\psi\phi^* + \sqrt{2}\,\bar\lambda\bar\psi\phi \right) \right] \tag{C.75}$$

where $D_\mu\phi \equiv \partial_\mu\phi - igqA_\mu\phi$ and $D_\mu\psi \equiv \partial_\mu\psi - igqA_\mu\psi$ are the covariant derivatives.

Returning to R-symmetry, we see that V being real does not transform. Thus since $\theta' = e^{-i\alpha}\theta$, $\bar\theta' = e^{i\alpha}\bar\theta$ and $D_\alpha \sim \partial/\partial\theta^\alpha$, $\bar D_{\dot\alpha} \sim \partial/\partial\bar\theta^{\dot\alpha}$, W_α has R-charge $+1$:

$$RW_\alpha(y,\theta) = e^{i\alpha}\,W_\alpha(y, e^{-i\alpha}\,\theta) \tag{C.76}$$

and $W^\alpha W_\alpha$ has R-charge $+2$: the action (C.71) is invariant under R-symmetry.

At the component field level, let us note that the gaugino λ_α being the lowest component has R-charge $+1$ as well:

$$R\,\lambda_\alpha(x) = e^{i\alpha}\,\lambda_\alpha(x). \tag{C.77}$$

The action (C.74) is obviously R-symmetric at the superfield level. One checks that as well, at the component field level, a term such as $\lambda\psi\varphi^*$ has R-charge $1+(r-1)-r = 0$.

For the case of nonabelian gauge symmetry, we refer the reader to the references below (see for example [362]) and only sketch the results. The vector superfield has a matrix structure $V = V^a t^a$, where t^a are the generators of the gauge group. The gauge transformation (C.66) takes the form

$$e^{-V'} = e^{i\Lambda^\dagger} e^{-V} e^{-i\Lambda} \tag{C.78}$$

where, similarly, $\Lambda \equiv \Lambda^a t^a$.

The supersymmetric field strength $\mathcal{W}_\alpha$ is now written[5]

$$\mathcal{W}_\alpha = -\frac{1}{4}\bar D_{\dot\alpha}\bar D^{\dot\alpha} e^{gV} D_\alpha e^{-gV}. \tag{C.79}$$

It transforms as

$$\mathcal{W}'_\alpha = e^{ig\Lambda}\mathcal{W}_\alpha e^{-ig\Lambda}. \tag{C.80}$$

We have

$$\frac{1}{4g^2}\int d^4x\, d^2\theta\, \mathcal{W}^{a\alpha}\mathcal{W}^a_\alpha = \int d^4x \left[-\frac{1}{8}F^{a\mu\nu}F^a_{\mu\nu} - \frac{i}{8}F^{a\mu\nu}\tilde F^a_{\mu\nu} + \frac{i}{2}\lambda^a\sigma^\mu D_\mu\bar\lambda^a + \frac{1}{4}D^a D^a \right] \tag{C.81}$$

where

$$\tilde F^a_{\mu\nu} = \tfrac{1}{2}\epsilon_{\mu\nu\rho\sigma}F^{a\rho\sigma}, \tag{C.82}$$

---

[5]Note that in the limit of abelian symmetry, $\mathcal{W}_\alpha = -gW_\alpha$. We have restored the gauge coupling in the nonabelian case because of the self-couplings of the gauge sector.

and[6]

$$D_\mu \lambda^a = \partial_\mu \lambda^a + gC^{abc} A_\mu^b \lambda^c, \tag{C.83}$$

g being the gauge coupling.

Finally, the coupling of the chiral field to the vector field is given as in (C.74) by

$$\int d^4x d^4\theta \, \Phi^\dagger e^{-2gV} \Phi = \int d^4x \left[D^\mu \phi^{i*} D_\mu \phi_i - i\psi_i \sigma^\mu D_\mu \bar\psi^i + F^{*i} F_i \right. \tag{C.84}$$

$$\left. + g \left(\phi^{i*} D^a t_i^{aj} \phi_j + \sqrt{2} \phi^{i*} \lambda^a t_i^{aj} \psi_j + \sqrt{2} \bar\lambda^a \bar\psi^i t_i^{aj} \phi_j \right) \right],$$

where

$$\begin{aligned} D_\mu \phi_i &= \partial_\mu \phi_i - ig A_\mu^a t_i^{aj} \phi_j \\ D_\mu \psi_i &= \partial_\mu \psi_i - ig A_\mu^a t_i^{aj} \psi_j. \end{aligned} \tag{C.85}$$

C.4   The linear superfield

We conclude this appendix by introducing another example of a superfield, which describes the supermultiplet associated with an antisymmetric tensor field $b_{\mu\nu}$ (known as a Kalb-Ramond field [244]). Such a field is present in string theory where it plays a central rôle in cancelling anomalies (see Section 10.4.2 of Chapter 10).

The linear superfield is a real superfield L which satisfies the covariant constraints:

$$D^2 L = 0, \quad \bar{D}^2 L = 0. \tag{C.86}$$

Its nonvanishing components are therefore:

$$\begin{aligned} L|_0 &= \ell(x), \\ D_\alpha L|_0 &= \Lambda_\alpha(x) \ , \quad \bar{D}^{\dot\alpha} L\big|_0 = \bar\Lambda^{\dot\alpha}, \\ [D_\alpha, \bar{D}_{\dot\alpha}] L\big|_0 &= \tfrac{1}{3} \sigma_{\lambda\alpha\dot\alpha} \epsilon^{\lambda\mu\nu\rho} h_{\mu\nu\rho}. \end{aligned} \tag{C.87}$$

In the last equation, $h_{\mu\nu\rho}$ is the field strength of the antisymmetric tensor

$$h_{\mu\nu\rho} = \partial_\mu b_{\nu\rho} + \partial_\nu b_{\rho\mu} + \partial_\rho b_{\mu\nu}, \tag{C.88}$$

invariant under the gauge transformation $\delta b_{\mu\nu} = \partial_\mu \Lambda_\nu - \partial_\nu \Lambda_\mu$.

The action simply reads

$$\begin{aligned} \mathcal{L} &= -\int d^4\theta L^2 = -\frac{1}{32} \left(D^2 \bar{D}^2 + \bar{D}^2 D^2 \right) L^2 \big|_0 \\ &= \frac{1}{12} h^{\mu\nu\rho} h_{\mu\nu\rho} + \frac{1}{2} \partial^\mu \ell \partial_\mu \ell - \frac{i}{2} \Lambda^\alpha \sigma^\mu_{\alpha\dot\alpha} \partial_\mu \bar\Lambda^{\dot\alpha} + \frac{i}{2} \partial_\mu \Lambda^\alpha \sigma^\mu_{\alpha\dot\alpha} \bar\Lambda^{\dot\alpha}. \end{aligned} \tag{C.89}$$

We note that none of the fields of the linear multiplet are auxiliary fields.

---

[6] The C^{abc} are the structure functions: $[t^a, t^b] = iC^{abc} t^c$.

We show in equation (10.120) of Chapter 10 that an antisymmetric tensor field is Hodge dual to a pseudoscalar field. This equivalence extends to the full supermultiplets and a linear superfield is dual to a chiral supermultiplet. Indeed, consider the following Lagrangian:

$$\mathcal{L} = -\int d^4\theta \left[X^2 + \sqrt{2}X(S + S^\dagger) \right],$$
(C.90)

where X is a real superfield, S is chiral (and thus $S^\dagger$ antichiral).

We may vary the action with respect to S, or better since S is a constrained (chiral) superfield, we may write it as $S = \bar{D}^2\Sigma$ with Σ unconstrained[7] and minimize with respect to Σ. Similarly for $S^\dagger$. This gives $D^2 X = \bar{D}^2 X = 0$ which shows that X is a linear superfield L. Thus the action (C.90) coincides with (C.89).

Alternatively, one may minimize with respect to X, which gives the superfield equation of motion

$$X = -\frac{1}{\sqrt{2}}(S + S^\dagger).$$
(C.91)

Then (C.90) is equivalent to the action

$$\mathcal{L} = \int d^4\theta S^\dagger S.$$
(C.92)

Thus (C.89) and (C.92) describe two theories equivalent on-shell (we have used equations of motion).

Further reading

- J. Wess and J. Bagger, *Supersymmetry and Supergravity*, Princeton series in physics.
- J.-P. Derendinger, *Lecture notes on globally supersymmetric theories in four and two dimensions*, preprint ETH-YH/90-21.

Exercises

Exercise 1 Show that

$$\varepsilon^{\alpha\beta}\frac{\partial}{\partial\theta^\beta} = -\frac{\partial}{\partial\theta_\alpha}, \qquad \varepsilon^{\alpha\beta}\frac{\partial}{\partial\theta^\alpha}\frac{\partial}{\partial\theta^\beta}\theta\theta = 4, \qquad \varepsilon_{\dot\alpha\dot\beta}\frac{\partial}{\partial\bar\theta_{\dot\alpha}}\frac{\partial}{\partial\bar\theta_{\dot\beta}}\bar\theta\bar\theta = 4.$$

Exercise 2 Derive from (C.29) and (C.39) the supersymmetry transformations and Lagrangian for a chiral superfield in four-component notation, *i.e.* (3.10) and (3.19) in Chapter 3: define $\phi = (A + iB)/\sqrt{2}$, $F = (F_1 + iF_2)/\sqrt{2}$, $\Psi = \begin{pmatrix} \psi_\alpha \\ \bar\psi^{\dot\alpha} \end{pmatrix}$ and $\epsilon = \begin{pmatrix} \eta_\alpha \\ \bar\eta^{\dot\alpha} \end{pmatrix}$.

---

[7]One may show that any chiral superfield may be written in this way. The unconstrained superfield Σ is known as a prepotential.

Exercise 3 Show that, in the Wess–Zumino gauge

$$V(y, \theta, \bar\theta) = \theta\sigma\bar\theta A_\mu(y) + \theta^2\bar\theta\bar\lambda(y) + \bar\theta^2\theta\lambda(y) - \tfrac{1}{2}\theta^2\bar\theta^2\left(D(y) - i\partial^\mu A_\mu(y)\right)$$

and prove (C.69).

Hints: Use $(\sigma_\mu\bar\sigma_\nu + \sigma_\nu\bar\sigma_\mu)_\alpha{}^\beta = 2\eta_{\mu\nu}\delta_\alpha^\beta$.

Exercise 4

(a) Using (C.69), show that (C.71) is written in terms of the component fields as (C.72).
(b) Prove similarly (C.75).
(c) Show that the four-component spinor versions of (C.72), (C.70), and (C.75) are given by the equations (3.38), (3.40), and (3.43) of Chapter 3.

Hints:

(a) Use $\mathrm{Tr}(\sigma^{\mu\nu}\sigma^{\rho\sigma}) = -\tfrac{1}{2}\left(\eta^{\mu\rho}\eta^{\nu\sigma} - \eta^{\mu\sigma}\eta^{\nu\rho} + i\epsilon^{\mu\nu\rho\sigma}\right)$ where $(\sigma^{\mu\nu})_\alpha{}^\beta = \tfrac{1}{4}(\sigma^\mu\bar\sigma^\nu - \sigma^\nu\bar\sigma^\mu)_\alpha{}^\beta$ as defined in (B.24) of Appendix B.
(b) Compute $\Phi^\dagger V\Phi\big|_{\theta^2\bar\theta^2}$ and $\Phi^\dagger V^2\Phi\big|_{\theta^2\bar\theta^2}$ in the Wess–Zumino gauge. Obviously $\Phi^\dagger V^n\Phi$ vanishes for $n \geq 3$.

Exercise 5 Prove, using (C.19), that the components of the full vector supermultiplet transform under supersymmetry as:

$$\begin{aligned}
\delta_s C &= i(\eta\chi - \bar\eta\bar\chi)\\
\delta_s \chi_\alpha &= \eta_\alpha(M + iN) - i\left(\sigma^\mu\bar\eta\right)_\alpha(A_\mu - i\partial_\mu C),\\
\delta_s A_\mu &= (\lambda\sigma_\mu\bar\eta + \eta\sigma_\mu\bar\lambda) + \eta\partial_\mu\chi + \bar\eta\partial_\mu\bar\chi,\\
\delta_s M &= i(\eta\lambda - \bar\eta\bar\lambda) - i(\eta\sigma^\mu\partial_\mu\bar\chi - \partial_\mu\chi\sigma^\mu\bar\eta),\\
\delta_s N &= -(\eta\lambda + \bar\eta\bar\lambda) + (\eta\sigma^\mu\partial_\mu\bar\chi + \partial_\mu\chi\sigma^\mu\bar\eta),\\
\delta_s \lambda_\alpha &= -\eta_\alpha D - \frac{i}{2}(\sigma^\mu\bar\sigma^\nu)_\alpha{}^\beta\eta_\beta F_{\mu\nu}\\
\delta_s D &= i(\eta\sigma^\mu\partial_\mu\bar\lambda + \bar\eta\bar\sigma^\mu\partial_\mu\lambda).
\end{aligned} \tag{C.93}$$

Exercise 6 Express the constraint (C.68) in terms of the component fields using the decomposition (C.69).

Exercise 7

(a) Express $D^2\bar D^2 K(\Phi_i, \Phi_j^\dagger)$ in terms of derivatives of K with respect to Φ_i and/or $\Phi_j^\dagger$ and covariant derivatives of Φ_i or $\Phi_j^\dagger$.
(b) Express $D_\alpha\bar D_{\dot\alpha}\Phi_i^\dagger\big|_0$, $D^\alpha\bar D^2\Phi_i^\dagger\big|_0$, and $D^2\bar D^2\Phi_i^\dagger\big|_0$ in terms of the component fields.
(c) Use (C.28) and the previous results to prove (C.50).

Hint: (b)

$$D_\alpha\bar D_{\dot\alpha}\Phi_i^\dagger\big|_0 = -2i\sigma^\mu_{\alpha\dot\alpha}\partial_\mu\phi_i^*, \quad D^\alpha\bar D^2\Phi_i^\dagger\big|_0 = 4i\sqrt{2}\left(\partial_\mu\bar\psi_i\bar\sigma^\mu\right)^\alpha, \quad D^2\bar D^2\Phi_i^\dagger\big|_0 = -16\Box\phi_i^*.$$

Exercise 8 *We discuss in this exercise the reparametrization invariance of the Kähler manifold associated with a supersymmetric nonlinear sigma model.*

Let us redefine the scalar fields according to $\phi = \phi(\phi')$, $\bar\phi = \bar\phi(\bar\phi')$. The kinetic term (C.54) is invariant if the Kähler metric transforms as:

$$g'_{m\bar n} = g_{i\bar j}\frac{\partial\phi^i}{\partial\phi'^m}\frac{\partial\bar\phi^{\bar j}}{\partial\bar\phi'^{\bar n}}.$$

(C.94)

(a) How does the inverse metric $g^{i\bar j}$ transform? Show that the Christoffel symbol (C.56) transforms as:

$$\Gamma'^i{}_{jk} = \frac{\partial\phi'^i}{\partial\phi^l}\left\{\Gamma^l{}_{mn}\frac{\partial\phi^m}{\partial\phi'^j}\frac{\partial\phi^n}{\partial\phi'^k} + \frac{\partial^2\phi^l}{\partial\phi'^j\partial\phi'^k}\right\}.$$

(C.95)

(b) Show that in order that the kinetic term for the fermion fields in (C.58) be invariant, one must impose

$$\psi^i = \psi'^j\frac{\partial\phi^i}{\partial\phi'^j}, \quad \bar\psi^{\bar i} = \bar\psi'^{\bar j}\frac{\partial\bar\phi^{\bar i}}{\partial\bar\phi'^{\bar j}}$$

(C.96)

i.e. as contravariant vectors, and similarly $\mathcal{D}_\mu\psi^i$, $\mathcal{D}_\mu\bar\psi^{\bar i}$.

(c) Using (a) show explicitly that, if ψ^i transforms as (C.96), then $\mathcal{D}_\mu\psi^i$, as defined in (C.59), transforms as a contravariant vector.

(d) Show that:

$$[\mathcal{D}_\mu, \mathcal{D}_\nu]\,\psi^i_\alpha = -g^{i\bar j}R_{j\bar j k\bar k}\left(\partial_\mu\phi^k\partial_\nu\bar\phi^{\bar k} - \partial_\nu\phi^k\partial_\mu\bar\phi^{\bar k}\right)\psi^j_\alpha.$$

(C.97)

where the Riemann tensor $R_{j\bar j k\bar k}$ is defined in (C.60).

Exercise 9 Consider a chiral superfield S with a Kähler potential $K = -\ln\left(S + S^\dagger\right)$. Show that the transformation:

$$S \to \frac{aS - ib}{icS + d}$$

(C.98)

with $ad - bc = 1$ amounts to a Kähler transformation (C.52). Such a chiral superfield appears in string theory: its scalar component is the string dilaton whose vacuum expectation value provides the gauge coupling g ($\langle S\rangle = 1/g^2$). Thus, the 'S-duality' transformation (C.98) with the choice $a = d = 0$ and $b = -c = 1$ corresponds to a strong/weak coupling duality (see Chapter 10).

Hints: $F(S) = \ln\left(icS + d\right)$.

<u>Exercise 10</u> Show that (C.81) and (C.84) are invariant under:

$$\delta_s \phi_i = \sqrt{2}\, \eta\, \psi_i$$

$$\delta_s \psi_{i\alpha} = \sqrt{2}\, \eta_\alpha F_i - i\sqrt{2}\, \sigma^\mu_{\alpha\dot\alpha}\bar\eta^{\dot\alpha}\, D_\mu \phi_i$$

$$\delta_s F_i = -i\sqrt{2}\, \bar\eta\bar\sigma^\mu D_\mu \psi_i - 2g\bar\eta\bar\lambda^a t_i^{aj}\phi_j$$

$$\delta_s A_\mu^a = \eta\sigma_\mu\bar\lambda^a - \bar\eta\bar\sigma_\mu\lambda^a$$

$$\delta_s \lambda_\alpha^a = -\eta_\alpha D^a - \frac{i}{2}(\sigma^\mu\bar\sigma^\nu)_\alpha{}^\beta \eta_\beta F_{\mu\nu}^a \tag{C.99}$$

$$\delta_s D^a = i\eta\sigma^\mu D_\mu\bar\lambda^a + i\bar\eta\bar\sigma^\mu D_\mu\lambda^a.$$

Hints: If one compares with the individual transformations of the chiral and vector supermultiplets (*i.e.* (C.29) and (C.70) with ordinary derivatives replaced by covariant derivatives), one identifies a single extra term (the second term in $\delta_s F_i$). To check its form, cancel all terms proportional to F^{*i} in the supersymmetry transformation of the Lagrangian. They originate from only two terms: $F^{*i}\delta_s F_i + g\sqrt{2}\bar\lambda^a\delta_s\bar\psi_i\, t_i^{aj}\phi_j$.

<u>Problem 1</u> *One of the problems of supersymmetry is the absence of conventions to which all authors would adhere. In this exercise, we will reformulate some of the expressions in the appendix to make explicit the conventions chosen here and in other textbooks. You may choose to prove part or all of the following equations, or just use them when you try to compare with other textbooks.*

The first convention is one of metric signature. We introduce here a sign ϵ which is $+1$ for a metric signature $(+,-,-,-)$ (our convention) and -1 for a metric signature $(-,+,+,+)$. In the computations necessary for what follows, the main consequence of the choice of metric signature is the formula:

$$\{\gamma^\mu, \gamma^\nu\} = 2\epsilon\eta^{\mu\nu} \tag{C.100}$$

which may be written

$$(\sigma^\mu\bar\sigma^\nu + \sigma^\nu\bar\sigma^\mu)_\alpha{}^\beta = 2\epsilon\,\eta^{\mu\nu}\delta_\alpha^\beta,$$

$$(\bar\sigma^\mu\sigma^\nu + \bar\sigma^\nu\sigma^\mu)^{\dot\alpha}{}_{\dot\beta} = 2\epsilon\,\eta^{\mu\nu}\delta^{\dot\alpha}_{\dot\beta}. \tag{C.101}$$

Thus, one has

$$\mathrm{Tr}\,(\sigma^\mu\bar\sigma^\nu) = 2\epsilon\eta^{\mu\nu}, \tag{C.102}$$

and, using (B.61) of Appendix B,

$$(\theta\sigma^\mu\bar\theta)(\theta\sigma^\nu\bar\theta) = \tfrac{1}{2}\epsilon\theta^2\,\bar\theta^2. \tag{C.103}$$

We start by constructing supersymmetry charges which satisfy:

$$\{Q_\alpha, \bar Q_{\dot\alpha}\} = 2\epsilon_Q\, \sigma^\mu_{\alpha\dot\alpha}\, i\partial_\mu \tag{C.104}$$

$$\{Q_\alpha, Q_\beta\} = 0 = \{\bar Q_{\dot\alpha}, \bar Q_{\dot\beta}\} \tag{C.105}$$

where ϵ_Q is ± 1.

The translation generator has the general form:

$$P_\mu = \frac{i}{\gamma_P} \partial_\mu, \tag{C.106}$$

with γ_P a normalization factor[8].

Following the method of Section C.1, we define the superspace translation by:

$$S\left(y^\mu, \eta, \bar\eta\right) = \exp i \left(\gamma_P \, y^\mu P_\mu + \gamma_Q \, \eta Q + \gamma_{\bar Q} \, \bar\eta \bar Q\right). \tag{C.107}$$

Then, one obtains, besides (C.106),

$$Q_\alpha = \frac{i}{\gamma_Q}\left[\frac{\partial}{\partial\theta^\alpha} + i\rho_Q \, \sigma^\mu_{\alpha\dot\alpha} \, \bar\theta^{\dot\alpha} \, \partial_\mu\right]$$

$$\bar Q_{\dot\alpha} = -\frac{i}{\gamma_{\bar Q}}\left[\frac{\partial}{\partial\bar\theta^{\dot\alpha}} + i\rho_Q \, \theta^\alpha \sigma^\mu_{\alpha\dot\alpha} \, \partial_\mu\right] \tag{C.108}$$

where we have defined

$$\rho_Q \equiv \epsilon_Q \gamma_Q \gamma_{\bar Q}, \tag{C.109}$$

which we will take to be real.

The supersymmetry transformation on a superfield $\mathcal{F}(x,\theta,\bar\theta)$ reads

$$\delta_S \, \mathcal{F}(x,\theta,\bar\theta) = \left[-i\gamma_Q \eta^\alpha Q_\alpha - i\gamma_{\bar Q}\bar\eta_{\dot\alpha}\bar Q^{\dot\alpha}\right]\mathcal{F}(x,\theta,\bar\theta). \tag{C.110}$$

The next step is to find covariant derivatives which satisfy:

$$\{D_\alpha, \bar D_{\dot\alpha}\} = 2\epsilon_D \, \sigma^\mu_{\alpha\dot\alpha} \, i\partial_\mu \qquad \{D_\alpha, D_\beta\} = 0 = \{\bar D_{\dot\alpha}, \bar D_{\dot\beta}\}$$

$$\{D_\alpha, Q_\beta\} = \{D_\alpha, \bar Q_{\dot\beta}\} = \{\bar D_{\dot\alpha}, Q_\beta\} = \{\bar D_{\dot\alpha}, \bar Q_{\dot\beta}\} = 0, \tag{C.111}$$

with $\epsilon_D = \pm 1$. One finds:

$$D_\alpha = \frac{\lambda_D}{\gamma_Q}\left(\frac{\partial}{\partial\theta^\alpha} - i\rho_Q \, \sigma^\mu_{\alpha\dot\alpha} \, \bar\theta^{\dot\alpha} \, \partial_\mu\right),$$

$$\bar D_{\dot\alpha} = -\frac{\lambda_{\bar D}}{\gamma_{\bar Q}}\left(\frac{\partial}{\partial\bar\theta^{\dot\alpha}} - i\rho_Q \theta^\alpha \sigma^\mu_{\alpha\dot\alpha} \, \partial_\mu\right), \tag{C.112}$$

where λ_D and $\lambda_{\bar D}$ are two normalization constants which satisfy:

$$\lambda_D \lambda_{\bar D} = \frac{\epsilon_D}{\epsilon_Q}. \tag{C.113}$$

We now turn to the chiral superfield. Defining the variable

$$y^\mu = x^\mu - i\rho_Q \, \theta\sigma^\mu\bar\theta, \tag{C.114}$$

---

[8]Note that, depending on the respective values of ϵ_Q and γ_P, one sometimes does not recover from (C.105) the standard anticommutation relations (C.7), *i.e.* if $\epsilon_Q \gamma_P \neq 1$. This is indeed the case of the most common convention, the one of Wess and Bagger [362].

we have

$$Q_\alpha = \frac{i}{\gamma_Q} \frac{\partial}{\partial\theta^\alpha},$$

$$\bar{Q}^{\dot\alpha} = \frac{i}{\gamma_Q} \left[\frac{\partial}{\partial\bar\theta_{\dot\alpha}} + 2i\rho_Q \, (\bar\sigma^\mu\theta)^{\dot\alpha} \, \frac{\partial}{\partial y^\mu} \right], \tag{C.115}$$

$$D_\alpha = \frac{\lambda_D}{\gamma_Q} \left[\frac{\partial}{\partial\theta^\alpha} - 2i\rho_Q \, (\sigma^\mu\bar\theta)_\alpha \, \frac{\partial}{\partial y^\mu} \right],$$

$$\bar{D}^{\dot\alpha} = \frac{\lambda_{\bar D}}{\gamma_{\bar Q}} \frac{\partial}{\partial\bar\theta_{\dot\alpha}}. \tag{C.116}$$

Thus, we can write a chiral superfield Φ which satisfies the condition $\bar{D}^{\dot\alpha}\Phi = 0$ as

$$\Phi = \phi(y) + \alpha_\psi \sqrt{2} \, \theta\psi(y) + \alpha_F \, \theta^2 \, F(y), \tag{C.117}$$

where α_ψ and α_F are real normalization constants.

The supersymmetry transformations (C.110) read in terms of component fields:

$$\delta_S\phi = \alpha_\psi \sqrt{2} \, \eta \, \psi$$

$$\alpha_\psi \delta_S\psi_\alpha = \sqrt{2} \, \alpha_F \, F\eta_\alpha - i\sqrt{2} \, \rho_Q \, (\sigma^\mu\bar\eta)_\alpha \, \partial_\mu\phi$$

$$\alpha_\psi \delta_S\bar\psi^{\dot\alpha} = \sqrt{2} \, \alpha_F \, F^*\bar\eta^{\dot\alpha} - i\sqrt{2} \, \rho_Q \, (\bar\sigma^\mu\eta)^{\dot\alpha} \, \partial_\mu\phi^*$$

$$\alpha_F \delta_S F = i\sqrt{2}\rho_Q\alpha_\psi \, \partial_\mu\psi\sigma^\mu\bar\eta = -i\sqrt{2}\rho_Q\alpha_\psi \, \bar\eta\bar\sigma^\mu\partial_\mu\psi \tag{C.118}$$

or in four-component notation (see Exercise 2 for notation and Appendix B for *our* conventions on gamma matrices):

$$\delta_S A = \alpha_\psi \, \bar\epsilon\Psi, \qquad \delta_S B = \alpha_\psi \, \bar\epsilon i\gamma_5\Psi,$$

$$\alpha_\psi \delta_S\Psi = \alpha_F \, (F_1 - i\gamma_5 F_2)\, \epsilon - i\rho_Q \, \gamma^\mu\partial_\mu \, (A + i\gamma_5 B) \, \epsilon, \tag{C.119}$$

$$\alpha_F \delta_S F_1 = -\rho_Q\alpha_\psi \, \bar\epsilon\gamma^\mu i\partial_\mu\psi, \quad \alpha_F \delta_S F_2 = -\rho_Q\alpha_\psi \, \bar\epsilon\gamma_5\gamma^\mu\partial_\mu\Psi.$$

As for the Lagrangian, we have

$$\Phi^\dagger(x)\Phi(x)\big|_{\theta^2\bar\theta^2} = \alpha_F^2 F^*(x)F(x) + \epsilon\rho_Q^2 \left[\tfrac{1}{2} \, \partial^\mu\phi^*(x)\partial_\mu\phi(x) - \tfrac{1}{4} \, \phi(x)\Box\phi^*(x) \right.$$

$$\left. - \tfrac{1}{4} \, \phi^*(x)\Box\phi(x) \right] + \frac{i}{2}\rho_Q\alpha_\psi^2 \left(\psi\sigma^\mu\partial_\mu\bar\psi - \partial_\mu\psi\sigma^\mu\bar\psi \right),$$

and

$$W\,(\Phi))\big|_{\theta^2} = \alpha_F \, F \, \frac{\partial W}{\partial\Phi} \, (\phi(y)) - \alpha_\psi^2 \, \frac{1}{2} \frac{\partial^2 W}{\partial\Phi^2} \, (\phi(y)) \, \psi^\alpha(y)\psi_\alpha(y). \tag{C.120}$$

Thus the action for the scalar field reads (up to total derivatives which do not contribute)

$$
\begin{aligned}
S &= \int d^4x\, d^2\theta\, d^2\bar\theta\; \Phi^\dagger(x,\theta,\bar\theta)\Phi(x,\theta,\bar\theta) + \left[\int d^4y\, d^2\theta\; W\left(\Phi(y,\theta)\right) + \text{h.c.}\right] \\
&= \int d^4x \left\{ \epsilon\rho_Q^2\, \partial_\mu\phi\partial^\mu\phi^* - \left|\frac{dW}{d\phi}\right|^2 \right. \\
&\quad \left. + \frac{1}{2}\alpha_\psi^2 \left[i\rho_Q\left(\psi\sigma^\mu\partial_\mu\bar\psi + \bar\psi\bar\sigma^\mu\partial_\mu\psi\right) - \left(\frac{d^2W}{d\phi^2}\psi^2 + \frac{d^2W^*}{d\phi^{*2}}\bar\psi^2\right)\right]\right\},
\end{aligned}
\tag{C.121}
$$

where we have solved for the auxiliary field:

$$
F = -\frac{1}{\alpha_F}\left(\frac{dW}{d\phi}\right)^*.
\tag{C.122}
$$

One may write the preceding action using four-component spinors (see Exercise 2 for notation). For example, in the case where

$$
W(\Phi) = \tfrac{1}{2}m\Phi^2 + \tfrac{1}{3}\lambda\Phi^3,
\tag{C.123}
$$

we have[9]

$$
\begin{aligned}
S = \int d^4x \left\{ \epsilon\rho_Q^2\, \partial_\mu\phi\partial^\mu\phi^* - \left|m\phi + \lambda\phi^2\right|^2 \right. \\
\left. + \alpha_\psi^2 \left[\frac{1}{2}i\rho_Q\,\bar\Psi\gamma^\mu\partial_\mu\Psi - \frac{1}{2}m\bar\Psi\Psi - \lambda\bar\Psi_R\phi\Psi_L - \lambda\bar\Psi_L\phi^*\Psi_R\right]\right\}.
\end{aligned}
\tag{C.124}
$$

This can be shown directly to be invariant under the supersymmetry transformations (C.119).

Finally, we write the abelian vector superfield decomposition as:

$$
\begin{aligned}
V &= C + i\theta\chi - i\bar\theta\bar\chi + \frac{i}{2}\theta^2\left[M + iN\right] - \frac{i}{2}\bar\theta^2\left[M - iN\right] \\
&\quad + \alpha_A\rho_Q\,\theta\sigma^\mu\bar\theta A_\mu - i\rho_Q\theta^2\bar\theta_{\dot\alpha}\left[\alpha_\lambda^*\bar\lambda^{\dot\alpha} + \frac{i}{2}\left(\bar\sigma^\mu\partial_\mu\chi\right)^{\dot\alpha}\right] \\
&\quad + i\rho_Q\bar\theta^2\theta^\alpha\left[\alpha_\lambda\lambda_\alpha + \frac{i}{2}\left(\sigma^\mu\partial_\mu\bar\chi\right)_\alpha\right] - \frac{1}{2}\epsilon\rho_Q^2\theta^2\bar\theta^2\left[D + \frac{1}{2}\Box C\right].
\end{aligned}
\tag{C.125}
$$

The corresponding gauge invariant superfield reads

$$
\begin{aligned}
W_\alpha = \epsilon_D\frac{\lambda_{\bar D}}{\gamma_{\bar Q}}\left[i\alpha_\lambda\lambda_\alpha - \rho_Q\left(\epsilon\,\delta_\alpha^\beta D + i\alpha_A(\sigma^{\mu\nu})_\alpha{}^\beta F_{\mu\nu}\right)\theta_\beta \right. \\
\left. + \rho_Q\alpha_\lambda^*\theta^2\sigma^\mu_{\alpha\dot\alpha}\partial_\mu\bar\lambda^{\dot\alpha}\right],
\end{aligned}
\tag{C.126}
$$

[9]Note that, for $\rho_Q = -1$ (as in Wess and Bagger [362]), we obtain an unusual form for the Dirac equation: $i\gamma_\mu\partial^\mu\Psi = -m\Psi$. It is to recover the standard Dirac equation that we have chosen $\rho_Q = +1$, and thus real γ_Q, $\gamma_{\bar Q}$ (see (C.109)).

whereas the complete supersymmetry transformations are

$$\delta_s C = i\,(\eta\chi - \bar{\eta}\bar{\chi})\,,$$

$$\delta_s \chi_\alpha = \eta_\alpha(M + iN) - i\rho_Q\,(\sigma^\mu\bar{\eta})_\alpha\,(\alpha_A A_\mu - i\partial_\mu C)\,,$$

$$\alpha_A\delta_s A_\mu = i\epsilon\,(\alpha_\lambda\lambda\sigma^\mu\bar{\eta} - \alpha_\lambda^*\eta\sigma^\mu\bar{\lambda}) + \eta\partial_\mu\chi + \bar{\eta}\partial_\mu\bar{\chi}\,,$$

$$\delta_s M = -\rho_Q\left[\alpha_\lambda\eta\lambda + \alpha_\lambda^*\bar{\eta}\bar{\lambda} + i\,(\eta\sigma^\mu\partial_\mu\bar{\chi} - \partial_\mu\chi\sigma^\mu\bar{\eta})\right]\,, \qquad \text{(C.127)}$$

$$\delta_s N = -i\rho_Q\left[\alpha_\lambda\eta\lambda - \alpha_\lambda^*\bar{\eta}\bar{\lambda} + i\,(\eta\sigma^\mu\partial_\mu\bar{\chi} + \partial_\mu\chi\sigma^\mu\bar{\eta})\right]\,,$$

$$\alpha_\lambda\delta_s\lambda_\alpha = i\epsilon\rho_Q\eta_\alpha D - \rho_Q\alpha_A\,(\sigma^{\mu\nu}\eta)_\alpha\,F_{\mu\nu}\,,$$

$$\delta_s D = \epsilon\,(\alpha_\lambda^*\eta\sigma^\mu\partial_\mu\bar{\lambda} - \alpha_\lambda\bar{\eta}\bar{\sigma}^\mu\partial_\mu\lambda)\,.$$

The conventions followed by various authors are summarized in the following table.

Authors:	HERE	Wess-Bagger [362]	Sohnius [342]	Derendinger [101]	YOUR CONVENTIONS
ϵ	$+1$	-1	$+1$	$+1$	
ϵ_5	$+1$	i	i		
ϵ_Q	$+1$	$+1$	$+1$	$+1$	
γ_P	$+1$	-1	$+1$	-1	
γ_Q	$+1$	i	$+1$	-1	
$\gamma_{\bar{Q}}$	$+1$	i	$+1$	-1	
ρ_Q	$+1$	-1	$+1$	$+1$	
ϵ_D	-1	-1	$+1$	-1	
λ_D	$+1$	i	$+1$	$+1$	
$\lambda_{\bar{D}}$	-1	i	$+1$	-1	
α_ψ	$+1$	$+1$	$\sqrt{2}$	$+1$	
α_F	$+1$	$+1$	-1	-1	
α_A	$+1$	$+1$	-1		
α_λ	$-i$	$+1$	$+1$		

Appendix D
An introduction to cosmology

Since the metric signature often chosen in general relativity is different from the one adopted in this book, we introduce in this appendix a sign ϵ which is $+1$ for a metric signature $(+,-,-,-)$ (our convention) and -1 for a metric signature $(-,+,+,+)$.[1]

D.1 Elements of general relativity

In general relativity, matter curves spacetime and the central dynamical object is the metric $g_{\mu\nu}(x)$ which depends on the spacetime coordinates $x^0, x^1, \ldots, x^{D-1}$ (to be general, we consider in this section D-dimensional spacetimes). Physics should not depend on the way we define the spacetime coordinates; in other words, it should be invariant under general coordinate transformations $x^\mu \to x'^\mu(x)$. Since the elementary line element

$$ds^2 = g_{\mu\nu}dx^\mu dx^\nu \tag{D.1}$$

should be invariant, the infinitesimal transformation

$$dx'^\mu = \frac{\partial x'^\mu}{\partial x^\nu}dx^\nu \tag{D.2}$$

implies the following transformation law for the metric tensor $g_{\mu\nu}$ and its inverse $g^{\mu\nu}$

$$g'_{\mu\nu}(x') = \frac{\partial x^\rho}{\partial x'^\mu}\frac{\partial x^\sigma}{\partial x'^\nu}g_{\rho\sigma}(x), \quad g'^{\mu\nu}(x') = \frac{\partial x'^\mu}{\partial x^\rho}\frac{\partial x'^\nu}{\partial x^\sigma}g^{\rho\sigma}(x), \tag{D.3}$$

which are the respective transformation laws of a covariant and a contravariant tensor.

One may define similarly the transformation laws of a contravariant vector V^ν or covariant vector V_ν:

$$V'^\nu = \frac{\partial x'^\nu}{\partial x^\rho}V^\rho, \quad V'_\nu = \frac{\partial x^\rho}{\partial x'^\nu}V_\rho, \tag{D.4}$$

However, the spacetime derivatives of vectors do not transform covariantly. For example,

$$\frac{\partial V'^\nu}{\partial x'^\mu} = \frac{\partial x'^\nu}{\partial x^\rho}\frac{\partial x^\sigma}{\partial x'^\mu}\frac{\partial V^\rho}{\partial x^\sigma} + \frac{\partial^2 x'^\nu}{\partial x^\rho \partial x^\sigma}\frac{\partial x^\sigma}{\partial x'^\mu}V^\rho.$$

---

[1] In the notation of Misner, Thorne and Wheeler [287], our conventions are as follows: $-\epsilon$ for g sign, $+1$ for Riemann sign, and $+1$ for Einstein sign.

The situation is reminiscent of the one encountered in gauge theories. An ordinary derivative $\partial_\mu \Psi$ of a gauge nonsinglet field Ψ does not transform covariantly under a gauge transformation. One has to introduce the covariant derivative $D_\mu \Psi = \partial_\mu \Psi - ig A_\mu^a t^a \Psi$ which has the same gauge transformation as Ψ (see Section A.1.4 of Appendix Appendix A).

Similarly here one defines the covariant derivatives

$$D_\mu V^\nu = \partial_\mu V^\nu + \Gamma^\nu_{\ \mu\lambda} V^\lambda, \quad D_\mu V_\nu = \partial_\mu V_\nu - \Gamma^\rho_{\ \mu\nu} V_\rho, \tag{D.5}$$

where $\Gamma^\rho_{\ \mu\nu}$, the analog of the gauge field, is called a Christoffel symbol or spin connection (it does not transform as a tensor under general coordinate transformations) and is defined in terms of the metric as:

$$\Gamma^\rho_{\ \mu\nu} = \tfrac{1}{2} g^{\rho\sigma} \left[\partial_\mu g_{\nu\sigma} + \partial_\nu g_{\mu\sigma} - \partial_\sigma g_{\mu\nu} \right]. \tag{D.6}$$

One can then check that $D_\mu V^\nu$ transforms as a (mixed) tensor:

$$(D_\mu V^\nu)' = \frac{\partial x^\rho}{\partial x'^\mu} \frac{\partial x'^\nu}{\partial x^\sigma} D_\rho V^\sigma. \tag{D.7}$$

In the same way that one defines the field strength by differentiating the gauge field ($F_{\mu\nu} = \partial_\mu A_\nu - \partial_\nu A_\mu - ig\left[A_\mu, A_\nu\right]$), one introduces the Riemann curvature tensor:

$$R^\mu_{\ \nu\alpha\beta} = \partial_\alpha \Gamma^\mu_{\ \nu\beta} - \partial_\beta \Gamma^\mu_{\ \nu\alpha} + \Gamma^\mu_{\ \alpha\sigma} \Gamma^\sigma_{\ \nu\beta} - \Gamma^\mu_{\ \beta\sigma} \Gamma^\sigma_{\ \nu\alpha}. \tag{D.8}$$

By contracting indices, one also defines the Ricci tensor $R_{\mu\nu}$ and the curvature scalar R

$$R_{\mu\nu} \equiv R^\alpha_{\ \mu\alpha\nu}, \quad R \equiv g^{\mu\nu} R_{\mu\nu}. \tag{D.9}$$

It is easy to show that

$$\left[D_\mu, D_\nu\right] V^\rho = R^\rho_{\ \sigma\mu\nu} V^\sigma, \tag{D.10}$$

which is reminiscent of a similar equation in gauge theories (see (A.51) of Appendix Appendix A).

An important notion connected with covariant differentiation is the notion of parallel transport. A vector is said to be parallel-transported along a curve parametrized by τ if its covariant derivative projected along the curve is zero:

$$\frac{DV^\mu}{D\tau} \equiv D_\lambda V^\mu \frac{dx^\lambda}{d\tau} = 0. \tag{D.11}$$

For example, the velocity $U^\mu = dx^\mu/d\tau$ (τ proper time: $d\tau^2 = \epsilon g_{\mu\nu} dx^\mu dx^\nu$) of a freely falling object is parallel-transported along the free fall trajectory (this generalizes the notion of zero acceleration in flat space). Such a path is called a geodesic. In other words, a geodesic is a curve that parallel transports its tangent vector along itself. On a sphere the geodesics are the great circles.

Let us now restrict to a $2n$-dimensional compact manifold K with metric g_{ij} and consider a small loop γ in the (x^i, x^j) plane at a point P. Take a vector V^k at P and parallel transport it along γ. If the space is flat, when one reaches P again the vector recovers its previous value. This is not true in curved space and V^k is changed by the amount (*cf.* (D.10)):

$$\delta V^k \propto R^k{}_{lij} V^l. \tag{D.12}$$

Let us restrict our attention to vectors tangent to K and note T_P the tangent space at point P. The holonomy group at point P is the set of transformations h_γ that associate to $V \in T_P$ the tangent vector $V' \in T_P$ obtained by parallel transporting V along the loop γ through P (each loop determines one element of H_P).

Using the equation of parallel transport (D.11) together with (D.5), one can write $(\delta V^k = -dx^i \Gamma^k{}_{ij} V^j)$, using Stokes' theorem,

$$V' = h_\gamma V = \left(\exp - \int_\gamma \Gamma dx \right) V = \left(\exp - \int_S R ds \right) V, \tag{D.13}$$

where S is a surface delimited by the loop γ. It is straightforward to check the group structure of H_P (take two loops γ_1 and γ_2 through P: $h_{\gamma_1} h_{\gamma_2} = h_{\gamma_1 \gamma_2}$). Let us note that (D.13) shows that the spin connection can be interpreted as the gauge field associated with the holonomy group $(\Gamma^k{}_{ij} = (A^i)^k{}_j)$.

Generally, for a complex manifold, $H_P \sim SO(2n)$, independently of P. Take the example of the sphere S_2 and consider first the loops γ which consist of a triangle made of (parts of) three great circles. Since these are geodesics it is easy to parallel transport a tangent vector V. By changing the angles of this triangle, one may generate any rotation in the tangent plane, hence any transformation of $SO(2)$.

The action of general relativity in D dimensions has the form

$$S = \int d^D x \sqrt{|g|} \left[-\frac{1}{2\kappa_D^2} (R - 2\epsilon\lambda) \right] + S_{\text{matter}} (g_{\mu\nu}, \Psi \cdots) \tag{D.14}$$

where $\kappa_D^2 = 8\pi G_{(D)}$ ($\kappa^2 = 8\pi G_N$ in the four-dimensional case), g is the determinant of $g_{\mu\nu}$, Ψ stands for the matter fields and λ is the cosmological constant. Using

$$\frac{2}{\sqrt{|g|}} \frac{\delta S_{\text{matter}}}{\delta g_{\mu\nu}} = T^{\mu\nu}, \tag{D.15}$$

one derives (see Exercise 1) from the Lagrangian (D.14) the equations of motion of the metric components, *i.e.* Einstein's equations:

$$G_{\mu\nu} \equiv R_{\mu\nu} - \tfrac{1}{2} g_{\mu\nu} R = \kappa_D^2 T_{\mu\nu} + \epsilon\lambda g_{\mu\nu}, \tag{D.16}$$

where the tensor $G_{\mu\nu}$ thus defined is called the Einstein tensor. In four dimensions,

$$G_{\mu\nu} = 8\pi G_N T_{\mu\nu} + \epsilon\lambda g_{\mu\nu}. \tag{D.17}$$

D.2 Friedmann–Robertson–Walker Universes

Under the assumption that the Universe is homogeneous and isotropic on scales of order 100 Mpc (1 pc $= 3.262$ light-year $= 3.086 \times 10^{16}$ m)^2, one may try to find an homogeneous and isotropic metric as a solution of Einstein equations. The most general ansatz is, up to coordinate redefinitions, the Robertson–Walker metric:

$$\epsilon \, ds^2 = c^2 dt^2 - a^2(t) \, \gamma_{ij} dx^i dx^j, \tag{D.18}$$

$$\gamma_{ij} dx^i dx^j = \frac{dr^2}{1 - kr^2} + r^2 \left(d\theta^2 + \sin^2 \theta d\phi^2 \right), \tag{D.19}$$

where $a(t)$ is the cosmic scale factor, which is time-dependent in an expanding or contracting Universe. The constant k which appears in the spatial metric γ_{ij} can be chosen to be equal to ± 1 or 0: the value 0 corresponds to flat space, *i.e.* usual Minkowski spacetime; the value $+1$ to closed space ($r^2 < 1$) and the value -1 to open space. Note that r is dimensionless whereas a has the dimension of a length. Physical distances (we will say proper distances) may be measured at time t along rays of constant θ and ϕ as

$$d(t) = \int_0^r \sqrt{|g_{rr}|} dr' = a(t) \int_0^r \frac{dr'}{\sqrt{1 - kr'^2}}. \tag{D.20}$$

The components of the Einstein tensor now read (see Exercise 2):

$$G_{00} = 3 \left(\frac{\dot{a}^2}{a^2} + \frac{k}{a^2} \right), \tag{D.21}$$

$$G_{ij} = -\gamma_{ij} \left(\dot{a}^2 + 2a\ddot{a} + k \right), \tag{D.22}$$

where we use standard notation: $\dot{a}$ is the first time derivative of the cosmic scale factor, $\ddot{a}$ the second time derivative.

For the energy–momentum tensor, we follow our assumption of homogeneity and isotropy and assimilate the content of the Universe to a perfect fluid:

$$T_{\mu\nu} = -\epsilon p g_{\mu\nu} + (p + \rho)U_\mu U_\nu, \tag{D.23}$$

where U_μ is the velocity 4-vector ($U^0 = 1, U^i = 0$). It follows from (D.23) that $T_{00} = \rho$ and $T_{ij} = a^2 p \gamma_{ij}$. The pressure p and energy density ρ usually satisfy the equation of state:

$$p = w\rho. \tag{D.24}$$

The constant w takes the value $w \sim 0$ for nonrelativistic matter (negligible pressure) and $w = 1/3$ for relativistic matter (radiation). In all generality, the perfect fluid consists of several components with different values of w.

---

^{2}To keep in mind orders of magnitude, the visible disk of a typical spiral galaxy has radius 10 kpc, a typical halo has radius larger than 50 kpc, and a typical intergalactic distance is 6 Mpc.

One now obtains from the $(0,0)$ and (i,j) components of the Einstein equations (D.17):

$$3\left(\frac{\dot{a}^2}{a^2} + \frac{k}{a^2}\right) = 8\pi G_N \rho + \lambda, \tag{D.25}$$

$$\dot{a}^2 + 2a\ddot{a} + k = -8\pi G_N a^2 p + a^2 \lambda. \tag{D.26}$$

D.2.1 Friedmann equation

The first of the preceding equations can be written as the Friedmann equation, which gives an expression for the Hubble parameter $H \equiv \dot{a}/a$ measuring the rate of the expansion of the Universe:

$$H^2 \equiv \frac{\dot{a}^2}{a^2} = \frac{1}{3}\left(\lambda + 8\pi G_N \rho\right) - \frac{k}{a^2}. \tag{D.27}$$

Note that the cosmological constant appears as a constant contribution to the Hubble parameter.

This equation should be supplemented by the conservation of the energy–momentum tensor which simply yields:

$$\dot{\rho} = -3H(p + \rho). \tag{D.28}$$

Hence a component with equation of state (D.24) has its energy density scaling as $\rho \sim a(t)^{-3(1+w)}$. Thus nonrelativistic matter (often referred to as matter) energy density scales as a^{-3}. In other words, the energy density of matter evolves in such a way that ρa^3 remains constant. Radiation scales as a^{-4} and a component with equation of state $p = -\rho$ ($w = -1$) has constant energy density. The latter case corresponds to a cosmological constant as can be seen from (D.26)–(D.27) where the cosmological constant can be replaced by a component with $\rho_\Lambda = -p_\Lambda = \lambda/(8\pi G_N)$.

The Friedmann equation allows us to define the Hubble constant H_0, i.e. the present value of the Hubble parameter, which sets the scale of our Universe at present time. Because of the troubled history of the measurement of the Hubble constant, it has become customary to express it in units of 100 km s^{-1} Mpc^{-1} which gives its order of magnitude. Present measurements give

$$h_0 \equiv \frac{H_0}{100 \text{ km s}^{-1} \text{ Mpc}^{-1}} = 0.7 \pm 0.1.$$

The corresponding length and time scales are:

$$\ell_{H_0} \equiv \frac{c}{H_0} = 3000 \, h_0^{-1} \text{ Mpc} = 9.25 \times 10^{25} \, h_0^{-1} \text{ m}, \tag{D.29}$$

$$t_{H_0} \equiv \frac{1}{H_0} = 3.1 \times 10^{17} \, h_0^{-1} \text{ s} = 9.8 \, h_0^{-1} \text{ Gyr}. \tag{D.30}$$

A reference energy density at present time t_0 is obtained from the Friedmann equation for vanishing cosmological ($\lambda = 0$) and flat space ($k = 0$):

$$\rho_c \equiv \frac{3H_0^2}{8\pi G_N} = 1.9\ 10^{-26}\ h_0^2\ \text{kg m}^{-3}. \tag{D.31}$$

This corresponds to approximately one galaxy per Mpc^3 or 5 protons per m^3. In fundamental units where $\hbar = c = 1$, this is of the order of $(10^{-3}\text{eV})^4$. In the case of a vanishing cosmological constant, it follows from (D.27) that, depending on whether the present energy density of the Universe ρ_0 is larger, equal or smaller than ρ_c, the present Universe is spatially open ($k > 0$), flat ($k = 0$) or closed ($k < 0$). Hence the name critical density for ρ_c.

It has become customary to normalize the different forms of energy density in the present Universe in terms of this critical density. Separating the energy density ρ_{M0} presently stored in nonrelativistic matter (baryons, neutrinos, dark matter, etc.) from the density ρ_{R0} presently stored in radiation (photons, relativistic neutrino, if any), one defines:

$$\Omega_M \equiv \frac{\rho_{M0}}{\rho_c}, \quad \Omega_R \equiv \frac{\rho_{R0}}{\rho_c}, \quad \Omega_\Lambda \equiv \frac{\lambda}{3H_0^2}, \quad \Omega_k \equiv -\frac{k}{a_0^2 H_0^2}. \tag{D.32}$$

The last term comes from the spatial curvature and is not strictly speaking a contribution to the energy density.

Then the Friedmann equation taken at time t_0 simply reads

$$\Omega_M + \Omega_R + \Omega_\Lambda + \Omega_k = 1. \tag{D.33}$$

Since matter dominates over radiation in the present Universe, we may neglect Ω_R in the preceding equation. Using the dependence of the different components with the scale factor $a(t)$, one may then rewrite the Friedmann equation at any time as:

$$H^2(t) = H_0^2 \left[\Omega_\Lambda + \Omega_M \left(\frac{a_0}{a(t)} \right)^3 + \Omega_R \left(\frac{a_0}{a(t)} \right)^4 + \Omega_k \left(\frac{a_0}{a(t)} \right)^2 \right], \tag{D.34}$$

where a_0 is the present value of the cosmic scale factor and all time dependences have been written explicitly. We note that, even if Ω_R is negligible in (D.33), this is not so in the early Universe because the radiation term increases faster than the matter term in (D.34) as one gets back in time (*i.e.* as $a(t)$ decreases). If we add an extra component X with equation of state $p_X = w_X \rho_X$, it contributes an extra term $\Omega_X \left(a_0 / a(t) \right)^{3(1+w_X)}$ where $\Omega_X = \rho_X / \rho_c$.

Important information about the evolution of the Universe at a given time is whether its expansion is accelerating or decelerating. This is obtained by studying the second derivative of the cosmic scale factor, which is easily extracted from (D.25) and (D.26):

$$\frac{\ddot{a}}{a} = -\frac{4\pi G_N}{3}(3p + \rho) + \frac{\lambda}{3}. \tag{D.35}$$

The acceleration of our Universe is usually measured by the deceleration parameter q which is defined as:

$$q \equiv -\frac{\ddot{a}a}{\dot{a}^2}. \tag{D.36}$$

Using (D.35) and separating again matter and radiation, we may write it at present time t_0 as:

$$q_0 = -\frac{1}{H_0^2}\left(\frac{\ddot{a}}{a}\right)_{t=t_0} = \frac{1}{2}\Omega_M + \Omega_R - \Omega_\Lambda. \tag{D.37}$$

Once again, the radiation term Ω_R can be neglected in this relation. We see that in order to have an acceleration of the expansion ($q_0 < 0$), we need the cosmological constant to dominate over the other terms. We can also write the deceleration parameter (D.36) at a given time t as

$$q(t) = \frac{1}{H^2(t)}\left[\frac{1}{2}\Omega_M\left(\frac{a_0}{a(t)}\right)^3 + \Omega_R\left(\frac{a_0}{a(t)}\right)^4 - \Omega_\Lambda\right]. \tag{D.38}$$

If we introduce an extra component X as above, it contributes a term $\Omega_X(a_0/a(t))^{3(1+w_X)}(1 + 3w_X)/2$: only components with equation of state parameter $w_X < -1/3$ tend to accelerate the expansion of the Universe.

The measurement of the Hubble constant and of the deceleration parameter allows us to obtain the behavior of the cosmic scale factor in the last stages of the evolution of the Universe:

$$a(t) = a_0\left[1 + \frac{t - t_0}{t_{H_0}} - \frac{q_0}{2}\frac{(t - t_0)^2}{t_{H_0}^2} + \cdots\right]. \tag{D.39}$$

D.2.2   Measure of distances

Measuring cosmological distances allows us to study the geometry of spacetime. Depending on the type of observation, one may define several distances.

In an expanding or contracting Universe, the light emitted by a distant source undergoes a frequency shift which gives a direct information on the time dependence of the cosmic scale factor $a(t)$. To obtain the explicit relation, we consider a photon propagating in a fixed direction (θ and ϕ fixed). Its equation of motion is given as in special relativity by setting $ds^2 = 0$ in (D.18):

$$c^2 dt^2 = a^2(t)\frac{dr^2}{1 - kr^2}. \tag{D.40}$$

Thus, if a photon (an electromagnetic wave) leaves at time t a galaxy located at distance r from us, it will reach us at time t_0 such that

$$\int_t^{t_0} \frac{c\,dt}{a(t)} = \int_0^r \frac{dr}{\sqrt{1-kr^2}}. \tag{D.41}$$

The electromagnetic wave is emitted with the same amplitude at a time $t+T$ where the period T is related to the wavelength of the emitted wave λ by the relation $\lambda = cT$. It is thus received with the same amplitude at the time $t_0 + T_0$ given by

$$\int_{t+T}^{t_0+T_0} \frac{c\,dt}{a(t)} = \int_0^r \frac{dr}{\sqrt{1-kr^2}}, \tag{D.42}$$

the wavelength of the received wave being simply $\lambda_0 = cT_0$. Since $T_0, T \ll t_0, t$, we obtain from comparing (D.41) and (D.42)

$$\frac{cT_0}{a_0} = \frac{cT}{a(t)}, \qquad \text{i.e.} \qquad \frac{\lambda_0}{\lambda} = \frac{a_0}{a(t)}. \tag{D.43}$$

Defining the redshift parameter z as the fractional increase in wavelength $z = (\lambda_0 - \lambda)/\lambda$, we have

$$1 + z = \frac{a_0}{a(t)}. \tag{D.44}$$

One may thus replace time by redshift since time decreases monotonically as redshift increases. For example, the expression of the Hubble parameter in (D.34) can be turned into

$$H^2(z) = H_0^2\left[\Omega_M(1+z)^3 + \Omega_R(1+z)^4 + \Omega_k(1+z)^2 + \Omega_\Lambda\right]. \tag{D.45}$$

Using

$$\int_t^{t_0} \frac{c\,dt}{a(t)} = \int_{a(t)}^{a_0} \frac{c\,da}{a\dot{a}} = \int_{a(t)}^{a_0} \frac{c\,da}{a^2 H} = \int_0^z \frac{c\,dz}{H(z)}$$

we may extract from (D.34) and (D.41) the proper distance (D.20) at time t_0:

$$a_0 \int_0^r \frac{dr}{\sqrt{1-kr^2}}$$

$$= a_0 \begin{cases} \sin^{-1} r & k = +1 \\ r & k = 0 \\ \sinh^{-1} r & k = -1 \end{cases} \tag{D.46}$$

$$= a_0 \int_t^{t_0} \frac{c\,dt'}{a(t')}$$

$$= \ell_{H_0} \int_0^z \frac{dz}{\left[\Omega_M(1+z)^3 + \Omega_R(1+z)^4 + \Omega_k(1+z)^2 + \Omega_\Lambda\right]^{1/2}}$$

where $\ell_{H_0} = cH_0^{-1}$. We can also write the deceleration parameter in (D.38) as

$$q = \frac{1}{H(z)^2} \left[\frac{1}{2}\Omega_M (1+z)^3 + \Omega_R (1+z)^4 - \Omega_\Lambda \right]. \tag{D.47}$$

This shows that the Universe starts accelerating at redshift values $1+z \sim (2\Omega_\Lambda/\Omega_M)^{1/3}$ (neglecting Ω_R), that is typically redshifts of order 1.

If a photon source of luminosity L (energy per unit time) is placed at a distance r from the observer, then the energy flux ϕ (energy per unit time and unit area) received by the observer is given by

$$\phi = \frac{L}{4\pi a_0^2 r^2 (1+z)^2} \equiv \frac{L}{4\pi d_L^2}. \tag{D.48}$$

The two powers of $1+z$ account for the photon energy redshift and the time dilatation between emission and observation. The quantity $d_L \equiv a_0 r(1+z)$ is called the luminosity distance.

If the source is at a redshift z of order one or smaller, then we can approximate the integral $\int_0^r dr/\sqrt{1-kr^2}$ in (D.41) by simply r and this equation gives

$$a_0 r \sim \int_t^{t_0} \frac{a_0 c\, dt}{a(t)} = \int_a^{a_0} \frac{a_0 c\, da}{a\dot{a}} \sim \ell_{H_0} \int_a^{a_0} \frac{da}{a\left[1 - q_0 H_0(t - t_0)\right]} \tag{D.49}$$

where we have used the development (D.39) with $t_{H_0} = \ell_{H_0}/c = H_0^{-1}$. Using $H_0(t - t_0) \sim (a - a_0)/a_0 \ll 1$ and $a = a_0/(1+z)$, we obtain for $z \ll 1$

$$a_0 r = \ell_{H_0} z \left(1 - \frac{1+q_0}{2} z + \cdots \right). \tag{D.50}$$

This result can be obtained directly from (D.46) (see Exercise 3). Thus, the luminosity distance reads, for $z \ll 1$,

$$d_L = \ell_{H_0} z \left(1 - \frac{1+q_0}{2} z + \cdots \right)(1+z) = \ell_{H_0} z \left(1 + \frac{1-q_0}{2} z + \cdots \right). \tag{D.51}$$

Hence measurement of deviations to the Hubble law ($d_L = \ell_{H_0} z$) at moderate redshift allow us to measure the combination $\Omega_M/2 - \Omega_\Lambda$ (see (D.37)).

Another distance is defined in cases where one measures the angular diameter δ of a source in the sky. If D is the diameter of the source, then D/δ would be the distance of the source in Euclidean geometry. In a Universe with a Robertson–Walker metric, it turns out to be $a(t)r = a_0 r/(1+z)$. This defines the angular diameter distance d_A

$$d_A = a(t)r = \frac{d_L}{(1+z)^2}. \tag{D.52}$$

Several distance measurements tend to point towards an evolution of the present Universe dominated by the cosmological constant contribution[3] and thus a late acceleration of its expansion.

The approach that has made the first strong case for such an hypothesis uses supernovae of type Ia as standard candles[4]. Two groups, the Supernova Cosmology Project [88] and the High-z Supernova Search [140] have found that distant supernovae appear to be fainter than expected in a flat matter-dominated Universe. If this is to have a cosmological origin, this means that, at fixed redshift, they are at larger distances than expected in such a context and thus that the Universe is accelerating its expansion.

More precisely, one uses the relation (D.48) between the flux ϕ received on Earth and the luminosity L of the supernova. Traditionally, flux and luminosity are expressed on a log scale as apparent magnitude m_B and absolute magnitude M (magnitude is $-2.5 \log_{10}$ luminosity + constant). The relation then reads

$$m_B = 5 \log(H_0 d_L) + M - 5 \log H_0 + 25. \tag{D.53}$$

The last terms are z-independent, *if one assumes that supernovae of type Ia are standard candles*; they are then measured by using low z supernovae. The first term, which involves the luminosity distance d_L, varies logarithmically with z up to corrections which depend on the geometry, more precisely on $q_0 = \Omega_M/2 - \Omega_\Lambda$ for small z as can be seen from (D.51). This allows us to compare with data cosmological models with different components participating to the energy budget, as can be seen from Fig. D.1.

This can be turned into a limit in the $\Omega_M - \Omega_\Lambda$ plane for the model considered here (see Fig. D.2).

Of course, such type of measurement is sensitive to many possible systematic effects (extra corrections besides the phenomenological stretch factor applied to the lightcurves, presence of dust, etc.), and this has fueled a healthy debate on the significance of supernova data, as well as a thorough study of possible systematic effects by the observational groups concerned.

Let us note that the combination $\Omega_M/2 - \Omega_\Lambda$ is 'orthogonal' to the combination $\Omega_M + \Omega_\Lambda$ measured in CMB experiments (see Section D.3.4). The two measurements are therefore complementary: this is sometimes referred to as 'cosmic complementarity'.

Other results come from gravitational lensing. The deviation of light rays by an accumulation of matter along the line of sight depends on the distance to the source (D.50) and thus on the cosmological parameters Ω_M and Ω_Λ. As q_0 decreases (*i.e.* as the Universe accelerates), there is more volume and thus more lenses between the observer and the object at redshift z. Several methods are used: abundance of multiply-imaged quasar sources [255], strong lensing by massive clusters of galaxies (providing multiple images or arcs) [356], and weak lensing [286].

---

[3]At least when analyzed in the framework of the model discussed in this section, *i.e.* including nonrelativistic matter, radiation and a cosmological constant. As discussed in Section 12.2.2 of Chapter 12, the cosmological constant may be replaced by a dynamical component.

[4]By calibrating them according to the timescale of their brightening and fading.

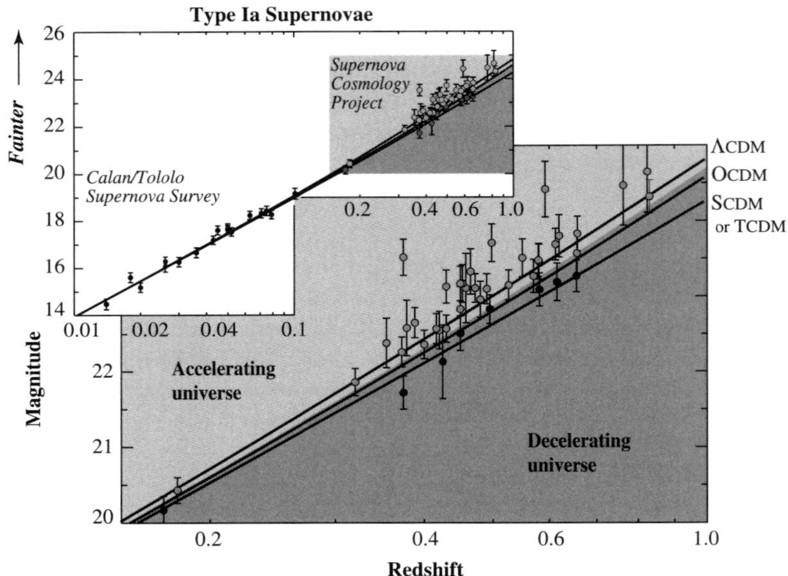

Fig. D.1 Hubble plot (magnitude versus redshift) for Type Ia supernovae observed at low redshift by the Calan–Tololo Supernova Survey and at moderate redshift by the Supernova Cosmology Project.

D.2.3   Age of the Universe

Since $1 + z = a_0/a(t)$, we can write (D.45) as

$$H^2(z) = \frac{1}{(1+z)^2}\left(\frac{dz}{dt}\right)^2 = H_0^2\left[\Omega_M(1+z)^3 + \Omega_R(1+z)^4 + \Omega_k(1+z)^2 + \Omega_\Lambda\right].$$

(D.54)

This is easily integrated to obtain the time–redshift relation. Sending z to infinity (or, for that matter, to the value of z corresponding to nucleosynthesis, up to which we think we understand the evolution of the Universe), we obtain the age of the Universe:

$$t_0 = t_{H_0}\int_0^\infty \frac{dz}{(1+z)\left[\Omega_M(1+z)^3 + \Omega_R(1+z)^4 + \Omega_k(1+z)^2 + \Omega_\Lambda\right]^{1/2}}$$

(D.55)

(we have set $t = 0$ at the Big Bang singularity; hence t_0 is the age of the Universe). One may neglect the radiation dominated era since it is short on the scale of the age of the Universe (see Table D.1 below). One can check, using (D.33) to express Ω_k, that t_0 increases with Ω_Λ at fixed Ω_M. Thus a nonvanishing cosmological constant helps to solve what was once known as the "age of Universe crisis": the Universe should be older than the oldest globular clusters in the Milky way whose age is estimated at 10 Gyr. Present data on $(\Omega_M, \Omega_\Lambda)$ favors an age of the Universe around 15 Gyr.

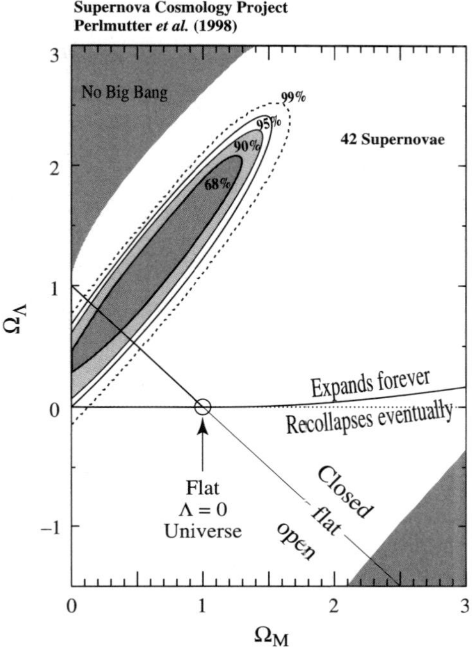

Fig. D.2 Best-fit coincidence regions in the $(\Omega_M, \Omega_\Lambda)$ plane, based on the analysis of 42 type Ia supernovae discovered by the Supernova Cosmology Project [88].

D.3   The hot Big Bang scenario

We summarize in Table D.1 the main stages of the evolution of the Universe which we will describe in this section. Before undertaking this task, let us recall a few facts about equilibrium distributions.

D.3.1   Equilibrium distributions

For a species in kinetic equilibrium, the phase space distribution functions are of the standard Fermi–Dirac or Bose–Einstein form[5]:

$$f(\mathbf{p}) = \frac{1}{\exp\left[E/(kT)\right] \pm 1}, \quad E = \sqrt{\mathbf{p}^2 + m^2}, \tag{D.56}$$

where the $+$ sign refers to Fermi–Dirac and the $-$ sign to Bose–Einstein. One can easily infer the number density n, energy density ρ, and pressure p of the corresponding species:

$$n = \frac{g}{(2\pi\hbar)^3} \int f(\mathbf{p}) \, d^3p,$$

$$\rho = \frac{g}{(2\pi\hbar)^3} \int E \, f(\mathbf{p}) \, d^3p, \tag{D.57}$$

$$p = \frac{g}{(2\pi\hbar)^3} \int \frac{\mathbf{p}^2}{3E} f(\mathbf{p}) \, d^3p,$$

---

[5]We disregard here any chemical potential.

t	kT_γ (eV)	z	
$t_0 \sim 15$ Gyr	2.35×10^{-4}	0	now
$\sim$ Gyr	$\sim 10^{-3}$	~ 4	formation of galaxies
$t_{\rm rec} \sim 4 \times 10^5$ yr	0.26	1100	recombination
$t_{\rm eq} \sim 4 \times 10^4$ yr	0.83	3500	matter-radiation equality
3 min	6×10^4	2×10^8	nucleosynthesis
1 s	10^6	3×10^9	e^+e^- annihilation
4×10^{-6} s	4×10^8	10^{12}	QCD phase transition
$< 4 \times 10^{-6}$ s	$> 10^9$		baryogenesis
			inflation
$t = 0$		∞	Big Bang

Table D.1 The different stages of the cosmological evolution in the standard scenario, given in terms of time t since the Big Bang singularity, the energy kT of the background photons, and the redshift z. The line following nucleosynthesis indicates the part of the evolution which has been tested through observation. The values ($h_0 = 0.7, \Omega_M = 0.3, \Omega_\Lambda = 0.7$) are adopted to compute explicit values.

where g is the number of internal degrees of freedom ($g = 1$ for a neutrino or an antineutrino and $g = 2$ for a massless vector field such as the photon, for an electron or a positron). We will need the explicit form in the relativistic limit ($kT \gg m$):

$$n = \left(\frac{3}{4}\right)_F \frac{\zeta(3)}{\pi^2} g \left(\frac{kT}{\hbar}\right)^3,$$

$$\rho = \left(\frac{7}{8}\right)_F \frac{1}{2} g\, a_{\rm BB} T^4, \tag{D.58}$$

$$p = \rho/3,$$

where the parenthesis $(\cdots)_F$ indicates the extra factor to be taken into account in the case of the Fermi–Dirac distribution and $a_{\rm BB}$ is the blackbody constant (we restore here the powers of c):

$$a_{\rm BB} \equiv \frac{\pi^2 k^4}{15 c^3 \hbar^3} = 7.56 \times 10^{-16}\ {\rm J\ m^{-3}\ K^{-4}} = 4.72\ {\rm keV\ m^{-3}\ K^{-4}}. \tag{D.59}$$

For example we have at present time ($T_0 = 2.725$ K) $n_{\gamma 0} = 411$ cm^{-3}.

In the nonrelativistic limit ($kT \ll m$), n and $\rho = mn$ are exponentially small:

$$n = g \left(\frac{mkT}{2\pi\hbar^2}\right)^{3/2} \exp\left[-m/(kT)\right]. \tag{D.60}$$

We conclude that, as long as some species remain relativistic, the matter energy density is dominated by this radiation and takes the form:

$$\rho_R = \tfrac{1}{2}g_* a_{\mathrm{BB}} T^4,$$

$$g_* = \sum_{\text{bosons } i} g_i \left(\frac{T_i}{T}\right)^4 + \frac{7}{8} \sum_{\text{fermions } i} g_i \left(\frac{T_i}{T}\right)^4, \tag{D.61}$$

where we have taken into account the possibility that the species i may have a thermal distribution at a temperature T_i different from the temperature T of the photons. If radiation dominates the energy density of the Universe, one obtains from (D.27)

$$H = \frac{2\pi}{3\hbar} \sqrt{\frac{\pi}{5}} \, g_*^{1/2} \frac{(kT)^2}{M_P}. \tag{D.62}$$

It is possible to show, using the second law of thermodynamics ($TdS = dE + pdV$), that the entropy per unit volume is simply the quantity

$$s \equiv \frac{S}{V} = \frac{\rho + p}{T} \tag{D.63}$$

and that the entropy in a covolume sa^3 remains constant. The entropy density is dominated by relativistic particles and reads

$$s = \tfrac{2}{3} g_s a_{\mathrm{BB}} T^3,$$

$$g_s = \sum_{\text{bosons } i} g_i \left(\frac{T_i}{T}\right)^3 + \frac{7}{8} \sum_{\text{fermions } i} g_i \left(\frac{T_i}{T}\right)^3, \tag{D.64}$$

where the sum extends *only* to the species in thermal equilibrium. Note that, when all species have the temperature of the photons T, which is true through most of the history of the Universe, $g_s = g_*$.

We deduce from the constancy of sa^3 that $g_s (aT)^3$ remains constant. Hence the temperature T of the Universe behaves as a^{-1} whenever g_s remains constant. This is so except when some species drop out of equilibrium. Indeed, a given species drops out of equilibrium when its interaction rate Γ drops below the expansion rate H. We will return to this question in Section D.3.3 and see for example that neutrinos decouple at temperatures below 1 MeV. Their temperature continues to decrease as a^{-1} and thus remains equal to T. However when kT drops below $2m_e$, electrons annihilate against positrons with no possibility of being regenerated and the entropy of the electron–positron pairs is transferred to the photons. Since $g_s|_{\gamma,e^\pm} = 2 + 4 \cdot 7/8 = 11/2$ and $g_s|_\gamma = 2$, the temperature of the photons becomes multiplied by a factor $(11/4)^{1/3}$. Since the neutrinos have already decoupled, they are not affected by this entropy release and their temperature remains untouched. Thus we have

$$\frac{T}{T_\nu} = \left(\frac{11}{4}\right)^{1/3} \sim 1.40. \tag{D.65}$$

We can now compute the value of g_s for temperatures much smaller than m_e: $g_s = 2 + (7/8)6(4/11) = 3.91$. We deduce that, at the present time ($T_0 = 2.725$ K), $s_0/k = 2890 \text{ cm}^{-3}$.

D.3.2   The evolution of the universe

It follows from the Friedmann equation (D.27) that if the Universe is dominated by a component of equation of state $p = w\rho$ (as we have seen following (D.28) $\rho(t) \sim a(t)^{-3(1+w)}$), then the cosmic scale factor $a(t)$ varies with time as $t^{2/[3(1+w)]}$. We start at present time t_0 with the energy budget discussed in Section D.2.2, say $\Omega_M = 0.3, \Omega_\Lambda = 0.7, \Omega_k \sim 0$. Radiation consists of photons and relativistic neutrinos:

$$\rho_R(t) = \rho_\gamma(t) \left[1 + \frac{7}{8} \left(\frac{4}{11} \right)^{4/3} N_\nu^{\mathrm{rel}}(t) \right], \tag{D.66}$$

where $N_\nu^{\mathrm{rel}}(t)$ is the number of relativistic neutrinos at time t. We have $\Omega_\gamma = \rho_\gamma(t_0)/\rho_c = 2.48 \times 10^{-5} \, h_0^{-2}$ and the mass limits on neutrinos imply $N_\nu^{\mathrm{rel}}(t_0) \leq 1$. In any case, $\Omega_R \ll \Omega_M$.

For redshifts larger than 1, the vacuum energy is subdominant and the Universe is matter-dominated ($a(t) \sim t^{2/3}$). However radiation energy density increases more rapidly (as $a(t)^{-4}$) than matter ($a(t)^{-3}$) as one goes back in time (as $a(t)$ decreases). At time t_{eq}, there is equality. This corresponds to

$$\frac{1}{1 + z_{\mathrm{eq}}} = \frac{a(t_{\mathrm{eq}})}{a_0} = \frac{1.68 \, \Omega_\gamma}{\Omega_M} = \frac{4.17 \times 10^{-5}}{\Omega_M h_0^2}, \tag{D.67}$$

where we have assumed three relativistic neutrinos at this time.

However, at $t_{\mathrm{rec}} > t_{\mathrm{eq}}$ i.e. still in the matter-dominated epoch, electrons recombine with the protons to form atoms of hydrogen and, because hydrogen is neutral, this induces the decoupling of matter and photon: from then on ($t_{\mathrm{rec}} < t < t_0$), the Universe becomes transparent[6]. This is the important recombination stage. After decoupling the energy density $\rho_\gamma \sim T^4$ of the primordial photons is redshifted according to the law

$$\frac{T(t)}{T_0} = \frac{a_0}{a(t)} = 1 + z. \tag{D.68}$$

One observes presently this cosmic microwave background (CMB) as a radiation with a blackbody spectrum at temperature $T_0 = 2.725$ K or energy $kT_0 = 2.35 \times 10^{-4}$ eV. The fact that the spectrum observed is very precisely a blackbody spectrum indicates that (D.68) applies even for periods earlier than recombination.

Since the binding energy of the ground state of atomic hydrogen is $E_b = 13.6$ eV, one may expect that the energy kT_{rec} is of the same order. It is substantially smaller because of the smallness of the ratio of baryons to photons $\eta = n_b/n_\gamma \sim 5 \times 10^{-10}$. Indeed, according to the Saha equation, the fraction x of ionized atoms is given by

$$\frac{n_p n_e}{n_H n_\gamma} = \frac{x^2}{(1 - x)} \eta = \frac{4.05 c^3}{\pi^2} \left(\frac{m_e}{2\pi kT} \right)^{3/2} e^{-E_b/kT}. \tag{D.69}$$

Hence, because $\eta \ll 1$, the ionized fraction x becomes negligible only for energies much smaller than E_b. A careful treatment gives $kT_{\mathrm{rec}} \sim 0.26$ eV. We return to CMB in a more detailed analysis in Section D.3.4 below.

---

[6]It is believed that hydrogen is later reionized by the photons produced by the first stars or quasars but the Universe is then sufficiently dilute to prevent recoupling.

D.3.3   Relics

In this book, we pay specific attention to the evolution of the abundance of stable or quasistable massive particles. They may provide good candidates for dark matter (see Chapter 5) or alternatively could become too abundant at present times to be consistent with observations (see Chapter 11). We have noted that the number densities of massive particles in thermal equilibrium become exponentially small for temperatures smaller than their mass. This is, however, under the assumption of thermal equilibrium. Because we are in an expanding Universe, particles may drop out of thermal equilibrium and their abundance be frozen at the corresponding temperature.

Let us consider the evolution of a given stable particle species of mass m_X. There are two competing effects which modify the abundance of this species: annihilation and expansion of the Universe. Indeed, the faster is the dilution associated with the expansion, the least effective is the annihilation because the particles recede from one another. This is illustrated by the following equation which gives the evolution with time of the particle number density n_X:

$$\frac{dn_X}{dt} + 3Hn_X = -\langle\sigma_{\mathrm{ann}}v\rangle\left(n_X^2 - n_X^{(\mathrm{eq})2}\right), \tag{D.70}$$

where $\langle\sigma_{\mathrm{ann}}v\rangle$ is the thermal average of the $X\bar{X}$ annihilation cross-section times the relative velocity of the two particles annihilating, $n_X^{(\mathrm{eq})}$ is the equilibrium density, as given in (D.57). The friction term $3Hn_X$ in (D.70) represents the effect of the expansion of the Universe: in the absence of the annihilation process, the number of particles in a covolume $n_X a^3$ or (since the temperature T behaves as a^{-1}) n_X/T^3 would be constant. Indeed, one can rewrite (D.70) as

$$\frac{d}{dt}\left(\frac{n_X}{T^3}\right) = -\langle\sigma_{\mathrm{ann}}v\rangle T^3\left[\left(\frac{n_X}{T^3}\right)^2 - \left(\frac{n_X^{(\mathrm{eq})}}{T^3}\right)^2\right]. \tag{D.71}$$

Because of annihilation, the evolution equation has two regimes:

- As long as the expansion rate H is smaller than the annihilation rate $\Gamma_X \equiv n_X\langle\sigma_{\mathrm{ann}}v\rangle$, one may solve (D.70) disregarding the expansion: $n_X/T^3 \sim n_X^{(\mathrm{eq})}/T^3 \sim \exp\left(-m_X/kT\right)$. Obviously, as time goes on, T decreases, the energy density n_X decreases and so does the annihilation rate.
- Once the expansion rate H becomes larger than the annihilation rate Γ_X, there is freezing of the number of particles in a covolume and n/T^3 becomes constant. This occurs at a freezing temperature T_f such that

$$n_X(T_f)\langle\sigma_{\mathrm{ann}}v\rangle \sim H(T_f). \tag{D.72}$$

The explicit form of $n_X(T_f)$ depends crucially on whether the species is still relativistic at the time of freezing. We start with the case of *cold relics*, which are nonrelativistic at the time of freezing. Let us define $x_f \equiv m_X/(kT_f)$. Using (D.60), with g_X number of degrees of freedom of the X particle, and (D.62), one obtains

$$x_f^{-1/2}\exp(x_f) = \frac{3}{4\pi^3}\sqrt{\frac{5}{2}}\frac{g_X}{g_*^{1/2}}\frac{m_X M_P\langle\sigma_{\mathrm{ann}}v\rangle}{\hbar^2},$$

or, neglecting a subleading term in $\ln x_f$,

$$x_f = \ln\left(0.038 \frac{g_x}{g_*^{1/2}} \frac{m_x M_P \langle\sigma_{\mathrm{ann}} v\rangle}{\hbar^2}\right). \tag{D.73}$$

We note that x_f is increasing with the mass m_x or the annihilation cross-section $\langle\sigma_{\mathrm{ann}} v\rangle$. As x_f increases, the freezing occurs later and the particle density follows longer the equilibrium distribution: the relic density is reduced. Thus, the more massive the particle, or the larger the annihilation cross-section , the smaller the relic density.

For $T < T_f$,

$$\frac{n_x(T)}{T^3} \sim \frac{n_x(T_f)}{T_f^3}. \tag{D.74}$$

We can deduce an estimate of the present X species abundance. Using (D.72), we have

$$n_x(T_0) \sim \frac{T_0^3}{T_f^3} \, n_x(T_f) \sim \frac{(kT_0)^3}{(kT_f)^3} \frac{H(T_f)}{\langle\sigma_{\mathrm{ann}} v\rangle}.$$

Thus, using (D.62) to express $H(T_f)$ and (D.64) to express $(kT_0)^3$, we may write the present X particle energy density $\rho_x(T_0) = n_x(T_0) m_x$ as

$$\rho_x(T_0) = \frac{15}{\pi} \sqrt{\frac{\pi}{5}} \frac{s_0}{k} \frac{x_f}{M_P \langle\sigma_{\mathrm{ann}} v\rangle} \frac{g_*^{1/2}}{g_s} \hbar^2, \tag{D.75}$$

where s_0 is the present entropy density of the Universe. Then

$$\Omega_x \equiv \frac{\rho_x(T_0)}{\rho_c} = \frac{8\pi\hbar}{3 M_P^2} \frac{\rho_x(T_0)}{H_0^2}$$

reads

$$\Omega_x h_0^2 = 40 \sqrt{\frac{\pi}{5}} \, x_f \frac{h_0^2}{H_0^2} \frac{s_0}{k} \frac{\hbar^3}{M_P^3 \langle\sigma_{\mathrm{ann}} v\rangle} \frac{g_*^{1/2}}{g_s} \sim 30 \text{ fbarn} \frac{x_f}{\langle\sigma_{\mathrm{ann}} v/c\rangle} \frac{g_*^{1/2}}{g_s}. \tag{D.76}$$

In the alternative case of *hot relics*, i.e. species which are relativistic at the time of freezing, we obtain from (D.58)

$$\frac{n_x(T_f)}{T_f^3} = \left(\frac{3}{4}\right)_F \frac{\zeta(3)}{\pi^2} g \frac{k^3}{\hbar^3} \tag{D.77}$$

which does not depend on the details of freezing. Then using (D.64) and (D.74), one obtains for $\Omega_x = n_x(T_0) m_x/\rho_c$:

$$\Omega_x h_0^2 = \frac{60}{\pi^3} \zeta(3) \frac{\hbar c}{M_P^2} \frac{h_0^2}{H_0^2} \frac{s_0}{k} m_x \frac{g}{g_s} \left(\frac{3}{4}\right)_F \sim 7.6 \times 10^{-2} \frac{m_x}{1 \text{ eV}/c^2} \frac{g}{g_s} \left(\frac{3}{4}\right)_F. \tag{D.78}$$

D.3.4 Cosmic microwave background

Before discussing the spectrum of CMB fluctuations, we introduce the important notion of a particle horizon in cosmology.

Because of the speed of light, a photon which is emitted at the Big Bang ($t = 0$) will have travelled a finite distance at time t. The proper distance (D.20) measured at time t is simply given as in (D.41) by the integral:

$$d_h(t) = a(t) \int_0^t \frac{c\,dt'}{a(t')} \tag{D.79}$$

$$= \frac{\ell_{H_0}}{1+z} \int_z^\infty \frac{dz}{[\Omega_M(1+z)^3 + \Omega_R(1+z)^4 + \Omega_k(1+z)^2 + \Omega_\Lambda]^{1/2}},$$

where, in the second line, we have used (D.46). This is the maximal distance that a photon (or any particle) could have travelled at time t since the Big Bang . In other words, it is possible to receive signals at a time t only from comoving particles within a sphere of radius $d_h(t)$. This distance is known as the particle horizon at time t.

A quantity of relevance for our discussion of CMB fluctuations is the horizon at the time of the recombination *i.e.* $z_{\rm rec} \sim 1100$. We note that the integral on the second line of (D.79) is dominated by the lowest values of z: $z \sim z_{\rm rec}$ where the Universe is still matter dominated. Hence

$$d_h(t_{\rm rec}) \sim \frac{2\ell_{H_0}}{\Omega_M^{1/2} z_{\rm rec}^{3/2}} \sim 0.3 \text{ Mpc.} \tag{D.80}$$

We note that this is simply twice the Hubble radius at recombination $H^{-1}(z_{\rm rec})$, as can be checked from (D.45):

$$R_H(t_{\rm rec}) \sim \frac{\ell_{H_0}}{\Omega_M^{1/2} z_{\rm rec}^{3/2}}. \tag{D.81}$$

This radius is seen from an observer at present time under an angle

$$\theta_H(t_{\rm rec}) = \frac{R_H(t_{\rm rec})}{d_A(t_{\rm rec})}, \tag{D.82}$$

where the angular distance has been defined in (D.52). We can compute analytically this angular distance under the assumption that the Universe is matter dominated (see Exercise 4). Using (D.133), we have

$$d_A(t_{\rm rec}) = \frac{a_0 r}{1 + z_{\rm rec}} \sim \frac{2\ell_{H_0}}{\Omega_M z_{\rm rec}}. \tag{D.83}$$

Thus, since, in our approximation, the total energy density Ω_T is given by Ω_M,

$$\theta_H(t_{\rm rec}) \sim \Omega_T^{1/2}/(2z_{\rm rec}^{1/2}) \sim 0.015 \text{ rad } \Omega_T^{1/2} \sim 1^\circ\, \Omega_T^{1/2}. \tag{D.84}$$

We have written in the latter equation Ω_T instead of Ω_M because numerical computations show that, in case where Ω_Λ is nonnegligible, the angle depends on $\Omega_M + \Omega_\Lambda = \Omega_T$.

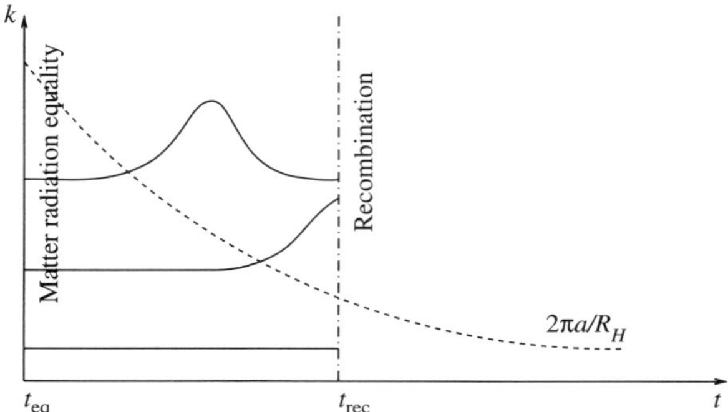

Fig. D.3 Evolution of the photon temperature fluctuations before the recombination. This diagram illustrates that oscillations start once the corresponding Fourier mode enters the Hubble radius (these oscillations are fluctuations in temperature, along a vertical axis orthogonal to the two axes that are drawn on the figure). They are frozen at t_{rec}.

We can now discuss the evolution of photon temperature fluctuations. For simplicity, we will assume a flat primordial spectrum of fluctuations: this leads to predictions in good agreement with experiment; moreover, as we will see in the next section, it is naturally explained in the context of inflation scenarios.

Before decoupling, the photons are strongly coupled with the baryons. In a gravitational potential well, gravity tends to pull this baryon–photon fluid down the well whereas radiation pressure tends to push it out. Thus, the fluid undergoes a series of acoustic oscillations. These oscillations can obviously only proceed if they are compatible with causality, *i.e.* if the corresponding wavelength is smaller than the horizon scale or the Hubble radius: $\lambda = 2\pi a(t)/k < R_H(t)$ or

$$k > 2\pi \frac{a(t)}{R_H(t)} \sim t^{-1/3}. \tag{D.85}$$

Starting with a flat primordial spectrum, we see that the first oscillation peak corresponds to $\lambda \sim R_H(t_{\mathrm{rec}})$, followed by other compression peaks at $R_H(t_{\mathrm{rec}})/n$ (see Fig. D.3). They correspond to an angular scale on the sky:

$$\theta_n \sim \frac{R_H(t_{\mathrm{rec}})}{d_A(t_{\mathrm{rec}})} \frac{1}{n} = \frac{\theta_H(t_{\mathrm{rec}})}{n}. \tag{D.86}$$

Since photons decouple at t_{rec}, we observe the same spectrum presently (up to a redshift in the photon temperature)[7].

Experiments usually measure the temperature difference of photons received by two antennas separated by an angle θ, averaged over a large fraction of the sky. Defining the correlation function

$$C(\theta) = \left\langle \frac{\Delta T}{T_0}(\mathbf{n}_1) \frac{\Delta T}{T_0}(\mathbf{n}_2) \right\rangle \tag{D.87}$$

---

[7]A more careful analysis indicates the presence of Doppler effects besides the gravitational effects that we have taken into account here (see for example [323]). Such Doppler effects turn out to be nonleading for odd values of n.

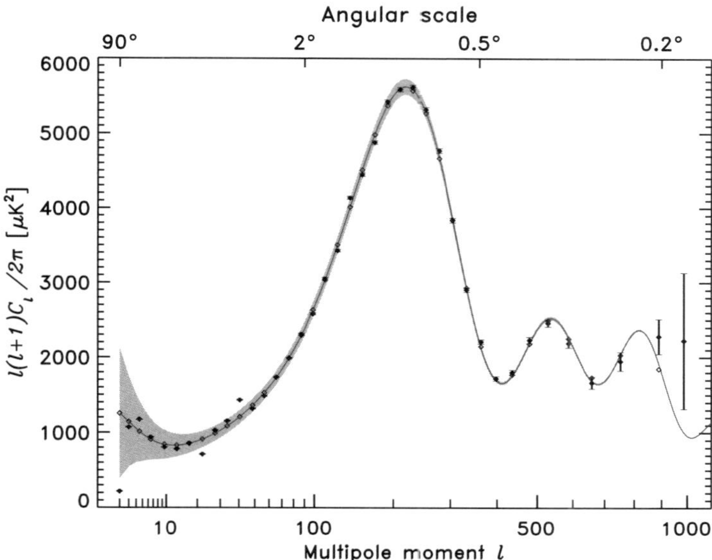

Fig. D.4 This figure compares the best-fit power law ΛCDM model to the temperature angular power spectrum observed by WMAP [225].

averaged over all $\mathbf{n}_1$ and $\mathbf{n}_2$ satisfying the condition $\mathbf{n}_1 \cdot \mathbf{n}_2 = \cos\theta$, we have indeed

$$\left\langle \left(\frac{T(\mathbf{n}_1) - T(\mathbf{n}_2)}{T_0}\right)^2 \right\rangle = 2\left(C(0) - C(\theta)\right). \tag{D.88}$$

We may decompose $C(\theta)$ over Legendre polynomials:

$$C(\theta) = \frac{1}{4\pi} \sum_{l}^{\infty} (2l + 1)C_l P_l(\cos\theta). \tag{D.89}$$

The monopole ($l = 0$) related to the overall temperature T_0, and the dipole ($l = 1$) due to the Solar system peculiar velocity bring no information on the primordial fluctuations. A given coefficient C_l characterizes the contribution of the multipole component l to the correlation function. If $\theta \ll 1$, the main contribution to C_l corresponds to an angular scale[8] $\theta \sim \pi/l \sim 200°/l$. The previous discussion (see (D.84) and (D.86)) implies that we expect the first acoustic peak at a value $l \sim 200\Omega_T^{-1/2}$.

The power spectrum obtained by the WMAP experiment is shown in Fig. D.4. One finds the first acoustic peak at $l \sim 200$, which constrains the ΛCDM model used to perform the fit to $\Omega_T = \Omega_M + \Omega_\Lambda \sim 1$. Many other constraints may be inferred from a detailed study of the power spectrum.

---

[8]The C_l are related to the coefficients a_{lm} in the expansion of $\Delta T/T$ in terms of the spherical harmonic Y_{lm}: $C_l = \langle |a_{lm}|^2 \rangle_m$. The relation between the value of l and the angle comes from the observation that Y_{lm} has $(l - m)$ zeros for $-1 < \cos\theta < 1$ and $\mathrm{Re}(Y_{lm})$ m zeros for $0 < \phi < 2\pi$.

D.3.5 Baryogenesis

It is obvious to note that the observed Universe contains much more matter than antimatter. This can be expressed in a more quantitative way. The success of the standard cosmological model predictions regarding the abundances of light elements (D, ^{3}He, ^{4}He and ^{7}Li), rests on the following hypothesis on the ratio of baryon density to photon density [142]:

$$\eta_B \equiv \frac{n_B}{n_\gamma} = (2.6 - 6.2) \times 10^{-10}. \tag{D.90}$$

We have seen earlier that the present value for n_γ is 411 cm^{-3}. Since no significant amount of antimatter has been detected in our Universe, we have

$$\eta_B = \frac{n_B - n_{\bar{B}}}{n_\gamma}. \tag{D.91}$$

We note that the photon is its own antiparticle, and thus this ratio is a measure of the unbalance between matter and antimatter. The precision on η_B has recently been improved by CMB experiments. The WMAP collaboration has measured it to be [141]:

$$\eta_B = \left(6.1^{+0.3}_{-0.2}\right) \times 10^{-10}. \tag{D.92}$$

A.D. [326] has listed the conditions necessary to dynamically generate a matter–antimatter asymmetry in an expanding Universe:

- baryon number violation;
- C and CP violation;
- deviation from thermal equilibrium.

The Standard Model fulfills in principle all these requirements. Indeed there are electroweak field configurations (sphalerons) [266] which violate B and L because the corresponding currents are anomalous in the presence of background W boson fields. The presence of sphalerons leads to effective interactions involving left-handed fields

$$\mathcal{O} = \sum_{i=1}^{3} q_{i_L} q_{i_L} q_{i_L} l_{i_L} \tag{D.93}$$

which violate B and L by three units each (and preserve $B - L$). This has no effect at zero temperature because of the small electroweak coupling but B and L-violating processes come into thermal equilibrium as one reaches the electroweak phase transition. Between the corresponding temperature (~ 100 GeV) and 10^{12} GeV, the processes (D.93) tend to wash out any nonzero value of $B + L$, except if there exists a nonvanishing $B - L$ asymmetry.

The last of the Sakharov conditions requires a first order phase transition. As the Universe undergoes this phase transition, bubbles of the true vacuum nucleate. Particles in the high temperature plasma are partially reflected when they encounter the bubble walls; these interactions can be CP violating. To avoid the washout of the baryons thus created inside the bubble, the mass of the sphaleron (given by the height of the barrier between two true vacua) must be large compared to the temperature.

The rate of sphaleron interactions inside the bubble goes as $e^{-\text{const.}(v_c/T_c)}$, where v_c is the Higgs vacuum expectation value at the critical temperature T_c when nucleation takes place (v at $T = 0$). The transition is thus sufficiently first order if

$$\frac{v_c}{T_c} \geq 1. \tag{D.94}$$

Since this ratio is inversely proportional to the quartic coupling, this can be translated into an upper value on the Higgs mass, which is incompatible with LEP limits: the electroweak phase transition is too weakly first order in the context of the Standard Model. We see in Chapter 11 that the situation is marginally improved in the MSSM model.

D.4 Inflationary cosmology

The inflation scenario has been proposed to solve a certain number of problems faced by the cosmology of the early Universe [214]. Among these one may cite:

- The flatness problem. If the total energy density ρ_T of the Universe is presently close to the critical density, it should have been even more so in the primordial Universe. Indeed, we can write (D.27) as

$$\frac{\rho_T(t)}{\rho_c(t)} - 1 = \frac{k}{\dot{a}^2}, \tag{D.95}$$

  where $\rho_c(t) = 3H^2(t)/(8\pi G_N)$ and the total energy density ρ_T includes the vacuum energy. If we take for example the radiation-dominated era where $a(t) \sim t^{1/2}$, then (D.95) can be written as ($\dot{a} \sim t^{-1/2} \sim a^{-1}$)

$$\frac{\rho_T(t)}{\rho_c(t)} - 1 = \left[\frac{\rho_T(t_U)}{\rho_c(t_U)} - 1\right] \left(\frac{a(t)}{a(t_U)}\right)^2 = \left[\frac{\rho_T(t_U)}{\rho_c(t_U)} - 1\right] \left(\frac{kT_U}{kT}\right)^2, \tag{D.96}$$

  where we have used (D.68) and taken as a reference point the epoch t_U of the grand unification phase transition. This means that, if the total energy density is close to the critical density at matter–radiation equality (as can be inferred from the present value), it must be even more so at the time of the grand unification phase transition: by a factor $\left(1\text{eV}/10^{16}\text{GeV}\right)^2 \sim 10^{-50}$! Obviously, the choice $k = 0$ in the spatial metric ensures $\rho_T = \rho_c$ but the previous estimate shows that this corresponds to initial conditions which are highly fine tuned.
- The horizon problem. We have stressed in the previous sections the isotropy and homogeneity of the cosmic microwave background and identified its primordial origin. It remains that the horizon at recombination is seen on the present sky under an angle of 2°. This means that two points opposite on the sky were separated by about 100 horizons at the time of recombination, and thus not causally connected. It is then extremely difficult to understand why the cosmic microwave background should be isotropic and homogeneous over the whole sky.
- The monopole problem. As we have seen in Section 4.5.3 of Chapter 4 on the example of the Georgi–Glashow model, monopoles occur whenever a simple gauge group is broken to a group with a $U(1)$ factor. This is precisely what happens

in grand unified theories. In this case their mass is of order M_U/g^2 where g is the value of the coupling at grand unification. Because we are dealing with stable particles with a superheavy mass, there is a danger to overclose the Universe, *i.e.* to have an energy density much larger than the critical density.

Indeed, if we assume the presence of at least one monopole per horizon at the time t_U of their formation, the number n_M of monopoles satisfies

$$n_M \geq d_h(t_U)^{-3} \sim H^3(T_U) \sim \left(\frac{T_U^2}{m_P}\right)^3 \tag{D.97}$$

where we have used the fact that in a radiation-dominated Universe $\rho \sim a^{-4} \sim T^4$. For n_M near the lower bound, monopoles are too scarce to annihilate among themselves. Realistic values of the parameters in (D.97) lead to a monopole energy density which is orders of magnitude larger than the critical density. We thus need some mechanism to dilute the relic density of monopoles.

Inflation provides a remarkably simple solution to these problems: it consists in a period of the evolution of the Universe where the expansion is exponential. Indeed, if the energy density of the Universe is dominated by the vacuum energy $\rho_{\rm vac}$, then (D.27) reads

$$H^2 = \frac{\dot{a}^2}{a^2} = \frac{\rho_0}{3m_P^2} - \frac{k}{a^2}. \tag{D.98}$$

If $\rho_{\rm vac} > 0$, this is readily solved as

$$a(t) = \begin{cases} H_{\rm vac}^{-1}\cosh H_{\rm vac}t & \text{if } k = +1 \\ H_{\rm vac}^{-1}e^{H_{\rm vac}t} & \text{if } k = 0 \\ H_{\rm vac}^{-1}\sinh H_{\rm vac}t & \text{if } k = -1 \end{cases} \quad \text{with } H_{\rm vac} \equiv \sqrt{\frac{\rho_{\rm vac}}{3m_P^2}}, \tag{D.99}$$

which corresponds to an exponential expansion at late times $(t \gg H_{\rm vac}^{-1})$. Such behavior is in fact observed whenever the magnitude of the Hubble parameter changes slowly with time, *i.e.* is such that $\left|\dot{H}\right| \ll H^2$.

Such a space was first proposed by [104, 105] with very different motivations and is thus called de Sitter space (described in Equations (6.40) and (6.41) of Chapter 6)[9].

An important property of de Sitter space is the fact that the particle horizon size is exponentially increasing in time. Indeed, we have for the horizon size, following (D.79),

$$d_h(t)\big|_{\text{de Sitter}} = a(t)\int_0^t \frac{cdt'}{a(t')} = \frac{c}{H_{\rm vac}}\, e^{H_{\rm vac}t} \quad \text{for } H_{\rm vac}t \gg 1. \tag{D.100}$$

Obviously a period of inflation will ease the horizon problem. It follows from the previous equation that a period of inflation extending from t_i to $t_f = t_i + \Delta t$ contributes to the horizon size a value $ce^{H_{\rm vac}\Delta t}/H_{\rm vac}$, which can be very large.

---

[9]It turns out that the three choices in (D.99) correspond to the same choice as can be shown by a redefinition of the spacetime coordinates (see Chapter 7 of [275]). In what follows we will take the flat $k = 0$ formulation when we discuss pure de Sitter space. This is obviously no longer true when one adds matter to de Sitter space.

We note that de Sitter space also has a finite event horizon. This is the maximal distance that comoving particles can travel between the time t where they are produced and $t = \infty$ (compare with (D.79)):

$$d_{\mathrm{eh}}(t) = a(t) \int_t^\infty \frac{c\,dt'}{a(t')}. \tag{D.101}$$

In the case of de Sitter space, this is simply

$$d_{\mathrm{eh}}(t)|_{\mathrm{de\ Sitter}} = \frac{c}{H_{\mathrm{vac}}} = R_H, \tag{D.102}$$

i.e. it corresponds to the Hubble radius (constant for de Sitter spacetime). This allows us to make an analogy between de Sitter spacetime and a black hole: a black hole of mass M has an event horizon at the Schwarzschild radius $R_g = 2G_N M$. For example, just as black holes evaporate by emitting radiation at Hawking temperature $T_H = 1/(4\pi R_g)$, an observer in de Sitter spacetime feels a thermal bath at temperature $T_H = H/(2\pi)$.

We see that it is the event horizon that fixes here the cut-off scale of microphysics. Since it is equal here to the Hubble radius, and since the Hubble radius is of the order of the particle horizon for matter or radiation-dominated Universe[10], it has become customary to compare the comoving scale associated to physical processes with the Hubble radius (we already did so in our discussion of acoustic peaks in CMB spectrum; see Fig. D.3, and D.5 below).

A period of exponential expansion of the Universe may also solve the monopole problem by diluting the concentration of monopoles by a very large factor. It also dilutes any kind of matter. Indeed, a sufficiently long period of inflation "empties" the Universe. However matter and radiation may be produced at the end of inflation by converting the energy stored in the vacuum. This conversion is known as reheating (because the temperature of the matter present in the initial stage of inflation behaves as $a^{-1}(t) \propto e^{-H_{\mathrm{vac}}t}$, it is very cold at the end of inflation; the new matter produced is hotter). If the reheating temperature is lower than the scale of grand unification, monopoles are not thermally produced and remain very scarce.

Finally, it is not surprising that the Universe comes out very flat after a period of exponential inflation. Indeed, the spatial curvature term in the Friedmann equation (D.98) is then damped by a factor $a^{-2} \propto e^{-2H_{\mathrm{vac}}\Delta t}$. For example, a value $H_{\mathrm{vac}}\Delta t \sim 60$ (one refers to it as 60 e-foldings) would easily account for the huge factor 10^{50} of adjustment that we found earlier.

Most inflation models rely on the dynamics of a scalar field in its potential. Inflation occurs whenever the scalar field evolves slowly enough in a region where the potential energy is large. The set up necessary to realize this situation has evolved with time: from the initial proposition of [214] where the field was trapped in a local minimum to "new inflation" with a plateau in the scalar potential, chaotic inflation [5, 274] where the field is trapped at values much larger than the Planck scale and more recently the hybrid inflation scenario [276] with at least two scalar fields, one allowing an easy exit from the inflation period.

---

[10]In an open or flat Universe, the event horizon (D.101) is infinite.

The equation of motion of a homogeneous scalar field $\phi(t)$ with potential $V(\phi)$ evolving in a Friedmann–Robertson–Walker Universe is:

$$\ddot{\phi} + 3H\dot{\phi} = -V'(\phi). \tag{D.103}$$

where $V'(\phi) \equiv dV/d\phi$. The term $3H\dot{\phi}$ is a friction term due to the expansion. The corresponding energy density and pressure are:

$$\rho = \tfrac{1}{2}\dot{\phi}^2 + V(\phi), \tag{D.104}$$

$$p = \tfrac{1}{2}\dot{\phi}^2 - V(\phi). \tag{D.105}$$

We may note that the equation of conservation of energy $\dot{\rho} = -3H(p + \rho)$ takes here simply the form of the equation of motion (D.103). These equations should be complemented with the Friedmann equation (D.98).

When the field is slowly moving in its potential, the friction term dominates over the acceleration term in the equation of motion (D.103) which reads:

$$3H\dot{\phi} \simeq -V'(\phi). \tag{D.106}$$

The curvature term may then be neglected in the Friedmann equation (D.98) which gives

$$H^2 \simeq \frac{\rho}{3m_P^2} \simeq \frac{V}{3m_P^2}. \tag{D.107}$$

Then the equation of conservation $\dot{\rho} = -3H(p + \rho) = -3H\dot{\phi}^2$ simply gives

$$\dot{H} \simeq -\frac{\dot{\phi}^2}{2m_P^2}. \tag{D.108}$$

It is easy to see that the condition $|\dot{H}| \ll H^2$ amounts to $\dot{\phi}^2/2 \ll \rho/3 \sim V(\phi)/3$, i.e. a kinetic energy for the scalar field much smaller than its potential energy. Using (D.106) and (D.107), the latter condition then reads

$$\varepsilon \equiv \frac{1}{2}\left(\frac{m_P V'}{V}\right)^2 \ll 1. \tag{D.109}$$

The so-called slow-roll regime is characterized by the two equations (D.106) and (D.107), as well as the condition (D.109). It is customary to introduce another small parameter:

$$\eta \equiv \frac{m_P^2 V''}{V} \ll 1, \tag{D.110}$$

which is easily seen to be a consequence of the previous equations[11].

---

[11] Differentiating (D.106), one obtains $\eta = \varepsilon - \ddot{\phi}/(H\dot{\phi})$.

An important quantity to be determined is the number of Hubble times elapsed during inflation. From some arbitrary time t to the time t_e marking the end of inflation (*i.e.* of the slow-roll regime), this number is given by

$$N(t) = \int_t^{t_e} H(t)dt. \tag{D.111}$$

It gives the number of e-foldings undergone by the scale factor $a(t)$ during this period (see (D.99)). Since $dN = -Hdt = -Hd\phi/\dot\phi$, one obtains from (D.106)) and (D.107)

$$N(\phi) = \int_{\phi_e}^{\phi} \frac{1}{m_P^2} \frac{V}{V'} d\phi. \tag{D.112}$$

During the inflationary phase, the scalar fluctuations of the metric may be written in a conformal Newtonian coordinate system as:

$$\epsilon ds^2 = a^2 \left[(1 + 2\Phi)d\eta^2 - (1 - 2\Phi)\delta_{ij}dx^i dx^j\right], \tag{D.113}$$

where η is conformal time $(ad\eta = dt = da/\dot a)$. We may write the correlation function in Fourier space $\mathcal{P}_S(k)$ by

$$\langle \Phi_{\mathbf{k}}\Phi^*_{\mathbf{k}'}\rangle = 2\pi^2 k^{-3}\mathcal{P}_S(k)\delta^3 (\mathbf{k} - \mathbf{k}'). \tag{D.114}$$

The origin of fluctuations is found in the quantum fluctuations of the scalar field during the de Sitter phase. Indeed, if we follow a given comoving scale $a(t)/k$ with time (see Fig. D.5), we have seen in Section D.3.4 that, some time during the matter-dominated phase, it enters the Hubble radius. Since the Hubble radius is constant during inflation, this means that at a much earlier time, it has emerged from the Hubble radius of the de Sitter phase. In this scenario, the origin of the fluctuations is thus found in the heart of the de Sitter event horizon: using quantum field theory in curved space, one may compute the amplitude of the quantum fluctuations of the scalar field; their wavelengths evolve as $a(t)/k$ until they outgrow the event horizon, *i.e.*

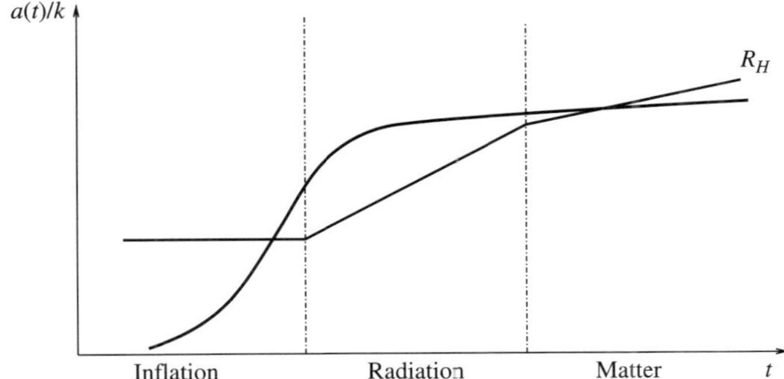

Fig. D.5 Evolution of a physical comoving fluctuation scale with respect to the Hubble radius during the inflation phase $(R_H(t) = H_{\mathrm{vac}}^{-1})$, the radiation-dominated phase $(R_H(t) = 2t)$, and matter-dominated phases $(R_H(t) = 3t/2)$.

the Hubble radius; they freeze out and continue to evolve classically. The fluctuation spectrum produced is given by

$$\mathcal{P}_S(k) = \left[\left(\frac{H^2}{\dot{\phi}^2} \right) \left(\frac{H}{2\pi} \right)^2 \right]_{k=aH} = \frac{128\pi}{3M_P^6} \left(\frac{V^3}{V'^2} \right)_{k=aH}, \tag{D.115}$$

where the subscript $k = aH$ means that the quantities are evaluated at Hubble radius crossing, as expected. We also note that H sets the scale of quantum fluctuations in the de Sitter phase.

The scalar spectral index $n_S(k)$ is computed to be:

$$n_S(k) - 1 \equiv \frac{d \ln \mathcal{P}_S(k)}{d \ln k} = -6\varepsilon + 2\eta. \tag{D.116}$$

Thus, because of the slow-roll , the fluctuation spectrum is almost scale invariant, a result that we have alluded to when we discussed the origin of CMB fluctuations.

Besides scalar fluctuations, inflation produces fluctuations which have a tensor structure, *i.e.* primordial gravitational waves. They can be written as perturbations of the metric of the form

$$ds^2 = a^2 \left[\eta_{\mu\nu} + h_{\mu\nu}^{TT} \right], \tag{D.117}$$

where $h_{\mu\nu}^{TT}$ is a traceless transverse tensor (which has two physical degrees of freedom, *i.e.* two polarizations). The corresponding tensor spectrum is given by

$$\mathcal{P}_T(k) = \frac{64\pi}{M_P^2} \left(\frac{H}{2\pi} \right)^2, \tag{D.118}$$

with a corresponding spectral index

$$n_T(k) \equiv \frac{d \ln \mathcal{P}_T(k)}{d \ln k} = -2\varepsilon. \tag{D.119}$$

We note that the ratio $\mathcal{P}_T/\mathcal{P}_S$ depends only on (V'/V) and thus on ε, which yields a consistency condition: $\mathcal{P}_T/\mathcal{P}_S = -8\,n_T$.

We conclude this discussion by reviewing briefly the three main classes of inflation models:

- Large field models $(0 < \eta < 2\varepsilon)$. This corresponds to the chaotic inflation scenario [274] where a field ϕ of value of the order of a few M_P slowly rolls down a power law potential typically $V(\phi) = \lambda\phi^4$. One drawback is the large value of the field which makes it necessary to include all nonrenormalizable corrections of order $(\phi/M_P)^n$.
- Small field models $(\eta < 0)$. In this class, illustrated first by the new inflation scenario, the field ϕ starts at a small value and rolls along an almost flat plateau (where $V''(\phi) < 0$) before falling to its ground state.
- Hybrid models $(0 < 2\varepsilon < \eta)$. The field rolls down to a minimum of large vacuum energy (where $V''(\phi) < 0$) from a small initial value. Inflation ends because, close to this minimum, another direction in field space takes over and brings the system to a minimum of vanishing energy [276]. We will see examples of such a scenario in Chapter 11.

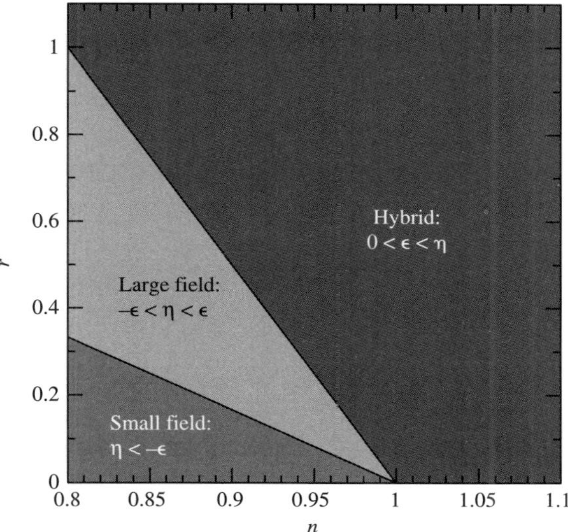

Fig. D.6 Regions corresponding to the different inflation models in the plot $r = C_2^{\text{tensor}}/C_2^{\text{scalar}}$ versus n_s [122].

These models make different predictions for scalar and tensor perturbations, as can be seen from Fig. D.6.

D.5 Cosmic strings

Among the topological defects, cosmic strings have the most interesting cosmological evolution: because of their interactions they may disappear with time and thus do not necessarily overclose the Universe.

Let us first illustrate what is a cosmic string on the simplest example of a quartic potential:

$$V(\Phi^\dagger\Phi) = \tfrac{1}{2}\lambda \left(\Phi^\dagger\Phi - \eta^2\right)^2, \tag{D.120}$$

which is invariant under the global $U(1)$ transformation: $\Phi \to e^{-i\alpha}\Phi$. Let us consider the field configuration (in cylindrical coordinates z, r, θ) away from the region where $r \sim 0$:

$$\Phi(z, r, \theta) = \eta e^{i\theta}. \tag{D.121}$$

Obviously, at $r = 0$, one must have $\Phi(z, 0, \theta) = 0$. Thus some potential energy $V \sim \lambda\eta^4$ is localized around $r = 0$, typically on a distance $\rho \sim m^{-1} = \lambda^{-1/2}\eta^{-1}$ (the Compton wavelength of the scalar field).

This finite energy configuration extending along the z axis is called a global string. Its energy per unit length is accordingly:

$$\mu(R) \sim \lambda\eta^4 \times \lambda^{-1}\eta^{-2} + \int_\rho^R \left|\frac{1}{r}\frac{\partial\Phi}{\partial\theta}\right|^2 2\pi r \, dr,$$

$$\sim \eta^2 + 2\pi\eta^2 \int_\rho^R \frac{dr}{r} \sim \eta^2 + 2\pi\eta^2 \ln(Rm), \tag{D.122}$$

where we have introduced a cut-off distance R which is the distance between two strings or the diameter of the loop, in the case of a closed string. We deduce that there is an interaction at long distance due to the presence of a Goldstone boson: the force per unit length is simply $d\mu/dR \sim \eta^2/R$. This is analogous to the case of vortex lines in liquid helium.

In the case of a gauge symmetry, the structure is richer since the solution (D.121) has to be associated with a gauge configuration. We take the example of an abelian gauge symmetry (identified with electrodynamics):

$$\mathcal{L} = -\tfrac{1}{4}F^{\mu\nu}F_{\mu\nu} + D^\mu\Phi^\dagger D_\mu\Phi - V(\Phi^\dagger\Phi), \tag{D.123}$$
$$D_\mu\Phi = \partial_\mu\Phi - igA_\mu\Phi,$$

where V is still given by (D.120).

At the center of the string, $\Phi = 0$ and $V = \tfrac{1}{2}\lambda\eta^4$. Far from the center, we have

$$\Phi = \eta e^{in\theta}. \tag{D.124}$$

In order to minimize also the kinetic energy, we choose a gauge configuration

$$A_\mu = -\frac{i}{g}\partial_\mu \ln\Phi = \frac{n}{g}\partial_\mu\theta = \frac{n}{g}\frac{1}{r}\delta^\theta_\mu \tag{D.125}$$

which ensures both $D_\mu\Phi = 0$ and $F_{\mu\nu} = 0$. Thus, the energy density vanishes outside the core of the string: this local string corresponds truly to a localized configuration of energy. The potential energy per unit length is then simply

$$\mu_\Phi \sim \lambda\eta^4 \times \left(m^{-1}\right)^2 \sim \eta^2. \tag{D.126}$$

Also, if $\mathbf{S}$ is a surface delimited by a curve $\mathcal{C}$ around the string, we have

$$\int_S \mathbf{B}\cdot\mathbf{dS} = \int_\mathcal{C} \mathbf{A}\cdot\mathbf{dl} = \frac{n}{g}\int_\mathcal{C} d\theta = \frac{2\pi n}{g}. \tag{D.127}$$

Hence the string carries n units of elementary magnetic flux $2\pi/g$. The size of the surface through which $\mathbf{B}$ is nonvanishing is typically its Compton wavelength $M_A^{-1} = (g\eta)^{-1}$. Thus, using (D.127), we have

$$B\left(M_A^{-1}\right)^2 \sim \frac{2\pi n}{g} \quad\text{or}\quad B \sim 2\pi ng\eta^2. \tag{D.128}$$

The corresponding energy per unit length is

$$\mu_A \sim B^2\left(M_A^{-1}\right)^2 \sim \eta^2, \tag{D.129}$$

hence of the same order as μ_Φ: the total energy per unit length μ scales as η^2.

Cosmic strings interact [249]: if two strings cross each other, the four ends have a probability (close to one for the structureless strings that we have considered so far) to change partners, leaving a pair of sharply kinked strings which straighten with time through gravitational wave emission. Occasionally, a string crosses itself, which leads to the formation of a loop. This loop shrinks until it vanishes. This has the effect of reducing the size of strings (and loops).

In realistic models, scalar fields couple to fermions, for example through Yukawa couplings:

$$\mathcal{L} = \bar{\Psi} i \gamma_\mu D^\mu \Psi + \lambda \bar{\Psi}_L \Phi \Psi_R + \text{h.c.} \tag{D.130}$$

Since Φ vanishes in the core of the string, the fermions become massless. We thus have zero modes which may move along the string: currents propagate along the string which is thus called a superconducting string [376]. Such currents may stabilize the loops: it remains then to check how the stabilized loops, sometimes called vortons, contribute to the energy density of the Universe. Supersymmetric models provide examples of such configurations, as we see in Section 11.4 of Chapter 11.

Further reading

- S. Weinberg, *Gravitation and cosmology: principles and applications of the general theory of relativity*, John Wiley and Sons 1972.
- J. Rich, *Fundamentals of Cosmology*, Springer 2001.

Exercises

<u>Exercise 1</u> We detail here the derivation of Einstein's equations by varying the action (D.14) with respect to the metric $g_{\mu\nu}$.

(a) Show that $\delta g^{\mu\nu} = -g^{\mu\rho} g^{\nu\sigma} \delta g_{\rho\sigma}$.

(b) Show that $\delta \sqrt{|g|} = \frac{1}{2} \sqrt{|g|} g^{\mu\nu} \delta g_{\mu\nu}$.

(c) Prove that $\sqrt{|g|} g^{\mu\nu} \delta R_{\mu\nu}$ is a total derivative.

(d) Vary the action (D.14) with respect to the metric $g_{\mu\nu}$ to obtain Einstein's equations (D.16).

Hints:

(a) $g^{\mu\nu} g_{\nu\rho} = \delta^\mu_\rho$ yields $\delta g^{\mu\nu} g_{\nu\rho} + g^{\mu\nu} \delta g_{\nu\rho} = 0$.

(b) Use $\det A = \exp(\text{Tr}\ln A)$ to obtain $\delta g = g g^{\mu\nu} \delta g_{\mu\nu}$.

(c) Check that $\delta R_{\mu\nu} = D_\alpha (\delta \Gamma^\alpha{}_{\mu\nu}) - D_\nu (\delta \Gamma^\alpha{}_{\mu\alpha})$. Then the metric tensor in $g_{\mu\nu} \delta R_{\mu\nu}$ can be written inside the covariant derivatives since $D_\alpha g_{\mu\nu} = 0$. Finally, use the property $D_\mu V^\mu = |g|^{-1/2} \partial_\mu \left(\sqrt{|g|} V^\mu \right)$ valid for any vector V^μ to prove that

$$\sqrt{|g|} g^{\mu\nu} \delta R_{\mu\nu} = \partial_\alpha \left(\sqrt{|g|} g^{\mu\nu} \delta \Gamma^\alpha{}_{\mu\nu} \right) - \partial_\nu \left(\sqrt{|g|} g^{\mu\nu} \delta \Gamma^\alpha{}_{\mu\alpha} \right).$$

<u>Exercise 2</u> In the case of the Robertson–Walker metric (D.18):

(a) compute the nonvanishing Christoffel symbols;

(b) using the fact that the Ricci tensor associated with the three-dimensional metric γ_{ij} is simply $R_{ij}(\gamma) = 2k\gamma_{ij}$, compute the components of the Ricci tensor and the scalar curvature;

(c) deduce the components of the Einstein tensor (D.25) and (D.26).

Hints:

(a) $\Gamma^i{}_{j0} = \delta^i_j \dot{a}/a$, $\Gamma^0{}_{ij} = a\dot{a}\gamma_{ij}$, $\Gamma^i{}_{jk} = \Gamma^i{}_{jk}(\gamma)$.

(b) $R_{00} = -3\ddot{a}/a$, $R_{ij} = \left(2k + \ddot{a}a + 2\dot{a}^2\right)\gamma_{ij}$, $R = -6\epsilon\left(k + \ddot{a}a + \dot{a}^2\right)/a^2$.

<u>Exercise 3</u> Prove (D.50) from (D.46).

Hint: Using (D.46) with (D.33), we obtain, for $z \ll 1$,

$$a_0 r \sim \ell_{H_0} \int_0^z dz/\left[1 + z(2 + \Omega_M + 2\Omega_R - 2\Omega_\Lambda)\right]^{1/2}.$$

Then integrate and use (D.37).

<u>Exercise 4</u> *We compute exactly the luminosity distance* $d_L = a_0 r(1 + z)$ *or angular distance* $d_A = a_0 r/(1 + z)$ *in the case of a matter-dominated Universe.*
    Defining

$$\zeta_k(r) \equiv \begin{cases} \sin^{-1} r & k = +1 \\ r & k = 0 \\ \sinh^{-1} r & k = -1. \end{cases} \tag{D.131}$$

use (D.46) which reads, in the case of a matter-dominated Universe,

$$a_0 \zeta_k(r) = \ell_{H_0} \int_0^z \frac{dz}{\left[\Omega_M(1 + z)^3 + (1 - \Omega_M)(1 + z)^2\right]^{1/2}} \tag{D.132}$$

to prove Mattig's formula [285]:

$$a_0 r = 2\ell_{H_0} \frac{\Omega_M z + (\Omega_M - 2)\left[\sqrt{1 + \Omega_M z} - 1\right]}{\Omega_M^2(1 + z)}. \tag{D.133}$$

Hints: For $k \neq 0$, change to the coordinate $u^2 = k(\Omega - 1)/\left[\Omega(1 + z)\right]$ in order to compute the integral (D.132). Using the last of equations (D.32), which reads $\ell_{H_0}^2/a_0^2 = k(\Omega - 1)$, one obtains

$$\zeta_k(r) = 2\left(\zeta_k\left[\sqrt{\frac{k(\Omega - 1)}{\Omega}}\right] - \zeta_k\left[\sqrt{\frac{k(\Omega - 1)}{(1 + z)\Omega}}\right]\right),$$

from which (D.133) can be inferred.

Appendix E

Renormalization group equations

We gather here a set of renormalization group equations [102] which are scattered in the book. The notations are the ones used in the main text. In the case of Yukawa couplings, soft scalar masses, and A-terms, we give the third family couplings. To obtain the corresponding formulas for the other two families, one should just neglect the Yukawa coupling terms on the right-hand side.

Throughout this appendix, we define $t \equiv \ln(\mu/\mu_0)$ where μ is the renormalization scale.

E.1   Gauge couplings

In the supersymmetric case,

$$16\pi^2 \frac{dg_i}{dt} = -b_i^{(1)} \, g_i^3 + O(g^5) \tag{E.1}$$

where, for N_F families and N_D Higgs doublets (the minimal model has $N_D = 2$), we have

$$b_1^{(1)} = -2N_F - \frac{3}{10}N_D, \tag{E.2}$$

$$b_2^{(1)} = 6 - 2N_F - \frac{1}{2}N_D, \tag{E.3}$$

$$b_3^{(1)} = 9 - 2N_F. \tag{E.4}$$

The $U(1)$ gauge coupling has the standard $SU(5)$ normalization ($g_1^2 = 5g'^2/3$).

E.2 μ parameter

We note that the μ parameter does not appear in the evolution equations of the other mass parameters.

$$16\pi^2 \frac{d\mu}{dt} = \mu \left(3|\lambda_t|^2 + 3|\lambda_b|^2 + |\lambda_\tau|^2 - 3g_2^2 - \frac{3}{5}g_1^2 \right). \tag{E.5}$$

E.3 Anomalous dimensions

At one loop, we have the following anomalous dimensions $\gamma_i = d\ln Z_i/dt$:

$$16\pi^2\gamma_Q = -2|\lambda_t|^2 - 2|\lambda_b|^2 + \frac{16}{3}g_3^2 + 3g_2^2 + \frac{1}{15}g_1^2, \tag{E.6}$$

$$16\pi^2\gamma_T = -4|\lambda_t|^2 + \frac{16}{3}g_3^2 + \frac{16}{15}g_1^2, \tag{E.7}$$

$$16\pi^2\gamma_B = -4|\lambda_b|^2 + \frac{16}{3}g_3^2 + \frac{4}{15}g_1^2, \tag{E.8}$$

$$16\pi^2\gamma_L = -2|\lambda_\tau|^2 + 3g_2^2 + \frac{3}{5}g_1^2, \tag{E.9}$$

$$16\pi^2\gamma_T = -4|\lambda_\tau|^2 + \frac{12}{5}g_1^2, \tag{E.10}$$

$$16\pi^2\gamma_{H_2} = -6|\lambda_t|^2 + 3g_2^2 + \frac{3}{5}g_1^2, \tag{E.11}$$

$$16\pi^2\gamma_{H_1} = -2|\lambda_\tau|^2 - 6|\lambda_b|^2 + 3g_2^2 + \frac{3}{5}g_1^2. \tag{E.12}$$

E.4 Yukawa couplings

For the third generation, we have, in the case of R-parity conservation,

$$16\pi^2\frac{d\lambda_t}{dt} = \lambda_t\left(6|\lambda_t|^2 + |\lambda_b|^2 - \frac{16}{3}g_3^2 - 3g_2^2 - \frac{13}{15}g_1^2\right), \tag{E.13}$$

$$16\pi^2\frac{d\lambda_b}{dt} = \lambda_b\left(|\lambda_t|^2 + 6|\lambda_b|^2 + |\lambda_\tau|^2 - \frac{16}{3}g_3^2 - 3g_2^2 - \frac{7}{15}g_1^2\right), \tag{E.14}$$

$$16\pi^2\frac{d\lambda_\tau}{dt} = \lambda_\tau\left(3|\lambda_b|^2 + 4|\lambda_\tau|^2 - 3g_2^2 - \frac{9}{5}g_1^2\right). \tag{E.15}$$

E.5 Gaugino masses

We have, for $i = 1, 2, 3$,

$$8\pi^2\frac{dM_i}{dt} = -b_i^{(1)}g_i^2 M_i, \tag{E.16}$$

where the one-loop coefficients are given in (E.2)–(E.4): $b_1^{(1)} = -33/5$, $b_2^{(1)} = -1$ and $b_3^{(1)} = 3$ for $N_F = 3$ and $N_D = 2$.

E.6   Soft scalar masses

We have for the Higgs soft scalar mass terms:

$$8\pi^2 \frac{dm^2_{H_1}}{dt} = 3|\lambda_b|^2 \left(m^2_Q + m^2_B + m^2_{H_1} + |A_b|^2\right) + |\lambda_\tau|^2 \left(m^2_L + m^2_T + m^2_{H_1} + |A_\tau|^2\right)$$

$$- \left(3g_2^2|M_2|^2 + \frac{3}{5}g_1^2|M_1|^2\right) - \frac{3}{5}g_1^2\xi_1, \tag{E.17}$$

$$8\pi^2 \frac{dm^2_{H_2}}{dt} = 3|\lambda_t|^2 \left(m^2_Q + m^2_T + m^2_{H_2} + |A_t|^2\right)$$

$$- \left(3g_2^2|M_2|^2 + \frac{3}{5}g_1^2|M_1|^2\right) + \frac{3}{5}g_1^2\xi_1, \tag{E.18}$$

and for the soft sfermion masses of the third family

$$8\pi^2 \frac{dm^2_Q}{dt} = |\lambda_t|^2 \left(m^2_Q + m^2_T + m^2_{H_2} + |A_t|^2\right) + |\lambda_b|^2 \left(m^2_Q + m^2_B + m^2_{H_1} + |A_b|^2\right)$$

$$- \left(\frac{16}{3}g_3^2|M_3|^2 + 3g_2^2|M_2|^2 + \frac{1}{15}g_1^2|M_1|^2\right) + \frac{1}{5}g_1^2\xi_1, \tag{E.19}$$

$$8\pi^2 \frac{dm^2_T}{dt} = 2|\lambda_t|^2 \left(m^2_Q + m^2_T + m^2_{H_2} + |A_t|^2\right)$$

$$- \left(\frac{16}{3}g_3^2|M_3|^2 + \frac{16}{15}g_1^2|M_1|^2\right) - \frac{4}{5}g_1^2\xi_1, \tag{E.20}$$

$$8\pi^2 \frac{dm^2_B}{dt} = 2|\lambda_b|^2 \left(m^2_Q + m^2_B + m^2_{H_1} + |A_b|^2\right)$$

$$- \left(\frac{16}{3}g_3^2|M_3|^2 + \frac{4}{15}g_1^2|M_1|^2\right) + \frac{2}{5}g_1^2\xi_1, \tag{E.21}$$

$$8\pi^2 \frac{dm^2_L}{dt} = |\lambda_\tau|^2 \left(m^2_L + m^2_T + m^2_{H_1} + |A_\tau|^2\right)$$

$$- \left(3g_2^2|M_2|^2 + \frac{3}{5}g_1^2|M_1|^2\right) - \frac{3}{5}g_1^2\xi_1, \tag{E.22}$$

$$8\pi^2 \frac{dm^2_T}{dt} = 2|\lambda_\tau|^2 \left(m^2_L + m^2_T + m^2_{H_1} + |A_\tau|^2\right)$$

$$- \frac{12}{5}g_1^2|M_1|^2 + \frac{6}{5}g_1^2\xi_1, \tag{E.23}$$

where[1]

$$\xi_1 \equiv \sum_{\text{scalar } j} y_j m^2_j \tag{E.24}$$

$$= \frac{1}{2}m^2_{H_2} - \frac{1}{2}m^2_{H_1} + \sum_{i=1}^{3}\left(\frac{1}{2}m^2_{Q_i} - m^2_{U_i} + \frac{1}{2}m^2_{D_i}\right) - \left(\frac{1}{2}m^2_{L_i} - \frac{1}{2}m^2_{E_i}\right).$$

---

[1] Note that similar terms for nonabelian gauge symmetry would be of the form $\xi^a = \text{Tr}\,(T^a m) = 0$. This is why such a term appears only for the abelian $U(1)$ symmetry.

E.7   *A*-terms

As for the Yukawa couplings, we give only the equations for the third family:

$$8\pi^2 \frac{dA_t}{dt} = 6\lambda_t^2 A_t + \lambda_b^2 A_b - \frac{16}{3} g_3^2 M_3 - 3g_2^2 M_2 - \frac{13}{15} g_1^2 M_1, \tag{E.25}$$

$$8\pi^2 \frac{dA_b}{dt} = \lambda_t^2 A_t + 6\lambda_b^2 A_b + \lambda_\tau^2 A_\tau - \frac{16}{3} g_3^2 M_3 - 3g_2^2 M_2 - \frac{7}{15} g_1^2 M_1, \tag{E.26}$$

$$8\pi^2 \frac{dA_\tau}{dt} = 3\lambda_b^2 A_b + 4\lambda_\tau^2 A_\tau - 3g_2^2 M_2 - \frac{9}{5} g_1^2 M_1. \tag{E.27}$$

Also, for $B \equiv B_\mu / \mu$, we have:

$$8\pi^2 \frac{dB}{dt} = 3\lambda_t^2 A_t + 3\lambda_b^2 A_b + \lambda_\tau^2 A_\tau - 3g_2^2 M_2 - \frac{3}{5} g_1^2 M_1. \tag{E.28}$$

E.8   **Dimensional reduction**

All preceding equations are understood in the so-called $\overline{DR}$ renormalization, where DR stands for Dimensional Reduction. Let us explain how this scheme is defined.

In usual gauge theories, the standard renormalisation scheme is based on dimensional regularization: divergent integrals are computed by performing an analytic continuation to $D = 4 - 2\epsilon$ spacetime dimensions. Since Ward identities do not depend on the dimension of spacetime, this procedure does not spoil gauge symmetries. The most remarkable consequence of such a continuation is that the gauge coupling is no longer dimensionless: one may write the bare gauge coupling g_0 in terms of the (dimensionless) renormalized one g_R

$$g_0 = \mu^\epsilon \frac{1}{Z^{1/2}} g_R \,, \tag{E.29}$$

where Z is the wave function renormalization constant and the scale μ plays the rôle of renormalization scale in dimensional regularization. When computing integrals over loop momenta, divergences appear as poles $1/\epsilon^k$.

In the minimal subtraction (noted MS), renormalized quantities are obtained by subtracting these poles (in the $\overline{MS}$ scheme, one also subtracts a generic additive constant $\ln 4\pi - \gamma$, where γ is the Euler constant).

It turns out that dimensional regularization is not compatible with the supersymmetry algebra. A straightforward way to see this is to note that, in $D \neq 4$ dimensions, the number of on-shell degrees of freedom of a gauge field $(D - 2)$ does not match the one of a gaugino Majorana field $(2^{(D-2)/2})$.

One illustration of the fact that the $\overline{MS}$ scheme does not respect supersymmetry is the fact that the renormalized gauge coupling g measured in gauge boson interactions no longer coincides with the coupling $\hat{g}$ of a gaugino to the fields of a chiral supermultiplet (cf. (3.43)):

$$\mathcal{L} = \hat{g} q \sqrt{2} \left(\bar{\lambda} \Psi_L \, \phi^* + \bar{\lambda} \Psi_R^c \, \phi \right) \,. \tag{E.30}$$

This has led to modify the regularization prescription in the following way: keep the algebra of fields four dimensional while still performing the dimensional continuation in loop momenta. This is the dimensional reduction scheme ($\overline{DR}$ corresponding to $\overline{MS}$) [85, 340].

Since $\overline{DR}$ is compatible with supersymmetry, we have as expected $g = \hat{g}$ in this scheme. On the other hand,

$$g_{\overline{MS}} = g_{\overline{DR}} \left[1 - \frac{g^2}{96\pi^2} C_2(G) \right] , \tag{E.31}$$

$$\hat{g}_{\overline{MS}} = g_{\overline{DR}} \left[1 + \frac{g^2}{96\pi^2} (C_2(G) - C_2(R)) \right] , \tag{E.32}$$

where $C_2(G)$ is defined in (9.18) and $C_2(R)\delta_{ij} = \sum_a (t^a t^a)_{ij}$ (R is the representation of the chiral supermultiplet of (E.30) under the gauge group G).

We also note the following relation for the gaugino mass parameters:

$$M_{\lambda\overline{MS}} = M_{\lambda\overline{DR}} \left[1 + \frac{g^2}{16\pi^2} C_2(G) \right] . \tag{E.33}$$

A complication arises in dimensional reduction: from the point of view of D dimensions, local invariance applies only to the components $A_0^a, \cdots, A_{D-1}^a$ of the vector fields. The remaining 2ϵ components $A_D^a, \cdots, A_3^a$ are scalars under the Lorentz group in D dimensions: they are called ϵ-scalars. They induce additional renormalization counter terms and care must be taken at higher orders to take these fields into account.

Finally, we note that, in the $\overline{DR}$ scheme, heavy thresholds effects are very simply taken into account in renormalization group analyses: they are properly described by a simple step function localized precisely at the mass of the heavy field [7].

Exercises

Exercise 1 Show that ξ_1 defined in (E.24) satisfies

$$8\pi^2 \frac{d\xi_1}{dt} = -b_1 g_1^2 \xi_1 \tag{E.34}$$

and solve this equation.

Hints: The equations for the first two families are read from (E.19)–(E.23) by neglecting the Yukawa couplings.

Solution: $\xi_1(\mu) = g_1^2(\mu)\xi_1(\mu_0)/g_1^2(\mu_0)$.

Bibliography

[1] I. Affleck, M. Dine, and N. Seiberg. Dynamical supersymmetry breaking in supersymmetric QCD. *Nucl. Phys.*, B241:493–534, 1984.

[2] I. Affleck, M. Dine, and N. Seiberg. Dynamical supersymmetry breaking in four dimensions and its phenomenological implications. *Nucl. Phys.*, B256:557–599, 1985.

[3] I. Affleck and M. Dine. A new mechanism for baryogenesis. *Nucl. Phys.*, B249:361, 1985.

[4] Y. Aharonov and D. Bohm. *Proc. Phys. Soc. (London)*, B62:8, 1949.

[5] A. Albrecht and P.J. Steinhardt. Cosmology for Grand Unified Theories with radiatively induced symmetry breaking. *Phys. Rev. Lett.*, 48:1220–1223, 1982.

[6] G. Altarelli and R. Barbieri. Vacuum polarization effects of new physics on electroweak processes. *Phys. Lett.*, B253:161–167, 1991.

[7] I. Antoniadis, C. Kounnas, and K. Tamvakis. Simple treatment of threshold effects. *Phys. Lett.*, 119B:377–380, 1982.

[8] T. Appelquist and J. Carazzone. Infrared singularities and massive fields. *Phys. Rev.*, D11:2856, 1975.

[9] N. Arkani-Hamed and S. Dimopoulos. Supersymmetric unification without low energy supersymmetry and signatures for fine-tuning at the LHC. *JHEP*, 0506:073, 2005.

[10] N. Arkani-Hamed, J. March-Russell, and H. Murayama. Building models of gauge-mediated supersymmetry breaking without a messenger sector. *Nucl. Phys.*, B509:3–32, 1998.

[11] C. Armendariz-Pico, V. Mukhanov, and P.J. Steinhardt. A dynamical solution to the problem of a small cosmological constant and late time cosmic acceleration. *Phys. Rev. Lett.*, 85:4438–4441, 2000.

[12] T. Asaka, K. Hamaguchi, M. Kawasaki, and T. Yanagida. Leptogenesis in inflationary universe. *Phys. Rev.*, D61:083512, 2000.

[13] J. Atick, L. Dixon, and A. Sen. String calculation of Fayet-Iliopoulos D-terms in arbitrary supersymmetric configurations. *Nucl. Phys.*, B292:109–149, 1987.

[14] M.F. Atiyah and I.M. Singer. The index of elliptic operators 1.3. *Ann. Math.*, 87:484–530,546–604, 1968.

[15] M.F. Atiyah and I.M. Singer. The index of elliptic operators 4. *Ann. Math.*, 93:119–138, 1971.

[16] B. Aubert, *et al.* [BABAR Collaboration]. Measurement of the $B \to X_s\gamma$ branching fraction and photon spectrum from a sum of exclusive final states. *Phys. Rev.*, D72, 052004, 2005.

[17] K.S. Babu, J.C. Pati, and F. Wilczek. Fermion masses, neutrino oscillations, and proton decay in the light of SuperKamiokande. *Nucl. Phys.*, B566:33–91, 2000.

[18] H. Baer and M. Brhlik. Neutralino dark matter in minimal supergravity: direct detection versus collider searches. *Phys. Rev.*, D57:567–577, 1998.

[19] J. Bagger, T. Moroi, and E. Poppitz. Anomaly mediation in supergravity theories. *JHEP*, 0004:009, 2000.

[20] T. Banks and M. Dine. Coping with strongly coupled string theory. *Phys. Rev.*, D50:7454–7466, 1994.

[21] T. Banks and A. Zaks. On the phase structure of vector-like gauge theories with massless fermions. *Nucl. Phys.*, B196:189–204, 1982.

[22] R. Barbieri, F. Caravaglios, and M. Frigeni. The supersymmetric higgs for heavy superpartners. *Phys. Lett.*, B258:167–170, 1991.

[23] R. Barbieri and G.F. Giudice. Upper bounds on supersymmetric particle masses. *Nucl. Phys.*, B306:63, 1988.

[24] R. Barbieri and G.F. Giudice. $b \to s$ gamma decay and supersymmetry. *Phys. Lett.*, B309:86–90, 1993.

[25] R. Barbieri and A. Strumia. About the fine tuning price of LEP. *Phys. Lett.*, B433:63–66, 1998.

[26] R. Barbieri and A. Strumia. The LEP Paradox. Fourth Rencontre du Vietnam: International conference on physics at extreme high energies, Hanoi, 2000.

[27] R. Barbier et al. R-parity violating supersymmetry. *Phys. Rep.*, 420:1–195, 2005.

[28] M. Beck et al. Searching for dark matter with the enriched Ge detectors of the Heidelberg-Moscow $\beta\beta$ experiment. *Phys. Lett.*, B336:141–146, 1994.

[29] G. Belanger, F. Boudjema, A. Pukhov, and A. Semenov. Micromegas: a program for calculating the relic density in the MSSM. *Comput. Phys. Commun.*, 149: 103–120, 2002.

[30] G.W. Bennett et al. [E821 Collaboration]. Measurement of the negative muon anomalous magnetic moment to 0.7 ppm. *Phys. Rev. Lett.*, 92:161–802, 2001.

[31] L. Bergstrom, P. Ullio, and J.H. Buckley. Observability of gamma rays from dark matter neutralino annihilation in the Milky Way halo. *Astropart. Phys.*, 9:137–162, 1998.

[32] S. Bertolini, F. Borzumati, A. Masiero, and G. Ridolfi. Effects of supergravity-induced electroweak breaking on rare B-decays and mixings. *Nucl. Phys.*, B353:591–649, 1991.

[33] J. Bijnens and C. Wetterich. Fermion masses from symmetry. *Nucl. Phys.*, B283:237, 1987.

[34] P. Binétruy, S. Dawson, M. K. Gaillard, and I. Hinchliffe. Supersymmetry breaking in superstring inspired models. *Phys. Rev.*, D37:2633, 1988.

[35] P. Binétruy, S. Dawson, and I. Hinchliffe. Gaugino masses in superstring models. *Phys. Rev.*, D35:2215, 1987.

[36] P. Binétruy and E. Dudas. Gaugino condensation and the anomalous $U(1)$. *Phys. Lett.*, B389:503–509, 1996.

[37] P. Binétruy and G. Dvali. D-term inflation. *Phys. Lett.*, B388:241, 1996.

[38] P. Binétruy, G. Girardi, R. Grimm, and M. Müller. Kähler transformations and the coupling of matter and Yang-Mills fields to supergravity. *Phys. Lett.*, B189:83–88, 1987.

[39] P. Binétruy, G. Girardi, and R. Grimm. Supergravity couplings: a geometric formulation. *Phys. Rep.*, 343:255–462, 2001.

[40] P. Binétruy, G. Girardi, and P. Salati. Constraints on a system of two neutral fermions from cosmology. *Nucl. Phys.*, B237:285, 1984.

[41] P. Binétruy, N. Irges, S. Lavignac, and P. Ramond. Anomalous U(1) and low-energy physics: the power of D flatness and holomorphy. *Phys. Lett.*, B403:38–46, 1997a.

[42] P. Binétruy, S. Lavignac, S. Petcov, and P. Ramond. Quasi-degenerate neutrinos from an abelian family symmetry. *Nucl. Phys.*, B496:3–23, 1997.

[43] P. Binétruy, S. Lavignac, and P. Ramond. Yukawa textures with anomalous horizontal abelian symmetry. *Nucl. Phys.*, B477:353–377, 1996.

[44] P. Binétruy and P. Ramond. Yukawa textures and anomaly. *Phys. Lett.*, B350: 49–57, 1995.

[45] P. Binétruy and C.A. Savoy. Higgs and top masses in a non-minimal supersymmetric theory. *Phys. Lett.*, B277:453–458, 1992.

[46] P. Binétruy. Models of dynamical supersymmetry breaking and quintessence. *Phys. Rev.*, D60:063502, 1999.

[47] E.B. Bogomol'nyi. The stability of classical solutions. *Sov. J. Nucl. Phys.*, 24:449–454, 1976.

[48] A. Bouquet, J. Kaplan, and C.A. Savoy. Low-energy constraints on supergravity parameters. *Nucl. Phys.*, B262:299–316, 1985.

[49] P. Brax and J. Martin. Quintessence and supergravity. *Phys. Lett.*, B468:40–45, 1999.

[50] A. Brignole, G. Degrassi, P. Slavich, and F. Zwirner. On the $\mathcal{O}(\alpha_t^2)$ two-loop corrections to the neutral higgs boson masses in the MSSM. *Nucl. Phys.*, B631: 195–218, 2002.

[51] A. Brignole, L. Ibánez, and C. Munoz. Toward a theory of soft terms for the supersymmetric Standard Model. *Nucl. Phys.*, B422:125–171, 1994. Erratum-ibid. B436 (1995) 747–748.

[52] A. Brignole. Radiative corrections to the supersymmetric neutral Higgs boson masses. *Phys. Lett.*, B281:284–294, 1992.

[53] J.D. Brown and J.C. Teitelboim. Dynamical neutralization of the cosmological constant. *Phys. Lett.*, B195:177–182, 1987.

[54] J.D. Brown and J.C. Teitelboim. Neutralization of the cosmological constant by membrane creation. *Nucl. Phys.*, B297:787–836, 1988.

[55] F. Buccella, J.P. Derendinger, S. Ferrara, and C.A. Savoy. Patterns of symmetry breaking in supersymmetric gauge theories. *Phys. Lett.*, 115B:375–379, 1982.

[56] W. Buchmuller and D. Wyler. CP violation and R invariance in supersymmetric models of strong and electroweak interactions. *Phys. Lett.*, B121:321, 1983.

[57] W. Buchmuller. Neutrinos, grand unification and leptogenesis. In *2001 European school of high-energy physics (Beatenberg, Switzerland)*, 2001.

[58] A.J. Buras, P. Gambino, M. Gorbahn, S. Jager, and L. Silvestrini. Universal unitarity triangle and physics beyond the standard model. *Phys. Lett.*, B500:161–167, 2001.

[59] A.J. Buras, M.E. Lautenbacher, and G. Ostermaier. Waiting for the top quark mass,... *Phys. Rev.*, D50:3433–3446, 1994.

[60] A.J. Buras, M. Misiak, M. Munz, and S. Pokorski. Theoretical uncertainties and phenomenological aspects of B $\rightarrow$ X$_s\gamma$ decay. *Nucl. Phys.*, B424:374–398, 1994.

[61] N. Cabibbo, L. Maiani, G. Parisi, and R. Petronzio. Bounds on the fermions and Higgs boson masses in grand unified theories. *Nucl. Phys.*, B158:295–305, 1979.

[62] N. Cabibbo and L. Maiani. The lifetime of charmed particles. *Phys. Lett.*, B79:109–111, 1978.

[63] R.R. Caldwell, R. Dave, and P.J. Steinhardt. *Phys. Rev. Lett.*, 75:2077, 1995.

[64] C.G. Callan, S. Coleman, and R. Jackiw. A new improved energy-momentum tensor. *Ann. Phys.(N.Y.)*, 59:42, 1970.

[65] C.G. Callan, D. Friedan, E.J. Martinec, and M. Perry. Strings in background fields. *Nucl. Phys.*, B262:593, 1985.

[66] P. Candelas, G. Horowitz, A. Strominger, and E. Witten. Vacuum configurations for superstrings. *Nucl. Phys.*, B258:46, 1985.

[67] M. Carena, M. Quiros, A. Riotto, I. Vilja, and C.E.M. Wagner. *Nucl. Phys.*, B503:404, 1997.

[68] S.M. Carroll. Quintessence and the rest of the world. *Phys. Rev. Lett.*, 81:3067–3070, 1998.

[69] J.A. Casas, J.R. Espinosa, and H.E. Haber. The Higgs mass in the MSSM infrared fixed point scenario. *Nucl. Phys.*, B526:3–20, 1998.

[70] J. A. Casas, A. Lleyda, C. Muñoz, and G.G. Ross. Hierarchical supersymmetry breaking and dynamical determination of compactification parameters by nonperturbative effects. *Nucl. Phys.*, B347:243–269, 1990.

[71] J. A. Casas, A. Lleyda, and C. Muñoz. Strong constraints on the parameter space of the MSSM from charge and color breaking. *Nucl. Phys.*, B471:3–58, 1996.

[72] N.P. Chang, S. Ouvry, and X. Wu. N=1 supergravity with nonminimal coupling: a class of models. *Phys. Rev. Lett.*, 51:327–330, 1983.

[73] M. Chanowitz and M.K. Gaillard. The Tev physics of strongly interacting W's and Z's. *Nucl. Phys.*, B261:379–431, 1985.

[74] K. Chetyrkin, M. Misiak, and M. Munz. Weak radiative B-meson decay beyond leading logarithms. *Phys. Lett.*, B400:206–219, 1997.

[75] P. Ciafaloni and A. Strumia. Naturalness upper bounds on gauge mediated soft terms. *Nucl. Phys.*, B494:41–53, 1997.

[76] CKMfitter group (J. Charles et al.). *Eur. Phys. J.*, 41:1–131, 2005. updated results available at http://ckmfitter.in2p3.fr.

[77] M. Claudson, L.J. Hall, and I. Hinchliffe. Low-energy supergravity: false vacua and vacuous predictions. *Nucl. Phys.*, B228:501–528, 1983.

[78] M. Claudson and M. Halpern. Supersymmetric ground state wave functions. *Nucl. Phys.*, B250:689–715, 1985.

[79] A. Cohen, D. Kaplan, and A. Nelson. The more minimal supersymmetric standard model. *Phys.Lett.*, B388:588–598, 1996.

[80] S. Coleman and J. Mandula. All possible symmetries of the S-matrix. *Phys. Rev.*, 159:1251–1256, 1967.

[81] S. Coleman, S. Parke, A. Neveu, and C.M. Sommerfield. Can one dent a dyon? *Phys. Rev.*, 15:544–545, 1977.

[82] S. Coleman, J. Wess, and B. Zumino. Structure of phenomeological lagrangians. *Phys. Rev.*, 177:2239–2250, 1969.

[83] S. Coleman. The invariance of the vacuum is the invariance of the world. *Journ. Math. Phys.*, 7:787, 1966.

[84] J. C. Collins, (1984). Renormalization, Cambridge University Press.

[85] D.M. Cooper, D.R.T. Jones, and P. van Nieuwenhuizen. *Nucl. Phys.*, B167:479, 1980.

[86] E.J. Copeland, A.R. Liddle, D. Lyth, E.D. Stewart, and D. Wands. False vacuum inflation with Einstein gravity. *Phys. Rev.*, D49:6410–6433, 1994.

[87] E.J. Copeland, A.R. Liddle, and D. Wands. Exponential potentials and cosmological scaling solutions. *Phys. Rev.*, D57:4686–4690, 1998.

[88] Supernova cosmology project. *Astrophys. J.*, 517:565, 1999.

[89] C.D. Coughlan, G. German, G.G. Ross, and G. Segré. The one loop effective potential and supersymmetry breaking in superstring models. *Phys. Lett.*, 198B:467, 1987.

[90] G.D. Coughlan, W. Fischler, E.W. Kolb, S. Raby, and G.G. Ross. Cosmological problems for the Polonyi potential. *Phys.Lett.*, 131B:59–64, 1983.

[91] E. Cremmer, S. Ferrara, L. Girardello, and A. van Proeyen. Yang-Mills theories with local supersymmetry: Lagrangian, transformation laws and super-Higgs effect. *Nucl. Phys.*, B212:413–442, 1983.

[92] E. Cremmer, S. Ferrara, C. Kounnas, and D.V. Nanopoulos. Naturally vanishing cosmological constant in N=1 supergravity. *Phys. Lett.*, 133B:61–66, 1983.

[93] G. D'Ambrosio, G.F. Giudice, G. Isidori, and A. Strumia. Minimal flavor violation: an effective field theory approach. *Nucl. Phys.*, B645:155, 2002.

[94] T. Damour and F.J. Dyson. The Oklo bound on the time variation of the fine structure constant revisited. *Nucl. Phys.*, B480:37–54, 1996.

[95] T. Damour and K. Nordtvedt. General relativity as a cosmological attractor of tensor-scalar theories. *Phys. Rev. Lett.*, 70:2217–2219, 1993.

[96] T. Damour and A.M. Polyakov. The string dilaton and a least coupling principle. *Nucl. Phys.*, B423:532–558, 1994.

[97] S.C. Davis, A.-C. Davis, and W.B. Perkins. Cosmic string zero modes and multiple phase transitions. *Phys. Lett.*, B 408:81–90, 1997.

[98] S.C. Davis, A.-C. Davis, and M. Trodden. N=1 supersymmetric cosmic strings. *Phys. Lett.*, B 405:257–264, 1997.

[99] S. Dawson. R parity breaking in supersymmetric theories. *Nucl. Phys.*, B261:297, 1985.

[100] G. Degrassi, P. Gambino, and G.F. Giudice. $B \to X_s\gamma$ in supersymmetry: large contributions beyond the leading order. *JHEP*, 0012:009, 2000.

[101] J.P. Derendinger. Lecture notes on globally supersymmetric theories in four and two dimensions. Unpublished, preprint ETH-YH/90-21, 1990.

[102] J.P. Derendinger and C.A. Savoy. Quantum effects and $SU2) \times U(1)$ breaking in supergravity gauge theories. *Nucl. Phys.*, B237:307–328, 1984.

[103] J. P. Derendinger, L. E. Ibáñez, and H. P. Nilles. On the low-energy D = 4, N = 1 supergravity theory extracted from the D = 10, N = 1 superstring. *Phys. Lett.*, 155B:65, 1985.

[104] W. de Sitter. *Proc. Kon. Ned. Akad. Wer.*, 19:217, 1917.

[105] W. de Sitter. *Proc. Kon. Ned. Akad. Wer.*, 20:229, 1917.

[106] M. Dimopoulos, S. Thomas, and J.D. Wells. Sparticle spectroscopy and electroweak symmetry breaking with gauge-mediated supersymmetry breaking. *Nucl. Phys.*, B488:39–91, 1997.

[107] S. Dimopoulos and H. Georgi. Softly broken supersymmetry and SU(5). *Nucl. Phys.*, B193:150, 1981.

[108] S. Dimopoulos, S. Raby, and F. Wilczek. Supersymmmetry and the scale od unification. *Phys.Rev.*, D24:1681–1683, 1981.

[109] S. Dimopoulos and S. Raby. Supercolor. *Nucl. Phys.*, B192:353, 1981.

[110] S. Dimopoulos and S. Thomas. Dynamical relaxation of the supersymmetric CP violating phases. *Nucl. Phys.*, B465:23–33, 1996.

[111] M. Dine, W. Fischler, and M. Srednicki. Supersymmetric technicolor. *Nucl. Phys.*, B189:575–593, 1981.

[112] M. Dine, A.E. Nelson, Y. Nir, and Y. Shirman. New tools for low energy dynamical supersymmetry breaking. *Phys. Rev.*, D53:2658–2669, 1996.

[113] M. Dine, A.E. Nelson, and Y. Shirman. Low energy dynamical supersymmetry breaking simplified. *Phys. Rev.*, D51:1362–1370, 1995.

[114] M. Dine, Y. Nir, and Y. Shirman. Variations on minimal gauge mediated supersymmetry breaking. *Phys. Rev.*, D55:1501, 1997.

[115] M. Dine, R. Rohm, N. Seiberg, and H.P. Nilles. Gluino condensation in superstring models. *Phys. Lett.*, 156B:55, 1985.

[116] M. Dine, N. Seiberg, and E. Witten. Fayet-Iliopoulos terms in string theory. *Nucl. Phys.*, B289:589–598, 1987.

[117] M. Dine and N. Seiberg. Nonrenormalization theorems in superstring theory. *Phys. Rev. Lett.*, 57:2625, 1986.

[118] P.A.M. Dirac. Quantised singularities in the electromagnetic field. *Proc. Roy. Soc.*, A33:60–72, 1931.

[119] P.A.M. Dirac. The cosmological constants. *Nature (London)*, 139:323, 1937.

[120] P.A.M. Dirac. New basis for cosmology. *Proc. Roy. Soc. London*, A165:199–208, 1938.

[121] L.J. Dixon, V.S. Kaplunovsky, and J. Louis. Moduli dependence of string loop corrections to gauge coupling constants. *Nucl. Phys.*, B355:649–688, 1991.

[122] S. Dodelson, W.H. Kinney, and E.W. Kolb. Cosmic microwave background measurements can discreminaate among inflation models. *Phys. Rev.*, D56:3207–3215, 1997.

[123] M. Drees and M. M. Nojiri. Neutralino-nucleon scattering reexamined. *Phys. Rev.*, D48:3483–3501, 1993.

[124] M. Drees and M. M. Nojiri. Neutralino relic density in minimal N=1 supergravity. *Phys. Rev.*, D47:376–408, 1993.

[125] E. Dudas, C. Grojean, S. Pokorski, and C.A. Savoy. Abelian flavor symmetries in supersymmetric models. *Nucl. Phys.*, B481:85–108, 1996.

[126] E. Dudas. Theory and phenomenology of type I strings and M-theory. *Class. Quantum Grav.*, 17:R41–R116, 2000.

[127] A. Einstein. *Sitzungsber. Preuss. Akad. Wiss.*, phus.-math. Klasse VI:142, 1917.

[128] J. Ellis, K. Enqvist, D.V. Nanopoulos, and F. Zwirner. Aspects of the superunification od strong, electroweak and gravitational interactions. *Nucl. Phys.*, B276:14, 1986.

[129] J. Ellis, J.F. Gunion, H.E. Haber, L. Roszkowski, and F. Zwirner. Higgs bosons in a nonminimal supersymmetric model. *Phys. Rev.*, D39:844, 1989.

[130] J. Ellis, C. Kounnas, and D. Nanopoulos. No-scale supersymmetric GUTs. *Nucl. Phys.*, B247:373–395, 1984.

[131] J. Ellis, C. Kounnas, and D. Nanopoulos. Phenomenological SU(1,1) supergravity. *Nucl. Phys.*, B241:406–428, 1984.

[132] J. Ellis, A.B. Lahanas, D. Nanopoulos, and K. Tamvakis. No-scale supersymmetric standard model. *Phys. Lett.*, 134B:429–435, 1984.

[133] J. Ellis, D.V. Nanopoulos, and M. Quiros. On the axion, dilaton, Polonyi, gravitino and shadow matter problems in supergravity and superstring models. *Phys. Lett.*, B 174:176–182, 1986.

[134] J. Ellis, G. Ridolfi, and F. Zwirner. Radiative corrections to the masses of supersymmetric Higgs bosons. *Phys. Lett.*, B 257:83–91, 1991.

[135] U. Ellwanger and M. Rausch de Traubenberg and C.A. Savoy. Phenomenology of supersymmetric models with a singlet. *Nucl. Phys.*, B492:21–50, 1997.

[136] B. Aubert et al. [BABAR Collaboration]. Observation of CP violation in the b^0 meson system. *Phys. Rev. Lett.*, 87:091801, 2001.

[137] K. Abe et al. [Belle Collaboration]. Observation of a large CP violation in the neutral B meson system. *Phys. Rev. Lett.*, 87:091802, 2001.

[138] Y. Hayato et al. [Super-Kamiokande Collaboration]. Search for proton decay through $p \to \bar{\nu}K^+$ in a large water cherenkov detector. *Phys. Rev. Lett.*, 83:1529, 1999.

[139] M. Bona et al [UTfit collaboration]. The UTfit collaboration report on the status of the unitarity triangle beyond the standard model I. Model-independent analysis and minimal flavor violation. *JHEP*, 0603:080, 2006.

[140] A.G. Riess et al. Observational evidence from supernovae for an accelerating universe and a cosmological constant. *Astron. J.*, 116:1009–1038, 1998.

[141] D.N. Spergel et al. First year WMAP observations: determination of cosmological parameters. *Astrophys. J. Supp.*, 148:175, 2003.

[142] K. Hagiwara et al. Review of particle physics. *Phys. Rev. D*, 66:010001, 2002.

[143] M. Ciuchini et al. δm_k and ϵ_k in SUSY at the next-to-leading order. *JHEP*, 9810:008, 1998.

[144] P.G. Harris et al. *Phys. Rev. Lett.*, 82:904, 1999.

[145] P. Fayet and J. Iliopoulos. Spontaneously broken supergauge symmetries and Goldstone spinors. *Phys. Lett.*, 51B:461–464, 1974.

[146] P. Fayet. Supergauge invariant extension of the Higgs mechanism and a model for the electron and its neutrino. *Nucl. Phys.*, B90:104–124, 1975.

[147] P. Fayet. Fermi-Bose hypersymmetry. *Nucl. Phys.*, B113:135–155, 1976.

[148] P. Fayet. Supersymmetry and weak, electromagnetic and strong interactions. *Phys. Lett.*, 64B:159, 1976.

[149] P. Fayet. Spontaneously broken supersymmetric theories of weak, electromagnetic and strong interactions. *Phys. Lett.*, 69B:489–494, 1977.

[150] P. Fayet. In S. Ferrara, J. Ellis, and P. van Nieuwenhuizen, editors, *Unification of the fundamental particle interactions*, page 587. Plenum, 1980.

[151] P. Fayet. Supersymmetric grand unification in a six-dimensional space-time. *Phys. Lett.*, B159:121, 1985.

[152] P. Fayet. *Nucl. Phys.*, B263:649, 1986.

[153] J.L. Feng, K.T. Matchev, and T. Moroi. Focus points and naturalness in supersymmetry. *Phys. Rev.*, D61:075005, 2000.

[154] J.L. Feng, K.T. Matchev, and T. Moroi. Multi-TeV scalars are natural in minimal supergravity. *Phys. Rev. Lett.*, 84:2322–2325, 2000.

[155] J.L. Feng and T. Moroi. Supernatural supersymmetry: phenomenological implications of anomaly-mediated supersymmetry breaking. *Phys. Rev.*, D61:095004, 2000.

[156] E. Fermi. An attempt of a theory of beta radiation 1. *Z. Phys.*, 88:161–177, 1934.

[157] E. Fermi. Trends to a theory of beta radiation. *Nuovo Cimento*, 11:1–19, 1934.

[158] S. Ferrara, J. Iliopoulos, and B. Zumino. Supergauge invariance and the Gell-Mann-Low eigenvalue. *Nucl. Phys.*, B77:413, 1974.

[159] S. Ferrara and E. Remiddi. Absence of the anomalous magnetic moment in a supersymmetric abelian gauge theory. *Phys. Lett.*, 53B:347–350, 1974.

[160] P. Ferreira and M. Joyce. Structure formation with a self-tuning scalar field. *Phys. Rev. Lett.*, 79:4740–4743, 1997.

[161] P. Ferreira and M. Joyce. Cosmology with a primordial scaling field. *Phys. Rev.*, D58:023503, 1998.

[162] R.P. Feynman and M. Gell-Mann. *Phys. Rev.*, 109:193, 1958.

[163] D. Finnell and P. Pouliot. Instanton calculations versus exact results in four-dimensional SUSY gauge theories. *Nucl. Phys.*, B453:225–239, 1995.

[164] K. Freese, J. Frieman, and A. Gould. Signal modulation in cold dark matter detection. *Phys. Rev.*, D37:3388, 1988.

[165] J.-M. Frère, D.R.T. Jones, and S. Raby. Fermion masses and induction of the weak scale by supergravity. *Nucl. Phys.*, B222:11–19, 1983.

[166] J. Frieman, C. Hill, A. Stebbins, and I. Waga. Cosmology with ultra-light Nambu-Goldstone bosons. *Phys. Rev.Lett.*, 75:2077–2080, 1995.

[167] C.D. Froggatt and H.B. Nielsen. Hierarchy of quark masses, Cabibbo angles and CP violation. *Nucl. Phys.*, B147:277–298, 1979.

[168] C.D. Froggatt and H.B. Nielsen. Statistical analysis of quark and lepton masses. *Nucl. Phys.*, B164:114–140, 1979.

[169] M. Fukugita and T. Yanagida. Baryogenesis without grand unification. *Phys. Lett.*, B174:45, 1986.

[170] F. Gabbiani, E. Gabrielli, A. Masiero, and L. Silvestrini. A complete analysis of FCNC and CP constraints in general SUSY extensions of the Standard Model. *Nucl. Phys.*, B477:321–352, 1996.

[171] M.K. Gaillard and B.W. Lee. Rare decay modes of the K mesons in gauge theories. *Phys. Rev.*, D10:897, 1974.

[172] M.K. Gaillard, B. Nelson, and Y.-Y. Wu. Gaugino masses in modular invariant supergravity. *Phys. Lett.*, B459:549–556, 1999.

[173] G. Gamberini, G. Ridolfi, and F. Zwirner. On radiative gauge symmetry breaking in the minimal supersymmetric model. *Nucl. Phys.*, B331:331–349, 1990.

[174] P. Gambino and M. Misiak. Quark mass effects in $\bar{B} \to X_s\gamma$. *Nucl. Phys.*, B611:338–366, 2001.

[175] R. Gatto, R. Sartori, and M. Tonin. Weak self masses, Cabibbo angles and broken $SU(2){\times}SU(2)$. *Phys. Lett.*, B28:128–130, 1968.

[176] M. Gell-Mann, R.J. Oakes, and B. Renner. Behavior of current divergences under $SU(3){\times}SU(3)$. *Phys. Rev.*, 175:2195–2199, 1968.

[177] M. Gell-Mann, P. Ramond, and R. Slansky. *Talk at the Sanibel symposium*, 1979. preprint CALT-68-709 (unpublished).

[178] H. Georgi. An almost realistic gauge hierarchy. *Phys. Lett.*, B108:283, 1982.

[179] H. Georgi and S.L. Glashow. Gauge theories without anomalies. *Phys. Rev.*, D6:429–431, 1972.

[180] H. Georgi and S.L. Glashow. Unified weak and electromagnetic interactions without neutral currents. *Phys. Rev. Lett.*, 28:1494–1497, 1972.

[181] H. Georgi and S.L. Glashow. Unity of all elementary particle forces. *Phys. Rev. Lett.*, 32:438–441, 1974.

[182] H. Georgi and C. Jarlskog. Soft breaking of supersymmetry. *Phys. Lett.*, 86B:297, 1979.

[183] H. Georgi, H.R. Quinn, and S. Weinberg. Hierarchy of interactions in unified gauge theories. *Phys. Rev. Lett.*, 33:451–454, 1974.

[184] T. Gherghetta, C.F. Kolda, and S.P. Martin. Flat directions in the scalar potential of the supersymmetric standard model. *Nucl. Phys.*, B468:37–58, 1996.

[185] S.B. Giddings, S. Kachru, and J. Polchinski. Hierarchies from fluxes in string compactification. *Phys. Rev.*, D66:106006, 2002.

[186] E. Gildener. Gauge symmetry hierarchies. *Phys. Rev.*, D14:1667, 1976.

[187] P. Ginsparg. Applied conformal field theory. In E. Brézin and J. Zinn-Justin, editors, *Fields, strings and critical phenomena (Les Houches, Session XLIX, 1988)*. Elsevier Science Publishers, 1989.

[188] L. Girardello and M.T. Grisaru. Soft breaking of supersymmetry. *Nucl. Phys.*, B194:65–76, 1982.

[189] G.F. Giudice, M. Luty, H. Murayama, and R. Rattazzi. Gaugino mass without singlets. *JHEP*, 9812:027, 1998.

[190] G.F. Giudice and A. Masiero. A natural solution to the μ problem in supergravity theories. *Phys. Lett.*, B206:480–484, 1988.

[191] G.F. Giudice, M. Peloso, A. Riotto, and I. Tkachev. Production of massive fermions at preheating and leptogenesis. *JHEP*, 9908:014, 1999.

[192] G.F. Giudice and A. Romanino. Split supersymmetry. *Nucl. Phys.*, B699:65, 2004.

[193] S.L. Glashow and J. Iliopoulos and L. Maiani. *Phys. Rev.*, D2:1285, 1970.

[194] F. Gliozzi, J. Scherk, and D. Olive. Supersymmetry, supergravity theories, and the dual spinor model. *Nucl. Phys.*, B122:253, 1977.

[195] H. Goldberg. Constraint on the photino mass from cosmology. *Phys. Rev. Lett.*, 50:1419, 1983.

[196] S.I. Goldberg. *Curvature and homology*. Dover, 1982.

[197] M. Goldhaber, L. Grodzins, and A.W. Sunyar. Helicity of neutrinos. *Phys. Rev.*, 109:1015–1017, 1982.

[198] J. Goldstone. Field theories with superconductor solutions. *Nuovo Cimento*, 19:154–164, 1961.

[199] A.S. Goncharov, A.D. Linde, and M.J. Visotsky. Cosmological problems for spontaneously broken supergravity. *Phys.Lett.*, 147B:279–283, 1984.

[200] P. Gondolo, J. Edsjo, P. Ullio, L. Bergstrom, M. Schelke, and E.A. Baltz. Darksusy: a numerical package for supersymmetric dark matter calculations. pages astro–ph/0211238.

[201] P. Gondolo and J. Silk. Dark matter annihilation at the galactic center. *Phys. Rev. Lett.*, 83:1719–1722, 1999.

[202] M. W. Goodman and E. Witten. Detectability of certain dark-matter candidates. *Phys. Rev.*, D31:3059–3063, 1985.

[203] T. Goto and T. Nihei. Effect of an RRRR dimension 5 operator on the proton decay in the minimal SU(5) SUGRA GUT model. *Phys. Rev.*, D59:115009, 1999.

[204] A. Gould. Resonant enhancements in Wimp capture by the Earth. *Astrophys. J.*, 321:571, 1987.

[205] M.B. Green and J.H. Schwarz. Anomaly cancellations in supersymmetric $d = 10$ gauge theory and superstring theory. *Phys. Lett.*, B149:117, 1984.

[206] K. Griest and D. Seckel. Cosmic asymmetry, neutrinos, and the Sun. *Nucl. Phys.*, B283:681, 1987.

[207] K. Griest and D. Seckel. Three exceptions in the calculation of relic abundances. *Phys. Rev.*, D43:3191–3203, 1991.

[208] B. Grinstein. A supersymmetric SU(5) gauge theory with no gauge hierarchy problem. *Nucl. Phys.*, B206:387, 1982.

[209] M.T. Grisaru, A. Karlhede, and M. Rocek. The superhiggs effect in superspace. *Phys. Lett.*, B120:110, 1983.

[210] M. Grisaru, M. Rocek, and W. Siegel. Improved methods for supergraphs. *Nucl. Phys.*, B159:429, 1979.

[211] J.-F. Grivaz. Status of susy searches. *Czech. J. Phys.*, 54:Suppl. A, 1984.

[212] D.J. Gross, J.A. Harvey, E. Martinec, and R. Rohm. Heterotic string theory (i). the free heterotic string. *Nucl. Phys.*, B256:253, 1985.

[213] D.J. Gross, J.A. Harvey, E. Martinec, and R. Rohm. Heterotic string theory (ii). the interacting heterotic string. *Nucl. Phys.*, B267:75, 1986.

[214] A.H. Guth. The inflationary universe: a possible solution to the horizon and flatness problems. *Phys. Rev.*, D23:347–356, 1981.

[215] R. Haag, J. Lopuszanski, and M. Sohnius. All possible generators of supersymmetries of the S-matrix. *Nucl. Phys.*, B88:257–274, 1975.

[216] H.E. Haber and R. Hempfling. Can the mass of the lightest Higgs boson of the Minimal Supserymmetric Model be larger than m_z? *Phys. Rev. Lett.*, 66:1815–1818, 1991.

[217] H.E. Haber. *Nucl. Phys. Proc. Suppl.*, 62:469, 1998.

[218] J. Halliwell. Scalar fields in cosmology with an exponential potential. *Phys. Lett.*, B185:341, 1987.

[219] L.J. Hall, V.A. Kostelecky, and S. Raby. New flavor violations in supergravity models. *Nucl. Phys.*, B267:415, 1986.

[220] L. Hall, R. Rattazzi, and U. Sarid. The top quark mass in supersymmetric SO(10) unification. *Phys. Rev.*, D50:7048–7065, 1994.

[221] E. Halyo. Hybrid inflation from supergravity D terms. *Phys. Lett.*, B387:43–47, 1996.

[222] J.A. Harvey. Magnetic monopoles, duality and supersymmetry. In E. Gava and A. Masiero and K.S. Narain and S. Randjbar-Daemi and Q. Shafi, *1995 Trieste Summer School in High Energy Physics and Cosmology*, pages 66–125. World Scientific, 1996.

[223] P. Hasenfratz, and G. 't Hooft, (1976). A fermion-boson puzzle in a gauge theory, *Phys. Rev. Lett.*, 36:1119, 1976.

[224] C.T. Hill. Quark and lepton masses from renormalization-group fixed points. *Phys. Rev.*, D 24:691–703, 1981.

[225] G. Hinshaw et al. Three year WMAP observations: temperature analysis, arXiv:astro-ph/0603451, 2006.

[226] Y. Hosotani. Dynamical mass generation by compact extra dimensions. *Phys. Lett.*, 126B:309–313, 1983.

[227] P. Hořava and E. Witten. Eleven-dimensional supergravity on a manifold with boundary. *Nucl. Phys.*, B475:94, 1996.

[228] P. Hořava and E. Witten. Heterotic and type I string dynamics from eleven dimensions. *Nucl. Phys.*, B460:506, 1996.

[229] P. Huet and A.E. Nelson. Electroweak baryogenesis in supersymmetric models. *Phys. Rev.*, D53:4578–4597, 1996.

[230] L.E. Ibáñez, C. López, and C. Muñoz. The low-energy supersymmetric spectrum according to $N = 1$ supergravity GUTS. *Nucl. Phys.*, B256:218–252, 1985.

[231] L.E. Ibáñez and C. López. N=1 supergravity, the weak scale and the low-energy particle spectrum. *Nucl. Phys.*, B233:511–544, 1984.

[232] L.E. Ibáñez and G.G. Ross. Discrete gauge symmetries and the origin of baryon and lepton number conservation in supersymmetric versions of the standard model. *Nucl. Phys.*, B368:3–37, 1992.

[233] L.E. Ibáñez and G.G. Ross. Fermion masses and mixing angles from gauge symmetries. *Phys. Lett.*, B332:100, 1994.

[234] L.E. Ibáñez. Computing the weak mixing angle from anomaly cancellation. *Phys. Lett.*, B 303:55–62, 1993.

[235] J. Iliopoulos and B. Zumino. Broken supergauge symmetry and renormalization. *Nucl. Phys.*, B76:310, 1974.

[236] R. Jackiw and C. Rebbi. Spin from isospin in a gauge theory. *Phys. Rev. Lett.*, 36:1116, 1976.

[237] R. Jackiw and P. Rossi. Zero modes of the vortex-fermion system. *Nucl. Phys.*, B190:681–691, 1981.

[238] C. Jarlskog. Commutator of the quark mass matrices in the standard electroweak model and a measure of maximal CP violation. *Phys. Rev. Lett.*, 55:1039, 1985.

[239] T.F. Jordan. Restrictions implied by Lorentz and spin invariance for scattering amplitudes. *Phys. Rev.*, 139:B149–B150, 1965.

[240] M. Joyce. Electroweak baryogenesis and the expansion rate of the universe. *Phys. Rev.*, D55:1875–1878, 1997.

[241] B. Julia and A. Zee. Poles with both magnetic and electric charges in non-abelian gauge theories. *Phys. Rev.*, D11:2227–2232, 1975.

[242] G. Jungman and M. Kamionkowski and K. Griest. Supersymmetric dark matter. *Phys. Lett.*, 267:195–373, 1996.

[243] S. Kachru, R. Kallosh, A. Linde, and S.P. Trivedi. De Sitter vacua in string theory. *Phys. Rev.*, D68:046005, 2003.

[244] M. Kalb and P. Ramond. Classical direct interstring action. *Phys. Rev.*, D9:2273–2284, 1974.

[245] T. Kaluza. On the problem of unity in physics. *Sitzungsber. Preuss Akad. Wiss. Berlin Math. Phys.*, 1921.

[246] G.L. Kane and S. F. King. Natural implications of LEP results. *Phys. Lett.*, B451:113, 1999.

[247] V.S. Kaplunovsky and J. Louis. On gauge couplings in string theory. *Nucl. Phys.*, B444:191, 1995.

[248] M.Yu. Khlopov and A.D. Linde. Is it easy to save the gravitino? *Phys. Lett.*, B138:265–268, 1984.

[249] T.W.B. Kibble. Topology of cosmic domains and strings. *J. Phys. A*, 9:1387–1398, 1976.

[250] J.E. Kim and H.P. Nilles. The μ problem and the strong CP problem. *Phys. Lett.*, 138B:150, 1984.

[251] O. Klein. Quantum theory and 5-dimensional relativity. *Z. Phys.*, 37:895, 1926.

[252] R. Kleiss. Derivation of the minimal Standard Model lagrangian. In A. Zichichi, editor, *1990 Erice Summer School: Physics up to 200 TeV*, pages 93–141, 1990.

[253] M. Knecht, A. Nyffeler, M. Perrottet, and E. de Rafael. Hadronic light by light scattering contribution to the muon $g - 2$: an effective field theory approach. *Phys. Rev. Lett.*, 88:071802, 2002.

[254] M. Kobayashi and M. Maskawa. *Prog. Theor. Phys.*, 49:652, 1973.

[255] C.S. Kochanek. Gravitational lensing limits on cold dark matter and its variants. *Astrophys. J.*, 453:545, 1995.

[256] C. Kolda and D. Lyth. Quintessential difficulties. *Phys. Lett.*, B458:197–201, 1999.

[257] C. Kolda and H. Murayama. The Higgs mass and new physics scales in the minimal Standard Model. *JHEP*, 0007:035, 2000.

[258] H. Komatsu. New constraints on parameters in the minimal supersymmetric model. *Phys. Lett.*, B215:323, 1988.

[259] Konishi, K. *Phys. Lett.*, 135B:439, 1984.

[260] E.J. Konopinski and H. Mahmoud. The universal Fermi interaction. *Phys. Rev.*, 92:1045–1049, 1953.

[261] Koppenburg, P. *et al.* [Belle Collaboration]. Inclusive measurement of the photon energy spectrum in $b \to s\gamma$ decays. *Phys. Rev. Lett.*, 93, 061803, 2004.

[262] N.V. Krasnikov. On supersymmetry breaking in superstring theories. *Phys. Lett. B.*, 193:37–40, 1987.

[263] L.M. Krauss and F. Wilczek. Discrete gauge symmetry in continuum theories. *Phys. Rev. Lett.*, 62:1221–1223, 1989.

[264] K. Kumekawa, T. Moroi, and T. Yanagida. Flat potential for inflaton with a discrete R invariance in supergravity. *Progr. Theor. Phys.*, 92:437–448, 1994.

[265] A. Kusenko, P. Langacker, and G. Segrè. Phase transitions and vacuum tunneling into charge- and color- breaking minima in the MSSM. *Phys. Rev.*, D54:5824–5834, 1996.

[266] V.A. Kuzmin, V.A. Rubakov, and M.E. Shaposhnikov. Classical direct interstring action. *Phys. Lett.*, 155:36, 1985.

[267] L.D. Landau, A.A. Abrikosov, and I.M. Khalatnikov. *Dokl. Akad. Nauk. SSSR*, 95:1177, 1954.

[268] B. Lee, C. Quigg, and H. Thacker. Weak interactions at very high energy: the role of the Higgs boson mass. *Phys. Rev.*, D16:1519, 1977.

[269] T.D. Lee and C.N. Yang. *Phys. Rev.*, 104:254, 1956.

[270] LEP Higgs working group, ALEPH, DELPHI, L3, and OPAL experiments.

[271] LEP Higgs working group, ALEPH, DELPHI, L3, and OPAL experiments. *LHWG-Note-2004-01*.

[272] M. Leurer, Y. Nir, and N. Seiberg. Mass matrix models. *Nucl. Phys.*, B398:319–342, 1993.

[273] M. Leurer, Y. Nir, and N. Seiberg. Mass matrix models: the sequel. *Nucl. Phys.*, B420:468–504, 1994.

[274] A.D. Linde. Chaotic inflation. *Phys. Lett.*, 129B:177–181, 1983.

[275] A.D. Linde. *Particle physics and inflationary cosmology*, volume 5 of *Contemporary Concepts in Physics*. Harwood Academic Publishers, 1990.

[276] A.D. Linde. Axions in inflationary cosmology. *Phys. Lett.*, 259B:38–47, 1991.

[277] M. Lindner. Implications of triviality for the Standard Model. *Z. Phys.*, C31: 295–300, 1986.

[278] M.A. Luty and W. Taylor IV. Varieties of vacua in classical supersymmetric gauge theories. *Phys. Rev.*, D53:3399–3405, 1996.

[279] G. Mack. All unitary ray representations of the conformal group SU(2,2) with positive energy. *Comm. math. Phys.*, 55:1–28, 1977.

[280] Z. Maki, M. Nakagawa, and S. Sakata. *Prog. Theor. Phys.*, 28:870, 1962.

[281] N. Manton. The force between 't Hooft-Polyakov monopoles. *Nucl. Phys.*, B126:525–541, 1977.

[282] W. Marciano, G. Valencia, and S. Willenbrock. Renormalization group-improved unitarity bounds on the Higgs boson and top quark masses. *Phys. Rev.*, D40:1725–1729, 1989.

[283] E. Martinec. Nonrenormalization theorems and fermionic string finiteness. *Phys. Lett.*, 171B:189, 1986.

[284] A. Masiero and D.V. Nanopoulos and K. Tamvakis and T. Yanagida. Naturally massless Higgs doublets in supersymmetric SU(5). *Phys. Lett.*, 115:380, 1982.

[285] S. Mattig. *Astron. Nachr.*, 284:109, 1958.

[286] Y. Mellier. Probing the universe with weak lensing. *Ann. Rev. Astron. Astrophys.*, 37:127–189, 1999.

[287] C.W. Misner, K.S. Thorne, and J.A. Wheeler. *Gravitation*. W.H. Freeman and company, 1972.

[288] C. Montonen and D.I. Olive. Magnetic monopoles as gauge particles? *Phys. Lett.*, 72B:117–120, 1977.

[289] B. Moore, T. Quinn, F. Governato, J. Stadel, and G. Lake. *Mon. Not. Roy. Astronom. Soc.*, 310:1147, 1999.

[290] H. Murayama and A. Pierce. Not even decoupling can save minimal supersymmetric SU(5). *Phys. Rev.*, D65:055009, 2002.

[291] H. Murayama. A model of direct gauge mediation. *Phys. Rev. Lett.*, 79:18–21, 1997.

[292] D.V. Nanopoulos and K. Tamvakis. Susy Guts:4 - Guts:3. *Phys.Lett.*, B113:151, 1952.

[293] J.F. Navarro, C.S. Frenk, and S.D.M. White. A universal density profile from hierarchical clustering. *Astrophys. J.*, 490:493–508, 1997.

[294] A. Neveu and J.H. Schwarz. Factorizable dual model of pions. *Nucl. Phys.*, B31:86, 1971.

[295] H.P. Nilles, M. Srednicki, and D. Wyler. Weak interaction breakdown induced by supergravity. *Phys. Lett.*, B120:346, 1983.

[296] Y. Nir and G. Raz. Quark-squark alignment revisited. *Phys. Rev.*, D66:035007, 2002.

[297] Y. Nir and N. Seiberg. Should squarks be degenerate? *Phys. Lett.*, B309:337–343, 1993.

[298] V. Novikov, M. Shifman, A. Vainshtein, and V. Zakharov. Exact Gell-Mann-Low function of supersymmetric Yang-Mills theories from instanton calculus. *Nucl. Phys.*, B229:381, 1983.

[299] Y. Okada, M. Yamaguchi, and T. Yanagida. Upper bound of the lightest Higgs boson mass in the Minimal Supersymmetric Standard Model. *Prog. Theor. Phys.*, 85:1–6, 1991.

[300] L. O'Raifeartaigh. Spontaneous symmetry breaking for chiral scalar superfields. *Nucl. Phys.*, B96:331–352, 1975.

[301] H. Osborn. Topological charges for N= 4 supersymmetric gauge theories and monopoles of spin 1. *Phys. Lett.*, 83B:321, 1979.

[302] J.C. Pati and A. Salam. Lepton number a the fourth color. *Phys. Rev.*, D10:275–289, 1974.

[303] R. Peccei and H. Quinn. *Phys. Rev. Lett.*, 38:1440, 1977.

[304] P.J.E. Peebles and B. Ratra. Cosmology with a time-variable cosmological constant. *Astrophys. J.*, 325:L17–L20, 1988.

[305] P.J.E. Peebles and A. Vilenkin. Quintessential inflation. *Phys. Rev.*, D59:063505, 1999.

[306] B. Pendleton and G.G. Ross. Mass and mixing angle prediction from infrared fixed points. *Phys. Lett.*, B98:291, 1981.

[307] M.E. Peskin and T. Takeuchi. A new constraint on a strongly interacting Higgs sector. *Phys. Rev. Lett.*, 65:964–967, 1990.

[308] M.E. Peskin and T. Takeuchi. Estimation of oblique electroweak corrections. *Phys. Rev.*, D46:381–409, 1992.

[309] J. Polchinski and M. Wise. The electric dipole moment of the neutron in low-energy supergravity. *Phys. Lett.*, B125:393, 1983.

[310] J. Polchinski. Renormalization and effective lagrangians. *Nucl. Phys.*, B231:269–295, 1984.

[311] J. Polchinski. Dirichlet-Branes and Ramond-Ramond charges. *Phys. Rev. Lett.*, 75:4724–4727, 1995.

[312] J. Polchinski. *String theory*, volume I and II. Cambridge University Press, 1998.

[313] J. Polonyi. Generalization of the massive scalar multiplet coupling to the supergravity. 1977. unpublished preprint KFKI-77-93.

[314] E. Poppitz and S. Trivedi. New models of gauge and gravity mediated supersymmetry breaking. *Phys. Rev.*, D55:5508–5519, 1997.

[315] M.K. Prasad and C.M. Sommerfield. Exact classical solutions for the 't Hooft monopole and the Julia-Zee dyon. *Phys. Rev. Lett.*, 35:760–762, 1975.

[316] R. Rajaraman. *Solitons and Instantons*. North-Holland, 1998.

[317] P. Ramond. Dual theory for free fermions. *Phys. Rev.*, D3:2415–2418, 1971.

[318] P. Ramond. Field theory, a modern primer, 2nd edition. *Frontiers in Physics*. Addison-Wesley, 1990.

[319] L. Randall and R. Sundrum. An alternative to compactification. *Phys. Rev. Lett.*, 83:4690–4693, 1999.

[320] L. Randall and R. Sundrum. Out of this world supersymmetry breaking. *Nucl. Phys.*, B557:79, 1999.

[321] W. Rarita and J. Schwinger. On a theory of particles of half-integral spin. *Phys. Rev.*, 60:61, 1941.

[322] B. Ratra and P.J.E. Peebles. Cosmological consequences of a rolling homogeneous scalar field. *Phys. Rev.*, D37:3406, 1988.

[323] J. Rich, Fundamentals of cosmology, Springer 2001.

[324] R. Rohm and E. Witten. The antisymmetric tensor field in supersymmetric theory. *Ann. of Phys.*, 170:454, 1986.

[325] G.G. Ross. Textures and flavour models. In S. Lavignac J. Orloff and M. Cribier, editors, *Seesaw 25*, pages 81–98. World Scientific, 2004.

[326] A.D. Sakharov. Violation of CP invariance, C asymmetry and baryon asymmetry of the universe. *JETP Lett.*, 5:24–27, 1967.

[327] A. Salam and J. Strathdee. Supersymmetry, parity and fermion-number conservation. *Nucl. Phys.*, B97:293, 1975.

[328] P. Salomonson and J.W. van Holten. Fermionic coordinates and supersymmetry in quantum mechanics. *Nucl. Phys.*, B196:509–531, 1982.

[329] J. Scherk and J.H. Schwarz. Dual models for non-hadrons. *Nucl. Phys.*, B81:118, 1974.

[330] J. Scherk and J.H. Schwarz. How to get masses from extra dimensions. *Nucl. Phys.*, B153:61–88, 1979.

[331] J. Scherk. Extended supersymmetry and supergravity theories. In *Advanced Study Institute on Gravitation: Recent developments, Cargèse*, 1978.

[332] J. Schwinger. A magnetic model of matter. *Science*, 165:757–761, 1969.

[333] N. Seiberg and E.Witten. Electric-magnetic duality, monopole condensation, and confinement in N=2 supersymmetric Yang-Mills theory. *Nucl. Phys.*, B426:19, 1994.

[334] N. Seiberg and E.Witten. Monopoles, duality and chiral symmetry breaking in N=2 supersymmetric QCD. *Nucl. Phys.*, B431:484, 1994.

[335] N. Seiberg. Exact results on the space of vacua of four-dimensional SUSY gauge theories. *Phys. Rev.*, D49:6857–6863, 1994.

[336] N. Seiberg. Electric-magnetic duality in supersymmetric nonabelian gauge theories. *Nucl. Phys.*, B435:129–146, 1995.

[337] M.A. Shifman and A.I. Vainshtein. Solutions of the anomaly puzzle in SUSY gauge theories and the Wilson operator expansion. *Nucl. Phys.*, B277:456–486, 1986.

[338] M.A. Shifman and A.I. Vainshtein. On holomorphic dependence and infrared effects in supersymmetric gauge theories. *Nucl. Phys.*, B359:571–580, 1991.

[339] M.A. Shifman. Beginning supersymmetry. Supersymmetry in quantum mechanics. *ITEP lectures on particle physics and field theory*, 1:301–344, 1995.

[340] W. Siegel. Supersymmetric dimensional regularization via dimensional reduction. *Phys.Lett.*, 84B:193–195, 1979.

[341] P. Sikivie, L. Susskind, Voloshin, and Zakharov. Isospin breaking in technicolor models. *Nucl. Phys.*, B173:189, 1980.

[342] M. Sohnius. Introducing supersymmetry. *Phys.Rep.*, C 128:39–204, 1985.

[343] D. Spergel and U. Pen. Cosmology in a string dominated universe. *Astrophys. J.*, 491:L67–L71, 1997.

[344] B. Spokoiny. Deflationary universe scenario. *Phys. Lett.*, B315:40–45, 1993.

[345] E.D. Stewart. Inflation, supergravity, and superstrings. *Phys. Rev.*, D51:6847–6853, 1995.

[346] E.C.G. Sudarshan and R.E. Marshak. Chirality invariance and the universal Fermi interaction. *Phys. Rev.*, 109:1860–1862, 1958.

[347] L. Susskind. Lattice models of quark confinement at high temperature. *Phys. Rev.*, D20:2610–2618, 1979.

[348] P.K. Townsend. Cosmological constant in supergravity. *Phys. Rev.*, D15:2802–2804, 1977.

[349] G. 't Hooft. Naturalness, chiral symmetry, and spontaneous chiral symmetry breaking. In *Recent developments in gauge theories (NATO ASI Series B: Physics Vol. 59)*, pages 135–157. Plenum Press, 1979.

[350] J.-Ph. Uzan. The fundamental constants and their variation: observational status and theoretical motivation. *Rev. Mod. Phys.*, 75:403, 2003.

[351] M. Veltman, *Acta Phys. Polon.*, B8:475, 1977.

[352] M. Veltman. *Acta Phys. Polon.*, B12:437, 1981.

[353] G. Veneziano and S. Yankielowicz. An effective lagrangian for the pure N=1 supersymmetric Yang-Mills theory. *Phys. Lett.*, 113B:231–236, 1982.

[354] G. Veneziano. Construction of a crossing-symmetric, Regge-behaved amplitude for linearly-rising trajectories. *Nuovo Cim.*, 57A:190, 1968.

[355] A. Vilenkin. String dominated universe. *Phys. Rev. Lett.*, 53:1016–1018, 1984.

[356] L. Wambsganss, R. Cen, J.P. Ostriker, and E.L. Turner. *Science*, 268:274, 1995.

[357] E. Weinberg. Parameter counting for multimonopole solutions. *Phys. Rev.*, D20:936–944, 1979.

[358] E. Weinberg. Index calculations for the fermion-vortex system. *Phys. Rev.*, D24:2669, 1981.

[359] S. Weinberg. Implications of dynamical symmetry breaking: an addendum. *Phys. Rev.*, D19:1277–1280, 1979.

[360] S. Weinberg. Cosmological constraints on the scale of supersymmetry breaking. *Phys. Rev. Lett.*, 48:1303–1306, 1982.

[361] S. Weinberg. The cosmological constant. *Rev. Mod. Phys.*, 61:1, 1989.

[362] J. Wess and J. Bagger. *Supersymmetry and Supergravity*. Princeton Series in Physics. Princeton University Press, Princeton, 1983. 2nd edition 1992.

[363] J. Wess and B. Zumino. Supergauge invariant extension of quantum electrodynamics. *Nucl. Phys.*, B78:1–13, 1974.

[364] J. Wess and B. Zumino. Supergauge transformations in four dimensions. *Nucl. Phys.*, B70:39–50, 1974.

[365] C. Wetterich. Cosmology and the fate of dilatation symmetry. *Nucl. Phys.*, B302:668, 1988.

[366] C.M. Will. *Theory and experiment in gravitational physics*. Cambridge University Press, 1993.

[367] K. Wilson. Renormalization group and critical phenomena. *Phys. Rev.*, B4:3174, 3184, 1971.

[368] E. Witten and D. Olive. Supersymmetry algebras that include topological charges. *Phys. Lett.*, 78B:97–101, 1978.

[369] E. Witten and D. Olive. Supersymmetry algenras that include topological charges. *Phys. Lett.*, 78B:97–101, 1978.

[370] E. Witten. Dyons of charge $e\theta/2\pi$. *Phys. Lett.*, 86B:283–287, 1979.

[371] E. Witten. Dynamical breaking of supersymmetry. *Nucl. Phys.*, B188:513, 1981.

[372] E. Witten. Mass hierarchies in supersymmetric theories. *Phys. Lett.*, B105:267, 1981.

[373] E. Witten. Constraints on supersymmetry breaking. *Nucl. Phys.*, B202:253–316, 1982.

[374] E. Witten. Introduction to supersymmetry. In A. Zichichi, *International School of Sunuclear Physics, Erice 1981*, page 305. Plenum Press, 1983.

[375] E. Witten. Dimensional reduction of superstring models. *Phys. Lett.*, 155B:151, 1985.

[376] E. Witten. Superconducting strings. *Nucl. Phys.*, B249:557–592, 1985.

[377] E. Witten. New issues in manifolds of SU(3) holonomy. *Nucl. Phys.*, B268:79, 1986.

[378] E. Witten. Strong coupling expansion of Calabi-Yau compactification. *Nucl. Phys.*, B471:135, 1996.

[379] L. Wolfenstein. Parametrization of the Kobayashi-Maskawa matrix. *Phys. Rev. Lett.*, 51:1945–1947, 1983.

[380] LEP SUSY working group, ALEPH, DELPHI, L3, and OPAL experiments.

[381] T.T. Wu and C.N. Yang. Concept of non-integrable phase factors and global formulation of gauge fields. *Phys. Rev.*, D12:3845–3857, 1975.

[382] C.S. Wu, *et al*. Experimental test of parity conservation in beta decay. *Phys. rev.*, 105:1413–1415, 1957.

[383] T. Yanagida. In *Workshop on Unified Theory and Baryon Number of the Universe, KEK, Japan, 1979*.

[384] C.N. Yang and R.L. Mills. Conservation of isotopic spin and isotopic gauge invariance. *Phys.Rev.*, 96:191–195, 1954.

[385] A. Zee. A theory of lepton number violation, neutrino majorana mass, and oscillation. *Phys. Lett.*, B93:389, 1980. Erratum-ibid.B95 (1980) 461.

[386] Ya.B. Zeldovich. *DAN SSSR*, 86:505, 1952.

[387] I. Zlatev, L. Wang, and P.J. Steinhardt. Quintessence, cosmic coincidence, and the cosmological constant. *Phys. Rev. Lett.*, 82:896–899, 1999.

[388] B. Zumino. Normal forms of complex matrices. *J. Math. Phys.*, 3:1055–1057, 1962.

[389] B. Zumino. Supersymmetry and Kähler manifolds. *Phys. Lett.*, 87B:203–206, 1979.

[390] D. Zwanziger. Quantum field theory of particles with both electric and magnetic charges. *Phys. Rev.*, 176:1489–1495, 1968.

Index

The suffix f following a page number indicates a figure, n indicates a footnote and t indicates a table. Greek letters are alphabetized as though spelled out.